WILDLIFE 2000

Modeling Habitat Relationships of Terrestrial Vertebrates

VOLUME EDITORS

Jared Verner, U.S. Department of Agriculture, Forest Service, Forestry Sciences Laboratory, 2081 East Sierra Avenue, Fresno, California 93710

Michael L. Morrison, Department of Forestry and Resource Management, 145 Mulford Hall, University of California, Berkeley, California 94720

C. John Ralph, U.S. Department of Agriculture, Forest Service, Redwood Sciences Laboratory, 1700 Bayview Drive, Arcata, California 95521

SYMPOSIUM SPONSORS

Pacific Southwest Forest and Range Experiment Station, Forest Service, U.S. Department of Agriculture

North Central Forest Experiment Station, Forest Service, U.S. Department of Agriculture

Bureau of Reclamation, U.S. Department of Interior

Division of Wildlife Research, Fish and Wildlife Service, U.S. Department of Interior

Pacific Gas and Electric Company

Bureau of Land Management, U.S. Department of Interior

Department of Forestry and Resource Management, University of California, Berkeley

U.S. Army Corps of Engineers

Southern California Edison Company

Region 5, Forest Service, U.S. Department of Agriculture

Cooperative Parks Studies Unit, University of California, Davis

SYMPOSIUM ORGANIZERS

San Francisco Bay Area Chapter, The Wildlife Society

Pacific Southwest Forest and Range Experiment Station, Forest Service, U.S. Department of Agriculture

Department of Forestry and Resource Management, University of California, Berkeley

Pacific Southwest Region, Forest Service, U.S. Department of Agriculture

WILDLIFE 2000

Modeling Habitat Relationships of Terrestrial Vertebrates

Edited by JARED VERNER
MICHAEL L. MORRISON
C. JOHN RALPH

Based on an International Symposium Held at
Stanford Sierra Camp, Fallen Leaf Lake, California
7–11 October 1984

The University of Wisconsin Press

Library of Congress Cataloging-in-Publication Data
Wildlife 2000.
Symposium sponsored by the Pacific Southwest Forest and Range Experiment Station, Forest Service, U.S. Dept. of Agriculture, et al. and organized by the San Francisco Bay Area Chapter of the Wildlife Society, et al.
Bibliography: pp. 433–454.
Includes index.
1. Habitat (Ecology)—Congresses. 2. Population biology—Congresses. 3. Biological models—Congresses. 4. Vertebrates—Habitat—Congresses. 5. Vertebrates—Ecology—Congresses. I. Verner, Jared. II. Morrison, Michael L. III. Ralph, C. John. IV. Pacific Southwest Forest and Range Experiment Station (Berkeley, Calif.) V. Wildlife Society. San Francisco Bay Area Chapter. VI. Title: Wildlife two thousand.
QL756.W53 1986 596′.05 85-40769
ISBN 0-299-10520-2

Published 1986

The University of Wisconsin Press
114 North Murray Street
Madison, Wisconsin 53715

The University of Wisconsin Press, Ltd.
1 Gower Street
London WC1E 6HA, England

First printing

Printed in the United States of America

Contents

Part II. Biometric Approaches to Modeling

Part III. When Habitats Fail as Predictors

Part IV. Predicting Effects of Habitat Patchiness and Fragmentation

Part V. Linking Wildlife Models with Models of Vegetation Succession

Part VI. Synopsis

Foreword

MARK F. DEDON and STEPHEN A. LAYMON

The international symposium "Wildlife 2000: Modeling Habitat Relationships of Terrestrial Vertebrates" was held 7–11 October 1984, at Stanford Sierra Camp, Fallen Leaf Lake, near South Lake Tahoe, California, USA. The idea for the symposium was conceived some 18 months earlier at a program-planning meeting of the San Francisco Bay Area Chapter of the Wildlife Society. Among topics discussed for possible exploration was a half-day workshop on the development and application of models intended to predict responses of wildlife to habitat changes. The topic was of particular interest because several of us were involved in research projects requiring the use of such models. As we explored the potential of this topic, we soon learned that other researchers and land managers throughout the world shared our interest. We then envisioned not a brief, half-day session but a full-scale, international symposium.

The Forest Service, U.S. Department of Agriculture, had previously sponsored several conferences with similar themes, beginning in 1975 in Tucson, Arizona, with the "Symposium on Management of Forest and Range Habitats for Nongame Birds." This was followed by regional workshops on the same theme in Portland, Oregon (1977), Atlanta, Georgia (1978), Minneapolis, Minnesota (1979), and Salt Lake City, Utah (1980). In 1981, the University of Vermont, in Burlington, hosted a major symposium on "The Use of Multivariate Statistics in Studies of Wildlife Habitat." Also in 1981, a symposium at the University of Florida, Gainesville, dealt with modeling of ecological communities.

Although none of these symposia or workshops dealt primarily with the development, application, and testing of wildlife-habitat models for all taxa of terrestrial vertebrates, they clearly spearheaded the current use of models in wildlife management. Therefore, the Wildlife 2000 Symposium was organized specifically to address the development and application of models intended to predict responses of wildlife to changes in their habitats. The response to our call for papers confirmed our belief in a major need for such a forum.

The objective of the symposium, as reflected in these proceedings, was to begin and to provide for the continuance of communication between researchers and land managers on the state of the art in wildlife-habitat modeling. Just as models form a foundation for testing hypotheses, we hope this volume will provide a foundation for future modeling technology.

MARK F. DEDON, *Program Chairman, San Francisco Bay Area Chapter, The Wildlife Society*

Department of Forestry and Resource Management, 145 Mulford Hall, University of California, Berkeley, California 94720. *Present address:* Department of Engineering Research, Pacific Gas and Electric Company, 3400 Crow Canyon Road, San Ramon, California 94583

STEPHEN A. LAYMON, *President, San Francisco Bay Area Chapter, The Wildlife Society*

Department of Forestry and Resource Management, 145 Mulford Hall, University of California 94720

Introduction

JARED VERNER, MICHAEL L. MORRISON, and C. JOHN RALPH

The challenge of wise resource management is the same today as it has always been, although history shows that we have not always met that challenge in appropriate ways. Much state and federal legislation in the United States, especially during the past two decades, addresses this challenge directly by requiring managers to assure long-term maintenance of all renewable resources. This requires a level of integration ("modeling") among resource priorities rarely seen before, partly because of the overwhelming quantity and complexity of the information normally needed to achieve such integration. The computer age is upon us, however, so this constraint no longer stands in the way of our implementing mandates for integrated resource management. What we need now are some appropriate ways to organize our information and get it into computer files, and suitable methods for analyzing the information in those files.

The various chapters in this book do not follow any specific modeling strategy. Our intent was to include many different methods for examining wildlife-habitat relationships. For this publication, modeling should thus be viewed in its broadest sense. We like the simple, yet revealing, definition of "model" given in the 1973 edition of the American Heritage Dictionary: "A tentative ideational structure used as a testing device." Models range from simple words to complex equations, and often (usually?) little relationship exists between the mathematical complexity of a model and its fit to reality. For this reason, our principal aim was to include chapters that evaluate the applicability of specific models to the user-defined "real world."

Some persons urged us to develop a unifying set of definitions for common terminology used in modeling wildlife-habitat relationships. Although we initially succumbed to this temptation, to the extent of considering a glossary, we quickly discovered that few terms are really exclusive to this arena, and many that we would have selected for a glossary deserve a prohibitively lengthy discussion. Furthermore, many of the common terms found in this volume (e.g., *habitat, niche, quality, optimality*) defy simple definition. We could not presume to accomplish what many before us have tried and apparently failed to do. Instead, we tried to make each contribution stand alone with its terminology, although we undoubtedly fell short of satisfying many readers. We stress, however, that terms common to different chapters should be viewed within the context of each—"optimal habitat quality" is not found in any dictionary.

Long-term maintenance of renewable resources assumes knowledge about long-term effects of habitat manipulation. Specifically, for wildlife resources, it assumes that we know (1) each wildlife species' needs for and responses to various habitat types and successional stages; (2)

the specific details of successional change in composition and structure of the vegetation of any given area following disturbance; and (3) all of the various interactions of these and still other environmental variables, such as competition, weather, disease, and predation. All of these variables result in considerable annual variation in measures of wildlife populations, but such variation, it is hoped, is around a habitat-mediated mean population value. Our knowledge of these variables is deficient, but it is increasing rapidly and will continue to do so into the foreseeable future. Meanwhile, the rate of human use of natural resources accelerates in proportion to population growth, regardless of whether or not we can manage those resources effectively. We have no alternative but to use all existing information to achieve the best possible assessments of short- and long-term impacts of habitat change on wildlife.

In addition to computers, models will be an essential component of any effective strategy for integrated, long-term management of renewable resources. As abstract analogs of the real world, most models include elements for only a subset of the interacting factors that exist in that real world. The object, therefore, is to develop useful models by including only the most important factors and to correctly allow for the ways in which those factors interact. No one can deny, however, that this procedure piles one assumption upon another, which inevitably means making mistakes, even very large ones. Probably no contributor to this volume would argue that his or her model is unflawed. But probably most would agree that a successful marriage between computer technology and modeling technology is an essential part of any effective resource-management plan of the future. This is a time to be bold, even though we know we will make mistakes. We must learn and profit from our mistakes to be assured that someday we can make reasonably accurate predictions of the results of our actions.

This book presents examples of the current generation of models used to predict responses of terrestrial vertebrates to changes in their habitats. It represents the proceedings of a symposium held at Stanford Sierra Camp, near South Lake Tahoe, California, between 7 and 11 October 1984. But it is much more than that. Each manuscript resulting from a presentation at the symposium went through several drafts, being revised according to comments from session chairpersons, summarizers, and the proceedings co-editors. As a result of this process, not all the symposium presentations have been included in these proceedings. In addition, some papers not given at the symposium were solicited, to give a more complete picture of modeling in the arena of wildlife-habitat relationships.

A central goal of the symposium was to provide a forum for wildlife managers and researchers to exchange views and share their experiences with modeling. In addition to the introductory presentations by F. Dale Robertson and Jack Ward Thomas, the book is divided into six parts, five technical sections and a synopsis. Each technical section includes a separate introduction and two summaries, one giving the manager's viewpoint and one giving the researcher's viewpoint. Finally, the manager's and researcher's perspectives of the entire proceedings are given in Part VI—the synopsis—by Hal Salwasser, and by Herman H. Shugart, Jr., and Dean L. Urban, respectively.

Part I—Development, Testing, and Application of Wildlife-Habitat Models—emphasizes underlying concepts and issues related to these important aspects of modeling. The chapters examine single- and multispecies models. Predictions can relate to single slices of a time frame or be extrapolated over time in relation to expected short-term or long-term changes in habitats. Models in this current generation require much optimism, because they are based on many untested assumptions. Some chapters describe the nature of these assumptions and attempt to evaluate how likely they are to be true, pointing out ways to test them. Other chapters describe improved ways to measure habitats or elements of vertebrate life histories.

Part II—Biometric Approaches to Modeling—describes various statistical tests and procedures used to develop species-habitat models. Chapters discuss assumptions of statistical methods and how well specific cases meet those assumptions. Appropriate sample sizes, sampling designs, and sampling procedures are also covered by some chapters.

Part III—When Habitats Fail as Predictors—includes chapters that challenge us not to rely on just our knowledge of the structure and composition of habitats to estimate some measure of animal abundance or reproductive success. The approach here is quite different from merely showing that a predictive model based on habitat elements did a poor job when tested, a result that could have occurred simply because the wrong elements of the habitat (or wrong values of correct elements) were used in the model. In other words, the model failed because it was flawed; a correct model could have predicted population responses to changes in the habitat. Instead, the goal of this section is to emphasize the fact that populations of some wildlife species are occasionally (or usually) so much influenced by other environmental factors (e.g., weather, predation, disease, or competition) that one is unable to consistently predict their response to habitat change. This is not to say, of course, that habitat is unimportant. Obviously a species will be absent from, or at least only transient in, unsuitable habitats. Beyond that, however, we need to know the extent to which we can use habitat factors alone to predict a species' abundance or its probability of successful reproduction. The key issue centers on the fact that practitioners must recognize when populations respond to something other than changes in their habitat.

Part IV—Predicting Effects of Habitat Patchiness and Fragmentation—is aimed primarily at identifying and predicting the effects of habitat fragmentation on wildlife populations. This is an important component of habitat—the sizes and juxtapositions of different patches. The extreme in the process of fragmentation is represented by forest remnants ("habitat islands") in the eastern United States, where patches exhibit a wide range of sizes, but all are more or less surrounded by cleared land now used for agriculture or urban development. Most forests of the western United States have not yet been so extremely modified. They represent, instead, the early beginnings of a process of habitat fragmentation, such as that begun in the eastern United States nearly two centuries ago. The contrast between eastern and western conditions gives us an opportunity to examine the process at two contrasting points, one near its beginning, the other much further along. Appropriate questions include (1) In what ways can island biogeographic theory be fruitfully applied to fragmented landscapes, and in what ways is it questionable or clearly wrong to do so? (2) What methods can be used to study efficiently and accurately the questions of patch size, interspersion of patches of different types, and the relationships between the diversity of patch types and the diversity of wildlife species in an area? (3) Can we establish general guidelines for the minimum patch size needed to maintain viable populations of selected species? (4) How does the provision of corridors of suitable habitat influence the minimum patch size required? (5) How should the dispersal behavior of a species affect a manager's decisions about patch size, patch interspersion, and the availability of corridors? Most of these questions are addressed, though not answered, in this section.

Part V—Linking Wildlife Models with Models of Vegetation Succession—represents the ultimate goal of most efforts to predict the effects of habitat change on wildlife species. Accurate linkage of wildlife and succession models would permit long-term forecasting of the effects of major land-management activities on wildlife. Several models now in use for this purpose are described in this section. They represent independent efforts to achieve the same goal, and, not surprisingly, they exhibit considerable convergence. Major issues center on (1) the assumptions of the various models and the extent to which biological systems meet those assumptions; (2) the accuracy of the models, both for wildlife and plant succession; (3) the cost-effectiveness of this approach; (4) the current status of testing and application of the models; (5) the suitability of various models in different situations; and (6) whether just a few, many, or all wildlife species should be modeled. Several chapters in this section show how managers are using models today to provide input into project planning. It is encouraging to learn from these that a more comprehensive and quantitative approach to evaluating project impacts on wildlife resources has resulted in greater consideration for wildlife in the planning process.

However inadequate the various models and their applications may be today, and regardless of how scanty our data may be to feed them, it is clear to many practitioners that models linking our knowledge of plant succession and wildlife-habitat relationships have enormous promise for the future of a diverse fauna on managed lands everywhere. To the extent that the frontiers explored by researchers are always ahead of developed technologies, and that those technologies are always ahead of their applications, this book is already out of date. It is the sincere wish of all of us involved in its making, however, that it will serve as a secure and multifaceted foundation for an accelerated effort to upgrade our knowledge and the state of our models to the point where we can better meet our goal of wise resource management. Although our efforts may be primitive, we have at least begun. We believe it was Lincoln who once said, "If we never try, we shall never succeed."

Acknowledgments

The symposium that provided the foundation for this book would not have occurred without the financial and logistical support of sponsors listed previously. But equally important were the persistence and dedication of individuals who took responsibility for key tasks. Section and volume summarizers and our board of consulting editors did yeoman service in reviewing and evaluating manuscripts. As chairman of the steering committee, Mark Dedon rode herd on all phases of planning and implementation. He also developed the computer software used to manage registration, room assignments, fees, and so on. Reginald Barrett served as chairman of the local arrangements committee. Stephen Laymon organized field trips and assisted with many other chores needed to make the symposium a success. William Laudenslayer made essential contacts with government agencies and private organizations. Lorraine Merkle handled most of the day-to-day business for registration, fees, and so on, maintaining a current computer file of all such information. Sherry Morrison managed room assignments, assembly of registration packets, and registration at the conference grounds. Lorraine and Sherry also spent many hours assisting us with final editorial preparation of the complete draft manuscript. Anthony Gomez assembled all audiovisual materials, including the boards used for mounting posters, and managed the audiovisual equipment throughout the conference. His operation of the slide and overhead projectors for all presentations was flawless and much appreciated. Linda Doerflinger did most of the work needed to compile a booklet of abstracts that was distributed to all persons attending the symposium. Irene Timossi, Ellen Woodward, and Janice Reid assisted with local arrangements, and Lawrence Davis and Robert Heald assisted with field trips. To all, we extend our sincere thanks.

Jared Verner, *Research Ecologist*
USDA Forest Service, Forestry Sciences Laboratory, 2081 East Sierra Avenue, Fresno, California 93710

Michael L. Morrison, *Assistant Professor*
Department of Forestry and Resource Management, 145 Mulford Hall, University of California, Berkeley, California 94720

C. John Ralph, *Research Ecologist*
USDA Forest Service, Redwood Sciences Laboratory, 1700 Bayview Drive, Arcata, California 95521

A Sponsor's Viewpoint

F. DALE ROBERTSON

I'm pleased to be here to welcome you to the Wildlife 2000 Symposium on behalf of the many sponsoring organizations and the symposium organizers who worked so hard to put this together. I especially welcome participants from other countries, who are adding an international dimension to this meeting. Wildlife management is a worldwide issue, and we in the United States are anxious to share progress with you. I understand that five nations are represented among you, as well as 42 universities, 10 federal agencies, 9 state agencies, and 10 private consulting firms. I suspect that the collective knowledge, wisdom, and experience of the people in this room are unmatched in the world of wildlife management and research.

The fact that the symposium is commencing on Columbus Day (and also Thanksgiving Day for the Canadian attendees) causes me to reflect on the forces that were set in motion nearly 500 years ago. When the first Europeans explored North America, the native human population was probably at or near the carrying capacity of the land—probably no more than 5 million persons. Much has happened since Columbus Day, 1492, and we can expect more of the same in the years ahead.

Through improved technologies, the United States supports 230 million people and feeds and clothes much of the rest of the world's population. This has been accomplished through intensive management and utilization of natural resources. Obviously such progress has had a tremendous effect on wildlife habitat. Technological advances have a tendency to be cumulative. They feed back on themselves to the point that change is now literally exploding in every dimension of our lives—political, economic, social, and particularly technological. Humans seem to be able to adapt to the change, though it is often painful for many of us. But what about wildlife?

Undoubtedly human progress will continue at an even faster pace. And the other side of this coin is that we can make bigger mistakes, and faster, than ever before. An ever-growing human population will continue to strive for higher standards of living, and that will put even more strain on natural systems and their flora and fauna. The challenge—or opportunity—for wildlife professionals is to find ways to blend fish- and wildlife-habitat needs into a "system" that is driven largely by other motives. The driving forces of this larger system will not stop while we do our research, refine our models, and go through all the usual gyrations that normally lead us, as natural resources professionals, to acceptable technical answers.

Fortunately we already know a great deal about wildlife and their habitat relationships. We have made much progress on this front, and we will make more. It is imperative that we take this knowledge and interject ourselves into the mainstream of resource decision making with the most persuasive information that we can muster. And we must never forget that the time schedule is set by someone else. Land-management decisions that affect wildlife resources *will* continue to be made by someone; I believe everyone would agree that those decisions should be based on input from wildlife specialists using the best information available at the time.

Developing and using wildlife-habitat-relationships models is one of the best ways to effect favorable decisions for wildlife. Proceedings of this symposium should serve as a platform from which we can be more effective with the current state of knowledge and reach new levels of expertise and success in conserving wildlife as a continuing part of our rich natural heritage.

F. Dale Robertson, *Associate Chief*
USDA Forest Service, P.O. Box 2417, Washington, D.C. 20013

Wildlife-Habitat Modeling—Cheers, Fears, and Introspection

JACK WARD THOMAS

I have a pecular mixture of reactions to the prospect of a conference on the state of the art in modeling wildlife-habitat relationships. Perusal of the program and the abstracts of the papers to be presented leads me to (1) cheer the progress that is being made toward understanding and describing wildlife habitat through modeling; (2) fear for our ability to mold these modeling approaches into tools comprehended by and useful to practicing resource-management professionals; and (3) become a bit introspective about where we are going with habitat modeling.

The application of modeling in forest management will serve throughout this paper as an example, but the principles and generalities could apply to other areas of natural resources management as well. How did we come to our present status in regard to habitat modeling? I conceive of this best in terms of an analogy involving watersheds. From these watersheds emerge streams called biology, politics, applied mathematics, and computer technology.

Studies in the realm of biology started with descriptive biology of individual species, then moved on to the examination of interspecific relationships, and from there to consideration of the entire community (i.e., ecology). This led to the recognition, as Barry Commoner (1971) put it, that (1) everything is connected to everything else; (2) everything has to go somewhere; (3) there is no free lunch; and (4) nature knows best. As these insights developed and gained widespread acceptance, pressure increased to see this knowledge applied to natural resources management.

At the same time, over in the political watershed, there was developing a tradition of "conservation" defined as "wise use." This movement emerged in forestry as a philosophy-policy known as "multiple-use." This has been called an "era of happy platitudes" filled with ethically loaded ambiguities (Zivnuska 1961). But the underlying assumption that a variety of products could somehow be simultaneously produced from appropriately managed forests was embodied in law with passage of the Multiple–Use Sustained Yield Act of 1960. At that time, there was, when viewed in retrospect, a certain naive reliance on the assumptions that people of my generation were taught in our professional training—i.e., "Good timber management is good wildlife management (or whatever)." Those who taught and those who learned turned those assumptions into articles of faith. After all, compared with "cut-out and get-out" practices that were all too common in the 1800s and early 1900s, these statements were relatively true. After 1960, there was increasing recognition that forest resources had limits in terms of availability and that forests, and other natural resources, should be rationally, carefully, and methodically exploited in a way that provided for multiple-use, protection of the resource base, and some degree of efficiency. The Forest and Rangeland Renewable Resources Planning Act of 1974 and the National Forest Management Act of 1976 marked the confluence of the biological and political streams.

As these streams intermingled, platitudes and assumptions began to give way to coldblooded analysis. It became increasingly obvious that, indeed, there was no free lunch. "Trade-off" emerged as the buzzword of the day. To meet the requirements of new laws and regulations, while making some sense of this puzzling and sometimes conflicting admixture of biological realities and political requirements, the natural resources planners moved unobtrusively from the shadows to the forefront of agency activites.

And, along with those planners, another stream entered the increasingly murky political flow. That stream emerged in full flow from two watersheds on the other side of the technological divide. In one, applied mathematics—including biometry, econometrics, and systems analysis—made advances. In the other, computer technology and data-handling techniques were developing. When the mathematics and computer streams joined, amazing things happened. Mathematical techniques and theories that were sound, but were unbearably cumbersome to apply, suddenly became applicable in the everyday world of biology, business, and natural resources management.

Within the past decade these four streams—biology, politics, mathematics, and computer technology—have intermingled to form a turbulent new river—a river not seen nor navigated until recently. But it is, I suspect, like the mighty Salmon—a River of No Return.

The name of this conference implies just that. "Wildlife 2000" clearly forecasts that the conference organizers believe that the discussions that will occur here are harbingers of what much of wildlife science and management will be in the year 2000—the modeling of wildlife-habitat requirements and the application of those models in land management.

Wildlife biologists were late, compared with some other resource-management professionals, in the common use of modeling. The first generation of resource-management plans, primitive as that art was then, developed to meet requirements of the Resources Planning Act, but did not benefit greatly from contributions of wildlife biologists. Aware of the inadequacies of those plans in their consideration of wildlife and wildlife habitat, biologists looked for reasons for those inadequacies and found some in perceived agency bias toward and disproportionate budget allocations for commodity production, and in shortage of personnel, among others. But when rationalizations were exhausted, it was still obvious that part of the problem was that wildlife biologists were, in general, ill disposed and poorly equipped to play effectively in planning and allocation games that required the expression of information and concepts in the form of models.

Although much information was available on the relationships of vertebrate wildlife species to habitat conditions, it was diffuse, diverse, and difficult to consider in land-use planning. The middle and late 1970s saw the first really effective efforts to organize and integrate available information on wildlife and habitat. Those efforts seem, as their developers fervently hoped they would do, to have triggered a dramatic increase in attention to how wildlife and its habitat can be modeled and used to predict effects on wildlife of changes in habitat over both short and long terms. The really important contribution of those early efforts was not the information presented or the insights, procedures, and formats produced. It was the recognition that it was possible for wildlife biologists to participate effectively in navigating the River of No Return.

The papers to be presented at this conference are an indication that wildlife biologists have come far in their ability to express knowledge about the relationship of wildlife to habitat in mathematical terms or in forms that can be considered in forest planning and management or both. Wildlife biologists are indeed making progress in that regard—more rapid progress than seemed possible even 10 years ago. Yet most recognize that this is just a beginning.

Those are the cheers—now for the fears. My greatest fear concerning modeling and its application to land management is best illustrated by the following story. Several years ago, I sat in a room filled with forest supervisors, district rangers, and assorted support staff of various disciplines. We were there to discuss the results of just completed analyses of timber harvests that could be anticipated under several combinations of constraints (financial, wildlife, soils, and so on). The tool that had been used to evaluate the situation was linear programming. The chief planner and his staff gave an excellent presentation of the results of the various analyses. But the presentation centered on results. Everyone in the room was impressed. There was no presentation of or questions about the procedures that produced the "answers" that would now be discussed. I was a bit befuddled and uneasy, but assumed that I was the only one—everyone else seemed at ease and confident. The discussion started. Obviously, projected outcomes were accepted by the audience. The ensuing discussion centered on things other than the procedure itself—personnel, staffing, political ramifications, the law, and time schedules.

I tried to engender some discussion about the assumptions of the linear programming model, the confidence limits around each of the variables considered, and so on. That caused a pause in the conversation, but it was only a brief pause. The conversation quickly returned to budgets, schedules, and deadlines. It slowly dawned on me that only three of the 21 persons in that room knew much, if anything, about the procedures that produced the information on which those participants were basing significant decisions. I felt a bit nervous. Things have improved since, but not much.

The moral of the story is not that administrators should suddenly become experts on the ins and outs of modeling. I do believe that people who produce and use models have the responsibility to explain processes, assumptions, strengths, and weaknesses to those who make decisions based on those models. Those who evaluate the results and make decisions have the responsibility to insist that those who produce and use the models explain what's going on—in plain English. Users should understand (at the very least) the assumptions inherent in the model(s); the sources and the variability associated with the data used; and the level of confidence that can be placed in the results. Blind acceptance of modeling results from the bowels of the computer can be as irrational as reliance on the honored and ancient skills used by the oracles in deciphering messages in the entrails of a sacrificial chicken.

Another fear concerning wildlife-habitat modeling is the "cubing problem.'' That is, we have a disturbing tendency to take rather simple observations about the relationship of wildlife to its habitat and endow, with little justification, that information with the appearance of more and more precision. Those data, with their newly enhanced precision, are then entered into a formula, replete with coefficients, that produces a numerical answer with several digits to the right of the decimal point!

A best guess (also called a best simulation or B.S.) concerning the relationship of wildlife to its habitat is an essential contribution to modeling the anticipated results from forest planning. There is, however, a constant pressure and temptation induced by the pressures of planning to go farther—to cube B.S. $B.S.^3$ can be a real problem.

Obviously there are extreme pressures—from the requirements of the law, from superiors in the agency hierarchy, from peers, and from ''competitors'' in the planning game who represent other interests—to provide more and more detail. One is sometimes reminded, when observing different specialty groups compete in producing and using models, of Shakespeare's line from *Macbeth* where Macbeth hurls the challenge: ''Lay on, Macduff, and damned be him that first cries, 'Hold, Enough!' ''

Yet, there is a time to cry out ''Hold, enough!'' Just where that point is, just when the exponent on B.S. becomes too large, is not clear. There seem to be no rules.

As an old white-water river runner, I know there are two ways to steer a drifting craft in fast-moving water. The oarsman either goes faster than the current (as with a canoe) or slower than the current (as with a raft or drift boat). My observations indicate that more canoes get wrapped around rocks than do rafts. I'm suggesting that perhaps we need to slow down a bit and steer a bit more carefully and methodically.

Models are merely the means to an end. They are not the end in and of themselves. Models are a formalized way of guiding adaptive management of our natural resources—no more, no less. Remember John Krebs's (1980:411) observation that ''a model is not meant to represent the complexity of Nature but to capture the essence of a phenomenon''; and Frank Egler's (Jenkins 1977:43) comment that ''ecosystems are not only more complex than we think, they are more complex than we can think.''

In the days of the Roman Empire, returning conquerors were given a triumphal parade. The conqueror rode ahead of his legions in a chariot to receive the accolades of the citizens of Rome. Behind him rode a slave holding a laurel wreath over the conqueror's head for all to see. The slave had an additional duty—to whisper over and over in the conqueror's ear, ''Remember thou art but a man.'' Those who produce, and certainly those who apply, models in natural resource management need to hear such a whisper saying, over and over, ''You are dealing only with an essence of what is—nature seen *through a glass darkly*.'' It is not real—it is but a shimmering image of the moment that will change as the viewer's perspective and need change.

I fear that as we move toward Wildlife 2000 we are apt to become so enthralled with our discovery of modeling and its possibilities that we forget that the vast majority of our colleagues neither understand nor highly value the approach. Most resource-management practitioners have been exposed to only basic statistical analysis. The grand finale of such a level of training in statistics was apt to have been the introduction to analysis of variance. Few have mathematical training beyond algebra and trigonometry. Reference to such procedures as canonical correlation and cluster, principal components, and discriminant function analyses can produce glazing of the eyes that can be matched only by a great gray owl closing its nictitating membranes. We need to keep in mind that what is being discussed at this conference is not yet the mainstream of wildlife biology. Knowing that will help keep our discussions in perspective.

I am told that recently trained wildlife biologists are better equipped than the "old hands" to deal with habitat modeling and its application to management. That doesn't help much in the short run. The old hands occupy the positions of power and influence. And, besides, most of the more recently trained biologists are pursuing careers other than working for state or federal resource-management agencies—most do not even find employment in the wildlife field. That seems unlikely to change in the foreseeable future.

We must ensure, therefore, that currently practicing wildlife biologists receive the training necessary to understand the advantages and limitations of modeling and how to "play the game" in a world of modelers and planners. I am most concerned about biologists who work for state wildlife agencies. Biologists employed by the federal land-management agencies will have at least some opportunity to learn, through training or experience or both, to deal with modeling and the application of modeling. Those federal agencies, for better or worse, are and will continue to be heavily engaged in planning and the modeling that goes with that planning. Wildlife biologists employed by those agencies will have to learn to deal with modeling or they will become ineffective and unappreciated. Their professional welfare and survival depend on what they must know about modeling. Survival is a most powerful incentive.

Those who work for state agencies that deal with federal land-management agencies may not have similar incentives. They will, however, be intimately involved in the creation, critique, and evaluation of land-management plans and management activities. If inability to understand or apply modeling exists and persists among state biologists, it must inevitably lead to their increasing ineffectiveness, frustration, alienation, and conflict when dealing with federal agencies.

We must plan and scheme now for appropriate and continuing training for wildlife biologists so that they can understand and apply modeling. High-priority candidates for such training include biologists—both federal and state—who deal with the management of public lands.

I fear, as well, the hesitation common among many who have the training and skill in modeling wildlife-habitat interactions to become involved in producing information useful to management. I wager that some people who present papers at this conference will add a caveat to their presentation that goes something like this: "I have spent several years and thousands of dollars [probably taxpayer dollars] studying and modeling the habitat associations of species *X* and can explain more of its habitat requirements than any other person. I recommend strongly that these results be viewed most cautiously and considered as preliminary. Additional research is required and more information will emerge in due time. Certainly, this information should not be applied in management." Now, that ought most emphatically to free the author from any possible responsibility for the application to management of the hard-won knowledge.

I have news to share. Managers will take information they need from whatever sources are available. They must, or at least will, make decisions and institute management action regardless of the amount or quality of the information available. The manager is, and should be, the one who decides what information to use and when and where to use it—it is the manager's career that is at risk, after all. Remember, if you publish information it is subject to being used by managers.

I recommend, instead, an attitude on the researcher's part expressed in some such statement as this: "It is an imperfect and uncertain world. But here is some information that you may be able to use. The information is not as good as I would like. I expect to have more and better informaton later. In the meantime, it is the best available. Its strengths are such and such and its weaknesses are so and so."

Now for a bit of introspection. My experience tells me that the natural course of things concerning the use of new technology in management goes through four distinct phases: (1) technology development—the slow, arduous development, often in widely diverse fields, of "pieces" that are ultimately combined to make the new technique; (2) stimulus—this usually takes the form of a first attempt, however crude, to combine the bits and pieces from technology development into a technique or process; (3) chaos—innovators, having seen the promise and possibilities of the new technique or process, leap into the arena with new approaches, innovations, and modifications; and (4) order—in the end order arises from chaos or the chaos exhausts itself, subsides, and dies.

Land-use planning, with all its subdisciplines, is probably the mechanism that demands that order arise from chaos. Further, it is the crucible in which the order produced will be tested.

I predict that those in attendance at Wildlife 2000 will feel by the conference's end that the area of wildlife-habitat modeling—particularly in terms of research—is indeed in chaos. How does order come from what some perceive as chaos? First, let us recognize that some continuing chaos is both inevitable and desirable. Some, perhaps most, researchers are not concerned with order, with synthesis, or with the needs of management. For them, chaos with its stimulation and constantly changing intellectual challenge is "where it's at." If managers need order, they are the ones who must request and encourage the emergence of order from perceived chaos. How can that be done? Let us examine one possible approach.

When biologists attempt to bring together information on species-habitat relationships for a particular area, they are apt to be impressed by the amount of germane information available and bewildered by how to integrate and present the information in some meaningful fashion. The "noise" and "clutter" inherent in the data and in any such system can be stunning, if not overwhelming.

If these modeling approaches and data banks are to help natural resource managers, if they are to contribute to larger data bases or models, if their usefulness is to be maximized, and if their "noise and clutter" are to be minimized, it may be well to develop and test the habitat-relationships models using standardized habitat variables, land units, and relationships. A perusal of the pertinent literature indicates that, for forest habitats at least, this could be done. To the extent possible, these variables should be expressed in a form that is or could be collected by forest-survey crews.

Many studies, and the resulting models from those studies, deal with individual species. Much scarce manpower and money have gone, and will go, into such efforts. That was acceptable, in the beginning at least, for much of that work was to develop approaches and techniques to produce wildlife-habitat models. But we do not have enough time, skilled people, or financial resources available to produce such models for all species, sometimes for several habitats. Obviously, there will be selection of the species to receive attention. This should involve a formal exercise to determine priorities for the species to be addressed in these model-development and testing exercises. For a start, federal land-management agencies should list the species, in priority order, on which such information is desired. These lists, at least, could guide the agency's own researchers and encourage others to work on high-priority species. If funding were forthcoming from those agencies for such efforts, so much the better and so much higher the likelihood of attention to those species. It's a trite but true statement—money talks.

JACK WARD THOMAS, *Chief Research Wildlife Biologist*
USDA Forest Service, Range and Wildlife Habitat Laboratory,
Route 2, Box 2315, LaGrande, Oregon 97850

I

Development, Testing, and Application of Wildlife-Habitat Models

Introduction: Development, Testing, and Application of Wildlife-Habitat Models

KRISTIN H. BERRY

Between 1969 and 1976, the U.S. Congress passed several major acts affecting management of fish and wildlife resources. Among the more important were the National Environmental Policy Act of 1969 (42 U.S.C. 4321–4347), Endangered Species Act of 1973 (16 U.S.C. 1531–1543), Forest and Rangelands Renewable Resources Planning Act of 1976 (16 U.S.C. 1601–1610), National Forest Management Act of 1976 (16 U.S.C. 1600–1614), and the Federal Land Policy and Management Act of 1976 (Public Law 94–579). These acts ushered in a new era and stimulated federal and state agencies to change their management of fish and wildlife resources. In response, wildlife biologists and managers first sought simple, rapid, and reliable methods to determine and predict the species and habitats present on lands within their jurisdictions. Their second step was to expand the data base for species and habitats, especially those identified as rare, threatened, endangered, or of special significance. Finally, managers attempted to predict the effects of various land-use actions on the species and their habitats.

Since the mid-1970s, government personnel and academicians have developed several techniques to accomplish the tasks. I group the techniques into three categories: (1) single-species models; (2) multiple-species or community models; and (3) habitat-analysis models. Single-species models include simple correlation or presence/absence models, statistical models, Habitat Suitability Index (HSI) models, Habitat Capability (HC) models, and Pattern Recognition (PATREC) models. Simple correlation models are common in many government environmental analysis and impact documents. Species may be displayed in matrices by the successional stages of plant communities or habitat types (see, e.g., Thomas et al. 1978; Thomas 1979; Verner and Boss 1980; Nelson and Salwasser 1982). Unlike simple correlation models, statistical models permit prediction of distribution and abundance. For example, Robbins (1978) used stepwise regression techniques to examine vegetative and environmental factors influencing distribution and abundance of birds on 80 census plots in deciduous and mixed woodland. He found significant correlations for at least five environmental factors among eight of 20 bird species tested. Turner and Medica (1982) used a similar method to predict abundance of a rare species, the flat-tailed horned lizard (*Phrynosoma mcallii*). Clawson et al. (1984) also used the technique to determine which habitat characteristics accounted for the most variability for 10 species of reptiles and amphibians occurring in old fields and upland forests.

The HSI models, developed by the U.S. Fish and Wildlife Service (Fish and Wildlife Service 1981a), have received more attention to date than other single-species models. The HSI is based on assessment of the physical and biological attributes of habitat for a particular animal species and is assumed to be proportional to carrying capacity. As of September 1984, the Fish and Wildlife Service had published models for 85 species in its HSI report series, and more are anticipated (A. Farmer, pers. comm.). Other agencies are also producing HSI models.

Similar to the HSI model is the Habitat Capability (HC) model, which was developed by the USDA Forest Service. The HC models are still evolving, but have been used to describe habitat conditions associated with or necessary to maintain different population levels of a species (Hurley et al. 1982). Sheppard et al. (1982) employed the HC model with a Habitat Capability Coefficient (HCC) to predict species' abundance. The HCC is an aggregated, weighted value based on habitat-capability ratings for each successional stage of vegetation used for reproduction, feeding, and resting.

The final single-species model is PATREC (Williams et al. 1978), which uses a set of hypothetical conditional probabilities for analyzing habitat for a species. Williams et al. (1978) chose the pronghorn (*Antilocapra americana*) as an example.

Multiple-species or community models are represented by (1) the U.S. Bureau of Land Management's Integrated Habitat Inventory and Classification (IHICS) System (Bureau of land Management 1982); (2) the Forest Service's life-form system (Thomas 1979); and (3) community guild models (Severinghaus 1981; Short 1983; Verner 1984). The IHICS, a system of data gathering, classification, and storage, has no capacity for predicting how animal species use habitat or how changes in habitat will affect a species. In contrast, the life-form and guild systems cluster species with similar habitat requirements for feeding and reproduction. The guild system can be used to estimate responses of species to alteration of habitat.

Habitat-analysis models overlap with some single-species models noted above, e.g., IHICS, statistical models, and PATREC. Other habitat-analysis models currently receiving attention are (1) the Wildlife and Fish Habitat Relationships (WFHR) program (Nelson and Salwasser 1982); (2) Habitat Evaluation Procedures (HEP) (Schamberger and Farmer

KRISTIN H. BERRY: Research, Studies, and Monitoring, U.S. Bureau of Land Management, California Desert District, 1695 Spruce Street, Riverside, California 92507

1978; Fish and Wildlife Service 1980a, 1980b); (3) simulation models, such as DYNAST; (4) optimization models, such as FORPLAN; and (5) economic-analysis models (Fish and Wildlife Service 1980c). See Nelson and Salwasser (1982) for a brief review.

Few models have been tested. The work of Lines and Perry (1978) is an example of a case in which the model was first described and then tested. Seitz et al. (1982) compared three habitat-evaluation procedures, the PATREC model (Williams et al. 1978), Suitability Index (SI) model (Giles 1978), and personal opinion approach. They found that PATREC and the SI models performed better than the personal opinion approach, but neither PATREC nor the SI models performed particularly well. Although dozens of HSI models have been constructed (primarily using the literature and opinions of professionals), few have been tested. Of the tested models, one required some refinements (Lancia et al. 1982), and several others were almost useless (Cole and Smith 1983). Applications of HEP for land-use planning (e.g., Rhodes et al. 1983; Urich and Graham 1983) are another source of concern because HEP usually relies on untested HSI models for several species. Validation, invalidation, and verification of models are critical processes and must receive more attention (Farmer et al. 1982; Marcot et al. 1983).

As of mid-1984, we have seen the promulgation of numerous species, community, and habitat models. The development of new models, some testing and modification of existing models, and increased use of modeling for land-use decisions is underway. On a smaller scale, biologists are reevaluating and reexamining model assumptions, e.g., the relationship between population density and habitat quality (Van Horne 1983), as well as problems with multivariate statistics (Johnson 1981a). The increased number of wildlife-habitat models being published in refereed journals also is indicative of the maturing process for this fledgling field.

Chapters in Part I will advance our knowledge and understanding of models significantly and offer major contributions to the literature. Several new types of species and habitat models are described, and numerous existing and new models are evaluated and tested. Several authors give warnings about model assumptions and applications. Other chapters describe models for predicting presence and abundance of species, for analyzing impacts, and for assisting with land-use decisions. The subjects cover the gamut of mesic wildlife habitats, from marshes and prairies to oak and coniferous forests. The range of species is considerably more limited, however. Lower vertebrates are poorly represented. Birds are obviously the group of choice for study.

The range of participants is another important and valuable feature of this work. Authors include academicians engaged in theoretical research, field-oriented wildlife biologists and managers, and land-use managers. The chapters represent an enormous scale of coverage, from the simple to the highly complex. In general, chapters describing the development and testing of models were prepared by academicians and wildlife biologists, whereas those focused on applications were written by wildlife and land-use managers. Although the different types of authors approach the problems of modeling with different viewpoints, they should strive to communicate with each other and to bridge the gaps between model theory and development, testing, and application. Improved communication, patience, and understanding are essential to advancing the state of the art. Part I offers a tremendous opportunity for growth and development of wildlife-habitat modeling, as well as enhanced communication of participants.

1

Concepts and Constraints of Habitat-Model Testing

MELVIN L. SCHAMBERGER and L. JEAN O'NEIL

Abstract.—Habitat models used in land-use-planning studies are based on assumed relationships between the animal and its environment, and between habitat suitability or capability and some measure of the animal. Attempts to validate habitat models have had mixed results. Some of the problems are related to the delineation of the components of habitat as contained in the model, and others are related to the difficulty of obtaining an independent measure that can serve as a suitable standard of comparison. This chapter briefly presents the conceptual basis for Habitat Suitability Index (HSI) models being developed by the U.S. Fish and Wildlife Service, discusses examples of model validations, and proposes methods for testing models as they relate to land-use applications.

The use of quantitative habitat models for fish and wildlife planning is relatively new, but is increasing rapidly for inventory, impact-assessment, mitigation, and wildlife-management studies (Thomas 1982; Urich and Graham 1983). Models provide a needed tool for studies where habitat is to be emphasized in the natural resources planning and decision-making process. Many types of models are collectively referred to as "habitat models," and most of these are used in an attempt to record or predict a species' response to its environment; this response may be described as occurrence, physiological condition, abundance, or other response of interest to the model user. The response of interest then becomes the model objective, and differing objectives must be recognized in model construction and testing.

Habitat has been selected as the basis for modeling efforts to assist in planning studies because habitat provides an integration between concepts of population and carrying capacity, and it can provide a consistent basis for baseline, impact-assessment, mitigation, and monitoring studies (Fish and Wildlife Service 1980a, 1980b). Habitat Suitability Index (HSI) models are being developed and used in the context of determining habitat quality (Fish and Wildlife Service 1981a). Such models usually are consistent with data needs for planning studies; they are fairly simple, can be applied in a timely manner with minimum cost, and the outputs are easily understood. The reliability of these planning models is not as high as we would like, yet alternatives have not been developed that are widely accepted. Science and planning are two separate realms (Romesburg 1981). Science requires high confidence levels, precision, and certainty. The standards of science are not applied to planning decisions that merely require that decisions be made, using the best available information. How can we integrate the concepts and rigor of science into the realm of planning? Perhaps habitat models can provide a link.

HSI model validation is difficult for several reasons. There are no standard methods for defining or measuring habitat quality. We often lack quantitative habitat data from which to develop models and data that exist are not available in a consistent format. Further, the models are developed around the concepts of habitat and carrying capacity, terms that have no commonly accepted definitions and that are difficult to quantify. Conversely, models are often tested around the concepts of population dynamics and habitat use, and testers have not always recognized the conceptual differences between habitat and population models.

This chapter provides a brief discussion of the conceptual relationships among habitat, carrying capacity, and habitat models as they relate to model testing. The focus is primarily on HSI models, although the concepts, constraints, and suggestions for testing are generally applicable to most types of habitat models. Several approaches to model testing are presented, with the overriding caution that models must be tested and applied within the context of model objectives and their intended use.

Habitat model concepts

Models are simplifications of the systems they depict, and they lose resolution in attaining simplicity. They require numerous assumptions and can never completely mimic the real world (Maynard Smith 1974; Hall and Day 1977). If the simplification is done properly, most system dynamics are preserved (Maynard Smith 1974) and extraneous information is removed from further consideration (Overton 1977). Models provide a framework around which qualitative habitat information can be structured for decision making and can structure qualitative and quantitative relationships into testable hypotheses.

Habitat models are based on the concepts of habitat and carrying capacity. Both can be defined either narrowly or broadly (e.g., Edwards and Fowle 1955; Pianka 1974; Giles 1978). Narrow definitions of habitat strictly limit the number

MELVIN L. SCHAMBERGER: Western Energy and Land Use Team, U.S. Fish and Wildlife Service, 2627 Redwing Road, Fort Collins, Colorado 80526

L. JEAN O'NEIL: Waterways Experiment Station, U.S. Army Corps of Engineers, P.O. Box 631, Vicksburg, Mississippi 39180

and type of considerations, such as the basic food, cover, or physical habitat attributes necessary for survival. Broad definitions include other factors, such as competition or abiotic interactions. The principal difference among definitions of carrying capacity is the number of factors perceived to be critical in determining population limits. The most comprehensive view is that carrying capacity is a function of all factors that interact to limit populations, including food, predators, inter- and intraspecific competition, disease, mortality, natality, weather, and habitat. Thus, habitat is not the only factor that determines animal presence or abundance, and habitat models may include only a few of the factors that determine population levels (Fig. 1.1). It is important to note that each habitat model contains restricted, operational definitions of habitat and carrying capacity, and those definitions must be considered when designing model tests.

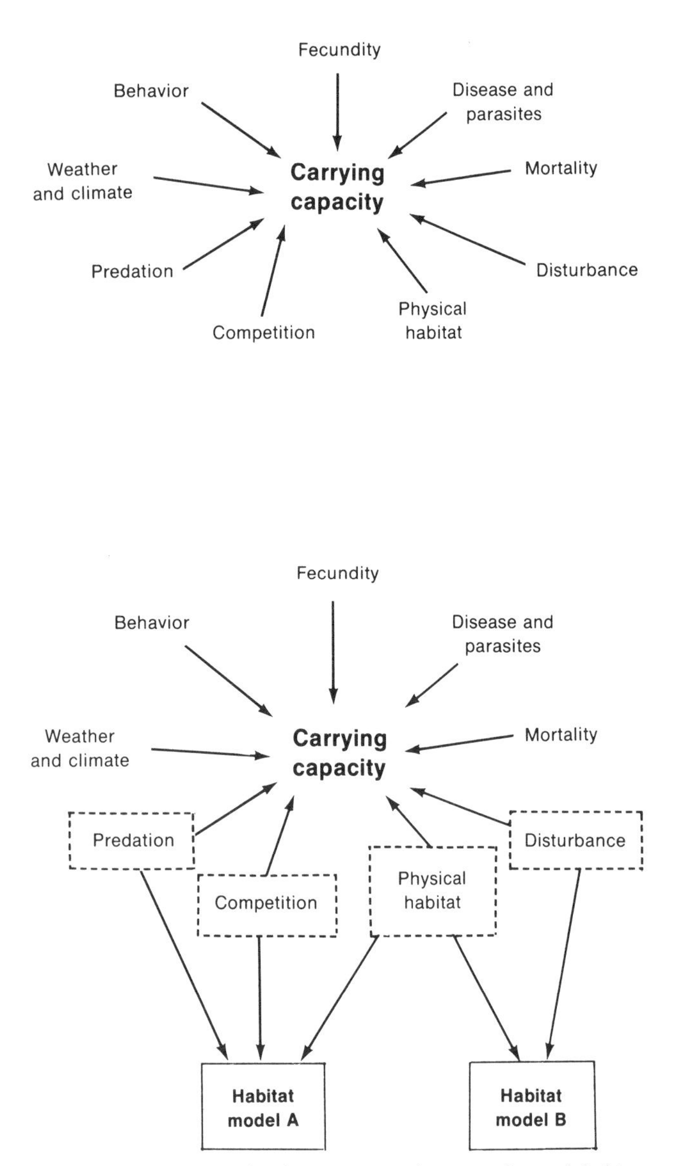

Figure 1.1. Relationships between carrying capacity and habitat models.

HSI model concepts

HSI models are specifically designed for use in planning and environmental impact assessment studies. They are meant to quantify habitat quality and to permit wildlife resources to be considered along with other aspects of project planning, such as engineering or economics. They are not models of carrying capacity because not all factors that influence animal abundance are included. They are constrained to basic habitat attributes thought to be important both to the wildlife species and to specific planning or management needs. HSI models are designed for use in situations where land use and, therefore, habitat condition are expected to change; they are intended to allow assessments of resultant changes in potential habitat quality and availability for selected wildlife species.

Variables included in the models are limited to those (1) to which the species responds; (2) that can be measured or estimated readily; (3) whose value can be predicted for future conditions; (4) that are vulnerable to change during the course of the project; and (5) that can be influenced by planning and management decisions. Many variables that are known to influence animal populations are excluded from HSI models if they cannot be readily measured (predation), managed (weather), or predicted for future conditions (competition). The result is a model that has a very restricted operational definition of habitat for a specific land-use study and for a specified geographic area.

Many characteristics of vegetation structure meet the above five conditions, e.g., basal area of mast-producing trees or density of the herbaceous layer. Work on multiple species such as the work of Elton and Miller (1954), MacArthur and MacArthur (1961), Hildén (1965), and Wiens (1969) has been important in forming the HSI concept, whereas more recent studies such as Nixon et al. (1980) have contributed to individual species models. In addition to vegetation structure, variables related to floristics (e.g., abundance of preferred foods), interspersion (distance to cover), landform (presence of standing water), and similar factors often meet the above conditions.

The HSI is determined through aggregation of one or more Suitability Index (SI) scores for life-requisite components, such as winter food. By definition, the SI and HSI provide a 0–1.0 index of habitat suitability. The HSI for a species at a given site is not intended to predict population levels, but an HSI of 0.8 should indicate better habitat quality than an HSI of 0.4 and should represent greater *potential* carrying capacity.

In summary, HSI models are not research models, carrying capacity models, population predictors, or comprehensive. HSI models are practical, operational planning models; designed to assess impacts of change; and based on a narrow definition of both habitat and carrying capacity. They also provide a bridge between the fields of planning and science, in that science is used to improve model performance in planning activities.

Model testing

Model testing serves two important purposes: (1) to provide information about model performance and reliability in specific applications; and (2) to provide data that can lead to model improvement for both the model tested and for similar models. This chapter focuses on the design of tests that will provide meaningful model-performance data. Model tests are intended to determine how well a model meets its stated objectives; therefore, if model objectives are not clearly stated, tests will be nearly impossible to design and results often will be inconclusive. Tests must be developed in consideration of the planned use of the model and should include evaluation of the variables in that model.

Results of model tests must be interpreted in the context of how the model will be applied, because a model may be adequate for one type of study and inadequate for another (Bella 1970). Marcot et al. (1983) supported this concept by referencing several criteria, including precision, accuracy, realism, and resolution. For this reason test results should report the level of confidence or significance attained; a model that performs poorly at the 95% confidence level may be acceptable for some types of operational studies if it performs well at the 75% confidence level.

One of the biggest problems we have encountered in HSI model-testing efforts is that of designing tests and locating either test sites or existing data that are consistent with model content and purpose. Because each model defines habitat in a slightly different manner, there is no standard approach to model testing. Each of the methods discussed below can be used, under certain circumstances, as the basis for model testing. However, considerable forethought is required to select the appropriate method. Habitat-use data, for example, may be appropriate in one situation, but inappropriate in another. Other considerations in test design are listed in Table 1.1 and discussed below.

Table 1.1. General considerations for HSI model testing

1. Identify model objectives, purposes, and performance levels before designing test. Model tests must be consistent with model objectives and model resolution.
2. Test models against a real-world measure of a response by the animal.
3. Long-term, multiyear data for both habitat variables and animal response should be collected, using stringent quality control of field data.
4. Models should be tested against data sets other than the ones used for model development.
5. Study design should use concepts of change.
6. Tests using exploited species require additional care in data analysis.
7. When variables not included in the model (e.g., predation and competition) are known to be limiting to the test population, correlations between model output and animal numbers probably will not occur.
8. It usually is preferable to test model variables and establish their validity before proceeding to tests of overall model output.
9. Shortcomings of the test data should be fully determined and recognized in reporting test results. Poor correlations between density and model output should not automatically lead to the conclusion that the model is incorrect. The data set may be inaccurate or inappropriate as test data.
10. Large sample sizes are essential; small sample sizes are difficult to interpret and may lead to erroneous conclusions.
11. The standard of comparison should be reflective of habitat quality.
12. Test sites should include the entire range of habitat quality.
13. Test sites should be several times the average home range size of the species being tested.
14. Understand the model before testing it.

Levels to test within a model

One of the first decisions in designing a model test is to determine what part of the model to test. Hypotheses can be formulated at any level (Fig. 1.2) within the model, including tests of assumptions (level A); variables (level B); components (level C); or overall output (level D). Nearly all HSI model tests to date have been at level D, which often provides little information for improving model performance because actual information about the species' response to changing habitat variables is lacking. Overall model performance can be evaluated, however, allowing earlier use of the models in the planning process. Much of the research on wildlife-habitat relationships in the last few years can contribute to tests of assumptions and variables and is a valuable source of information for model construction.

Testing individual model variables (level B) or assumptions (level A) is one approach that provides information for determining and improving model reliability. For example, the species depicted in Figure 1.2 feeds on insects in the tree canopy, but measurement of insect abundance in the tree canopy is difficult. A surrogate measure, foliage density, was substituted for insect abundance in the model, on the assumption that increased amounts of foliage would increase insect habitat, thereby increasing insect abundance. This assumption is a testable hypothesis and was shown by Blenden (1982) and Blenden et al. (Chapter 2) to be correct. Testing at this level also benefits other models that use the same assumption. Level B tests evaluate the relationship between an individual variable and a response by the species (see, e.g., Scott and Oldemeyer 1983), whereas interactions among variables can be tested at level C, in which sensitivity analysis is essential.

The most comprehensive tests should proceed from a testing of assumptions at level A before proceeding to level B, and so forth throughout the model, with the level D test occurring last. Such a thorough test of a model is time-consuming and costly (Caswell 1976b; Overton 1977), but may provide the most productive results in terms of model improvement.

Tests of change

The most logical test is one that evaluates the model within the conditions and purposes for which that model is intended to function. In the case of HSI models, these are planning situations. Because HSI models are designed to help assess habitat changes anticipated from alteration of environmental features, the best HSI model tests are those that evaluate the model or model variables under conditions

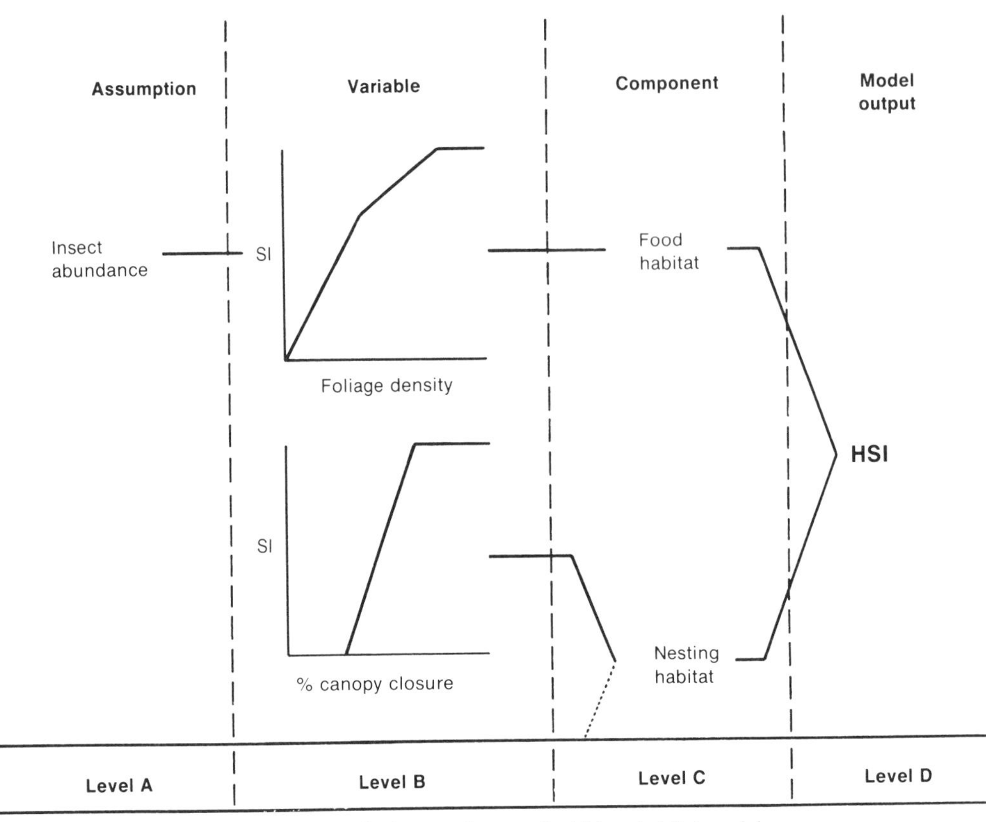

Figure 1.2. Levels that can be tested within a habitat model.

in which variable values are changed. In such studies, variations in the animal's response (e.g., changes in population levels or reproductive success) can be predicted and directly related to differing values of the variables.

Tests employing change can use a variety of test methods; the experimental design of the test is the critical element. There are three general categories of tests of change. The first includes laboratory or field tests where, under controlled conditions, different variable values are related to animal response (e.g., Wiens and Rotenberry 1981b). For example, grass composition and height might be manipulated and the response by nesting waterfowl determined.

The second category encompasses situations in which one uses different existing vegetation conditions to represent change in the values of interest. For example, O'Neil is presently testing a pine warbler (*Dendroica pinus*) model on sites that display a range of variability in tree density, one of the model variables. Current work on vegetation-succession models (e.g., Smith et al. 1981a; Raedeke and Lehmkuhl, Chapter 53) is expected to improve the sophistication and usefulness of this type of test. An alternative approach is to select sites with a range of values for the animal response (e.g., sign, density) and then compare observed scores with scores predicted by the model. A drawback to tests that simulate change is that usually there is insufficient control to determine cause-and-effect relationships and variable interactions, and extreme care must be used in data interpretation.

The third category of tests of changes uses model application and evaluation in operational settings where land-use changes are proposed and habitat gains or losses are predicted. The changes are implemented and species' responses over time are monitored and compared to model predictions. We advocate greater use of this type of test.

Standard of comparison

Once the structure of a test has been determined, an appropriate standard, such as habitat use, density, or condition factors, must be selected for comparing the output of the model or model component. The standard of comparison is selected after consideration of characteristics of the species; model variables, objectives, and assumptions; field conditions and season; data used in model construction; and the reliability of the potential standard.

Theoretically, a model should be developed and tested using two or more different sets of the same type of data. Realistically, most terrestrial HSI models are developed from several sources using different types of data, such as habitat preference, reproductive success, or lethal condition (e.g., winter temperature). The model, by use of suitability-index graphs (Fig. 1.3), converts these data to a common 0–1.0 index format. However, when model output is tested, often only one standard is used for comparison; inadequate consideration is given to matching the test data with the data used for curve and model development. For example, if lethal-condition data were used to develop a model, perhaps

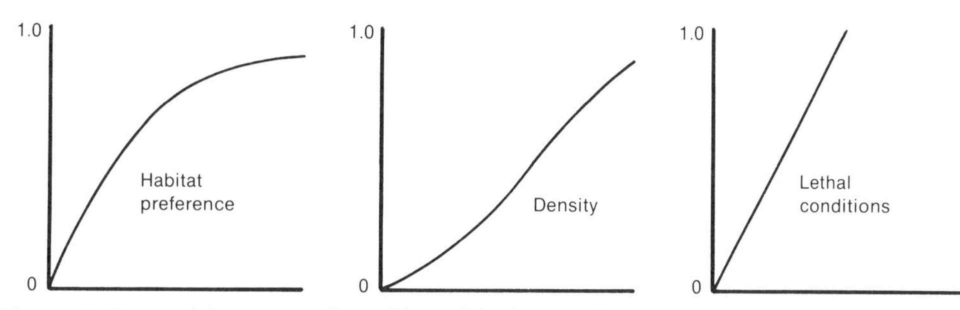

Figure 1.3. Types of data commonly used in models that must be converted to a 0–1.0 scale for model output.

presence/absence data could be used as a test statistic, whereas continuous-measurement data such as animal density might not be appropriate. Conversely, a model designed to assess potential presence/absence conditions should not be expected to predict population densities over a wide range of habitat values.

There is no simple solution to this problem, although we can urge our colleagues to gather and report species' habitat data in a common and consistent format that can be translated into suitability curves for habitat models. In the interim, it is incumbent upon model builders to identify clearly the type(s) of data used as the basis for curve development. In this way, model testers will know the assumptions used in the model and can design tests accordingly. Until consistent data are available, we may need several models for the same species, each based on different types of habitat data and designed for a particular use. The tests must recognize these differences in models.

Tests addressed in this chapter are those that relate model output to measures of the "real world" and can be categorized into (1) habitat-use information; (2) abundance data, such as standing crop or population indices; or (3) measures of well-being, such as growth, reproductive success, or endocrine levels. Regardless of the measure used, the measure itself must reflect habitat quality; otherwise that standard of comparison is inappropriate. Model tests are complicated because the independent measures used to define habitat quality have unique sets of assumptions that create problems in data use, reliability, and interpretation (see Miller 1984).

Habitat use

Habitat-use data document the species' use of or preference for particular areas within its range. It is assumed that (1) a species will select and use areas that are best able to satisfy its life requirements; and (2) as a result, greater use will occur in higher-quality habitat. Numerous techniques are available to assess habitat use, including direct observation, animal tracking, call count, and radio telemetry. For example, Lancia et al. (1982) used radio-tagged bobcats in the Southeast to determine whether habitat use by tagged bobcats correlated with HSI model predictions. The two assumptions mentioned above are not always valid, because factors other than habitat characteristics may affect an animal's use of a site or our perception of that use. The researcher must demonstrate that a measure of use and habitat quality are related. Further, species that have low population levels unrelated to habitat limitations are difficult to test with these techniques (see Part III of this volume). Species with low mobility, specific habitat requirements, and high detectability are more suited to this type of habitat-quality measure. Laymon and Barrett (Chapter 14) discuss some particular problems with this approach.

Animal-abundance data

Animal-abundance data include complete animal counts, indirect counts, and various indices. Three factors may confound the use of abundance data as a standard of comparison: (1) density may not be a function of habitat quality in the study area; (2) density is difficult to measure accurately; and (3) not all factors limiting the population are incorporated in HSI models.

Population levels do not necessarily reflect habitat quality, as Van Horne (1983) showed. Population determinants such as those in Figure 1.1 may override habitat features, or variation in animal numbers may be explained by considering the scale of measurement or stochastic factors (Rotenberry, Chapter 31). Van Horne (1983) provided examples in which density may be higher in low-quality habitat and vice versa because of social interactions. Population levels of many species are often determined at locations or times of the year other than those that are the subject of a model (e.g., Fretwell 1970). Further, point-in-time or short-term population studies reflect only the recent past and may reflect long-term abundance inadequately. Van Horne (1983) recommended consideration of several factors in combination, including density, immigration, and reproductive success, to better link relationships between habitat quality and populations.

The second confounding factor is that the reliability of all population data is often low or uncertain. For example, some individuals or species have varying responses to capture or observation attempts, such as "trap-happy" small mammals or wary small birds. Harvest data are subject to vagaries such as a change in hunting effort, weather, or market prices. Some species experience cyclic changes in population densities, both over seasons and over years, and such cycles are not always habitat-related. In addition, established techniques for gathering population data may be unreliable or may be applied in an unreliable manner. Sources of error include observer ability and consistency, weather conditions, animal detectability, gear efficiency, and other factors (Miller 1984).

The third confounding factor relates to the fact that HSI models do not incorporate all birth- and death-rate factors that influence abundance. Habitat models usually do not include the direct impact that competitors, predators, parasites, prey, or exploitation may have on populations, even though these factors may strongly influence animal abundance. For example, exploitation is not usually included as a habitat variable in HSI models, yet the results of exploitation may reduce population levels, alter standing crop estimates, change behavior patterns, or force animals into or out of specific habitats.

HSI tests using population data include Clawson et al. (1984) in Missouri, Cole and Smith (1983) in West Virginia, and Clark and Lewis (1983) in Georgia. In the latter studies, the authors used data from a single year and found it inadequate, a point well illustrated by Rice et al. (1983b). Clark and Lewis also found a mismatch between variables in the model and the limiting factors; hence, the model did not predict animal density. Relative abundance is best used as test data when (1) a large number of sample sites are included; (2) the data represent long-term abundance; (3) unreliable data can be screened, as, for example, when unusual weather locally affects one site but not others; (4) field methods are consistent over sites, techniques, observers, seasons, and times of day; and (5) the model and test data are for the same geographic area. This discussion is not intended to conclude that density or animal-use data should not be used as model-test data. Rather, it is intended to point out that density or use data alone will not always provide a valid standard of comparison for HSI model testing. Alternative measures could include reproductive success or condition factors.

Measures of well-being

Measures of well-being or condition factors reflect the state or health of an individual or population and are assumed to reflect habitat quality. Van Horne (1983) felt that, in some instances, factors such as reproductive success or mean body weight may be better indicators than density of habitat quality. These measures pertain to vigor, growth, reproductive success, stored energy, or endocrine system products that affect metabolism or other features such as blood chemistry, blood cell counts, serum proteins, and enzyme levels. Measurements of well-being have not been used for HSI model tests to date, but they have been used in wildlife management (Kie et al. 1983); Norman and Kirkpatrick 1981; Weber et al. 1984) and are discussed here to stimulate thought regarding their use as standards of comparison. Measures of well-being or condition factors should be used in conjunction with population data. Additional research is needed to identify parameters that correlate with habitat suitability (e.g., Grue et al. 1983; Weber et al. 1984).

Acknowledgments

This chapter is a condensation of a longer manuscript that has been under development for some time. Individuals that contributed to the thoughts and concepts in the larger document include Adrian Farmer, Carl Armour, Carroll Cordes, Jim Terrell, and Mike Bordy. Reviewers of this manuscript have provided thoughtful comments; we express our appreciation to C. John Ralph, Jake Rice, Bill Laudenslayer, Tom Roberts, and Bill Krohn. Editorial comments were provided by Paul Opler and Cathy Short, illustrations by Jennifer Shoemaker, and manuscript preparation by Carolyn Gulzow and Dora Ibarra.

2

Evaluation of Model Assumptions: The Relationship Between Plant Biomass and Arthropod Abundance

MICHAEL D. BLENDEN, MICHAEL J. ARMBRUSTER, THOMAS S. BASKETT, and ADRIAN H. FARMER

Abstract.—Evaluation of individual model assumptions is a critical first step in achieving a credible tool for land-use planning. We quantified arthropod abundance in the ground and shrub layers of six successionally different sites in central Missouri, and then related our findings to easily measured habitat variables commonly used to represent arthropod food potential in habitat models for selected vertebrates. Arthropod sample weights decreased with maturing seral stage, as did herbaceous biomass. Percent cover of forbs, vegetation height, and forb weight were important variables in various seasonal regression models explaining both Malaise-trap and sweep-net samples. Temperature was also a good predictor variable, but this was probably a function of arthropod activity as well as abundance. Testing of individual hypotheses is feasible, and we offer some suggestions for future work.

Habitat-assessment models are mechanisms for synthesizing existing information about a species and its habitat requirements (Thomas 1982). They are designed for land-use planning and attempt to identify ranges of suitability for a few critical variables associated with the supply of food, cover, water, and spatial needs of the species of interest (Farmer et al. 1982). These same variables often are subject to alteration by land-management activities. Variables are identified from reported research that identifies habitat characteristics assumed to be important in explaining the presence, absence, or relative abundance of selected species. Aggregation of variables into models often incorporates untested assumptions relating species to perceived habitat requirements.

Validation of habitat models is now receiving increased attention at both the managerial and academic levels of wildlife biology. We view validation as an attempt to determine the level of agreement between model output and some system measurement (Farmer et al. 1982). The most common approach to validation involves a comparison of total model output with some estimate of species' density (Cole and Smith 1983). However, habitat models are not population models. They do not contain birth- and death-rate components and make no pretext of being estimators of population densities at any single point in time. It is not surprising, therefore, that validation efforts that purport to test the relationships between model estimates of habitat potential and short-term estimates of population densities often fail to exhibit statistical significance or to yield biological insight. Such efforts must assume, first, that population density is estimated accurately, and second, that the estimate represents the potential of the site at that particular time. The first assumption is suspect, given the difficulty of obtaining meaningful estimates of population size (Johnson 1981a; Verner 1985). Without an understanding of the demographic history of the population (Van Horne 1983), the second assumption is unacceptable.

We believe the first step in efforts to validate habitat models should focus on the evaluation of individual model assumptions. These assumptions often pertain to the use of easily measured variables as proxies for site-specific habitat conditions. For example, habitat models commonly use characteristics of vegetation as variables to predict a site's potential food supply (Baskett et al. 1980). Arthropods are an important food source for many vertebrates, and variables commonly assumed to represent arthropod abundance include percentage of herbaceous cover and average height of herbaceous vegetation (Schroeder and Sousa 1982; Sousa 1982, 1983). We were interested in evaluating assumed relationships between vegetation measurements and arthropod abundance on sites commonly altered by resource-management activities in central Missouri. This chapter presents (1) an example directed at model-assumption validation that uses relationships between measures of vegetation and arthropod abundance; and (2) suggestions for validation of model components.

Study area and methods

Six study plots, from 0.76 to 1.72 ha, were located on or adjoining the University of Missouri's Ashland Wildlife Research Area, Boone County, Missouri (Blenden 1982). Plots were selected to represent various seral stages along a successional gradient from open old-field to mature oak-hickory (*Quercus-Carya*) forest.

The four old-field plots (plots 1–4) varied from a site

MICHAEL D. BLENDEN: School of Forestry, Fisheries and Wildlife, University of Missouri, Columbia, Missouri 65211

MICHAEL J. ARMBRUSTER and ADRIAN H. FARMER: Western Energy and Land Use Team, U.S. Fish and Wildlife Service, 2627 Redwing Road, Fort Collins, Colorado 80526

THOMAS S. BASKETT: Missouri Cooperative Wildlife Research Unit, 112 Stephens Hall, University of Missouri, Columbia, Missouri 65211

Table 2.1. Selected characteristics of six study plots

	Old-field plots				Forest plots	
Character	1	2	3	4	5	6
Area (ha)	1.21	1.72	0.90	1.53	0.76	0.98
Tree canopy (%)	0	12	28	37	89	91
Woody stems/ha						
Class I = <2.5 cm	843	1801	3414	2376	1554	2232
Class II = 2.5–15.2 cm	0	129	1753	966	3150	1715
Class III = >15.2 cm	0	14	5	41	477	291
Herbaceous biomass (gm/m^2)	924	522	422	337	66	86

mowed 3 years previously (0% canopy cover) to a site supporting 37% canopy cover from trees >1 m tall (Table 2.1). Woody stems <2.5 cm dbh varied in density from 843 to 3414/ha and were often clumped. Peak standing green-herbaceous biomass ranged from 337 to 924 gm/m^2, with the predominant species being Missouri goldenrod (*Solidago missouriensis*), sedges (*Carex* spp.), panic grasses (*Panicum* spp.), and poverty grass (*Danthonia spicata*). Common woody species were American plum (*Prunus americana*), common persimmon (*Diospyros virginiana*), smooth sumac (*Rhus glabra*), eastern red cedar (*Juniperus virginiana*), elm (*Ulmus* spp.), and multiflora rose (*Rosa multiflora*).

The two forested plots (plots 5 and 6) had canopy closures of 89% and 91% with associated peak standing green-herbaceous biomass of 66 gm/m^2 and 86 gm/m^2, respectively (Table 2.1). Plot 5 was dominated by an overstory of white oak (*Quercus alba*) and shagbark hickory (*Carya ovata*) with a combined density of 531 stems/ha. An eastern red cedar understory was present at 170 stems/ha. Thirty percent of the woody vegetation was <2.5 cm dbh, 60% was 2.5–15.2, and 10% was >15.2 cm dbh. The largest tree on this site was 38.1 cm dbh. Plot 6 also was dominated by shagbark hickory and white oak at densities of 503 and 372 stems/ha, respectively. Sugar maple (*Acer saccharum*) was the common understory species (1631 stems/ha). Fifty-three percent of the woody vegetation was <2.5 cm dbh, 40% was 2.5–15.2, and 7% was >15.2 cm. The largest tree was 53.3 cm dbh. Dominant herbaceous plants on both sites were sedges, tick trefoils (*Desmodium* spp.), and panic grasses.

Arthropod populations were sampled on each plot during May–October 1979 and March–May 1980. Malaise traps (Marston 1965) were centrally located on each plot and operated continuously. Traps were checked at midday every 3 days. Each plot was also sampled with a standard sweep net at approximately 9-day intervals. One 50-sweep transect was randomly located in each of four quadrants within each plot. All plots were sampled on the same day after dew had dried from vegetation. All arthropod samples (Malaise and sweep net) were dried for 3 days at 60°C, counted, and weighed.

Weather data were obtained from a National Weather Service station located approximately 6.4 km north, at Columbia Regional Airport. Mean daily temperature, precipitation, wind speed, and minutes of sunlight/day were averaged over either a 3-day period encompassing each Malaise sample date or a 2-day period surrounding each sweep-net sample date.

Data were collected on percent cover of grasses, forbs, and woody growth >1 m (Daubenmire 1959); litter depth, vegetation height, species composition, and percent canopy cover of larger trees were sampled four times between May and August. Eight estimates of green-herbaceous biomass were obtained from each plot during the growing season. Woody stems >1 m tall were counted once on all plots and lumped into three dbh categories (Table 2.1).

Because initial comparisons between arthropod numbers and weight for both Malaise and sweep-net samples indicated a strong relationship ($r = 0.78$, $P = 0.001$; $r = 0.63$, $P = 0.001$, respectively), only sample weight means were subsequently analyzed. A two-way analysis of variance was used to test for differences in sample weight versus vegetation characteristics, or weather conditions, between plots and time periods, respectively. Statistical Analysis Systems' (SAS) general linear model procedure with least square of means option (Helwig and Council 1979) was used to test for significant differences between sample weight means. A stepwise multiple regression (SAS maximum R^2 improvement option) was performed to determine the relative importance (partial F-value) of vegetation and weather variables in explaining variation in both Malaise and sweep-net samples. All independent variables included in the models were significant at $P < 0.05$ and were used only if they increased R^2 by 0.05 or more.

Results

Arthropod samples

Malaise-trap and sweep-net captures generally exhibited decreasing sample weights with maturing successional stages of study plots (Table 2.2). The season-long mean Malaise sample weights varied from 9.66 gm on plot 1, the most open old-field site (Table 2.1), to 4.62 gm on plot 6, the mature forest site. Sweep-net samples showed a similar trend with mean sample weight decreasing from 4.53 gm on plot 1 to 0.28 gm on plot 6. Sample weights from plot 1 were significantly larger ($P < 0.05$) than samples from any other plot, and old-field sample weights were larger ($P < 0.05$) than forest sample weights. Plots 2–4 exhibited similar ($P > 0.05$) sample weights, and samples from forested plots were not significantly different from one another.

Capture data were also examined by season: spring = 24 March to 7 June, summer = 10 June to 23 August, and fall = 24 August to 31 October. Plot 1 exhibited a significantly higher ($P < 0.05$) mean spring Malaise sample weight (7.51 gm) than any other site except plot 3 (Table 2.2). Old-field Malaise summer samples were similar ($P > 0.05$), but were significantly larger ($P < 0.05$) than forest samples. Fall Malaise samples were smaller than summer means. Although sample weights from forested plots were not significantly different within sample periods, they differed significantly ($P < 0.05$) from season to season, with the lowest weights in fall and the highest in summer.

Table 2.2. Mean Malaise-trap and sweep-net sample weights in gm/sample for the entire season, spring, summer, and fall. (Standard deviations shown in parentheses.)

	Old-field plots				Forest plots	
	1	2	3	4	5	6
MALAISE-TRAP SAMPLE						
Entire season	9.66[a] (7.77)	7.34[b] (5.69)	7.74[bc] (6.03)	7.83[bc] (6.27)	5.07[d] (3.23)	4.62[d] (3.16)
Spring	7.51[e] (7.49)	4.07[f] (4.44)	5.79[ef] (5.29)	4.18[f] (3.97)	4.93[f] (2.98)	3.50[f] (2.60)
Summer	13.97[g] (8.34)	12.25[g] (5.53)	12.87[g] (5.77)	13.91[g] (5.96)	7.62[h] (2.75)	7.54[h] (2.25)
Fall	7.49[i] (5.60)	5.71[ij] (3.42)	4.55[jk] (2.52)	5.40[ij] (3.39)	2.66[k] (1.63)	2.80[k] (2.33)
SWEEP-NET SAMPLE						
Entire season	4.53[l] (3.94)	1.25[m] (1.03)	0.93[mn] (0.85)	0.69[no] (0.47)	0.42[nop] (0.44)	0.28[nop] (0.21)
Spring	0.48[q] (0.57)	0.16[q] (0.19)	0.22[q] (0.20)	0.21[q] (0.20)	0.36[q] (0.37)	0.27[q] (0.29)
Summer	7.12[r] (3.10)	1.95[s] (0.86)	1.18[st] (0.54)	1.06[st] (0.37)	0.64[t] (0.53)	0.36[t] (0.21)
Fall	6.00[u] (3.47)	1.64[v] (0.78)	1.39[vw] (1.06)	0.73[vw] (0.27)	0.28[w] (0.22)	0.21[w] (0.10)

Note: Similar superscript letters denote no significant differences ($P > 0.05$), within a given season and sample method.

No significant between-plot differences were detected for spring sweep-net samples (Table 2.2). Plot 1 yielded larger ($P < 0.05$) mean summer (7.12 gm) and fall (6.00 gm) sample weights than any other plot. Plots 2–4 and 3–6 exhibited similar ($P > 0.05$) summer and fall sample weights.

Table 2.3. Independent variables of Malaise-trap regression models for the entire season, spring, summer, and fall (Partial *F*-values indicate relative importance of each variable.)

Independent variable	Slope	Partial F	Combined R^2
ENTIRE SEASON			
Temperature	+	257.43	
Percent cover of forbs	+	138.79	0.54
SPRING			
Temperature	+	112.43	
(Vegetation weight)^2	+	49.05	0.66
SUMMER			
Percent cover of forbs	+	98.22	
Log of temperature	+	32.34	
Density of class I trees	+	22.99	
Woody vegetation weight	–	10.64	0.46
FALL			
Vegetation height	+	45.89	
Log of temperature	+	38.11	0.41

Table 2.4. Independent variables of sweep-net regression models for the entire season, spring, summer, and fall. (Partial *F*-values indicate relative importance of each variable.)

Independent variable	Slope	Partial F	Combined R^2
ENTIRE SEASON			
Vegetation height	+	174.37	
Density of class I trees	–	32.18	0.64
SPRING			
(Forb weight)^2	+	64.26	
Temperature	+	11.98	
Density of class III trees	+	10.24	
Woody vegetation weight	+	7.16	0.84
SUMMER			
(Vegetation height)^2	+	87.21	
Density of class I trees	–	50.48	
Litter depth	–	31.05	0.83
FALL			
(Forb weight)^2	+	53.79	
(Grass weight)^2	–	20.16	0.80

Vegetation and weather variables

Fifty-four percent of the variation in season-long Malaise samples was explained by mean daily temperature and percent cover of forbs (Table 2.3). Temperature was the best predictor variable in the spring Malaise regression model ($R^2 = 0.66$), followed by the square of vegetation weight. Forty-six percent of the variation in summer Malaise sample weight was explained by percent cover of forbs, the log of temperature, density of class I trees, and weight of woody vegetation. Vegetation height and log of temperature explained 41% of the variation in fall Malaise samples.

Vegetation height and density of class I trees explained 64% of the variation in season-long sweep-net sample weights (Table 2.4). The square of forb weight, temperature, density of class III trees, and weight of woody vegetation explained 84% of the variation in spring sweep-net samples. The square of vegetation height, density of class I trees, and litter depth explained 83% of summer sweep-net samples. The fall regression model contained the square of forb weight and the square of grass weight. It explained 80% of the variation in sweep-net sample weight.

Discussion

We quantified arthropod abundance in the ground and shrub layers of what we viewed as six successionally different sites in central Missouri. We then related our findings to easily measured habitat variables (including weather variables) commonly used to represent arthropod food potential in habitat models for selected vertebrates. Both Malaise traps and sweep-net transects probably sampled only arthropods in the first 1 m above ground. In this stratum, both arthropod sample weights and herbaceous biomass decreased with maturing seral stage.

Several variables characteristic of the sampled vegetation were statistically important in explaining arthropod catches. Percent cover of forbs, vegetation height, and forb weight were important variables in various seasonal regression models for both Malaise and sweep-net samples. Temperature was also identified as an important variable explaining Malaise catches, but this relationship may have been confounded by the effect of temperature on arthropod activity. Temperature was included only in the sweep-net regression model for spring samples.

Extrapolation of these apparent relationships to the building, use, and testing of habitat models must be done in a realistic manner. Both the Malaise-trap and sweep-net approaches to arthropod sampling have inherent shortcomings. Topographic and vegetative heterogeneity can create natural flyways and influence Malaise-trap catches within a sample site (Blenden 1982). Height of surrounding vegetation (Townes 1962) and amount of sunlight striking the collecting jar (Matthews and Matthews 1970) may affect Malaise-trap samples. Trap efficiency may vary for different insect orders (Nicholls 1960). Sweep-net efficiency can be influenced by vegetation characteristics such as size, density, and flexibility (Whittaker 1952; Southwood 1966) and can change with arthropod activity patterns, taxa, stage of life cycle, and varying weather conditions (DeLong 1932; Romney 1945; Southwood 1966). Given these potential problems, however, we believe the techniques were adequate for our purposes, i.e., determining relative abundance of arthropods associated with different cover types in central Missouri.

Additional considerations must be addressed in efforts directed at model-assumption validation. First, the study sites should represent a range in values for the variables being evaluated. A range of conditions is important so that the findings can be extrapolated to other areas. Results from single-site studies do not provide enough data to allow generalizations about the model assumption tested. Our experiences indicate that locating study sites that exhibit an appropriate range of values for the variables of interest may be one of the most difficult tasks in designing validation efforts. Cook et al. (1984) found it necessary to locate study sites in four western states in order to obtain a range of values for evaluation of a pronghorn (*Antilocapra americana*) habitat model.

Size of sample plots is also an important consideration during validation. The size of our plots was probably adequate for evaluating the relationship between arthropod abundance and vegetation. However, our plots would not be large enough to evaluate relationships between habitat resources and larger animals such as deer (*Odocoileus* spp.). Size and juxtaposition of evaluation sites to specific cover types were thought to be important factors determining herpetofaunal use of sites during an earlier study at this location (Clawson et al. 1984). Telemetry studies, such as those of Lancia et al. (1982), offer promise for solving many interpretative problems associated with study-plot size and cover-type usage by directing sampling efforts to areas actually used by the animal.

Most exploratory studies (including the present one) that attempt to characterize habitat look at a variety of variables that may or may not include those important to habitat occupancy and use (Johnson 1981a). Such studies stop with the generation of hypotheses (Romesburg 1981). Without experimental manipulation of identified variables, followed by careful observation of responses, it is often difficult to determine whether or not the identified variables are causal or are simply examples of covariance.

Potential concerns about covariance suggest that the next step in the work presented here would be to manipulate experimentally those variables explaining arthropod abundance and to record responses. Such an approach is applicable to a variety of habitat-model variables. For example, habitat models which attempt to characterize nesting cover for various waterfowl species often use some measure of visual obscurity (Kirsch et al. 1978), as well as ratios of emergent vegetation to open water (Weller and Spatcher 1965), for pair- and brood-habitat potential. On the basis of manipulative studies, Dwernychuk and Boag (1973) and Kaminski and Prince (1981) have suggested that waterfowl respond to visual obscurity when selecting nesting cover or to emergent : open water ratios for pair usage.

In summary, we recommend that efforts to test habitat models focus on individual model assumptions. Our results demonstrate the feasibility of testing individual hypotheses, and we believe that the following benefits can be obtained: (1) tests can be designed and controlled much more easily at the level of individual assumptions; (2) test results may be applicable to a range of habitats and species wider than the actual test situations (e.g., our general relationships between vegetation and arthropods); and (3) tests of the output from a habitat model may be difficult or impossible to interpret unless one first has information concerning individual model assumptions and the situations under which any single assumption is likely to be invalid.

When all model assumptions have been validated (i.e., we failed to invalidate them [see Holling 1978]), we can proceed to explore the functional relationships between variables and ultimately to gain an understanding of the model's reliability in various land-use-planning situations.

Acknowledgments

This research was funded in part by Research Agreement #USDI 14-16-0008-2014 between the U.S. Fish and Wildlife Service and the University of Missouri. Our appreciation goes to J. N. Burroughs, J. A. Ellis, and F. B. Samson for their assistance in project design. P. A. Opler, S. J. Chaplin, and F. B. Samson offered valuable suggestions for improvements to an earlier draft. Word processing was provided by C. Gulzow and D. Ibarra. This chapter is a contribution from the Missouri Cooperative Wildlife Research Unit (U.S. Fish and Wildlife Service; Missouri Department of Conservation; Wildlife Management Institute; and School of Forestry, Fisheries, and Wildlife, University of Missouri, cooperating), and from Missouri Agricultural Experiment Station Projects 182 and 184, Journal Series No. 9704.

3

A Habitat Model for Ruffed Grouse in Michigan

JAMES H. HAMMILL and RICHARD J. MORAN

Abstract.—A Habitat Suitability Index model for ruffed grouse (*Bonasa umbellus*) was developed in draft form by the U.S. Fish and Wildlife Service. Two critical life requisites, fall-to-spring cover and a winter food source, provided the foundation for five habitat variables: (1) equivalent stem densities of deciduous trees, coniferous trees, and shrubs; (2) height of deciduous trees; (3) lower canopy height of conifers; (4) height of deciduous shrubs; and (5) interspersion of fall-to-spring cover with the winter food source. We modified the draft HSI model to better fit observations of ruffed grouse habitat use in Michigan and Wisconsin. Modifications of the model included adjustment of equivalent stem densities for regenerating shrubs and conifers, denial of the penalty for a conifer component in the stand, and assumption that the winter food source (1% canopy of mature aspen) is always available. Tests of the model's ability to predict breeding densities of ruffed grouse on five study areas differed from field population estimates by 1% to 34%. Application of the model to evaluate three forest-management alternatives indicated a 41% increase in breeding territories when the Standard Practices alternative was used and an additional 28% with specific guidelines for ruffed grouse.

The Habitat Evaluation Procedures (HEP) Group of the Western Energy and Land Use Team, U.S. Fish and Wildlife Service, developed a Habitat Suitability Index (HSI) model for ruffed grouse (*Bonasa umbellus*) in 1981, based largely on data from long-term studies of the species in Minnesota (Gullion and Svoboda 1972; Gullion 1977). This model has not been published; we refer to it hereafter as the "reference model." Modifications of the reference model, based on our observations in Michigan, were incorporated into a model that we subsequently tested in Michigan and Wisconsin. This model is referred to hereafter as the "Michigan model." This chapter reports results of our independent tests of that model's ability to predict the density of territorial male ruffed grouse at selected sites in Michigan and Wisconsin.

Model development

We adjusted the habitat variables of the reference model to meet conditions that we felt were important to grouse in Michigan. The determination of a HSI for ruffed grouse involves consideration of habitat variables independently, as well as the interspersion of these components. Breeding male ruffed grouse require 2.4–4.0 ha (6–10 acres) of suitable habitat to establish a territory (Gullion 1977). Optimum density (HSI = 1.0) was taken as occupancy by one territorial male (breeding pair) per 4 ha (10 acres). Optimal habitat for this species consists of forests that provide two life requisites: fall-to-spring cover and winter food.

James H. Hammill: Michigan Department of Natural Resources, P.O. Box 300, Crystal Falls, Michigan 49920

Richard J. Moran: Michigan Department of Natural Resources, P.O. Box 158, Houghton Lake Heights, Michigan 48630

Habitat components

Fall-to-spring cover

Suitable fall-to-spring cover for ruffed grouse is characterized by dense vertical stems of regenerating deciduous trees, shrubs, or coniferous trees in pure or mixed stands. These stands provide protection from predation for breeding birds, particularly the drumming males, and are used as summer brood range and winter cover. The efficacy of cover provided by regenerating aspens (*Populus grandidentata* and *P. tremuloides*) is well documented (Gullion and Svoboda 1972). Stem density is the basis for comparing vertical structure provided by various growth forms of woody vegetation in both models. However, the growth form of certain components of some stands is considerably different from the growth form of deciduous trees. Thus it is necessary to equate shrub and conifer stem densities to an *equivalent stem density* (ESD) value.

Many shrubs common to the Lake States exhibit a multistemmed form originating from parent root stock or a stump. The reference model used an ESD factor of 0.25, assuming that four shrub stems would occupy the same space as one deciduous sapling. We used a shrub conversion factor of 0.5 for the Michigan model.

Conifers have a growth form that provides more protective cover because of the typically dense crown, which occupies more space in the stand. To equate the stem densities of conifers with those of deciduous saplings, both models used an ESD factor of 4.0.

The ESD was variable 1 (V_1) in the determination of the HSI and was calculated from equation (1):

$$\text{ESD} = d + 4c + 0.50s, \qquad (1)$$

in which d = number of deciduous saplings/ha, c = number of conifer stems/ha, and s = number of shrub stems/ha.

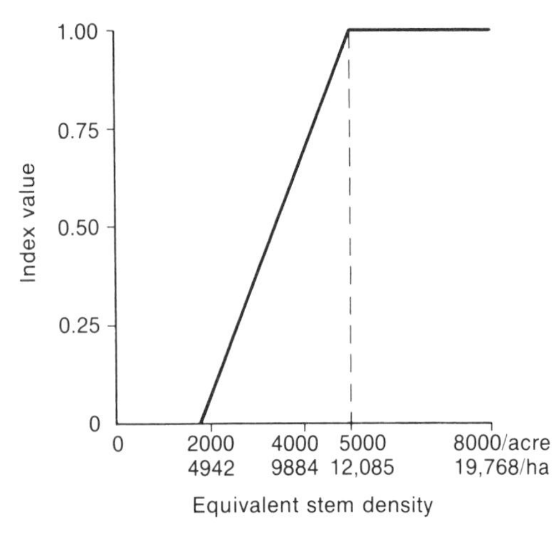

Figure 3.1. Index value for equivalent stem density (V_1) in the Michigan model.

According to Gullion (1977), optimum cover for wintering and breeding ruffed grouse develops at 14,000–20,000 stems/ha (Gullion 1977). The Michigan model assumes that ESD remains optimum at any value in excess of 12,000 stems/ha (5000/acre) (Fig. 3.1).

Adequate height of overhead cover (V_2) adds to the security of fall-to-spring cover for ruffed grouse. Deciduous saplings do not provide suitable fall-to-spring cover when they are less than 4.6 m (15 feet) in height. Optimal cover is provided when these stems average 6.1 m (20 feet) in height (Gullion 1977) (Fig. 3.2). Although Figure 3.2 does not show this, the index declines over time with increasing stem height. The decline is long-term and highly variable, depending upon understory and site index of the stand. Because self-thinning occurs with the aging of stands, the ideal combination of density and height is transitory in nature. With aging, stand height increases and stem density decreases, resulting in thinning below the required density for optimal fall-to-spring cover and decreasing the value of such stands for fall-to-spring cover. This, indeed, is the crux of managing aspen for ruffed grouse. The aging stand does, however, provide the critical winter food component. In the model, decline due to increase in height of stems is reflected in the lower stem densities (V_1) that accompany aging.

Gullion (1977) considered conifer cover to be detrimental to ruffed grouse. However, coniferous cover is optimal for ruffed grouse when both cover and long-range visibility are provided in the stand. In Michigan we have found a significant number of active territories (drumming males) within lowland conifer cover types. Consequently, the quality of habitat in the Michigan model was not penalized by the percentage of conifers present, as is the case in the reference model.

In coniferous cover, the height factor was applied to the level of the lowest branches (V_3). Conifers can provide cover from terrestrial and avian predators throughout the various age-classes. Live boughs extending to the ground increase the opportunity for terrestrial predators to capture grouse. Also, when the lowest live conifer branches are higher than 0.9 m (3 feet), the value of the stand component decreases, because it increases the ability of avian predators to pursue and capture grouse. When the lowest live branches on conifers are higher than 4.6 m (15 feet), this element is assumed to be unsuitable for ruffed grouse (Fig. 3.3).

Gullion (1977) recognized the value of the understory shrub component as fall-to-spring cover. In Michigan, shrubs also provide additional overhead cover and contribute to the total stem density available as protective cover in the stand (V_4). The reference model assumed that the minimum shrub height useful to ruffed grouse was 1.5 m (5 feet) and that a shrub component averaging 1.8 m (6 feet) or higher provided optimum shrub conditions (Fig. 3.4).

Winter food source

Ruffed grouse eat the buds of several species of deciduous shrubs and trees during winter, including white birch (*Betula papyrifera*), ironwood (*Ostrya virginiana*), red maple (*Acer rubrum*), hazelnut (*Corylus* spp.), juneberry (*Amelanchier* spp.), and apple (*Malus* spp.). The flower buds of mature

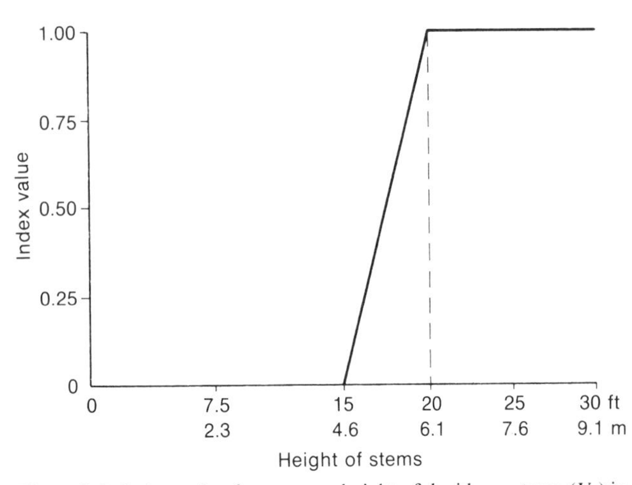

Figure 3.2. Index value for average height of deciduous trees (V_2) in the Michigan model.

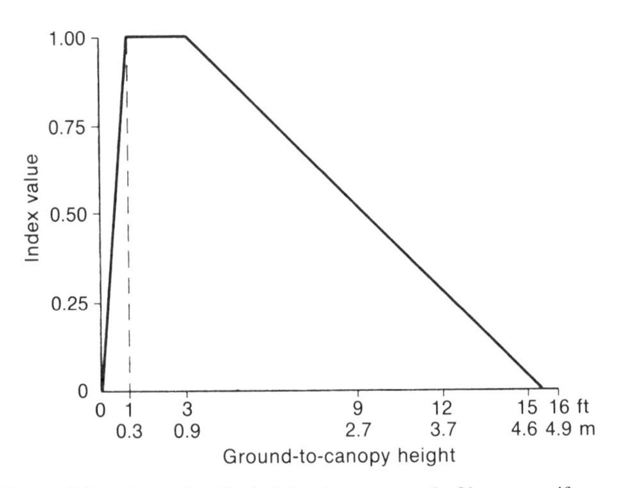

Figure 3.3. Index value for height above ground of lower coniferous branches (V_3) in the Michigan model.

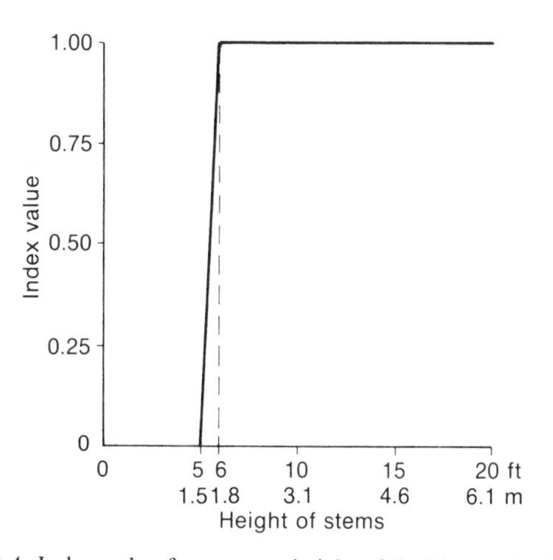

Figure 3.4. Index value for average height of deciduous shrubs (V_4) in the Michigan model.

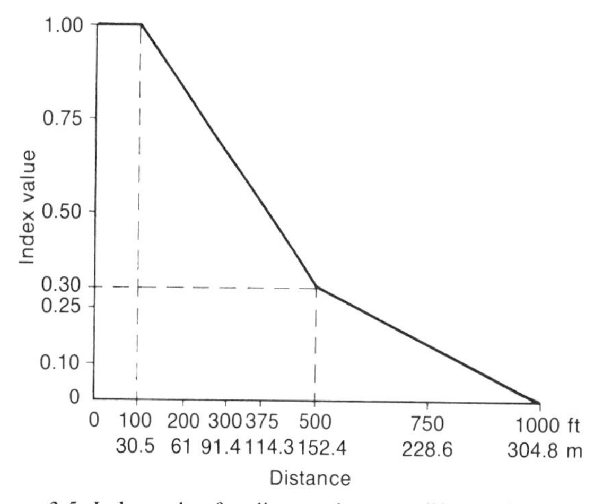

Figure 3.5. Index value for distance between life requisites (V_5) in the Michigan model.

male aspens, however, constitute the staple winter food of ruffed grouse in the Upper Midwest and Lake States (Gullion and Svoboda 1972). Other food sources are generally of lower nutritional value and are eaten only when aspen buds are unavailable. The Michigan model assumes that mature aspen are available within the area being evaluated. If a stand contained more than 1% canopy cover of well-distributed mature aspen, the reference model assumed that the life requisite for winter food was satisfied. Thus, the Michigan model applies only to stands with at least a 1% canopy of mature aspen. The 1% composition figure is based on Gullion's (1977) recommendation that life requisites, including mature aspen, be found within a 2.4- to 4.0-ha (6- to 10-acre) activity center. This could be supplied by a single aspen clone of 0.04 ha (0.10 acre).

Interspersion

Optimum habitat conditions were assumed in the reference model when winter food could be found within 30.5 m (100 feet) of fall-to-spring cover. As the distance between the two life requisites increased, habitat suitability decreased. If the two life requisites were separated by 305 m (1000 feet) or more, they were not suitable for ruffed grouse. To calculate the interspersion index, random points were selected on a map in fall-to-spring stands that were devoid of a winter food source. The average distance between these points and the nearest winter food source was then calculated to find the interspersion index (V_5) (Fig. 3.5). Note that only those stands useful to ruffed grouse, and only those lacking a winter food source, were considered.

Calculation of HSI in the Michigan Model

The HSI of a particular stand was calculated according to equation (2):

$$\mathrm{HSI} = V_1\left[\frac{d(V_2) + 4c(V_3) + 0.50s(V_4)}{d + 4c + 0.50s}\right]V_5, \qquad (2)$$

in which d = density of deciduous trees in stems/ha, c = density of coniferous trees in stems/ha, s = density of deciduous shrubs in stems/ha, and V_n = the suitability index of habitat variables, as determined from Figures 3.1–3.5. The rationale for integrating these variables, as in equation (2), followed that in the reference HSI model. The result was then converted directly to a predicted number of breeding male grouse/ha for comparison with study-area surveys of drumming males.

Methods

The model was applied to stands in five study areas located in four counties in Michigan and one in Wisconsin (Table 3.1). Breeding grouse populations on test areas were determined by locating all drumming males annually on each study area. In Michigan, surveys were done for 6 years (1975, 1976, 1978, 1979, 1981, and 1982) at a site in Gladwin County, for 2 years (1983 and 1984) at one site each in Osceola and Montmorency counties, and only during 1981 at a site in Iron County. Surveys were done in 1978, 1979, and 1980 at a site in Wood County, Wisconsin.

The estimated densities of active territories (drumming males), based on field surveys, were then compared with predictions of the Michigan model. The latter values did not vary between sampling years for a given area, because habitat variables were constant over the relatively short span of years studied.

Table 3.1. Test results of Michigan HSI model for ruffed grouse on five study areas in Michigan (MI) and Wisconsin (WI)

Location of study areas by county and state	Size (ha)	Years surveyed	Breeding male territories: Model prediction	Field survey	SE	% error
Gladwin, MI	151.0	6	78.0	101.0	1.49	+29.5
Iron, MI	158.6	1	35.5	35.0	—	− 1.0
Montmorency, MI	136.4	2	26.0	32.0	1.00	+23.1
Osceola, MI	64.8	2	8.8	7.0	1.50	−20.5
Wood, WI	66.4	3	45.7	30.0	0.58	−34.4

Results and discussion

The measured error in these tests ranged from 1% in Iron County, Michigan, to 34% in Wood County, Wisconsin (Table 3.1). The mean error of prediction for all five sites was 6% below actual field estimates. Errors may have resulted from a lack of fine-tuning of model variables to fit the specific characteristics of the individual areas, or the differences could have resulted from errors in estimating numbers of males, or both. In any case, we believe the results indicate that our Michigan HSI model was a remarkably good predictor of grouse numbers.

Forest-management practices are the key to Michigan's forest-wildlife-management program. In 1979, aspen constituted 35% of all forest products harvested in northern Michigan; the state now administers about 25% of the 1,379,100 ha (3,407,709 acres) of aspen forest type in Michigan, and private individuals administer another 46% (Raile and Smith 1983). The aspen type has been recognized as critical, if not obligatory, to the abundance and distribution of ruffed grouse in North America (Gullion and Svoboda 1972; Gullion 1977). Most aspen stands on state forest lands in Michigan exceed 20 ha (50 acres) (Hammill and Visser 1984). The tendency to maintain large stands results in low interspersion of life requisites for ruffed grouse.

The Michigan model is currently being used on a trial basis to evaluate past, present, and future forest-management practices on public lands. It may also be used to develop forest-management guidelines for production of ruffed grouse. To evaluate past, present, and proposed land-management schemes further, a habitat-assessment project for ruffed grouse was begun in 1983. Wildlife biologists were asked to use the Michigan HSI model to evaluate 38 randomly selected land-management compartments on public land. These compartments comprised 25,256 ha (62,406 acres), or about 2% of all state-owned lands. The units were to be evaluated for three management alternatives:

1. *Base Productivity*, which assumes that the compartment is not manipulated.
2. *Standard Practices*, which assume that the compartment is treated to produce an optimal mix of wildlife, aesthetics, and so on.
3. *Ruffed Grouse Guidelines*, which assume that the compartment will be treated as follows:
    a. aspen acreage not capable of being held for 10 years will be harvested to preserve type; cutting widths narrow;
    b. all treatments (other than as in *a*) in upland types capable of producing aspen regeneration for fall-to-spring cover will not be more than 201 m (660 feet) wide;
    c. treatments in lowland conifer types will not be more than 100 m (330 feet) wide;
    d. the same acreage must be harvested in each type as was originally prescribed in Standard Practices.

State biologists predicted 447 breeding territories of ruffed grouse under the Base Productivity alternative. As a result of Standard Practices, they predicted 631 breeding territories, a 41% increase over Base Productivity. This increase would be the result of forest management for other products like pulpwood, and habitat improvement for white-tailed deer (*Odocoileus virginianus*). When ruffed grouse guidelines were applied to the same compartments to increase interspersion while providing for similar wood product output (alternative 3), 805 breeding territories were predicted, providing an additional 28% increase.

The accuracy of the Michigan HSI model for predicting breeding male ruffed grouse densities is encouraging. More testing is needed, especially in stands with more diverse habitat conditions than were available on study areas for which we had population data. The model has been used to evaluate quality of habitat both prior to cover manipulation and as an artifact of manipulation. The cost-effectiveness of management decisions can also be monitored by employing these relatively simple standards. Implementing guidelines for ruffed grouse will mean additional costs in timber sales, depending on the intensity of management. For example, improving interspersion requires more working hours for cutting lines. In some instances, aspen will be sacrificed at less than optimum price to preserve the type. With the model, results of several forest-management options can be tested prior to choosing a course of action.

Acknowledgments

P. Sousa of the HEP Group, U.S. Fish and Wildlife Service, gave us instruction and assistance with the reference HSI model. G. Gullion gave valuable advice about modifying habitat variables for the Michigan model. J. F. Kubisiac, K. R. McCaffery, and W. A. Creed provided population and habitat data from Wisconsin for model testing. T. F. Carlson, T. R. Prawdzik, and L. R. Smith provided population and habitat data for study areas in Michigan. H. R. Hill assisted with sampling for habitat assessments, and L. G. Visser and J. W. Urbain reviewed the manuscript. W. N. Bronner and J. W. Urbain gave us encouragement and support. K. Douglas typed the draft manuscripts and G. Lehman prepared the figures. Finally, many wildlife biologists in Regions I and II of the Michigan Department of Natural Resources performed habitat assessments. We sincerely thank all of these persons, whose contributions were essential for the successful completion of this work.

4

Simulating the Roosting Habitat of Sandhill Cranes and Validating Suitability-of-Use Indices

DOUGLAS C. LATKA and JAMES W. YAHNKE

Abstract.—The objective of this study was to develop habitat-suitability indices for roosting sandhill cranes (*Grus canadensis*) in flowing water along the Platte River in central Nebraska. We quantified water depth, water velocity, and distance from roosting birds to the nearest high bank or large island, using field measurements from transects across roosting areas, together with predawn aerial photographs of roosting cranes. A roosting-habitat model was developed from these measurements at one location, and results were used to predict the distribution of roosting cranes at a separate, independent location. The predicted and actual distributions of cranes at the second location were significantly correlated.

A wildlife-habitat model can be used to describe the response of a species to changes in its habitat. One use of such models has been to describe the impact of future actions on the habitats of target species (Fish and Wildlife Service 1980b). For a model to be useful, its variables must represent the species' preference for those physical characteristics that are components of its habitat (Hardy et al. 1982). Therefore, the first goal of wildlife-habitat modeling should be to verify that the variables perceived to constitute suitable habitat are related to the species' use of an area. Ideally, such models should be applicable over a substantial geographic area, which means that they need to be tested in several independent locations. The objective of this study was to measure the water depth, water velocity, and a location factor variable of the roosting habitat of the sandhill crane (*Grus canadensis*) and use these variables to develop a predictive model of crane roosting habitat. These variables were measured using techniques developed for the Instream Flow Incremental Methodology (IFIM).

Study area

The Platte River begins at the confluence of the North and South Platte rivers at North Platte, Nebraska, and flows 500 km (311 miles) east to Plattsmouth, Nebraska. The 143 km (89 miles) of the Big Bend section of the Platte River, from Lexington to Chapman, supports the majority of the midcontinent population of sandhill cranes during migration (Lewis 1974; Fish and Wildlife Service 1981b).

The south channel at Mormon Island, in the eastern end of the Big Bend area, was selected for study because it supports the highest densities of roosting sandhill cranes on the Platte River (Fish and Wildlife Service 1981b). Study site 1 was approximately 3.2 km (2.0 miles) west of Highway 281, and study site 2 was 1.6 km (1.0 miles) west of site 1. Each site encompassed one meander in the river. Islands at both study sites were previously cleared of vegetation to improve roosting (Currier 1984). The study sites were located in a private preserve with no permanent human disturbances nearby. One of the largest native grassland and wet meadow complexes remaining in the valley was on Mormon Island. Cornfields were numerous on the south side of the river. Therefore, the food source component and the disturbance component were optimal within the study area.

Methods

Sandhill cranes roosted in shallow, slow-moving water in the middle of the Platte River. To describe this habitat, suitability-of-use indices were developed at study site 1: one each for water depth, water velocity, and distance to the nearest bank or island wider than 46 m (150 feet). Depths and velocities were estimated using data collected for the Water Surface Profile (WSP) hydraulic simulation program (Milhouse et al. 1984). Islands more than 46 m (150 feet) wide were considered significant because they represented permanent stuctures in the stream with effects similar to those of the streambank.

We developed habitat-use information from aerial photographs taken during the peak of the migration season. Suitability indices for water depth and velocity and the distance from a roosting crane to the nearest riverbank or significant island were developed from study site 1 and used to predict roosting habitat for study site 2. The predicted habitat was then compared with actual use of study site 2. We assumed that the study sites were at or close to saturation with cranes at the time of sampling.

We placed transects across the study site to represent the stream reach, following the methods of Bovee and Milhouse (1978). The physical habitat model, which is part of the Instream Flow Incremental Methodology (IFIM), assumes that the measurements taken along each transect are projected halfway to the adjacent transects upstream and downstream (Milhouse et al. 1984). However, to control for an absence of

DOUGLAS C. LATKA: U.S. Bureau of Reclamation, P.O. Box 1607, Grand Island, Nebraska 68802
JAMES W. YAHNKE: U.S. Bureau of Reclamation, P.O. Box 25007, Denver, Colorado 80225

this condition, weighting factors were assigned. For example, if the characteristics of a transect extended only 30% of the distance upstream to the next transect, 0.3 was assigned as the weight and the measurements were all extended 30% of the distance upstream (weights range from 0 to 1.0). The next upstream transect was then assumed to project back 70% of the previous transect. In this study, large sandbars and the channel were used as characteristics for assigning weights.

Measurements of depth, velocity, and distance from the transect headstake were taken across the transect wherever changes in bed elevation occurred (Bovee and Milhouse 1978). The points where these measurements were taken defined the individual cell boundaries (Milhouse et al. 1984). The average of the hydraulic measurements was the value for the characteristics of the individual cell. The right-hand cell boundary (facing upstream) was used to identify the distance to the nearest stream bank or significant island. The distance was measured to the nearest 0.3 m (1 foot) and grouped in 3-m (9.8-foot) intervals. The distance the cell was projected upstream was determined by its weight. Each cell was assigned an average depth, an average velocity, an area, and a distance to the nearest bank or significant island.

To document crane use at study site 1, low-altitude aerial photographs were taken before dawn on 22 March 1983, the morning before the hydraulic measurements were taken. A staff gauge and stage recorder were used to measure changes in river stage between the times the photographs and hydraulic measurements were taken.

We overlaid the cell grid created by the measurements on the photographs and counted the cranes in each cell. Crane use was summed into depth, velocity, and distance classes. A frequency distribution of the area available was determined by summing the areas of the cells in each class. A density estimate, in cranes/100 m^2 (1076 ft^2), for each class was derived by dividing the crane use by the area available for that physical characteristic. We derived polynomial equations using multiple regression techniques for the depth and velocity vs density relationships; linear regression was used to describe the distance vs crane density relationship, as described in the Results section.

We used the curve created from the regression techniques to create the HSIs for the three physical characteristics. The interval with the highest density on the predicted curve was assigned a weight of 1.0. The density for each of the other intervals was divided by the highest density to get its weight.

Transects were then established at study site 2. Photographs and field measurements were taken in the same manner as for study site 1. A habitat value for each cell in study site 2 was computed by the following equation, using the suitability indices created from study site 1 (Bovee 1982):

$$\mathrm{WUA} = \mathrm{Area} \times S(D) \times S(V) \times S(Ds),$$

in which WUA = Weighted Usable Area or Habitat Value;
Area = area of the cell;
S(D) = suitability of the depth of the cell determined by the index developed from site 1;
S(V) = Suitability of the velocity of the cell;
S(Ds) = suitability of the distance of the cell from the bank or a significant island.

The grid was then overlaid on the photograph for study site 2, and cranes were counted in each cell. A Spearman correlation between the number of cranes per cell and the predicted WUA per cell was run to test for significant association.

Results

On study site 1, 7895 sandhill cranes were counted—about 1.6% of the total estimated population that migrates through the Platte Valley (Fish and Wildlife Service 1981b). Most cranes were found in depths between 0 and 12 cm (0 and 4.7 inches) and velocities between 0 and 40 cm/sec (0 and 1.31 feet/sec) (Fig. 4.1). No cranes were found less than 15 m (49 feet from the high bank or significant islands, although suitable depths and velocities occurred there. Only six cranes were found in depths greater than 24 cm (9.5 inches). Cranes were not found in velocities greater than 68 cm/sec (2.15 feet/sec). However, there were no suitable depths available with velocities greater than 68 cm/sec (2.15 feet/sec). The available depth within a cell increased with the velocity within a cell ($r = 0.64$; $P < 0.0001$). Therefore, it was assumed that only two physical characteristics could be viewed independently by cranes in their selection of a cell. That is, cranes rejected a cell if either the depth (which would include velocity) or the distance was unsuitable.

Third-order polynomial equations were derived for the density estimates by water depth and velocity, using multiple regression techniques (Table 4.1). Depth explained 68% of the variation in the density estimate, and velocity explained 65% of the variation. Higher-order equations were also derived but rejected, because they did not fit the data as well.

A linear regression of crane density on distance was based on distances ranging from 0 to 56 m (0 to 184 feet) from the bank. This regression explained about 80% of the variation in crane density over this range of distances. At 56 m the estimated density from the regression was 31 cranes/100 m^2 (1076 ft^2); this was assigned a weight of 1.0, as were all distances greater than 56 m (184 feet).

Cranes were not observed at a distance less than 15 m (49 feet) (Fig. 4.1) from the bank. The density of cranes increased gradually from 15 m (49 feet) to about 40 m (131 feet), with a dramatic increase to about 37 cranes per 100 m^2 (1076 ft^2) at 45 m (148 feet). The densities between 45 and 72 m (148 and 236 feet) averaged about 30 cranes/100 m^2 (1076 ft^2). This represented a plateau where the distance was optimal. Beyond 72 m (236 feet) crane densities fell, and we obtained mostly zero densities. A review of the data shows

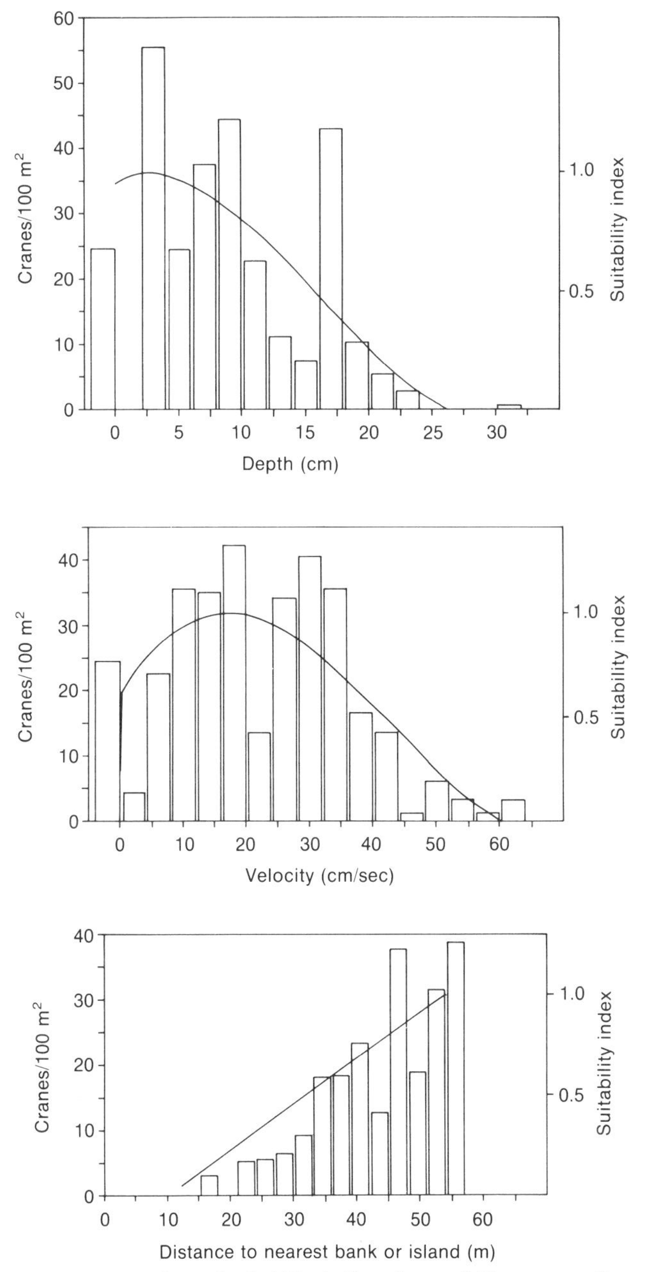

Figure 4.1. Density and suitability indices for sandhill crane roosting habitat characteristics at Platte River, Nebraska, 22 March 1983.

that coincidental to this decline in density there was a decline in the number of observations per distance interval; most observations were five or fewer, although only five of the cells had unsuitable depth or velocity. All of these were factors in the decline or absence of cranes at the greater distances. We believe that the decline in density at greater distances was an anomaly due to the sampling method rather than a real reaction by the cranes to greater distance. Distances less than 72 m represent cells that were found on either side of the middle of the channel. Distances greater than 72 m and up to 100 m were represented only by the very middle of the channel and at only two transects. To represent accurately the preference for distances greater than 72 m, a study site with a wider channel (and, therefore, a greater distance to the middle) would have to be measured. We did not, therefore, have enough samples to define crane use at the greater distances accurately. A linear regression, rather than a polynomial equation, was used to represent the distance data.

We found a significant positive relationship when we used the habitat characteristics developed from site 1 to predict crane use at study site 2 ($r = 0.76$, $P < 0.0001$) (Fig. 4.2).

Discussion

This study demonstrates how aerial photography can be used to test the performance of a wildlife-habitat model for a particular area. The result of the one test flight illustrates the potential for a much wider application of this technique for testing wildlife-habitat models. The result from the one flight cannot be used as a conclusive test of the overall performance of the model; however, it does indicate that for one independent study site the model was successful in predicting usable sandhill crane roosting habitat. Additional aerial photographs in different geographic areas would reveal further variation in sandhill crane habitat use. This could improve the overall performance of the model and make it more applicable to different geographic areas.

The suitability-of-use indices created from study site 1 agreed with observations made by others (Frith 1974; Lewis 1974; Fish and Wildlife Service 1981b): sandhill cranes roost in shallow depths in the middle of the river channel. On the basis of the results of the correlation, the HSI model developed from site 1 was adequate for predicting usable roosting areas at site 2 when all other habitat requirements were provided.

If the suitability-of-use indices were accurate, the assumption that the study sites were saturated with cranes was not correct. Some cells with simulated habitat at study site 2 had no cranes. Despite these empty cells, we obtained significant results at study site 2.

The fact that the habitat was not saturated also affected the development of the suitability curve for distance. All of the distance intervals greater than 84 m (276 feet) and about half of those between 75 and 84 m (246 and 276 feet) were on two transects. Both of those transects were devoid of birds despite the presence of good roost sites. Roosting cranes may have vacated these transects in the predawn hours, before sampling. Although these factors also affected the goodness of fit of the model, they were not significant enough for us to reject our original premise.

By providing the known variables associated with roosting sandhill crane habitat, we were able to isolate the water-flow variables and to evaluate their importance in providing roosting habitat. To compare simulated habitat with actual crane use of areas of the Platte River where other variables are not optimal, the effect of these variables on roosting sandhill cranes must be quantitatively determined. For instance, Krapu et al. (1984) reported that cranes avoided

Table 4.1. Polynomial or linear regressions of crane density at site 1

Variable or Source	Partial regression coefficient	R^2	F	P	95% CI of partial regression coefficient	
Depth (D)	1.068	0.617	0.250	0.624	−3.487	5.623
D^2	−0.183	0.627	1.867	0.192	−0.469	0.103
D^3	3.55×10^{-3}	0.676	2.266	0.153	-1.48×10^{-3}	8.58×10^{-3}
D_0	34.397	—	13.698	0.002	14.588	54.206
Regression	—	0.676	10.420	0.001	—	—
Velocity (V)	2.346	0.292	13.137	0.002	0.981	3.712
V^2	-6.66×10^{-2}	0.400	15.269	0.001	−0.103	−0.0306
V^3	4.40×10^{-4}	0.653	12.517	0.003	1.78×10^{-4}	7.02×10^{-4}
V_0	7.900	—	1.520	0.234	−5.617	21.411
Regression	—	0.653	10.641	<0.0005	—	—
Distance (Ds)	0.696	0.806	70.476	<0.0005	0.521	0.871
$(Ds)_0$	−7.747	—	8.068	0.011	−13.502	−1.993
Regression	—	0.806	70.476	<0.0005	—	—

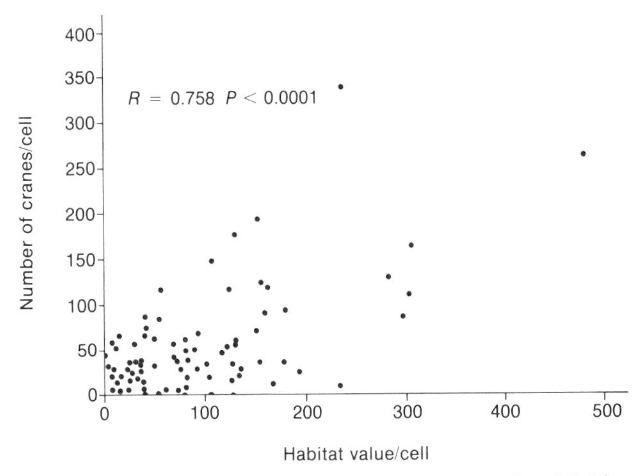

Figure 4.2. Correlation of number of cranes to predicted habitat values at study site 2 at Platte River, Nebraska, 24 March 1983.

channels less than 50 m (165 feet) wide and areas with high vegetation. These factors, however, were evaluated independently of depth. In Indiana, Lovvorn and Kirkpatrick (1981) found that when suitable depths or bare soil was available for landing, cranes did not avoid vegetation. They concluded that in riverine situations the availability of unvegetated shoreline or bars for landing is critical. The use of narrow channels during migration may be limited not by width but rather by the deep water that flows through these channels. Future studies should be done to determine the usability of these other channels, with the depth variable considered.

Further study could also use the techniques developed for the IFIM to measure other flow rates. The other flow rates would change the availability of suitable depths and velocities, changing the habitat value for each new flow. If this were done for a variety of flow rates, a habitat vs flow-rate curve could be developed (Bovee 1982). Aerial photography could then be used to test whether crane use changes with a change in habitat value.

Acknowledgments

The Platte River Whooping Crane Critical Habitat Maintenance Trust provided access to its land for data collection. The U.S. Fish and Wildlife Service, Nebraska Game and Parks Commission, and the Trust helped collect the hydrologic data. C. Stalnaker and the staff at the Instream Flow Group made valuable suggestions and comments on the study. G. Krapu and T. Hardy reviewed the manuscript. M. Pucherelli and the staff of the Remote Sensing Section at the Engineering and Research Center, Bureau of Reclamation, made and interpreted the early-morning photographs. The Hydrology Section of the Grand Island Project Office, especially C. Schwieger, E. Kouma, and D. Woodward, provided useful guidance in all phases of the data collection and analysis. J. Ziewitz of the U.S. Fish and Wildlife Service helped with the data analysis, and Donna Janisch typed the manuscript. To all we express our sincere appreciation.

5

Evaluation of a Mallard Productivity Model

DOUGLAS H. JOHNSON, LEWIS M. COWARDIN, and DONALD W. SPARLING

Abstract.—A stochastic model of mallard (*Anas platyrhynchos*) productivity has been developed over a 10-year period and successfully applied to several management questions. Here we review the model and describe some recent uses and improvements that increase its realism and applicability, including naturally occurring changes in wetland habitat, catastrophic weather events, and the migrational homing of mallards. The amount of wetland habitat influenced productivity primarily by affecting the renesting rate. Late snowstorms severely reduced productivity, whereas the loss of nests due to flooding was largely compensated for by increased renesting, often in habitats where hatching rates were better. Migrational homing was shown to be an important phenomenon in population modeling and should be considered when evaluating management plans.

The mallard (*Anas platyrhynchos*) has the most extensive breeding range, and is the most common and most heavily harvested of North American waterfowl. There has been widespread concern in recent years about the status of the mallard population, which seems to be declining. Despite considerable study, however, a thorough understanding of mallard population dynamics has eluded waterfowl biologists, possibly because their investigations have been either extensive but superficial or detailed but narrowly focused.

The model we describe originated in 1972 when a team of researchers at Northern Prairie Wildlife Research Center recognized the potential value of coordinating their research activities. They identified simulation modeling as a powerful tool for integrating their findings and identifying areas in which further research would be most productive. This stochastic model is now a central component of a modeling system designed to aid decision making for management and acquisition of habitat for mallard production (Cowardin et al. 1983). Other components are various data-handling and display routines. The system has recently been applied to a number of actual management situations, such as evaluating land-acquisition options, comparing habitat-management practices, and projecting future impacts of wetland drainage (Cowardin et al. 1983). This model differs from many others presented in this volume in that it describes the process of mallard recruitment, as influenced by habitat and other variables, rather than merely predicting the presence/absence or the abundance of the species. In this sense, it more closely characterizes the quality of the habitat for survival and reproduction of mallards.

Douglas H. Johnson, Lewis M. Cowardin, and Donald W. Sparling: U.S. Fish and Wildlife Service Northern Prairie Wildlife Research Center, P.O. Box 2096, Jamestown, North Dakota 58402

The basic model

General overview

In the model, mallards are followed from arrival on the breeding grounds, in the spring, through the summer reproductive season, until the onset of hunting in the fall. The survival and reproduction of individual females are formulated as functions of wetland habitat conditions, nesting habitats available (their area, vegetative features, and the security from destruction of nests), length of breeding season, population age structure, and physical condition of individual birds. Many of these modeled features vary dynamically throughout the season, and all can be manipulated by the investigator. A detailed description of the model is forthcoming, but the following will provide an overview. Some parameter values were based on field investigations, others were developed during the modeling process.

Each simulated hen follows the flow-chart path shown in Figure 5.1 on each day of a 120-day breeding season. A nonnesting hen can begin nesting with a probability that depends on (1) her physical condition as reflected by her body weight, which initially is a normal random variate but declines after repeated nesting efforts; (2) wetland availability, which normally declines during the season; and (3) the date within the season (a mallard hen being unlikely to initiate a nest late in the season regardless of her condition or the quality of the habitat).

A nesting hen selects a habitat from those available; usually we restrict her to a 4-legal-section block (10.36 km^2). The probability that a particular habitat is selected is a function of the area of that habitat multiplied by its attractiveness. Attractiveness to mallards is indexed by a measurement of the height and density of nesting cover (Robel et al. 1970; Kirsch et al. 1978). After nesting commences, the hen's weight is reduced by 10 gm for each egg laid and by 2 gm for each day spent incubating her clutch.

The clutch must survive about 34 days of egg laying and incubation to hatch. On each of those days, the clutch and

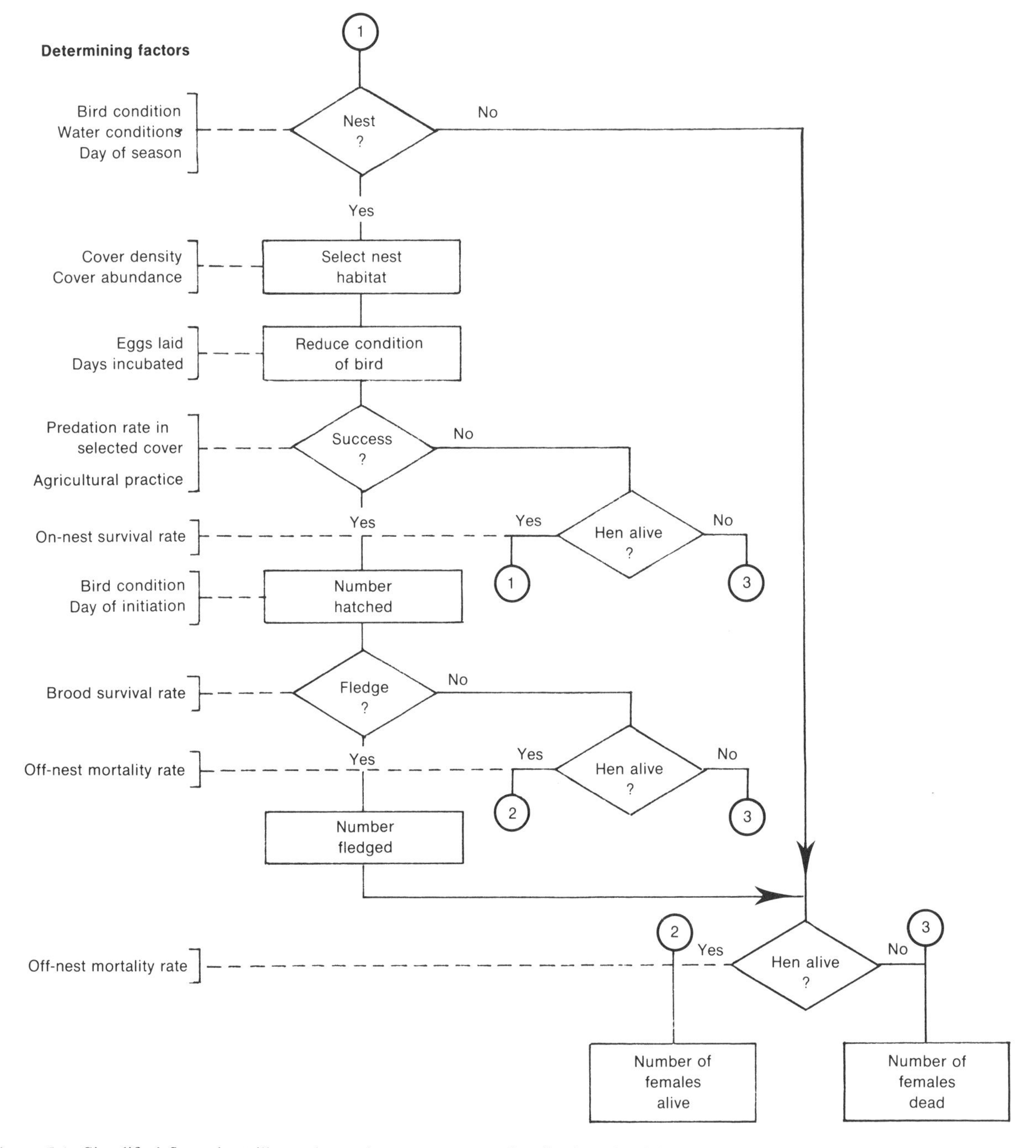

Figure 5.1. Simplified flow chart illustrating major components of mallard productivity model. (See text for explanation of numbers.)

the hen are at risk to a variety or predators. Certain habitats pose additional hazards, e.g., tillage of cropland and mowing of hayland and roadsides. Survival of the nest varies among habitats and can also be altered by certain management practices, such as predator reduction or habitat manipulation.

If a nest is destroyed, the hen may or may not suffer the same fate, with probability of death equal to 0.06. In addition, hens have a small chance (0.001) each day of dying in other ways, such as predation away from the nest, accidents, and disease. Once a simulated hen dies, her death is tallied and she no longer plays a role.

Should the simulated clutch survive the requisite 34 days, the eggs hatch. The number of eggs hatched equals the number laid, which was a random variable dependent on the hen's condition and the initiation date.

Not all ducklings survive to fledging. Some broods succumb completely, and nearly half the ducklings are lost from remaining broods. Those ducklings that survive all the hazards are tallied as fledged recruits, the ultimate product of the reproductive season.

Model output

The standard FORTRAN listing resulting from the model includes values of the key input variables, the random number seed used, and the available nesting habitats and their characteristics, including changes that occur during the breeding season. Results of the simulation are succinctly summarized on one page showing the percentage of clutches that hatch, the percentage of hens that are successful in hatching a clutch, the average number of nests per hen, the summer mortality rate of hens, and the recruitment rate (number of fledged females per adult female in the spring population). For each type of nesting habitat, the program lists the number of nesting attempts initiated, the number of successful attempts, and the number of recruits that were hatched in that habitat.

An optional SAS (SAS Institute 1982a) routine produces several detailed tables and figures that facilitate a close inspection of the model's operation. One listing is of all nests, indicating the hen involved, her age, her weight at nest initiation, the habitat, dates on which the nest was begun and terminated, the size of the clutch, and the fate of the nest. Summary tables display the total number of nests according to habitat and fate by age of hen. Another insightful display (Fig. 5.2) shows the history of each hen throughout the nesting season, including the dates, habitat, and fate of each nest. The success of nests according to date initiated is shown in both a table and a figure (Table 5.1 and Fig. 5.3). These data are also presented for each habitat type. Final tables (Tables 5.2 and 5.3) give the necessary statistics for the calculation of Mayfield (1961) estimates of nest success rates (Johnson 1979).

The variety of outputs and the flexibility of viewing the simulation process have been valuable aids in debugging the program and in modifying it to conform to biological findings. In addition, many of the displays were used to analyze biases in an intensive field study of radio-telemetered mallards (Cowardin et al. 1985).

Validation

Complete validation of the model would require the comparison of results from the model with measurements of

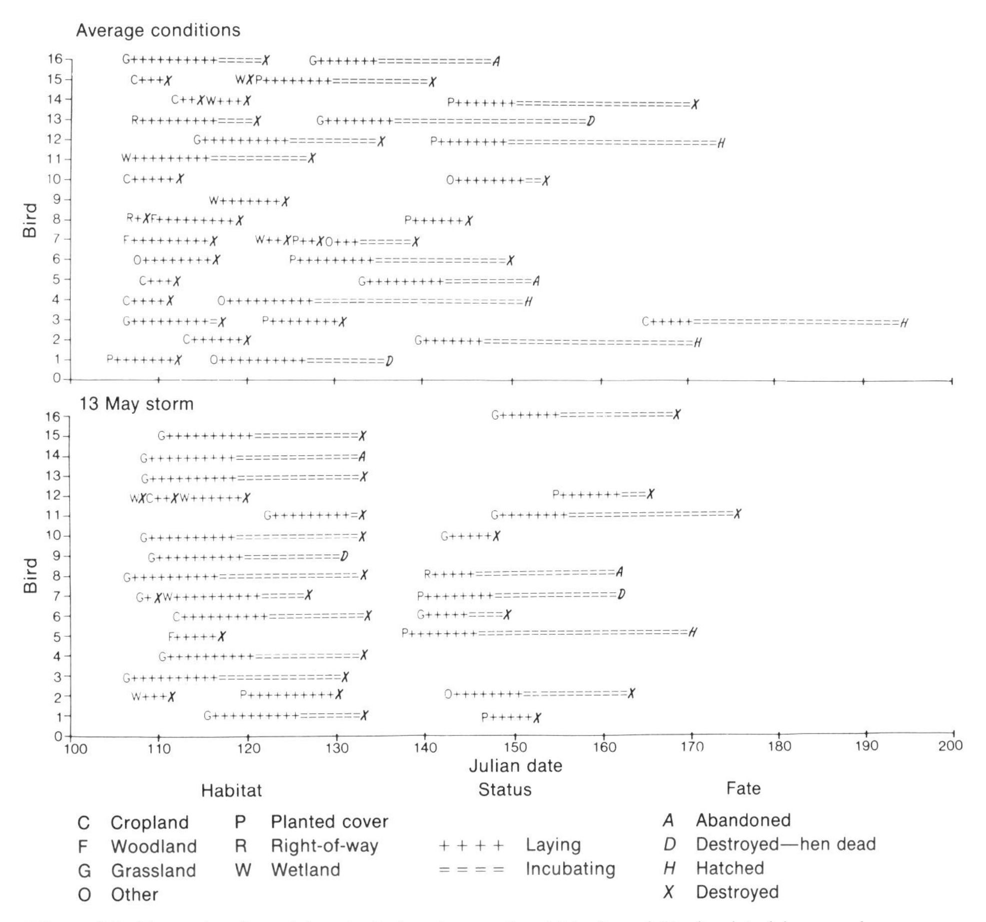

Figure 5.2. Example of model output showing nesting histories of 16 simulated hens under average conditions (*above*) and with 13 May snowstorm (*below*).

Table 5.1. Mallard productivity parameters under a situation with average water conditions and under five other climatological situations, as predicted by a simulation model

	Average conditions	Dry year	Wet year	Snow-storm, 3 May	Snow-storm, 13 May	Flood, 21 May
Nests per hen	1.97	1.18**	2.14**	2.33**	2.31**	2.11**
Nest success rate	0.12	0.13	0.12	0.08**	0.05**	0.11
Hen success rate	0.23	0.14**	0.25	0.20**	0.13**	0.24
Summer survival rate	0.76	0.76	0.75	0.78	0.81*	0.77
Recruitment rate	0.42	0.24**	0.45	0.33**	0.21**	0.42
Population change	0.94	0.84**	0.95	0.86**	0.77**	0.95

Note: Significance level for difference from average situation, based on *t*-test with five simulations each of 500 hens: * $P < 0.05$, ** $P < 0.01$.

Table 5.2. Distribution by habitat of nest initiations and successful nests in an average situation and under a simulated 21 May flood

	Percent of initiated nests		Percent of successful nests	
Habitat	Average	Flood	Average	Flood
Cropland	6.2	6.3	3.5	4.4
Grassland	35.6	35.0	51.8	56.3
Hayland	8.5	8.6	8.7	6.7
Odd areas	6.2	6.3	2.3	3.2
Planted cover	21.9	23.4	19.5	21.2
Right-of-way	2.6	2.4	0.5	1.2
Wetland	13.1	11.6	10.8	4.9
Scrubland and woodland	5.9	6.4	3.0	2.2

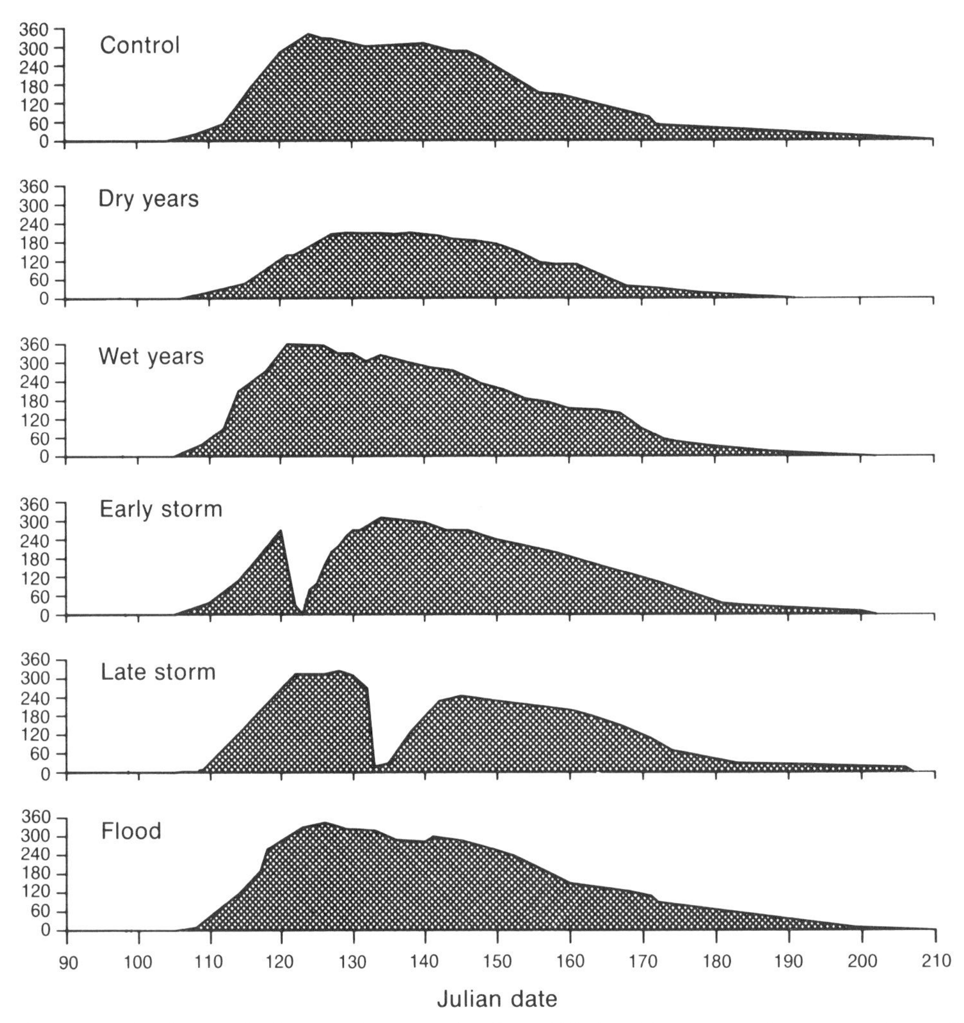

Figure 5.3. Number of simulated active nests, by day within breeding season, for average (control) situation and five modifications.

Table 5.3. Mallard productivity parameters simulated by models without and with homing

	Results of model	
	Without homing	With homing
Breeding population (females)	21.0	22.6
Initiated nests		
Inside fence	3.5	5.9
Outside fence	35.9	34.7
Successful nests		
Inside fence	1.9	3.3
Outside fence	4.2	4.5
Recruitment rate	0.537	0.633
Production (females)	11.3	14.3

those parameters determined from field studies. This poses a difficult problem for mallard production, because there is no known way to actually count the young fledged from a breeding population. In practice, only indices of mallard production are obtained. It is possible, however, to measure many of the intermediate parameters calculated by the model and compare them with field estimates. Although we have not done studies specifically designed to validate the model, we did apply the model to one area where detailed field studies were done (Cowardin et al. 1983). Model predictions compared with actual estimates were 13.2 vs 13.8 for number of successful nests, 68.9 vs 74.3 for number of nests initiated, 1.78 vs 1.55 for average number of initiated nests per hen, and 0.22 vs 0.19 for the nest success rate.

We also found that this model gave results consistent with other models. Cowardin and Johnson (1979) presented a simple deterministic model relating hen success to nest success. Although this model is completely independent of the stochastic model described here, results have been consistently similar. We also used output from the stochastic model as input to a different deterministic model (Cowardin and Johnson 1979) designed to predict the change in population size from one year to the next. The results suggested a slowly declining population, a conclusion that agrees with continental counts of mallards. The consistency of these results suggests that the model is sufficiently realistic to serve as a tool to assist in the making of management decisions.

Modifications of the model

Wetland conditions

Wetland conditions markedly influence the number of mallards attracted to breeding areas; they also affect the probability of nesting, the length of the nesting season, and the number of nests attempted (Krapu et al. 1983; Cowardin et al. 1985). The relation between wetland conditions and reproductive effort has been recognized for some time (Crissey 1969) and is currently incorporated in a model used by the U.S. Fish and Wildlife Service for predicting the size of the fall flight and for setting hunting regulations (Martin et al. 1979). Models currently in use are simple and do not account for the complexity of wetland conditions, which vary in time and space, or for the interactions of wetland conditions with other factors that determine reproductive success.

Our model allows for daily changes in wetland conditions throughout the breeding season and for interactions among factors that influence recruitment rate. We can simulate the impact of various wetland conditions that occur in nature and predict reproductive parameters such as nest success, hen success, summer hen survival, and recruitment.

We simulated dry, average, and wet water conditions during the breeding season. For each set of conditions, we used a data set containing an index to water conditions for each day of the season. The index value approximates the percentage of semipermanent wetland basins (defined by Stewart and Kantrud 1971) containing water on that day. These data sets reflect the normal drying of wetland basins during the season; differences in percentages at the beginning of the breeding season and rates of loss distinguish the dry, average, and wet years. These indices can be entered as data without modifying the structure of the model.

Simulated dry conditions had considerable influence on recruitment, reducing the total number of nests and the nesting activity. Nests per hen declined from 1.97 under average situations to 1.18 in dry years (Table 5.1). Nesting activity was of diminished intensity and reduced duration (Fig. 5.3). Nest survival rates were similar under all wetland conditions. Reduced nesting affected the recruitment rate, which dropped from 0.42 under average conditions to 0.24 under dry conditions.

Changes due to wet-year conditions were noticeable in the number of nests per hen (Table 5.1), which reflected increased nesting persistence due to improved wetlands. The net results showed slightly higher rates of hen success and recruitment.

Catastrophic events

Reproductive success is affected by catastrophic weather events, which can be either local or widespread across the breeding range. The net effect of these events on recruitment rate is difficult to isolate in field studies because a hen may renest after her nest is destroyed. The probability of renesting depends on the hen's physical condition, wetland conditions, and the time of the season when nest destruction occurs. Our model furnished an opportunity to investigate the impact of catastrophic events that would be difficult, if not impossible, to assess through field study.

We simulated three events that are fairly common in the northern prairies. The first two were snowstorms, one on 3 May and a later one on 13 May. Storms of this type bury nests under deep snow and cause their destruction or abandonment. They also blanket the nesting habitat and temporarily prevent further nest initiation. Such storms, although frequently severe, are of short duration, and nest initiation resumes as soon as habitat becomes available.

The simulations were accomplished by setting the daily nest mortality rate at 0.90 on the day of the simulated storm. The input data for wetlands were then set to zero to prevent any nest initiations for a period of 3 days.

The third catastrophic event simulated was a midseason

flood. Mallards nest primarily in uplands, but some individuals nest in emergent vegetation over water and in wet meadows surrounding wetlands. These nests may be destroyed by rapid rises in water level occurring after heavy summer rainstorms. The simulation was conducted by setting the daily nest mortality rate to 0.5 on the day of the simulated rainstorm for all nests located in wetland. We assumed that half the wetland nests would survive the flood. We also reduced by half the height-density measurement for wetland nests during the remainder of the season. This change simulated the reduced attractiveness of wetland sites for nest initiations following the flood. Also, the data representing wetland conditions were modified to simulate improved water levels, which occur after storms and extend the length of the nesting season.

Both the 3 May and the 13 May snowstorms substantially reduced the survival rate of simulated nests (Table 5.1). Virtually all nests were destroyed during the 13 May storm. The reduction in nest survival was somewhat compensated for by increased renesting, as new nests were initiated a few days after the storm (Fig. 5.2). The average number of nests per hen increased by about 0.35, because destruction of many nests early in the season (Fig. 5.3) allowed the birds to renest. Nests destroyed by the storm did not result in the associated loss of the hen, which we assume occurs in 6% of the nests destroyed by predators. The simulated hens were precluded from nesting for a period of time, which decreased their risk to predation and resulted in a slight increase in summer survival (Table 5.1). The 13 May storm was disastrous and resulted in the lowest recruitment rate and smallest index to population change of all simulated events. This storm caused a greater reduction in nest success because more birds were nesting at that time. Moreover, these birds had expended more resources and had a lower probability of renesting. Inclement weather could also affect real populations by reducing the food supply available to nesting hens, thereby further reducing the probability of renesting.

The simulated flood had little impact on the population because relatively few nests were affected. The decrease in number of successful nests in wetland was from 10.8% to 4.9% (Table 5.2). The loss of wetland nests was compensated for by increased nesting associated with improved wetland conditions. Most renests were in habitats more secure than wetland. The percentage of nests initiated in planted cover, for example, increased from 21.9 to 23.4.

Homing

Migrational homing, the predilection of an adult bird to return to the area where it spent the previous nesting season or a yearling bird to come back to the area where it was reared, is recognized in several species of ducks, including mallards (Bellrose 1976). This characteristic may be a critical determinant in certain nesting areas, such as islands, where densities of nesting birds grow far larger than would be anticipated on the basis of wetland habitat. Such population trajectories may be fueled by homing, which is particularly common among adults that successfully produce young (Doty and Lee 1974). Successful hens are hypothesized to favor the same habitat for subsequent nesting.

One way to assess the importance of homing is to follow individual birds from one year to the next. Our current model does not identify birds in subsequent years, however, so we approximated a solution. First, we modified the probability of selecting a particular habitat for nesting so that it depended on the nest success rate, as well as the area and attractiveness of the habitat. This change reflected the propensity of successful hens to select the same habitat for subsequent nests. In the original model, the probability of a hen choosing a habitat for nesting was proportional to the product of the area of that habitat times the height and density measurement of vegetation in that habitat. To these factors we added a function of the nest success rate (P) in the habitat. The function was 1 at $P = 0$, increased linearly to 3 at $P = 0.5$, and was constant thereafter.

Second, we allowed the number of breeding hens in an area to depend not only on the number and area of wetlands, but also on the hen success rate of that area in the previous year. This change permitted population growth due to homing by successful hens and their female offspring. We assumed that 75% of successful hens return to the breeding area in the next year, as well as 30% of their female offspring. In addition, a certain number of hens pioneer into the area because of the wetlands there. We assumed that the number of pioneers would be reduced by one-half for each homing hen. Simulations suggested that the population would be stable if 28% of hens are successful, each producing an average of two young females. The population would increase if the hen success rate exceeded 28% and would decline if it fell below 28%.

We explored the effects of these changes by simulating production on a 10.36-km^2 area typical of fairly good habitat in North Dakota. The field contained a 7-ha plot of planted nesting cover, protected by a predator-resistant fence. This habitat is an attractive one that has a high hen success rate and yields many offspring. Wetlands on the area attracted an estimated 21 mallard hens during the initial year.

Under the original model (based on three replications), 3.5 of the 39.4 nests initiated were in the fenced plot (Table 5.3). Total production of females was (21)(0.537) = 11.3.

From the model with homing, we projected the breeding population to be 22.6 females, and 5.9 of the 40.6 initiated nests were in the fenced plot. This difference from the original model appears modest, but it caused increases in the hen success rate and in the recruitment rate. Total production of females was (22.6)(0.633) = 14.3, a 27% increase compared with the former model.

Discussion

The mallard has been the subject of several other modeling efforts primarily designed to estimate its fall flight or to develop exploitation strategies (Geis et al. 1969; Hammack and Brown 1974; Anderson 1975a, 1975b; Hochbaum and Caswell 1978; Martin et al. 1979). These models, most of which covered the entire North American range, typically treated reproduction as a function involving only the breeding population size and the number of wetlands containing water. Our model is much more restricted in scope, pertaining to discrete areas within the glaciated prairie. In part because of this restriction, and in part because of the greater biological content of the model, it incorporates far more of the factors now known to influence productivity.

Models such as the one presented here allow interactions among variables to arise naturally. These interactions often account for behavior that appears contrary to intuition. One such example from the present application is the effect of flooding, a weather calamity that destroyed many wetland nests. Despite this loss, however, total productivity was unaffected, because hens that lost nests renested—often in more secure habitats—and improved wetland conditions increased renesting among the population in general.

The extensions of the model described here illustrate several facets of the population dynamics of mallards. The analyses contrasting wet, average, and dry years were consistent with expectations based on field experience and showed the marked effect of drought. Spring storms, another common phenomenon in much of the mallard's breeding range, were also influential, but their severity depended on when they occurred in relation to the breeding season of the species. The destruction of nests by spring flooding was offset by increased renesting, often in more secure habitats. The homing phenomenon was shown to exert considerable influence on productivity, and it should therefore be considered in further research and be incorporated in future management plans.

The value of the model described here as a research tool lies in its ability to incorporate new biological findings and to allow ''experimenting'' with the simulated population. This may shed light on how investigations of the real population should be most profitably designed. The management applications are diverse, ranging from assisting with the interpretation of operationally gathered productivity indices, to exploring potential results of selected habitat manipulations, and to evaluating strategies for acquisition of further habitat.

Despite these virtues, the model is no more than a collection of equations reflecting our beliefs about mallard population dynamics. Actions taken in response to predictions from the model should be monitored to assess the model's performance. It can be continually improved as we learn more about the processes involved.

Acknowledgments

We are grateful to T. J. Dwyer, R. J. Greenwood, G. L. Krapu, A. B. Sargeant, and G. A. Swanson, all members of the original mallard research team that conceived the model discussed here. R. D. Crawford and D. S. Gilmer also contributed to that effort, as did all biologists who have developed knowledge about mallard reproductive dynamics. A. M. Frank assisted with computer programming. This report benefitted from reviews by W. E. Grenfell, A. T. Klett, G. L. Krapu, M. G. Raphael, T. L. Shaffer, and the editors of the proceedings.

6

Developing a Practical Model to Predict Nesting Habitat of Woodland Hawks

JAMES A. MOSHER, KIMBERLY TITUS, and MARK R. FULLER

Abstract.—Some problems associated with developing habitat models include large required sample sizes, high cost of data collection, and applicability of data collected in one region to another. Our studies of raptor nesting habitat in the eastern woodlands of North America showed that it is possible to develop models that should provide an acceptable level of accuracy for management application. The variables required for the model can be measured economically and are often included in data bases generally used by forest managers. The models can be applied with a small reduction in accuracy to a large portion of the species' breeding range.

Three general types of ecological models exist: frameworks to test hypotheses about the functioning of ecological systems, detailed descriptions of ecological processes, and aids for land managers in the conservation and/or management of species or communities. Recently, the last type has become common. Several rather elaborate systems are now in use. Models must be used with caution because they simplify complex ecological processes and because incomplete data bases are often used for model parameterization. For example, an early model we reviewed for the "Habitat Evaluation Procedures" (Fish and Wildlife Service 1981a) predicted that red-tailed hawks (*Buteo jamaicensis*) would not breed in continuous forest habitat, when in fact they do, quite regularly. That model relied on data from red-tailed hawk nests in small woodlots, lone trees, or on cliffs. Thus, that general model was not useful for much of the northeastern United States.

We have identified the following six criteria for ideal management models, some of which are also shared by other model types. A model should (1) predict accurately (>80%); (2) be economical to parameterize; (3) contain variables measurable by managers; (4) have variables compatible with existing forest-management systems; (5) use variables that can be demonstrated to be important for identifying areas used by the species of concern; and (6) be applicable over all or most of a species' range. Any of these criteria can be difficult to achieve singly, and the problems are compounded when we attempt to satisfy all criteria in a single model.

Our purposes are to (1) illustrate some problems encountered when developing models for woodland raptor habitat; and (2) demonstrate ways of alleviating some of these problems by using sampling and by measuring variables that reduce time spent in the field without sacrificing all the useful information. We developed the model by using data for many variables to achieve the highest accuracy possible; then we simplified that model to make it more practical as a management tool.

JAMES A. MOSHER: University of Maryland, Appalachian Environmental Laboratory, Frostburg, Maryland 21532. *Present address:* Savage River Consulting, P.O. Box 71, Frostburg, Maryland 21532.

KIMBERLY TITUS: University of Maryland, Appalachian Environmental Laboratory, Frostburg, Maryland 21532.

MARK R. FULLER: U.S. Fish and Wildlife Service, Patuxent Wildlife Research Center, Laurel, Maryland 20708.

Methods

STUDY SITES

We measured habitat characteristics (Table 6.1) at the nests of broad-winged (*B. platypterus*) and red-shouldered (*B. lineatus*) hawks and at random plots during six breeding seasons in western Maryland (1978–1983) and three seasons in northeastern Wisconsin (1980–1982). Study sites and field techniques have been previously described (Titus and Mosher 1981; Titus 1984).

The western Maryland study area was within the Appalachian Province in Garrett and Allegany counties. Relief was oriented northeast to southwest and elevation varied from 200 to 900 m. The area was extensively forested (72%). The dominant deciduous tree species were white oak (*Quercus alba*), northern red oak (*Q. rubra*), chestnut oak (*Q. prinus*), red maple (*Acer rubrum*), and sugar maple (*A. saccharum*).

The northeastern Wisconsin study area was within the Lakewood district of the Nicolet National Forest (Oconto and Forest counties), which is 94% forested. Elevation varied from 300 to 450 m. This area lies within the Superior Upland section of the hemlock–white pine–northern hardwoods region as defined by Braun (1950). Dominant deciduous tree species included white birch (*Betula papyrifera*), aspen (*Populus tremuloides* and *P. deltoides*), red maple, and gray oak (*Q. borealis*).

We collected habitat data at active nest sites (i.e., nests in which at least one egg was laid) of red-shouldered and broad-winged hawks from middle to late summer. If a nest tree was used more than once by the same species, only the first sample was included in these analyses. We also made similar habitat measurements at randomly selected sites on the study areas (Titus and Mosher 1981).

Methods of measurement included those of James and Shugart (1970) as well as point-centered-quarter (Roth 1976) and density board (Noon 1981a) techniques for describing the structure and composition of the vegetation. In addition, we measured physiographic variables, including slope of sample plot and distance to water and forest openings. Details of habitat-measurement procedures are described in more detail elsewhere (Titus and Mosher 1981).

Statistical methods

We calculated minimum sample sizes separately for each hawk species for each habitat variable. The criterion of acceptable sample size was that estimates were within 20% of the mean with 95% accuracy (Cochran 1963; DeVos and Mosby 1969). The two distance measures had lognormal distributions and were transformed using X' = natural log $(X+1)$.

Two types of models were developed from 33 variables (Table 6.1) for each hawk species. The "ecological model" included all available variables for which we had complete data, except that only one of a pair of correlated ($r \geq 0.7$) variables was retained (Green 1979). We also eliminated variables that required sample sizes larger than the pooled samples from both states (Table 6.1). We reduced the set of variables in the ecological model to yield the "management model" by eliminating those not readily available to or measurable by forest managers (Table 6.2).

Direct discriminant analyses (DDA) (Nie et al. 1975) were done for each hawk species and for the random data from each study area with two sets of habitat variables. All discriminant analyses were based on prior probabilities of the group sample sizes. The DDA approach was used to allow a comparison between models because a stepwise analysis would probably result in different variables in each model.

The DDAs were used to determine differences between the next site and the randomly selected site. For each hawk species, we conducted two analyses incorporating, first, the ecological-model variables, then, separately, the management-model variables. These models were used to predict what portion of available habitat might provide suitable nesting sites. We used the classification portion of discriminant function analysis to determine what percentage of known samples were placed in the correct group (Rice et al. 1983a; Williams 1983). The ecological and management models were compared, and the DDA equation from each region was also used to predict the nest-site and random-habitat data from the other region. Thus, independent data sets were used to examine how well the model predicted nest-site habitat in another region.

Results

Thirty-three variables describing vegetation structure and composition, physiography, and nest-tree characteristics were available for analysis (Table 6.1). The variable-elimination procedures resulted in ecological and management models that incorporated 10 and 13 ecological variables and eight and six management variables for the red-shouldered and broad-winged hawks, respectively (Table 6.2).

Table 6.1. Minimum sample sizes needed for variables measured at woodland hawk nest sites

	Minimum sample size needed[a]	
Variables measured (units)	Broad-winged hawk	Red-shouldered hawk
Percent slope	47 (133)	118 (58)
Log distance to water (m)	5 (133)	8 (58)
Log distance to forest opening (m)	8 (133)	11 (58)
Canopy height (m)	2 (133)	2 (58)
Shrub density (total of four transects)	65 (133)	96 (58)
Shrub index (total of four transects)	53 (133)	54 (58)
Number of tree species	12 (122)	14 (55)
Number of shrub species	11 (122)	19 (55)
Number of trees/0.04 ha	28 (133)	24 (58)
Number of snags/0.04 ha	750 (133)	306 (58)
Evergreen canopy cover (%)	1280 (133)	791 (58)
Total canopy cover (%)	3 (133)	3 (58)
Evergreen understory cover (%)	1030 (133)	496 (58)
Total understory cover (%)	11 (133)	13 (58)
Evergreen ground cover (%)	1919 (133)	703 (58)
Total ground cover (%)	24 (133)	20 (58)
Point-centered-quarter coefficient of variation (%)	20 (60)	11 (33)
Point-centered-quarter mean tree distance (m)	14 (60)	22 (33)
Number of understory stems dbh 1–4 cm	42 (133)	74 (58)
Number of understory stems dbh 5–8 cm	56 (133)	31 (58)
Number of understory stems dbh ≥9 cm	86 (133)	110 (58)
Number of trees dbh ≤20 cm	87 (133)	90 (58)
Number of trees dbh 21–40 cm	25 (133)	24 (58)
Number of trees dbh ≥41 cm	125 (133)	41 (58)
Basal area (m^2/ha)	9 (133)	5 (58)
Density board-level 0–0.3 m ($\bar{x}$ of four measures)	35 (67)	36 (37)
Density board-level 0.3–1.0 m ($\bar{x}$ of four measures)	71 (67)	70 (37)
Density board-level 1.0–2.0 m ($\bar{x}$ of four measures)	73 (67)	98 (37)
Density board-level 2.0–3.0 m ($\bar{x}$ of four measures)	42 (67)	80 (37)
Height of nest tree (m)	2 (133)	2 (58)
Height of nest (m)	7 (133)	4 (58)
Percent nest height (m)	4 (133)	3 (58)
Dbh of nest tree (cm)	11 (133)	8 (58)

[a] Sample size given in parentheses.

The DDAs comparing nest sites with random sites provided an acceptable level of overall accuracy (>80% correctly classified, meeting an initial criterion) for both species (Table 6.3). This accuracy (correct classification rate) refers to the proportion of nest sites or random sites that were statistically assigned to the group with which they were most associated on the basis of the discriminant function equation and a given probability. The correct classification of nest sites ranged from a low for red-shouldered hawks of 69%, for all study areas combined, to a high of 83% for the Maryland ecological model. For broad-winged hawks, the range was from 56% for the Wisconsin management model to 86% for

Table 6.2. Hawk habitat-selection model variables incorporated in discriminant function analysis models (indicated by X)

Variable	Ecological model: Broad-winged hawk	Ecological model: Red-shouldered hawk	Management model: Broad-winged hawk	Management model: Red-shouldered hawk
Percent slope	X		X	
Log distance to water	X	X	X	X
Log distance to forest opening	X	X	X	X
Canopy height	X	X	X	X
Shrub index	X	X		
Total canopy cover	X	X		
Total understory cover	X	X		
Total ground cover	X	X		
Understory stems dbh ≥9 cm	X			
Number of trees dbh ≤20 cm	X		X	
Number of trees dbh 21–40 cm	X	X	X	X
Number of trees dbh ≥41 cm	X	X	X	X
Basal area	X	X	X	X

the Maryland ecological model. The mean correct classification rate for the random samples (91.5%) was higher than for the nest-site samples (73.7%) in all models.

DDAs run on the reduced set of variables produced management models nearly as accurate as the ecological models (Table 6.3). For red-shouldered hawks, the mean correct classification of nest sites was 73% for both models, while this percentage decreased from 77% (ecological model) to 71% (management model) for broad-winged hawks. Except for the Wisconsin broad-winged hawks (management model), the region-specific models yielded more accurate classification results than when regions were combined. Thus, accuracy was lost by combining data from the two regions and by eliminating some ecological variables. Also, the level of accuracy differed between the two species.

Table 6.3. Differences between actual nest sites and randomly selected sites for each hawk species, using discriminant function analysis. (Ecological model shown first; management model in parentheses.)

	Red-shouldered hawk: Nest site	Red-shouldered hawk: Random	Broad-winged hawk: Nest site	Broad-winged hawk: Random
MARYLAND				
Nest site	29 (27)	6 (8)	85 (82)	14 (17)
Random	3 (2)	97 (98)	13 (12)	87 (88)
Percent correct	93.3 (92.6)		86.4 (85.4)	
WISCONSIN				
Nest site	16 (17)	7 (6)	23 (19)	11 (15)
Random	4 (5)	69 (68)	9 (8)	64 (65)
Percent correct	88.5 (88.5)		81.3 (78.5)	
COMBINED				
Nest site	39 (40)	19 (18)	103 (101)	30 (32)
Random	11 (11)	162 (162)	25 (27)	148 (146)
Percent correct	87.1 (87.4)		82.0 (80.7)	

Linear discriminant functions created with the data from one state classified the samples from the other state less accurately (Table 6.4). The decrease averaged 9.6% for broad-winged hawks and 12.6% for red-shouldered hawks.

Discussion

The development of a practical and statistically sound model for predicting woodland hawk nesting habitat depended on several factors. First, a priori selection of habitat variables was not possible. Many potentially useful variables had to be measured at many nest sites in two separate geographic regions. Second, it was necessary to determine (1) if the measurements were amenable to statistical testing; (2) if the variables could be interpreted in an ecological context; and (3) if they were likely to be usable by land managers. Finally, we had to determine if the characteristics we had chosen had good predictive value.

The variance associated with some variables was so high that adequate sample sizes were not realistic to achieve. Examples are certain evergreen characteristics and the number of snags (Table 6.1). For certain other variables (e.g., distance measurements), data transformations were necessary for proper statistical treatment. Because some variables, such as those obtained from density boards and other measurements of the understory (e.g., shrub density), were correlated with other variables (Table 6.1), we used only the most easily measured variable in the models (Table 6.2). Therefore, developing the models required careful evaluation of variables, resulting in selection of those that would be both manageable for statistical analyses and feasible to measure.

Many habitat characteristics were useful for differentiating beween nest sites and the habitat available elsewhere (our random samples). Distances from nests to water and forest openings, canopy height, and the number of medium- and large-sized trees were variables found to be useful in the predictive models. By contrast, other variables (e.g., ground cover, altitude) were not useful predictors. Overall, the models accurately classified samples in nest-site or random-habitat categories in 78% to 92% of the cases (Table 6.3) when applied within the region from which the data had been gathered.

The process of predicting habitat use in one region on the basis of data from another produced mixed results. For example, predicting hawk nest sites in Maryland using the Wisconsin management model was successful for 44% of the broad-winged hawk samples and 83% of the red-shouldered hawk samples (Table 6.4). Predicting Wisconsin nest sites from the Maryland model, however, was successful for 85% of the broad-winged hawk samples and 65% of the red-shouldered hawk samples. Thus, there was variation due both to species and to direction of the predictive process. This occurred for statistical reasons (Lachenbruch and Mickey 1968; Daruna and Karrer 1981) and because of geo-

Table 6.4. Number of samples classified in each category by ecological and management models, using discriminant function analysis, when predicting classification of samples from one state by models created from the other. (MD = Maryland; WI = Wisconsin.)

	Red-shouldered hawk			Broad-winged hawk		
	n	Nest site	Random	*n*	Nest site	Random
ECOLOGICAL MODEL						
Nest sites						
MD predicting WI	23	12	11	34	25	9
WI predicting MD	35	30	5	99	58	41
Random samples						
MD predicting WI	73	12	61	73	15	58
WI predicting MD	100	24	76	100	12	88
Overall percent correctly classified						
MD predicting WI		76.0			76.6	
WI predicting MD		78.5			73.4	
MANAGEMENT MODEL						
Nest sites						
MD predicting WI	23	15	8	34	29	5
WI predicting MD	35	29	6	99	44	55
Random samples						
MD predicting WI	73	13	60	73	24	49
WI predicting MD	100	21	79	100	6	94
Overall percent correctly classified						
MD predicting WI		78.1			72.9	
WI predicting MD		80.0			69.3	

graphic variation in habitat use by birds (see, e.g., Noon et al. 1980; Collins 1983b; Titus 1984).

Some of the geographic variation might have been due to differences in available habitat between regions. Titus (1984) found significant differences between regions (univariate and multivariate tests) in both habitat availability and nest-site habitat used by each hawk species. However, when he rescaled nest-site data with a z-score transformation using the regional habitat availability (random) data, tests on the rescaled data showed that differences in raptor habitat use between regions were due to differences in available habitat in nearly all cases. Similar results were noted by Vassalo and Rice (1982) between insular and mainland avifaunas. Results using the number of overstory trees in the plot illustrate the commonest outcome of the univariate tests (Fig. 6.1). The means for the random data differed between regions, as did the means for the nest-site data ($P < 0.01$). However, there was no difference ($P = 0.39$) when the rescaled data were tested. In essence, although habitat gradient scales differed between regions, hawks selected the same portion of the gradient.

Our results from cross-regional model application indicate that forest managers may be able to use available data from their inventories and nest-site data from other areas to identify potential hawk nesting habitat. To verify the accuracy of the model, those areas classified as potential nest sites should be searched for the presence of nesting raptors. More accurate results should be obtained if nest-site data are available from the forest, or an area nearby.

The major cost of developing our models was the travel and manpower required to locate an adequate number of nests. Many methods have been used to find raptor nests (Fuller and Mosher 1981). Managers can learn of nest sites from foresters, local birdwatchers, ornithologists and wildlife biologists. Additional expenses include the cost of instruments and the time required to measure the variables at each site. Fortunately, the instruments (e.g., dbh and meter tape, clinometer, or altimeter) are usually inexpensive and often might be on hand. By minimizing the number of variables to be measured, the time at each nest site is kept short. The collection of data for 33 variables by two experienced people

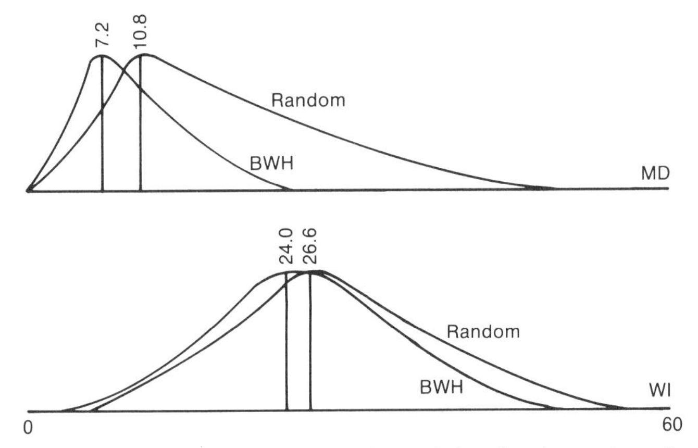

Figure 6.1. Stylized distributions of actual data for the number of overstory trees/0.04 ha with dbh ≤20 cm. Random = random habitat samples, BWH = broad-winged hawk habitat samples, MD = Maryland, and WI = Wisconsin.

required an hour, while about one-half man-hour is required at each nest to gather data for the management models presented in this chapter.

We believe the methods used to gather data are efficient, and that the variables measured are useful for identifying habitat associated with red-shouldered and broad-winged hawk nest sites. If additional data are gathered in a similar manner, from a diversity of areas, it should be possible to develop more generally applicable models for a variety of woodland raptors.

Acknowledgments

We are grateful to B. Haug, J. Devereux, J. Coleman, D. Lyons, M. Kopeny, J. Partelow, M. Mahaffy, and D. Hurley for the time and effort they devoted to finding nests and measuring habitat. We appreciate the cooperation of the personnel of the Nicolet National Forest and the Maryland State Forest Service. Computer services were provided by the University of Maryland Computer Science Center. Funds for this research were provided through USFWS contracts (FWS–14–16–0009–80–007 and FWS–14–16–0009–79–007) to J. A. Mosher and a data analysis contract (FWS 85800–5336–3) to K. Titus. We thank G. A. Feldhamer, R. Drobney, W. F. Laudenslayer, Jr., C. J. Ralph, J. C. Rice, R. P. Morgan II, M. J. Conroy, M. A. Howe and C. S. Robbins for their comments on drafts of this chapter. This is Contribution No. 1604–AEL of the Appalachian Environmental Laboratory and Technical Report-11 of the Central Appalachian Raptor Ecology Program.

7

Determining Avian Habitat Preference by Bird-Centered Vegetation Sampling

DIANE L. LARSON and CARL E. BOCK

Abstract.—A common method for determining avian habitat preference is to estimate bird numbers across sites and correlate them with average characteristics of those sites. An alternative is to measure habitat variables on small plots centered on individual birds and compare these results with data from randomly situated plots. We estimated numbers of four breeding bird species on nine shrubsteppe sites in New Mexico, and measured major habitat characteristics on 25-m radius plots located randomly or centered on individual birds. Bird abundance estimates were correlated with average characteristics of the nine sites. We compared attributes of random vs bird-centered plots across and within the sites using analysis of variance. Correlation and bird-centered analysis (BCA) never gave conflicting results, but BCA permitted detection of many more statistically significant habitat preferences. BCA allowed us to determine when species used different criteria for general site selection than for selection of specific places within occupied sites. We were also able to construct habitat-suitability-index models using BCA data. Advantages of the BCA method include (1) reduction of certain problems associated with estimating bird numbers; (2) generation of larger sample sizes for a similar amount of field work, at least in species-poor habitats; and (3) determination of avian habitat preferences and predictability at a variety of geographic scales using the same data set.

Habitat selection may occur at a number of spatial scales and need not be based on the same criteria at each (Wiens 1981b). Correlating average habitat characteristics with abundances of birds across a number of sites cannot reveal the selection of specific kinds of habitat features within a given occupied area. Bird-centered vegetation sampling, such as that developed and tested by James (1971), Whitmore (1975), Roth (1979), and Karr and Freemark (1983), might better enable us to determine the spatial scale at which individual birds select areas with various habitat features.

This study had three objectives: (1) to compare the correlation method of habitat analysis with the bird-centered analysis (BCA) method, to determine which provided the most insight into avian habitat selection; (2) to compare habitat-selection information at several spatial scales to see which habitat characteristics appeared necessary for occupation of general areas, whether the same characteristics also predicted locations of birds within occupied sites, and whether habitat selection by a species was predictable from site to site; and (3) to use BCA data to construct a habitat-suitability-index model as an example of the utility of the method.

DIANE L. LARSON: Department of Environmental, Population, and Organismic Biology, Box B-334, University of Colorado, Boulder, Colorado 80309. *Present address:* Department of Biology, University of Iowa, Iowa City, Iowa 52242

CARL E. BOCK: Department of Environmental, Population, and Organismic Biology, Box B-334, University of Colorado, Boulder, Colorado 80309

Study area and species

We analyzed habitat preferences of four species of breeding birds—horned lark (*Eremophila alpestris*), sage sparrow (*Amphispiza belli*), Brewer's sparrow (*Spizella breweri*), and western meadowlark (*Sturnella neglecta*)—on nine shrubsteppe sites in the Rio Puerco basin, Sandoval County, northwestern New Mexico. Data were collected in June and July 1983. The nine study sites fell into two broad categories: lowland areas with scattered saltbush (*Atriplex canescens*) and cholla (*Opuntia imbricata*), and upland areas where dominant shrubs were rabbitbrush (*Chrysothamnus nauseosus*) and especially big sagebrush (*Artemisia tridentata*). Sites were delineated on the basis of general uniformity of habitat; each was large enough to accommodate two or three 500-m line transects at least 400 m apart, or approximately 80–120 ha.

Methods

DATA COLLECTION

Bird species' abundances were estimated along the 500-m transects, using the method of Emlen (1971) as applied to shrubsteppe habitat by Wiens and Rotenberry (1981b, and pers. comm.). We walked each transect six times over the 2-month period (except for two transects on site 6, which became inaccessible in July and were therefore walked only three times). Results were pooled to give a single abundance estimate for each species at each site.

Habitat characteristics were measured on 10 randomly located, 25-m radius plots on each of the nine study sites. At 50 sample points per plot, located on lines oriented in the four cardinal compass directions, we recorded presence of bare ground and/or presence (by canopy) and maximum height (cm) of each species of forb, grass, and shrub. These data were pooled to give estimates of major habitat characteristics of each plot, including (1) percent bare ground; (2) percent canopy cover of forbs, grass, and/or shrubs; (3) mean maximum height of all vegetation; (4) mean maximum grass-forb height; (5) mean maximum shrub height; and (6) percentages of the total shrub canopy consisting of the various shrub species.

Vegetation was sampled in an identical manner around locations of individual birds at each study site. Birds usually were located from a distance of at least 20 m to reduce observer influence on them. We did this for most of the species present (Larson 1984) but report here only on four of the most common species. We walked along bird transects at each site for 1 hour per morning on 4 consecutive days each month and marked locations where individuals were first seen. We located and subsequently sampled 144 bird-centered vegetation plots for the four species (n = 37, 33, 41, and 33 for horned larks, sage sparrows, Brewer's sparrows, and western meadowlarks, respectively). We made no effort to locate a predetermined number of birds; rather, we assumed that birds were located in proportion to their abundances at each site. This method is similar to that used by James (1971), except that we sampled all adult birds encountered, rather than just singing males, to avoid a bias in vegetation height. Between half and two-thirds of the birds sampled were singing males. Vegetation data from bird-centered plots were grouped into those points $\leq$5 m (n = 12), $\leq$11 m (n = 24), and $\leq$25 m (n = 50) from the plot centers.

Data analysis

Pearson product moment correlation coefficients were computed between estimates of species' abundances and habitat characteristics, averaged over the 10 randomly located 25-m radius plots on each site. This was done for each species across all sites (n = 9 in all cases) and across occupied sites only. Horned larks, sage sparrows, Brewer's sparrows, and western meadowlarks occurred at five, three, five, and five of the nine sites, respectively. The small sample size across occupied sites made it difficult to achieve statistical significance with the correlation method; however, the time required to increase the sample size substantially would have been prohibitive. Therefore, to be as conservative as possible in our comparison of the two methods, we report significant correlations across occupied sites at both the $P \leq 0.01$ and $P \leq 0.05$ levels. For all other statistical tests, alpha was set at $P \leq 0.01$, because of the large number of comparisons being made.

Characteristics of bird-centered analysis (BCA) plots were compared with those of random plots through one-way analysis of variance. BCA plots were compared with random plots from all sites to investigate criteria for site occupancy. This analysis was repeated for each species, using data from occupied sites only. Finally, BCA plots and random plots were compared for each bird on each site individually, in an attempt to learn if habitat characteristics that determined site occupancy consistently predicted locations of birds within occupied sites. We also examined predictability of the characteristics of species' BCA plots by comparing the distributions of habitat variables around their median values.

Results and discussion

Comparison of the methods

Seven habitat variables either correlated with the abundances of one or more of the four species or differed significantly on random vs bird-centered plots occupied by at least one species (Table 7.1). Although some tests could not be considered statistically independent (e.g., bare ground vs grass cover, occupied sites vs all sites), our results clearly indicated the value of the BCA method in the present case. Results of the different analyses were never contradictory, but bird-centered analyses identified more significant habitat associations for the four species. Using data from all sites and species, correlation analysis revealed three significant associations, but 5-, 11-, and 25-m radius BCA detected 14, 13, and 13 significant habitat preferences, respectively (Table 7.1). Restricting the analysis for each species to occupied sites only, correlation analysis showed two significant associations at the 0.01 level and four additional associations at the 0.05 level, but BCA revealed 12, 7, and 7 habitat preferences for the three plot sizes.

Larger sample sizes necessarily increased the power of BCA over correlation analysis, even though BCA involved less field time. Collection of data from random plots was necessary for both analyses. In addition, it took about 15 min to collect vegetation data for each BCA plot ($\times$144 = 36 hours) and approximately 30 hours to locate the birds (a total of 66 hours of field time). There were 24 transects on the nine sites, each walked six times at about 0.5 hour/walk (= 72 hours field time). Restricting the BCA analysis to 5-m radius plots would make this comparison even more favorable, and the smaller plots were at least as powerful at revealing habitat preferences as the larger plots (Table 7.1).

Scale of habitat selection

BCA revealed much about the geographic scale at which birds selected certain habitat features. For example, the proportion of the shrub canopy consisting of sagebrush was a powerful predictor of occurrence of the four species across all nine study sites (Table 7.1). Brewer's and sage sparrows are considered sagebrush-dependent in the western United States (Braun et al. 1976; Rotenberry and Wiens 1980); our data showed that they selected sites where it was common. On the other hand, western meadowlarks and horned larks were absent from sagebrush-dominated sites. None of the four species occurred on BCA plots across occupied sites where the proportion of sagebrush differed from random (Table 7.1). It appears that for most birds, dominance by sagebrush was associated with habitat selection only at the broadest geographic scale we measured—namely, whether or not a species occurred at a particular site at all. Western meadowlarks were an exception, however, in that they were negatively correlated with sagebrush across occupied sites (Table 7.1). The lack of evidence of such an association on

Table 7.1. Significant associations between birds and habitat variables

	Horned lark				Sage sparrow				Brewer's sparrow				Western meadowlark			
		BCA plot radius[a]				BCA plot radius[a]				BCA plot radius[a]				BCA plot radius[a]		
Variable	Correlation[b]	5 m	11 m	25 m	Correlation[b]	5 m	11 m	25 m	Correlation[b]	5 m	11 m	25 m	Correlation[b]	5 m	11 m	25 m
Shrub cover (%)																
All sites	−0.46				0.36	+	+	+	0.83**	+	+	+	−0.18			
Occupied sites	−0.74				0.99**	+	+	+	0.81*	+	+	+	−0.40			
Bare ground (%)																
All sites	−0.23				0.27	+			−0.60	−	−	−	−0.64	−	−	−
Occupied sites	0.44	+	+	+	0.85				−0.87*	−	−	−	−0.15	−		
Grass cover (%)																
All sites	0.40				−0.31	−	−	−	−0.04				0.57	+	+	+
Occupied sites	0.13	−			−0.58	−	−	−	0.35				0.43	+		
Maximum height of vegetation																
All sites	−0.03				−0.01	+	+	+	0.80**	+	+	+	0.29	+		
Occupied sites	−0.52				0.93	+	+	+	0.90*	+	+	+	0.02			
Maximum grass/forb height																
All sites	0.28				−0.35				−0.58			−	0.27		+	+
Occupied sites	0.41				0.72				−0.72				0.76			
Maximum shrub height																
All sites	0.22				−0.02				0.38	+			0.35			
Occupied sites	−0.07				0.81	+			0.90*	+			−0.18			
Sagebrush (%)																
All sites	−0.35	−	−		0.83**	+	+	+	0.12	+	+	+	−0.37		−	−
Occupied sites	−0.39				−0.28				−0.50				−0.88**			

[a] For random bird-centered analysis (BCA) plots, $n = 90$ for all sites; 50 for horned larks, Brewer's sparrows, and western meadowlarks across occupied sites; and 30 for sage sparrows across occupied sites. Symbols indicate significant ($P \leq 0.01$) positive (+) and negative (−) associations.

[b] For all species across all sites, $n = 9$; across occupied sites, $n = 5$, 3, 5, and 5 for horned larks, sage sparrows, Brewer's sparrows, and western meadowlarks, respectively.

** r values $P \leq 0.01$.

* r values $P \leq 0.05$ (occupied sites only; see Methods).

meadowlark BCA plots suggests that some other habitat characteristic (e.g., grass cover) may actually have been more important at a local level.

Brewer's and sage sparrows selected areas within occupied sites with above average shrub cover and height (Table 7.1). However, site-by-site analysis suggested that these two species actually were selecting local areas based on very different criteria. Brewer's sparrows occurred on those five sites with the highest average shrub canopy cover (Fig. 7.1*a*). They also selected areas within each site with greater than average shrub canopy, although these differences were significant in only two of five cases. By contrast, although sage sparrows occupied BCA plots with greater than average shrub canopy across all occupied sites combined, they were missing from site 2, which had the most shrub canopy (Fig. 7.1*b*). The dominant shrub at site 2 was rabbitbrush; this was also the most abundant shrub at site 7, where Brewer's sparrows again occurred in the absence of sage sparrows. Sagebrush was the dominant shrub at sites 1, 8, and 9, which were the only places we found sage sparrows.

These results suggest that Brewer's sparrows selected areas with the most shrubs regardless of species, but sage sparrows selected areas where sagebrush was the dominant ground cover. BCA revealed no preference among sage sparrows for areas dominated by sagebrush within occupied sites (Table 7.1), because they occurred only where it grew in nearly monotypic stands.

Western meadowlarks were missing from the three sites with the least grass cover and selected local areas with greater than randomly available grass cover across the five occupied sites combined (Fig. 7.1*c*). However, the amount of grass cover was highly variable on BCA plots centered on western meadowlarks from site to site, which is consistent with the observation that this species is a habitat generalist (Wiens and Rotenberry 1981b).

Horned larks were unique among the four species, in that it was easier to predict where they would occur within an occupied site than to predict which sites they would occupy. Horned larks generally are considered bare-ground species (Cannings and Threlfall 1981). BCA plots centered on horned larks included significantly more bare ground than random plots across all occupied sites (Table 7.1), yet horned larks were missing from the three sites with the most bare ground (Fig. 7.1*d*). Selection of areas with large amounts of bare ground apparently occurred at a local level

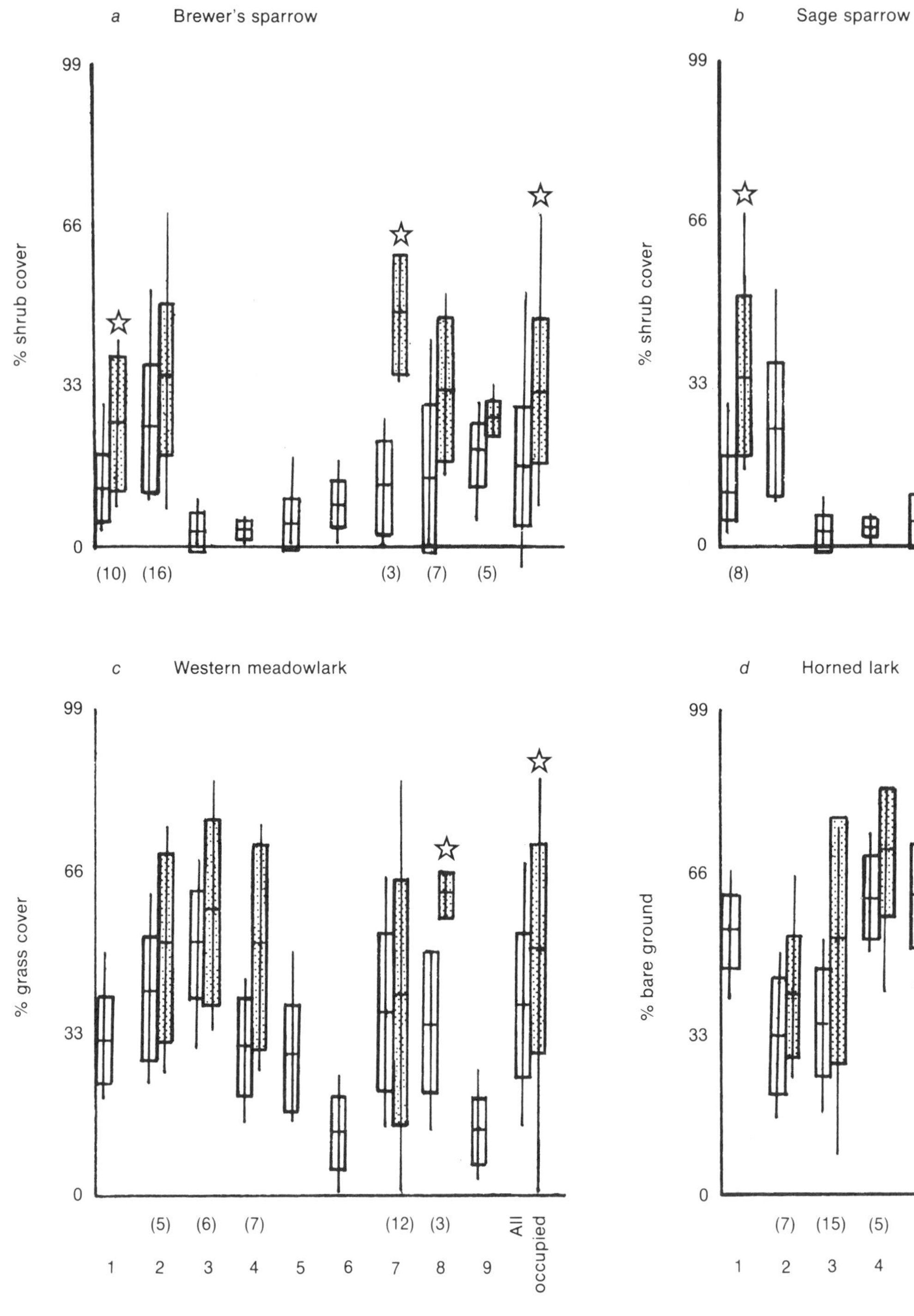

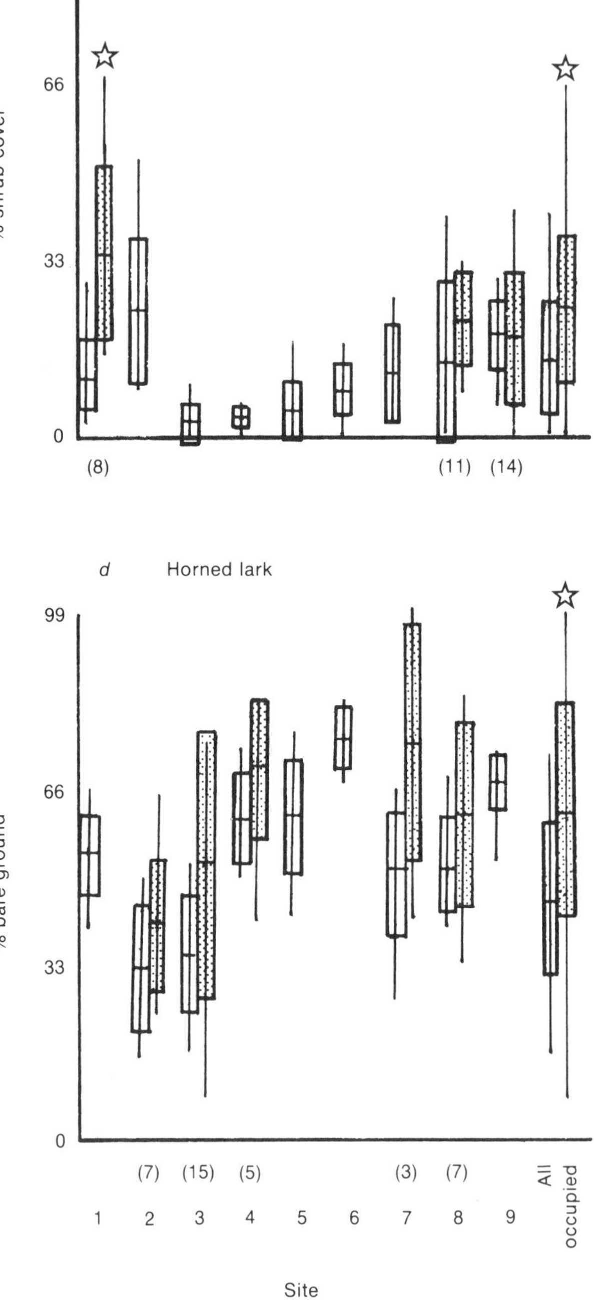

Figure 7.1. Bird-centered (shaded) and random (open) values for selected plot characteristics based on 5-m BCA. The horizontal line is the mean, the vertical line is the range, and the rectangle encloses ± 1 *SD*. Stars indicate a significant difference between bird-centered and random plots using one-way ANOVA. Numbers in parentheses are sample sizes of bird-centered plots.

in the Rio Puerco basin, after general site selection already had taken place.

Predictability of habitat selection

For a species to be a valuable indicator of environmental conditions, the variation of occupied habitats should be small, or at least should be understood. We examined this by plotting the distribution of major habitat variables for each species around their median values (Fig. 7.2). We expected variables differing from random on BCA plots to be the most consistent, because the birds presumably selected for those attributes. This was not always the case. Sage sparrows selected areas with tall vegetation both across all sites and across occupied sites (Table 7.1), and they did occupy BCA plots that were relatively predictable for this characteristic (Fig. 7.2). However, they also chose areas with sparse grass cover and above-average shrub canopy, characteristics for which their BCA plots were comparatively variable (Fig. 7.2). Western meadowlark BCA plots were relatively unpredictable for all major attributes we measured, which again may reflect meadowlarks' overall habitat generalization (Wiens and Rotenberry 1981b). Brewer's sparrow BCA plots

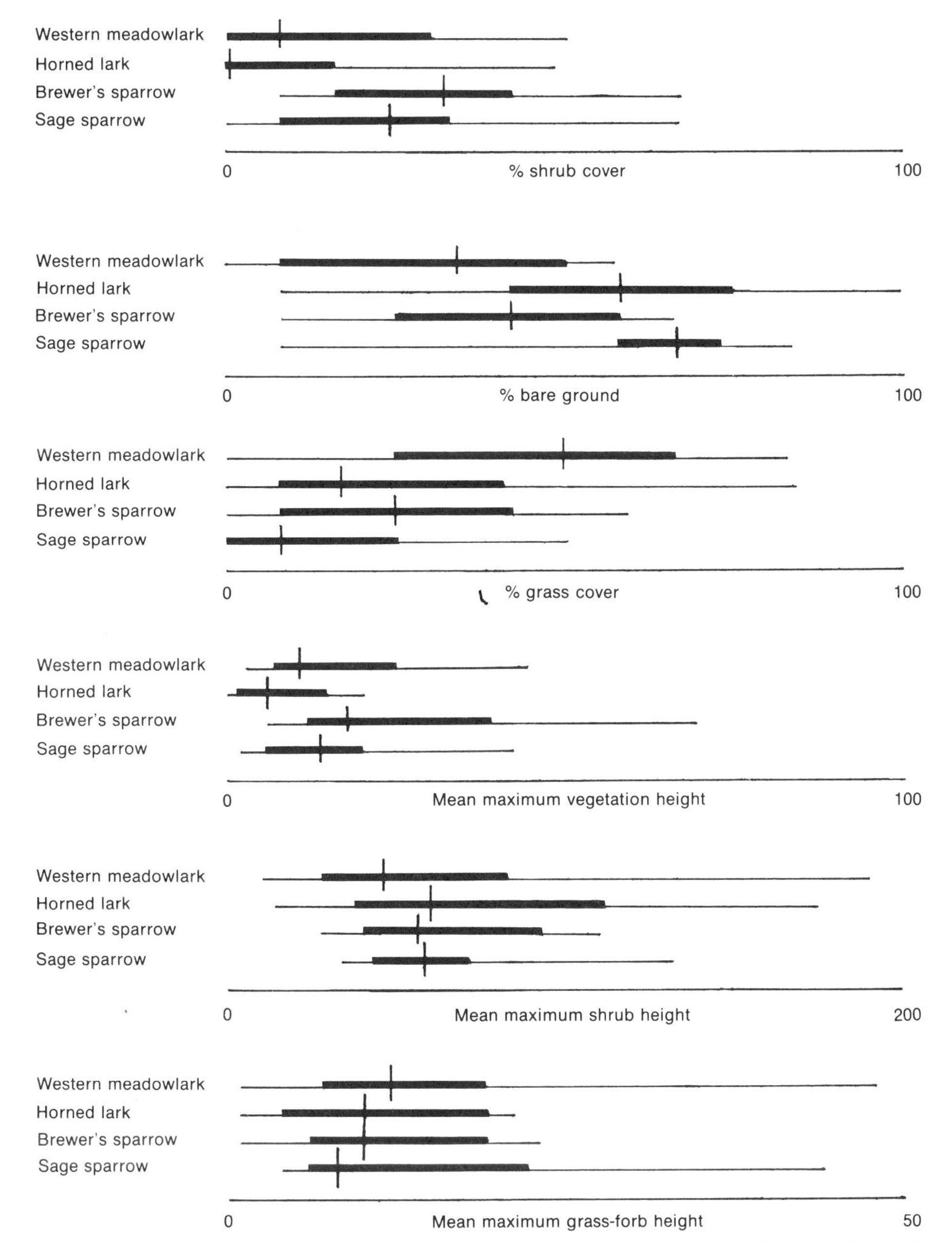

Figure 7.2. Variability of bird-centered plots for six habitat characteristics. Vertical lines are the medians, horizontal lines the ranges. Bars encompass two-thirds of all plots, centered on the median.

were most variable in terms of amounts of bare ground and grass cover (Fig. 7.2), but bare ground was significantly lower on BCA than on random plots (Table 7.1).

Horned larks selected areas with less grass and more bare ground than average in occupied sites (Table 7.1; Fig. 7.1*d*). However, shrub cover was the least variable attribute of their BCA plots (Fig. 7.2). Bock and Webb (1984) also found low shrub canopy cover to be the best predictor of horned larks in southeastern Arizona. It may be that average shrub cover in our New Mexico study areas was near the ideal for horned larks, so that BCA and random plots did not differ on average, even though this is an important variable for the species.

In some cases, variability in selection characteristics may be a result of a consistent pattern or *direction* of habitat selection, rather than the result of a consistent habitat "search image" (Klopfer 1963). For example, the average amount of bare ground on horned lark BCA plots was very different from site to site (Fig. 7.1*d*), leading to a high overall variability for this characteristic (Fig. 7.2). However, in each of the five occupied sites, horned lark BCA plots had greater than average bare ground (Fig. 7.1*d*; runs test, $n = 5$, $P = 0.03$). If a species consistently chooses areas with the extremes for a given habitat characteristic, rather than from some predetermined range of values, then the species' predictability for that characteristic will not reflect the characteristic's importance to the bird. BCA can reveal this sort of habitat selection, as well as selection for areas with absolute habitat characteristics.

Habitat suitability modeling

Data from BCA can be used to construct habitat suitability index (HSI) models. Three pieces of information must be used to construct the HSI model: (1) the values of the habitat characteristic in question on sites where the species did not occur (HSI = 0); (2) the values on sites where the species did occur but selected plots differing from random (connects HSI = 1 with HSI = 0); and (3) the values on sites where the species occurred and selected plots not differing from random (HSI = 1).

To illustrate, we have drawn a suitability-index model showing the relationship between Brewer's sparrows and shrub cover on our study sites in northwestern New Mexico (Fig. 7.3). Brewer's sparrows did not occupy sites with 8% or less shrub cover, resulting in an HSI value of 0. When shrub cover exceeded 8% but was less than 13%, they selected plots with significantly more shrub cover than was available randomly, suggesting that habitat there was acceptable but not ideal. On sites having 13% or more shrub cover, plots occupied by Brewer's sparrows did not differ from random for this characteristic (Fig. 7.1), suggesting an HSI value of 1. Because none of our sites exceeded 30% shrub canopy cover, we cannot specify an upper limit on this model. It may be that when shrub cover exceeds a certain value, suitability for Brewer's sparrows declines. If an upper limit could be identified, the range of values over which suitability was equal to 1 could be used as a measure of predictability.

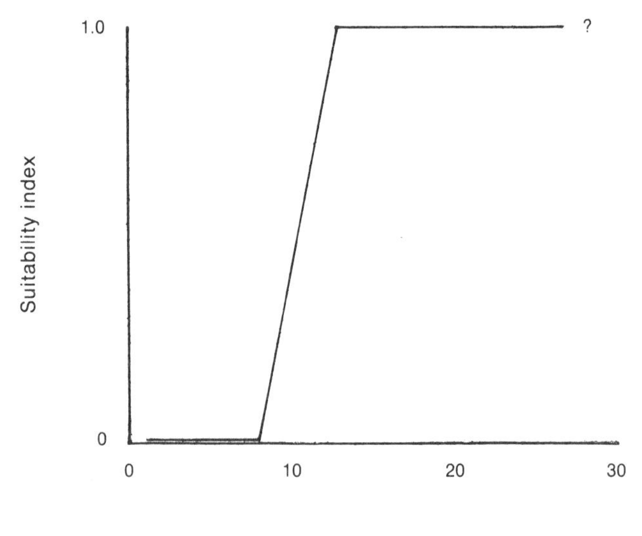

Figure 7.3. Habitat-suitability-index model for Brewer's sparrows and shrub cover in the Rio Puerco Basin, northwestern New Mexico.

HSI models cannot describe directional habitat-selection patterns such as we found in the horned lark relationship with bare ground. However, because this pattern did not influence the birds' general site selection, it is probably of little importance in terms of management considerations.

Conclusions

The four species we studied exhibited definite habitat preferences, both at the site-occupancy level and within occupied sites. No overall trends in habitat predictability were evident, although our results suggest that habitat selection occurred in two ways—either for absolute characteristics or for extremes from among those configurations locally available.

There is much debate about "structure" in shrubsteppe breeding bird assemblages, or, conversely, the lack of structure (e.g., Wiens and Rotenberry 1981b). Our study was too limited in temporal or geographic extent to add much to this debate. It does suggest, however, that bird-centered habitat analysis is a more powerful tool for examining habitat relationships than other traditional methods.

The primary advantages of BCA are its precision and efficiency. Results suggest that plots as small as 5 m in radius showed most major habitat relationships in shrubsteppe communities, although larger plots might be necessary in forested areas. We sampled vegetation on only 12 points in each 5-m plot, which usually involved 5–10 min of field work. The large sample sizes associated with BCA made resulting statistical analyses more powerful. Data could be pooled at various geographic scales to examine habitat preferences within individual stands of vegetation, among a series of occupied stands, or among all stands in a region.

The BCA method has certain limitations. Roth (1979)

noted that if a species selects areas with intermediate characteristics and avoids extremes, occupied vs random plots may have the same average values. This may have been the case in the present study for horned larks and shrub canopy. This sort of event, however, should be detectable through examination of the consistency of habitat selection (Fig. 7.2).

BCA cannot be used to describe correlation between bird species' abundances and habitat variables. However, annual abundances, at least for shrubsteppe birds, are so highly variable and so weakly related to habitat structure (Wiens and Rotenberry 1981b) that such correlations may not be meaningful in any event.

In BCA, birds must be sampled randomly and in approximate proportion to their abundance at the various sites. Detectability may vary with habitat, sex, and nesting phenology, requiring the introduction of a correction factor such as that used in many transect methods (Ralph and Scott 1981). However, various other problems inherent in estimating bird numbers are avoided or minimized because vegetation characteristics are measured only where a bird actually was present. Although statistical power is increased by the large sample size associated with BCA, stochastic variation is also increased because locations of individual birds are used. A bias may be introduced toward those individuals occupying more obvious perches, although this did not appear to be the case in our study.

Finally, whole-community analysis may be difficult in species-rich habitats because of the large number of BCA plots that would have to be sampled. We found BCA an efficient and insightful method for assessing avian habitat relationships in our study, where we dealt with species-poor bird assemblages in structurally simple and similar habitats.

Acknowledgments

This study was supported in part by the USDA Forest Service, Rocky Mountain Forest and Range Experiment Station (Cooperative Agreement No. 28–C2–218), and by the National Audubon Society. We thank E. Aldon, D. Armstrong, D. Bodine, A. Cruz, T. Eicher, M. Grant, W. Grenfell, B. Knapp, S. Larson, M. Raphael, T. Strong, and J. Verner for advice and assistance.

8

Building Predictive Models of Species Occurrence from Total-Count Transect Data and Habitat Measurements

KIMBERLY G. SMITH and PETER G. CONNORS

Abstract.—We present a method combining total counts of birds (i.e., categorical data) with habitat measurements to predict occurrence of individual species. We used two statistical programs to achieve these models. We first used logistic regression with a stepwise procedure to identify potentially important habitat variables. We then used a functional analysis of categorical data (FUNCAT) to build predictive models. Significance of both the predictive model and predictor variables within each model is calculated directly in FUNCAT in a manner analogous to analysis of variance. We demonstrate this technique for eight postbreeding species of waterbirds, using habitat measurements and total-count data (i.e., total individuals for each species) from 25 transects censused 10 times each in late summer 1975–1978 near Barrow, Alaska. Although significant FUNCAT models based on one habitat variable were found for most species in most years, usefulness of the technique appeared better for some species (e.g., Baird's sandpiper [*Calidris bairdii*]) than for others (e.g., ruddy turnstone [*Arenaria interpres*], semipalmated sandpiper [*C. pusilla*]). Reasons for this are discussed, as are criteria used to select the best FUNCAT models: (1) likelihood ratio probabilities, which are lack-of-fit tests of model adequacy; (2) variable probabilities, which tell the significance of variables in the models; and (3) number of empty cells.

A recent trend in avian ecology has been to quantify habitat data in conjunction with avian population studies. To date, most of these studies have been descriptive with only a few attempts to build predictive models of avian habitat relationships, notable exceptions being Robbins (1978), Rice et al. (1984), and Rotenberry (Chapter 31). But, as Noon et al. (1980) aptly pointed out, this predictive value will ultimately determine the usefulness of any habitat model.

In many avian population studies, transect censuses are used to determine the number of individuals (i.e., total-count data) (see Verner and Ritter 1985). Habitat variables are also measured along each transect. Managers, who must combine these data sets to make meaningful decisions, face a dilemma as to how these data should be analyzed because of the categorical nature of total-count data. Because count data are not continuous, they are not analyzed appropriately by many of the statistical procedures familiar to managers. Here we discuss a method that combines habitat measurements with total-count (i.e., categorical) data collected over several census periods to produce models that predict individual species' occurrences based on habitat variables. We illustrate this technique using data on some common post-breeding birds and habitat measurements gathered on transects in middle to late summer near Barrow, Alaska.

KIMBERLY G. SMITH: Department of Zoology, University of Arkansas, Fayetteville, Arkansas 72701

PETER G. CONNORS: Bodega Marine Laboratory, University of California, Bodega Bay, California 94923

Methods

STUDY AREA AND SAMPLING PROCEDURES

Birds were censused in summers of 1975–1978 near Barrow, Alaska (71°17′N, 156°46′W), within the coastal area extending 22 km southwest and 6 km southeast of Point Barrow. Data were collected in conjunction with investigations to assess dependence of shorebirds on arctic habitats potentially susceptible to perturbations from offshore oil development activities (Connors et al. 1984). A variety of habitats were studied, including tundra-backed ocean beach, gravel spit beach, lagoon estuary, ocean estuary, closed brackish lagoon, saltmarsh, mudflats, and brackish storm-flooded pools.

Each summer we censused along permanent transects marked with stakes at 50-m intervals. In narrow shoreline areas, stakes defined a single row of square census subplots, 50 m on a side. In areas of more extensive continuous habitat, such as mudflats, stakes defined a double parallel row of 50-m subplots. Estimates of 50-m distances were made by eye to one side of the stakes or to each side of the stakes in double-row subplots. Eleven transects were censused in 1975, 22 in 1976 and 1977, and 23 in 1978. Most transects were 1 km long, and 19 transects encompassed the same area (50,000 m^2). Five transects were half of the above area because of limited available habitat, and one was about three times that area. In most studies concerned with presence or absence, census areas should be the same on all transects (Verner and Ritter 1985). Because we are only focusing on the eight most common bird species (see below), any bias due to different census areas is probably minimal. (For more details concerning the characteristics of each habitat, see Connors et al. [1984].)

All transects were censused once within each 5-day interval, beginning in mid-July and ending in late August or early September, yielding 10 census periods per year. Here we analyze data for the eight most common species: ruddy turnstone (*Arenaria interpres*), pectoral sandpiper (*Calidris melanotos*), Baird's sandpiper (*C. bairdii*), dunlin (*C. alpina*), semipalmated sandpiper (*C. pusilla*), western sandpiper (*C. mauri*), red phalarope (*Phalaropus fulicaria*), and glaucous gull (*Larus hyperboreus*).

Within each 50-m subplot on all transects, we characterized the littoral habitat in either 1976 or 1978 by estimating or measuring the following seven variables: (1) distance from shore (SHORE) = distance from the center of 50-m subplot to nearest major shoreline ($\bar{x}$ = 19.6 m, range = – 16.6–275.0 m);(2) width of normal flood zone (NFLD) = distance from mean water level to tide limit of annual storm inundation ($\bar{x}$ = 57.6 m, range = 10.0–400.0 m); (3) width of maximum flood zone (MFLD) = distance from mean water level to highest water level as indicated by farthest inland driftwood line ($\bar{x}$ = 163.6 m, range = 0–800.0 m); (4) water cover (H2OCOV) = percent of plot covered by water ($\bar{x}$ = 7.5%, range = 0–50.5%); (5) salinity (SAL) = salt concentration of water measured on the subplot ($\bar{x}$ = 23.0 ppt, range = 4.0–35.0 ppt); (6) substrate particle size (SUB) = nonlinear size particle gradient from mud (0) to gravel (5) ($\bar{x}$ = 3.86, range = 0.7–5.0); and (7) vegetation cover (VEGCOV) = percent of exposed area on plot covered by plants ($\bar{x}$ = 14.2%, range = 0–95.2%). Although habitat measurements were collected in each subplot, means and ranges represent average values for each transect. No transformations were performed on any variables because no statistical assumptions were associated with LOGIST or FUNCAT concerning the underlying distribution of individual variables (other than that the variables were multivariately normally distributed). We assumed that these variables were nearly constant over the census period in all 4 years and therefore used the same habitat measurements for each census.

Statistical analyses

Building predictive models for each species was accomplished by using two computer procedures in the SAS statistical program package: logistic stepwise regression (PROC LOGIST [SAS Institute 1983]) and functional analysis of categorical data (PROC FUNCAT [SAS Institute 1982b]). These statistical analyses were performed on a data matrix formed by combining the total number of each species observed in each census on a transect with habitat variables for that particular transect. For example, in 1978 this translated into a 15 (eight bird species and seven habitat variables) by 230 (10 censuses on each of 23 transects) matrix.

FUNCAT is the statistical analysis of interest, but it is difficult to use initially because the program does not discriminate statistically unimportant from important variables. Rather, FUNCAT uses all variables supplied to build predictive models, regardless of whether a particular variable is a good predictor or not. To determine potentially important variables for FUNCAT analysis, the entire data matrix was first analyzed with a logistic regression analysis using a stepwise procedure. LOGIST, like any stepwise analysis, builds regression models by adding, in order of importance, variables that meet some criterion of probability (e.g., $P = 0.10$) and ignores the other variables. In our example, we used LOGIST to identify a few habitat variables for each bird species to be examined in the subsequent FUNCAT analyses. LOGIST uses two states for the response variable—presence (1) or absence (0)—and does not take into account multiple sampling, as FUNCAT does. LOGIST also assumes continuous variables, although categorical variables can be handled by creation of dummy variables.

Like most regression analyses, a problem with using LOGIST is that it is difficult to assess the statistical adequacy of the models accurately. In LOGIST, a residual chi-square value can be calculated (using a PRINTQ statement) for all variables not included in a LOGIST model. This statistic represents a lack-of-fit test such that a significant chi-square value (e.g., $P < 0.05$) means that the model does not adequately fit the data. Thus, the smaller the chi-square value (and the higher the associated P-value), the "better" the model (see Table 8.1). A high residual chi-square value would be due either to error or to a variable that was not included. One would like, however, to have a direct test of the model's adequacy; this is the rationale for continuing the analysis with FUNCAT.

Unlike LOGIST, FUNCAT uses repeated observations on each transect to construct models of species' presence based on frequency—the number of censuses in which a species was present divided by the total number of censuses. Statistical adequacy of these models can be tested directly by calculating a maximum likelihood ratio probability in a manner analogous to an analysis of variance (ANOVA) table. (For a more detailed discussion, see SAS Institute 1982b.) Likewise, as in ANOVA, significance of each predictor variable can also be calculated in each analysis. Thus, for any given FUNCAT model, statistical adequacy of the model, plus the significance of each variable within the model, can be determined.

In our example, the probability of only two states (presence and absence) was estimated, although models that estimate probabilities for more than two states are possible. Statistical assumptions associated with these models are (1) independence of response variables among transects; (2) independence of response variables within transects, i.e., each census is independent; and (3) a binomial distribution between probability of presence and absence such that the sum is 1.00. In our study of postbreeding and migrating birds, we assumed that a 5-day interval between censuses was sufficient to satisfy the second assumption concerning independence within transects. This approach also assumes that different 5-day censuses are equivalent samples of species' occurrence on a transect, thereby ignoring the inherent seasonality of these postbreeding bird populations. Provided that no seasonal shifts occur in habitat use during census periods, however, seasonal fluctuations are constant across all habitat variables. Further, the effect of population seasonality is dampened by the use here of presence/absence data rather than total numbers. FUNCAT assumes categorical variables but will accept continuous variables if a DIRECT statement is used.

Table 8.1. Summary of logistic stepwise regression analyses of bird-habitat relationships, showing significant habitat variables, the probability (*P*) associated with the residual chi-square (see text), and whether subsequent FUNCAT analyses produced a significant model using the same habitat variables. (Dashes indicate that no significant models were found.)

Species	Year	Variable in model	*P*	Significant FUNCAT model
Ruddy turnstone	1976	—	—	—
	1977	SAL, SUB, VEGCOV	0.03	No
	1978	SAL	0.16	No
Pectoral sandpiper	1976	SAL	0.79	Yes
	1977	SUB, SHORE, VEGCOV	0.97	No
	1978	SAL	0.18	Yes
Baird's sandpiper	1976	NFLD, VEGCOV	0.74	Yes
	1977	SUB	0.63	Yes
	1978	SUB	0.65	Yes
Dunlin	1976	H2OCOV	0.02	Yes
	1977	H2OCOV, VEGCOV	0.35	Yes
	1978	SUB	0.00	No
Semipalmated sandpiper	1976	SAL, VEGCOV	0.56	No
	1977	SUB, SAL	0.18	No
	1978	SAL	0.03	No
Western sandpiper	1976	SAL	0.55	Yes
	1977	SUB, SAL	0.54	Yes
	1978	SUB, H2OCOV	0.06	Yes
Red phalarope	1976	NFLD	0.99	Yes
	1977	SAL, SUB, NFLD, H2OCOV	0.47	Yes
	1978	SAL, VEGCOV	0.54	Yes
Glaucous gull	1976	NFLD	0.54	Yes
	1977	SAL	0.02	No
	1978	SAL	0.78	Yes

Unlike LOGIST analyses, in which data for each species were analyzed in one computer run, the best FUNCAT models were determined by performing several computer runs, but still examining only the variable (or variables) deemed important for each species from LOGIST analyses. This included examining individual habitat variables, combinations of variables, interactions between variables, and curvilinear (i.e., squared) relationships.

FUNCAT builds models based on the number of unique values measured for a particular variable. For example, 14 different values were measured for salinity in 1976 and 1977, so a FUNCAT model based on salinity alone in those 2 years would have 14 cells. If more than one variable is used in a FUNCAT model, the number of cells equals the number of unique variable combinations. Sample sizes must be large enough so that enough degrees of freedom remain when a model is constructed based on unique variable values. In our example, censusing only 11 transects in 1975 resulted in too few degrees of freedom to construct meaningful models for most bird species. We therefore focus on the other 3 years in the remaining sections of this chapter.

Results

Results of the LOGIST analyses are presented in Table 8.1. Statistically adequate ($P < 0.10$) logistic regression models (total of 23) were found for all species in all years with the exception of ruddy turnstone in 1976. Thirteen of these models contained only one variable, seven contained two variables, two contained three variables, and the model for red phalarope in 1977 contained four variables. Salinity was a significant variable in 13 models, followed by substrate (9), vegetation cover (6), water cover (4), normal flood zone (4), and distance to shore line (1). Maximum flood zone did not enter into any of the models. Twelve of the 23 models had probabilities above 0.50, and 11 had probabilities below 0.50. Ten of the 12 "better" models also produced adequate FUNCAT models (Table 8.1) using exactly the same variables selected by LOGIST, but only four of the 11 "poorer" models produced adequate FUNCAT models. This suggests that initial screening of habitat variables using LOGIST can assist in developing adequate FUNCAT models.

In 20 of 24 FUNCAT analyses, good models were found for the eight species in the 3 years of study (Table 8.2). Of these, 18 models were based on one predictor variable (or its square), one (red phalarope in 1978) was based on two variables, and one (red phalarope in 1977) was based on three variables. Salinity was a significant predictor in 10 FUNCAT models, followed by substrate (6), water cover (3), vegetation cover (1), and normal flood zone (1). Distance to shore and maximum flood zone did not appear in any of the final FUNCAT models.

For several bird species, the best FUNCAT model was not based on the LOGIST habitat variable (or variables) for that particular year, but on habitat variables that were significant in 1 of the other 2 years. For example, although

Table 8.2. Summary of FUNCAT analyses of bird-habitat relationships, showing likelihood probability associated with each model (model *P*), intercept and predictor variable(s) with associated weights (*b*) for that model, probability that a predictor variable is a significant predictor (variable *P*), and number of empty cells/total cells possible (see text). (Dashes indicate that no significant models were found.)

Species/Year	Model *P*	Intercept	*b* (variable)	Variable *P*	Empty cells/ total cells
Ruddy turnstone					
1976	—	—	—	—	—
1977	0.33	1.893	−0.457(SAL)	0.001	4/14
			0.012(SAL)(SAL)	0.001	
1978	—	—	—	—	—
Pectoral sandpiper					
1976	0.83	0.120	−0.207(SAL)	0.001	9/14
1977	0.07	−4.320	0.543(SAL)	0.13	10/14
			−0.025(SAL)(SAL)	0.06	
1978	0.08	0.973	−0.333(SAL)	0.02	8/15
			0.006(SAL)(SAL)	0.14	
Baird's sandpiper					
1976	0.23	−0.001	−0.502(SUB)	0.001	0/7
1977	0.62	0.393	−0.663(SUB)	0.001	0/7
1978	0.72	1.080	−0.798(SUB)	0.001	0/7
Dunlin					
1976	0.35	−1.529	0.068(H2OCOV)	0.001	0/6
1977	0.42	−1.301	0.260(H2OCOV)	0.001	0/6
			−0.004(H2OCOV)(H2OCOV)	0.001	
1978	0.10	−1.174	0.035(H2OCOV)	0.001	0/6
Semipalmated sandpiper					
1976	—	—	—	—	—
1977	0.54	−0.384	0.814(SAL)	0.27	0/7
			−0.275(SAL)(SAL)	0.03	
1978	—	—	—	—	—
Western sandpiper					
1976	0.40	0.662	−0.159(SAL)	0.001	5/14
1977	0.23	0.856	−0.888(SUB)	0.001	0/7
1978	0.17	−1.732	1.115(SUB)	0.13	0/7
			−0.268(SUB)(SUB)	0.03	
Red phalarope					
1976	0.81	0.190	−0.005(NFLD)	0.03	0/16
1977	0.25	−1.284	0.091(SAL)	0.001	0/15
			−0.314(SUB)	0.05	
			0.024(H2OCOV)	0.08	
1978	0.60	−3.082	0.079(SAL)	0.001	1/17
			0.021(VEGCOV)	0.01	
Glaucous gull					
1976	0.42	0.586	−0.158(SAL)	0.05	0/14
			0.004(SAL)(SAL)	0.05	
1977	0.06	−0.025	−0.161(SAL)	0.06	1/14
			0.006(SAL)(SAL)	0.005	
1978	0.85	−4.477	0.123(SAL)	0.001	5/15

normal flood zone and vegetation cover were significant habitat variables in the 1976 Baird's sandpiper LOGIST model, and although these two variables produced an adequate FUNCAT model, substrate, the variable most important in the 1977 and 1978 LOGIST models, also produced the best FUNCAT model in 1976.

A single habitat variable was a significant predictor in all 3 years for pectoral sandpiper, Baird's sandpiper, and glaucous gull (Table 8.2). Of these, models for Baird's sandpiper were superior because all three were based on simple linear relationships. The probability of predicting the occurrence of a Baird's sandpiper varied somewhat among years at the low (mud) end of the substrate gradient. For example, the probability of observing a Baird's sandpiper on a transect with a substrate value of 1.0 was about 40% in 1976, 50% in 1977, and 60% in 1978 (Fig. 8.1). The probability of observing a Baird's sandpiper was fairly consistent at the upper (gravel) end of the gradient, fluctuating between 5% and 10% (Fig. 8.1).

In 2 of 3 years, models for glaucous gull contained an interaction term of salinity squared (Table 8.2), suggesting that glaucous gulls are to be expected more commonly on transects with low or high salinities than on transects with intermediate values (Fig. 8.2). This may have some biological reality, or it may be a statistical artifact based on high probability of occurrence at 9 and 35 ppt salinities (Fig. 8.2). In any event, the high variability of observed probabilities around the predicted probability curve from the FUNCAT

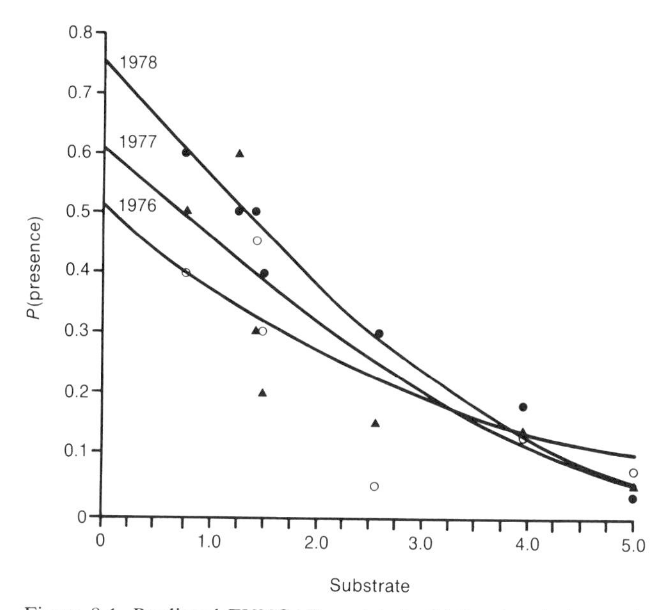

Figure 8.1. Predicted FUNCAT models (solid lines) and observed (symbols) probabilities of presence for postbreeding Baird's sandpiper as a function of substrate in 1976 (open circles), 1977 (triangles), and 1978 (solid circles).

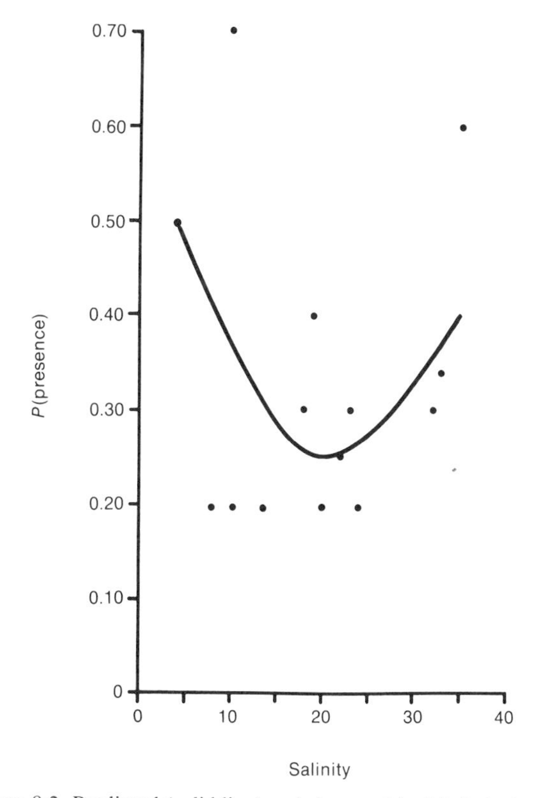

Figure 8.2. Predicted (solid line) and observed (solid circles) probabilities of presence for glaucous gull as a function of salinity in 1976.

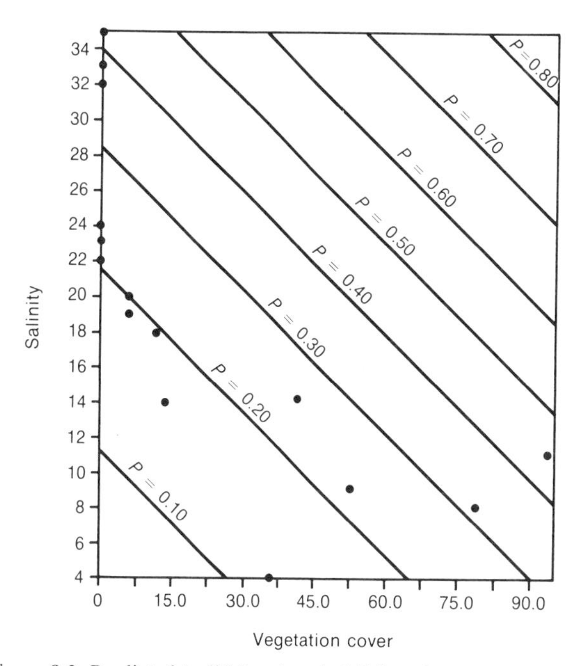

Figure 8.3. Predicted (solid lines) probabilities of presence for postbreeding red phalaropes as a function of salinity and vegetation cover in 1978. The solid circles show the combinations of salinity and vegetation cover measured. Actual probabilities of occurrence are not shown.

model suggests a limited applicability of this model in management decisions.

Models containing two variables can also be easily depicted. For example, two variables influenced red phalaropes in 1978. The model contained both salinity and vegetation cover, and the probability of occurrence ranged from 20% to 40% when vegetation cover was absent (Fig. 8.3). However, the probability of occurrence also increased to about 45% as vegetation cover approached 100%. Note that this model predicts high probability of occurrence when both salinity and vegetation cover are high, a situation which may not occur in nature.

Discussion

For six of eight species, the FUNCAT procedure produced predictive models of species' occurrence based on one or a few habitat variables, suggesting its potential usefulness in management decisions. From a management point of view, however, usefulness of a model could vary considerably, depending on the management question being asked. For example, models derived for Baird's sandpiper would seem to be useful because there is a consistent, year-to-year relationship between one habitat variable and the probability of occurrence of that species. Models based on the same variable among years, but in which the relationship switched from linear to curvilinear, as occurred in models for four species (Table 8.2), may not be as useful as those in which the mathematical relationship is the same for all years.

Models that exhibit no consistency in important variables among years, as shown by those for red phalarope, although still biologically interesting, may have limited use in management decisions. In our own work on oil development and postbreeding shorebird habitat selection, determination of bird-habitat relationships would allow assessment of habitat suitability during a longer period of summer and fall prior to bird migration, rather than during peak bird migration.

In our analyses, we purposely attempted to find single-variable FUNCAT models. We did this for several reasons, including simplicity in interpretation, simplicity in year-to-year comparisons both within and between species, and ease of depicting models graphically. It may be unrealistic to expect to predict species' occurrence with a high probability based on only one variable, because occurrence of most species will be affected by a host of factors. It is possible that we sacrificed some predictive power by examining only simple models. The most useful models, of course, would be those that have few predictor variables and consistently high predictive power.

Careful selection and sampling of variables is critical to the success of a FUNCAT analysis. In an analysis of the type presented here, variables should be selected that potentially relate to presence of the species under study. It is then important that replicate samples be taken throughout the range of each variable. In our study, for example, salinity ranged from 0 to 35 ppt, and we sampled at 14–15 sites within this range. Replicate sampling along lower and higher ends of the salinity gradient could have helped determine whether the curvilinear relationship between salinity and glaucous gull occurrence was real or was a statistical artifact based on a few outlying data points (Fig. 8.2). We suggest that extremes and median values for a particular variable should be sampled intensively to produce adequate FUNCAT models.

FUNCAT was developed to analyze data sets where multiple observations have been made on a categorical response variable (presence or absence of bird species in our example) in conjunction with either categorical or continuous predictor variables. (Continuous predictor variables are recommended because categorical predictor variables use up degrees of freedom rapidly. This problem, however, can be overcome by appropriate experimental design prior to sampling.) Response variables should be sampled enough times to reflect accurately the response to predictor variables. In our example, we believed that 10 censuses over the fall migration period accurately estimated distribution of bird species on the transects. Predictor variables, on the other hand, should be sampled as often as deemed necessary. We sampled habitat variables once in 4 years (one sample for each transect). We could have sampled each year (four samples) or sampled after each census (40 samples). Choice of predictor variables will, in most cases, determine the sampling regime.

Three criteria can be used to select the "best" FUNCAT models. First, the likelihood ratio probability, which is a lack-of-fit test of model adequacy, should be examined. Because the null hypothesis is that there is no significant difference between observed data points and those calculated from the FUNCAT model, the higher the likelihood ratio probability, the better the model fits the data. Using this criterion, 20 of 24 FUNCAT models in this study were adequate ($P > 0.05$) fits to the observed data (Table 8.2).

The second criterion is to examine variable probabilities. That is, given that a model is a good fit to the observed data, are the variables in the model significant (e.g., $P < 0.05$) predictors of the observed data? It does not necessarily follow that a good model must have significant predictor variables, as shown by the pectoral sandpiper FUNCAT model in 1977 (Table 8.2). The likelihood ratio P is > 0.05, signifying an adequate fit to the observed data, but neither variable—(SAL) or (SAL) (SAL)—is a significant predictor ($P > 0.05$). In some cases the curvilinear term, which had to be included to obtain a model $P > 0.05$, turned out to be a nonsignificant predictor variable (e.g., pectoral sandpiper in 1978). The opposite was also true: a significant curvilinear term coupled with a nonsignificant linear term (e.g., semipalmated sandpiper in 1977, western sandpiper in 1978).

The third criterion is to examine the number of empty cells. An empty cell refers to the situation in which, for a unique predictor value, no individual of a particular bird species was present. As the introduction to FUNCAT cautions (SAS Institute 1982b), models with many empty cells may lead to spurious conclusions concerning important predictor variables. In our example, only pectoral sandpiper models are suspect, with high numbers of empty cells in all 3 years (Table 8.2). However, this reflects absence of pectoral sandpipers on transects with high salinities, so there is some biological reality to these models. The large number of empty cells is due to the infrequency of pectoral sandpipers on certain transects, but not to sampling problems (e.g., all transects were sampled an equal number of times). Empty cells that occur from inadequately sampling the entire range of a particular variable, or from an inadequate number of samples, could cause problems in interpretation.

Acknowledgments

Financial assistance was provided by the National Oceanic and Atmospheric Administration as part of the Outer Continental Shelf Environmental Assessment Program, by the J. William Fulbright College of Arts and Sciences, University of Arkansas, and grant BSR–8408090 from the National Science Foundation. Computer funds were furnished by the Department of Zoology, University of Arkansas. Initial conversations with J. E. Dunn were helpful. Rita Berg designed the computer analyses, helped interpret results, and made many helpful suggestions concerning description of statistical analyses. William Laudenslayer and Jake Rice offered many suggestions on the entire work.

9

Refinement of the Shugart-Patten-Dueser Model for Analyzing Ecological Niche Patterns

DOUGLAS A. JAMES and M. JOSEPH LOCKERD

Abstract.—Old-field rodent communities in northwestern Arkansas were investigated, using the niche pattern model developed by Shugart, Patten, and Dueser. We refined this model by modifying all three axes: (1) species' abundance; (2) realized niche; and (3) niche breadth. The results showed that the hispid cotton rat (*Sigmodon hispidus*) filled its available habitat, the woodland vole (*Microtus pinetorum*) did not, and habitat for the fulvous harvest mouse (*Reithrodontomys fulvescens*) was scarce. The modified model is useful in predicting future trends in community structure.

The success of a species in a community depends on the prevalence of its habitat requirements there (realized niche), its degree of resource exploitation (niche breadth), and its population level (abundance). These three factors form a three-dimensional niche pattern model proposed by Shugart and Patten (1972), modified first by Dueser et al. (1976), and again by Dueser and Shugart (1979). We present still another refinement of this Shugart-Patten-Dueser model: one that removes interdependencies among the three axes of the model, giving each a unique property, and one that permits the prediction of results due to ecological change. We confine our evaluation to the habitat niche, as did the previous investigators, but ours is based on small-mammal communities occurring in old-field habitats in northwestern Arkansas. Determining niche patterns of species in a community is important because the model identifies factors contributing to species' abundances and allows predictions of population changes.

Methods

Mammal populations

Data were collected between 29 September and 7 November 1978 in four old fields at the Buffalo National River in Arkansas, two at Buffalo Point, one each at Steel Creek and Pruitt. Trap grids were established using rat-sized Sherman noncollapsible livetraps (25 × 8 × 8 cm) placed at 10-m intervals. Grid size and length of trapping sessions varied in each field (see Results, Table 9.2). Captured mammals were ear-tagged and released, and capture locations were noted.

Species' population levels, including 95% confidence limits (Giles 1969), were estimated, using a modification (Lockerd et al. 1979) of the Overton (1965) iterative extension of the Schnabel (1938) method,

$$\hat{N}_{j+1} = \hat{N}_j + \frac{\sum_{i=1}^{k} \frac{M_i n_i}{\hat{N}_j - Z_i} - \sum_{i=1}^{k} x_i}{\sum_{i=1}^{k} \frac{M_i n_i}{(\hat{N}_j - Z_i)^2}}$$

$$j = 0,1,\ldots \text{until convergence},$$

where

$$\hat{N}_0 = \sum_{i=1}^{k} \frac{M_i n_i}{\sum_{i=1}^{k} x_i},$$

in which

$\hat{N}$ = population estimate,
n_i = total number caught in the ith trapping session,
x_i = number of marked individuals captured in the ith trapping session,
Z_i = total number of individuals removed previous to the ith trapping session,
M_i = number of marked individuals at risk of capture in the ith trapping session, and
k = total number of trapping sessions.

Total rodent populations for each plot were estimated from combined species capture-recapture data. Abundance of each species on each plot was calculated as a percentage of the total based on the number of individuals handled, excluding recaptures. This percentage approach, although reducing accuracy, was necessary because of uncommon species in various fields.

Habitat analysis

Traps in which small mammals were captured were used as sample centers for habitat characterization, with not more than two trapping sites used in cases of multiple catches of the same individual. Approximately 20% of the traps in each grid were selected at random to determine overall habitat

Douglas A. James: Department of Zoology, University of Arkansas, Fayetteville, Arkansas 72701

M. Joseph Lockerd: Computer Sciences Corp./EPA, 1860 Lincoln Street, Denver, Colorado 80295

Table 9.1. Correlations between habitat factors and each of the two discriminant functions. (An asterisk identifies the factors used to describe the particular discriminant function; "dicot" and "monocot" refer to dicotyledenous and monocotyledonous plants, respectively.)

Habitat factor	First discriminant function	Second discriminant function
1. Dicot contacts (20–30 cm high)	0.005	0.332
2. Dicot contacts (30–40 cm high)	0.105	0.352
3. Dicot contacts (50–70 cm high)	0.010	0.427*
4. Dicot contacts (70–90 cm high)	0.133	0.393
5. Monocot contacts (20–30 cm high)	−0.012	−0.207
6. Monocot contacts (30–40 cm high)	0.014	−0.135
7. Dicot vertical height diversity	0.197	0.331
8. Monocot biomass in 0.5 m^2	0.106	−0.093
9. Dicot biomass in 0.5 m^2	0.090	0.050
10. Number of dicot species/m^2	0.658*	0.112
11. Average height of tallest sapling	0.084	0.525*
12. Average distance to tallest sapling	0.206	0.159
13. Woody stem density at waist height	0.121	0.113
14. Average vegetation height	0.159	0.176
15. Dicot foliage density (0–120 cm)	0.114	0.452*
16. Monocot foliage density (0–120 cm)	0.133	−0.224

availability in the region. When different species' points and/or random points coincided, the single-habitat sample was used in each data set. Some random points designated traps that had not caught mammals.

A suite of 46 environmental factors for habitat characterization was selected (Lockerd 1980), which was reduced to the 16 vegetational factors named in Table 9.1. These were the factors most highly correlated with each of the first two principal components of rodent-habitat analysis and which thus accounted for the greatest variance in the system. The factors were correlated with each other by a correlation coefficient as high as +0.88 and as low as +0.02. The rest were evenly distributed between $< \pm 0.1$ and ± 0.6. Thus, they were associated strongly with the two components, but were not primarily overly associated with each other. The detailed descriptions of each factor appear in Lockerd (1980). To assist understanding here, it should be explained that the term *contacts* (Table 9.1) refers to total leaves touching a vertical rod in each of various height zones over multiple sampling points around a trap. Vertical height diversity was calculated using the Shannon index (Pielou 1975) applied to the leaf-contact data across nine height categories. Stem density was a count of woody stems along waist-high transects, and total foliage density was the sum of leaf contacts through nine vertical strata from 0 to 120 cm at 12 sampling points. The remaining factors (Table 9.1) should be self-evident.

Variance-covariance structure of the 16 factors was stabilized using a simultaneous power transformation routine of the form $Y = (X^\lambda - 1)/\lambda$, in which Y is the transformed variable, X the original variable, and λ the power transformation. The criterion was to maximize the multivariate normal likelihood function. Homogeneity of variance-covariance matrices of original and transformed data for vegetational factors was evaluated using the Box test statistic (Cooley and Lohnes 1971). Species' differences in habitat use were examined using multivariate analysis of variance (MANOVA). A nested model was adopted to adjust for differences in study sites (fields) that could obscure differences between species. The test statistic was the F-approximation of Wilks's lambda criterion (Cooley and Lohnes 1971).

The overall MANOVA produced characteristic vectors that were used in discriminant function analysis to calculate species' means in discriminant space. A graphics routine produced 95% confidence ellipses around species' means with respect to the first two discriminant functions. The ellipses were based on the standard deviations of the multivariate sample point solutions for each species.

Niche pattern models

Niche pattern diagrams were constructed using the model of Dueser and Shugart (1979) and the alternative methods we adopted. In our model the species' abundance axis was the average population level for each rodent over the four study fields, based on capture-recapture data. Scores for habitat samples on the first two discriminant function axes were used in determining niche size. Thus, niche breadth for each species was the area of the 95% confidence ellipse circumscribing its sample values in the two-dimensional discriminant space. The random habitat samples were used in calculating realized niche availability for each species. The random samples, therefore, represented the availability of various habitats in the study fields, and the species' habitat samples formed the multivariate classification for categorizing each random sample. A random sample was considered to represent habitat available to a particular rodent species if it had a probability greater than 0.75 of fitting the multivariate habitat characterization for that species. The species' values for the realized niche axis were calculated as the percentage of all random samples that were classified as fitting each species. (Among the 105 random samples, 30% did not meet the $P > 0.75$ criterion for any rodent species.)

Computer analyses involving principal components analysis (FACTOR), multivariate analysis of variance (GLM), and classification procedure (DISCRIM) were from the Statistical Analysis System (Barr et al. 1976). Testing for equal variance-covariance matrices (EQCOVA), providing variance stabilization transformations (VARSTB), and producing confidence ellipses (CONFEL) were library programs of Computing Services, University of Arkansas.

Results

Population levels

Sigmodon hispidus (hispid cotton rat) dominated two of the four fields, and *Microtus pinetorum* (woodland vole) dominated the other two (Table 9.2), both showing high variability in numbers among fields. *Reithrodontomys fulvescens* (fulvous harvest mouse) constituted a small but fairly consistent percentage of the total population in fields where it occurred.

Table 9.2. Trapping results in each of the four study fields

	Study field					
	Buffalo Point I	Buffalo Point II	Steel Creek	Pruitt	Combined total	Average
Number of captures (percent of total)						
Sigmodon	114 (98)	36 (74)	0 (0)	2 (18)	152 (70)	
Microtus	2 (2)	0 (0)	34 (83)	4 (37)	40 (18)	
Reithrodontomys	0 (0)	11 (22)	5 (12)	2 (18)	18 (8)	
Other species	0 (0)	2 (4)	2 (5)	3 (27)	7 (3)	
Total captures	116	49	41	11	217	
Density (number/40 ha)						
Sigmodon	7987	1151	0	78		2304
Microtus	163	0	1350	155		417
Reithrodontomys	0	342	199	78		115
Other species	0	62	82	117		65
Total population	8150	1555	1631	428		2941
95% confidence limits[a]						
Upper	9691	2030	2662	739	3780	
Lower	6854	1192	961	247	2313	
Number traps/nights	92/5	118/9	163/13	120/8		
Trapping area (ha)	0.62	1.20	1.35	0.98		

[a]Confidence limits are given only for the total rodent population estimates.

Habitat differences

Over all fields, 196 total habitat characterizations were made: 105 random points, plus 45 *Sigmodon*, 28 *Microtus*, and 18 *Reithrodontomys* points. This represented all *Reithrodontomys* capture points, nearly all for *Microtus*, but less than a third of the *Sigmodon* points (because of repeats). The resulting data were significantly heterogeneous with respect to species' variance-covariance matrices (F = 2.444, df = 408 and 13,345, $P \simeq 0.0$), but variance stabilization transformations reduced F considerably (F = 1.306, $P < 0.05$). Although still heterogeneous, the reduction was considered adequate because MANOVA is robust under slight departures from homogeneity in matrices (Cooley and Lohnes 1971). (If the matrices were stabilized, species' niche breadths based on confidence ellipses in discriminant space would be equal. Our matrices were not completely stabilized, and thus species' confidence ellipses were not equal in size. However, it would be best to use unmodified variance-covariance matrices in generating the ellipses or, alternatively, using sizes of confidence ellipses around species' principal components scores, as did Smith [1977] and White and James [1978].)

There were significant differences in rodent habitats based on MANOVA (F = 1.55, df = 80 and 326, P = 0.004), and paired comparisons showed that *Sigmodon* differed from *Microtus* (F = 1.60, df = 32 and 104, P = 0.04) but not from *Reithrodontomys* (P = 0.28); and *Microtus* and *Reithrodontomys* habitats were significantly different (F = 2.49, df = 32 and 50, P = 0.002).

Correlations of habitat factors with discriminant functions (Table 9.1) identified ecological clines in old-field environs that were important in separating rodents. The first discriminant function was strongly correlated only with the number of herbaceous dicotyledenous species present (r = 0.658) and represents the degree of dicotyledenous species richness. Three factors were strongly correlated with the second discriminant function: average height of the nearby tallest saplings (r = 0.525), total dicotyledenous foliage density (r = 0.452), and dicotyledenous foliage density between 50 and 70 cm tall (r = 0.427). This represents old-field succession where dicotyledenous foliage density and stature increase with sapling development.

With respect to the first discriminant function (Fig. 9.1), typical *Microtus* habitat contained the fewest dicotyledenous species, whereas *Reithrodontomys* was found in portions of fields that were richer in these plants and *Sigmodon* was intermediate. The rodents showed considerable overlap in their habitats along this environmental axis; *Microtus* was the least selective (widest ellipse dimension on this axis), while *Reithrodontomys* was the most specialized (narrowest). Although the three rodents were similar in stage of field succession (Fig. 9.1), the habitat ellipse mean for *Reithrodontomys* was positioned in an earlier successional stage than for *Sigmodon* and *Microtus*, and the mean for *Microtus* was in the most successionally advanced position. Again *Microtus* was the most generalized and *Reithrodontomys* the most specialized. Differences in habitat use by *Sigmodon* compared to *Reithrodontomys* were also noted in Texas (Kincaid et al. 1983), using a model for evaluating realized habitat niches (Kincaid and Bryant 1983).

Niche patterns

Relationships between species' values for niche pattern parameters calculated according to Dueser and Shugart (1979) (Fig. 9.2) differed from those produced by our model (Fig. 9.3) only on the niche breadth axis.

The measure of species' abundance ($\hat{a}$, Fig. 9.2) by Dueser and Shugart (1979) was a prediction of species' abundance in

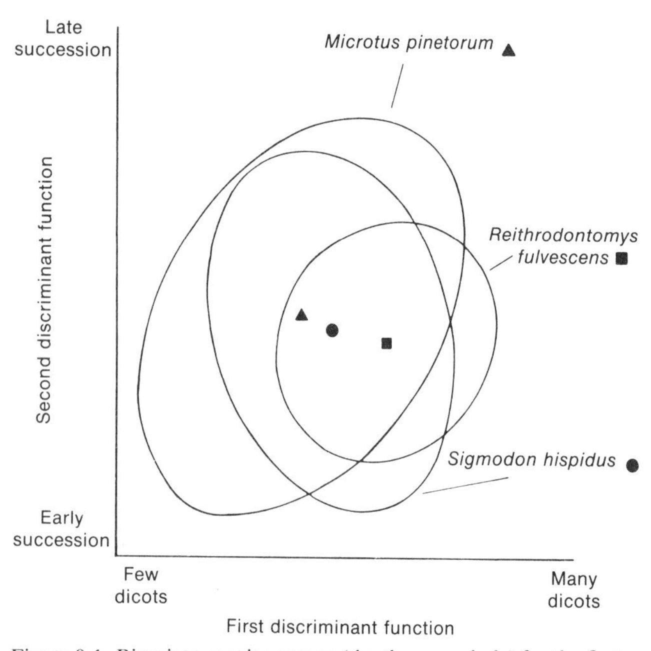

Figure 9.1. Bivariate species means (the three symbols) for the first two discriminant axes, circumscribed by 95% confidence ellipses based on the standard deviation of individual sample points for each species. The discriminant axes represent niche dimensions along which species exhibit habitat selection. ("Dicots" refer to dicotyledonous plants.)

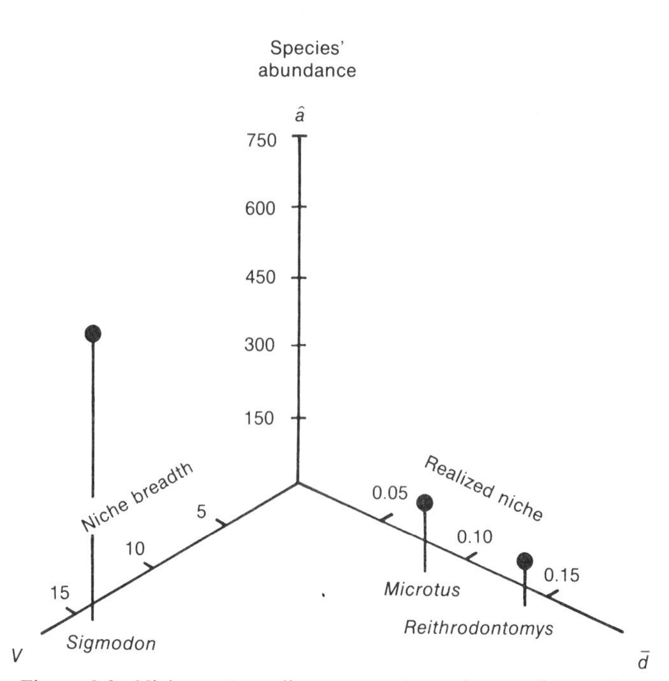

Figure 9.2. Niche pattern diagram constructed according to the methods used by Dueser and Shugart (1979). Increasing values on the realized niche axis represent lesser abundance of preferred habitat. Increasing values on the niche breadth axis represent less habitat specialization. Values on the species' abundance axis are estimates of abundance in optimal habitat.

optimal habitat, whereas the alternative measure we propose ($\bar{a}$, Fig. 9.3) was an empirical determination of average population levels existing in the study areas. The two measures produced nearly identical results on a relative scale. Both models showed *Sigmodon* to be the most abundant species, while both *Microtus* and *Reithrodontomys* had smaller population levels in the region. The approximately equal population heights in Figures 9.2 and 9.3 indicate that the actual populations were near levels expected in optimal habitat.

The realized niche measure ($\bar{d}$, Fig. 9.2) of Dueser and Shugart (1979) was an inverse function in which smaller values indicate greater abundance of preferred microhabitat conditions, whereas our presentation of this parameter ($\hat{h}$, Fig. 9.3) was the percentage of available habitat, and therefore increased as habitat availability increased. Thus, when comparing the two models, one must remember that small values in Figure 9.2 have the same meaning as large values in Figure 9.3. The ordering of the species along the two realized niche axes was the same. Typical *Sigmodon* habitat was most abundant, but less pronounced, in Figure 9.3 than in Figure 9.2, and *Reithrodontomys* habitat was the least common, especially as shown in Figure 9.2. Availability of *Microtus* habitat was intermediate.

In both niche pattern diagrams the niche breadth measures (V, Fig. 9.2; V', Fig. 9.3) were inversely related to specialization by the species, where a high value identified habitat generalization. The biggest difference in the outcome of the two models was in this axis. One showed that *Sigmodon* was an extreme habitat generalist with respect to the other species (Fig. 9.2), but the other indicated that *Microtus* was the greatest generalist (Fig. 9.3).

Discussion

Shugart and Patten (1972) introduced a niche pattern model consisting of three orthogonal axes representing the structure and function in biotic communities. The reliability of the model to depict community relationships accurately and to predict future community structure depends on the theoretical criteria for using one measure of a niche pattern parameter over another. We present below the theoretical basis for developing alternative methods for niche pattern modeling.

Aspects of the original models have been criticized recently (Carnes and Slade 1982; Van Horne and Ford 1982), charging (1) that random habitat samples would represent habitat availability better than combined-species samples; (2) that variance in species' canonical scores would best depict niche breadth; and (3) that unequal sample sizes between species produce large biases. Early recognition of the first two problems (Lockerd 1980) led to our use of random samples to define realized niches and to construct confidence ellipses based on species' canonical scores for niche breadth. The problem of unequal samples remains and is always a difficulty with small-mammal communities, which characteristically have both abundant and rare species. However, Dueser and Shugart (1982) contend that important

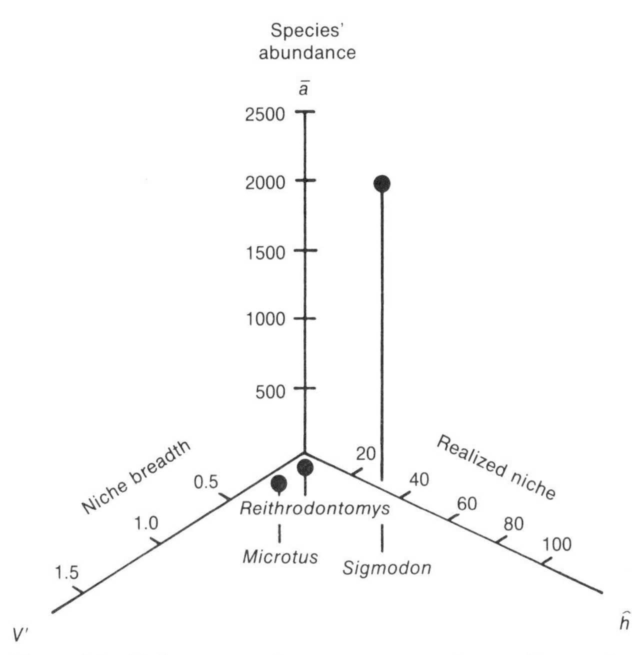

Figure 9.3. Niche pattern diagram constructed according to the methods proposed in the present study. Increasing values on the realized niche axis represent increased abundance of preferred habitat. Increasing values on the niche breadth axis represent less habitat specialization. The third axis shows levels in population abundance.

information is lost in artificially attempting to equalize samples across species. In our study, we adjusted sample sizes somewhat by using all habitat samples of all individuals for rare species and only a portion of those for common ones. Also, our results showed that niche breadth did not have the same rank order as relative abundance (Fig. 9.3), which is what a sampling bias should produce.

Species' abundance

This model axis was first proposed by Shugart and Patten (1972) as a measure of species' abundance in optimal habitat. Dueser et al. (1976) departed from this and used an index of abundance based on trapping results. Dueser and Shugart (1979) used a frequency-of-occurrence function that was based on the variability in their measure of optimal habitat availability ($\bar{d}$, Fig. 9.2). This function was a good predictor of relative population levels in the small-mammal species we studied, because their frequency-of-occurrence function ($\hat{a}$) from Figure 9.2 and our empirically determined abundance measure ($\bar{a}$) from Figure 9.3 produced similar results on a relative scale.

Granting this similarity in results, two factors recommend our empirically derived population estimator ($\bar{a}$) over their frequency-of-occurrence function ($\hat{a}$). First, representing niche pattern parameters with orthogonal axes implies that the three axes are independent. The frequency-of-occurrence function ($\hat{a}$) contains a variance measure that is also a parameter of the niche breadth estimator (V). Thus these axes are not independent. Dependencies between axes make accurate interpretation of results difficult because the axes concerned partly represent common factors and partly different ones. Second, using a predictor of abundance in optimal habitat rather than estimating existing population levels could miss cases in which favored habitat is abundant but the species is actually uncommon or, alternatively, where a species is abundant even though its habitat is limited. These situations could arise among species having cyclic population fluctuations, or at times when recent climatic or other factors have affected population levels.

Realized niche

This axis is a measure of the realized niches (Hutchinson 1957) of the species in the community and implies an index to regional availability of preferred microhabitat conditions. Shugart and Patten (1972), Dueser et al. (1976), and Dueser and Shugart (1979) used distance measures in discriminant space ($\bar{d}$, Fig. 9.2), which assumed that the distance from individual species' means to the combined-species mean (grand mean) represented relative availability of favored habitat. This assumption was based on the premise that the grand mean approximated average regional habitat. Because the data used to compute the grand mean were collected at points where animals were encountered, however, the grand mean was only a representation of average habitat for the species concerned, not of average regional habitat.

Use of this function has an inherent potential of biasing the result toward species that are most abundant. Because more data are generally collected for those species, they therefore contribute more information to the discriminant analysis. This characteristically places the common species' mean scores closest to the grand mean. It should also be noted that because the niche breadth measure (V) contains $\bar{d}$ as a parameter, the V axis and the $\bar{d}$ axis are not independent (Fig. 9.2), thus producing further problems in interpretation.

Our alternative method of deriving this niche dimension ($\hat{h}$, Fig. 9.3) has the advantage of sampling random points in the region. The measure results in an unbiased estimate of the proportion of regional habitat that a particular species could be expected to occupy. This measure considers regional habitat that is not occupied by any of the species concerned, allowing speculation about population expansion and potential immigration into the area. Because $\hat{h}$ is a percentage, its value increases as habitat availability increases; thus all three axes bear the same orientation to the origin, which prevents confusion and misinterpretation of results. (Anderson and Shugart [1974] and Shugart et al. [1974] used principal components to represent realized habitat niches of birds in a way that agrees with our recommendations on the matter.)

Niche breadth

The niche breadth axis is an indicator of resource exploitation or degree of specialization by a species. Dueser and Shugart (1979) discussed various possible functions, including lengths of axes of confidence ellipses in habitat space (Green 1974), but used the variability of distances in discriminant space from individual species' points to the grand mean

Table 9.3. Possible niche pattern conditions and expectations for future changes

	Present condition				
Case	Abundance[a]	Realized niche[b]	Niche breadth[c]	Characterization[d]	Future expectations[e]
I	High	Much	Wide	High carrying capacity, peak numbers, ecological space filled.	Will either stay abundant or decline gradually.
II	High	Much	Narrow	Peak numbers, could be overcrowded.	No increase; decrease likely.
III	High	Little	Wide	Peak numbers, using marginal habitats.	Could decline if realized niche not increased.
IV	High	Little	Narrow	Overpopulated, cyclical species at peak numbers.	Will crash if realized niche stays low.
V	Low	Much	Wide	Ecological space unfilled.	Will increase.
VI	Low	Much	Narrow	Not saturated.	Will increase somewhat.
VII	Low	Little	Wide	Ecological space probably filled.	Increase possible if realized niche increased.
VIII	Low	Little	Narrow	Low carrying capacity, ecological space filled.	Stable unless realized niche increased, never abundant unless cyclical.

[a] High: peak numbers, little increase expected; low: sparse population, increase possible.
[b] Much: species' habitat abundant; little: species' habitat scarce.
[c] Wide: generalist, occupying many microhabitats; narrow: specialist, utilizing few microhabitats.
[d] Present conditions based on abundance, realized niche, and niche breadth.
[e] Predicted changes based on present abundance, realized niche, and niche breadth.

(V, Fig. 9.2). Besides producing dependencies between this axis and both the $\bar{d}$ and $\hat{a}$ axes, the V function has an interpretational problem. The meaning of species variability around the grand mean representing combined-species average habitat is uncertain. The slopes of habitat-density functions as used by Shugart and Patten (1972) for niche width have these same drawbacks. A measure of variability around separate species' means representing average habitat utilization of that species would better depict within-species flexibility along niche dimensions.

Our use of areas of 95% confidence ellipses for species' means in discriminant space is a good indicator of relative niche breadth (V', Fig. 9.3). The ellipses are large for generalized species and small for those with specialized habitat requirements. Our method could overestimate niche breadth when observations of species are clustered in a portion of the ellipse, because the entire ellipse is drawn regardless of the dispersion of points within it. The problem is minimized when sample sizes are large and species' variance-covariance matrices are stabilized.

Community niche pattern

Niche pattern analysis designates the relative positions of species with respect to major community properties and is useful for predicting future changes. It has been used with birds and mammals with respect to vegetation, yet it has a much wider application. Conditions can be characterized given a knowledge of species' abundances, their realized niche availabilities, and the niche breadths; and recognizing that some variation in interpretation exists, expected changes can be predicted from these three properties (Table 9.3). This kind of analysis probably should be confined to evaluations of member species in a common guild. Note, however, that the predictive property is possible only with our version of the model, in which species' abundance is the present population level, and is not the extrapolation to abundance in optimal habitats used in the antecedent versions.

Using the results from our version of the model, *Sigmodon* would correspond most closely to case I in Table 9.3, except for having only a moderately wide niche breadth, thus making a further increase even more unlikely. It could increase, however, as ecological succession produces more of its late-stage habitat. At the opposite extreme, *Reithrodontomys* fits case VIII perfectly. It is low in numbers and has little expectation for increase, accentuated because it is an early successional species. *Microtus* is fairly close to case V except for having only a moderate amount of its habitat present. Nevertheless, its ecological space still is not saturated, it is low in abundance, it is an ecological generalist, and thus an increase is imminent. Only cyclical species like *Microtus* become low in abundance when the opposite would be expected based on niche availability and ecological breadth.

Acknowledgments

This study was conducted at the Buffalo National River in Arkansas under contract no. CX–702980013 with the National Park Service. James E. Dunn developed the mathematics for determining population levels, and both he and Jackie D. Tubbs assisted with statistical matters. Herman H. Shugart, Jr., provided essential advice, and Donald W. Kaufman, Raymond D. Dueser, and Kimberly G. Smith commented on a preliminary draft of this chapter. Support was provided by the J. William Fulbright College of Arts and Sciences, University of Arkansas, and under NSF grant no. BSR–8408090. We gratefully acknowledge all of these contributions.

10

Using Food Abundance to Predict Habitat Use by Birds

TIMOTHY BRUSH and EDMUND W. STILES

Abstract.—Many studies have used foliage volume or foliage height diversity to describe habitat use by birds. We tested predictions of habitat use and bird density based on arthropod biomass in the New Jersey Pine Barrens. Predictions of habitat use by insectivorous birds were accurate in most cases and were significantly better than predictions based on foliage volume or diversity. Birds were more abundant in either pine- or oak-dominated habitats, depending on which had a higher arthropod biomass. Densities of cavity-nesting birds showed the same trends as open-nesters, and species were not limited by the availability of nest sites. Densities of insectivores were closely associated with changes in arthropod biomass occurring in their preferred tree taxa. Phenological characteristics of dominant plant species can be used to predict times of arthropod abundance and scarcity for foliage-gleaning birds.

Habitat associations of bird species have been assessed mainly by using foliage measurements such as volume or foliage height diversity (MacArthur and MacArthur 1961; James 1971; Anderson and Shugart 1974; Smith 1977; Swift et al. 1984). Although these indices of habitat suitability have sometimes been accurate, they may be indirect measures of habitat quality if animal prey is the limiting factor. Prey abundance per se has seldom been used to predict the occurrence or abundance of forest birds.

In this chapter we develop a set of predictions relating density of foliage-gleaning birds to arthropod biomass. These predictions incorporate measures of foliage volume and (foliage) arthropod biomass at the plant taxon and habitat levels. We compared bird densities in different habitats at different times of year in the New Jersey Pine Barrens and contrasted predictions based on arthropod biomass with those based on foliage volume alone.

Methods

Study sites

The New Jersey Pine Barrens are dominated by oaks (*Quercus* spp.) and pines (*Pinus* spp., mainly pitch pine, *P. rigida*), with Atlantic white-cedar (*Chamaecyparis thyoides*) and red maple (*Acer rubrum*) important in lowland habitats (Forman 1979). Heaths (Ericaceae) and shrubby oaks are the most frequent understory plants. Canopy cover is seldom complete, but the shrub layer is usually dense. There are no major macroclimatic differences among Pine Barrens habitats, except that cedar swamps are cooler in summer than other habitats (Havens 1979).

Timothy Brush: Department of Biological Sciences, Rutgers University, Piscataway, New Jersey 08854. *Present address:* Manomet Bird Observatory, Box 936, Manomet, Massachusetts 02345

Edmund W. Stiles: Department of Biological Sciences and Bureau of Biological Research, Rutgers University Piscataway, New Jersey 08854

We established nine study sites in eight representative habitats, mainly in pine- and oak-dominated habitats (Table 10.1). The study sites were located in Lebanon and Penn state forests near Chatsworth, Burlington County, New Jersey. Fire history and soil moisture determine the dominant plant species, height, and density of vegetation on these otherwise similar sites (Forman and Boerner 1981). Foliage volume and foliage height diversity were estimated, using the technique of MacArthur and MacArthur (1961). Distances between the sites ranged from 0.1 to 17 km, but most were within 2 km of another site.

Census methods

Arthropods

We collected samples of arthropods from foliage to determine the biomass of arthropods available as food for birds. Each sample involved insertion of 50 randomly chosen terminal foliage clusters 0.3 m (1 foot) in length into a standard cloth insect net with an opening 32 cm (12.6 inches) wide. Each branch was shaken vigorously to dislodge arthropods into the net. Like Gula (1977), we observed few arthropods escaping or failing to be dislodged. Samples were taken at midday (11:00–14:00) on sunny days to reduce weather-related variability. Each individual foliage cluster was taken from a different tree or shrub, one which was at least 3.7 m (12 feet) from the next nearest plant from which arthropods were collected. On subsequent visits to each site, we chose new locations for sampling. After collection, arthropods were frozen, then dried at 60°C (140°F) for 24 hours, or until a constant weight was reached. We weighed each sample on a Mettler A30 balance accurate to 0.0001 gm.

We pooled samples from pines—pitch pine and shortleaf pine (*P. echinata*)—and oaks—black oak (*Q. velutina*), blackjack oak (*Q. marilandica*), scrub oak (*Q. ilicifolia*), white oak (*Q. alba*), and chestnut oak (*Q. prinus*)—because we found similar arthropod taxa in similar abundances within these plant genera. We also pooled samples from heaths (blueberries [*Vaccinium* spp.] and huckleberries [*Gaylussacia* spp.]) and

Table 10.1. Descriptions of study sites censused in the New Jersey Pine Barrens, March–September 1983

Habitat	Size (ha)	Foliage density (m^2/m^3)	Canopy height (m)	Dominant species	Brief habitat description
Oak	24.8	13.8	18–21	Oak	Tall oaks, open understory
Oak-Pine	44.7	14.9	15–18	Oak	Oak and pine, dense shrubs
West Plains–Unburned	24.8	5.2	1.5	Oak	Unburned, shrubby oaks and pines
West Plains–Burned	24.8	4.0	0.6	Oak	Burned, shrubby oaks and pines
Pine-Oak	49.6	8.7	12–15	Pine	Pine with open understory
Penn Pine–Oak	44.7	12.7	12–15	Pine	Pine with dense understory
Pine–Oak Burn	64.5	7.7	6–9	Pine	Burned April 1982
Pine Swamp	24.8	17.0	15–18	Pine	Dense deciduous understory
Cedar Swamp	16.1	9.7	21	Cedar	Tall, dense cedars

from deciduous understory plants (Pine Swamp site) because it was impossible to collect arthropods separately from plants growing so close together.

Arthropods were collected during five seasonal periods—March (3 days of sample collection), April (5 days), early breeding season (1 May through 15 June; 8 days), late breeding season (16 June through 31 July; 6 days), and September (3 days). Periods were chosen to match important phenological events (pers. obs.; Gula 1977). From the beginning of May to mid-June (1 May–15 June), oaks leafed out, while pines showed little needle growth. From mid-June through July (16 June–31 July), oak leaves toughened, while pines completed their annual growth. These events are correlated with major changes in the physical and chemical suitability of foliage as arthropod food (Feeny 1970; Van Balen 1973; Gula 1977; Schultz 1983).

Although our modified sweep-sampling technique was probably not an accurate way to determine actual numbers of arthropods, it provided good estimates of their relative abundance (Folse 1982). This was especially the case because only three arthropod taxa dominated most of the samples: caterpillars (Lepidoptera larvae), beetles (Coleoptera), and spiders (Arachnida). However, we excluded aposematic, probably unpalatable arthropods (small numbers of coccinellids, hemipterans, and homopterans) from our analyses.

The arthropod biomass for each plant taxon was combined with the relative foliage volume of each taxon to develop an index of relative arthropod biomass. For each period, we multiplied mean arthropod biomass per plant taxon by the proportional foliage volume of that plant taxon. We then added the resulting numbers for all plant taxa on each site. Site-level values were multiplied by total foliage volume at each site to arrive at relative arthropod biomass. For example, $[(35AO)(0.62FO) + (40AP)(0.38FP)]14.94FT = 551.3$ (relative arthropod biomass in OP); where AO = arthropod biomass (mg) in oaks, FO = foliage volume (m^2/m^3) of oaks, AP = arthropod biomass (mg) in pines, FP = foliage volume of pines, and FT = total foliage volume.

Birds

Brush censused birds by using the Emlen (1971, 1977) line-transect technique, and densities are reported as birds/40 ha (100 acres). Transects were established along narrow firebreaks or sand roads and were 650–1000 m (2132–3280 feet) long. The use of firebreaks and sand roads was not a problem because the Pine Barrens are characterized by frequent disturbances, and most plants and animals present are edge or successional species. Plants and birds along the edges of firebreaks and sand roads are similar to those away from firebreaks or roads. One or two transects were censused per habitat, and four to eight censuses were done per habitat per period. Although uneven sampling intensity may affect the number and identity of species found in different habitats, those were not our main areas of interest. Density estimates on individual transects in study areas with two transects were similar to overall density estimates for those study areas, such that reasonably accurate comparisons of bird densities in different habitats could be made.

Censusing commenced at first light and continued for no more than 2 hours (except in March and April, when censusing continued until 11:00). The order and direction of censusing were varied randomly within each period. Censuses were done on 4 days in March, 9 in April, 17 in May to mid-June, 17 from mid-June through July, 7 in August, and 4 in September.

PREDICTIONS AND STATISTICAL TESTS

We used foliage volume, foliage height diversity, and relative arthropod biomass separately to predict relative patterns of bird density. We used correlations and G-tests (Sokal and Rohlf 1973) to compare predicted and observed patterns of bird density.

Results

ARTHROPOD BIOMASS

Plant taxa

Arthropods sampled in pines peaked in biomass in mid-June through July, after increasing from very low levels in March and April (Fig. 10.1). Oak arthropods were present mainly at low levels, except for a peak in May to mid-June and intermediate biomass from mid-June through July. Arthropod biomass peaked in May to mid-June in white-cedars,

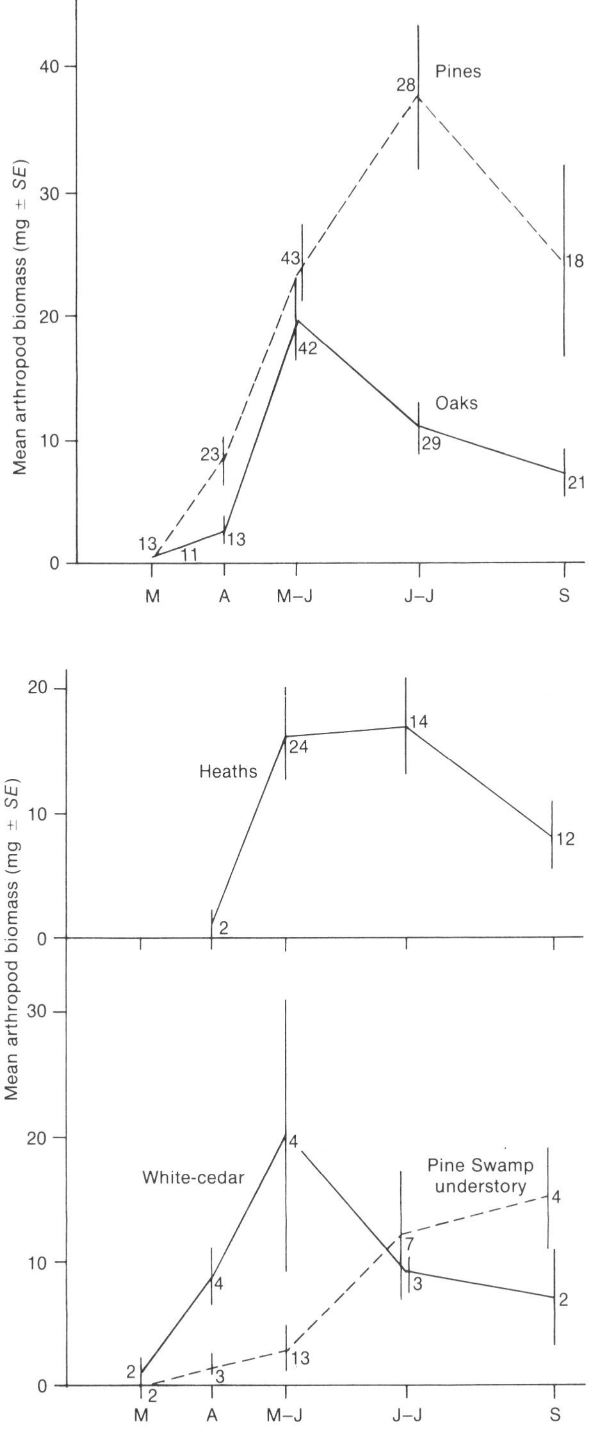

Figure 10.1. Arthropod biomass in different plant taxa in the New Jersey Pine Barrens, March–September 1983. Solid and dashed lines connect means; vertical lines = *SE*. M = March, A = April, M–J = May to mid-June, J–J = mid-June through July, S = September.

but was more variable than in oaks or pines. In the deciduous understory (Pine Swamp), arthropods reached their highest abundance in late summer. Arthropods in heaths were more variable but generally were low in biomass. Spiders (Arachnida) dominated in March and April in all plant taxa, and beetles (Coleoptera) and Lepidopteran larvae were prevalent in most samples taken from May to September (Fig. 10.1).

Study sites

Multiplying arthropod biomass by the proportional foliage volume of each taxon elevated the importance of trends in the dominant taxa, oaks and pines (Fig. 10.2). Pine-dominated sites (listed in Table 10.1) showed peak arthropod biomass from mid-June through July, but two oak-dominated sites (Oak and Oak-Pine) had their highest biomass in May to mid-June. The difference between the oak and pine sites was significant ($G = 459.4$, $P < 0.001$). In the other two oak-dominated sites (West Plains–Unburned and West Plains–Burned), a late freeze (10–11 May) resulted in the destruction of young oak leaves and their associated arthropods. The oaks leafed out again by middle to late June, but the arthropod peak was delayed, coinciding with the peak in pines from mid-June through July. The differences in arthropod biomass between oak sites affected and those unaffected by the freeze was significant ($G = 169.0$, $P < 0.0001$). The pines were not affected by the freeze. Biomass in the cedar-dominated site peaked in May to mid-June but was significantly lower than on the oak and pine sites ($G = 409.6$, $P < 0.001$; and $G = 180.0$, $P < 0.001$, respectively, Fig. 10.2).

Bird abundance related to arthropod abundance

Trends in pine- and oak-dominated forests

Seventy-three bird species were seen on transects during the study, and densities ranged from 0 to 207 per 40 ha. The number of species per study site per period ranged from 0 to 43, and 48 species were present in densities of at least 0.5 birds/40 ha on at least one study site. These species occurred on a mean of 4.6 ($SD = 2.5$) study sites. The 13 most common species made up 88% of the bird community by density and occurred on a mean of 6.9 ($SD = 1.8$) sites (Table 10.2). We concentrated on these species because they were common and relatively widespread. Rufous-sided towhees, for example, constituted 57% of the bird community.

Birds were most abundant at times and places of highest arthropod abundance. On oak-dominated sites, bird densities peaked in May to mid-June, while densities were highest on pine-dominated sites from mid-June through July (different patterns in the two habitats, $G = 26.1$, $P < 0.001$; Fig. 10.3). Bird density and relative arthropod biomass were highly correlated for the March–September period ($r^2 = 0.55$, $P < 0.05$) and for May–July ($r^2 = 0.57$, $P < 0.05$). Percentages of 12 of the 13 most common bird species in oak-dominated habitat declined between May to mid-June and mid-June through July (Table 10.2; average decline 18.4% [$SD = 5.0$]—a significant overall decline; sign test performed on arcsin-transformed data, $P = 0.003$). Chipping sparrows were found only on pine-dominated sites.

Site-by-site variation

Bird density generally varied in concert with arthropod biomass also at the site level (Fig. 10.2). Correlation coefficients for the sites ranged from 0.52 to 0.98 ($\bar{x} = 0.91$) and were significant except for West Plains–Unburned and West Plains–Burned. Individual bird species showed a close association with the plant taxa in which they did most of their foraging (pers. obs.). Densities of rufous-sided towhees closely followed habitat-level food abundance in both the OP and PO sites (Fig. 10.4). Pine warblers followed changes in arthropod biomass in pines. Similarly, black-throated green warblers (*Dendroica virens*) were closely associated with arthropod biomass fluctuations in cedars.

The relationship of bird abundance to other predictors

Foliage volume was a poor predictor of bird density. For the entire period (March–September), and for early summer only (May–July), bird density was not significantly correlated with foliage volume ($r^2 = 0.06$, and $r^2 = 0$, respectively; $P > 0.05$ in both cases). Birds were less common than expected in oak-dominated habitats, except in May to mid-June (predictions based on foliage volume; May to mid-June, $G = 0.5$, $P > 0.25$; mid-June through July, $G = 12.0$, $P < 0.001$; August, $G = 5.6$, $P < 0.025$; September, $G = 12.5$, $P < 0.001$; see Fig. 10.3). Foliage height diversity and bird density were not significantly correlated in summer ($r^2 = 0.11$, $P > 0.05$). Foliage height diversity was significantly correlated with bird species diversity (BSD), but relative arthropod biomass was not ($r^2 = 0.47$, $P < 0.01$; and $r^2 = 0.11$, $P > 0.05$, respectively).

Cavity-nesting birds showed the same trends as open-nesters. Population changes of the two groups in pine- and oak-dominated habitats between May to mid-June and mid-June through July were not significantly different from each other (overall, $G = 0.6$, $P > 0.1$; pine-dominated, $G = 0.2$, $P > 0.5$; oak-dominated, $G = 0.2$, $P > 0.5$). Only 9.3% of 247 apparently suitable snags contained cavities. Densities of Carolina chickadees, the most abundant cavity-nesters, did not change following removal or addition of cavities (Brush, pers. obs.).

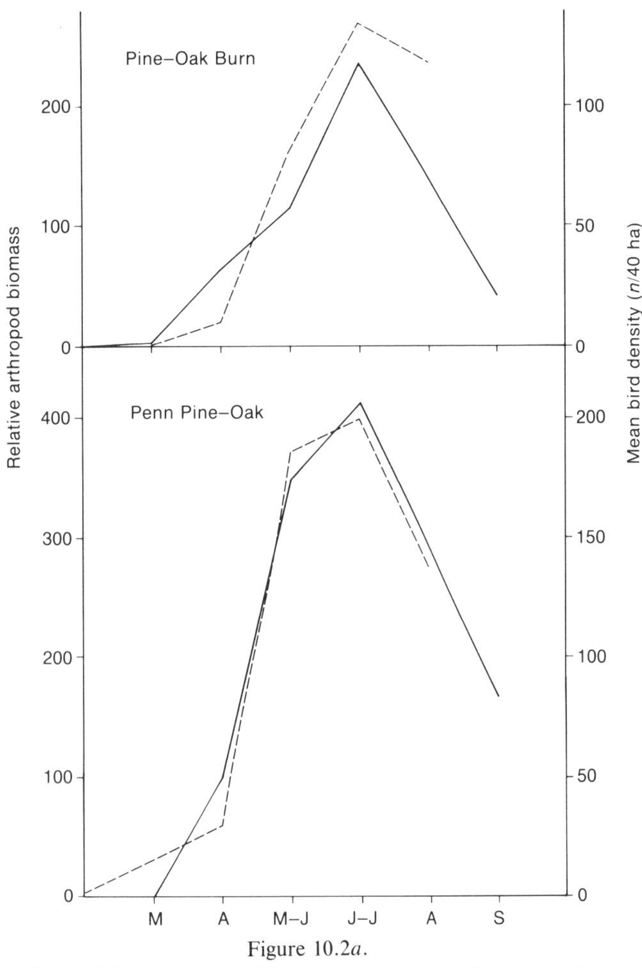

Figure 10.2*a*.

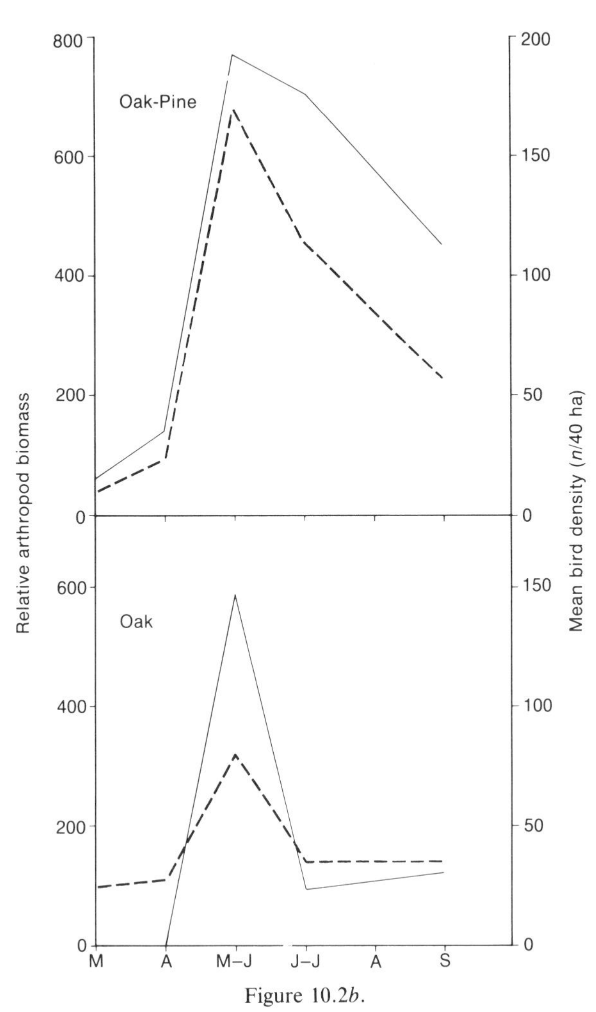

Figure 10.2*b*.

Figure 10.2*a–e*. Relationship between arthropod biomass and bird density in nine study sites in the New Jersey Pine Barrens, March–September 1983. Solid lines = relative arthropod biomass; dashed lines = bird density. See Table 10.1 for description of study sites.

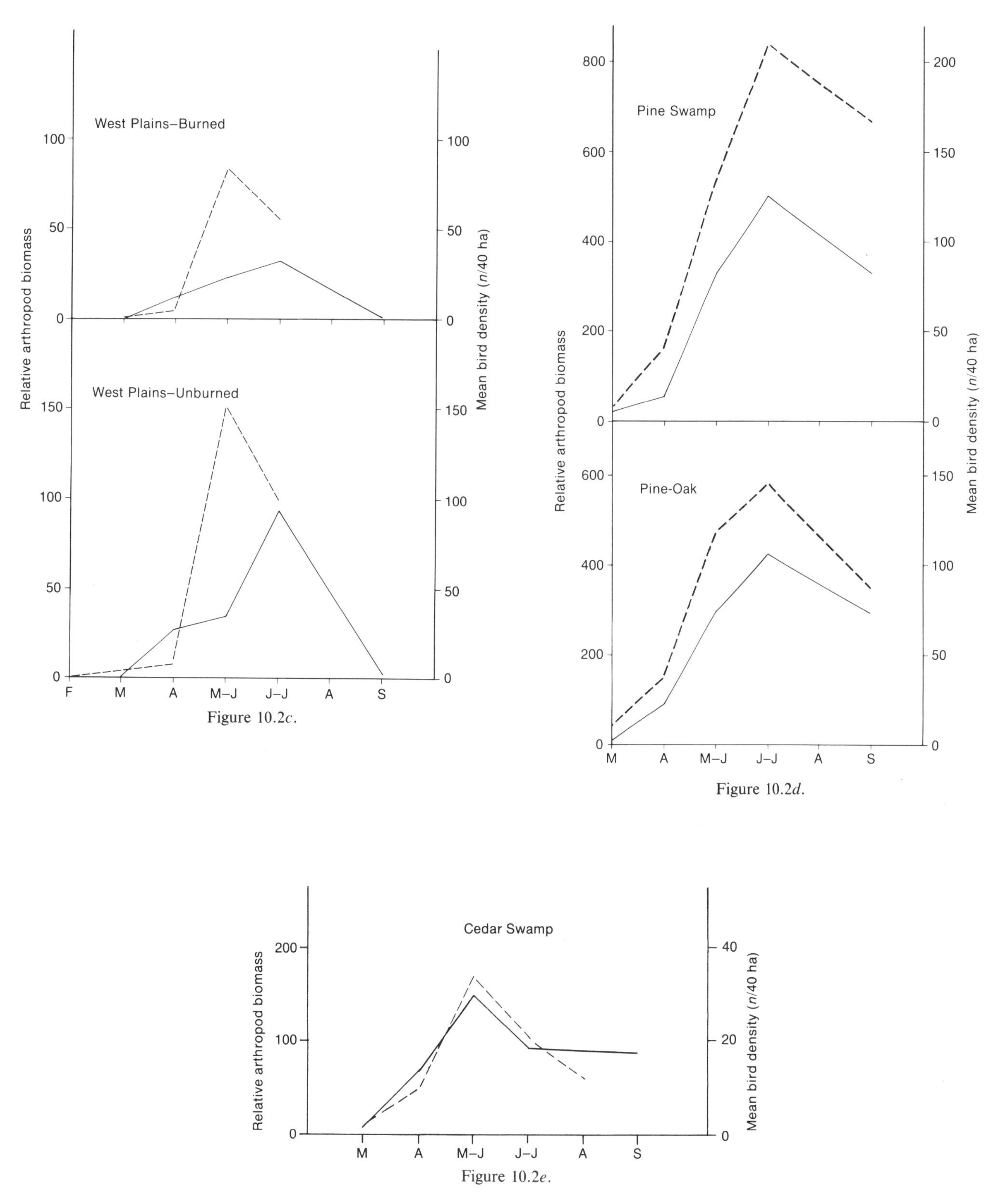

Figure 10.2*c*.

Figure 10.2*d*.

Figure 10.2*e*.

Table 10.2. Mean densities (*n*/40 ha) in oak- and pine-dominated sites during the breeding season, May–July 1983. (See Table 10.1 for list of study sites.)

Species	May to mid-June			Mid-June through July		
	Oak	Pine	% Oak[a]	Oak	Pine	% Oak[a]
Rufous-sided towhee (*Pipilo erythrophthalmus*)	74	61	55	53	86	38
Pine warbler (*Dendroica pinus*)	3	12	19	2	32	6
Common yellowthroat (*Geothlypis trichas*)	3	18	13	1	14	7
Carolina chickadee (*Parus carolinensis*)	2	6	28	3	9	27
Prairie warbler (*Dendroica discolor*)	9	7	58	3	6	34
Ovenbird (*Seiurus aurocapillus*)	6	8	44	<1	1	15
Brown thrasher (*Toxostoma rufum*)	4	1	75	6	5	57
Chipping sparrow (*Spizella passerina*)	0	5	0	0	5	0
Black-and-white warbler (*Mniotilta varia*)	1	2	38	0	2	0
Blue jay (*Cyanocitta cristata*	2	1	71	1	1	53
Field sparrow (*Spizella pusilla*)	<1	1	33	1	3	26
Tufted titmouse (*Parus bicolor*	1	<1	91	1	2	26
Brown-headed cowbird (*Molothrus ater*)	1	1	57	1	<1	54

[a] Percentage of population in oak-dominated habitat out of total population in oak- and pine-dominated habitats.

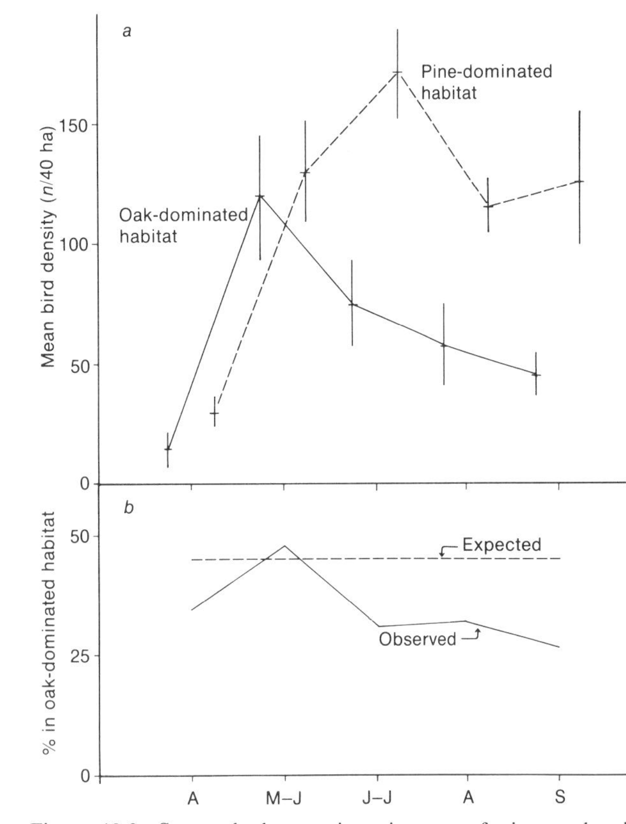

Figure 10.3. Seasonal changes in avian use of pine- and oak-dominated habitats in the New Jersey Pine Barrens, April–September 1983. (*a*) Bird density in oak- and pine-dominated habitats (horizontal bars = means, vertical lines = *SD*); (*b*) percentage of the bird community in oak-dominated habitat out of the total in both habitat types (expected value based on relative foliage volume in oak and pine sites).

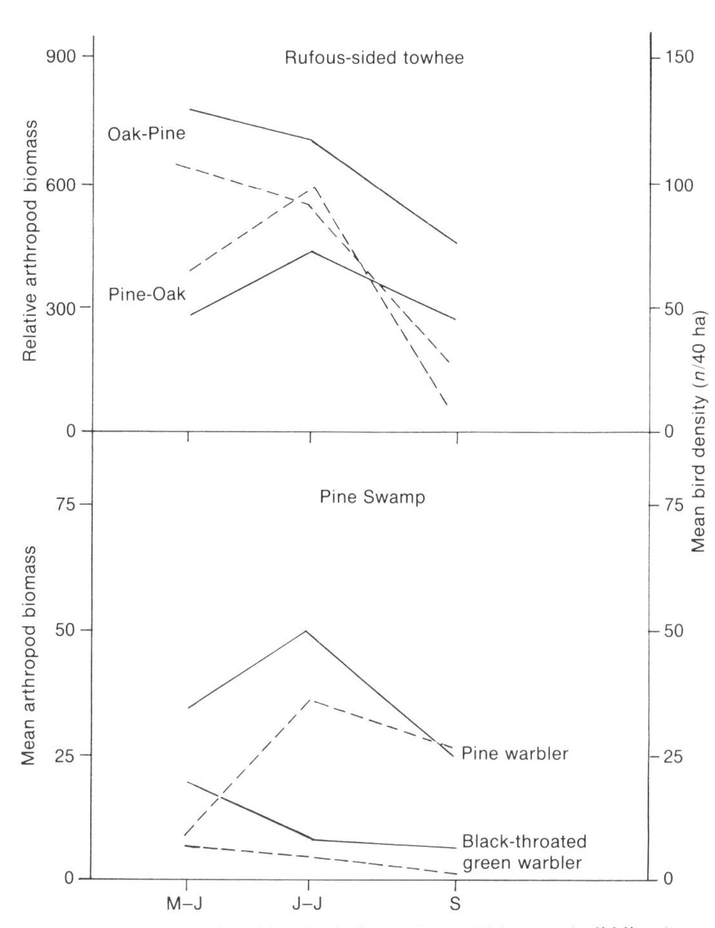

Figure 10.4. Relationship of relative arthropod biomass (solid lines) to densities of selected bird species (dashed lines) in the New Jersey Pine Barrens, May–September 1983. Biomass values are shown for plant taxa preferred by foraging pine warblers, which prefer pines, and for black-throated green warblers, which prefer white-cedars. Because rufous-sided towhees used both oaks and pines, arthropod abundance at the habitat level was compared with towhee densities. Abbreviations as in Table 10.1.

Discussion

Predictors of bird abundance

Prey abundance best predicted densities of insectivores at several levels. At the community level, bird density was highly correlated with arthropod biomass (Fig. 10.2) but not with foliage volume or foliage height diversity. Birds responded to changing arthropod biomass by shifting to areas with a greater food supply. The lower, nonsignificant correlations on two study sites (West Plains–Unburned and West Plains–Burned) were probably caused by the unusually late freeze. However, in the study area as a whole, the bird community (which was composed of basically the same species on all sites) closely followed food peaks in pine- and oak-dominated sites (Figs. 10.1 and 10.3). For species with clear foraging preferences, within- and between-habitat differences in arthropod biomass were excellent predictors of changes in bird density (Fig. 10.4). Generalist foragers such as rufous-sided towhees (Brush and Stiles, pers. obs.) were responsive to arthropod fluctuations in both pines and oaks, and specialists such as black-throated green and pine warblers tracked arthropod fluctuations in their preferred plant genera. The patterns of changing habitat use during the summer could not have been predicted from knowledge of foliage volume alone because the latter changed only slightly during the summer. Cavity-nesting birds made up a small fraction of the bird community, but they were just as food-responsive as were open-nesting species and did not appear to be limited by the number of nest sites.

Our findings agree with and extend the results of other studies in oak (Gibb 1950; Van Balen 1973; Gula 1977; Graber and Graber 1983b) and pine forests (Gibb and Betts 1963; Van Balen 1973; Gula 1977; Tagashi and Takahashi 1977; Larsson and Tenow 1979). Many of these studies (e.g., Gibb and Betts [1963]; Van Balen [1973]) showed strong relationships among tree phenology, arthropod abundance, and the timing of bird foraging and breeding activities.

Oaks and pines had fairly short periods of leaf and needle growth (Kozlowski and Clausen 1966), thus providing relatively predictable times of arthropod abundance and scarcity. In contrast, both deciduous understory shrubs and heaths had later and more variable peaks in arthropod biomass. This pattern is consistent with the longer, indeterminate growth period for those taxa (pers. obs.) and consequently less predictable leaf chemistry (Niemela and Haukioja 1982; Schultz 1983). These patterns of arthropod abundance are probably due to variability in leaf or needle chemistry, which determines suitability for herbivorous arthropods (Feeny 1970; Schultz 1983). Although we did not measure leaf chemistry, our data show that birds responded seasonally to changes in arthropod abundance. The strength and reciprocity of observed changes in bird abundance in oak- and pine-dominated sites suggest that birds moved from one habitat to another, on the basis of changing arthropod abundance. All study sites were near enough to other habitats for birds to sample and move among habitats dominated by oaks, pines, or cedars.

Management implications

The approach outlined here can be used by managers wishing to evaluate habitat suitability for insectivorous birds. Fairly simple correlations of prey and predator abundance can reveal major temporal and spatial differences in habitat suitability. This method should be useful in many wooded habitats because many are dominated by only a few plant and arthropod taxa. One modification of this approach might involve taking ground samples of frass-fall from arboreal caterpillars, which is less time-consuming and has been used successfully by others (Gibb 1950; Van Balen 1973). Also, for some widespread tree genera, such as oaks and pines, sufficient background information exists on growth phenology (Kozlowski and Clausen 1966) or arthropod fluctuations (Niemela and Haukioja 1982; Schultz 1983) for managers to make predictions of bird-community dynamics in habitats dominated by those genera without having to sample arthropods. Further information could be gained by studying the diet and food supply of individual bird species, although a study using dietary guides and overall food abundance may prove useful and less time-consuming.

11

Temporal and Spatial Aspects of Species-Habitat Models

RICHARD A. LANCIA, DAVID A. ADAMS, and EDWARD M. LUNK

Abstract.—We used a temporal model to predict impacts of a proposed peat-mining operation and spatial models to assess habitat quality for three bird species, and we attempted to validate the spatial models. The former simulated land-use changes over time, predicted their impact upon habitat quality for 10 wildlife species, and accrued annualized index values by species. The latter utilized time-static, geobased, environmental data to develop spatial arrays of environmental factors and derived habitat-quality indices. Habitat-quality index validation, using the spatial approach, was more successful for common, range-restricted, more specialized species than for rare, wide-ranging, and/or generalized forms. The temporal approach is more applicable to broad-scale strategic analyses, and the spatial approach is of more value in small-scale tactical situations.

Evaluation of habitat suitability, defined as the ability of habitat to provide life requisites, may involve determining a numerical index of suitability for a wildlife species (see, e.g., Adams 1980; Army Corps of Engineers 1980; Fish and Wildlife Service 1980b; Adams et al. 1983) and assigning that value to a given land-use category or vegetation type at a given point in time. This temporally static approach assumes that all instances of a given cover type have the same habitat suitability for a given species, and it can be employed with little reference to locations on the ground. However, wildlife habitat is a temporally and spatially dynamic resource; its character is constantly molded by succession and disturbance. Thus, habitat suitability should be assessed by integrating the effects of ongoing processes through time.

Furthermore, wildlife habitat and the animal populations obtaining life's necessities from it are arranged in spatial patterns in such a way that associations between cover types—their interspersion and juxtaposition—may be important for determining habitat suitability. Because the character of habitat is largely determined by its distribution and variation on the ground, assessing habitat suitability with spatially related components depicted in map form can be a valuable addition to the habitat evaluation process. Also, validation of habitat-suitability models may be facilitated by using spatial methods that involve patterns of use by animal populations. These measures of the distribution of animals probably are tied more directly to habitat conditions than is population density, because density varies in response to a multiplicity of factors other than habitat conditions.

Grid cells are well suited to spatial modeling and validation, because each cell provides values for environmental and population measurements independent of habitat boundaries, thereby preserving environmental gradients. We present an example that includes a spatial dimension in species-habitat models for three bird species. To validate the models, we compared grid-cell maps of habitat suitability to similar maps of observed frequency of use. The approach is similar to one used in modeling bobcat habitat suitability by Lancia et al. (1981, 1982).

Example with temporal emphasis

Methods

The project area encompassed the "Pamlimarle" peninsula of eastern North Carolina—low-lying, poorly drained land east of the Suffolk Scarp between Pamlico and Albemarle sounds. Much of the peninsula is underlain by peat deposits up to 4.9 m (16 feet) thick and totaling about 150,953 ha (373,000 acres). Recent interest in developing energy resources resulted in four proposals to use the peat as fuel and to reclaim the mined land to a combination of agriculture, residential property, and lakes.

Three 20-year peat-mining scenarios and a no-change alternative were developed. Scenario 1 assumed that all permitted areas would be mined in accordance with permit conditions (7026 ha [17,360 acres]). Scenario 2 assumed that all reserves (deposits 1.2 m [4 feet] or greater in depth) would be mined (33,995 ha [84,000 acres]). Scenario 3 assumed that development would begin as in scenario 1, but peat mining would collapse economically at year 10 with only 3448 ha (8520 acres) mined.

Each scenario dictated the preproject land-use/habitat-type inventory, mining rates, and reclamation plans. Wildlife biologists representing the North Carolina Wildlife Resources Commission and the U.S. Fish and Wildlife Service supplied a list of 10 species representing those characteristic of forested areas—pine warbler (*Dendroica pinus*) and hairy woodpecker (*Picoides villosus*); those associated with large areas of native vegetation—black bear (*Ursus americanus*) and bobcat (*Felis rufus*); those requiring pocosin—marsh rabbit (*Sylvilagus palustris*), or marsh and water—muskrat

RICHARD A. LANCIA, DAVID A. ADAMS, and EDWARD M. LUNK: Department of Forestry, North Carolina State University, Raleigh, North Carolina 27695

(*Ondatra zibethicus*); and those tolerant of disturbance and agricultural development—white-tailed deer (*Odocoileus virginianus*), northern bobwhite (*Colinus virginianus*), and mourning dove (*Zenaida macroura*). The average of five independent Habitat Suitability Indices (HSIs) estimated by wildlife biologists with experience in the project area for each of 16 land-use/habitat types was used in the simulations. Rates of habitat-type conversions resulting from the project and successional change were expressed as linear, exponential, sinusoidal, or phased functions of time dictated by the conditions of each scenario.

A program was developed for the Apple II microcomputer which, with user-specified input of project duration, initial land inventory, indicator species, HSIs, and type conversions, would (1) accrue annualized habitat-suitability units by species and type (units = (HSI)(area)); (2) compute the mean HSI (weighted by land-use/habitat-type areas) for each species with and without project alternatives; and (3) compute a sum of habitat units over species and habitats. The program provided tabular comparisons of each scenario with the "no-change" alternative and each scenario with all others.

Results

Land-use conversions

Without the project, land-use conversions reflect natural succession and management typical of the study area today plus conditions imposed by failure of prospective mining to develop (Table 11.1). Land in agriculture is retained; disturbed land succeeds into pocosin and forest; prospective mining areas become flooded following abandonment; and normal timber rotations are imposed on forest land.

Table 11.1. Land-use and habitat-type conversions (in hectares) associated with proposed peat mining, Pamlimarle Peninsula, North Carolina

Land use or cover type	Scenarios						
			Postproject				
	Preproject		Without project		With project		
	1 and 3	2	1 and 3	2	1	2	3
Residential	0	0	0	0	0	981	0
Roads and rights-of-way	0	0	0	0	0	0	0
Agriculture	1,633	1,733	1,633	1,733	3,969	18,178	1,021
Evergreen forest	387	500	478	375	582	3,215	225
White-cedar swamp	0	54	0	69	0	9	0
Mixed swamp	29	3,979	48	262	11	470	17
Disturbed land	4,845	5,742	144	694	1,212	0	3,028
High pocosin	770	10,496	1,968	14,315	313	1,120	461
Low pocosin	1,639	11,750	5,033	16,979	768	1,550	1,027
Bay forest	0	379	0	206	0	65	0
Water and marsh	132	242	263	367	645	2,995	3,788
Active peat mine	0	0	0	0	362	1,750	0
Inactive peat mine	132	125	0	0	1,705	4,667	0
Total	9,567	35,000	9,567	35,000	9,567	35,000	9,567

Mining and subsequent reclamation increased the area of agricultural land and evergreen forest (scenarios 1 and 2) and residential development (scenario 2); reduced the area of high and low pocosin vegetation (all scenarios, but less so in scenario 3); expanded water and marshland (all scenarios); and left considerable acreage in the mining cycle (scenarios 1 and 2).

Species effects

Cumulative impacts during the project period reflected diverse habitat requirements of the indicator species (Table 11.2). Compared to without-project conditions, habitat suitability for white-tailed deer and northern bobwhite, species tolerant of disturbance and agriculture, generally changed less than $\pm 10\%$. Muskrat and mourning dove habitat improved under all scenarios; nearly 70% in scenario 3 for muskrat as a result of flooding following abandonment and in scenario 2 for mourning dove as a result of extensive agricultural development.

Species dependent upon forests and native vegetation fared less well. Habitat suitability for pine warblers and hairy woodpeckers declined generally more than 40% as evergreen forests were cycled through mining operations; but, at the end of the project, more evergreen forest existed under with-project than without-project conditions (scenarios 1 and 2) because of conversions from pocosins to pine plantations. Marsh rabbit habitat suitability declined about 30% in scenarios 1 and 2 and about 13% in scenario 3 because there was proportionately more water and marsh as pocosins were converted to other types. Perhaps the most significant losses were for bobcats where habitat suitability diminished 20–30% in scenarios 1 and 2 and 17% in scenario 3, and for black bear where reductions were >30% in all scenarios.

Validation

We did not attempt to validate the HSIs or successional rates in the field—an obvious weakness; however, the re-

Table 11.2. Accrued annualized habitat-suitability units (in hectares) with and without the proposed peat-mining project

Species	Scenario					
	1		2		3	
	With	Without	With	Without	With	Without
Black bear	8,746	13,658	36,442	58,505	9,304	13,658
White-tailed deer	11,931	12,590	39,374	47,336	11,888	12,590
Marsh rabbit	8,199	11,581	33,176	47,566	10,043	11,581
Muskrat	5,138	4,098	17,317	11,225	6,939	4,098
Northern bobwhite	11,810	10,149	33,907	31,727	10,924	10,149
Mourning dove	10,199	7,181	28,199	16,866	8,609	7,181
Great horned owl	9,176	8,898	30,399	32,173	9,024	8,898
Pine warbler	3,841	7,427	19,163	31,361	3,524	7,427
Hairy woodpecker	3,845	7,352	19,708	33,203	3,773	7,352
Bobcat	10,313	13,152	37,105	53,266	10,886	13,152
Total	83,198	96,086	294,790	363,228	84,914	96,086

sults of the simulations could be logically interpreted, and hence appeared reasonable to us.

Example with spatial emphasis

Methods

Habitat

The study area (572 ha [1413 acres]) was located on the Croatan National Forest in the Coastal Plain of North Carolina. Vegetation is typical of the southeastern evergreen region with pine forests and plantations, pocosins, bottomland hardwood forests along drainages, and agricultural fields. We modified existing species-habitat models for three bird species—prairie warbler (*Dendroica discolor*) (Sheffield 1981), pine warbler (Schroeder 1982), and pileated woodpecker (*Dryocopus pileatus*) (Schroeder 1983)—that were indicator species or species of special concern on national forests in the Southeast. Environmental factors (Lancia and Adams, pers. obs.) required in the models were measured at 67 sample points (located at 1000 m [3281 feet] UTM grid intersections) in March 1983. Resulting values were assigned to every (N = 5200) grid cell (0.11 ha) using SYMAP, a grid-cell-based, computer-mapping program (Dougenik and Sheehan 1979), and were combined within each cell, using HSI functions described in the species-habitat-relationships models to produce a grid-cell map of HSI values.

Habitat use

Two bird surveys were done, once each in March and April 1983, by recording the number and species of birds heard calling at each sample point during a 5-min period. We calculated a frequency index as the average number of individuals of a given species heard per sample point and used SYMAP to assign a frequency index value to each cell. Estimation of an index permitted relative comparisons among points, assuming that the same proportion of birds is detected at each point (Dawson 1981). Bart and Schoultz (1984), however, showed that a greater proportion of birds is missed as density increases, resulting in underestimates. Nevertheless, the relative ranking among points should not be affected.

Validation

We used two methods to validate the bird models. The first was a contingency table analysis of the relationship between maps of habitat suitability and frequency indices. We assumed that use was evidence that an area fulfilled some life requisites and that, as more life requisites were met, frequency of use increased. Three interpretive zones were identified: zone I—model performance good (coincident or nearly coincident HSI and use); zone II—less than fully occupied habitat or model errors (high HSI and low use); and zone III—model error (low HSI and high use).

The second method converted occurrence, calculated as the percentage of the grid cells in a given HSI quartile in which a particular species was heard (i.e., frequency index >0), to density of singing males. The frequency-density conversion fit a Poisson distribution, assuming that all individuals had an equal probability of occurring at each sample point and were equally observable (Caughley 1977). Clumped distributions lead to underestimates of density; thus the conversion provided a conservative estimate of density.

Means of the HSI quartiles were plotted against corresponding density estimates from the frequency-density conversion. If the models performed as expected, the relationship should be linear, pass through the origin, and have a positive slope. These relationships were tested with linear regression analysis; however, because only four points per species were used, the validation attempt is crude. Statistical significance is not presented because the choice of significance levels should be contingent on how the models are applied.

Results

Habitat-suitability distribution

Mean HSIs for all grid cells for the prairie warbler, pine warbler, and pileated woodpecker were 0.35 ± 0.26 ($\bar{x}$ ± SD), 0.40 ± 0.18, and 0.28 ± 0.24, respectively. Figure 11.1 shows the distribution of habitat suitability for the prairie warbler as an example of spatially explicit HSIs. In this case higher values were associated with a young pine plantation near the center of the map and lower values with older stands around the periphery.

Validation

Contingency table analyses showed 60–74% of the grid cells in zone I for both warblers, but only about 49% in zone I for the pileated woodpecker. The relationship between HSI and relative density (based on the frequency-density conversion) for both warblers was nearly linear and passed through the origin (Fig. 11.2). No linear relationship was evident for the pileated woodpecker, probably because our tests were inappropriate for a species that could be heard for long distances, traveled over wide areas to use nonuniformly distributed resources, and was relatively uncommon. Thus, both validity tests suggest better performance of the warbler models than the woodpecker model.

Discussion

Comparison of temporal and spatial approaches

The temporal simulation permitted evaluations of several alternative project configurations at minimal cost in terms of computer time and manpower. Simulated land-use/habitat conditions conformed to those specified in the scenarios, but validity of results rested heavily upon the degree to which HSIs reflected suitability as perceived by the indicator species. HSIs should have been validated in the field, but funding and time constraints prevented this effort. Thus, the accuracy of the model predictions is only as good as the

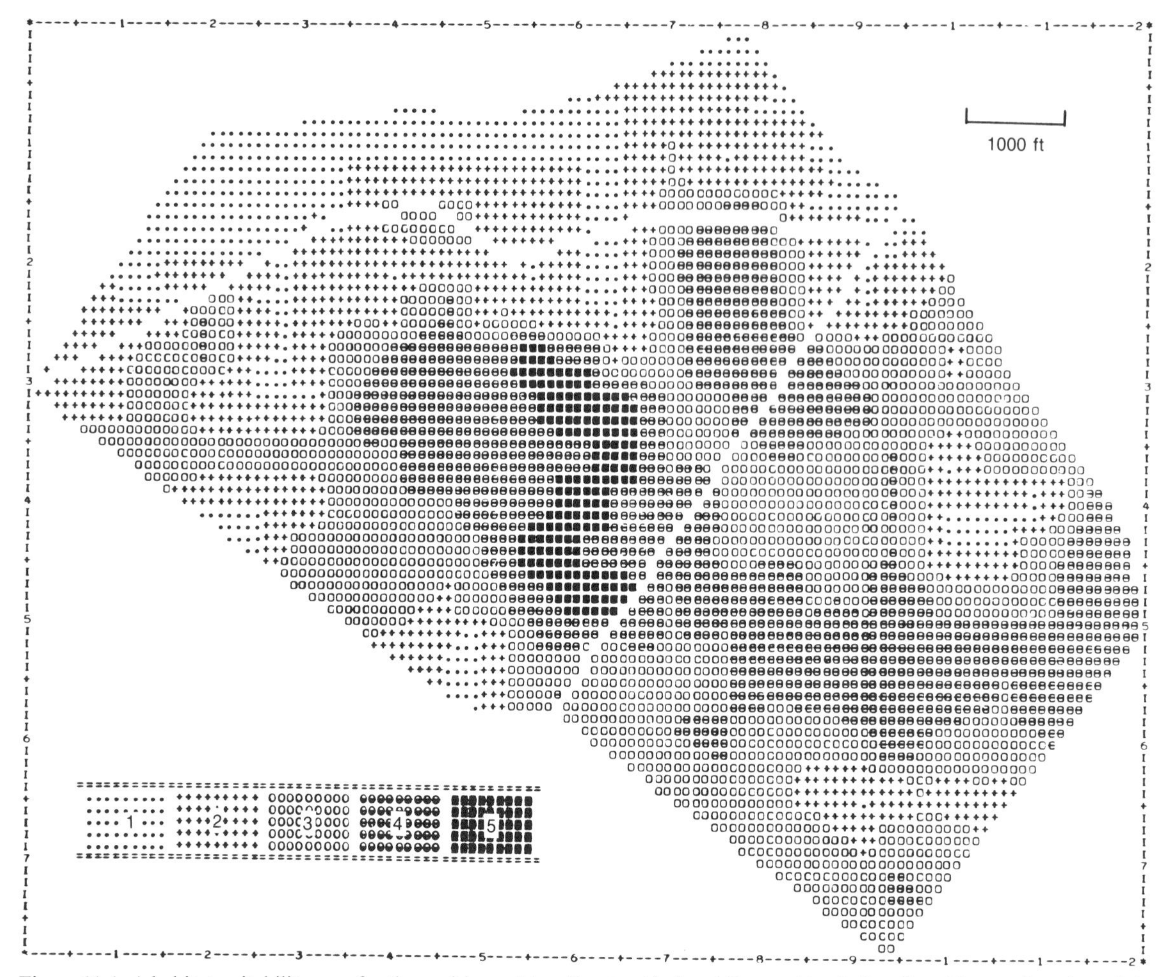

Figure 11.1. A habitat-suitability map for the prairie warbler, Croatan National Forest, North Carolina. The gradient from light to dark represents a continuum from low (1) to high (5) suitability.

accuracy of the HSIs and the conversion rates among land-use/habitat types.

The simulation program permitted great latitude in modeling land-use conversion functions, but required an assumption that all examples of a given land-use/habitat type exhibited the same suitability for a given species. The approach could not explicitly consider effects of type juxtaposition, edge, and size of individual tracts, except as these considerations were implicit in the HSIs. The model reflected temporal considerations well, but spatial considerations poorly.

Advantages of assessing habitat suitability spatially and developing a map of habitat suitability are (1) evaluation of model performance can be facilitated because both habitat use and populations vary spatially; (2) managers can evaluate options with respect to the spatial arrangement of types; and (3) the dependence of HSIs on arbitrary decisions about habitat types and boundaries can be reduced. A current disadvantage of this approach is the requirement of a large mainframe computer with sophisticated mapping, data management, and statistical software.

Recommendations

Strategic versus tactical considerations

Although both approaches discussed have obvious limitations, each has attractive characteristics in particular situations. For evaluating strategic considerations involving large areas over a long time, the temporal approach is superior. It permits consideration of a large number of options at a low cost, providing an array of results from which the manager can select. Our experience indicates that errors inherent in HSI validity and land-use conversion functions are usually tolerable for strategic decision making.

For tactical considerations, a manager is concerned with decisions on the ground; location, shape, size, and juxtaposition of cover types become important. Techniques that integrate spatial factors in HSIs can provide habitat-suitability maps depicting distribution as well as magnitude of impacts. Our approach for spatial analysis currently requires a mainframe computer, and the cost of iterations precludes testing many alternatives or considering temporal aspects. An obvi-

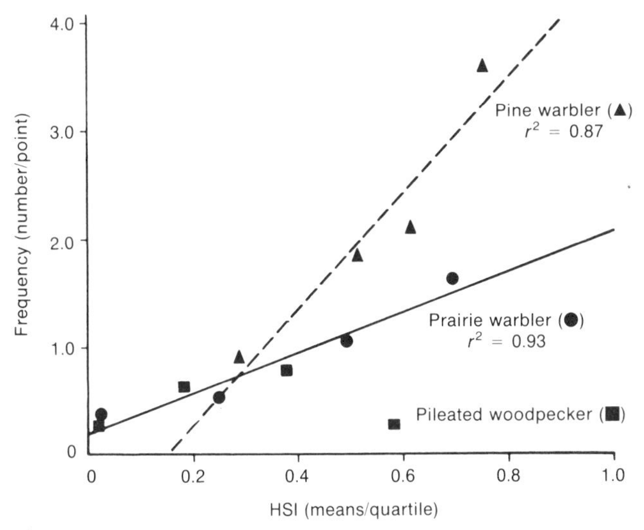

Figure 11.2. Relationship between the habitat-suitability index and frequency for three bird species, Croatan National Forest, North Carolina.

ous goal is a spatially explicit, temporal model that can be run inexpensively on a microcomputer.

Validation

Choosing an appropriate means to validate species-habitat models is important. Although the ability of animals to maintain viable populations in a habitat may be the ultimate measure of that habitat's suitability, this is extremely difficult to measure. Habitat productivity may be inferred from carrying capacity and the intrinsic rate of natural increase, but estimating these demographic parameters requires a sequence of population estimates and/or controlled field experiments. Also, factors not encompassed by habitat undoubtedly influence population levels, which may confound efforts to establish significant relationships between habitats and populations (Van Horne 1983). Thus, only models for common species that are relatively easy to locate, capture, or observe are likely to be validated with demographic or population-density data.

For many species, population analysis may require several years of intensive trapping, capture, and observation to obtain even rough estimates of demographic parameters. Therefore, frequency-of-use validation may be the most feasible approach. If frequency of use is related directly to satisfaction of life requisites, then use can be an indicator of habitat suitability. However, this assumption has not been tested.

Acknowledgments

Support for this study was provided in part by the USDA Forest Service, the U.S. Fish and Wildlife Service, the North Carolina Wildlife Resources Commission, the Water Resources Research Institute of the University of North Carolina, and the School of Forest Resources, North Carolina State University. R. H. Barrett, W. E. Grenfell, M. R. Lennartz, B. G. Marcot, M. L. Morrison, M. G. Raphael, and D. L. Stewart reviewed earlier drafts. Also, special thanks go to Hal Salwasser.

12

Effects of Habitat Type and Sample Size on Habitat Suitability Index Models

DEAN F. STAUFFER and LOUIS B. BEST

Abstract.—We evaluated the similarity of Habitat Suitability Index (HSI) models developed from data for different habitat types and different sample sizes. HSI curves were developed for 12 bird species and for three habitat variables (densities of trees, snags, and low vegetation) from data collected in streamside habitats in Iowa. Models were developed for each bird-variable combination from data sets composed of 1349, 1034, 780, 446, 254, and 240 sampling points in both upland and floodplain forests. Similarity of models representing bird relationships to vegetation characteristics was evaluated with an overlap index. Only 11 of 180 curve pairs compared for different sample sizes and habitats were dissimilar for low-vegetation density. Sixty-nine percent (of $n = 36$) of the tree-density and 53% of the snag-density curves were different between floodplain and three upland forest data sets. The similarity of curve pairs decreased with smaller data sets. We conclude that bird relationships to habitat variables can vary with habitat type. Thus HSI models for deciduous forests may not reflect these differences if floodplain and upland hardwood types are combined in model development. We also conclude that models derived from small data sets may not adequately represent the true relationship of species to habitat variables.

Much effort currently is being devoted to developing wildlife-habitat models. Some uses of these models include evaluating management plans, assessing current conditions, and predicting the effect of future habitat manipulations upon wildlife communities or individual species. Various methods have been used to develop models of wildlife-habitat relationships. In this chapter we address aspects of developing Habitat Suitability Index (HSI) models (Fish and Wildlife Service 1980a, 1980b, 1981a) used in the Habitat Evaluation Procedures.

A fundamental step in evaluating habitat involves developing HSIs that relate the perceived suitability of an area for the species of concern to habitat characteristics of the area. Habitat variables selected to develop models should have ecological relevance to the species or guilds for which the models are being designed. Often models of wildlife-habitat relationships are not readily available, so HSI models for a target species must be developed by synthesizing information available in the literature (Fish and Wildlife Service 1981a). Data available for individual species vary in quality and quantity and may have been obtained from a variety of habitat types. Moreover, sample sizes on which results are based often differ greatly among studies.

Because HSI models synthesized from literature reviews are based on data from diverse sources, it is necessary to know how variation in the data base influences HSI models. Our objectives in this chapter are (1) to evaluate the similarity of HSI models derived from data for different habitat types (floodplain and upland woodlands); and (2) to assess the effects of sample size on HSI models. We develop models for 12 bird species and three habitat variables. We use HSI models because they are currently being promoted by the U.S. Fish and Wildlife Service and have been applied by a variety of resource-management agencies. We acknowledge that this modeling system may not necessarily be the best for our data base or for studies in general, but our aim is to investigate the relationship between empirical models derived from different habitat types or sample sizes, not to present definitive models of bird-habitat relationships.

Study area

Twenty-eight study plots were selected in Guthrie County, Iowa, to represent a gradient of streamside habitats from hayfields to closed-canopy woodlands. Common tree species of the floodplain woodlands were American elm (*Ulmus americana*), silver maple (*Acer saccharinum*), boxelder (*A. negundo*), willow (*Salix* spp.), black walnut (*Juglans nigra*), and ash (*Fraxinus* spp.). The upland woodlands were dominated by shagbark hickory (*Carya ovata*), white oak (*Quercus alba*), northern red oak (*Q. rubra*), black walnut, eastern hop hornbeam (*Ostrya virginiana*), and American basswood (*Tilia americana*). Additional vegetative characteristics of the study area are detailed in Stauffer and Best (1980).

Methods

Field work

Study plots consisted of parallel transects, marked at 25-m intervals and positioned 50 m apart (Fig. 12.1). The length (350–500 m) and number (1–5) of transects per plot were

Dean F. Stauffer: Department of Fisheries and Wildlife Sciences, Virginia Polytechnic Institute and State University, Blacksburg, Virginia 24061

Louis B. Best: Department of Animal Ecology, Iowa State University, Ames, Iowa 50011

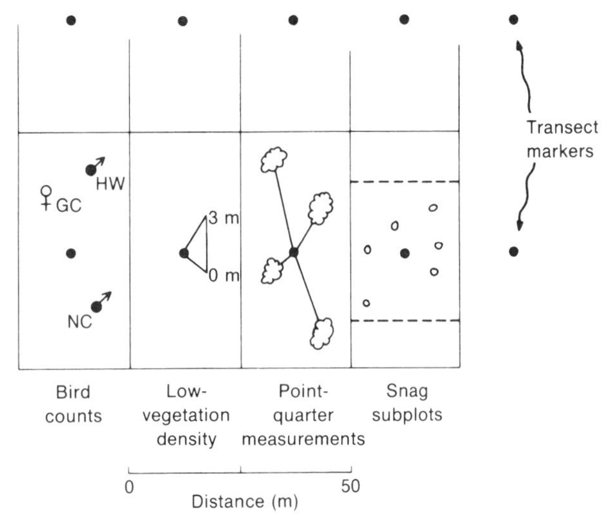

Figure 12.1. Relationship of the 25×50-m microplots to the transects, as used for this study. Each variable was recorded for each microplot. See the text for a description of the variables used.

determined by the extent of relatively homogeneous habitat. Each plot was gridded into 25×50-m microplots, centered on the transect markers, that served as sampling units (n = 1349).

Breeding birds were censused by conducting 12 counts on each plot from mid-April through mid-July. The number of observations of each species in each microplot was computed for comparison to selected habitat characteristics of the microplot.

We measured three habitat characteristics typical of those used in wildlife-habitat analyses (James 1971; Anderson and Shugart 1974). Tree density was calculated for each microplot based upon a point-centered-quarter sample (Cottam and Curtis 1956) taken at each transect marker. We recorded presence of live vegetation within each of three height classes (0–0.5, >0.5–1.5, and >1.5–3.0 m) at each transect marker. From these data an index (percent of maximum possible) of low-vegetation density was calculated. We also recorded the number of snags in a 30×25-m subplot centered within each microplot.

Data analysis

We used 12 species (Table 12.1) and designed HSI models for the three habitat variables. These species were selected because they occurred in at least 400 of the 1349 microplots and they represent open- and cavity-nesting and edge and interior woodland species.

To relate relative abundance of the bird species to the selected habitat variables, we first identified classes of each habitat variable such that at least 50 microplots fell within each category. Fourteen classes of tree density were used: 0, 1–50, 51–100, 101–200, 201–300, 301–400, 401–500, 501–600, 601–700, 701–900, 901–1100, 1101–1300, 1301–1600, and >1600 trees/ha. Low-vegetation density had four levels: 0, 33, 67, and 100%. Seven levels of snag density were identified: 0, 1, 2, 3, 4, 5–6, and >6 snags per 30×25-m sample plot.

To generate HSI curves, we calculated the mean number of birds seen per microplot for each class of each habitat variable. Means for each species-variable combination were scaled to range from 0 to 1.0 by dividing the mean values by the maximum mean value for that combination. The scaled values (suitability-index values) were then plotted against classes of the habitat variable to produce an HSI curve that related relative bird abundance to the habitat variable (e.g., see Fig. 12.2). This procedure assumed that where the curve reaches its maximum (i.e., 1.0), habitat conditions are optimal for that species (Fish and Wildlife Service 1980b).

To evaluate the effects of habitat type and sample size, HSI curves were generated for six data sets: all data (ALL [code used in tables and figures]), 1349 sample points from 28 study plots; woodland data only (WD), 1034 points from 15 plots; upland woodland data (UP), 780 points from 10 plots; first subset of upland data (UP1), 446 points from five plots; second subset of upland data (UP2), 240 points from four plots (plots used in UP1 and UP2 were randomly selected

Table 12.1. Median similarity between species' HSI curves for floodplain woodlands and curves for three upland woodland data sets

Species	Median similarity (n = 3)		
	Tree density	Snag density	Low-vegetation density
Red-headed woodpecker (*Melanerpes erythrocephalus*)	0.65	0.92	0.97
Downy woodpecker (*Picoides pubescens*)	0.85	0.91	0.86
Great crested flycatcher (*Myiarchus crinitus*)	0.65	0.88	0.97
Eastern wood-pewee (*Contopus virens*)	0.78	0.77	0.92
Blue jay (*Cyanocitta cristata*)	0.86	0.87	0.89
Black-capped chickadee (*Parus atricapillus*)	0.86	0.90	0.98
White-breasted nuthatch (*Sitta carolinensis*)	0.90	0.94	0.96
House wren (*Troglodytes aedon*)	0.94	0.98	0.97
Gray catbird (*Dumatella carolinensis*)	0.90	0.83	0.98
Northern cardinal (*Cardinalis cardinalis*)	0.92	0.94	0.98
Rose-breasted grosbeak (*Pheucticus ludovicianus*)	0.91	0.94	0.96
Indigo bunting (*Passerina cyanea*)	0.61	0.77	0.98

from those constituting UP); and floodplain woodland data (FP), 254 points from five plots. HSI curves were developed for each combination of species, variable, and data set, for a total of 540 curves. Note that not all microplots and data sets used were independent of one another and did not represent true statistical replicates (see, e.g., Hurlbert 1984). A result of this interdependence is that some of the perceived differences between HSI curves may represent spurious relationships rather than real differences.

We evaluated the similarity of HSI curves by means of an index (Steinhorst 1979):

$$S = \frac{2\,(x'y)}{x'x + y'y},$$

where x represents a vector of suitability index values for one combination of species, variable, and data set; and y represents the corresponding values for a second data set. This index ranges from 0 to 1.0. We calculated the similarity between HSI curves for all pairwise combinations of data sets (15 curve pairs for each species-variable combination).

Results

Similarity of bird relationships to tree density between different data sets ranged from 0.47 to 0.99; indices for snag density took values from 0.49 to 0.99; and relationships of birds to low-vegetation density ranged from 0.79 to 0.99 in similarity. The similarity index we used is conservative, and pairs of HSI curves had to diverge substantially to generate index values <0.90 (see Figs. 12.2, 12.3, and 12.4). Thus, we selected a similarity value of 0.90 as the dividing line between similar and dissimilar HSI curves.

Of 540 pairs of bird-habitat relationships evaluated, 129 (24%) were dissimilar. Sixty-four of the pairs of tree-density curves were dissimilar; of these, 15 pairs of relationships had similar maxima on the tree-density gradient (e.g., indigo bunting, Fig. 12.2) and 49 pairs had different maxima (e.g., eastern wood-pewee for $S = 0.78$). Of the 54 dissimilar pairs of snag-density curves, 20 had similar maxima (e.g., great crested flycatcher, Fig. 12.3) and 34 did not (e.g., red-headed woodpecker). Only 11 pairs of bird relationships to low-vegetation density were dissimilar; all had different maxima (e.g., indigo bunting, Fig. 12.4).

Habitat effects

To evaluate HSI curves developed from data for different habitat types, we compared similarity of bird-habitat relationships derived from floodplain woodland data to those for three sets of upland woodland data (Table 12.2). The similarity of curves for tree density was relatively low, with 25 (69%) of the pairs having a similarity <0.90. Nineteen (53%) of the pairs examined for bird relationships to snag density were dissimilar, as were seven (19%) of the pairs for density of low vegetation. These percentages of dissimilar curves

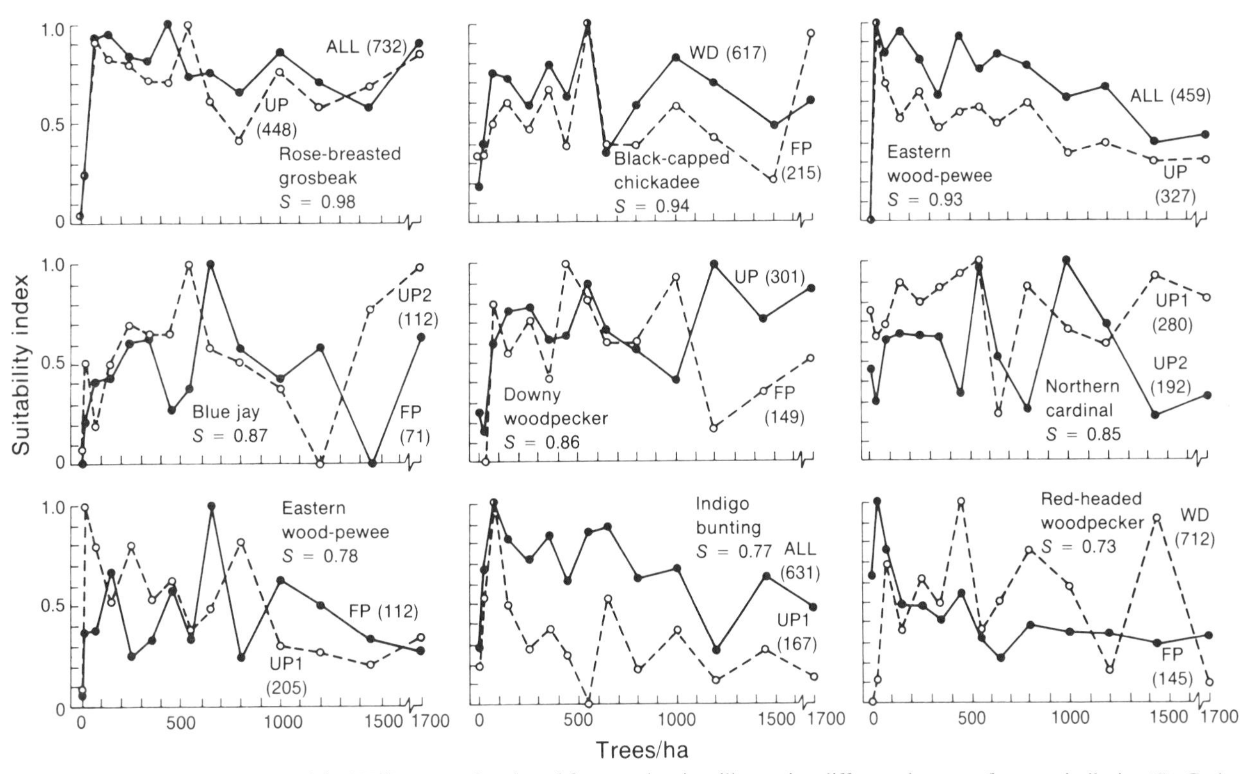

Figure 12.2. Representative empirical HSI curves developed for tree density, illustrating different degrees of curve similarity (S). Codes for the data sets for which the curves were developed are in the text. Values in parentheses are the number of bird observations for the particular species and data set.

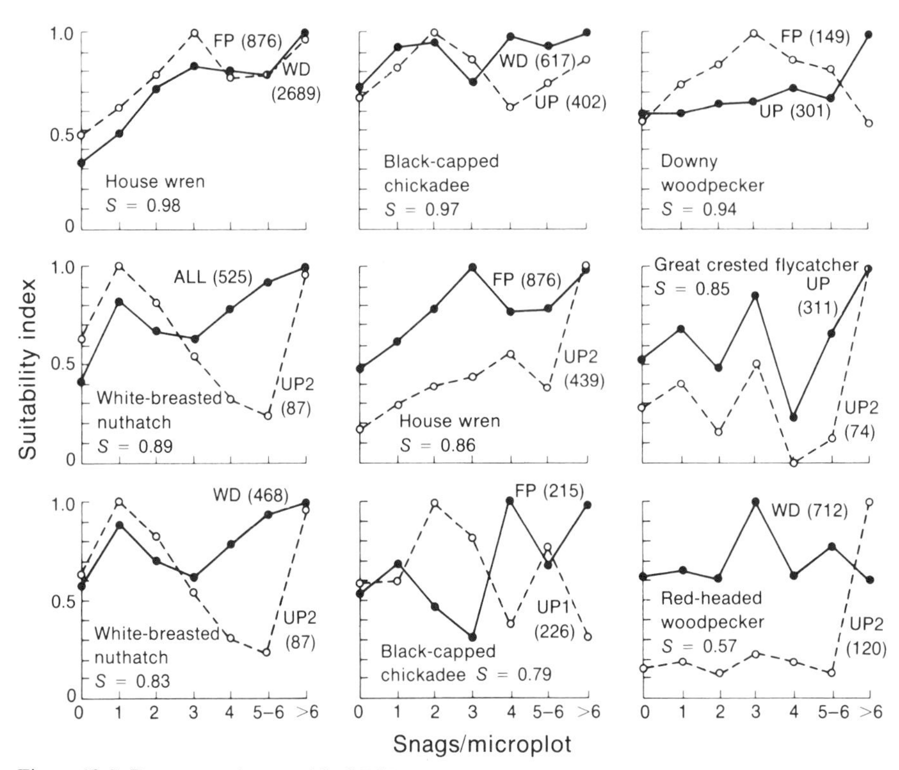

Figure 12.3. Representative empirical HSI curves developed for snag density, illustrating different degrees of curve similarity (S). Codes for the data sets for which the curves were developed are in the text. Values in parentheses are the number of bird observations for the particular species and data set.

were higher (z-test for proportions, z = 4.59, 2.61, and 3.64, respectively, for the three variables, $P < 0.01$) than the percentages of dissimilar curves for the other comparisons made between data sets. Median similarity of bird-habitat relationships between floodplain and subsets of the upland data tended to decrease with smaller samples of the upland data for tree and snag density; there was no comparable trend for the density of low vegetation (Table 12.2).

We found considerable variation among the 12 species in median similarity of floodplain curves compared with those of upland data sets (Table 12.1). For the rose-breasted grosbeak, white-breasted nuthatch, northern cardinal, and house wren, median similarity of the relationship to all three habitat variables was relatively high (>0.90). Median similarity was <0.90 for at least one variable for every other species. Only two species had a median similarity <0.90 for low-vegetation-density curves, whereas five and seven species had medians <0.90 for snag and tree density, respectively (Table 12.1).

Space limitations preclude presenting all data; here we present curves for only three species-variable combinations. The floodplain HSI curve of the indigo bunting for low-vegetation density was similar to two upland data sets (UP and UP2) (Fig. 12.5). For these three curves, the interpretation is that suitability increases as low-vegetation density increases. The floodplain curve was dissimilar from that of UP1, which had a maximum suitability at a density of 33% and declined at higher densities.

The relationship of house wrens to snag density was fairly similar between the floodplain and UP and UP1 data sets (Fig. 12.5). Although the maximum value for the relationship in floodplains occurred at 3 snags/plot, the curve also was high at >6 snags/plot, where the UP and UP1 curves reached their maxima. The UP2 curve was very dissimilar from the other three curves, taking relatively low suitability values for all snag densities until the maximum was reached at >6 snags/point.

The relationship of great crested flycatchers to tree density in upland habitats was dissimilar from that for floodplain (Fig. 12.5). Based on these curves, optimal tree density for flycatchers in upland habitats is 1–100 trees/ha, whereas for floodplain it is 600–700 trees/ha.

Sample size effects

To assess the influence of sample size on HSI curves, we calculated the similarity between bird-habitat relationships derived from the data sets ALL, WD, UP, UP1, and UP2 for each of the 12 bird species. Median similarity for curves based on any data-set pair tended to decrease as the sample size of the data sets used decreased (Table 12.3). We calculated Spearman's rank correlations between median similarity (based on 12 species) and the ratio of the smaller data set n divided by the larger data set n for each variable. For tree density, r_s = 0.79, for snag density, r_s = 0.79, and for low-vegetation density, r_s = 0.86 ($P < 0.05$, n = 10 for each

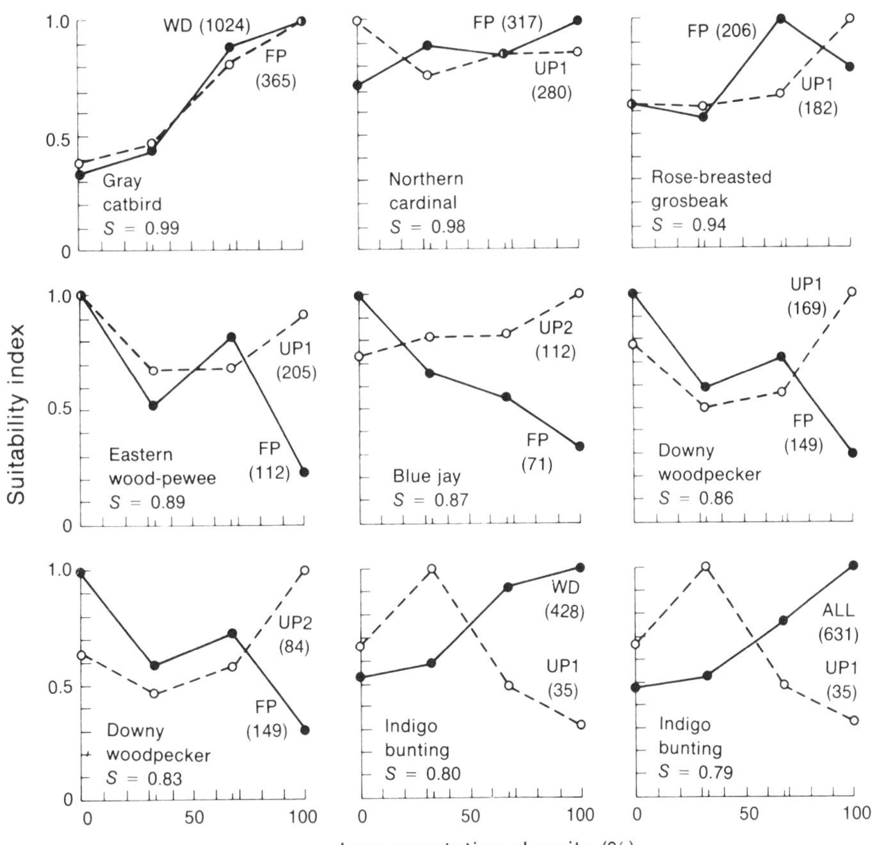

Figure 12.4. Representative empirical HSI curves developed for low-vegetation density, illustrating different degrees of curve similarity (S). Codes for the data sets for which the curves were developed are in the text. Values in parentheses are the number of bird observations for the particular species and data set.

r_s). Thus, as sample size between any pair of data diverges, the similarity of resultant HSI curves declines.

Overall, similarity was high between relationships derived for low-vegetation density, regardless of sample size (Table 12.3). The indigo bunting was the only species for which S was <0.90; in all four instances, the curve for the UP1 data set was dissimilar to those for the other data sets (Fig. 12.4).

Of 120 pairs of bird-tree density relationships evaluated, 32 (27%) were dissimilar (Table 12.3). Of these, 27 were for comparisons of the smallest data set, UP2, with the larger data sets. Nine of 10 curve pairs considered for the indigo bunting were <0.90; the median of the 10 pairs was 0.75. For the red-headed woodpecker, blue jay, and eastern wood-pewee, all comparisons between UP2 and the other data sets had $S < 0.90$ (medians for the four comparisons made were 0.82, 0.85, and 0.83 for the three bird species, respectively). Three of the similarities between UP2 and the other four data sets were dissimilar for the great crested flycatcher, and the median similarity was 0.86. Two or fewer comparisons of tree-density curves had $S < 0.90$ for the other seven species.

Snag density is presumably most important for cavity-nesting species, and only those will be considered here, although snag density can be used as an index of habitat characteristics important to open-nesting species (Marzluff and Lyon 1983). Of 60 comparisons of bird-snag relationships for cavity-nesters, 21 (35%) were dissimilar. Of these, 17 (81%) were comparisons of UP2 to the other data sets. Median similarities ($n = 4$) of the UP2 curves with the other data sets were: red-headed woodpecker, 0.56; downy woodpecker, 0.88; house wren, 0.92; black-capped chickadee, 0.88; great crested flycatcher, 0.85; and white-breasted nuthatch, 0.90.

For all three variables, similarity between bird-variable

Table 12.2. Median similarity among HSI curves derived from upland and floodplain woodland data sets[a]

Habitat variable	Median similarity (n = 12)		
	UP	UP1	UP2
Tree density	0.87 (7)	0.86 (9)	0.84 (9)
Low-vegetation density	0.97 (1)	0.96 (4)	0.96 (2)
Snag density	0.93 (4)	0.88 (7)	0.85 (8)

[a] Figures in parentheses in the body of the table indicate number of species (up to a maximum of 12) for which similarity was <0.90.

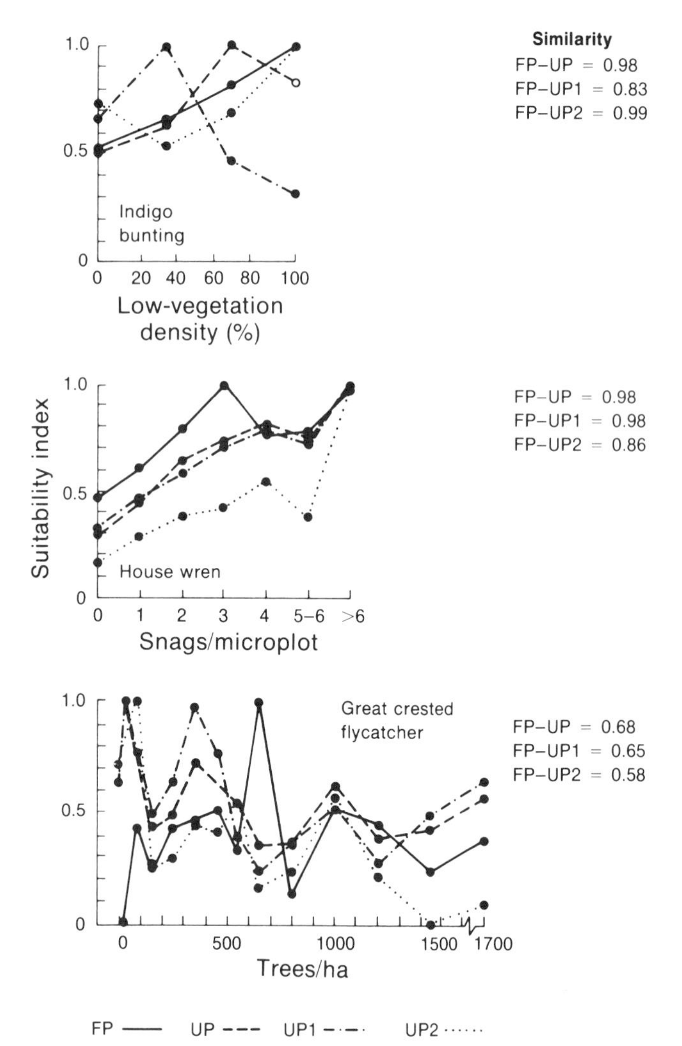

Figure 12.5. HSI curves from floodplain woodland data and three upland woodland data sets for the three species and the three variables considered. Codes for data sets are found in the text.

relationships derived from the larger data sets (ALL, WD, UP) was relatively strong (Table 12.3). The smallest data set, UP2, accounted for 53 (75%) of the 71 similarity values <0.90.

Discussion

Overall, HSI curves from data sets representing different habitats or sample sizes were similar. However, concern may be warranted for those pairs of bird-habitat relationships (24%) that were dissimilar. We found two general patterns in dissimilarity between curves: (1) either the curves differed at all levels of the habitat variable under consideration; or, less commonly, (2) the maxima (representing the most suitable habitat) were the same, but the remainder of the curves diverged. When the maxima are the same, either curve would depict the most suitable habitat similarly, but

Table 12.3. Median similarity between HSI curves derived from data sets of different sizes[a]

Variable/Data set	Median similarity (n = 12) WD	UP	UP1	UP2
Tree density				
ALL	0.98 (0)	0.97 (1)	0.94 (1)	0.88 (7)
WD		0.99 (1)	0.97 (1)	0.90 (6)
UP			0.97 (1)	0.91 (4)
UP1				0.85 (10)
Low-vegetation density				
ALL	0.99 (0)	0.99 (0)	0.98 (1)	0.96 (0)
WD		0.99 (0)	0.98 (1)	0.97 (0)
UP			0.99 (1)	0.98 (0)
UP1				0.96 (1)
Snag density				
ALL	0.99 (0)	0.97 (1)	0.94 (3)	0.88 (8)
WD		0.99 (1)	0.97 (3)	0.89 (7)
UP			0.97 (2)	0.90 (3)
UP1				0.85 (7)

[a]Data set sample sizes were: ALL, 1349; WD, 1034; UP, 780; UP1, 445; and UP2, 240. Data set codes are in the text. Figures in parentheses in the body of the table indicate number of species (up to a maximum of 12) for which similarity was <0.90.

assessments of habitat quality above and below the maxima may not agree.

Concern is justified when models for a particular species and habitat variable from different data sets are dissimilar and have different maxima. Our models may have differed for several reasons. First, the perceived differences may be spurious and may result from sampling error. The fact that some of our data were not independent of one another and the method by which we generated the HSI curves (based on means of species counts that had a Poisson distribution) may have led us, in some instances, to conclude erroneously that differences existed when in fact they did not. Conversely, some HSI models we assumed to be similar may actually have been different. Second, habitat variables selected for model development must be relevant to the species under consideration. If population numbers are not related to a particular habitat characteristic, it is unlikely that a consistent pattern of animal numbers to levels of the variable will be observed. Lastly, the observed differences may be real and must be considered when integrating and applying habitat models.

An animal's response to a particular habitat characteristic may depend on habitat type. More than 50% of the tree- and snag-density HSI curve pairs differed between floodplain and the three upland woodland data sets. This implies different habitat-selection patterns within a relatively small geographic area; in contrast, Noon et al. (1980) found relatively little variation in patterns of habitat selection by birds over a large geographic expanse. Although some of our observed differences may be spurious, many probably represent different responses of birds to the same habitat variable in different habitat types. Thus, when developing HSI models, it may be necessary to consider floodplain and upland woodlands separately. The category of "deciduous forest" as proposed for developing HSI models (Fish and Wildlife Service 1981a) may be too inclusive in some cases and could obscure real relationships of birds to habitat characteristics.

Data on any particular species often are dispersed throughout the literature and commonly represent intensive studies conducted on a few study sites. We found that HSI curves derived from data collected from four or five study plots in upland woodlands often diverged from analogous curves for larger data sets based on 10–28 study plots. To evaluate these differences, the investigator must decide which data set represents the "true" underlying relationship between the bird species and the habitat variable. If, as seems reasonable, HSI curves based upon larger data sets represent the real bird-variable relationship, then curves developed from the smaller data sets in some cases would not provide valid assessments of habitat quality.

On the basis of our results, we recommend that HSI curves be developed from the largest data sets available for the different habitat types (within gross categories such as deciduous forest). The similarity of HSI curves from smaller data sets or different habitats can be assessed, and, if the curves are found to be sufficiently alike, the data can be combined to develop the HSI curve to be used for habitat assessment. If the curves differ enough to cause concern, we suggest that different models be used for different habitat types.

There are trade-offs involved with using the largest available data sets. Large sets of data often represent a relatively superficial sampling of a large number of points and may include considerable among-site variation. In these cases, models developed may not provide a good fit of animal abundances to the habitat characteristics. Alternatively, smaller data sets representing intensive study of a particular species on a few sites may supply data to develop models with a good fit to the data, but these models may not be applicable over the range of the species. Perhaps the best situation is one in which intensive data are collected for a variety of sites that encompass the variation of the species-habitat relationships over the species' range. Our results are based upon data collected by an intensive study within a relatively small geographic area. It would be worthwhile to conduct a similar analysis for data collected less intensively over a larger geographic expanse to see if the trends we noted apply.

Determining the minimum acceptable sample size is difficult and will vary with species and habitat type. We recommend, on the basis of our results, that data from at least five sampling sites represented by at least 300–400 microplots, or their equivalent, be used when developing HSI models. We realize, however, that data are often sparse for the species of greatest concern and models must be based upon the best available information rather than waiting for the ideal data sets to be collected.

Acknowledgments

T. Rosburg, R. Deitchler, J. Vogler, C. Beckert, R. Fitton, and K. Spangler aided in collecting the field data. Reviews of earlier drafts by D. J. Orth, M. R. Ryan, R. H. Giles, Jr., W. F. Laudenslayer, B. A. Garrison, C. J. Ralph, E. P. Smith, R. G. Oderwald, and J. R. Rice are appreciated. This project was funded by the U.S. Fish and Wildlife Service, Office of Biological Services, administered through the Cooperative Wildlife Research Unit, Iowa State University, Ames, Iowa. Computer time was provided by the School of Forestry and Wildlife Resources, Virginia Polytechnic Institute and State University.

13

Limits in a Data-Rich Model: Modeling Experience with Habitat Management on the Colorado River

JAKE C. RICE, ROBERT D. OHMART, and BERTIN W. ANDERSON

Abstract.—We describe a model for predicting the status by season of all birds expected in riparian habitats along the lower Colorado River. We illustrate ways the model can be used in land planning and assessments. For two independent test sites, the model predicted avian statuses correctly for 90% of all bird species. These high rates of correct prediction are compared to accuracies of two alternative models: one based solely on breadths of species' distributions and one based solely on naturally occurring assemblages of species. Alternative models predicted less accurately than our habitat-based model but were correct in 60–80% of predictions. From these comparisons we conclude that our model does represent bird-habitat relationships and is appropriate for management advice. In general, models must be correct much more often than two-thirds of the time before one may conclude that true bird-habitat relationships are reflected by predictions.

We have spent more than a decade studying bird-habitat relationships in lower Colorado River riparian habitats. From the outset, one of our primary objectives has been to provide improved management information on desert riparian communities. Management needs the ability both to assess suitabilities of various habitats for all species of birds occurring in the region and to manipulate habitats to enhance bird species composition, richness, or densities in an area. Our management information had to be appropriate for all types of riparian vegetation, for all bird species in these riverine habitats, and for a realistic range of climatic conditions in all seasons.

Our study design included replicate transects in all types of desert riparian habitats. These transects were studied monthly for several years. As a result of our large spatial and temporal scale, conclusions and management tools we have devised are widely applicable. Because of the size of the data set, the tasks of data analysis, model construction, and model testing were extensive, and there were few precedents to follow. We had to resolve a number of difficulties that other workers can expect to encounter in similar large-scale modeling and analysis projects; thus, our experiences provide some useful guidelines for modelers and managers.

First we shall briefly review our overall study design, then we will outline the structure of our bird community–habitat management model. We will illustrate several problems encountered in developing the model, indicating the reasoning used in their resolution and our solutions. Because much of this work has been published elsewhere, our treatments will be brief. However, our final section will address in detail our tests of the model and the substantive question of what actually constitutes a test of a community model.

Methods

THE DATA BASE

At the heart of our riparian study were 72 line transects, established in stands of homogeneous riparian vegetation. Transects were 1.6 km (1 mile) or 0.8 km (0.5 mile) long, depending on the size of the stand. Transects were placed in 23 types of riparian vegetation, with replication proportionate to the occurrence of each type of vegetation in the entire lower Colorado River valley from Davis Dam to the Mexican border (Anderson et al. 1983).

For each transect, two types of data were collected: abundances of birds and attributes of vegetation. To estimate bird abundances, each transect was censused three times monthly for 4 years, using the variable-distance line-transect method (Emlen 1971; Anderson et al. 1977). Bird observations of all censuses within each season were combined to give a single estimate of density (birds/40 ha [100 acres]) of every bird species each season and each year. Vegetation measurements included counts of the six dominant tree species, the presence of mistletoe (*Phoradendron californicum*), the density of foliage in three strata (0–0.6 m [0–2 feet]; 1–3 m [3–10 feet]; ≥4.5 m [≥15 feet]), and foliage height diversity (Anderson et al. 1983). There were no significant changes among years in vegetation measures for any of the transects (Rice et al. 1984).

ANALYSES PRIOR TO MODEL CONSTRUCTION

Initially we assumed that abundance of each bird species on a transect would reflect suitability of the habitat on that transect for that bird species. Thus, regressions of abundance on the measurements of habitat should provide a simple quantitative model to predict expected abundance of

JAKE C. RICE: Northwest Atlantic Fisheries Center Box 5667, St. John's, Newfoundland, Canada A1C 5X1

ROBERT D. OHMART and BERTIN W. ANDERSON: Center for Environmental Studies, Arizona State University, Tempe, Arizona 85287

each species from vegetation measurements on new or altered sites. These regression equations, one for each species in each season, would provide a comprehensive model to predict the suitability of any habitat we might evaluate or develop in the future for all birds in the community.

We knew, however, that these regression equations would not be appropriate for all species of birds in all riparian habitats. Few species were encountered on all (or even nearly all) transects. For species of intermediate or narrow breadth of distribution, abundance (the dependent variable in the regressions) would be zero for a range of values of the habitat attributes (the independent variables). Hence, linear (or nonlinear) regressions would be inappropriate (Fig. 13.1*a*). Alternatively, more specialized models, such as stagewise regression, might be suitable (Draper and Smith 1981:337), but such models immediately present the question of where to partition the used from the unused sites. A higher-order model is required to determine whether any specific site falls in the used (open) or not-used (shaded) category of habitats for each species (Fig. 13.1*b*). Only in the first case would the regression equation be applied.

This problem prompted us to perform discriminant function analyses (DFA) between transects used and not used by each species. If the accuracy with which we could separate the groups using these discriminant functions was high, we would have an empirical tool for partitioning all possible habitats into those used and those not used by each species. The point on the canonical axis separating scores of used transects from scores of not-used transects (hereafter the "cutpoint" between used and unused sites) would provide the reference point for the stagewise regressions. Then, for habitats or transects predicted to be suitable for a species, regression models would be appropriate for predicting expected abundance. In these two steps we would have a comprehensive model for differentiating quality among all the habitats that might be used by any species.

The DFAs for each species each season were supposed to clarify quantitative relationships of bird occurrence to vegetation. Instead, the analyses exposed an entirely different problem. Initially, for each species in each season, the 72 transects were sorted into two groups: those used and those not used in 1976. Across all species, the large majority of DFAs were statistically significant; that is, used and not-used habitats differed. However, on the average only 65% of transects were correctly partitioned as used or not used (i.e., classification accuracy = 65%). We desired greater accuracy; therefore, we searched for causes of the high number of classification errors. We found that variation in occurrence among years was substantial (Rice et al. 1981). Thus, our problem was interpreting the meaning of a single season's record of abundance or occurrence. We were able to use information on the reliability of occurrence of a species over the years at specific sites to adjust our grouping criteria for the DFAs. If the two groups to be discriminated were transects consistently used for 2 years or consistently not used for 2 years, classification accuracy increased to 88% (Rice et al. 1983a). If 3 years of occurrence data were used in forming the two groups (present in at least 2 of 3 years,

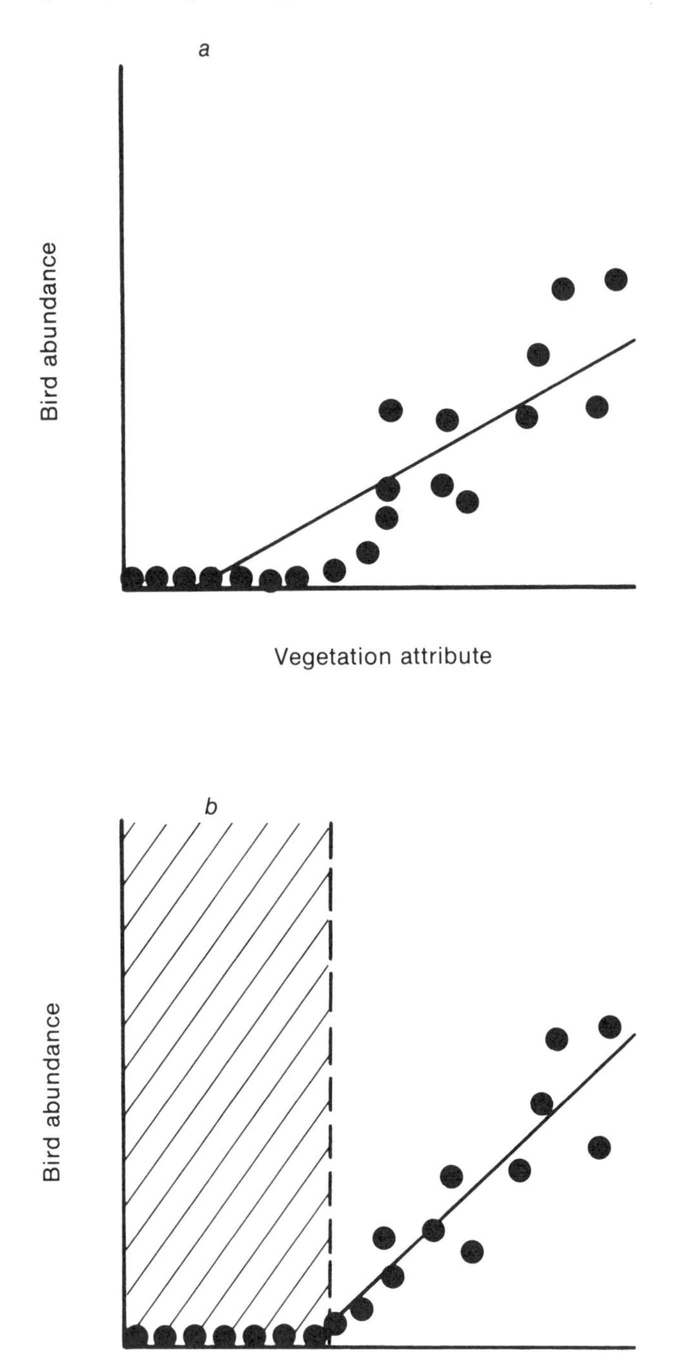

Figure 13.1. (*a*) Linear regression of bird abundance on vegetation attribute for all cases, including all zero bird abundances. Suitability overestimated at intermediate values on vegetation axis, underestimated at high values. (*b*) Predicted abundance is zero in the not-used portion. Regression for all cases in the used portion of the vegetation axis, performed after using discriminant classification function to separate used (clear) and not-used (shaded) areas on vegetation axis.

absent in all years), accuracy increased to 94% (Rice et al. 1984). Reassuringly, transects used in only 1 of the 2 or 3 years were ordinated in intermediate positions on the canonical axes, based on their values for habitat attributes. Sites of intermediate value on our measure of occurrence were also of intermediate position on our scale of habitat quality.

STRUCTURE OF OUR RIPARIAN MODEL

The large rate of turnover of most birds in our study sites (Rice et al. 1983b) led to a change in our modeling approach. Consistency of occurrence rather than abundance was used to indicate the suitability of habitats. We decided to manage for habitats where species of interest were sure to be present, no matter how poor the particular year might be for the species (a minimax modeling objective—aiming for the best possible result under the worst conditions) rather than to manage for habitats where the species might reach maximum abundance under some circumstances, but might be completely absent in other years (a least-square minimization objective—doing as well as possible under average conditions, regardless of how poor the outcome might be under some possible conditions). In practice, this change of modeling objective was not drastic; transects with the most reliable presences of species were often ones with the high abundances as well. We substituted a quantitatively tractable problem for a less tractable one, without losing a biologically interpretable objective for the model.

The canonical functions from the DFA became the core of the predictive component of our model, rather than merely a preliminary step prior to regression-based predictions. Categories of occurrence (Present, Likely, Irregular, Unlikely, Absent; Fig. 13.2) were determined by the position of consistently used, intermediate, and consistently unused transects along the canonical axis for each species. The model then predicted statuses of all species using a matrix of canonical functions, one per species per season, and the cutpoints separating each of the categories of occurrence along each function. For any specific site, the values of the vegetation attributes can be multiplied by the matrix of canonical function coefficients. The resulting discriminant scores (one for each species each season) are compared to the cutpoints, and the predicted status of each species is tabulated each season (Fig. 13.3).

The model also contains simulation components. Users may specify alternative habitat-management regimes, changing plant species composition and foliage density. The model simulates the effects of proposed changes on the vegetation. The new vegetation values are used to predict changes in the bird community expected to result from the vegetation manipulations. Thus, in a single series of iterations, the model can serve as an assessment tool, predicting the avian community of an area, and as a planning tool, investigating alternative management plans for that area. Predictions are available for every riparian species individually, rather than just for a few key or indicator species, or for some property such as richness or diversity. Predictions are provided for all seasons, not just for summer, and reflect expected status over several years.

Results

MODEL TESTING

Before any model is implemented, it should be tested rigorously. Tests should be both independent and unambiguous. We had no difficulty with the first requirement. Several transects from earlier years had not been used in the model development; neither bird occurrence nor vegetation data had been used in any of the analyses for model parameterization. Because of limitations on the length of this report, we will present only the results of two test transects in one season (May–July). Furthermore, we assess the accuracy of our predictions yearly, to allow comparisons between performance of our model and alternative models which do not make predictions over several years.

One test transect was vegetationally simple: a structurally homogeneous stand of saltcedar (*Tamarix chinensis*) approximately 2–3 m (7–10 feet) tall. The other was complex: a mature stand of cottonwood (*Populus fremontii*) and Gooding willow (*Salix gooddingii*) with dense understory, broken canopy, and some honey mesquite (*Prosopis glandulosa*) present. For the simple transect, our rate of predicting bird occurrence statuses correctly was 89% for each single year. For the complex transect, accuracy ranged from 88% to 92% (Table 13.1). (Note that rates of correct prediction can only be calculated using Present and Absent predictions. Predictions of Intermediate status cannot be tested with observations from a single year because either observed status would be correct.)

It was more difficult to meet the second criterion for a good test, that of being unambiguous. Given high rates of interannual variation in species' occurrences, little can be concluded from either a match or difference between the observed and the predicted status of a species in a single

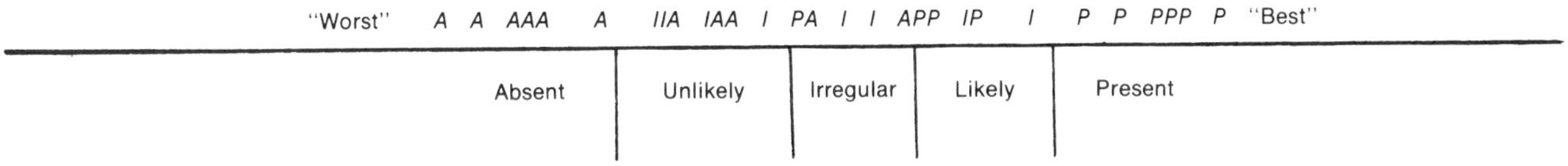

"Worst" A A AAA A IIA IAA I PA I I APP IP I P P PPP P "Best"

Absent | Unlikely | Irregular | Likely | Present

Figure 13.2. Determination of five categories of occurrence of avian species. Line represents canonical axis from used/not-used discriminant analysis. *A*s are scores of transects where the species was consistently absent; *P*s are scores of transects where the species was present at least 2 of 3 years; *I*s are scores of transects where the species was present in only 1 year.

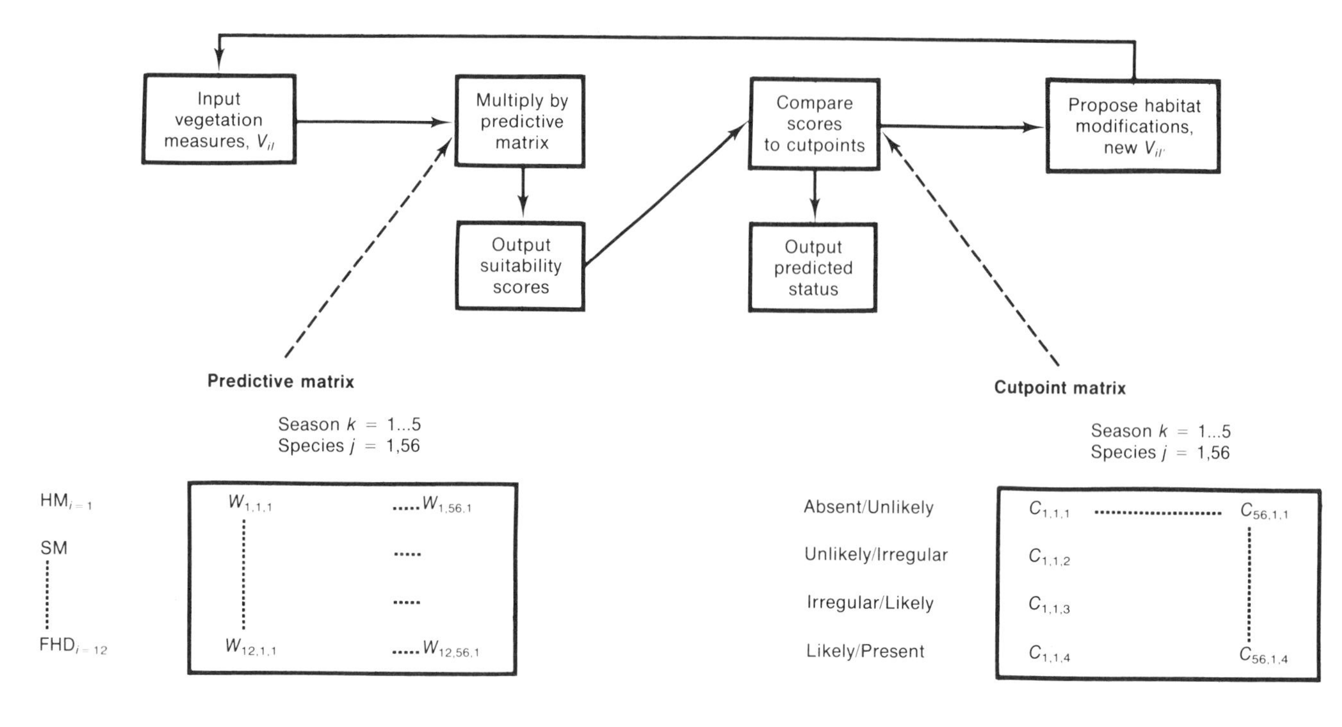

Figure 13.3. Flow diagram of steps in desert riparian model, illustrating sequence of steps leading to model predictions. Predictive matrix and cutpoint matrix are "exploded" below the flow chart. Subscripts: i refers to vegetation measure, i = 1–12 (honey mesquite to FHD); j refers to species, j = 1,56 (*Agelaius phoeniceus* [red-winged blackbird] to *Zonotrichia leucophrys* [white-crowned sparrow]); k refers to season, k = 1,5 (summer to spring); l refers to site, for any site to be used with model; m refers to category limits (Absent/Unlikely, Unlikely/Irregular, Irregular/Likely, Likely/Present). Subscripts for DFA canonical weights W are vegetation measure, species, season; cutpoints C are species, season, status category (1–4); vegetation measures V are specific measure and site.

year. To illustrate, consider the rates of correct prediction for a series of single years on the two test transects. When the 3 years are combined into pairs or triplets of years, the number of consistently incorrect predictions decreases. Although there were three prediction errors on the simple transect for 1977 and three more for 1978, only two species were predicted to have incorrect status in both years, and only one species was predicted incorrectly in all 3 years. For the complex transect, in single years there were three, four, and three errors; however, only two species were predicted incorrectly in each sequential pair of years, and again the status of only one species was predicted incorrectly in all 3 years. Our model was based on bird occurrences over several years and performed well when implemented on such time scales.

Clearly, the more data examined, the better the model appears to function. This is only true for Present and Intermediate predictions, however. For each species the number of transects usually stabilized after 3 or 4 years (Fig. 13.4*a*). For Absent transects, after 4 years of censusing, species are still appearing on transects not used previously (Fig. 13.4*b*). The correct number of years for an adequate test of model predictions has to be a compromise between including enough years to obtain realistic estimates of areas used by each species without including so many records that atypical species' occurrences obscure truly unsuitable habitats.

Alternative hypotheses for our model predictions

Although a single year's data are unlikely to be the best standard for testing a model, our model predictions were accurate even at that time scale. However, we have no guideline for comparison. In other words, what is our null hypothesis; how well should we expect to predict species' occurrences, knowing nothing of bird-vegetation relationships? For the rest of this chapter we will address the neglected question of how accurate model predictions must be before modelers can conclude that significant bird-habitat relationships are represented. We shall examine two alternative hypotheses.

Table 13.1. Number of species cross-tabulated by actual status on test transects and status predicted by riparian-habitat model. (Correct predictions of model are matches of Predicted and Observed statuses; Intermediate status includes species predicted in Likely, Irregular, or Unlikely categories [see Fig. 13.2].)

	Observed status			
	Simple transect		Complex transect	
Predicted status	Absent	Present	Absent	Present
	1977			
Absent	11	2	8	2
Intermediate	5	4	0	2
Present	1	13	1	23
	1978			
Absent	12	1	8	2
Intermediate	5	4	0	2
Present	2	12	2	22
	1979			
Absent	12	1	9	1
Intermediate	6	3	1	1
Present	0	14	2	22

Breadth of occurrence

Our first alternative hypothesis is that breadth of occurrence across transects is the determinant of avian occurrences. Breadth is used here simply to reflect the number of different habitats a species will occupy and does not consider evenness of the species' abundance across various habitats. An alternative model based on such breadths would simply predict that most common, widespread species will occur on a site and that most rare, narrowly distributed species will not. Habitat affinities play no role. We tested this alternative in the following manner. Seasonally, across 72 transects and 4 years, there are 288 opportunities for a species to occur. Breadth was defined as the proportion of 288 opportunities where a species was present. Breadths ranged from 0.007 (indigo bunting [*Passerina cyanea*]; individuals were detected on two opportunities) to 1.000 (brown-headed cowbird [*Molothrus ater*] and mourning dove [*Zenaida macroura*] on all 288 opportunities). Expected communities were constructed using sets of random numbers, bounded on the interval of 0–1.000. A species was assigned the status Present if the random number generated on its turn was less than or equal to the breadth calculated for the species. Otherwise, it was assigned the status Absent. A series of such communities were constructed and compared to the observed communities on the test transects.

For the simple transect, the statuses of 68% to 80% of the observed species matched those predicted by the breadth model. For the complex transect, the correspondence ranged from 64% to 67% (Table 13.2). These results are not fortuitous cases. Breadth communities were simulated 50 times for each transect and matched to the bird communities observed in 1977. The modal and median match rates were 78% for the simple transect (5–13 errors) and 58% for the complex transect (10–20 errors). These accuracies of prediction are less than those observed from our habitat-based model, but they are still high. They are as high, for example, as the average classification accuracy of our DFAs when based on a single year's records of occurrence.

Community-matching model

Comparing observed communities with ones predicted solely on breadths of occurrences of species is a type of neutral model analysis, where computers are used to simulate communities expected under certain hypothesized conditions (Caswell 1976a). These types of analyses have been criticized for being too neutral; many processes, in addition to the particular ones in question, have often been deleted (Diamond and Gilpin 1982). In our application, we wished to delete habitat affinities from community predictions. Predictions based solely on breadth do eliminate specific habitat affinities but also may eliminate interspecific interactions as well. Breadth-based models, at most, partly reflect the effects of occurrences where some species are strongly influenced (either positively or negatively) by the occurrences of other species. We do not know exactly how or by how much the occurrence of any one species is determined by the occurrence of other species; the role of competition in community structure is still hotly debated (Connell 1983; Roughgarden 1983; Simberloff 1983). Nonetheless, whatever processes function interspecifically, real avian communities must include the outcomes of such processes. Thus, we can use naturally occurring avian communities as targets for comparison with communities on our test transects.

In this series of tests, using our community-matching model we randomly chose several transects from our core set of 72. We compared the bird communities on each randomly chosen transect to the communities on our test transects. The average similarity over a number of such random pairings reflects the accuracy of a model that contains no true bird-habitat relationships but simply predicts that a typical community should be present. Strongly competing species would not be predicted to occur simultaneously, and if the sizes of guilds are truly regulated, our predicted communities would have guilds of the naturally occurring sizes.

When the bird communities on the simple transect were compared to the bird communities on a series of randomly chosen transects, median accuracy was 75% (Table 13.3). For the complex transect, median accuracy was 61%. Exhaustive iterative tests were not done for these comparisons because the finite number of observed transects would quickly be depleted and statistical independence between tests would be lost. Again, the accuracies of prediction in these tests were lower than rates of correct prediction by our habitat-based model. We may conclude that our model does contain true bird-habitat relationships and is a suitable model for planning habitat management. Again, however, these rates of correct prediction are fairly high, indicating that a model must be right considerably more than half the time before it can be concluded that true bird-habitat relationships are present.

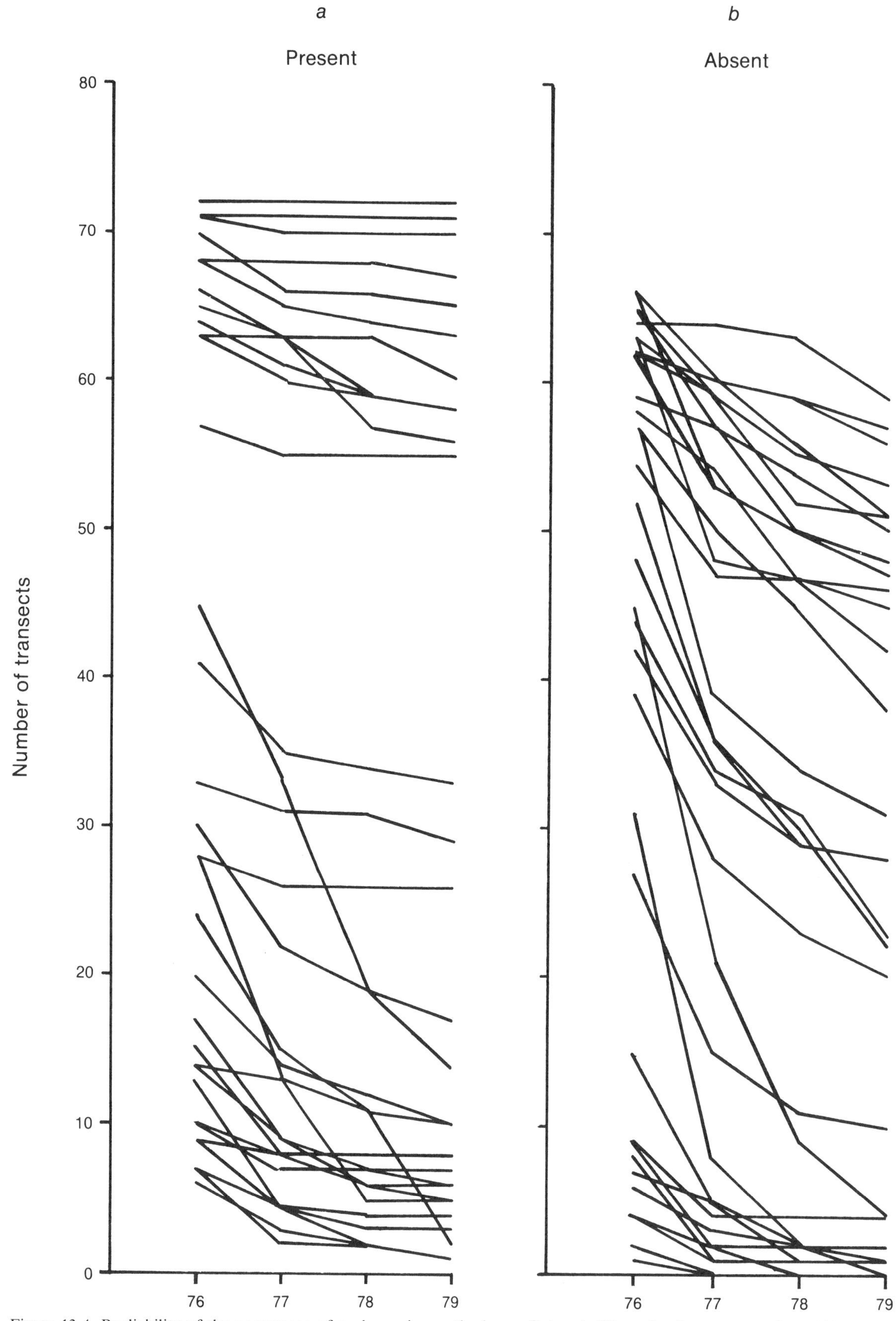

Figure 13.4. Realiability of the occurrence of each species on the lower Colorado River riparian transects from 1976 to 1979 (summer). Entries for 1976 are the number of present (*a*) or absent (*b*) transects for each species in that year. For subsequent years (intervals on the *x*-axis), the number of transects are shown where the species was present (or absent) in that year and all previous years back to 1976. Note that lines can only decrease or remain the same, because only patterns of consistent occurrence on transects are tabulated. As the number of consistently present and consistently absent transects decreases for a species, the number of transects in the Intermediate category would increase correspondingly.

Table 13.2. Number of species cross-tabulated by actual status on test transects and status predicted by breadth-of-occurrence model. (Correct predictions of model are matches of Predicted and Observed statuses. For each individual test, a separate breadth community was predicted. All predictions based on occurrence data from 1976 to 1979. Observations were for the year specified in the table entry.)

	Observed status			
	Simple transect		Complex transect	
Predicted status	Absent	Present	Absent	Present
	1977			
Absent	10	5	7	9
Present	7	14	3	17
	1978			
Absent	15	4	9	10
Present	4	13	2	15
	1979			
Absent	14	3	8	9
Present	4	15	4	15

Table 13.3. Number of species cross-tabulated by actual status on test transects and status predicted by community-matching model. (Correct predictions of model are matches of Predicted and Observed statuses. For each test, a transect was randomly selected from the 72 core transects and a year was chosen from the three available years. Thus, bird communities being compared were always from the same year, but were from different sites.)

	Observed status			
	Simple transect		Complex transect	
Predicted status	Absent	Present	Absent	Present
	TEST 1			
Absent	14	3	5	8
Present	4	15	4	19
	TEST 2			
Absent	16	6	6	10
Present	3	11	4	16
	TEST 3			
Absent	16	3	7	8
Present	2	15	5	16
	TEST 4			
Absent	4	4	8	11
Present	15	13	3	14
	TEST 5			
Absent	13	3	6	10
Present	5	15	6	14

Discussion

It is clear from our investigations that we have an accurate model for relating habitat characteristics to bird occurrences. Our rates of correct predictions on our test transects are consistently higher than the rates of correct predictions using alternative hypotheses. Furthermore, our model is robust across diverse habitats, seasons, and years. Therefore we can recommend implementation of our model in land-use planning in riparian habitats along the lower Colorado River.

It is also clear from our investigations that when models are being tested, it is necessary to have rigorous criteria for determining what constitutes a successful prediction by the model. Predictions that are correct in two of every three trials are possible without any reference at all to the habitat specific to a site. Both the breadth model and the community-matching model performed that well.

How accurate is accurate enough? There is no universal answer to that question. Accuracy of both the breadth and the community-matching models depends on characteristics of the avifauna under study. For regional avifaunas with few widespread species, the breadth model will make fewer correct predictions. For regional avifaunas with many locally distinct communities of species, the community-matching model will make fewer correct predictions. Nonetheless, these alternatives require consideration when models are being tested.

Our alternative predictive models are crude. For example, the breadth model would give the same breadth to a species that occurred on exactly six transects in each of 4 years as to a species that occurred on 24 transects in 1 year and was absent in the region in the other 3 years. For our community-matching model, all observed communities were treated as equally valid products of whatever interspecific interactions, if any, influenced community organization. Critics could argue that more refined alternative models were necessary to produce valid alternative hypotheses. Such refinements are possible, and we are currently exploring such avenues. Certainly no conceptual difficulties are presented by modifying the breadth measure to reflect year-to-year variation in the distribution of the birds. From our existing data base we have calculated many of the traditional measures of ecological overlap and competition among species. Construction of models that adjust the probabilities of occurrence of each species by the presence or absence of other species is in progress. Parameterization of alternative models is being kept independent of model tests.

That such refined models are possible does not detract from our overall point, however. Refined alternative models would probably make even fewer errors of prediction. Hence, models of bird-habitat relationships would have to perform even more accurately before the habitat model could be accepted.

Critics could also argue that our findings on the accuracy of nonhabitat-based models result from our decision to predict reliability of occurrence across years rather than using density as the predictive strategy. Redefining breadth to reflect proportional abundance by each species or changing the competition coefficient to reflect joint abundance would allow construction of corresponding alternative models for predicting abundances instead of reliability of occurrences. We feel our past research has shown that reliability of occurrence can be predicted with greater accuracy and precision than can abundance (and hence a prediction of occurrence can be more readily falsified, if wrong, than can abundance models). Furthermore, the variation among years in abundance and in distribution is an even greater impediment to the rigorous testing of abundance models than it was to testing our models which predicted reliability of occurrence.

Having said a number of pessimistic things about the feasibility of testing models once they are constructed, we conclude with some positive points. No one expects a model developed for application in a large region to be both simple and accurate. For example, even within the lower Colorado River valley some species display range limitations. Bewick's wren (*Thryomanes bewickii*) occurs only in the northern half of the study area, so when the model is applied to habitats in the southern portion of the region, occasionally its presence is erroneously predicted. Such special cases are unavoidable, so a priori extra rules will be necessary and legitimate.

In addition to errors due to special cases, genuine errors of prediction can also be informative. Careful testing of model predictions may pinpoint cases where our understanding of factors determining bird distributions is poor. Such diagnoses can lead to better biology. Certain patterns of model errors may appear as well. For example, our model seems to predict a few too many species in species-poor sites (our simple transect) and a few too few species in species-rich sites (our complex transect). Seeing those patterns may help us to improve our model, leading to better modeling as well.

Few of us believe that habitat relationships are the sole determinant of species' occurrences. In this chapter we have illustrated how other factors can make the investigation of the habitat relationships more difficult, and we have examined some approaches that can be used to distinguish habitat relationships that are susceptible to management activities from others that interfere with the ability to assess management tools.

Acknowledgments

We would like to thank the numerous field workers whose efforts provided the data base for this study, and A. Kurt Webb, John Murnane, and Linda C. Richardson for coordinating computer work. Thanks also to Susan M. Cook, Jane R. Durham, Cristie Roe, and Cindy D. Zisner for assistance with the preparation of the manuscript. Reviews by John Rotenberry, Barry Noon, C. J. Ralph, and William Laudenslayer improved the presentation of material in this manuscript. Funding for the Colorado River project was provided by the U.S. Bureau of Reclamation Contract No. 7–07–30–V0009, support that is gratefully acknowledged.

14

Developing and Testing Habitat-Capability Models: Pitfalls and Recommendations

STEPHEN A. LAYMON and REGINALD H. BARRETT

Abstract.—Predictive models of the relationships between wildlife populations and their habitats are becoming widely used by wildlife managers. The U.S. Fish and Wildlife Service Habitat Suitability Index (HSI) model series is a good example of this process. In tests of HSI models for the spotted owl (*Strix occidentalis*), marten (*Martes americana*), and Douglas' squirrel (*Tamiasciurus douglasii*), we experienced poor results for the three models even though they were based on what was believed to be good information. Problem areas included geographic resolution of the model, conflicting test results, unexpected variation in life history parameters, presence of predators, and difficulty in obtaining accurate habitat data. We present possible solutions to the problems we encountered, including a discussion of proper study design for HSI model tests. For example, we suggest incorporating predator presence into the HSI of the prey and relevant habitat information into geographic information systems. We strongly discourage the use of untested models because they lack credibility. We stress the need for a systematic research effort in concert with ongoing management programs.

Recent developments in the area of wildlife-management planning involve attempts to (1) quantify wildlife-habitat relationships (Fish and Wildlife Service 1981a; Nelson and Salwasser 1982); (2) establish computerized data bases of this information (Salwasser 1982); (3) integrate wildlife-habitat-relationships models into dynamic vegetation models (Barrett and Salwasser 1982); and (4) integrate such models with geographic information systems incorporating relevant habitat characteristics (Davis 1980). The development of habitat-capability models that predict species' occurrence is a critical link in the planning process. Habitat Suitability Index (HSI) models (Fish and Wildlife Service 1981a) are based on the premise that habitat suitability can be linked to individual habitat variables with primarily linear relationships, and that these variables can be combined into a meaningful index using a variety of variable weighting and averaging procedures. This single index is intended to represent the suitability of a given patch of habitat for a particular wildlife species.

HSI models have been proposed for many species, but their accuracy and geographic applicability is generally unknown because they have not been field tested. Inaccurate predictions by HSI models can arise for many reasons. Some assumptions used in HSI models have been questioned in the literature and in some cases have been shown to be false, such as (1) linearity of relationships between wildlife density and individual habitat parameters (Meents et al. 1983); (2) superiority of simple indices to those based on multivariate analysis (Green 1979); (3) invariability of habitat use regardless of life stage or season (Patterson 1976); (4) adequacy of a species' observed density as an indicator of habitat quality (Van Horne 1983); and (5) minimal effect of predators and other interspecific interactions on the abundance of their prey (Morin 1981). On the basis of a series of field tests in northern California of HSI models for the spotted owl (*Strix occidentalis*) (Laymon et al. 1985; Laymon and Reid, Chapter 15), marten (*Martes americana*) (Spencer et al. 1983; W. D. Spencer and R. H. Barrett, pers. obs.), and Douglas' squirrel (*Tamiasciurus douglasii*) (E. Hird and R. H. Barrett, pers. obs.), we illustrate a number of pitfalls that can arise in the development and validation of HSI models. We also discuss some procedures that may help avoid these pitfalls.

Spotted owl

Methods

The study area was located on the Eldorado National Forest, on the west slope of the Sierra Nevada between 1000 and 2000-m elevation. A map of the 170,000-ha study area was overlain with a 16-ha grid. Each grid cell was rated for three habitat variables: (1) average dbh of canopy trees; (2) percent canopy closure; and (3) stand structure. Ratings were derived from 1:24,000 color aerial photographs and orthophoto maps. The variables were suggested by results of research on spotted owls in Oregon (Forsman et al. 1984).

Average dbh of canopy trees was rated 0.1 if it was at least 40 cm and increased linearly up to 1.0 when the average dbh was >90 cm. Dbh was estimated to the nearest 10-cm increment from crown width estimated from the aerial photography. Canopy closure was rated 0.1 if 20% of the ground was covered by tree foliage. The rating increased linearly up to

STEPHEN A. LAYMON and REGINALD H. BARRETT: Department of Forestry and Resource Management, 145 Mulford Hall, University of California, Berkeley, California 94720

1.0 when the site had 80% closure. Approximately 10% increments were used. Forest structure was a dichotomous variable, receiving a rating of 0.2 when the stand was single-layered and 1.0 if it was multilayered.

The variables were combined using a geometric mean based on the assumption that they were partially compensatory and that a zero rating for any variable resulted in the habitat being unsuitable. The resulting HSI values for each grid cell were plotted on 1:24,000 overlay maps.

Test 1: Historical locations

The locations of 70 sites where spotted owls had been found in the course of owl surveys and normal USDA Forest Service operations were plotted on the 16-ha grid maps. These sites were believed to represent nonoverlapping territories of paired owls. The HSI values of the center 16-ha cell and the eight surrounding cells were recorded. The six cells with the highest HSI values were averaged to obtain an HSI for the 144 ha around each owl location. The three cells with the lowest HSI were excluded to keep clearcuts, brushfields, and south-facing slopes from depressing the mean HSI. Seventy additional cells were chosen randomly from the study area. HSI values were calculated for this sample as for the owl locations.

We used a t-test (Zar 1974:105) to explore differences in mean HSI values between the used and the random locations. A preference index (Strauss 1979) was used to quantify owl selectivity for four habitat-suitability categories: (1) unsuitable (HSI = 0–0.24); (2) marginal (HSI = 0.25–0.47); (3) suitable (HSI = 0.48–0.64); and (4) optimal (HSI = 0.65–1.0).

Test 2: Call count locations

Fourteen locations, each composed of nine 16-ha grid cells, were randomly selected from each of the four previously defined HSI suitability strata. These 56 sites were visited once in mid-July within a 7-night period having clear skies and no wind. From two to 12 sites were surveyed each night. At each site, the researcher imitated a male spotted owl giving a series of territorial calls. Ten calls were given over a 10-min period at each of three calling points, located 400 m apart, within each sample location.

A test for a monotonically increasing trend in calling response rate (proportion of 14 sites with owl response) from the unsuitable to the optimal habitat categories was made within the framework of the log-linear model (Marascuilo and Levin 1983:419–433). The strength of association between calling rate and increased habitat suitability was calculated using V^*, the log-linear analog to r (Marascuilo and Levin 1983:433).

Test 3: Telemetry locations, 16-ha resolution

Six owls were equipped with radio transmitters and tracked for 5 months. Nocturnal fixes were determined by triangulation. We plotted all owl locations on 1:24,000 maps and delineated a home range for each pair by connecting the outermost points. Owl use of a cell, regardless of frequency, was the dependent variable. The three habitat variables were combined in all possible, equally weighted combinations using arithmetic and geometric means. The best-fitting combination was determined by the Friedman test (Marascuilo and McSweeney 1977:355–367) and a preference test (Strauss 1979). The Strauss index (L) was also used to quantify the degree of selectivity by habitat-suitability class. In test 3 the classes used were (1) marginal (HSI = 0–0.2); (2) suitable (HSI = 0.3–0.6); and (3) optimal (HSI = 0.7–1.0) (Laymon and Reid, Chapter 15).

Test 4: Telemetry locations, 4-ha resolution

Similar procedures were followed for the 4-ha grid-cell resolution as outlined for the 16-ha resolution data (test 3). Habitat variables were resampled with a 4-ha grid on the six owl home ranges.

Test 5: Frequency of occurrence vs presence data

In this test the telemetry data described above (tests 3 and 4) were used to estimate the number of times an owl used each grid cell within a home range, rather than simple presence. Correlation analysis (Zar 1974:198–214) was used to detect relationships between predicted HSI values and frequency of cell use by owls.

Results

Test 1: Historical locations

The mean HSI value of locations where owls were found was 0.53. This was significantly higher by 42% than the mean HSI value of 0.31 for the random points (n = 140, $t_{\alpha=0.95}$ = 7.53, $P < 0.0001$).

The analysis of use vs availability data by habitat-suitability class showed that the owls avoided the unsuitable class ($L = -0.32$, $P_{used} = 0.04$, $P_{available} = 0.36$). They showed a significant positive preference for the suitable class ($L = 0.22$, $P_{used} = 0.43$, $P_{available} = 0.21$). The optimal class had a positive preference index that was not significantly greater than zero ($L = 0.19$, $P_{used} = 0.21$, $P_{available} = 0.02$). This was a result of the small number (two) of optimal sites in the random sample of available habitats.

Test 2: Call count locations

The proportion of sites where owls responded to calls increased from the unsuitable to the optimal habitat classes ($P_{unsuitable} = 0$, $P_{marginal} = 0.14$, $P_{suitable} = 0.36$, $P_{optimal} = 0.86$). The log-linear analysis yielded a monotonic trend contrast of 8.37 and a z-value of 6.95 ($P < 0.0001$), indicating a significantly increasing owl response rate with increasingly suitable habitat. V^* was 0.61, indicating a relatively strong relationship and thus substantiating the results of test 1.

Test 3: Telemetry locations, 16-ha resolution

The HSI model with the highest predictive power for the 16-ha resolution data was the arithmetic mean of percent canopy closure and forest structure. However, only two of the six owls showed significant positive preference for the optimal-habitat class. Two other owls showed significant positive preferences for the suitable-habitat class. No other indices were significantly different from zero, suggesting that, with a few exceptions, once owls choose a home range they utilize the habitats within its boundaries in proportion to their availability.

Test 4: Telemetry locations, 4-ha resolution

The HSI model with the highest predictive power for the 4-ha resolution data was the geometric mean of average dbh of canopy trees and percent canopy closure. All six owls sampled showed a significant, positive preference for optimal habitat. However, two owls showed significant, negative preference for the suitable-habitat class. No other significant preferences were found. Thus the higher resolution of habitat assessment was somewhat better able to document a tendency for these owls to focus their activity in relatively small but dense stands of large trees within their home ranges.

Test 5: Frequency of occurrence vs presence data

The frequency of owl use of individual cells showed no significant correlation with cell HSI value. The average correlation coefficient for six owls was 0.20 (r^2 = 0.04) at the 16-ha resolution and 0.05 (r^2 = 0.003) at the 4-ha resolution. These are very low values showing no significant relationships.

Marten

Methods

A marten HSI model based on published literature and expert opinion (Allen 1982a) was field tested on eight sites distributed among three 4000-ha study areas in northeastern California: (1) the Medicine Lake Highlands area on the Modoc National Forest; (2) the Swain Mountain Experimental Forest in Lassen National Forest; and (3) the Sagehen Creek watershed on the Tahoe National Forest. All areas include varying amounts of California red fir (*Abies magnifica*), white fir (*A. concolor*), wet meadow, lodgepole pine (*Pinus contorta*), Jeffrey pine (*P. jeffreyi*), and brushfield habitats. Previous research suggested that marten prefer the above habitats in the order listed, red fir being optimal and brushfields being unsuitable (Spencer et al. 1983).

Allen's (1982a) model incorporates four habitat variables: (1) percent tree canopy cover; (2) percent of canopy composed of fir; (3) stand age; and (4) percent of ground covered by downfall >8 cm diameter. They are combined in a geometric mean. Optimal marten habitat was assumed to have over 50% canopy cover, more than half of which was fir in mature or older age classes, and downfall covering 25–50% of the ground. The first three variables were obtained from 1:24,000 vegetation-type maps or aerial photographs provided by the Forest Service. Ground cover was visually estimated along transects, with calibrated photos used as a guide (Blonski and Schramel 1981). Predicted HSI values were calculated for each of the eight sample sites by using mean values for each habitat variable. Means were derived by weighting values for each mapping unit by its area.

Smoked aluminum track-plates (Barrett 1983) were used to field-test the marten HSI model by recording marten use at each site during the winter and spring of 1982. Track-plates were in cubbies, plywood or aluminum boxes (20 × 20 × 50 cm) with one open end through which marten could enter; the cubbies were attached to tree boles 1–2 m above snow level. An aluminum plate blackened with kerosene soot was placed in each cubby, along with frozen trout as bait. Animals stepping on the smoked plate left clear tracks which could be lifted with cellophane tape. A commercial marten lure (S. Stanley Hawbaker and Sons, Fort Loudon, PA) was placed on the cubby to enhance its attractiveness. The cubby protected the carboned plate from precipitation and minimized the number of nontarget animals stepping on the plate.

Eight clusters of 18–20 cubbies placed in a 400 × 600 m grid were allocated over the range of habitats involved. Each cluster sampled a relatively homogeneous site of approximately 800 ha. Cubbies were checked weekly for 5–10 weeks. Any number of marten tracks detected on a track-plate after a week's exposure was considered a visit. Results were summarized as the mean visitation rate per cubby for each cluster. Since marten have not been harvested in California for many years and populations are presumed limited by winter cover (Spencer et al. 1983), this method provided an index of marten use that was assumed to be correlated with marten carrying capacity. The results were tested for a significant trend with the Kruskall-Wallis Trend Test for nonparametric data (Marascuilo and McSweeney 1977:309).

Results

Of the eight sites sampled, three were rated marginal (HSI = 0.1–0.3), one was rated suitable (HSI = 0.4–0.6), and three were rated optimal (HSI = 0.7–1.0). Mean marten visitation rates increased with increasing habitat ratings ($P_{marginal}$ = 0.27, $P_{suitable}$ = 0.37, $P_{optimal}$ = 0.82), but the trend was not statistically significant (χ^2 for linear trend, 1 *df* = 3.66, P = 0.06). The visitation rates in both the marginal and optimal classes ranged over 40%, but it is possible that a larger sampling effort would have yielded a statistically significant finding.

Douglas' squirrel

Methods

An HSI model was constructed for the Douglas' squirrel based on information in the literature (Hatt 1929; Koford 1979; Woods 1981) using the graphic method outlined by the Fish and Wildlife Service (1981a). Three habitat variables (percent tree canopy cover, stand age, percent of canopy as Jeffrey pine) were combined by their geometric mean to index Douglas' squirrel carrying capacity as set by winter cover.

A 2000-ha portion of the Sagehen Creek watershed was censused for Douglas' squirrels during August and September 1982, using the line-transect method (Burnham et al. 1980). Squirrels were stimulated to call by the playing of prerecorded squirrel territorial calls at 50-m intervals. Pilot surveys indicated that optimum hours for detecting squirrels were within the 3 hours following sunrise, and that by recording only squirrels within 100 m of a transect, squirrel counts per 50-m interval could be equated to squirrel density (E. Hird and R. H. Barrett, pers. obs.). A total of 201 4-ha sample units (each including a 200-m transect segment) were surveyed for the three habitat variables required by the HSI model as well as for squirrel density.

HSI values were computed for each sample unit. Simple and multiple regression methods (Green 1979) were used to relate individual habitat variables and HSI values to squirrel density (d) (transformed by $\ln[d + 1]$). Analyses were made on subsets of the data for each major habitat (red fir, $n = 53$; lodgepole pine, $n = 28$; Jeffrey pine, $n = 110$), as well as the entire data set ($n = 201$). Finally, to facilitate comparison of predicted vs observed values, HSIs were transformed into squirrel-density estimates by assuming that the highest density recorded in the study for a 4-ha sample unit was equivalent to an HSI of 1.0. We assumed a linear relationship between the two variables.

Results

Despite the relatively large sample size, no statistically significant relationships could be detected between habitat variables of HSI and Douglas' squirrel density (no $r > 0.1$). The discrepancy between predicted and observed mean densities for major habitat types was greatest for the red fir habitat (−83%) and least for the Jeffrey pine habitat (−30%). The predicted density for the red fir type was about 1.5 squirrels/ha, whereas the observed density was only 0.25/ha (E. Hird and R. H. Barrett, pers. obs.).

Discussion

The three models that we developed and tested were all believed to be good ones. All were developed after a comprehensive literature search, yet none performed as well as one would hope, even after model variables were changed or variable weights readjusted. Problems we encountered are discussed below.

Geographic resolution

The geographic scale at which HSI models are applied and tested is important. It should reflect (1) the size of the animal's home range; (2) the degree of habitat specialization by the animal; (3) the heterogeneity of the habitat; and (4) the intended use of the model.

The effectiveness of the spotted owl models was dependent on the intended use and the size of the grid cell used during analysis. A resolution of 144 ha (nine 16-ha cells) was adequate to predict presence/absence over a broad landscape (tests 1 and 2). However, increasing the resolution to 4 ha was necessary for predicting the location of activity centers within home ranges. The poor test results for the Douglas' squirrel model may have been the result of using sampling units that were too large (4 ha) in relation to a typical squirrel home range (<1 ha).

We suggest that a rule of thumb might be to use grid cells of about one quarter the area of the species' normal home range. A species with highly specialized microhabitat requirements living in a heterogeneous landscape might require a higher resolution for models predicting preference. Presence/absence models for habitat generalists in homogeneous landscapes would require a much lower resolution. In general, the most important factors determining the optimum resolution will be home range size and landscape heterogeneity.

Conflicting test results

Conflicting results or a lack of statistically significant results are particularly bothersome. In the case of the spotted owl, test 1 was marred by a small sample of optimal habitats. That test was then used as a pilot sample to design test 2, which allocated equal but adequate sample sizes to each habitat-suitability class. Radio-telemetry location data were compared with HSI cell ratings at both 16-ha and 4-ha cell sizes (tests 3–5) with relatively poor and somewhat conflicting results. It was easier to model presence of territorial males at a resolution of 144 ha than to model habitat preference of individual owls at a resolution of 4 ha.

A good study design is critical in validating HSI models. A sound study design would include the following: (1) study areas that represent the range of habitats in which the animal is found; (2) a sufficient number of randomly selected individuals or sample plots to avoid bias from aberrant individual behavior and to determine the range of individual behavior; (3) more than one test method; and (4) data from more than a single year to avoid bias from atypical weather patterns or year-to-year population fluctuations that might influence habitat selection. The number of study areas would depend on the number of habitat types in which the animal occurs and the geographic range for which the model is designed. A minimum of three study areas should be considered. The number of samples (individuals or sample plots) obtained at each site would depend on the variation observed in each case, preferably by a pilot sample. For radio-tracking studies of predictable species, five individuals per study area might be sufficient; a behaviorally plastic species might require 20 or more subjects. The optimum number of sample plots, trap stations, or transects would vary just as widely. Whenever possible, it is wise to use more than one test method. Agreement among methods is probably the best way to assure that a poor model is not being depended on for management decisions. We stress the importance of careful study design. For further information on study design, we refer the reader to Green (1979), Romesburg (1981), and Hurlbert (1984).

Imperfect life history information

Accurate life history information for local populations is necessary to develop a good HSI model. This is especially true for species that have high behavioral plasticity. In developing the spotted owl model, it was assumed that habitat use would be similar throughout the year for this presumably resident species. However, radio-tracking data did not support this assumption. August and early September foraging locations were in more open habitats, where the owls were feeding on large insects, than were either early summer or fall locations (Laymon, pers. obs.). In addition, the daytime roost site locations were dependent on current weather patterns and were in much denser and more shaded locations during the summer than in other seasons. Marten habitat preferences also varied seasonally (Zielinski et al. 1983; Spencer et al. 1983).

The spotted owl in Oregon (Forsman 1984) and in northwestern California (D. Solis and C. Sisco, pers. comm.) is apparently nonmigratory. However, in the western Sierra Nevada a migratory movement of 12–25 km may occur in at least some years. In the late fall of 1983 all five radio-collared owls moved downslope 1000 m in elevation and established winter home ranges. The winter home ranges were in oak woodland, while the summer home ranges were in mixed-conifer forest.

The most appropriate season to sample Douglas' squirrel habitat use is probably not during the late summer when young are dispersing into less suitable habitats, even though it is easier to stimulate these squirrels to call during this time. Correlations between squirrel density and habitat variables might be more apparent if sampling occurred during the spring, even though mean spring density might be substantially lower than that observed in the late summer.

Studies quantifying seasonal habitat needs have been carried out for few wildlife species. Many HSI models are based on natural history information generated by casual observations. Even in the case of species for which systematic studies have been done, HSI model builders may try to generalize too far. Much ecological literature indicates that from one subspecies, ecotype, or local population to another, important behavioral differences in habitat use can occur. The spotted owl is one example, but such variability is most likely the norm.

Presence of predators

The effect of predators on habitat occupancy, and therefore on model test results, may be important in some cases. However, it is typically difficult to obtain information about the importance of predation to local prey populations. Douglas' squirrels are a major food source for marten in red fir habitat (Zielinski et al. 1983). The presence of higher densities of marten in mature red fir habitat at Sagehen Creek apparently limited the number of squirrels there below the density expected if food or cover were limiting. The density of squirrels was directly proportional to the number of large logs in habitats where marten were uncommon or absent, but it was inversely proportional to the number of large logs in red fir habitat (E. Hird and R. H. Barrett, pers. obs.). Marten find ready access to sleeping squirrels in winter dens located in or under large logs despite deep snow cover (Spencer et al. 1983). We hypothesize that Douglas' squirrels would be more common in red fir habitat devoid of marten.

Where predation can be shown to be a factor limiting a prey population, a linking of predator and prey HSI models may be appropriate. In the case of the Douglas' squirrel, marten density could be used as a variable in the squirrel HSI model. If high marten populations are found to be dependent on the presence of large logs in certain habitat types, those habitat types would receive a lower suitability rating for Douglas' squirrels when large logs were present. The effect of predators on habitat suitability is a topic requiring more research.

Unavailability of accurate habitat data

The ideal HSI model is one that accurately predicts habitat suitability, either using variables derived from existing vegetation-type maps or those derived quickly by remote-sensing methods. Unfortunately, for some species of wildlife the habitat variables most highly correlated with carrying capacity are not available in data bases of land-management agencies. Examples are understory vegetation, snag density, dead-and-down woody material, snow depth, and soil moisture. Models that require field sampling of habitat variables will not be used for large-scale project planning because of the high cost of obtaining the information. The cost of obtaining the data on the percentage of ground covered by woody downfall for the marten HSI test, for example, was greater than that of gathering all the other variables combined.

Certain variables useful in wildlife modeling could be included in compartment inventories periodically made by the Forest Service, or similar field surveys, thus increasing the usefulness of such efforts. The ultimate answer to this problem will be the development of computerized, geographic information systems (GISs) as proposed by Davis (1980). GISs will incorporate information needed by all wildland-resource managers. GISs may not be available on a large scale in the near future, but, in the meantime, detailed habitat information being collected by various resource inventories should be preserved as soon as possible for use in future studies.

Concluding remarks

It is dangerous to rely on untested HSI models. On the basis of our experiences, what we thought were good models proved in field tests to be poor ones. While we approve of the general approach of wildlife-habitat modeling, we argue that more, not less, field research is needed to successfully implement predictive models for wildlife-management planning. Untested models are not credible, and therefore are of little use. Perhaps the greatest value of such models at this point is not in management application, but rather in stimulating a more concerted and systematic research effort coordinated with ongoing management programs.

Acknowledgments

We thank W. D. Spencer and E. Hird for the use of their unpublished data, and the numerous field assistants who have contributed their time to gathering data for this study. Financial support was provided by the USDA Forest Service and the California Agricultural Experiment Station, Project 3501MS. We appreciate the helpful comments of C. J. Ralph, W. E. Grenfell, M. G. Raphael, and M. F. Dedon on previous drafts of this chapter.

15

Effects of Grid-Cell Size on Tests of a Spotted Owl HSI Model

STEPHEN A. LAYMON and JANICE A. REID

Abstract.—Habitat Suitability Index (HSI) models are receiving wide use in land-management planning. We present a test of a spotted owl (*Strix occidentalis*) model by comparing HSI values of radio-telemetry locations of owls with HSI values of random points. We used grid-cell sizes of 4 ha and 16 ha. The Friedman test and the Strauss Preference Index were also calculated to determine the best-fitting model. Owl locations corresponded better to high HSI values in the 4-ha than in the 16-ha cell size. The majority of averages of the four 4-ha cells had higher HSI values than the corresponding 16-ha cell, suggesting that the larger cell size masked small habitat pockets of high value. Selecting the proper cell size is important when validating HSI models.

The Habitat Suitability Index (HSI) model approach to wildlife-habitat-relationships modeling is an integral part of the U.S. Fish and Wildlife Service Habitat Evaluation Procedures (HEP) (Fish and Wildlife Service 1981a). HSI models for many species have been proposed, but few have been tested and even fewer have been tested using more than one method.

Laymon et al. (1985) have presented an HSI model for the spotted owl (*Strix occidentalis*) which is based on the geometric mean of three variables: (1) percent canopy closure; (2) vertical structure; and (3) mean dbh of the canopy trees. These variables were found to be important in determining spotted owl habitat in Oregon (Forsman et al. 1984). The model's ability to predict occupied spotted owl habitat in the Sierra Nevada of California was tested by comparing HSI values of 70 sites where owls were known to occur ($\bar{x}$ = 0.53) with HSI values of 70 randomly located sites ($\bar{x}$ = 0.31) (t = 7.53, CV of t = 1.96, $P < 0.001$). A second test, a field survey using vocal imitations of spotted owl calls, was conducted at 56 sites, 14 in each of four HSI categories: (1) unsuitable (HSI = 0–0.23); (2) low (HSI = 0.24–0.47); (3) moderate (HSI = 0.48–0.64); and (4) high (HSI = 0.65–1.00). Owls responded at none of the unsuitable sites, 14% of the low sites, 36% of the moderate sites, and 86% of the high sites. The results showed an increasing trend in owl response with an increase in HSI value (z = 6.95, CV of z = 1.96, $P < 0.001$).

In this chapter we report an additional test of the model by comparing predicted HSI values of the owls' foraging locations, determined by radio telemetry, with HSI values of a random sample of points in the owls' home range. The purposes of this chapter are (1) to determine the usefulness of the model in predicting within-territory habitat use by spotted owls; (2) to determine the best-fitting model combination for both 16- and 4-ha grid-cell resolutions; (3) to determine the effect of grid-cell resolution on model performance; and (4) to determine the causes of differing results caused by grid-cell resolution.

Study area and methods

The study area was located on the Eldorado National Forest in El Dorado and Placer counties, California, on the west slope of the Sierra Nevada between 1100 and 1700-m elevation. We equipped three pairs of spotted owls (referred to as the Ramsey, Bacchi, and Gaddis pairs) with radio transmitters and tracked them from 10 June to 1 October 1983. We determined the owls' locations (fixes) by triangulation. No fixes with an error polygon >0.5 ha were used in the analysis. Most fixes were taken at night and therefore represent foraging locations.

Each fix was plotted on a map. A home range was delineated for each pair of owls, using the minimum convex polygon method. We rated each 16-ha block and the four corresponding 4-ha blocks within a home range for three habitat variables—(1) forest structure; (2) percent tree canopy closure; and (3) average dbh of canopy trees—using estimates obtained directly from aerial photographs (scale = 1:24,000). Forest structure was a dichotomous variable, which received an index rating of 0.2 if the forest was even-aged, showing a single-story effect on the aerial photograph, and a 1.0 if uneven-aged, showing a multiple-story effect on the photograph. Percent canopy closure was rated 0.1 if the site had 20% crown closure from all trees. The rating increased linearly, in increments of approximately 0.15, up to 1.0 when the site had >80% canopy closure. Average dbh of canopy trees, estimated from crown width, was rated 0.1 if the mean dbh of the canopy trees was at least 40 cm and increased linearly up to 1.0 when the mean dbh was >90 cm. Grid cells were rated for the habitat type that made up the largest portion (usually >50%) of the cell.

HSI values, rounded to one decimal point, of all 16- and 4-ha cells within the owls' home ranges were determined for each of the three habitat variables and for the eight possible equally weighted two- and three-way arithmetic and

STEPHEN A. LAYMON and JANICE A. REID: Department of Forestry and Resource Management, 145 Mulford Hall, University of California, Berkeley, California 94720

geometric mean combinations. These were plotted in the form of 11 HSI maps for each pair. The HSI values for each owl were grouped into three suitability categories: (1) low (HSI = 0–0.2); (2) moderate (HSI = 0.3–0.6); and (3) high (HSI = 0.7–1.0). Cell use and availability data of each owl were generated by suitability categories for the three variables and eight combinations. This was done by superimposing a template of used and random locations over each HSI map. Random cells were selected until the sample was approximately equal to used cells.

We used the Strauss Preference Index (Strauss 1979) to quantify the preference of each owl for the three suitability categories, using the three individual variables and eight model combinations. The Strauss Preference Index (L) represents use minus availability within each category. Values for L can range from -1.0 (absolute avoidance) to 1.0 (absolute preference). The L values are interdependent across categories (i.e., an increase in preference in one category requires a decrease in another) but are independent of sample size.

We compared the L value for each category, using three pairwise comparisons: (1) $L_{moderate\ suitability} - L_{high\ suitability}$; (2) $L_{low\ suitability} - L_{high\ suitability}$; and (3) $L_{low\ suitability} - L_{moderate\ suitability}$. We did this to determine the degree of discrimination for each model between suitability categories, under the assumption that the highest degree of consistent discrimination should represent the best-fitting model. We made these comparisons for each owl and each variable and model combination (198 comparisons). The best-fitting model for both the 16- and 4-ha grid-cell resolutions was determined using the Friedman test (Marascuilo and McSweeney 1977:355–367). We applied this test to two 18 × 11 tables (one each for 4- and 16-ha resolutions) of L-difference scores for each of the three L-difference comparisons for the six owls (18 rows) by three individual variables and eight combinations (11 colums). We then ranked each L-difference comparison for each bird, across the three individual variables and eight combinations. We determined the average rank for each column. The omnibus Friedman test was used to determine whether differences in model capability existed. Post hoc, pairwise comparisons were made to determine the model variable or combination which had the greatest ability to describe the relationship of spotted owls to their habitat.

Statistical comparisons of frequency of owl use of cells versus HSI values of cells were performed using least-square linear regression (Zar 1974:198–214). We performed the calculations for all analysis on an IBM–PC/XT microcomputer using a standard spreadsheet program.

Results

Model test results

16-ha grid-cell resolution

None of the model combinations or single variables was successful in predicting within-home-range spotted owl habitat use when we used a 16-ha grid-cell resolution. The results of the Strauss Preference Index tests showed consistently significant positive preference for high-suitability habitat for only two of the six owls (Ramsey pair). Only 33% of these 66 model combinations for the six owls were significant. Twenty-one of the 22 significant preferences were for the Ramsey pair. A further examination of the 44 nonsignificant L-values revealed that 32 were negative, indicating an aversion by the remaining four owls to what the models described as high-suitability habitat. The average L-value for all 66 high-suitability comparisons was close to zero ($\bar{x}$ = 0.07). The average 95% confidence interval of L was 0.17. We found consistent negative preference for low-suitability habitat for only one owl and for only 15% of the 66 model combinations.

The omnibus Friedman test showed a significant difference in model performance (Table 15.1). The model combination that had the most power in describing the relationship between spotted owls and their habitat was the arithmetic mean of the percent canopy closure (C) and forest structure (S) (Table 15.2). The single variable with the highest predictive power was forest structure. This value was only slightly higher than the mean ranks for percent canopy closure and average canopy tree dbh (Table 15.2). The best-fitting model ($C + S/2$) was the only model combination that was much higher than the single variable, forest structure. We found no model combination to be statistically better than any other when we tested the mean ranks on a post hoc, pairwise basis. A complex, and probably uninterpretable, contrast was responsible for the statistical significance in the omnibus test. We can conclude that all model combinations are equally poor at predicting spotted owl habitat at the 16-ha resolution.

4-ha grid-cell resolution

When we used the 4-ha grid-cell resolution in the analysis, all model combinations, for all six owls, performed better at predicting within-home-range habitat use. The Strauss Preference Index analysis showed consistent significant preference for the high-suitability category for all six owls and on 95% of the 66 model combinations. No high-suitability L-values were negative, and 98% of them increased from the corresponding L-value in the 16-ha case. The average L-value increased more than fourfold ($\bar{x}$ = 0.29). The sample sizes were also larger than in the 16-ha case (Table 15.3). This increase in sample size reduced the average 95% confidence interval (CI) approximately 50% (mean CI = 0.09). To determine the cause of the increase in significant

Table 15.1. The Friedman test statistics for comparisons of mean rank of Strauss Preference Index L-values

	Grid-cell resolution	
Suitability comparison	16-ha	4-ha
$L_{moderate}-L_{high}$	18.68*	25.43*
$L_{low}-L_{high}$	10.77	19.08*
$L_{low}-L_{moderate}$	6.56	21.36*
Overall	22.55*	40.06*

* Significant: chi-square with 10 *df* and 95% confidence interval = 18.31.

Table 15.2. Mean Friedman test ranks of suitability-class comparisons of Strauss Preference Index values of L for the three spotted owl model variables and eight model combinations for 4- and 16-ha grid-cell resolution

Suitability-class	Variables			Model combinations							
L-comparisons	Canopy (*C*)	Dbh (*D*)	Structure (*S*)	$D + S/2$	$C + D/2$	$C + S/2$	$C + D + S/3$	$(D \times S)^{1/2}$	$(C \times D)^{1/2}$	$(C \times S)^{1/2}$	$(C \times D \times S)^{1/3}$
16-HA GRID-CELL RESOLUTION											
Moderate–high	6.5	3.3	8.4	5.5	4.5	8.5	7.3	5.4	3.0	6.9	6.5
Low–high	5.0	6.3	4.4	8.0	3.3	7.0	5.8	7.3	5.2	6.8	6.9
Low–moderate	5.6	7.3	6.8	6.2	6.4	7.8	5.0	6.4	5.8	4.4	4.4
Overall rank	5.7	5.6	6.5	6.6	4.8	7.8	6.0	6.4	4.6	6.1	6.5
4-HA GRID-CELL RESOLUTION											
Moderate–high	6.5	8.8	5.0	4.1	7.3	3.7	7.3	4.9	9.7	2.8	5.8
Low–high	7.7	7.8	4.5	5.5	6.2	2.5	4.6	5.8	8.6	8.0	4.8
Low–moderate	4.8	8.2	6.0	3.3	6.0	9.2	6.0	3.6	8.6	6.2	5.2
Overall rank	6.3	8.3	5.2	4.3	6.5	5.1	6.0	4.8	8.9	5.9	5.0

differences, we applied the 4-ha resolution confidence intervals to the 44 nonsignificant 16-ha resolution *L*-values. Twenty-two of these were then judged significant; however, all were negative. This showed that most of the increase in predictive power was caused by the increase in *L*-values rather than by the reduction of the confidence intervals. An example of this increase in predictive power of the model is shown in Figure 15.1.

Only 6% of the 66 model combinations showed significant negative preference in the low-suitability category. This was a slight reduction in number from the 16-ha resolution. The model did not perform well in either the 16- or 4-ha case with regard to predicting use of low-suitability spotted owl habitat.

The omnibus Friedman test showed that a difference existed in at least one of the potential contrasts. The model combination best describing the relationship between the owls and their habitat was the geometric mean of percent canopy closure and average diameter of canopy trees (Table 15.2). This combination was only slightly better than the mean rank of average diameter of canopy tree. The forest-structure variable had the lowest rank, and all model combinations that included this variable were lower than the two that did not (Table 15.2). Again, none of the post hoc, pairwise comparisons were significantly different. In the 4-ha

Table 15.3. Number of used and random cells within the six spotted owl home ranges for 4- and 16-ha grid-cell resolution

	16-ha resolution		4-ha resolution	
Individual owl	No. points used	No. random points	No. points used	No. random points
Ramsey male	11	24	22	30
Ramsey female	13	24	22	30
Bacchi male	21	32	26	36
Bacchi female	28	32	56	36
Gaddis male	24	33	33	34
Gaddis female	35	33	55	34
Mean cells/owl	22.0	29.7	35.7	33.3

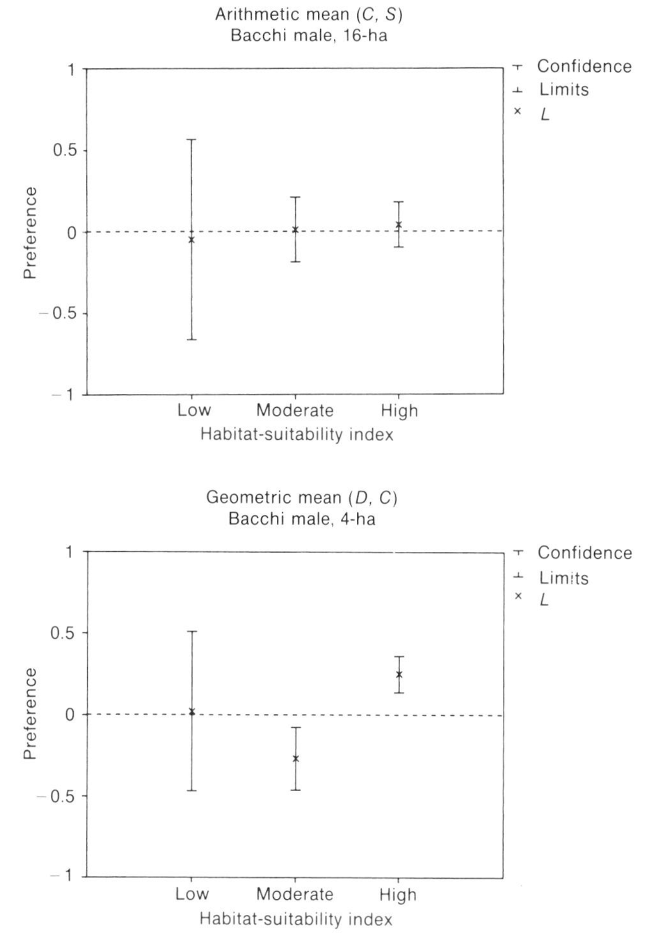

Figure 15.1. Example of Strauss Preference Index *L*-value and confidence interval of the best-fitting model for both 16- and 4-ha grid-cell resolutions, showing increased model ability to predict spotted owl habitat use of high-suitability habitat resulting from a decrease in grid-cell size.

case, the results of the Friedman test show that all model combinations are performing equally well in describing the relationship between owls and their habitat.

FREQUENCY OF CELL USE

We conducted an additional test using correlation analysis to compare frequency of spotted owl cell use with HSI values of used cells. All relationships were weak (Table 15.4). This weak relationship held for both 16- and 4-ha cell sizes and all model combinations.

EFFECT OF GRID-CELL SIZE

Because we rated each grid cell independently, it was possible to compare ratings for the two grid-cell sizes. We took the rating of the individual habitat variables of the 16-ha cells and compared them to the average of the corresponding four 4-ha cells. Comparison showed a weighting toward larger values of the mean of the four 4-ha cells (Fig. 15.2). This trend was similar for all three variables: forest-structure ratings were larger in 51% of the cells; average-diameter ratings were larger in 70% of the cells; and percent canopy closure ratings were larger in 54% of the cells. This indicates that small pockets of habitat ranging in size from 2 to 7 ha, with uneven-aged characteristics, large average dbh, and high canopy closure were masked by the 16-ha resolution ratings.

Discussion

Grid-cell resolution is important in testing HSI models. Choice of a grid-cell size should reflect the expected use of the model, the size of the animal's home range, and the homogeneity of the habitat. For an animal with a home range of 800–1600 ha (e.g., a spotted owl), a cell size of 16-ha, representing 1–2% of the home range, should be adequate. This was true when presence/absence of owls was being tested over a large geographic area. However, when the model was used to represent habitat preference within an owl's home range, 16-ha cells were no longer adequate because of the heterogeneity of the habitat. When a 4-ha grid was used on the home ranges of the owls, small patches of high-suitability habitat that the owls were selecting for foraging became evident. The usefulness of the model for predicting within-home-range habitat use was dramatically increased with the decrease in grid-cell size.

When frequency of owl use was compared to HSI values, the lack of model fit for both 16- and 4-ha resolutions indicated that factors other than habitat quality were also important. For example, an owl might choose how often to forage in a particular area based on past hunting success, as well as on vegetation characteristics.

Table 15.4. Correlation coefficients (*r*) of frequency of owl cell use to HSI values of cells of the best-fitting model combination for 4- and 16-ha cell resolutions

Individual owl	Best-fitting model combination	
	16-ha: (canopy + structure)/2	4-ha: $[(\text{canopy})(\text{dbh})]^{1/2}$
Ramsey male	0.22	−0.21
Ramsey female	0.21	−0.09
Bacchi male	0.30	0.36
Bacchi female	0.35	0.10
Gaddis male	−0.06	0.12
Gaddis female	0.20	0.00

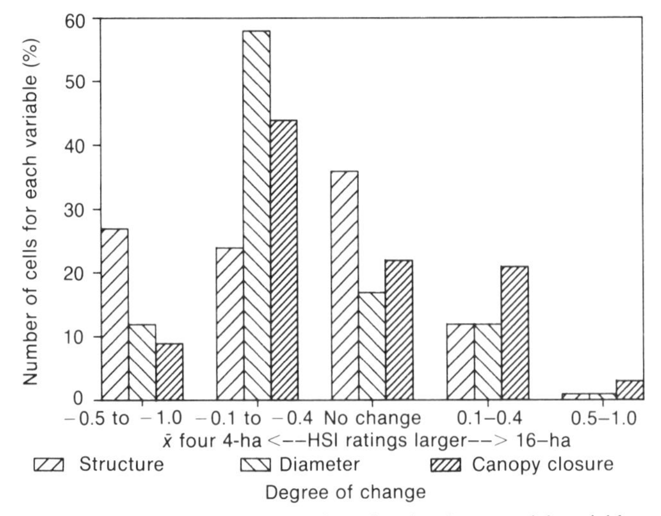

Figure 15.2. Change of HSI ratings for the three model variables resulting from a comparison of the mean of the four 4-ha cells with the corresponding 16-ha cell.

The use of the Strauss Preference Index to determine model fit, followed by the Friedman test to find the best-fitting model, is a novel approach to habitat model testing. The approach appears to be especially useful when sample sizes of the random or use points are small. The Friedman test is a powerful nonparametric, repeated measures design, which is suited to small sample sizes (small numbers of individuals), as is often the case in radio-telemetry studies. The lack of statistical differences in post hoc, pairwise comparisons in our study is the result of (1) the small number of individuals in the study; (2) the large number of model combinations examined; and (3) the lack of strong, consistent differences between the model combinations' predictive ability. The model combination with the highest mean rank should still be preferred, even though it is not statistically different from the others.

The comparison of the 16-ha grid cells with the corresponding four 4-ha cells was an important tool in determining the reason for the increased predictive power of the smaller resolution. Two processes were simultaneously in operation: (1) small pockets of habitat in the 2–7 ha range, predicted by the model to have high suitability, were being masked by the 16-ha cell ratings; and (2) the owls were choosing these small, high-quality sites for foraging. The use of the smaller cell resolution made these high-quality sites more evident. We recommend that researchers and managers take a critical look at habitat heterogeneity, intended model use, and grid-cell resolution before models are tested and implemented.

Acknowledgments

We are most grateful for the careful reviews and helpful comments on earlier drafts of this chapter by R. H. Barrett, M. F. Dedon, W. E. Grenfell, B. G. Marcot, C. J. Ralph, and M. G. Raphael.

16

Evaluating the Structure of Habitat for Wildlife

HENRY L. SHORT and SAMUEL C. WILLIAMSON

Abstract.—Methods for evaluating habitat structure that are based on the interpretation and ground truthing of aerial photographs to obtain land-use and surface-cover information are described. Information about habitat structure is considered pertinent to wildlife management because of the assumption that wildlife species are associated with the structure of habitat. Methods described herein can be used to (1) measure habitat structure for inventory and assessment work; (2) describe the relative structural complexity of different land units; (3) describe the direction and rate of change in habitat structure over time; and (4) describe the potential distribution of those wildlife species having particular dependencies on the specialized structure of habitats. It seems likely that information about habitat layers coupled with information about cover types will enhance wildlife-habitat management in the future.

Future wildlife management will almost certainly make increased use of remotely sensed data about land cover; this chapter describes a novel way to use aerial photography. Aerial photographs are frequently used to describe the distribution of vegetation cover types and physical features, including man-made developments, within the environment (Fig. 16.1). These data may be difficult to interpret because the use of habitats by wildlife may not be correlated with cover type. This chapter demonstrates that aerial photographs can describe the structure of vegetative cover types (Fig. 16.1), and that this information can be used to (1) compare the structure of habitats between study areas and time periods; and (2) develop maps of habitat structure that can be used to predict where particular wildlife species might be distributed within a study area. Such information is useful to managers because it provides a way to inventory the structure of wildlife habitat at a point in time, describe the rate and direction of change in habitat structure across time, and indicate the direction of change in habitat structure needed to provide habitat for particular wildlife species.

The following four assumptions are basic to the mapping of habitat structure: (1) the structure of cover types can be considered as habitat layers (Fig. 16.2), and habitats with more layers can be considered structurally more complex; (2) layers of habitat have definable limits (Table 16.1); (3) wildlife species can be associated with the structure of terrestrial habitats (see, e.g., Martin 1960; MacArthur et al. 1966; Karr 1971; Rabenold 1978; Geibert 1979; Maser et al. 1981; Heatwole 1982) by positioning their foraging and breeding activities within one or more layers of habitat. Wildlife groups can be formed that consist of vertebrate species whose foraging and breeding activities occur in the same habitat layer or layers (Short and Burnham 1982; Short 1983). Structurally complex habitats accommodate more wildlife groups and wildlife species; and (4) habitats can be compared and evaluated on the basis of the number of habitat layers present and the area and distribution of each of those habitat layers within a bounded area.

HENRY L. SHORT: Habitat Evaluation Procedures Group, Western Energy and Land Use Team, U.S. Fish and Wildlife Service, 2627 Redwing Road, Fort Collins, Colorado 80526

SAMUEL C. WILLIAMSON: Western Energy and Land Use Team, U.S. Fish and Wildlife Service, 2627 Redwing Road, Fort Collins, Colorado 80526

Methods

Examples described in this chapter were developed from aerial photographs of the Big Stone National Wildlife Refuge (NWR) and its immediate vicinity in southwestern Minnesota. The major cover types on the study area are upland and bottomland hardwood, upland shrubland, grassland, cropland, riverine and lacustrine areas, and emergent and submergent wetlands.

Color infrared aerial photography, at a scale of 1:12,000, was acquired in July 1983 as part of a refuge planning effort. Cover maps were developed and intensive on-site ground verification occurred during the photo-interpretation process. Minimum map units of 0.2 ha were used on the 8000-ha study area. Transparent map overlays were constructed via zoom-transfer. Individual map overlays were digitized into a computer-compatible vector-based format. Map-based information was converted into 0.1-ha cell format via the Map Overlay and Statistical System (Lee et al. 1985). Individual cover types were reclassified and aggregated using an octal number to represent the presence (1) or absence (0) of each habitat layer (Table 16.1). The eight-digit binary number allowed for the representation of up to 256 habitat layer combinations, although many fewer combinations are actually observed in nature.

A series of workshops were conducted with U.S. Fish and Wildlife Service personnel, and wildlife species breeding on the Big Stone NWR were associated with specific habitat strata. This data base was used to classify groups of species having similar dependencies on habitat structure.

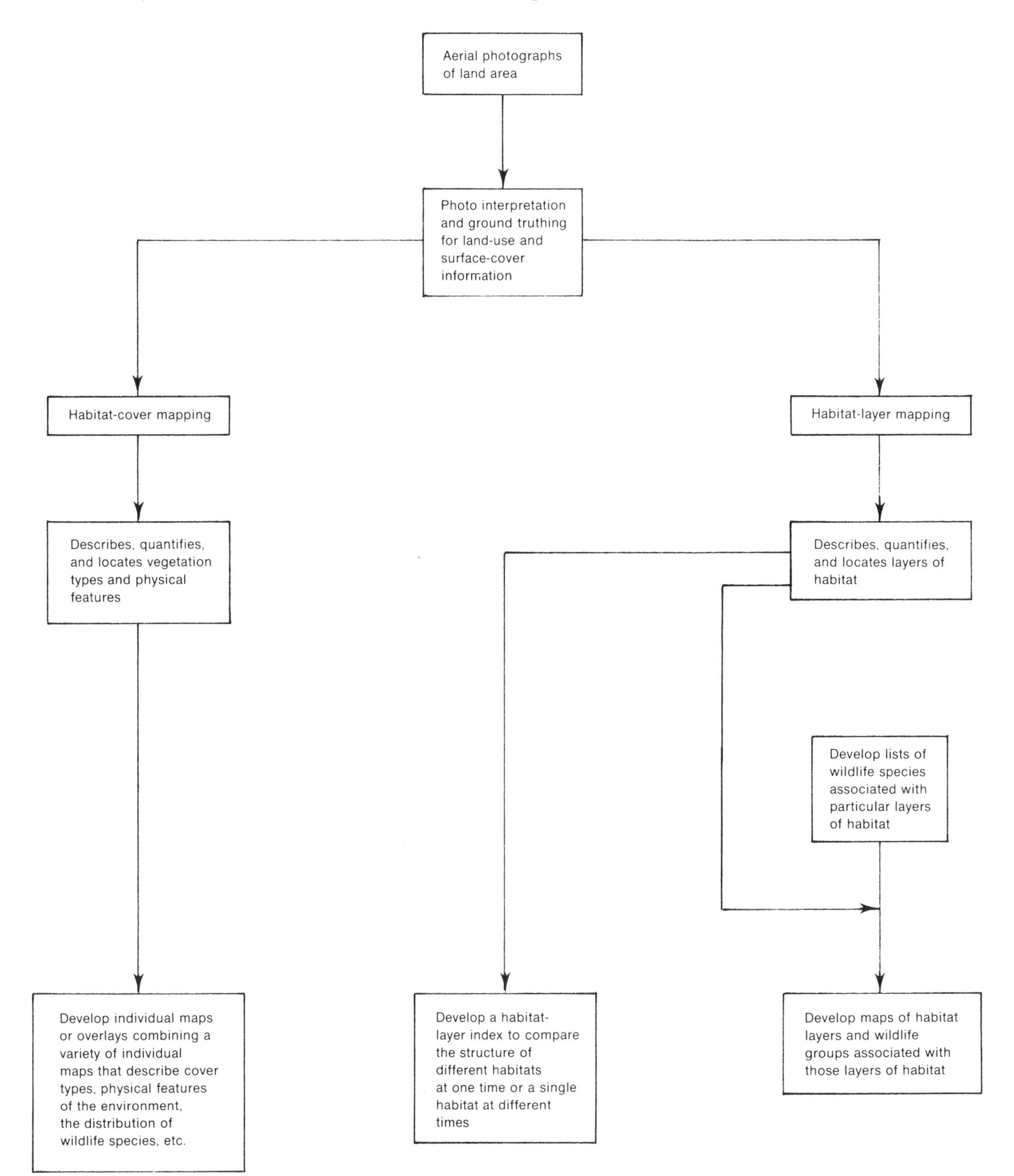

Figure 16.1. Use of aerial photographs for mapping habitat cover and habitat layers.

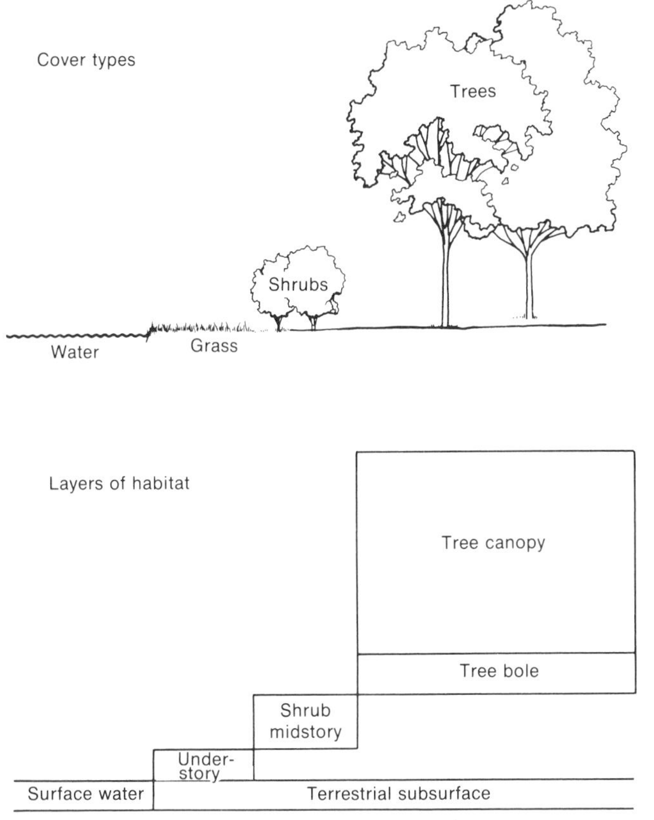

Figure 16.2. Abstract representation of vegetation cover types as layers of habitat.

Table 16.1. Criteria for determining the presence of different layers of habitat

Layer	Criteria
1. Bottom of water column	Water–terrestrial surface interface under more than 25 cm (10 inches) of water.
2. Water column	Water between the surface and the bottom of the water column.
3. Surface water	Land surface–water interface and shallow water up to 25 cm (10 inches) deep.
4. Terrestrial subsurface	Extends from 10 cm (4 inches) below the ground surface downward.
5. Understory	Vegetation extends from 10 cm (4 inches) below the ground surface up to 50 cm (20 inches) above the ground surface and provides at least 20% cover when projected to the surface.
6. Shrub midstory	Vegetation height from 50 cm (20 inches) up to 8 m (25 feet), which provides at least 20% cover when projected to the surface.
7. Tree bole	Tree trunks have a dbh ≥ 20 cm (8 inches) and occur at a density ≥ 12/ha (5/acre).
8. Tree canopy (overstory)	Vegetation, other than tree trunks, that is 8 m (25 feet) or more above the terrestrial or aquatic surface and provides at least 20% cover when projected to the surface.

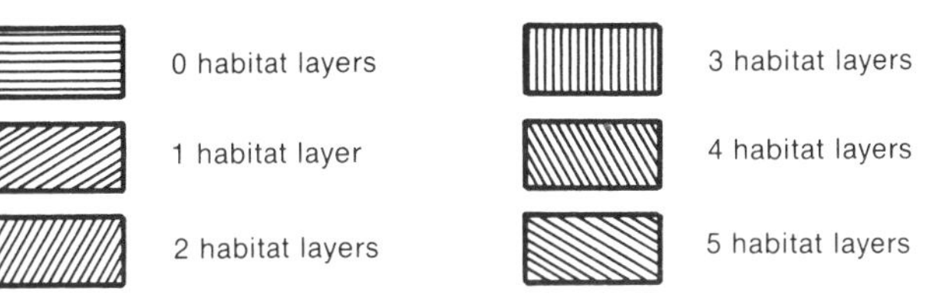

Figure 16.3. Computer-generated map of the portion of the Big Stone National Wildlife Refuge described in Table 16.2.

Results

A portion of the Big Stone NWR is mapped in Figure 16.3. This map portrays cells that contain 0, 1, 2, 3, 4, or 5 layers of habitat. Numbers of habitat layers on a land unit can be considered a map feature even though the individual layers of habitat for a cell are not identified. This representation distinguishes simple and structurally complex habitats, regardless of the vegetation on an area. Thirteen different combinations of habitat layers occurred in individual cells (Table 16.2) in the portion of the Big Stone NWR represented in Figure 16.3. Combination 1 in Table 16.2, for example, represents land units covered by gravel pits, quarries, roads, and buildings, and was interpreted as providing no layer of habitat for wildlife. Combination 11 in Table 16.2 represents habitats with hydric soils and dense stands of bottomland hardwoods, which we have interpreted as providing understory, tree bole, and tree canopy layers of habitat.

Table 16.2. Characteristics of areas and the presence (1) or absence (0) of habitat layers (from Table 16.1) in the portion of the Big Stone NWR shown in Figure 16.3

Combination no.	Example	Area (ha)	Percent total area	Habitat layers						
				2	3	4	5	6	7	8
1	Quarries, roads	37.5	11.7	0	0	0	0	0	0	0
2	Lakes, ponds	10.5	3.3	1	1	0	0	0	0	0
3	Shallow wetlands without emergents	1.1	0.3	0	1	0	0	0	0	0
4	Wet meadow	31.4	9.8	0	0	0	1	0	0	0
5	Shallow wetlands with emergents	32.1	10.0	0	1	0	1	0	0	0
6	Grasslands, croplands	113.3	35.3	0	0	1	1	0	0	0
7	Lowland shrubs (hydric soils)	7.5	2.3	0	0	0	1	1	0	0
8	Upland shrubs	36.1	11.2	0	0	1	1	1	0	0
9	Flood-killed trees (hydric soils)	2.1	0.7	0	0	0	1	0	1	0
10	Flood-killed trees in water	4.0	1.3	0	1	0	1	0	1	0
11	Mature trees (hydric soils), no midstory	3.2	1.0	0	0	0	1	0	1	1
12	Mature trees (upland), no midstory	30.7	9.5	0	0	1	1	0	1	1
13	Mature trees (upland), with midstory	11.5	3.6	0	0	1	1	1	1	1
		321.0	100.0							
Total area (ha) of individual habitat layer				10.5	47.7	191.6	271.9	55.1	51.5	45.4

Several analyses can be performed with the data listed in Table 16.2, including (1) indexing habitat structure; (2) graphically or mathematically interpreting the structure of habitat to describe differences between study areas or differences across time for one study area; and (3) producing maps that display the occurrence of particular habitat structures and the wildlife species dependent on those habitat structures.

The Habitat Layer Index

The Habitat Layer Index (HLI) was developed for use with Habitat Evaluation Procedures (HEP) (Fish and Wildlife Service 1980b). The HLI measures the relative structural diversity of habitats and differs from most HEP models which evaluate the potential usefulness of habitats for individual wildlife species.

The HLI measures the number and areas of habitat layers present on a terrestrial study area and compares that measure to a maximum value representing the number of habitat layers and the maximum area of those habitat layers that could occur on certain riparian study areas. The index is a ratio estimator that ranges from 0 (no structural habitat layers are available to wildlife, e.g., a paved parking lot) to 1 (all layers of habitat are present and extend throughout the study area, e.g., some riparian areas). Advantages of using the HLI are that a numerical value representing the structure of habitat on a study area can be quickly developed and the numerical value can become a data point in comparisons of different habitats or of the same habitat at different times. A HLI does not describe a specific habitat condition, nor does it predict the usefulness of a habitat for an individual wildlife species. However, species requiring the presence of several habitat layers are more likely to occur in areas with a high HLI than in areas with a low HLI.

The formula for calculating the HLI for an area like that pictured in Figure 16.3 is

$$\text{HLI} = \frac{l \sum_{i=1}^{l} A_i}{(6)(5) \sum_{j=1}^{n} B_j},$$

where l = the number of layers of habitat present within some bounded area, A_i = the area of layer of habitat i within the bounded area, B_j = the surface area of cover type j within the bounded area, and n = the number of different cover types present within the bounded area. The numerator is the product of the number of different layers of habitat that occur on a study area and the total area of those layers of habitat. The denominator is the product of the maximum number of layers of habitat (six) that could occur on the study area and the maximum area of those habitat layers (five times the area). The analysis of information from interpreted aerial photography and ground truthing indicated that six layers of habitat (surface water, terrestrial subsurface, understory, shrub midstory, tree bole, and tree canopy) are present on the portion of the refuge pictured in Figure 16.3 (see Table 16.2). Using the data summarized in Table 16.2, the HLI is calculated as follows:

$$\text{HLI} = \frac{6(47.7 + 191.6 + \ldots + 45.4)}{(6)(5)(37.5 + 10.5 + \ldots + 11.5)} = \frac{6(663.2)}{30(320.8)} = 0.41.$$

Therefore, the structural diversity of this area is 41% of the maximum structural diversity that could occur on the study area.

GRAPHIC REPRESENTATION

The HLI calculations consider the numbers of habitat layers and the total area of those different habitat layers. The percent total area of the 13 different combinations of layers of habitat represented in Table 16.2 is also useful information. A bar graph, for example, can be used to visually compare the limited presence of unique habitats, for example, combinations 9–11 compared to the extensive amount of habitat combinations 1, 6, and 8 (Table 16.2).

HABITAT LAYERS–WILDLIFE GROUP MAPS

Computer-generated maps of the cells containing the 13 different combinations of habitat layers were developed so that the locations of the different habitat combinations could be visualized. Two examples of these maps are provided (Figs. 16.4 and 16.5), along with lists of the wildlife species that could occur within the habitat structure represented on each map (Table 16.3).

Wildlife groups included for the habitat layers–wildlife group maps are nonfish vertebrate species that have similar dependencies on the physical structure of habitats (Table 16.3). These groups, which were developed from a list of species that breed on the Big Stone NWR, may contain species with dissimilar food habits, nutritional physiologies, and reproductive requirements. They are species, however, that may respond similarly to changes in the structure of habitats.

Groups were formed following the general procedure developed by Short and Burnham (1982), as modified by Short (1983), which is based on positioning the foraging and reproduction activity of species within the structure of habitats. A species-habitat matrix is developed with layers of habitat used for reproduction (columns) and layers of habitat used for foraging (rows). A wildlife species is positioned within the species-habitat matrix by identifying one or more elements of the matrix that represent where that species breeds and feeds within the structure of a habitat. The volume of space represented by these elements of the matrix broadly defines the habitat dependencies of that species. The appropriate elements for 153 wildlife species (35 mammals, 107 birds, and 11 reptiles and amphibians) that breed on the Big Stone NWR were determined and sorted with a computer routine to determine groups of wildlife species whose niche space occurs within the same elements. Only the groups of primary consumers are listed and described in this chapter (Table 16.3).

The following examples illustrate the association of wildlife species with particular layers of habitat. Figure 16.4 is a map of habitat with the structure of combination 8 (Table 16.2). These cells contain upland shrubs, dry meadows with scattered shrubs, and a variety of young upland forest types less than 8 m in height. Terrestrial subsurface, understory, and shrub midstory layers of habitat are present. A variety of primary consumer species occur in these habitats. These species are identified in the list in Table 16.3 of wildlife groups dependent on this habitat structure. Species in groups 11–14, 17–19, 23–26, 31–35, and 40 breed in habitats like those represented in Figure 16.4. Several species—such as the indigo bunting (*Passerina cyanea*), northern cardinal (*Cardinalis cardinalis*), brown thrasher (*Toxostoma rufum*), gray catbird (*Dumetella carolinensis*), and American goldfinch (*Carduelis tristis*)—breed only in the midstory layer, so they may be limited by the presence of habitats with a shrub midstory layer.

Combination 10 (Table 16.2) is uncommon in the study area (Fig. 16.5). It exists where the surface water, understory, and tree bole layers of habitat occur together, which happens when large trees have been killed by high water. These snags are located on flooded wetlands with an understory of mixed emergents. Wildlife species in groups 1–8, 10, 15, 21, 27–28, and 36–37 (Table 16.3) are expected to breed in this habitat, which is in short supply.

Discussion

This chapter describes a conceptual framework for analyses of wildlife habitat that can be accomplished following the interpretation of land use and surface cover from aerial photography. These analyses provide the manager and planner with a simplification and abstraction of habitat that may reduce the difficulty in associating wildlife species with cover types. Most cover-type classification systems, even extremely complex ones, can be collapsed into this structural model of habitat layers.

The Habitat Layers Index (HLI) describes the relative structural complexity of habitat on a study area by comparing the number of habitat layers present and the total area of those habitat layers with the most complex habitat structure that could occur on that area. The HLI may be useful to managers who are charged with measuring the cumulative impacts of land-use change when those changes affect structurally complex habitats. The HLI may also help describe the direction and rate of change in habitat structure through time. Moreover, it may be useful in comparing the habitat structure of different land units for purposes such as land acquisition, mitigation, and planning for the development of sites.

The representation of habitat structure in terms of layers of habitat can also provide a statistical statement of the impacts of change on wildlife habitat. Lists of acreages can be compared for baseline conditions and expected conditions after a proposed land-use change in order to predict whether or not the structure of the habitat will change significantly with development. The effect of changes in structure on the wildlife community can be predicted because wildlife species can be associated with layers of habitat using a species-habitat matrix.

Habitat layers–wildlife group maps locate habitats with specialized structures and list the wildlife species that seem dependent on those structures. Information about the distribution of habitat structure can help describe animal distribution patterns throughout a study area and the impacts on the wildlife community when the quantity and distribution of cells providing particular habitat layers are changed.

Table 16.3. Layers of habitat (from Table 16.1) used for feeding and breeding by groups of primary consumers that breed on the Big Stone NWR

	Feeding layers								Breeding layers							
Group no. and members	1	2	3	4	5	6	7	8	1	2	3	4	5	6	7	8
1																
Canvasback (*Aythya valisineria*)	1	2	3								3					
Redhead (*Aythya americana*)	1	2	3								3					
Ring-necked duck (*Aythya collaris*)	1	2	3								3					
Ruddy duck (*Oxyura jamaicensis*)	1	2	3								3					
2																
Snapping turtle (*Chelydra serpentina*)	1	2	3									4	5			
3																
Eastern painted turtle (*Chrysemys picta*)	1	2	3										5			
4																
American coot (*Fulica americana*)		2	3								3					
5																
Beaver (*Castor canadensis*)			3	4	5	6					3	4				
6																
Muskrat (*Ondatra zibethicus*)			3		5						3	4				
7																
American wigeon (*Anas americana*)			3		5						3		5			
Blue-winged teal (*Anas discors*)			3		5						3		5			
Northern shoveler (*Anas clypeata*)			3		5						3		5			
Canadian goose (*Branta canadensis*)			3		5						3		5			
Northern pintail (*Anas acuta*)			3		5						3		5			
Red-winged blackbird (*Agelaius phoeniceus*)			3		5						3		5			
Green-winged teal (*Anas crecca*)			3		5						3		5			
Gadwall (*Anas strepera*)			3		5						3		5			
Mallard (*Anas platyrhynchos*)			3		5						3		5			
8																
Yellow-headed blackbird (*Xanthocephalus xanthocephalus*)			3		5						3					
9																
Killdeer (*Charadrius vociferus*)			3		5								5			
10																
Wood duck (*Aix sponsa*)			3		5										7	
11																
Deer mouse (*Peromyscus maniculatus*)				4	5	6						4	5			
12																
Coyote (*Canis latrans*)				4	5	6						4				
13																
Striped skunk (*Mephitis mephitis*)				4	5							4	5			
14																
Plains pocket gopher (*Geomys bursarius*)				4	5							4				
15																
Raccoon (*Procyon lotor*)					5	6		8				4	5		7	
16																
Red squirrel (*Tamiasciurus hudsonicus*)					5	6		8				4		6	7	8
17																
Song sparrow (*Melospiza melodia*)					5	6		8					5	6		
18																
American robin (*Turdus migratorius*)					5	6		8						6		8
19																
Indigo bunting (*Passerina cyanea*)					5	6		8						6		
Northern cardinal (*Cardinalis cardinalis*)					5	6		8						6		
20																
Eastern fox squirrel (*Sciurus niger*)					5	6		8							7	8
Eastern gray squirrel (*Sciurus carolinensis*)					5	6		8							7	8
21																
Black-capped chickadee (*Parus atricapillus*)					5	6		8							7	
White-breasted nuthatch (*Sitta carolinensis*)					5	6		8							7	
22																
Blue jay (*Cyanocitta cristata*)					5	6		8								8
23																
Gray fox (*Urocyon cinereoargenteus*)					5	6						4	5			
24																
Clay-colored sparrow (*Spizella pallida*)					5	6							5	6		
25																
White-tailed deer (*Odocoileus virginianus*)					5	6							5			

Table 16.3. *(Continued)*

Group no. and members	Feeding layers								Breeding layers							
	1	2	3	4	5	6	7	8	1	2	3	4	5	6	7	8
26																
Gray catbird (*Dumetella carolinensis*)					5	6								6		
Brown thrasher (*Toxostoma rufum*)					5	6								6		
27																
European starling (*Sturnus vulgaris*)					5	6									7	
Red-bellied woodpecker (*Melanerpes carolinus*)					5	6									7	
Eastern bluebird (*Sialia sialis*)					5	6									7	
28																
Yellow-bellied sapsucker (*Sphyrapicus varius*)					5		7								7	
29																
Eastern chipmunk (*Tamias striatus*)					5							4	5		7	
30																
Red fox (*Vulpes vulpes*)					5							4				
Thirteen-lined ground squirrel (*Spermophilus tridecemlineatus*)					5							4				
Meadow jumping mouse (*Zapus hudsonius*)					5							4				
Richardson's ground squirrel (*Spermophilus richardsonii*)					5							4				
Franklin's ground squirrel (*Spermophilus franklinii*)					5							4				
Woodchuck (*Marmota monax*)					5							4				
31																
Mourning dove (*Zenaida macroura*)					5								5	6		8
Brown-headed cowbird (*Molothrus ater*)					5								5	6		8
32																
Field sparrow (*Spizella pusilla*)					5								5	6		
Dickcissel (*Spiza americana*)					5								5	6		
33																
Gray partridge (*Perdix perdix*)					5								5			
House mouse (*Mus musculus*)					5								5			
Ring-necked pheasant (*Phasianus colchicus*)					5								5			
Eastern cottontail (*Sylvilagus floridanus*)					5								5			
Grasshopper sparrow (*Ammodramus savannarum*)					5								5			
Western meadowlark (*Sturnella neglecta*)					5								5			
White-tailed jackrabbit (*Lepus townsendii*)					5								5			
Northern bobwhite (*Colinus virginianus*)					5								5			
Vesper sparrow (*Pooecetes gramineus*)					5								5			
Prairie vole (*Microtus ochrogaster*)					5								5			
Norway rat (*Rattus norvegicus*)					5								5			
Horned lark (*Eremophila alpestris*)					5								5			
Bobolink (*Dolichonyx oryzivorus*)					5								5			
Rock dove (*Columba livia*)					5								5			
Savannah sparrow (*Passerculus sandwichensis*)					5								5			
34																
Chipping sparrow (*Spizella passerina*)					5									6		8
Common grackle (*Quiscalus quiscula*)					5									6		8
35																
American goldfinch (*Carduelis tristis*)					5									6		
36																
House sparrow (*Passer domesticus*)					5										7	8
37																
Hairy woodpecker (*Picoides villosus*)					5										7	
Downy woodpecker (*Picoides pubescens*)					5										7	
Red-headed woodpecker (*Melanerpes erythrocephalus*)					5										7	
Northern flicker (*Colaptes auratus*)					5										7	
38																
American crow (*Corvus brachyrhynchos*)					5											8
39																
Rose-breasted grosbeak (*Pheucticus ludovicianus*)						6		8								8
40																
Ruby-throated hummingbird (*Archilochus colubris*)						6								6		8
41																
Northern oriole (*Icterus galbula*)								8								8
Orchard oriole (*Icterus spurius*)								8								8

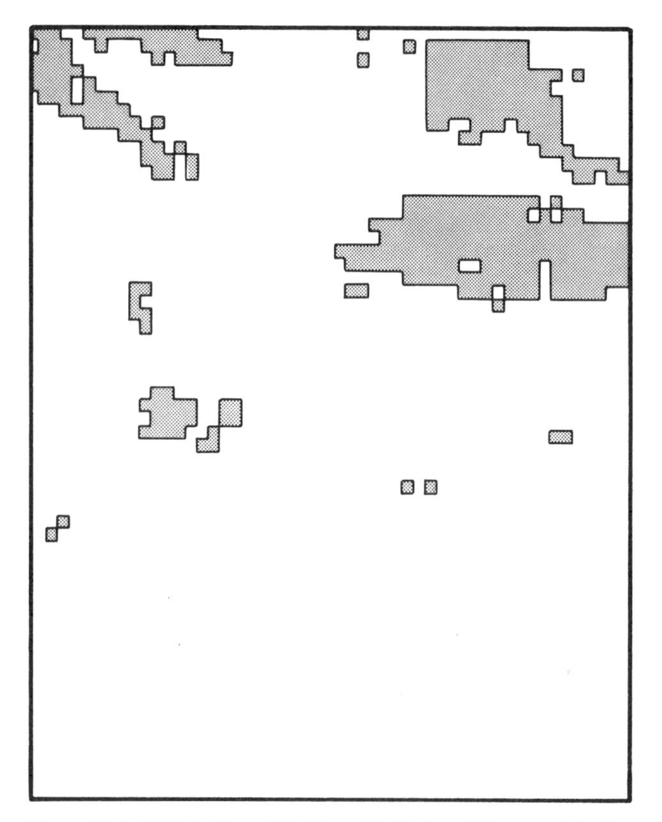

Figure 16.4. Distribution of 0.1-ha cells that contain a combination of a terrestrial subsurface layer, an understory layer, and a shrub midstory layer (combination 8 in Table 16.2).

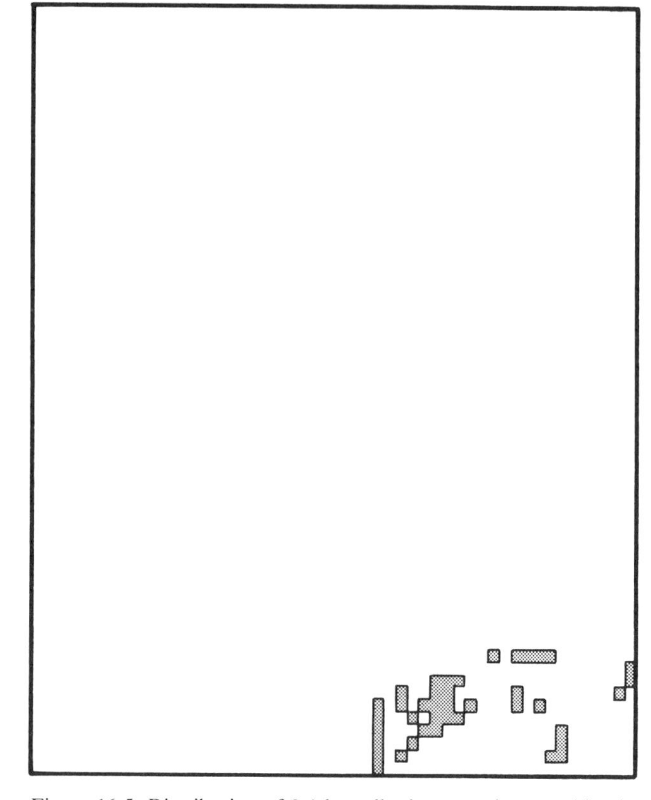

Figure 16.5. Distribution of 0.1-ha cells that contain a combination of a surface water layer, an understory layer, and a tree-bole layer (combination 10 in Table 16.2).

The development and the use of these methods suggest a logical procedure for the management of wildlife resources on public and private lands. One should (1) determine the wildlife species of management interest; (2) determine, from simple word models, the habitat requirements for the selected species; (3) summarize the habitat-structure and cover-type requirements for the selected species; (4) determine the quantity and distribution of the required combinations of habitat layer and cover type within a bounded area; and (5) formulate any necessary management plans to modify habitat structure and cover types to favor the species of interest.

Acknowledgments

We appreciate the help of Art Allen, Bruce Bury, Brad Harper, John Oldemeyer, Virgil Scott, Pat Sousa, and Kathy Stose in the development of the data base and in formulating lists of wildlife groups. Carolyn Gulzow and Dora Ibarra provided word processing, Jennifer Shoemaker prepared the graphics, and Cathy Short provided an editorial review. We also appreciate the assistance and information provided by Jim Heinecke and Rich Papasso of the Big Stone NWR and photography and user guidance provided by Jim Lutey and Mike Marxen. Carl Armour, Adrian Farmer, and Charles Segelquist provided technical reviews of earlier drafts of the manuscript.

17

Species Selection for Habitat-Evaluation Procedures

MICHAEL E. FRY, ROLAND J. RISSER, HARRISON A. STUBBS, and JEFFREY P. LEIGHTON

Abstract.—The U.S. Fish and Wildlife Service's Habitat Evaluation Procedures (HEP) contain only cursory guidelines for the selection of wildlife species to be considered in an evaluation. However, the number of species and the particular ones selected can strongly influence the outcome of a HEP analysis. The California State–Federal Interagency Wildlife Working Group has developed a data base of wildlife-habitat relationships for California. We present a logic structure for selecting evaluation species based on this information. We show that stratified random selection of evaluation species by taxonomic class is generally preferable to selecting species according to guild membership or by simple random selection. Finally, the number of species chosen to represent a habitat type is inversely related to the maximum percent of error in the sample mean Habitat Suitability Index (HSI). Hence, one should maximize the number of evaluation species within the constraints of a project's budget.

California leads all other states in numbers of Habitat Evaluation Procedures (HEP) studies completed or under way since 1980 (Fish and Wildlife Service Western Energy and Land Use Team, Fort Collins, Colorado, pers. comm.). Still, HEP is a new methodology that is evolving rapidly. In this chapter we assume that the reader is familiar with HEP (Fish and Wildlife Service 1980a, 1980b, 1981a).

Species selection

It is at present impractical to consider impacts to all species found in a proposed project area. When one's objective is to assess impacts at a community level (i.e., to consider wildlife species that are representative of all habitats being studied) the Fish and Wildlife Service (1980b) and Short and Burnham (1982) recommend a guild approach to species selection. In theory, guilds offer a way of representing large groups of species by using selected indicator species (Szaro and Balda 1982).

To date, no specific guidelines have been published on how to choose those wildlife species that will accurately represent a study area on the community level. This raises certain questions: (1) Is a truer estimate of a habitat's value to wildlife obtained when all guilds known to occur there are represented by the evaluation species? (2) What is the effect of over- or underrepresenting a particular taxonomic class in the species array? Existing HEP guidelines do not ensure that all taxonomic classes are represented. The implied assumption is that one taxonomic class (e.g., birds) can adequately represent impacts occurring within all other classes of wildlife.

We present an approach to species selection designed to achieve broad ecological representation when less than the entire wildlife community can be considered because of cost constraints. Our approach relies on the use of a wildlife-habitat-relationships data base, such as that of Verner and Boss (1980). This data base provides information which can be used to structure selection criteria for a given locale (i.e., habitat type, successional stage, habitat value [qualitative], habitat uses, and season of use).

The principal assumptions underlying our species-selection process are as follows: (1) it is impractical to consider each and every species when assessing impacts of a proposed project; (2) best representation of the wildlife in a given habitat is achieved by selecting species that are strongly associated with that habitat (i.e., the habitat type satisfies all life requisites); (3) information contained in the data base is complete and factual; (4) selecting species from each taxonomic class (in the same proportions as they are known to occur) assures better overall representation; and (5) within a taxonomic class, guilds offer an acceptable means of assessing impacts to many species by studying only a few.

We evaluated assumptions (4) and (5) using a computer simulation model. The model was also used to calculate the sampling precision expected for a given number of evaluation species.

Methods

Species selection

The following steps demonstrate the decision process for selecting evaluation species from a wildlife-habitat-relationships data base.

Step 1.—Identify all habitat types within the study area. Include information on successional stage and tree cover canopy class where appropriate.

Step 2.—For all habitat types identified in step 1, use the data base to list all species that make optimal or suitable use

Michael E. Fry, Roland J. Risser, and Jeffrey P. Leighton: Department of Engineering Research, Pacific Gas and Electric Company, 3400 Crow Canyon Road, San Ramon, California 94583

Harrison A. Stubbs: 6425 Girvin Drive, Oakland, California 94611

of the habitat type for all life requisites (i.e., breeding, feeding and resting) at least seasonally.

Step 3.—Screen the candidate lists for species whose geographic range does not include the project area and for species that require special habitat features not found within the project area.

Step 4.—Arrange the species on each candidate list by taxonomic class (i.e., amphibians, reptiles, birds, and mammals). Determine for each list the percent composition of each class.

Step 5.—If guild diversification is desired, give each species in each list a guild descriptor. An appropriate level of detail for guild designation could include foraging location and strategy (e.g., aerial insectivore or ground/shrub herbivore).

Step 6.—The complete lists would be the ideal array of species to use in a HEP analysis. Typically, however, these lists are long. Decide upon the number of species to use in evaluating each habitat, given cost and time constraints. Then determine how many species are required from each taxonomic class for optimal allocation of species. Make the proportions among taxonomic classes in the sample equal to those in the candidate list. Small numbers of species in some taxonomic classes make true proportional allocation difficult. Those classes with small numbers of species can be overrepresented in the initial sample list (e.g., 1 of 10, when the class actually represents <5% of the total) and then weighted back to their true proportion after Habitat Suitability Indexes (HSIs) are assigned.

Step 7.—Select individual species to fill the taxonomic class categories, with or without guild diversification. For guild diversification, either choose species from the stratified candidate list that represent as broad an array of guilds as possible, or randomly select species, rejecting any whose guild has been previously selected if there are still unrepresented guilds within the class. If guild diversification is not desired, randomly select species within taxonomic classes without considering guilds.

Sampling and statistical methods

Analyses were conducted on data sets from two habitat types found on the west slope of the Sierra Nevada: lodgepole pine (*Pinus contorta*) forest (types 4a and 4b) and wet mountain meadow (Verner and Boss 1980). HSI values for some species were determined at Pacific Gas and Electric Company project sites in the central and southern Sierra Nevada by multiagency HEP teams using verbal HSI models (Fish and Wildlife Service 1980b, 1981a). We determined additional values for the remaining species in the two habitat types, using the same methods and the same study areas.

We compared the two approaches to species selection described in step 7 above. In the first approach, species were sampled from taxonomic strata (classes) without replacement and regardless of guild membership. In the second approach, species were sampled within strata without replacement, but once a species from a particular guild was selected no other species from that guild were included until all guilds in the taxonomic class were represented. This latter constraint was intended to maximize the number of guilds represented in the sample and is hereafter referred to as "sampling with guild diversification."

We attempted to achieve proportional allocation of species by taxonomic class while including at least one species from each class. Best proportional allocation of species consisted of the set of numbers of species per stratum that yielded the minimum chi-square goodness-of-fit value. (A detailed discussion of the mathematical framework for application of the model is available from the authors on request.)

Computer simulation of the sampling procedure

We wrote a computer program for selecting a sample of species with or without guild diversification. This program also calculates the mean HSI and its standard error for a specified sample size. We ran 1000 independent simulations of the process, with and without guild diversification, for species lists of 10 through 20 and 25 species for each habitat type. The output from each simulation run consisted of a single mean HSI for each habitat type.

Results

Candidate lists of 58 and 28 species were produced for the lodgepole pine and meadow habitats, respectively (Table 17.1). The mean HSI values varied in each habitat by taxonomic class. Differences were most pronounced in the lodgepole pine habitat. Here the mean HSI value for birds (0.56) was greater than that for mammals (0.36) and for all species combined (0.45). The mean HSI values in the meadow habitat were somewhat higher for mammals (0.55) than for birds (0.46). The mean HSI values for reptiles and amphibians were similar within each habitat but differed between habitats. The HSI values were most variable among taxonomic classes in the lodgepole pine habitat (Table 17.1).

Sampling without guild diversification

Examination of the standard errors of mean HSI values for the two habitats (Table 17.2) disclosed a monotonic decrease with increasing number of species in the sample. As the estimator of the overall HSI is a sample mean, its sampling distribution is approximately normal. Therefore, the coefficient of variation can be multiplied by a z-statistic for a given confidence level to determine the approximate range of values expected for the sample HSI as a percentage of the true HSI for each sample size. There was less variation in the sample mean HSIs in the meadow habitat. The coefficients of variation for this habitat were approximately one-half of those for the lodgepole pine habitat (Table 17.2).

Sampling with guild diversification

As HSI values within individual guilds tend to be more homogeneous than values among guilds, sampling with guild diversification should generally result in samples that are internally more heterogeneous than those arising from sampling without guild diversification. However, our results (Table 17.3) indicate that sample mean HSIs based on sampling with guild diversification are biased. In the lodgepole

Table 17.1. Numbers of species, guilds, and HSI values by taxonomic class for two habitat types

Taxonomic class	Species		Guilds		HSI values			
	Number	%	Number	%	Min	Max	Mean	*SD*
LODGEPOLE PINE								
Amphibians	1	1.7	1	3.1	0.20	0.20	0.20	0.00
Reptiles	2	3.5	2	6.3	0.20	0.25	0.23	0.03
Birds	30	51.7	16	50.0	0.20	0.97	0.56	0.26
Mammals	25	43.1	13	40.6	0.10	0.70	0.36	0.16
All species	58	100.0	32	100.0				
MEADOW								
Amphibians	5	17.9	3	17.6	0.50	0.80	0.67	0.13
Reptiles	2	7.1	1	5.9	0.53	0.70	0.62	0.08
Birds	10	35.7	6	35.3	0.20	0.70	0.46	0.15
Mammals	11	39.3	7	41.2	0.34	0.80	0.55	0.17
All species	28	100.0	17	100.0				

pine habitat all of the sample means for this method were 0.8–2.6% larger than the true mean of 0.45. In the meadow habitat the sample means were 1.7–3.4% below the true mean of 0.54.

The skewness and kurtosis statistics for the lodgepole pine data indicate that the shapes of the sampling distributions are approximately normal. However, these distributions for the meadow habitat appear slightly skewed and more peaked than expected for a normal distribution (Table 17.3).

The variability (*SE*) of the mean HSI with guild diversification was approximatley 10% less than without guild diversification for the lodgepole pine habitat and 6–40% less for the meadow habitat (Table 17.3). However, the gain in precision of the estimator with guild diversification was less than the loss due to the bias described above.

Table 17.2. Statistical properties of the sampling distribution of the mean HSI for sampling without guild diversification

Number of species	Lodgepole pine				Meadow			
			Maximum percent error				Maximum percent error	
	SE	*CV*	α = 0.05	α = 0.10	*SE*	*CV*	α = 0.05	α = 0.10
5	0.116	25.6	50.1	42.1	0.075	13.8	27.0	22.7
10	0.071	15.6	30.6	25.7	0.043	7.8	15.4	12.9
11	0.064	14.1	27.7	23.2	0.039	7.2	14.1	11.9
12	0.062	13.7	26.8	22.5	0.036	6.6	12.9	10.8
13	0.057	12.6	24.6	20.7	0.033	6.2	12.1	10.1
14	0.053	11.7	22.9	19.2	0.031	5.7	11.1	9.3
15	0.051	11.3	22.2	18.6	0.029	5.3	10.5	8.8
16	0.048	10.6	20.8	17.5	0.027	5.0	9.7	8.2
17	0.047	10.3	20.2	17.0	0.025	4.5	8.9	7.5
18	0.044	9.7	19.1	16.0	0.023	4.2	8.3	7.0
19	0.043	9.5	18.6	15.6	0.021	3.9	7.6	6.3
20	0.041	9.0	17.6	14.7	0.020	3.6	7.1	5.9
25	0.033	7.3	14.3	12.0	0.011	2.0	3.9	3.3

Table 17.3. Statistical results of computer simulation of species selection based on 1000 independent trials with and without the guild diversification criterion

Number of species	Without guild diversification				With guild diversification			
	$\bar{X}$	*SE*	Skewness	Kurtosis	$\bar{X}$	*SE*	Skewness	Kurtosis
LODGEPOLE PINE (μ = 0.4534)								
10	0.4526	0.070	0.18	−0.04	0.4592	0.066	−0.05	−0.11
11	0.4530	0.066	−0.04	−0.34	0.4567	0.060	−0.01	−0.12
12	0.4527	0.065	0.09	0.02	0.4603	0.056	0.15	−0.19
13	0.4553	0.058	0.01	−0.20	0.4572	0.053	0.07	0.02
14	0.4518	0.053	0.19	0.08	0.4623	0.049	0.04	−0.26
15	0.4539	0.051	0.08	−0.18	0.4625	0.047	0.01	0.02
16	0.4533	0.048	0.08	−0.35	0.4626	0.044	0.08	−0.24
17	0.4514	0.047	−0.09	0.00	0.4644	0.041	−0.15	−0.09
18	0.4540	0.045	0.13	−0.33	0.4621	0.036	−0.01	0.13
19	0.4534	0.042	0.02	0.16	0.4650	0.035	−0.07	−0.14
20	0.4548	0.042	0.09	−0.08	0.4627	0.034	0.09	0.06
MEADOW (μ = 0.5432)								
10	0.5445	0.043	0.07	−0.24	0.5332	0.040	−0.13	−0.38
11	0.5427	0.037	0.08	−0.34	0.5338	0.036	−0.02	−0.25
12	0.5432	0.036	−0.01	−0.12	0.5340	0.032	−0.10	−0.26
13	0.5430	0.032	−0.01	−0.26	0.5311	0.029	−0.10	−0.14
14	0.5438	0.031	0.07	−0.26	0.5317	0.026	−0.18	−0.04
15	0.5438	0.028	−0.03	−0.19	0.5292	0.022	−0.11	−0.47
16	0.5434	0.027	−0.00	−0.18	0.5252	0.020	−0.20	−0.46
17	0.5438	0.025	−0.04	−0.28	0.5249	0.015	−0.16	−0.58
18	0.5422	0.023	−0.00	−0.31	0.5255	0.016	−0.05	−0.44
19	0.5427	0.021	−0.05	−0.01	0.5290	0.016	0.06	−0.17
20	0.5431	0.020	−0.05	−0.24	0.5317	0.016	0.11	−0.37

A comparison of the precision of the sampling procedure without guild diversification with that of simple random sampling without replacement (i.e., ignoring taxonomic groups) showed only small differences between the standard errors of the two methods for our data (Table 17.4). Selection based upon stratification by taxonomic class appears to result in lower values for percent error of the sample mean HSI. However, our results indicate that this procedure does not provide substantial gains in statistical precision.

Discussion

Stratification by Taxonomic Class

Bias on the part of investigators is minimized by requiring proportional allocation of all taxonomic classes of wildlife

Table 17.4. Precision ($SE(\overline{HSI})$) of the estimator of overall HSI, using two sampling methods for two habitat types

Number of species in sample	Lodgepole pine				Meadow			
	Stratified random sampling[a]		Simple random sampling		Stratified random sampling[a]		Simple random sampling	
	$SE(\overline{HSI})$	Percent error[b]	$SE(\overline{HSI})$	Percent error[b]	$SE(\overline{HSI})$	Percent error[b]	$SE(\overline{HSI})$	Percent error[b]
10	0.07	15.6	0.07	15.3	0.04	7.8	0.04	8.1
15	0.05	11.3	0.05	11.8	0.03	5.3	0.03	5.6
20	0.04	9.0	0.04	9.6	0.02	3.6	0.02	3.8
25	0.03	7.3	0.04	8.1	0.01	2.0	0.01	2.1

[a]Without guild diversification.
[b]Percent error = $[(SE(\overline{HSI})/\text{true mean})]100$.

represented within the habitat area. Mean HSI values varied considerably between taxonomic groups in this study (Table 17.1), indicating that the taxonomic composition of the species used in an evaluation can greatly affect the outcome of a HEP study. Therefore, if the management objective is to maintain species diversity within the terrestrial vertebrate community, stratification by taxonomic class is recommended. Furthermore, while the differences are small, stratification by taxonomic class resulted in lower percent error values compared with simple random selection (Table 17.4).

Use of guilds

Variability in sample mean HSI values for species selected with guild diversification is generally lower than for selection without guild diversification (Table 17.3). However, mean HSI values obtained with guild diversification for both habitats considered in this study were consistently biased. Furthermore, the small gains in precision were negated by the loss of accuracy. For this reason we recommend stratified random selection (without replacement) over selection based upon guild diversification. Although the use of guilds is intuitively appealing, it appears that their use is not warranted in this context.

Number of species needed

Sample-size requirements will always depend upon the precision and level of confidence one wishes to achieve, as well as on the inherent variability of the data. Ten species/habitat has been suggested (Fish and Wildlife Service 1980b) as a generally adequate number, although no justification for this was given.

We have shown that the number of species used to assess a habitat is inversely related to the maximum percent of error in the sample mean HSI, such that the greater the number of species used the closer the estimated values will be to the true mean (Table 17.2). Furthermore, our two habitats are quite different with respect to the magnitude of this inverse relationship (i.e., acceptable results could be obtained with fewer species in the meadow habitat than in the lodgepole pine habitat).

Further analysis of other habitats may eventually allow the construction of reliable guidelines for determining the number of species to use in HEP studies. Until then the old adage, "More is better," would seem to apply.

Acknowledgments

We thank Mark F. Dedon, Stephen G. Granholm, and Ellen H. Yeoman for their review of our draft manuscript. We also thank Pacific Gas and Electric Company for its support of this research.

18

The Use of Guilds and Guild-Indicator Species for Assessing Habitat Suitability

WILLIAM M. BLOCK, LEONARD A. BRENNAN, and R. J. GUTIÉRREZ

Abstract.—We evaluated the use of a guild of ground-foraging birds and a guild-indicator species from this guild for assessing habitat suitability. We inferred that species in the guild used different microhabitats. Thus, monitoring habitat suitability for a guild-indicator species may not reflect habitat suitability for other species in the guild. We found that it was more economical and statistically less variable to monitor the population of the guild as a unit than to monitor the population of any single species. Results from a preliminary test of the guild-indicator concept suggested that investigators cannot infer the presence of other species in the guild based solely on the presence of the guild-indicator species.

Recent federal laws mandate that National Forests maintain populations of native vertebrate species and enhance the habitats of selected management indicator species (Salwasser et al. 1983). Management indicator species are species whose populations can be monitored to measure the effects of resource-management practices on their populations and those of other species with similar habitat requirements (Salwasser et al. 1983). Consequently, resource managers require accurate, cost-effective methods to monitor vertebrate populations and to assess the suitability of the habitat for those species. Monitoring populations of single species entails considerable sampling and is neither time- nor cost-effective (Verner 1983, 1984). Severinghaus (1981) and Verner (1984) proposed alternative approaches to monitoring vertebrate populations based on the guild concept (sensu Root 1967).

A guild is a group of species that exploit a resource in a similar fashion (Root 1967). Root used a guild to study coexistence among birds that potentially competed for food. He determined that each species in the guild had a unique foraging niche; consequently, competition was minimized and coexistence was possible. Thus, as Root and others (e.g., Eckhardt 1979; Noon 1981b; Mannan et al. 1984) have demonstrated, species that are similar in a general mode of resource exploitation may be dissimilar in certain aspects of obtaining that resource.

In contrast to the guild concept, the guild-indicator concept assumes that members of the guild use identical rather than similar resources (Severinghaus 1981; Verner 1984). Thus, if the impact of environmental change is determined for one species from the guild, the remaining species should be affected similarly. The guild-indicator concept is appealing because it is potentially a cost-effective approach. Theoretically, agencies could monitor populations of a few guild-indicator species to index population levels of all birds, mammals, reptiles, and amphibians.

Verner (1983, 1984) questioned the economic benefit of monitoring indicator species and found that an inordinate number of species counts were required to detect changes in population numbers. Conversely, Verner showed that if species in a guild were combined and monitored as a single unit, it took fewer censuses to detect changes in total intraguild numbers. This guild-unit approach may be untenable if populations of species within the guild respond differently to environmental perturbations (Mannan et al. 1984).

We evaluate the use of guilds and guild-indicator species as wildlife-management tools by examining a guild of ground-foraging birds. We base these evaluations on three criteria. First, we compare the statistical and economic efficiencies of monitoring populations of individual species of the guild with those of monitoring the population of the guild as a single unit. Second, we examine the habitat relationships of the species within this guild to test the hypothesis that species use similar habitats. Finally, we test whether the presence of a guild-indicator species can be used to predict the presence of other species in the guild.

Study sites and methods

A guild of ground-foraging birds was studied on two Douglas-fir (*Pseudotsuga menziesii*) clearcuts (Kinsey Ridge and Hogback Ridge) in the Coast Range, Humboldt County, California, approximately 45 km east of Arcata. We chose these clearcuts because they were similar in aspect, slope, and in composition and structure of the vegetation. Topographic similarity was determined from topographic maps and aerial photographs, and vegetative similarity was assessed by visually comparing the composition and spatial

WILLIAM M. BLOCK: Department of Wildlife, Humboldt State University, Arcata, California 95521. *Present address:* Department of Forestry and Resource Management, 145 Mulford Hall, University of California, Berkeley, California 94720

LEONARD A. BRENNAN: Department of Wildlife, Humboldt State University, Arcata, California, 95521. *Present address:* P.O. Box 1144, Laytonville, California 95454

R. J. GUTIÉRREZ: Department of Wildlife, Humboldt State University, Arcata, California 95521

arrangement of vegetation between the two sites. Both sites were mixed brushfields of deerbrush (*Ceanothus integerrimus*), blue elder (*Sambucus cerulea*), beaked hazel (*Corylus cornuta*), willow (*Salix* spp.), and berries (*Rubus* spp.), with residual Douglas-fir, tanoak (*Lithocarpus densiflorus*), Pacific madrone (*Arbutus menziesii*), and bigleaf maple (*Acer macrophyllum*) providing a sparse tree canopy.

Inclusion of species to the guild was based on our field observations and on published accounts (Grinnell and Miller 1944; Pitelka 1951; Sumner and Dixon 1953; Davis 1957; Norris 1968; Bock and Lynch 1970; Gutiérrez 1980) which indicated that these species used ground as a foraging substrate. We censused and collected habitat-use data at Kinsey Ridge for frequently detected species (those detected >40 times during censuses).We restricted our analyses to these species to ensure adequate sample sizes for valid comparisons. Species richness data were collected at both Hogback and Kinsey ridges. We listed all ground-foraging species detected, regardless of frequency of occurrence.

During the 1983 breeding season we censused Kinsey Ridge for five species of ground-foraging birds: mountain quail (*Oreortyx pictus*), lazuli bunting (*Passerina amoena*), green-tailed towhee (*Pipilo chlorurus*), rufous-sided towhee (*P. erythrophthalmus*), and dark-eyed junco (*Junco hyemalis*). We used the variable-width line-transect method (Burnham et al. 1980) to survey the area and a Fourier series estimator from the computer program TRANSECT (Laake al. 1979) to estimate bird densities. We conducted six surveys to accumulate the minimum number of 40 detections per species suggested by Burnham et al. (1980). From each survey we pooled all detections and calculated densities for the entire guild.

Approximately 12 km of transects were established along former logging trails situated throughout the habitat. We walked these fixed transects and also random transects to locate species in the guild. Although placement of these transects was biased, they allowed us to sample the entire area and to minimize effects of our movement through the vegetation on bird activity. The location of each sighting corresponded to the center of a 0.02-ha circular plot where a series of structural and floristic habitat variables were measured. Structural habitat characteristics were first compared among species by using one-way analysis of variance (ANOVA) (Sokal and Rohlf 1969). If an F-ratio was significant ($P < 0.05$), specific groupings were determined by using Duncan's multiple range comparisons (Steel and Torrie 1960). Stepwise discriminant function analysis was used to ordinate microhabitats of species along a set of habitat gradients. F-ratios were calculated on the Mahalanobis' D^2 (Green 1978) between species to test if microhabitat characteristics between species were significantly different. To determine if microhabitats of species differed floristically, we compared the rankings of the relative cover of shrub species contributing >5% cover among the microhabitats of the species in the guild, using Spearman rank-order correlation analyses (Conover 1971).

We used the following design to determine whether presence of one species could be used to predict presence of ecologically similar species. We chose mountain quail as the guild-indicator species because it met two of the four criteria given by Salwasser et al. (1982) for management indicator species. In particular, mountain quail are valued by the public because they are commonly hunted (Gutiérrez 1975; Leopold et al. 1981), and they also require a specific habitat configuration consisting of dense shrub cover and free water (Gutiérrez 1975; Brennan 1984). We then listed all ground-foraging birds found at Kinsey Ridge. Aerial photographs were used to locate Hogback Ridge on the basis of its topographic and vegetative resemblance to Kinsey Ridge. Hogback Ridge was surveyed for all ground-foraging birds. Eight working days were spent surveying during July and September 1983 along approximately 8 km of transects. Transects were placed along remnant logging trails and sampled the entire area. A Jaccard coefficient was calculated to measure guild resemblance between the two study sites (Jaccard 1912; Legendre and Legendre 1983). This coefficient measures the proportion of species which were detected at both sites and ranges from 0 (no similarity) to 1 (total similarity). Thus, a value of 1 would support the guild-indicator concept, and a value less than 1 would fail to support the assumption that the presence of the guild-indicator species predicts the presence of particular species from the guild.

Table 18.1. Density estimates of ground-foraging birds and a guild-unit found in the Coast Range, Humboldt County, California

Species[a]	No. surveys	No. birds detected	Density (*n*/ha)[b]	% *CV*
Mountain quail	6	70	0.43	22
Lazuli bunting	6	60	0.40	22
Green-tailed towhee	6	43	0.42	24
Rufous-sided towhee	6	71	0.51	24
Dark-eyed junco	6	85	1.43	16
Guild-unit	1	27	4.39	25
Guild-unit	2	90	3.76	14
Guild-unit	3	138	3.37	11
Guild-unit	4	203	3.22	10
Guild-unit	5	248	3.08	9
Guild-unit	6	329	3.03	8

[a] In addition to these ground-foraging birds, the scrub jay and chipping sparrow were present at Kinsey Ridge. On nearby Hogback Ridge, mountain quail, rufous-sided towhee, dark-eyed junco, scrub jay, chipping sparrow, and fox sparrow were present. The use of guild-units groups detections of the species censused.

[b] Densities were calculated with a Fourier series estimator from the computer program TRANSECT (Laake et al. 1979).

Results

Dark-eyed junco was the most frequently detected species; the other four species were present in comparable numbers ($\chi^2 = 14.9$, $df = 4$, $P < 0.01$). Coefficients of variation of the density estimates ranged from 16% to 24% (Table 18.1). When detections of all species were pooled, only two surveys were required to obtain a density estimate that was more precise (i.e., a lower % *CV*) than those obtained for any single species after six surveys.

At least one species in the guild differed from the others

Table 18.2. Comparisons of the physical characteristics of the habitats used by ground-foraging birds at Kinsey Ridge, Humboldt County, California. (Sample size of 25 habitat plots/species.)

Variable	*F*-ratio[a]	Mountain quail	Dark-eyed junco	Green-tailed towhee	Rufous-sided towhee	Lazuli bunting
Slope (degrees)	3.27	A	B	A,B	A,B	A,B
Distance to cover (m)	4.28	A	B	A	A	A
Distance to water (m)	ns					
Distance to edge (m)	ns					
Minimum shrub height (m)	ns					
Maximum shrub height (m)	ns					
Canopy cover (%)	ns					
Shrub cover (%)	4.43	A	B	A	A	A
Dead woody cover (%)	ns					
Herbaceous cover (%)	4.94	A	B	B,C	A,C	B,C
Litter depth (cm)	3.60	A	A	A,B	B	A,B
Shrub species richness	ns					
Tree-to-shrub ratio	2.76	A	B	B,C	B,C	B,C
Shrub species diversity	ns					

Note: Within the same row, means with the same letter are not significantly different ($P > 0.05$, Duncan's multiple-range test).
[a] One-way analysis of variance; $df = 4, 119$; *F*-ratio is significant at $P < 0.05$ unless noted as nonsignificant (ns).

for six of the 14 variables (Table 18.2). Dark-eyed juncos used areas of less slope and shrub cover, more trees and herbaceous cover, and further from cover than other species used. Mountain quail habitat consisted of steeper slopes, greater shrub cover, little herbaceous cover, and few trees. Green-tailed towhees were found on moderate slopes with extensive shrub and herbaceous cover but few trees. Rufous-sided towhees used areas of moderate slope, abundant shrub cover, and little tree canopy cover. Lazuli bunting habitat was near water with shrubs interspersed among herbaceous open areas.

The first three functions derived from discriminant function analysis explained 95% of the total variation and maximized interspecific habitat differences (Fig. 18.1). Classification success ranged from 48% to 68% (Table 18.3). Wilks' lambda ($\Lambda = 0.284$; $\chi^2 = 143.4$, $df = 68$, $P < 0.001$) showed significant differences between species' microhabitats.

Mahalanobis' D^2 between all possible pairs of species showed that the structural characteristics of mountain quail habitat differed significantly from the habitats of all sympatric guild associates (*F*-ratios, $P < 0.05$; Table 18.4). Dark-eyed junco habitat was structurally different from the habitats of rufous-sided towhee and lazuli bunting (*F*-ratios, $P < 0.05$; Table 18.4).

Some species within the guild used habitats that exhibited different floristic components (Table 18.4). Rankings of coverage by shrub species were not significantly correlated between green-tailed towhees and rufous-sided towhees ($r_s = 0.524$, $df = 14$, $P > 0.05$), green-tailed towhees and dark-eyed juncos ($r_s = 0.503$, $df = 14$, $P > 0.05$), and dark-eyed juncos and lazuli buntings ($r_s = 0.522$, $df = 14$, $P > 0.05$).

The Jaccard coefficient (0.63) between the two sites suggested only moderate resemblance among the species of ground-foraging birds present at each site. This measure is positively biased because the indicator species, mountain quail, was present at both sites by study design.

Discussion

Our results suggested that the population of a guild-unit is more efficiently monitored than the population of a guild-indicator species. We found that less effort was required to obtain a more precise density estimate for a guild-unit than for a single species. If managers use population measures to assess habitat quality, then estimates must be precise to detect changes. Pooling detections of all species effectively increases sample size and the statistical ability to detect population change (Verner 1983, 1984). If fewer censuses are required to obtain a density estimate of a desired precision, field costs are minimized. Unfortunately, density estimates for a guild may provide little insight into population trends of individual species.

Problems arise when species are assigned to a guild a priori or on the basis of the literature. Many species differ in patterns of resource use from those reported in the literature. Jaskić (1981) recommended that guilds be based on site-specific information. We initially based our guild on the literature and later refined the list of species according to our observations. Our guild might have been improved if the inclusion of species had been based on quantified foraging data obtained at the study sites.

We measured significant microhabitat differences among species in the guild, suggesting that each species used different resources. Gleason (1939) showed that each vegetative community was a unique arrangement of resources. Thus, if birds select habitats because they provide the resources they require (Hildén 1965), one vegetated community might have the resources required by a bird, but a nearby community might not. Our observations that certain species did not occur at a site might be partly explained by the absence of suitable arrangements of resources selected by those species.

Caution is necessary when using guilds and guild-indicator species in management applications. Although species in a

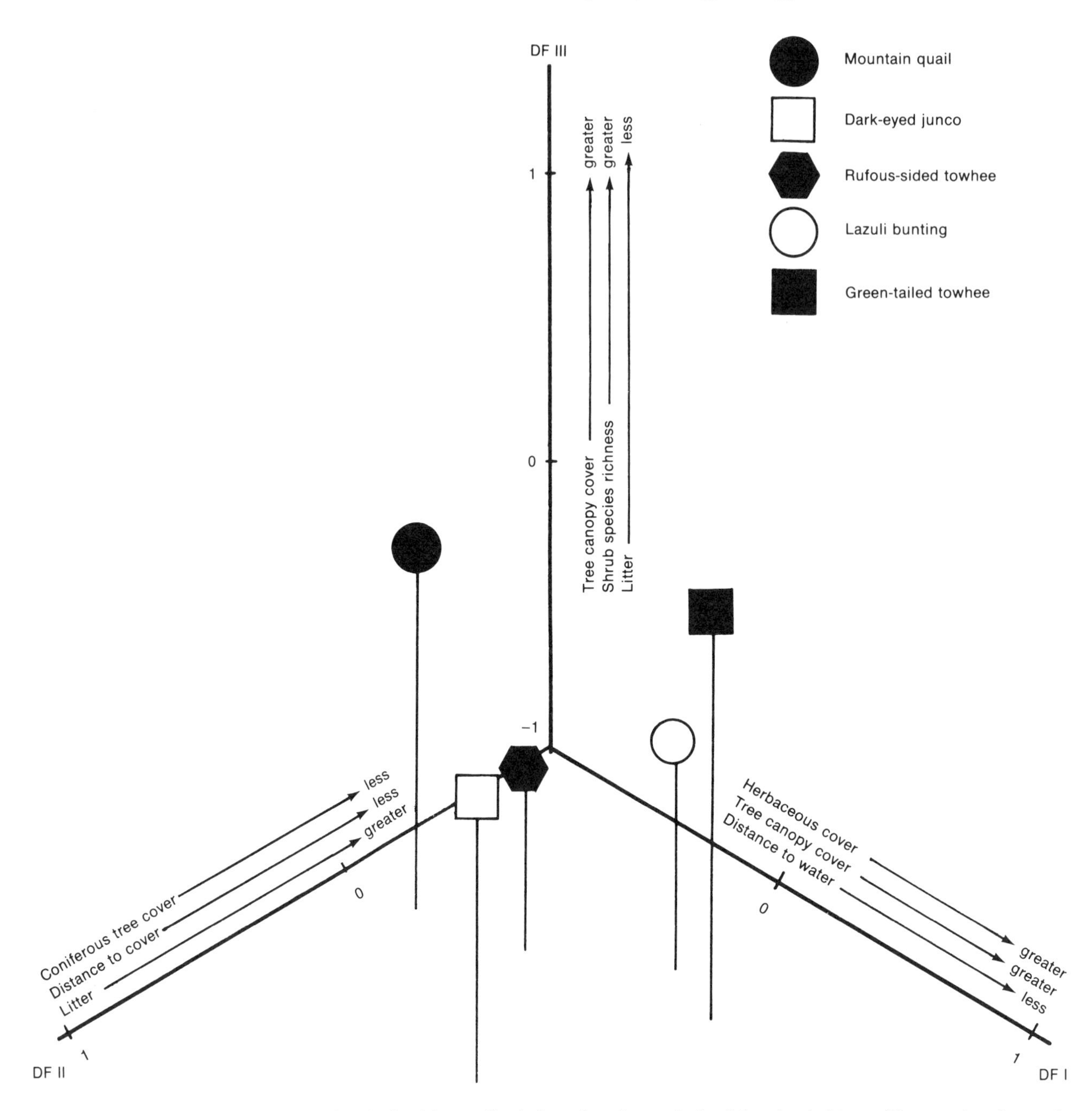

Figure 18.1. Three-dimensional ordination obtained from a discriminant function analysis of the microhabitats of five species of ground-foraging birds found in the Coast Range, Humboldt County, California.

Table 18.3. Classification results from a discriminant function analysis of the habitats of five species of ground-foraging birds found at Kinsey Ridge, Humboldt County, California

	Predicted group				
Actual group	Mountain quail	Dark-eyed junco	Green-tailed towhee	Rufous-sided towhee	Lazuli bunting
Mountain quail	68	4	8	16	4
Dark-eyed junco	12	64	8	16	0
Green-tailed towhee	4	0	60	12	24
Rufous-sided towhee	20	4	8	64	4
Lazuli bunting	4	12	20	16	48

Note: Prior probability of correct classification is equal (0.20) for each species based on relative sample size of the group. Value given is percentage of habitat plots for each species classified to a predicted group. Sample size is 25 for each species.

Table 18.4. Summary of microhabitat comparisons of five species of ground-foraging birds found in the Coast Range, Humboldt County, California

	Mountain quail	Lazuli bunting	Green-tailed towhee	Rufous-sided towhee	Dark-eyed junco
Mountain quail					
Lazuli bunting	A				
Green-tailed towhee	A	nmd			
Rufous-sided towhee	A	nmd	B		
Dark-eyed junco	A	A,B	B	A	

Note: A = significant Mahalanobis' D^2 (F-ratio; $P < 0.05$) between structural characteristics of species' microhabitats; B = nonsignificant Spearman correlation coefficient ($P > 0.05$) between floristic components of species' microhabitats; nmd = no microhabitat difference was determined.

guild may overlap considerably in the use of a particular resource, they may also differ greatly in other aspects of their respective ecologies. Managing a habitat for an indicator species may not meet the needs of other species in the guild. Thus, monitoring the population of a guild or guild-indicator may not reflect population fluctuations of individual species because species have different ecologies and respond differently to environmental change (Mannan et al. 1984).

Acknowledgments

C. W. Barrows, K. H. Berry, R. T. Golightly, B. G. Marcot, E. C. Meslow, and M. L. Morrison critically reviewed earlier drafts of this chapter. The California Department of Forestry Forest Resource Assessment Program and the International Quail Foundation provided funding for this research.

19

Evaluating Models of Wildlife-Habitat Relationships of Birds in Black Oak and Mixed-Conifer Habitats

MARK F. DEDON, STEPHEN A. LAYMON, and REGINALD H. BARRETT

Abstract.—We illustrate a method of testing wildlife-habitat-relationships (WHR) models using data from a bird and habitat survey conducted in California black oak (*Quercus kelloggii*) and mixed-conifer habitats in the Sierra Nevada of northern California. To construct a contingency table, we categorized bird species by the predicted degree of habitat suitability and by their sampled presence. We then examined the proportions of detected species for their relationship to predicted habitat suitability, using a nonparametric test for binomial trends. We found that the proportion of detected species increased significantly with increasing predicted habitat suitability at the black oak site but not at the mixed-conifer site. We also show that the predictive power of WHR models for some species is poor, and we urge users of WHR models to employ these or other methods to test the models in projects involving field surveys.

To meet the challenge of managing the hundreds of wildlife species in forest ecosystems, the USDA Forest Service developed several types of models that predict the suitability of habitats for supporting wildlife species. Wildlife-habitat-relationships (WHR) models are relatively simple (Salwasser 1982) and use a standardized set of habitat descriptors to predict habitat suitability for many species. WHR models are attractive to forest planners as a cost-effective means of predicting the effects of habitat alteration on a large number of species. The models consist of a matrix portion and a narrative portion containing species notes, range maps, and special habitat requirements. The matrix portion consists of suitabilities organized in a two-dimensional table of habitat descriptors by species. The models were designed conservatively, in the sense that if a species could possibly inhabit a given habitat it should be identified in the matrix so that a WHR user would be prompted to at least consider its presence. Use of the narrative portion of the model could result in the species being subsequently dropped from consideration. Hence, the role of the matrix is to provide a list of all species potentially using a habitat. A serious deficiency in the matrix would be a prediction that a habitat is unsuitable when in fact it is at least marginally suitable.

WHR models are developed using the literature and the expert opinions of biologists. Because information on species varies, so does the accuracy of the models, and without field verification this accuracy is unknown. The modelers admit to limitations in model accuracy, yet they claim that their models provide "the best single source available for permitting forest managers to assess the effects of habitat modification on wildlife species" (Verner and Boss 1980:10). The danger lies in the widespread application of models that may have poor predictive power.

Very few studies have tested WHR models. Marcot et al. (1983) reviewed some WHR validation processes and Raphael and Marcot (Chapter 21) demonstrated a validation of WHR models in Douglas-fir (*Pseudotsuga menziesii*) habitats. We present an approach for using field data to test the accuracy of predictions made by WHR models. The method is illustrated by evaluating the Verner and Boss (1980) WHR models for California black oak (*Quercus kelloggii*) and mixed-conifer habitats. Finally, we discuss some implications of errors found in these models.

Methods

The WHR models

Verner and Boss (1980) published WHR models for amphibians, reptiles, birds, and mammals found in the western Sierra Nevada, California. They defined 70 habitat stages by dividing 13 forest cover types into one to four seral stages and one to three canopy cover classes. The suitability was rated for each habitat stage and provided the feeding, breeding, and resting requirements of each terrestrial vertebrate. The season of major use is also noted. For some species, special habitat requirements necessary for their occurrence are given. Maps show the expected range of each species, and a narrative provides relevant life history information.

Bird sampling and habitat typing

Field data for this study were a subset of information gathered following the methods described by Dedon and Barrett (1982). We conducted the surveys on 80 30-m radius

Mark F. Dedon: Department of Forestry and Resource Management, 145 Mulford Hall, University of California, Berkeley, California 94720. *Present address:* Department of Engineering Research, Pacific Gas and Electric Company, 3400 Crow Canyon Road, San Ramon, California 94583

Stephen A. Laymon and Reginald H. Barrett: Department of Forestry and Resource Management, 145 Mulford Hall, University of California, Berkeley, California 94720

plots in the northern Sierra Nevada from May through July 1980. Forty plots were placed in a black oak habitat (with minor amounts of ponderosa pine [*Pinus ponderosa*] and Douglas-fir) located in the Shasta-Trinity National Forest, Shasta County, at about 1100-m elevation. The remaining 40 plots were placed in a mixed-conifer habitat (an association of mixed conifer-pine and tanoak [*Lithocarpus densiflorus*]–Pacific madrone [*Arbutus menziesii*] communities [Parker and Matyas 1978]) located in the Plumas National Forest, Butte County, at about 850-m elevation. We distributed the plots evenly over an area of about 35 ha in each habitat. The closest distance between plot centers was 70 m.

During the breeding season (May and June), one observer sampled each plot once for a period of 200 min, beginning 30 min after sunrise, and recorded all birds seen or heard within the plot. The lengthy sampling period was designed to increase the likelihood of detecting all bird species, except nocturnal birds, with a breeding home range encompassing the plot.

We also sampled botanical composition, canopy closure, and canopy height so that we could assign each plot a habitat stage, as defined by Verner and Boss (1980). Botanical composition was sampled as percent cover of each tree species which was used to determine the dominant forest type. Canopy closure was calculated on each plot as the average of four visual estimates, using a spherical densiometer. Canopy height was calculated as the average of 20 measurements, using an optical rangefinder (Dedon and Barrett 1982). The canopy closure and height estimates were compared directly with the criteria Verner and Boss (1980) used to designate cover and seral stage classes.

WHR predictions and sampled presence

After categorizing the study sites according to the WHR-defined habitat stages, we reviewed the WHR matrix, range maps, and species notes to compile a list of species predicted to occur in each study site in the season and locale sampled by our survey. We included species for which the model specified special habitat requirements only if we believed those requirements were available. We also noted the predicted habitat suitability as optimal, suitable, marginal, or unsuitable (Verner and Boss 1980:6). Nocturnal birds were not surveyed and thus are excluded from this list.

We tallied predicted and detected species in a contingency table by habitat-suitability categories. Some species were detected in habitats that the model predicted to be unsuitable for breeding, feeding, or resting. To include detected and predicted species in an unsuitable category required a tally of species finding the habitats in the study area unsuitable. This list could potentially include species in any habitat not found in the study areas, even habitats found outside the Sierra Nevada. However, it is of little interest to determine rigorously whether the Sierrian WHR models successfully exclude species from non-Sierrian habitats. Since the finest resolution of the WHR-model matrix for distinguishing habitat suitabilities is the habitat stage, the most rigorous analysis would limit the species expected in unsuitable habitats to those found in habitat stages not occurring at the study sites but still within the same forest cover types (mixed-conifer and black oak). We also excluded species predicted to occur in herb and shrub stages, because they do not represent any particular forest type as much as they do open or brushy habitats in general.

Trend analyses

For well-functioning WHR models, we would expect the proportion of detected species to increase in habitats predicted to be more suitable. We tested whether this relationship deviated from chance using a trend analysis for binomial proportions (Marascuilo and Levin 1983:402–405). While chi-square (χ^2) or G^2 (Fienberg 1980) statistics would reveal interactions between species' occurrence and predicted habitat suitability, trend analysis takes advantage of our expectation that a larger proportion of species should be detected with increasing habitat suitability.

Trend analysis is performed by calculating a contrast (y) for the linear component by assigning orthogonal polynomial weights to the proportion of species detected in each habitat suitability category:

$$y_{linear} = 3p_{optimal} + 1p_{suitable} - 1p_{marginal} - 3p_{unsuitable},$$

where $p_{optimal} \ldots p_{unsuitable}$ are the proportions of species detected in habitats predicted to be optimal . . . unsuitable. The associated standard error (SE) is defined as:

$$(SEy)^2_{linear} = pq(3^2/n_{optimal} + 1^2/n_{suitable} + 1^2/n_{marginal} + 3^2/n_{unsuitable}),$$

where p is the proportion of species detected in all suitability categories, q is the proportion of species predicted but undetected in all suitability categories, and $n_{optimal} \ldots n_{unsuitable}$ are the total numbers of species in the optimal . . . unsuitable categories. As the trend approaches linearity and as the slope approaches unity, y_{linear} increases; $(SEy)^2_{linear}$ increases as the sample size decreases and as p and q approach equality.

Because the direction of a meaningful trend is known (a larger proportion of species detected in more suitable habitats), a one-tailed test is appropriate and is performed with a z-test where:

$$z = \sqrt{y^2_{linear}/(SEy)^2_{linear}}.$$

The z-test can be compared with the normal distribution to determine the significance of the trend indicated by y_{linear}.

Results

At the black oak and mixed-conifer sites we detected 38 and 35 bird species, respectively (Table 19.1). In the black oak site we categorized 53% of the plots as black oak woodland 3B and 38% as black oak woodland 3C. The WHR matrix assumes that any black oak stands with canopy closure >70% would be included with class B stands, so we chose stage 3B (6–15-m canopy height and >40% closure) to represent this site. In the mixed-conifer site, we categorized 85% of the plots by the WHR habitat stage of mixed-conifer 4C (>15-m canopy height and >70% closure), and we used

Table 19.1. Detected and undetected bird species predicted to be in California black oak 3B (6–15-m canopy height and 40–69% canopy closure) and mixed-conifer 4C (>15-m canopy height and >70% canopy closure) habitats (Verner and Boss 1980). (Predicted habitat suitability is coded as O = optimal, S = suitable, M = marginal, and U = unsuitable. Unsuitable category refers to species predicted to occur in the same forest type as the study sites, but in different habitat stages [excluding herbaceous and shrub stages].)

	Black oak								Mixed-conifer							
	Detected				Undetected				Detected				Undetected			
Species	O	S	M	U	O	S	M	U	O	S	M	U	O	S	M	U
Turkey vulture (*Cathartes aura*)							X				X					
Sharp-shinned hawk (*Accipiter striatus*)						X								X		
Cooper's hawk (*Accipiter cooperii*)						X									X	
Northern goshawk (*Accipiter gentilis*)						X				X						
Red-tailed hawk (*Buteo jamaicensis*)			X									X				
Golden eagle (*Aquila chrysaetos*)							X								X	
Blue grouse (*Dendragapus obscurus*)			X												X	
Wild turkey (*Meleagris gallopavo*)						X								X		
California quail (*Callipepla californica*)								X								
Mountain quail (*Oreortyx pictus*)							X					X				
Band-tailed pigeon (*Columba fasciata*)						X					X					
Mourning dove (*Zenaida macroura*)			X									X				
Vaux's swift (*Chaetura vauxi*)							X							X		
Black-chinned hummingbird (*Archilochus alexandri*)							X								X	
Anna's hummingbird (*Calypte anna*)						X										X
Calliope hummingbird (*Stellula calliope*)							X									X
Allen's/Rufous hummingbird (*Selasphorus* sp.)				X								X				
Lewis's woodpecker (*Melanerpes lewis*)							X									X
Acorn woodpecker (*Melanerpes formicivorus*)		X														
Red-breasted sapsucker (*Sphyrapicus ruber*)						X					X					
Nuttall's woodpecker (*Picoides nuttallii*)						X										
Downy woodpecker (*Picoides pubescens*)		X										X				
Hairy woodpecker (*Picoides villosus*)		X									X					
White-headed woodpecker (*Picoides albolarvatus*)						X								X		
Northern flicker (*Colaptes auratus*)			X								X					
Pileated woodpecker (*Dryocopus pileatus*)								X	X							
Olive-sided flycatcher (*Contopus borealis*)								X							X	
Western wood-pewee (*Contopus sordidulus*)		X												X		
Hammond's/Dusky flycatcher (*Empidonax* sp.)					X				X							
Western flycatcher (*Empidonax difficilis*)	X											X				
Say's phoebe (*Sayornis saya*)								X								
Ash-throated flycatcher (*Myiarchus cinerascens*)		X														
Western kingbird (*Tyrannus verticalis*)							X									
Violet-green swallow (*Tachycineta thalassina*)		X														
Steller's jay (*Cyanocitta stelleri*)			X							X						
Scrub jay (*Aphelocoma coerulescens*)								X								
Mountain chickadee (*Parus gambeli*)						X								X		
Chestnut-backed chickadee (*Parus rufescens*)			X						X							
Plain titmouse (*Parus inornatus*)							X									
Bushtit (*Psaltriparus minimus*)							X					X				
Red-breasted nuthatch (*Sitta canadensis*)							X		X							
White-breasted nuthatch (*Sitta carolinensis*)	X														X	
Pygmy nuthatch (*Sitta pygmaea*)																X
Brown creeper (*Certhia americana*)		X							X							
Bewick's wren (*Thryomanes bewickii*)								X								
House wren (*Troglodytes aedon*)			X													X
Winter wren (*Troglodytes troglodytes*)													X			
Golden-crowned kinglet (*Regulus satrapa*)													X			
Ruby-crowned kinglet (*Regulus calendula*)							X									X
Blue-gray gnatcatcher (*Polioptila caerulea*)								X								X
Western bluebird (*Sialia mexicana*)		X														X
Mountain bluebird (*Sialia currucoides*)																X
Townsend's solitaire (*Myadestes townsendi*)															X	
Hermit thrush (*Catharus guttatus*)						X			X							
American robin (*Turdus migratorius*)		X								X						
Wrentit (*Chamaea fasciata*)								X								X
Cedar waxwing (*Bombycilla cedrorum*)		X													X	

Continued on following page

Table 19.1. (*Continued*)

	Black oak								Mixed-conifer							
	Detected				Undetected				Detected				Undetected			
Species	O	S	M	U	O	S	M	U	O	S	M	U	O	S	M	U
Solitary vireo (*Vireo solitarius*)		X									X					
Hutton's vireo (*Vireo huttoni*)				X								X				
Warbling vireo (*Vireo gilvus*)		X													X	
Orange-crowned warbler (*Vermivora celata*)		X										X				
Nashville warbler (*Vermivora ruficapilla*)		X										X				
Yellow-rumped warbler (*Dendroica coronata*)							X									
Black-throated gray warbler (*Dendroica nigrescens*)	X											X				
Townsend's warbler (*Dendroica townsendi*)			X												X	
Hermit warbler (*Dendroica occidentalis*)							X			X						
MacGillivray's warbler (*Oporornis tolmiei*)		X										X				
Wilson's warbler (*Wilsonia pusilla*)			X													
Western tanager (*Piranga ludoviciana*)		X								X						
Black-headed grosbeak (*Pheucticus melanocephalus*)			X								X					
Lazuli bunting (*Passerina amoena*)				X												X
Green-tailed towhee (*Pipilo chlorurus*)								X								X
Rufous-sided towhee (*Pipilo erythrophthalmus*)			X									X				
Chipping sparrow (*Spizella passerina*)		X														X
Fox sparrow (*Passerella iliaca*)								X								X
Dark-eyed junco (*Junco hyemalis*)		X								X						
Brewer's blackbird (*Euphagus cyanocephalus*)								X								X
Brown-headed cowbird (*Molothrus ater*)							X				X					
Northern oriole (*Icterus galbula*)								X								
Purple finch (*Carpodacus purpureus*)							X		X							
Cassin's finch (*Carpodacus cassinii*)																X
Pine siskin (*Carduelis pinus*)				X											X	
Lesser goldfinch (*Carduelis psaltria*)			X									X				
Lawrence's goldfinch (*Carduelis lawrencei*)						X										
Evening grosbeak (*Coccothraustes vespertinus*)			X											X		

4C to represent the entire mixed-conifer site. This site contained an abundance of young tanoak and black oak in the understory. The remaining mixed-conifer plots were categorized as 3C or 4B stages, even though they approached stage 4C, as determined by canopy height and crown closure. Because the WHR models were intended to be applied over broad areas, we felt justified in representing each study site by the habitat stage matched by the majority of the plots.

At the black oak site, the WHR models predicted that 63 bird species would find the habitat marginal or better. Thirty-eight species were actually detected by the survey, of which four were predicted to find the habitat unsuitable (Table 19.2). Trend analysis revealed a significant, monotonic increase of the proportion of detected species in habitats predicted to be increasingly more suitable (y_{linear} = 1.65, z = 1.95, P = 0.025). At the mixed-conifer site, 41 bird species were predicted to find the habitat marginal or better, and 35 species were detected, of which the WHR models predicted 14 to find the habitat unsuitable (Table 19.2). Trend analysis failed to detect a significant trend in the proportion of detected species (y_{linear} = 0.97, z = 1.63, P = 0.051). Thus, on the basis of the sampled presence of bird species, the WHR models developed for the black oak site appear to be more accurate than those for the mixed-conifer site.

Discussion

Sampling and model errors

When a species is detected in habitats predicted to be unsuitable (Type I errors), or when a species remains undetected in habitats predicted to be suitable (Type II errors), either the survey or the model must be in error. The two major sources of survey failure are (1) sampling in time or space other than that intended by the model (which produces Type I survey errors); and (2) inadequate sampling effort (which produces Type II survey errors).

Table 19.2. Contingency table of detected and undetected bird species in California black oak 3B and mixed-conifer 4C habitats (see Table 19.1 for habitat definitions). (Predicted habitat suitability is coded as O = optimal, S = suitable, M = marginal, and U = unsuitable.)

	Black oak				Mixed-conifer			
	O	S	M	U	O	S	M	U
Detected	3	18	13	4	7	6	8	14
Undetected	1	12	16	12	2	7	11	16
Proportion detected	0.75	0.60	0.45	0.25	0.78	0.46	0.42	0.47

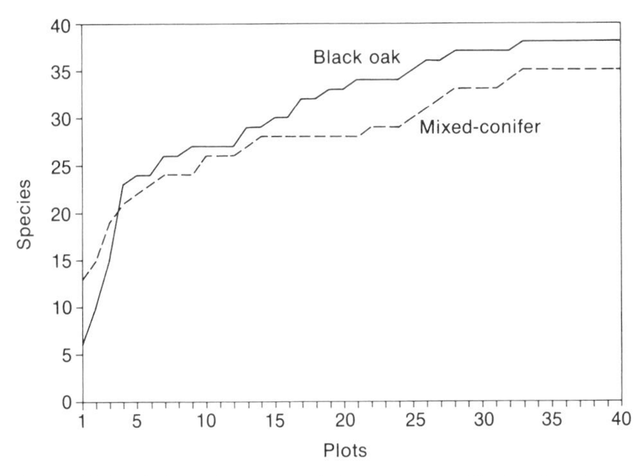

Figure 19.1. Bird species accumulation curves at the black oak and mixed-conifer study sites. Plots are ordered by sampling date.

To substantiate Type I model errors, the survey must follow the intended model application closely. Model parameters, including season, species' range, habitat stage, and special habitat requirements, must be accurately determined by the survey in the manner intended by the WHR model.

Adequate sampling effort is required to substantiate Type II model errors. To determine the adequacy of our survey for detecting all bird species using the study sites, we constructed species accumulation curves for each site. Species accumulation was rare by the end of the sampling period at both sites (Fig. 19.1), which supported the assumption that the survey was adequate to substantiate most Type II errors.

The importance of special habitat components

Eighteen bird species were detected for which the matrix portion of the WHR models predicted the habitat to be unsuitable (Table 19.1). These errors are noteworthy because they defeat the intention that the models should provide a list of all terrestrial vertebrate species potentially found in a habitat. Most of the discrepancies between the WHR predictions and the results of our field studies were probably due to the failure of the WHR models to recognize important understory components. For example, the WHR model for the black-throated gray warbler predicted that mixed-conifer 4C was unsuitable habitat. Earlier successional stages were predicted to be of marginal or suitable value only with the presence of "oaks; trees/shrubs" as special habitat requirements (Verner and Boss 1980:89). This warbler was detected on 50% of the mixed-conifer 4C plots. All these plots contained oaks as an important understory component. Therefore, this WHR model should be revised to denote mixed-conifer 4C habitat as suitable or marginal for this species when the special habitat requirements are met.

Conclusions

Little effort has yet been spent validating WHR models. The results of the present study suggest that even in habitats where the requirements of bird species are believed to be well understood, the predictive value of the applicable WHR models may be poor. If WHR models are to be of value in making land-management decisions, model testing will require a considerably greater research effort than has been envisioned to date.

The techniques we describe are useful for both land managers and modelers. Managers and decision makers need to know how well WHR models are working and whether the information they provide is reliable. Modelers need to know where specific revisions should be made to improve model performance. This study focused solely on birds in only two of the 70 habitats defined by Verner and Boss (1980). Similar studies for other habitats, and for these same habitats at other study sites, will be necessary before WHR models can be applied with confidence.

Acknowledgments

We thank Erica Dedon, Brad Hardenbrook, Sandra Martin, and Keityln Watson for their assistance in the field surveys. We appreciate reviews by Michael L. Avery, Kristen H. Berry, Henry W. Elliot III, Michael E. Fry, Bruce G. Marcot, Charles E. Meslow, Michael L. Morrison, Peter M. Meyer, Barry R. Noon, and C. John Ralph. Funding was jointly provided by the Agricultural Experimental Station, Project 3501MS, the USDA Forest Service, Pacific Southwest Region, and Erica Dedon.

20

The Consistent Characteristics of Habitats: A Question of Scale

PAUL B. HAMEL, NOEL D. COST, and RAYMOND M. SHEFFIELD

Abstract.—As part of validation studies of a Wildlife and Fish Habitat Relationships data base we tested the assumption that habitats possess consistent characteristics. We compared vegetative and faunal characteristics in the South Carolina Piedmont at the regional and stand scales. At the regional scale, control and test samples were similar. At smaller scales, samples of vegetative and bird communities were less similar to regional norms. Bird-community models were reasonably successful at predicting community composition at the regional scale and less successful at smaller scales. These findings suggested that habitats defined by forest type and stand condition may represent a mesoscale phenomenon intermediate between the scale of the individual stand and that of the biome. We concluded that (1) a small number of general parameters could be used to predict a larger array of quantitative structural characteristics; and (2) the assumption of consistency of characteristics of habitats was valid in this instance.

Wildlife managers have developed many data-base systems relating wildlife species to their habitats (e.g., Hoekstra and Cushwa 1979; Thomas 1979; Verner and Boss 1980; Lehmkuhl and Patton 1982). Each data base is structured as a predictive model whose success depends on underlying assumptions. Perhaps most fundamental is the assumption that a habitat, as defined in the data base, is an entity with a set of consistent, quantitative characteristics. A corollary is that the habitat exists in similar form across the geographic region to which the model applies. Frequently neither this assumption nor the corollary mentioned above has been examined, despite the serious consequences for modeling and management efforts should these assumptions be invalid.

Our research was designed to assess the validity of concepts and the accuracy of predictions concerning birds in the habitat-relationships data base maintained by the USDA Forest Service, Southern Region (Hamel and Efird 1985). Our objectives were to test the assumptions that (1) a habitat, defined from a small number of general parameters, possessed consistent quantitative vegetative characteristics; (2) these vegetative characteristics were consistent at different scales of examination; (3) bird-community composition was consistent across the geographic range of the habitat; (4) bird-community composition was consistent at different scales of examination; and (5) composition of the bird community could be predicted reliably from general habitat parameters. Our desired outcome was to provide a statement for managers of the limits of applicability of the data base.

Paul B. Hamel: Department of Biological Sciences, Clemson University, Clemson, South Carolina 29631. *Present address:* Ecological Services Division, Tennessee Department of Conservation, 701 Broadway, Nashville, Tennessee 37219

Noel D. Cost and Raymond M. Sheffield: USDA Forest Service, Southeastern Forest Experiment Station, 200 Weaver Boulevard, Asheville, North Carolina 28804

Methods

Study design

We tested the above assumptions by (1) designating a habitat from five physical and vegetative parameters; (2) selecting sample sites, using specific values of each of these parameters plus an additional areal constraint; (3) predicting bird-community response from five models of habitat relationships; (4) developing regional summaries of plant and bird communities at three hierarchical scales; (5) comparing regional summaries for plant and bird communities with test samples at three hierarchical scales; and (6) comparing predicted with observed bird communities.

Habitat definition and site selection

Logistics dictated choice of the South Carolina Piedmont as the region of the test. Funding constraints necessitated examination of a single habitat in a single year. We chose oak-hickory (*Quercus-Carya* spp.) forest because it is the most extensive type in North America (Forest Service 1977a) and is widespread in the South Carolina Piedmont.

Study stands met six criteria: (1) oak-hickory forest; (2) sawtimber, or mature-condition class; (3) rolling upland topography; (4) canopy cover >60%; (5) ground cover >15%; and (6) stand area >40 ha (100 acres) to enable intrastand as well as interstand comparisons and to control for area sensitivities of some bird species (Whitcomb et al. 1981; Temple, Chapter 43). Test stands were located in five different counties to sample the breadth of the Piedmont physiographic province in South Carolina.

Vegetation sampling

We used two vegetation samples (Table 20.1). The regional norm, or control sample, included 146 stands measured by the Forest Inventory and Analysis (FIA) unit of the Southeastern Forest Experiment Station in 1977. Vegetation

Table 20.1. Allocation of plant- and bird-community samples among hierarchical scales

Sample and scale of examination	Plant-community data[a]	Bird-community data[b]
Control sample: all scales	One sample in each of 146 stands	36 censuses on 13 stands in the southeastern United States
Predicted community: all scales	—	Northeastern United States (DeGraaf et al. 1980), southeastern United States (Dickson et al. 1980), Ozark Mountains (Evans and Kirkman 1981), South Atlantic coastal plain (Hamel et al. 1982).
Test sample		
Regional scale	55 samples in five stands	10 censuses on five stands in South Carolina, 24 censuses on six stands in Kansas, one combined census on two stands in Arkansas
Forest-stand scale	11 samples in each of five stands	Two 10-ha samples in each of five stands in South Carolina, four annual 8-ha samples in each of six stands in Kansas
Intrastand tract or mapping-grid scale	Five samples in each of 10 10-ha tracts	Single sample in each of 10 10-ha mapping grids (tracts) in South Carolina

[a]Control sample measured by personnel of the Southeastern Forest Experiment Station in 1977–1978; test samples measured for this study in 1982; all samples taken from stands in South Carolina Piedmont.

[b]Control sample of 36 censuses of 13 stands in the South Atlantic coastal plain (H. E. LeGrand and P. B. Hamel, pers. obs.); test samples in South Carolina measured in this study; test samples in Arkansas measured by Wooten (1981); test samples in Kansas measured by Cink and Boyd (1979, 1981), Boyd and Cink (1980), and Boyd et al. (1982).

measurements on the control sample consisted of 16 quantitative measures of vegetation structure (Forest Service 1977b) and 48 quantitative measures of vegetation composition (three measures each of 16 species). The basic sampling unit was a cluster of variable-radius sampling points (Forest Service 1977b), and a single observation consisted of a set of 10 such points in the control sample, but only five in the test sample.

The test sample consisted of measurements made in five stands. Five sets of measurements were made in each of two 10-ha (25-acre) tracts within each stand. An additional set of measurements was made at a point located at random in each stand. Vegetation measurements consisted of those of the FIA and two others from James and Shugart (1970). Our study sites represented three hierarchical levels of geographic scale: (1) the individual 10-ha (25-acre) tract; (2) the forest stand of at least 40 ha (100 acres); and (3) the region of at least 10^3–10^4 ha (2500–25,000 acres).

Univariate summaries of the measurements made in a tract (n = 5), in a stand (n = 11), or in all five stands (n = 55) constituted our test-sample estimates of the structure and composition of the vegetation at the tract, stand, and regional scales. These summaries were compared with the summary of the control sample to determine the extent to which regional characteristics were consistent at different scales of examination of the test sample. In all cases, we tested for significance at P = 0.05, using either t-tests or confidence intervals about test sample means.

Bird sampling

Bird-community data

Primary test samples (Table 20.1) consisted of spot-map censuses (Robbins 1970) made on tracts (mapping grids) of the vegetation test sample. Each 10-ha (25-acre) tract was visited weekly from late April until July 1982 (Hamel 1983). To permit calculations on all spot-map data, species with densities below 0.5 pair/plot were given a density of 1 pair/40 ha (100 acres); visitors were assigned a density of 0.1 pair/40 ha. Additional test samples for bird-community analyses came from censuses in northwestern Arkansas and in northeastern Kansas (Table 20.1). We selected four synthetic models and one empirical data set that associate birds with oak-hickory forests as sources for predictions of bird-community composition (Table 20.1).

Comparison of observed with predicted bird communities

We compared the three empirical data sets to the five sets of predictions at the regional, stand, and tract scales. Four possible outcomes of each comparison existed for each species: (1) predicted and found—a success; (2) predicted but not found—a miss; (3) found but not predicted—a surprise; and (4) neither predicted nor found. We ignored category (4) because of the difficulty involved in determining the full species pool for all prediction sets. Each of the other categories was treated as follows: (1) the success rate was the proportion of predicted species that were found; and (2) the surprise rate was the proportion of observed species that were not predicted. In this sense, the more effective models simultaneously maximized the success rate and minimized the surprise rate. For comparing the models we also calculated accuracy rate as successes/(successes + surprises + misses).

Results

Plant-community analyses

Site selection

All stands and nine of the 10 10-ha (25-acre) intrastand tracts of the test sample met all selection criteria. On the tenth tract, sawtimber was not the dominant size-class (poletimber basal area was slightly greater than sawtimber basal area). All criteria were also met by the tracts measured by Cink and Boyd (1979) in Kansas, except that sawtimber was not the dominant size-class on two tracts. Approximately 10% of the total area of the test stands consisted of narrow stream margins, a physiographic position distinct from the rolling upland criterion.

Table 20.2. Mean values (±1 *SE*) of measures of vegetation structure of control and test stands compared at regional, stand, and tract scales

		Test stands			
				Number of comparisons[b] significantly different at	
Variable compared	Control stands[a] (n = 146)	Regional scale (n = 55)	Probability at regional scale	Stand scale (n = 5 stands)	Tract scale (n = 10 tracts)
Total tree density (stems > 7.6 cm dbh/ha)	844.8 ± 27	826.5 ± 52	0.69	4	6
Total basal area (m²/ha, stems > 7.6 cm dbh)	21.6 ± 2.7	22.0 ± 0.71	> 0.9	3	4
Basal area, saplings (m²/ha, stems 7.6–12.7 cm dbh)	2.9 ± 1.1	3.3 ± 0.3	0.69	4	7
Basal area, poletimber (m²/ha, pine stems 12.8–23.0 cm, hardwood stems 12.8–28.0 cm dbh)	8.8 ± 0.3	8.1 ± 0.6	0.23	4	4
Basal area, sawtimber (m²/ha, pine stems > 23.0 cm, hardwood stems > 28.0 cm dbh)	9.9 ± 0.5	11.2 ± 0.7	0.17	2	2
Height of tallest tree (m)	26.0 ± 0.3	27.0 ± 0.7	0.17	4	6
Mean canopy height (m)	19.7 ± 0.2	21.1 ± 0.5	0.007	3	2
Canopy cover (%)	—	82.7 ± 2.4	—	—	—
Shrub density (stems/ha)	—	80,552 ± 10,052	—	—	—
Number of tree species recorded on a cluster	8.0 ± 0.2	6.4 ± 0.3	< 0.001	3	5
Dead tree density (stems > 7.6 cm dbh/ha)	—	20.1 ± 2.7	—	—	—
Basal area of dead trees (m²/ha, stems > 7.6 cm dbh)	—	10.3 ± 1.2	—	—	—
Clusters with water on sample acre (%)	27	16	0.10	—	—
Age (years)	48.7 ± 1.1	72.3 ± 2.8	< 0.001	2	2
Total space occupied by live vegetation (m³/ha)	108,579 ± 2,775	129,714 ± 4,378	< 0.001	3	1
Space occupied by live vegetation, by layers (m³/ha)					
0–0.6 m (0–2 feet)	1,507 ± 85	2,235 ± 148	< 0.001	3	3
0.7–2.1 m (3–7 feet)	2,364 ± 134	2,885 ± 240	0.06	2	3
2.2–6.1 m (8–20 feet)	10,241 ± 557	11,813 ± 827	0.12	3	5
6.2–16.5 m (21–54 feet)	63,799 ± 1,646	66,597 ± 2,528	0.35	2	3
16.6+ m (55+ feet)	31,537 ± 1,695	46,184 ± 3,763	< 0.001	2	1

[a]Each control stand was sampled with one cluster of 10 variable-radius sampling points. Each sample in the test stands consisted of one cluster of five variable-radius sampling points. No rarefaction or other technique has been applied to these data to correct the number of species for this difference in sampling methodology.

[b]Entries in stand- and tract-scale columns are those whose mean value of the variable was significantly different from the mean value of control stands. The mean value for a stand or tract was judged significantly different from the control value when the 95% confidence interval about the stand (or tract) mean did not include the mean value of the control stands.

Control vs test sample

Regional scale.—Ten of 16 comparisons of vegetation structure on control and test samples were not significantly different at $P = 0.05$ by t-tests (Table 20.2). Six of the seven significant differences, all except stand age, may have been due to diffrences in sampling, as discussed below. Samples taken in narrow stream margins (10% of test sample but excluded from control sample) had taller trees and much denser understory vegetation than did their counterparts on rolling uplands.

Foliage height profiles (Fig. 20.1) of the two samples had overlapping 95% confidence intervals over about half of their height. Overstory composition of the two samples was similar: 14 of the 15 most common tree species were the same in both samples (Table 20.3).

Stand and tract scales.—Comparison of vegetation structure and composition of the control sample to summaries of the test sample at stand and tract scales revealed a pattern superficially similar to that at the regional scale (Tables 20.2, 20.3). At both stand and tract scales, nonsignificantly different comparisons outnumbered significant differences. These results must be viewed with some caution, however, because the smaller sample sizes produced longer confidence intervals which, in turn, masked greater variation in these comparisons than in those at the regional scale. Mean values of individual stands and intrastand tracts departed succes-

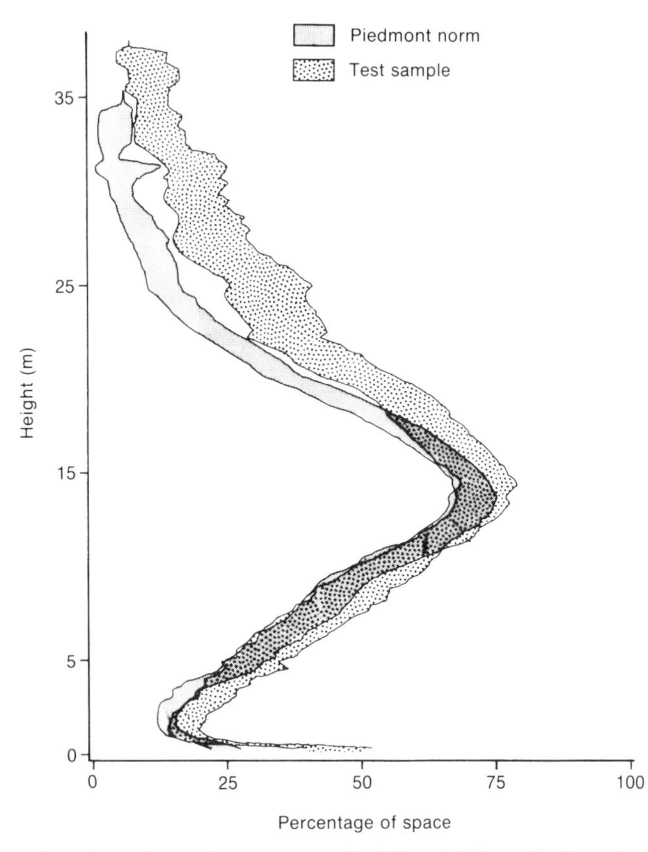

Figure 20.1. Comparison of composite foliage height profile from the test sample with that from the control sample representing norm of oak-hickory forest in South Carolina Piedmont. Shaded areas are 95% confidence intervals about mean percent of space occupied by living vegetation, measured in 0.3 m (1 foot) intervals.

sively farther from the mean values of the control sample than did the mean values of the test sample taken as a whole. Further analyses at these scales (not presented here) compared sample values of randomly located clusters to the mean values of the tracts or stands in which the clusters were located. Results of those comparisons indicated uniformly that the values observed on a randomly located vegetation cluster were not good predictors of the stand or tract mean values.

Bird-community analyses

Comparison of empirical data sets

Regional scale.—Eighty bird species were recorded in at least one of the three study localities—South Carolina, Kansas, and Arkansas. Of those species, 39% were recorded in all three localities and 65% were recorded in two. Mean density of those species recorded in only one locality was a low 0.5 pair/40 ha (100 acres).

Density and frequency values were available for species recorded in the South Carolina and Kansas studies but not for the Arkansas study. Thirty-seven species were recorded on one of those two studies at either (1) high density, indicated by an average density of at least 1 pair/plot; or (2) high frequency, indicated by a frequency >0.60. Thirty-four of these 37 common species were recorded by both studies (70% of the common species in each study). Eleven of 19 high-density species were recorded at high density on both lists (76% of high-density species of South Carolina and 70% of high-density species of Kansas). Nineteen of 37 high-frequency species were recorded at high frequency on both studies (70% of high-frequency species in each study).

Stand and tract scales.—Fifty-nine species were recorded in the South Carolina study and 64 in the Kansas study; 45

Table 20.3. Mean values (± 1 *SE*) of density, basal area, and frequency of tree species of control and test stands compared at regional, stand, and tract scales

Species/Variable[a]	Control stands (n = 146)	Test stands: Regional scale (n = 55)	Test stands: Probability at regional scale	Test stands: Number of comparisons[b] significantly different at Stand scale (n = 5 stands)	Test stands: Number of comparisons[b] significantly different at Tract scale (n = 10 tracts)
White oak (*Quercus alba*)					
Basal area	3.3 ± 0.2	3.8 ± 0.5	0.40	4	6
Density	118.4 ± 14.2	95.1 ± 13.7	0.24	1	1
Frequency	0.84	0.76	0.21	—	—
Scarlet oak (*Quercus coccinea*)					
Basal area	1.0 ± 0.2	3.6 ± 0.5	< 0.001	1	3
Density	31.6 ± 7.5	80.1 ± 14.9	0.006	2	2
Frequency	0.38	0.62	0.004	—	—
Hickory (*Carya* sp.)					
Basal area	2.0 ± 0.2	1.8 ± 0.3	0.56	1	1
Density	88.2 ± 12.4	72.0 ± 15.0	0.41	1	2
Frequency	0.62	0.53	0.25	—	—

Table 20.3. (*Continued*)

Species/Variable[a]	Control stands (*n* = 146)	Test stands: Regional scale (*n* = 55)	Probability at regional scale	Number of comparisons[b] significantly different at: Stand scale (*n* = 5 stands)	Tract scale (*n* = 10 tracts)
Yellow-poplar (*Liriodendron tulipifera*)					
Basal area	2.1 ± 0.3	1.9 ± 0.5	0.62	3	3
Density	40.6 ± 6.8	26.0 ± 8.2	0.18	2	2
Frequency	0.60	0.47	0.10	—	—
Sweetgum (*Liquidambar styraciflua*)					
Basal area	2.0 ± 0.2	0.5 ± 0.1	< 0.001	0	1
Density	85.4 ± 12.2	25.2 ± 9.2	< 0.001	1	1
Frequency	0.52	0.24	< 0.001	—	—
Southern red oak (*Quercus falcata*)					
Basal area	1.5 ± 0.2	1.2 ± 0.2	0.32	2	4
Density	45.6 ± 8.0	37.7 ± 9.8	0.52	2	2
Frequency	0.54	0.45	0.26	—	—
Shortleaf pine (*Pinus echinata*)					
Basal area	1.1 ± 0.1	2.0 ± 0.4	0.03	2	1
Density	42.8 ± 6.7	76.8 ± 19.4	0.10	1	1
Frequency	0.53	0.45	0.32	—	—
Northern red oak (*Quercus rubra*)					
Basal area	1.1 ± 0.2	0.4 ± 0.1	0.004	0	0
Density	29.1 ± 6.0	12.8 ± 5.0	0.04	0	0
Frequency	0.38	0.18	0.007	—	—
Post oak (*Quercus stellata*)					
Basal area	0.9 ± 0.1	0.7 ± 0.2	0.38	1	1
Density	33.5 ± 7.5	20.4 ± 9.6	0.37	0	0
Frequency	0.38	0.25	0.08	—	—
Red maple (*Acer rubrum*)					
Basal area	0.8 ± 0.1	1.0 ± 0.3	0.57	1	1
Density	39.8 ± 8.7	59.9 ± 19.7	0.36	1	0
Frequency	0.36	0.31	0.50	—	—
Black oak (*Quercus velutina*)					
Basal area	0.8 ± 0.1	0.6 ± 0.1	0.30	2	2
Density	31.8 ± 7.2	18.4 ± 5.4	0.14	0	1
Frequency	0.35	0.33	0.72	—	—
Sourwood (*Oxydendrum arboreum*)					
Basal area	0.5 ± 0.1	1.2 ± 0.2	0.001	1	2
Density	34.8 ± 6.4	118.0 ± 20.2	< 0.001	2	1
Frequency	0.27	0.54	< 0.001	—	—
Chestnut oak (*Quercus prinus*)					
Basal area	0.3 ± 0.1	1.5 ± 0.4	0.008	0	1
Density	7.4 ± 3.8	47.0 ± 16.2	0.02	0	1
Frequency	0.06	0.29	< 0.001	—	—
Water oak (*Quercus nigra*)					
Basal area	0.7 ± 0.1	0.2 ± 0.1	0.005	1	0
Density	22.6 ± 5.0	5.5 ± 3.8	0.009	0	0
Frequency	0.24	0.09	0.01	—	—
Flowering dogwood (*Cornus florida*)					
Basal area	0.4 ± 0.1	0.6 ± 0.2	0.36	3	2
Density	49.3 ± 10.5	74.7 ± 21.0	0.28	2	2
Frequency	0.20	0.31	0.11	—	—
Blackgum (*Nyssa sylvatica*)					
Basal area	0.2 ± 0.1	0.2 ± 0.1	0.64	0	0
Density	13.2 ± 5.2	10.7 ± 4.3	0.65	0	0
Frequency	0.19	0.14	0.40	—	—

[a] Basal area in m^2/ha, density as stems 7.6 cm dbh/ha, frequency as proportion of sample clusters on which the species was recorded.
[b] Entries in columns are the numbers of stands (or tracts) whose mean basal area or density was significantly different from the mean value of control stands. The mean value for a stand or tract was judged significantly different from the control value when the 95% confidence interval about the stand (or tract) mean did not include the mean value of the control stands.

were recorded by both studies. Mean species richness for single censuses was 35 (range = 28–38, n = 10) in South Carolina and 31 (range = 23–41, n = 24) in Kansas. Thus, an average tract in a single year had 48% (Kansas) to 59% (South Carolina) of the respective oak-hickory forest bird species.

At a somewhat longer temporal scale, the average tract in the Kansas study recorded 41 species (range = 35–51, n = 6) in at least 1 of the 4 years of the study. At a larger spatial scale, the average South Carolina stand, estimated from two 10-ha (25-acre) tracts in a single year, supported 40 species (range = 37–42, n = 5).

Comparison of observed with predicted bird-community composition

Regional scale.—At this scale, the several models performed similarly (Table 20.4). Each was most successful at predicting the South Carolina bird community and least successful with the Arkansas bird community. Success rates and surprise rates were correlated inversely to the number of species predicted. The existing censuses (H. E. LeGrand and P. B. Hamel, pers. obs.) produced the minimum surprise rates at the expense of low success rates. Dickson et al. (1980) used a small number of published censuses; their success rate was very high at a cost of a large number of surprises. Predictions by other models were between these extremes. The model of DeGraaf et al. (1980) for old-growth northern red oak was uniformly less successful than the other models.

When predictions were compared with observations of the most abundant species, success rates of all prediction sets except DeGraaf et al. (1980) were 0.94 or better. Success rates for the most frequent species in those comparisons were between 0.80 and 0.96, again with the exception of DeGraaf et al. (1980).

Stand, tract, and annual scales.—Success rates declined as the extent of the area sampled decreased, either in space or in time (Table 20.5 shows results for the Kansas study; results of the South Carolina study were similar). This was the product of the comparison of a fixed number of predictions to a declining number of species recorded (species-area effect: Catzeflis 1978; Engstrom and James 1981). Surprise rates changed less than success rates. As at the regional scale, the rank order of prediction models by success rate or surprise rate changed little as the scale changed, or from one empirical data set to another.

Discussion

Forest land managers usually possess general inventory information about the stands in their care. Wildlife biologists who attempt to assess the opportunities for wildlife in the context of land-use planning must be able to associate species with habitats as defined in the inventory. Any model that provides a common focus for the forester and for the wildlife biologist on the relationships of wildlife to their habitats has utility (Salwasser, Chapter 59). The extent of that utility will increase as the accuracy of model predictions increases. The validity of assumptions underlying the model is an important determinant of the accuracy of model predictions.

Frequently the habitats defined for land-management in-

Table 20.4. Comparison of observed and predicted bird communities in oak-hickory forests, at regional scale

	Empirical data sets[a]		
Prediction source and number of species predicted	South Carolina (59 species found)	Kansas study (64 species found)	Arkansas study (41 species found)
Hamel et al. (1982) predicted 66 species for sawtimber stands of oak-hickory forest	0.80/0.14 (0.72)	0.74/0.23 (0.60)	0.56/0.10 (0.53)
Evans and Kirkman (1981) predicted 56 species for old-growth stands of oak-hickory forest	0.85/0.20 (0.71)	0.75/0.34 (0.56)	0.62/0.15 (0.55)
DeGraaf et al. (1980) predicted 51 species for old-growth stands of northern red oak forest	0.75/0.37 (0.54)	0.65/0.48 (0.44)	0.51/0.36 (0.38)
Dickson et al. (1980) predicted 37 species for sawtimber stands of oak-hickory forest	0.94/0.44 (0.66)	0.78/0.55 (0.51)	0.73/0.34 (0.54)
96 species were recorded on 36 censuses of 13 stands in the Southeast (H. E. LeGrand and P. B. Hamel, pers. obs.)	0.61/0.10 (0.56)	0.59/0.11 (0.59)	0.42/0.02 (0.41)

Note: Figures are expressed as success rate/surprise rate (accuracy rate), in which the success rate is the proportion of predicted species that were found, the surprise rate is the proportion of observed species that were not predicted, and the accuracy rate = successes/(successes + surprises + misses).

[a] Bird censuses of the test sample described for Table 20.1.

Table 20.5. Comparison of predicted (from Table 20.4) and observed bird species composition at tract, year, and tract-year scales in oak-hickory forests in northeastern Kansas, 1978–1981

	Scale of examination (mean and range)					
	Year[a]		Tract[b]		Tract-year[c]	
Prediction source	Success rate	Surprise rate	Success rate	Surprise rate	Success rate	Surprise rate
Hamel et al. (1982)	0.63 0.58–0.68	0.18 0.16–0.20	0.53 0.44–0.67	0.15 0.14–0.17	0.41 0.32–0.56	0.11 0.07–0.17
Evans and Kirkman (1981)	0.62 0.55–0.68	0.32 0.28–0.33	0.54 0.45–0.64	0.27 0.20–0.29	0.45 0.38–0.61	0.18 0.07–0.29
DeGraaf et al. (1980)	0.52 0.49–0.59	0.47 0.46–0.49	0.44 0.37–0.55	0.46 0.41–0.51	0.36 0.29–0.53	0.41 0.28–0.48
Dickson et al. (1980)	0.74 0.70–0.78	0.46 0.43–0.48	0.65 0.57–0.78	0.42 0.35–0.45	0.57 0.45–0.65	0.32 0.13–0.41
American Birds censuses 1947–1979	0.49 0.47–0.52	0.07 0.02–0.11	0.40 0.33–0.48	0.07 0.03–0.10	0.31 0.23–0.38	0.05 0–0.10

[a]Summarizes results of six 8-ha (20-acre) tracts in each of 4 years ($n = 4$).
[b]Summarizes results of 4-year censuses on each of six 8-ha (20-acre) tracts ($n = 6$).
[c]Each tract-year combination is a single census on a single 8-ha (20-acre) tract in a single year ($n = 24$).

ventory purposes are discrete categories such as forest-type and stand-condition combinations. However, a considerable body of ecological theory and research, from Gleason (1926) and Whittaker (1967) to the present (Wiens 1983), suggests that communities exist as continua, not as discrete entities, and that species are distributed independently of each other. The tests reported here were designed to determine whether continuous ecological communities could be treated, at some scale, as discrete entities for management purposes.

We conducted our test in a habitat in which the assumptions were probably true. If we were able to refute them under such circumstances, then validity in other cases would be less likely. Our test of five assumptions revealed areas in which assumptions appeared to be valid and areas where validity was questionable, and it also yielded implications for managers concerning the applicability of habitat-relationships data bases. The assumption that a habitat defined in general terms possesses consistent, quantitative characteristics appeared at first to be of questionable validity. Closer examination of the data convinced us that a number of the differences we observed were the result of sampling differences or were in some way unrelated to the validity of the assumption. Restriction of the control sample to measurements made only on rolling upland sites is such an example. Areas of other physiographic position within stands of the control sample were not examined. Stands of the test sample were measured systematically without regard to physiographic position, and 10% of our measurements were made in narrow stream margins. This, in our opinion, was sufficient to cause the 1.4-m (4.4-foot) difference in mean canopy height and the differences noted in foliage volume and foliage height distributions. The observed difference of 1.6 species in number of tree species recorded from the average cluster reflects the examination of 10 points in the clusters of the control sample and five points in those of the test sample. Thus only one of the statistically significant differences between test and control samples, that of stand age, cannot be related directly to differences in sampling. From these considerations we concluded that the observed consistency of the test and control samples was an underestimate of the actual consistency of habitat characteristics at the regional scale.

The assumption that habitat characteristics were consistent at different scales of examination was not valid. At smaller scales the structure and composition of sawtimber oak-hickory forest was not consistent with that at the regional scale. Single sample measurements were sufficient to identify a stand in terms of the selection criteria but not to describe the quantitative characteristics of the stand.

Our third assumption, that bird-community composition was consistent across the range of the habitat, was not completely true. More than 90% of the species in the small Arkansas sample were also recorded from the South Carolina study, and 76% of the species in the Arkansas sample were also found in the Kansas study. About 70% of the species recorded in the Kansas study were also found in South Carolina, and vice versa. An even higher proportion of species noted as common in one study were found in both Kansas and South Carolina. Only half of the common species were common in both studies, however. This consistency was less than that observed for the plant community at the regional scale.

The fourth assumption, that bird-community composition was consistent at different scales, was suspect. Similarities existed, as noted above, between the samples from different parts of the range. Individual census samples were variable, however. Single samples of either plant- or bird-community composition were not adequate to characterize the respective community.

The fifth assumption was that bird-community composition could be predicted reliably from general habitat parameters. The success rates we found (Table 20.4) at first appeared to be in the range achieved by null models (Rice et al., Chapter 13). However, our calculations did not include the "neither predicted nor found" portion of the species pool and were thus underestimates of the success rates of Rice et al. When the full species pool of Hamel et al. (1982) was used in the computations, accuracy rates for that model exceeded 0.90. We therefore concluded that bird-community composition in sawtimber oak-hickory forest could be predicted reliably at the regional scale.

Comparison of different predictions (Evans and Kirkman 1981; Hamel et al. 1982) for oak-hickory sawtimber revealed that models designed for one part of the range of the type worked as well elsewhere. On the other hand, a model designed for northern red oak forests (DeGraaf et al. 1980) did not work as well in oak-hickory forests; it may have been inappropriate for oak-hickory forests of the Southeast (R. M. DeGraaf, pers. comm.). This demonstrates the importance of using models only within the region for which they were developed. Models designed for oak-hickory forest worked well through 1600 km (1000 miles) of longitude but across much less latitude. Our results here were probably conservative, because the probability of recording species that occur at low density is a function of sampling intensity (Haila, Chapter 45). Our maximum sampling intensity was 100 ha (250 acres)/year.

Two considerations affected interpretation of our results. First, the geographic scale of examination had an overwhelming effect on these results, as Wiens (1981b) observed in his review. At the regional scale, estimated by an area of at least 10^3–10^4 ha (2500–25,000 acres), oak-hickory forest and the avifauna associated with it had consistent characteristics of composition and structure. At smaller scales this consistency was less apparent. Predictions of avian community composition were also scale-dependent in their success. Other approaches, such as those of Haefner (1981) or Robbins (1978), may be required at smaller scales; the approach used here applied well at the regional scale.

Second, the importance of the context in which results were to be interpreted (Järvinen and Väisänen 1981) should not be underestimated. The assumption that habitats possessed consistent characteristics was valid for applications involving the average stand of oak-hickory sawtimber. But the assumption was certainly not valid for all stands of oak-hickory sawtimber. From this we inferred that a habitat defined in terms of forest type and stand-condition class was an entity at a mesoscale, intermediate between the small scale of an individual stand and the large scale of the biome. We believe that managers and others interested in applications at the regional scale may cautiously treat this assumption as valid. However, when their applications involve particular stands, managers may not necessarily assume that those stands possess characteristics that conform to regional norms.

Finally, we have conducted these tests on only one habitat. We have not examined a successional continuum in a single forest type, nor have we examined the response of wildlife communities other than birds. The assumption that habitats possess consistent characteristics was valid in this instance, at a particular scale. Validity of the assumption awaits determination under other conditions.

Acknowledgments

This work was supported by the USDA Forest Service. S. Gauthreaux, M. Lennartz, and H. LeGrand helped design the study. J. McClure aided in plot selection. N. Brunswig, E. Graves, A. Schenck, C. Dachelet, and R. Sims helped with the field work. The manuscript has benefitted from critical reviews by D. Adams, K. Evans, T. Hoekstra, R. DeGraaf, H. Salwasser, D. Durham, C. Nicholson, D. Eagar, S. Pearsall, and E. Bridges.

21

Validation of a Wildlife-Habitat-Relationships Model: Vertebrates in a Douglas-fir Sere

MARTIN G. RAPHAEL and BRUCE G. MARCOT

Abstract.—We assessed the reliability of a wildlife-habitat-relationships (WHR) model in four seral stages of mixed-evergreen forest in northwestern California. We sampled terrestrial vertebrates on 191 study sites and compared patterns of species' occurrence and abundance with those predicted by the WHR model. Species richness among stages differed significantly from model predictions for birds but not for amphibians, reptiles, and mammals. Overall, 11% of the species predicted to occur were not observed, whereas 14% were observed but not predicted. The model predicted changes in relative abundance between stages less successfully. Of 650 between-stage predictions, 43% were in error, but only 10% were serious (i.e., involving unexpected declines in abundance). Error rates were constant across seral stages but varied among taxa. We discuss reasons for model failure, effects of sampling effort on comparing predicted and observed occurrences of species, and appropriateness of seral stage classifications.

Following the work of Patton (1978) and Thomas (1979), many resource agencies have developed information systems and models for relating occurrence of wildlife species to habitats. The California Wildlife Habitat Relationships (WHR) Program is one such effort (Salwasser et al. 1980). WHR matrix models are a component of this system and are used to organize existing wildlife-habitat information systematically and to serve as tools for land-use planning (Ohmann 1983).

Despite the extensive use of WHR models, especially for forest-management applications, their reliability generally has not been evaluated. Studies in California have compared wildlife species lists of occurrence generated by WHR models to those from field observations in oak (Verner 1980; Dedon 1982) and mixed-conifer (Dedon et al., Chapter 19) habitats. To date, we are not aware of any studies that compare predicted and observed patterns of abundance of species with the objective of validating a WHR model.

Our study was designed to evaluate reliability of an untested WHR matrix for mixed-evergreen forest of California's North Coast-Cascades Zone (Marcot 1979). This model relates four levels of habitat suitability (species is absent, habitat is marginal, suitable, or optimum) to each of four seral stages, the two oldest of which are further divided into three classes of canopy cover.

MARTIN G. RAPHAEL: Department of Forestry and Resource Management, 145 Mulford Hall, University of California, Berkeley, California 94720. *Present address:* USDA Forest Service, Rocky Mountain Forest and Range Experiment Station, 222 South 22nd Street, Laramie, Wyoming 82070

BRUCE G. MARCOT: Cooperative Wildlife Research Unit, Department of Fisheries and Wildlife, Oregon State University, Corvallis, Oregon 97331. *Present address:* USDA Forest Service, Fish and Wildlife Staff, 319 SW Pine Street, Box 3623, Portland, Oregon 97208

Our validation study is based on a comparison of an extensive wildlife survey within these four seral stages with a set of predictions generated by the model. Our objectives were to (1) compare predicted and observed patterns of species' occurrence; (2) compare predicted and observed abundance of species between pairs of seral stages; and (3) assess the appropriateness of seral stages defined for the model.

The WHR matrix model

Verner and Boss (1980) described the development and content of the WHR system. To summarize, the California WHR Program is divided into five geographic zones. Within each zone, and for the state as a whole, working groups developed matrices showing the suitability of available habitat (vegetation) types and seral stages for each vertebrate species (except fishes and accidentals). For each species, the WHR documents also include a distribution map, notes on seasonal occurrence, any special habitat requirements (such as cliffs or snags), and a habitat-suitability rating for each seral stage of each habitat type.

The habitat-suitability ratings for the North Coast–Cascades Zone form the basis of our study. The suitability of each habitat stage (combination of habitat type and seral stage) for each species was rated by contract specialists (one each in herpetology, mammalogy, and ornithology) on the basis of their literature review, personal experience, consultation with others, and best professional judgment. Each specialist was provided with a detailed set of instructions outlining the characteristics of appropriate habitat stages and criteria by which to assign suitability ratings. Habitat stages used by species were rated as (1) optimum (best-quality habitat as judged by high breeding density of the species or high frequency of use for feeding or resting cover); (2) suitable (good habitat but not among best, as judged by moderate breeding density or frequency of use for feeding or resting); or (3) marginal (habitat used by the species, but it does not contribute over time to the maintenance of any population) (Verner and Boss 1980:6). These ratings were compiled into a draft volume (Marcot 1979). Included for all 18 habitat types and their seral stages are ratings for 25 amphibian, 27 reptile, 246 bird, and 97 mammal species. For this study, we extracted only the ratings for the mixed-evergreen habitat type.

Methods

Plot selection

Our study sites were located on the Klamath and Trinity River drainages of the Six Rivers, Shasta-Trinity, and Klamath national forests within a 50-km radius of the town of Willow Creek in the mixed-evergreen zone of northwestern California. Forest cover was dominated by Douglas-fir (*Pseudotsuga menziesii*) in association with tanoak (*Lithocarpus densiflorus*) and Pacific madrone (*Arbutus menziesii*) at elevations between 427 and 1220 m.

We selected potential study areas from timber maps or aerial photographs. Final selections were based on ground examinations to verify seral stage designations. We sampled at least 20 stands in each of four WHR seral stages (Table 21.1). Stand areas varied from 5 to 455 ha and averaged 68 ha.

The four WHR seral stages (Table 21.1) are broader than those used as timber strata by the Pacific Southwest Region of the USDA Forest Service. In particular, the two oldest WHR stages are each comparable to aggregates of two Forest Service timber strata. In addition, the WHR shrub-sapling stage covers a broad range of structural conditions from seedlings and brush < 10 cm tall to saplings up to 6 m tall. Therefore, we split the shrub-sapling and the two oldest WHR stages into pairs of substages to test for differences in wildlife species' abundance. Thus, for analysis we considered a total of seven seral stages (Table 21.1). Because all our study stands had canopy coverage >70% (considering both hardwood and coniferous tree species), we did not include the stages represented by <70% cover in our model tests.

Table 21.1. Characteristics of seral stages of Douglas-fir forest

Designation in present study	WHR code[a]	Description
Grass-forb	1	Annual and perennial grasses and forbs, with or without scattered shrubs and seedlings
Early shrub-sapling	2	Mixed or pure stands of shrubs and seedlings or saplings, usually <2 m tall
Late shrub-sapling	2	Mixed or pure stands of shrubs and seedlings or saplings, usually >2 m tall
Pole	3c	Trees ranging from 6 to 15 m tall with crowns 2–4 m in diameter; stand age >50 years; crown closure >70%
Sawtimber	3c	As above, but tree crowns 4–8 m in diameter and stand age 50–150 years
Mature	4c	Trees generally >15 m tall with crowns 8–15 m in diameter; stand age 150–250 years; crown closure >70%
Old-growth	4c	As above but crowns generally >12 m in diameter and age usually >250 years

[a] Pairs of stages with same code were not distinguished by the WHR matrix model.

Vertebrate sampling methods

We sampled birds in each of the seven habitat stages and amphibians, reptiles, and mammals in all but the grass-forb stage. Sampling the wide array of species required using a variety of sampling methods.

Variable-radius circular plots

We counted birds from each plot center, using the variable–radius circular-plot technique (Reynolds et al. 1980). Twelve 10-min counts were made from each plot center in each of two winters (1982, 1983) and one spring (1982). We estimated the distance to each bird seen or heard and used these distances to test for differences in detectability of birds among seral stages. With the exception of the grass-forb stage, birds were equally detectable among stages (Raphael and Marcot, pers. obs.). Therefore, we used the numbers of individuals tabulated over the 12 spring or 24 winter counts as estimates of relative abundance in each season. We reduced counts for the grass-forb stage by half to adjust for the higher detectability of birds in this type. Raphael (pers. obs.) found that the bird counts were linearly related to estimated densities. Thus, we consider our counts for each species as linear indices of their densities among stages.

Pitfall traps

We used pitfall arrays to capture small mammals (especially insectivores), reptiles, and salamanders. Each array, composed of 10 2-gallon plastic buckets buried flush with the ground and covered by plywood lids, was arranged in a 2 × 5 grid with 20-m spacing. We placed one array at each plot center and checked traps at weekly-to-monthly intervals from December 1981 (sawtimber, mature, old-growth; n = 27, 56, and 52 sites in each stage, respectively) or August 1982 (early shrub-sapling, late shrub-sapling, pole; n = 10 sites each) until October 1983. We recorded 9928 captures in 898,431 trap-nights; results for each species were expressed as captures/1000 trap-nights on each plot. We then calculated mean capture rates for all plots within each stage. Raphael and Rosenberg (1983) demonstrated that abundance estimates (capture rates) had stabilized by the time the traps were run for 15 months.

Drift fence arrays

Results of the 1981–1982 sampling demonstrated that our pitfalls were not effective for capturing snakes. Workers in the eastern United States found that drift fences, fitted with funnel traps, were quite effective for capturing snakes (Campbell and Christman 1982; Vogt and Hine 1982). Therefore, in May 1983 we installed drift fence arrays on 60 randomly selected plots (10/stage, except that eight mature sites and 12 old-growth sites were sampled). An array consisted of two 5-gallon buckets placed 7.6 m apart and connected by an aluminum fence 7.6 m long by 50 cm tall, with funnel traps 20 cm in diameter by 76 cm long, one on each side of the fence. These traps captured 890 animals, but only 15 snakes, in 5 months of operation. All capture data were added to those from the pitfalls for analysis, after we verified that doing so introduced no significant biases.

Track stations

We recorded tracks of squirrels and other larger mammals on each site, using a smoked aluminum plate baited with tuna pet food (Barrett 1983; Raphael and Barrett 1984). After a pilot study to determine optimum sampling duration (Raphael and Barrett 1981), we monitored each station for 8 days in August or September in 1982 and 1983. We sampled 20 stations in each of the three early stages and 81, 168, and 157 stations in the sawtimber, mature, and old-growth stages, respectively, and used the proportion of stations in each seral stage on which a species occurred as an index of species' abundance.

Livetrap grids

Because some mammal species avoid or escape from pitfalls, we established 27 livetrap grids (three in each of the earliest stages and five, seven, and six in the three later stages), each of which usually consisted of 100 25-cm Sherman livetraps arranged in a 10 × 10 grid with 20-m spacing. We used other grid sizes or shapes when the plot configuration would not contain the standard grid. Traps were checked each day for 5 days (based on pilot studies, Raphael and Barrett 1981) during July in 1981 (late stages only), 1982, and 1983 (all stages). Results for each species were expressed as mean number of captures/500 trap-nights.

Surface search

Some salamander species, especially those with small home ranges, are undersampled using pitfalls. To sample these species better we conducted time- and area-constrained searches (Bury and Raphael 1983) on a subset of sites in 1981 (late stages), 1982, and 1983 (all stages). A two-person team searched under all movable objects and within logs on three randomly located 0.04-ha circular subplots (fall 1981, 1982) or within a 1-ha area for 4 working hours (spring 1983). We sampled 20 sites in each of the early stages and 29, 39, and 48 sites in the three late stages. We captured 1636 animals; results were expressed as mean captures/site and were averaged within each stage.

Opportunistic observations

We recorded the presence of any vertebrates or any identifiable sign observed incidental to the above procedures. These observations were tallied to calculate frequency of occurrence of rarer species within each stage.

Data analysis

To compare model predictions and field observations, we first built a matrix that contained a row for each species, either predicted or observed, and columns indicating (1) codes for suitability levels of each habitat stage as provided by the WHR matrix (0 = absent, 1 = marginal, 2 = suitable, 3 = optimum); and (2) observed mean and standard deviation of abundance (count, capture rate, or frequency) in each stage derived from the most appropriate sampling method. We then compared the model predictions and observed values at four different levels: (1) species richness; (2) species similarity; (3) between-stage differences; and (4) within-stage differences.

Species richness

For this level of analysis, we simply compared the number of species predicted to occur in each stage (regardless of suitability rating) with the number we actually observed using a chi-square goodness-of-fit test. We also tallied the number of species observed in the field but not predicted to occur by the model, and the number predicted but not observed in each stage. We further recognized two types of errors, minor and severe. Minor errors included (1) cases where a species was absent in a stage, but its predicted suitability was "marginal" (e.g., western bluebird [*Sialia mexicana*] in the mature–old-growth stage during spring); (2) species with a limited geographic distribution within the sampled area (e.g., plain titmouse [*Parus inornatus*]); (3) species that occurred in specialized habitats such as streams or cliffs that were not sampled (e.g., river otter [*Lutra canadensis*] and canyon wren [*Catherpes mexicanus*]); and (4) species that could not be sampled reliably by our methods, including some of the wide-ranging mammalian carnivores and raptorial birds. Severe errors (i.e., errors that require model correction) were those where discrepancies could not be attributed to any of the preceding causes. For example, if a species was predicted to occur in a particular stage, was easily sampled, was not associated with a special habitat, and yet was not detected, we tallied a severe error.

Species similarity

Measures of species richness address neither differences of species composition nor differences of relative abundance. To evaluate similarities of amphibian, reptile, bird, and mammal communities among stages, we applied an agglomerative, hierarchical clustering algorithm to both our observations and the model patterns. To give each species equal weight, we divided the mean abundance of each species in each stage by the greatest mean for that species in any stage. Thus, we transformed the values for each species to a scale varying from 0 to 1.0. Our analysis of the model was based on the code values that varied from 0 (absent) to 3 (optimum). These values were also rescaled to vary from 0 to 1.0. We used the Bray-Curtis index (Bray and Curtis 1957) to compute the dissimilarity of each pair of stages based on the scaled abundance of each species among stages. The Lance-Williams Flexible Method (Lance and Williams 1967) was used to perform a cluster analysis on the similarity matrix calculated above. All calculations and the construction of dendrograms were performed using the program CLUSTER

(Keniston, unpublished, Marine Science Center, Oregon State University).

Between-stage comparisons

One major use of the WHR model is for comparison of wildlife populations among seral stages, such as the response of a species to conversion from an old-growth stand to a clearcut. The model distinguishes four levels of habitat suitability; thus 16 possible outcomes of any pairwise comparison of stages exist. To limit the number of predictions we examined, we simplified these to three: the population could remain the same, increase, or decrease. For these three predictions, there are also three results of field observations, yielding nine possible results of tests. For three of these, the model is correct (i.e., both remain the same, both decrease, both increase). The remaining six outcomes were tallied as errors (e.g., the model indicated increase, but field data showed no difference or a decrease). For most species, differences in abundance were tested using Student's t. For a sample of 10 species, we also evaluated use of a chi-square test to compare observed numbers of captures with expected values between pairs of stages. In general, the results were comparable to the t-tests except that two comparisons yielded significant differences under the chi-square model but not for the t-tests. Examination of these cases revealed that large counts on one or two plots were responsible and that the t-test results seemed more realistic.

Where abundance was estimated by frequency of occurrence (Caughley 1977:20), we computed Cohen's test based on differences between arcsine transformed proportions (Marascuilo and McSweeney 1977:147). We did not include in any tests species that were (1) too rare for analysis; (2) had limited distributions in the study area; or (3) were associated with specialized habitats. Table 21.2 provides an example of the determination of model error.

We estimated expected between-stage error rates under two different assumptions. First, we assumed that each of the four habitat-suitability ratings was assigned randomly to each stage. If each rating was assigned with equal probability, the proportions of predictions from pairwise comparisons would be 0.250 that the suitability rating would be the same (4 of 16), 0.375 that the suitability would increase, and 0.375 that it would decrease (6 of 16 for each). Second, we assumed that each stage was equally suitable, that is, we assigned the same rating for each stage for every species. Under this assumption, the probability would be 1.0 that any two stages would be rated the same for a species and 0 that the suitability rating would decrease or increase. For each of these assumptions, the proportions of species predicted to be the same, to increase, or to decrease between stages were used as marginal probabilities for the WHR model. We also tallied outcomes under each of the three predictions from the field data and calculated those marginal probabilities. To calculate the probability of each combination of model and field prediction, we multiplied the marginal probabilities of each outcome. For example, if the probability of the model predicting an increase was 0.375 and the probability of finding an increase for the field data was 0.189, the joint probability of both the model and field data predicting an increase would have been $(0.375)(0.189) = 0.071$. We filled all cells in the 3×3 matrix of outcomes in this manner and then summed the values on the diagonal (where both the model and field data make matching predictions) to calculate the expected proportion of correct predictions. These calculations were repeated for each of the two model assumptions.

Table 21.2. Example of process used to test predictions generated from the WHR matrix for the deer mouse (*Peromyscus maniculatus*) in the North Coast–Cascades zone of California

Stage	Habitat suitability[a]	Observed mean abundance (*SE*)[b]
Early shrub-sapling	Optimum	5.1 (1.1)
Late shrub-sapling	Optimum	3.1 (0.8)
Pole	Suitable	0.4 (0.1)
Sawtimber	Suitable	0.6 (0.1)
Mature	Marginal	1.0 (0.1)
Old-growth	Marginal	1.3 (0.1)

Contrast	Prediction[c]	Outcome[d]	Error?
Within-stage			
Early vs late shrub-sapling	Same	Different	Yes
Pole vs sawtimber	Same	Same	No
Mature vs old-growth	Same	Same	No
Between-stage			
Shrub-sapling vs pole-sawtimber	Decrease	Decrease	No
Shrub-sapling vs mature–old-growth	Decrease	Decrease	No
Pole-sawtimber vs mature–old-growth	Decrease	Increase	Yes

[a]From the WHR matrix (Marcot 1979).

[b]Based on captures of 1127 individuals during 898,431 pitfall trap-nights on 165 sites. Values are mean numbers of captures per 1000 trap-nights.

[c]Change in abundance between stages predicted by matrix ratings.

[d]Results of t-tests comparing observed mean abundance between stages or substages.

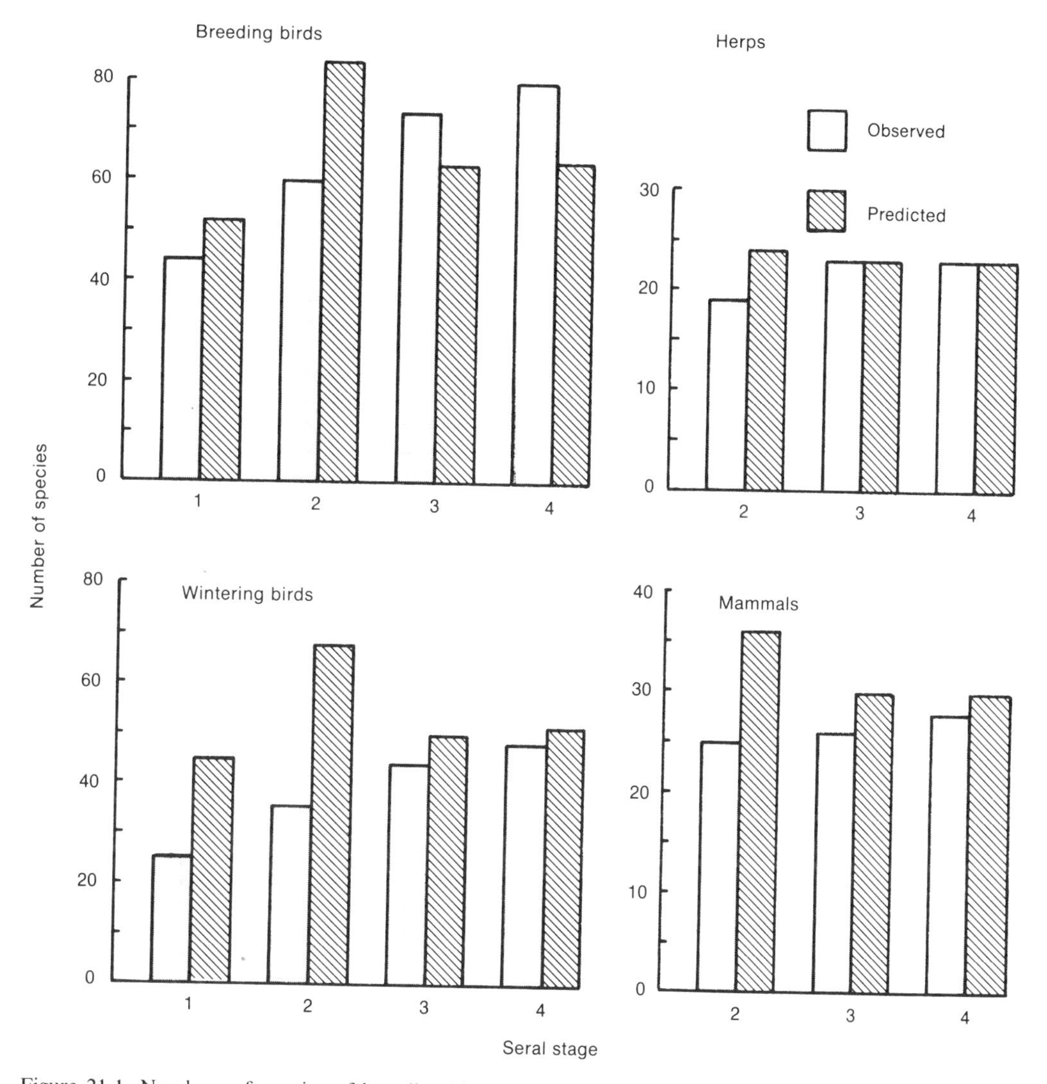

Figure 21.1. Numbers of species of breeding birds, wintering birds, reptiles and amphibians (herps), and mammals observed in the field and predicted by the WHR model in seral stages of Douglas-fir forest. See Table 21.1 for a description of each stage.

Within-stage comparisons

The WHR matrix is based on broadly defined seral stages. One source of model error may lie in an oversimplification of these stages, that is, species may respond differently to different substages. To examine this possibility, we tested for differences in species' abundance between two substages within each of three WHR stages. As above, differences in abundance were tested using Student's t or Cohen's test. Any significant difference ($P < 0.05$) of a species' abundance between substages of a given seral stage was tallied as an error because the model does not recognize substages (Table 21.2).

Results

Species richness

The numbers of breeding and wintering bird species observed in the four seral stages differed from the numbers predicted by the model ($P < 0.01$, Fig. 21.1). In both seasons, the greatest discrepancy occurred in the shrub-sapling stage. In the grass-forb and shrub-sapling stages, we observed fewer bird species than the model predicted. By contrast, the observed numbers of reptile, amphibian, and mammal species did not differ significantly from the numbers predicted among the three stages ($P > 0.50$, Fig. 21.1).

To better evaluate comparisons of species richness in each stage, we also tallied numbers of species observed but not predicted (ONP) and predicted but not observed (PNO) (Fig. 21.2). Overall, we evaluated 1322 combinations of species and stages; of these, 20% were in error. When minor errors were eliminated, the error rate was reduced to 12%. Over all stages and taxa, 11% of the species were PNO and 14% were ONP. Among taxa and stages, these rates varied from 3% (breeding birds PNO in mature–old-growth) to 35% (breeding birds ONP in grass-forb) (Fig. 21.2). PNO rates declined from early to late seral stage, whereas ONP rates were more constant (Fig. 21.2). Correct predictions (numbers both ob-

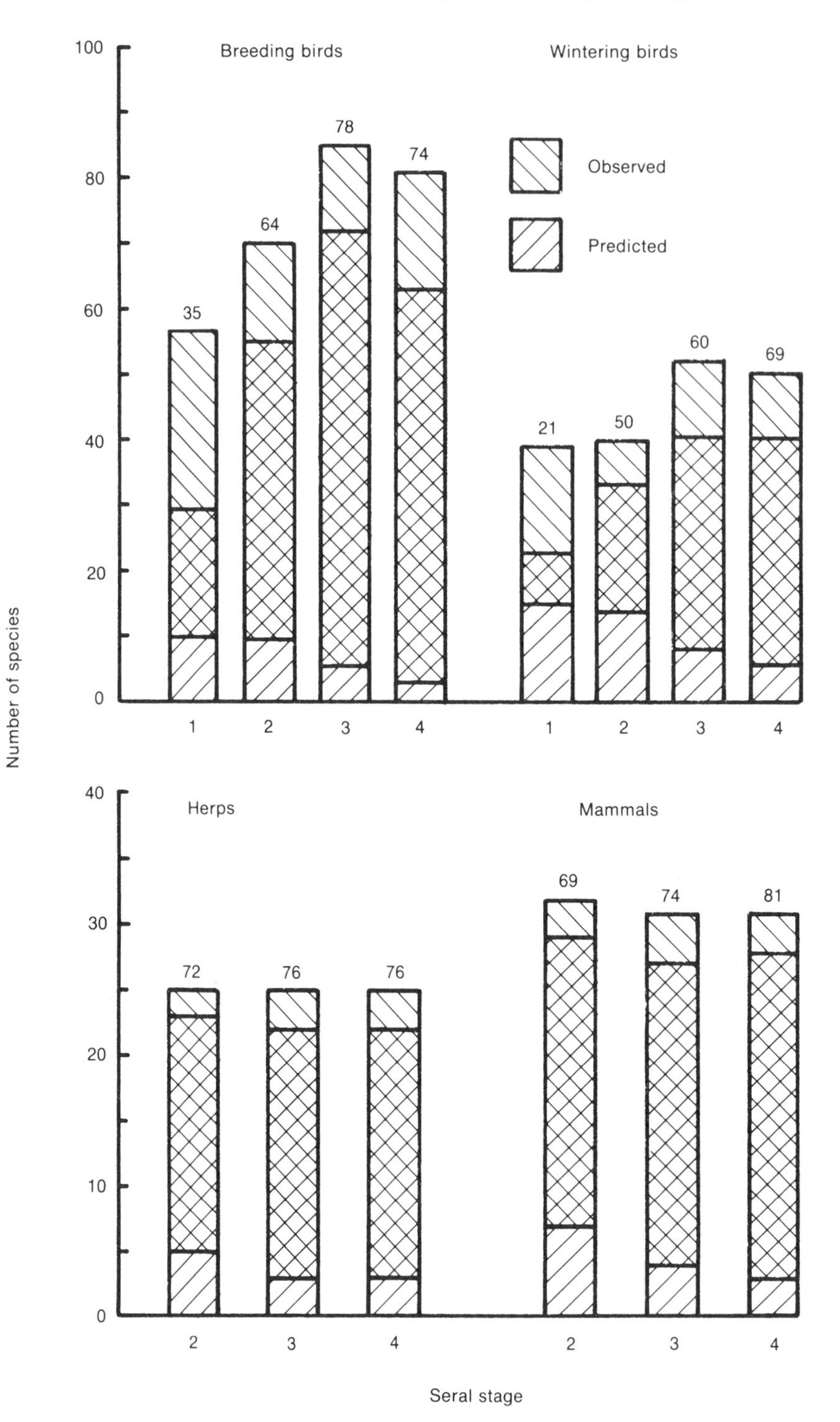

Figure 21.2. Numbers of all species of breeding birds, wintering birds, reptiles and amphibians (herps), and mammals that were either observed in the field but not predicted by the WHR model, predicted but not observed, or both predicted and observed (crosshatching) in seral stages of Douglas-fir forest. Numbers above histograms are the percentages of species that were both observed and predicted. See Table 21.1 for a description of each stage.

served and predicted) varied from a low of 21% for wintering birds in the grass-forb stage to 81% for mammals in mature–old-growth (Fig. 21.2).

BETWEEN-STAGE COMPARISONS

In general, the model often incorrectly predicted change of abundance for pairwise comparisons of stages (Table 21.3); overall, 43% of the 650 predictions were in error. The expected error rate, calculated under the assumption that suitability ratings were assigned randomly, was 70% (Table 21.4), a value significantly greater than that actually observed (Rohlf and Sokal 1981:156). However, if we assumed that the same suitability rating was assigned to each stage, the error rate would have been only 38%. The error rate was lowest (31%) between the two oldest stages and highest (52%) between the two youngest stages (Table 21.3).

Some errors should be considered more serious than others. For example, a land manager would be more concerned when a species actually decreased although the model predicted it to be equally abundant (e.g., brush mouse [*Peromyscus boylii*] in stage 3 vs stage 4) or to have increased (e.g., northern flicker [*Colaptes auratus*] in stage 2 vs stage 3). Overall, of 650 tests, only 61 outcomes (10%) were errors of these types (Table 21.3), a value close to the expected error rate of 11%, assuming random assignment of suitability ratings (Table 21.4). Assuming equal ratings for all stages, this rate would increase to 19%.

Table 21.3. Results of between-stage comparisons[a] of predicted and observed abundance estimates of vertebrates in seral stages of Douglas-fir forest, North Coast–Cascades Zone, California. (Values are number of species tallied under each outcome.)

Model prediction	Field observation			Total errors (%)
	Same	Decrease	Increase	
Grass-forb vs shrub-sapling[b] (*n* = 110 spp.)				
Same	37	6	18	52
Decrease	2	0	0	
Increase	30	1	16	
Grass-forb vs mature–old-growth[b] (*n* = 110 spp.)				
Same	34	3	1	44
Decrease	16	7	0	
Increase	26	2	21	
Shrub-sapling vs pole-sawtimber (*n* = 143 spp.)				
Same	37	16	7	46
Decrease	29	27	1	
Increase	11	2	13	
Shrub-sapling vs mature–old-growth (*n* = 143 spp.)				
Same	33	13	9	47
Decrease	25	28	5	
Increase	10	5	15	
Pole-sawtimber vs mature–old-growth (*n* = 144 spp.)				
Same	94	13	12	31
Decrease	4	2	2	
Increase	13	0	3	
All comparisons (*n* = 650 spp.)				
Same	236	51	47	43
Decrease	76	64	8	
Increase	90	10	68	

[a]No comparison was made for grass-forb vs pole-sawtimber.
[b]Comparisons include birds only.

Table 21.4. Expected probabilities of each combination of model prediction and field observation of vertebrate abundance among all seral stage comparisons, assuming random assignment of suitability values

Model prediction	Field observation			
	Same	Increase	Decrease	Totals[a]
Same	0.155[b]	0.047	0.048	0.250
Increase	0.232	0.071	0.072	0.375
Decrease	0.232	0.071	0.072	0.375
Totals[c]	0.618	0.189	0.192	1.000

[a]Proportions of 16 possible outcomes falling under prediction, assuming random assignment of suitability classes to each pair of seral stages.
[b]Each value is the product of the two associated marginal totals.
[c]Calculated from totals given in Table 21.3. Marginal totals may differ slightly from sums of cells because of rounding error.

Error rates were fairly consistent among taxa but were higher for mammals other than rodents in the two shrub-sapling comparisons and lower for rodents in the same comparisons (Table 21.5). We recorded no errors among reptiles and nonpasserine birds (spring) for the pole-sawtimber vs mature–old-growth comparison (Table 21.5).

The cluster classification (Fig. 21.3) also can be used to judge model performance, although at a much broader level. For example, model predictions indicated that the pattern of abundance of mammals in the shrub-sapling stages was relatively distinct from that in the other four stages. This general pattern was upheld by our observations (Fig. 21.3). We found similar results for birds in both seasons (Fig. 21.3); however, the stages were not grouped similarly for the herpetofauna. For most species groups, the model correctly predicted general patterns of community similarity among seral stages.

WITHIN-STAGE COMPARISONS

Among the 146 species that could be tested for differences in abundance between substages, 28% differed significantly within the shrub-sapling stage, 13% differed in the pole-sawtimber stage, and 9% differed in the mature–old-growth stage (Table 21.6). The greatest proportion of errors (42%) occurred with passerine birds in winter in the shrub-sapling stage, including hermit thrush (*Catharus guttatus*) and scrub jay (*Aphelocoma coerulescens*). Error rates for amphibians also were large compared to other taxa in the two older stages (Table 21.6). Examples included rough-skinned newt (*Taricha granulosa*) and western toad (*Bufo boreas*).

Cluster classification of stages based on similarity coefficients of species' abundance among substages rein-

Table 21.5. Summary of between-stage errors[a] among groups of species for a WHR matrix model, North Coast–Cascades Zone, California

Species group	No. spp. tested	Percent errors for each comparison: Grass-forb vs shrub-sapling	Grass-forb vs mature–old-growth	Shrub-sapling vs pole-sawtimber	Shrub-sapling vs mature–old-growth	Pole-sawtimber vs mature–old-growth
Amphibians	8	—[b]	—	38	63	50
Reptiles	5	—	—	40	40	0
Rodents	10	—	—	30	30	50
Other mammals	11	—	—	82	82	36
Passerine birds						
Spring	49	49	31	41	47	31
Winter	31	61	42	52	42	39
Nonpasserine birds						
Spring	19	47	53	42	42	0
Winter	11	45	73	45	36	27

[a]Errors were counted when differences in mean abundance as estimated from field samples did not match differences predicted by the WHR model (see Table 21.2).
[b]Amphibians, reptiles, and mammals were not sampled in the grass-forb stage.

forced these results (Fig. 21.3). In most cases, substages were grouped together. For observed amphibians and reptiles, however, the early shrub-sapling substage was more similar to the sawtimber substage than to the late shrub-sapling substage (Fig. 21.3).

Discussion

Species' abundance

Marcot et al. (1983) outlined 23 criteria for model validation. Our study explicitly addressed two of these, usefulness and reliability. A model is considered useful if at least some model predictions are empirically correct (Marcot et al. 1983). Clearly, such is the case for this model. Reliability, however, is a measure of the fraction of model predictions that are empirically correct. For the predictions we made, we found that the model was correct fairly often. Nonetheless, substantial revision will be necessary to improve model performance, particularly for between-stage comparisons. The overall error rate of 43% for such comparisons is quite high, given the potential importance of these models in land-use planning.

We recognized several limitations of our approach. First, we had no objective criteria by which to link absolute changes in abundance to changes in habitat suitability. Our tests were designed to measure statistically significant changes in abundance, but because statistical significance is a complex function of several factors (absolute difference between means, magnitude of variance, and sample size), we did not always know the biological significance of observed differences. Second, we treated all predictions with equal weight, regardless of the actual suitability ratings being compared. For example, we recorded a predicted decrease across two stages whether the suitability rating changed from optimum to suitable, optimum to marginal, or optimum to absent. Certainly the first of these is a more subtle change than the last. Unfortunately we had no way to relate the magnitude of abundance to the suitability ratings and, hence, could not distinguish among the various levels of increase or decrease.

Species' occurrence

Discrepancies in numbers of species observed vs predicted can be attributed in part to differences in sampling effort (number of plots sampled) among stages. In general, the greatest differences occurred in early stages (Fig. 21.1) where sampling effort was smallest. Differences in species richness primarily involved rarer species that required greater sampling effort. Furthermore, the number of species that were predicted to occur but not observed decreased from 36 in the less intensively sampled shrub-sapling stage (23 plots) to 15 in the intensively sampled mature–old-growth stage (109 plots). In contrast, numbers of species observed but not predicted increased from 11 to 16, respectively, in these two stages.

Because species predicted to occur but not observed may have been missed because of unknown sampling inadequacies, we are most concerned with those species observed but not predicted. For example, the Del Norte salamander (*Plethodon elongatus*) was predicted as absent in each sampled stage, yet it was abundant in all but the pole-sawtimber stages. In such cases, an adjustment of the matrix clearly is in order. Where we failed to observe predicted species, decisions regarding model adjustment are more difficult.

Identification of seral stages

Seral stages are intervals along a continuum of vegetative change. Our analysis of within-stage patterns of abundance was meant to evaluate the possibility of overly broad seral-stage definitions. Additional seral stages could be defined to increase model realism; because some species showed significantly different responses to substages, subdividing existing stages might be appropriate and worth the cost. The shrub-sapling stage, in which 28% of the tested species differed in abundance between substages (Table 21.2), definitely should be considered for subdivision. For ex-

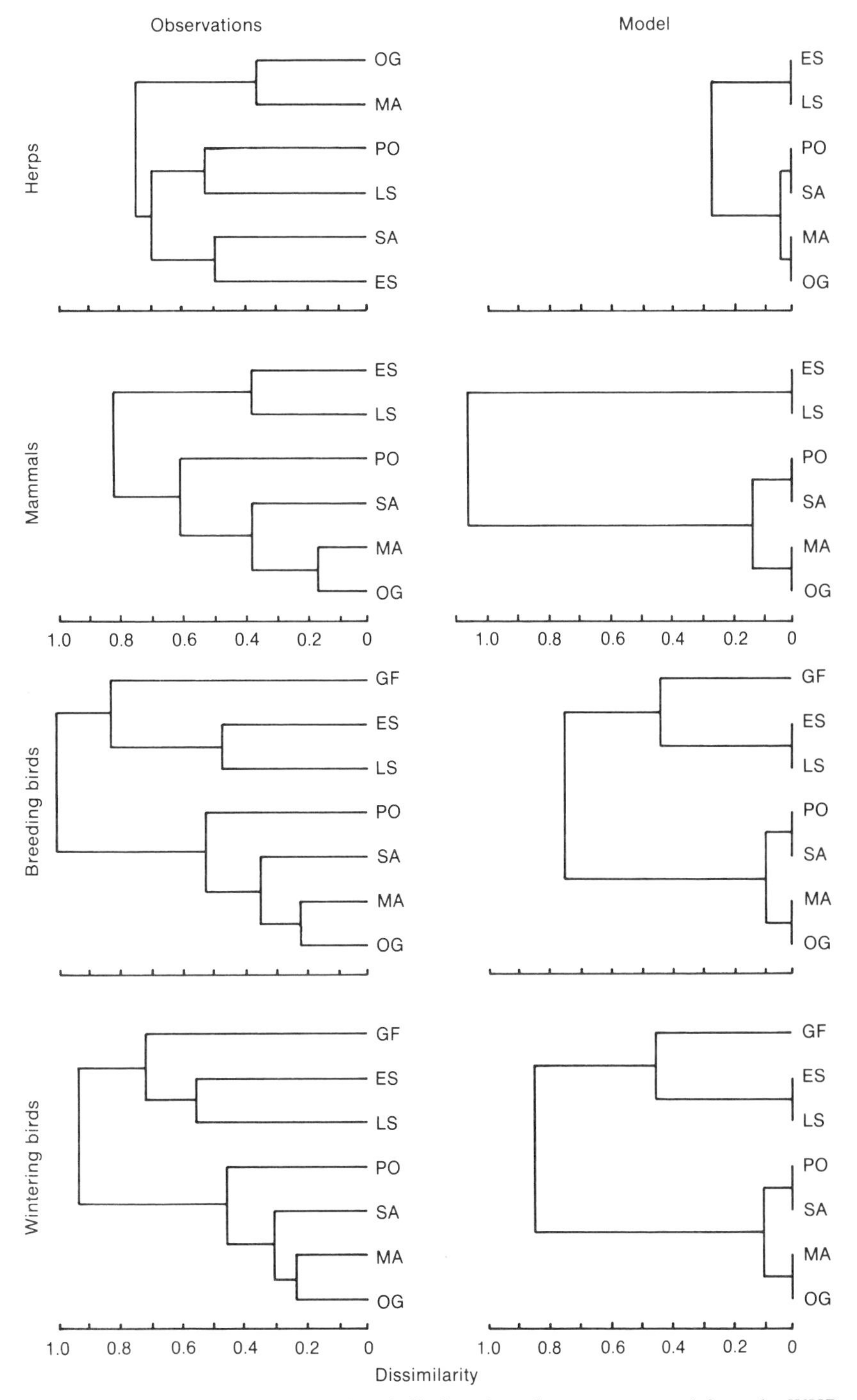

Figure 21.3. Dendrograms comparing similarity of seral stages computed from the WHR model (based on habitat-suitability codes for each species, *right*), and from the observed relative abundances in each stage (*left*) for reptiles and amphibians (herps), mammals, breeding birds, and wintering birds. Seral stage codes are: GF, grass-forb; ES, early shrub–sapling; LS, late shrub–sapling; PO, pole; SA, sawtimber; MA, mature; OG, old growth. These stages are described in Table 21.1.

Table 21.6. Percentage of species that differed in abundance between two substages within seral stages defined by the WHR matrix for the North Coast–Cascades Zone of California

	Seral stage[b]		
Species group[a]	Shrub-sapling	Pole-sawtimber	Mature–old-growth
Amphibians (8)	25	38	25
Reptiles (5)	20	0	0
Rodents (11)	27	0	0
Other mammals (11)	0	27	18
Passerine birds			
Spring (50)	28	14	12
Winter (31)	42	16	6
Other birds			
Spring (19)	32	5	5
Winter (11)	18	0	0
All species (146)	28	13	9

[a]Values in parentheses are the numbers of species tested.
[b]See Table 21.1 for description of stages.

ample, the sagebrush lizard (*Sceloporus graciosus*) was four times as abundant in early vs late shrub stages. Because many species responded differently to substages of each WHR stage, we suggest that the WHR model should be redesigned to incorporate substages.

Scale problems

WHR models are more reliably applied over large geographic areas (watershed or larger). Because of structural and compositional heterogeneity typical of mixed-evergreen forests of California and the large home-range size of many wildlife species, WHR models cannot always be expected to perform well at the single project (e.g., timber sale) level. Thus, at this phase of WHR model development, we suggest that predicted changes in wildlife communities resulting from timber management are best analyzed by combining individual timber sales or other treatments over at least the watershed level.

It is apparent from our analysis that this WHR model should be used with caution. Managers need to realize that model predictions are associated with a rather large degree of uncertainty. The model performed well when predicting the occurrence of species, but did not predict changes in species' abundance between stages with any more accuracy than if all stages had been rated equally suitable. It is true that results of our field study can be used to update and improve the model, but we studied only one of the model's 18 habitat types. The field work and subsequent analyses for our project cost about $600,000. It is unlikely that such extensive validation studies will be repeated for many other habitat types, so managers will continue to rely on untested models. Our estimates of model error can probably be extended to the WHR models for most coniferous habitats of the North Coast–Cascades. Until these models can be improved, the challenge will be to recognize model uncertainty and somehow to incorporate this uncertainty into model applications.

Acknowledgments

Field work was supported by the Pacific Southwest Region and by the Pacific Southwest Forest and Range Experiment Station of the USDA Forest Service, and by the University of California, Department of Forestry and Resource Management, AES project 3501–MS. We particularly wish to thank all of our field assistants for their (sometimes heroic) efforts. Our analysis benefitted from discussions with B. R. Noon and D. A. Sharpnack. We also thank R. H. Barrett, E. C. Meslow, C. J. Ralph, H. Salwasser, and J. Verner for institutional support; D. A. Airola, R. G. Anthony, K. H. Berry, D. M. Finch, W. F. Laudenslayer, Jr., C. J. Ralph, and J. Rice for comments on the manuscript; and L. J. Kelly and M. L. Armstrong for their patience in typing it.

22

A Management Strategy for Habitat Diversity: Using Models of Wildlife-Habitat Relationships

EDWARD F. TOTH, DAVID M. SOLIS, and BRUCE G. MARCOT

Abstract.—We describe a management strategy for habitat diversity to meet the habitat requirements for the majority of wildlife species in a mixed forest of Douglas-fir (*Pseudotsuga menziesii*) and hardwoods. We used a California Wildlife Habitat Relationships Program model to assess three parameters associated with diversity of wildlife habitat: presence of three special habitat components (dead-and-down wood, snags, and hardwoods), habitat patch size, and spatial configuration of seral stages. To accommodate most species (1) special habitat components should be maintained in all habitat stages; (2) areas of individual patches of shrub-seedling-sapling and forested stages should be 5–10 ha (12–25 acres) and 16–24 ha (40–60 acres), respectively; and (3) shrubfields or open forest stands should be juxtaposed with dense, mature forest stands.

The general problems of habitat diversity and patterns of habitat patch configuration directly relate to habitat fragmentation and to the isolation of habitat patches in forest ecosystems (Whitcomb et al. 1981). Development of specific-criteria patterns of habitat diversity helps avoid the creation of widespread, monotypic timber stands and the excessive fragmentation or isolation of habitat types, which may cause extinction of species locally (see Samson 1980). The problems of habitat fragmentation and isolation are addressed in insular biogeography. Some principles of island biogeography may be applicable to the question of configuration of habitat patches (Goeden 1979; McCoy 1982).

Individual components of habitat diversity have been empirically and theoretically studied by Kelker (1964), Galli et al. (1976), Rudis and Ek (1981), and others. Thomas et al. (1979) described habitat diversity as consisting of horizontal and vertical components. Horizontal components included the size, shape, composition, and relative spatial arrangement of habitat patches. Vertical components included the number, relative density, composition, and absolute height of different vegetation layers.

The National Forest Management Act (NFMA) of 1976 provided strong and explicit direction to manage for habitat diversity on National Forest lands. Specifically, the act mandated National Forests to maintain viable populations of existing native and desired non-native vertebrate species. Further, the subsequent 1982 federal regulations for implementation of the act (36 CFR Part 219, Section 219.19) stated that habitats should be "well distributed." However, the regulations do not provide direction or guidelines for this distribution of habitat. Thus, it is the responsibility of National Forests to further define what constitutes habitat diversity and to develop quantitative and qualitative methods for its assessment.

The Six Rivers National Forest (SRNF) is a major producer of commercial timber in California. Land-management activities on SRNF have changed habitat characteristics on approximately 60,000 ha (150,000 acres) or 21% of the forested area within the last 30 years. Habitat fragmentation has increased, and the abundance of some forest components (snags and fallen dead-and-down woody material) has decreased as a result. Thus, land managers on SRNF had an immediate need for a strategy for habitat management which would temper the effects of these trends and contribute to the long-term maintenance of the distribution and viability of wildlife populations.

However, the development of any habitat-management strategy requires a comprehensive understanding of the habitat requirements of wildlife species in an area. Detailed knowledge of wildlife-habitat requirements and species-specific habitat-management models exist for few species. The wildlife-habitat-relationships (WHR) models (Thomas et al. 1979; Marcot 1979; Verner and Boss 1980) presently available provide land managers with a conceptual framework for the development of habitat-management strategies.

WHR models depict relationships between wildlife (amphibians, reptiles, birds, and mammals) and their associated habitats (vegetation types and successional stages) and are based both on literature review and expert opinion (Salwasser et al. 1980, 1982; Raphael and Marcot, Chapter 21). WHR models also summarize biological information (e.g., home

EDWARD F. TOTH: USDA Forest Service, Six Rivers National Forest, 507 F Street, Eureka, California 95501. *Present address:* USDA Forest Service, Forestry Support Program, Room 1208-RPE, Box 2417, Washington, D.C. 20013

DAVID M. SOLIS: Orleans Ranger District, USDA Forest Service, Drawer B, Orleans, California 95556. *Present address:* USDA Forest Service, Six Rivers National Forest, 507 F Street, Eureka, California 95501

BRUCE G. MARCOT: Cooperative Wildlife Research Unit, Department of Fisheries and Wildlife, Oregon State University, Corvallis, Oregon 97331. *Present address:* USDA Forest Service, Fish and Wildlife Staff, 319 SW Pine Street, Box 3623, Portland, Oregon 97208

range sizes, diet, use of breeding and foraging substrates) of wildlife and illustrate trade-offs associated with management options for habitat alteration.

Our objective is to show how an intensive analysis of a WHR model provides the preliminary step in the development of a generalized management approach for habitat diversity specific to the SRNF. To assure maintenance of a semblance of the naturally occurring forest habitat mosaic and viable populations of the majority of wildlife species, we assessed a WHR model for the size, configuration, and composition of habitat patches.

Study area and methods

The SRNF extends approximately 224 km (140 miles) south from the Oregon-California border and contains about 382,000 ha (956,000 acres) in the Coast Range and Klamath mountains. Although SRNF contains a number of forest vegetation types, our analysis focused on the mixed forest of Douglas-fir (*Pseudotsuga menziesii*) and hardwood that occurs at 600–1200 m (2000–4000 feet) and that constitutes 70% of the forest land area. The majority of forest-management activities on SRNF occur in the Douglas-fir–hardwood forest. Habitat patch sizes typically vary from 24 to 60 ha (60–150 acres). Stands are vegetatively complex with hardwoods, snags, and down woody material contributing to structural characteristics of forest stands.

We used information derived from the North Coast–Cascades Zone of the California Wildlife Habitat Relationships Program of the USDA Forest Service (Marcot 1979, 1980; Salwasser et al. 1980). The WHR model recognizes eight habitat conditions in the Douglas-fir–hardwood forest type. Four seral stages are described: grass-forb (seral stage 1), shrub-seedling-sapling (seral stage 2), pole-medium tree (seral stage 3), and mature–old-growth tree (seral stage 4). The last two stages, in turn, are divided into three structural types: 0–39% (cover class A), 40–69% (cover class B), and 70–100% (cover class C) total overstory canopy closure (Marcot 1979). The WHR model describes the use of these eight habitat conditions for breeding and feeding by each wildlife species on an ordinal, four-class scale: optimal, suboptimal, marginal, and absent. Although information contained in the WHR model is based on studies conducted over a wide geographic area, we assumed that the model could be used to give a preliminary assessment of wildlife species richness associated with specific sizes, configurations, and compositions of habitat patches on the SRNF. Further field study would be needed to validate our conclusions.

The Wildlife Habitat Inventory Matrix Program (WHIMP) is a computerized data storage and retrieval system used with a Tektronic 4051 minicomputer to access the WHR data base (Marcot 1979, 1980). WHIMP produces a list of species which meet user-specified habitat and/or niche characteristics.

The influence of habitat patch size on wildlife species richness was assessed by querying the number of wildlife species with a specified mean or maximum breeding territory or home range that breed at optimal or suboptimal levels in each habitat condition (Fig. 22.1). We assumed that wildlife species would occur in patches of habitat whose areas were equal to or greater than the specified mean or maximum territory or home range values. If territory or home range needs of a species were met by habitat patch size, the species would occupy the habitat patch. The WHR data base does not (1) distinguish between core area and minimum home range; (2) address the heterogeneity and edge values of habitat patches; (3) consider the effects of intra- or interspecific competition or stochastic environmental events; or (4) indicate the minimum number of breeding pairs or the total area of habitat needed to maintain viable populations of wildlife.

The influence of the configuration of habitat stages on wildlife species richness was assessed by querying which of the 304 wildlife species on SRNF use a combination of two or more habitat conditions to meet their breeding and feeding requirements. We specified in the model query that species must breed at optimal or suboptimal levels in a habitat condition different from that used for feeding at optimal or suboptimal levels. We excluded the grass-forb stage from the analysis because of its relatively short duration in a managed mixed Douglas-fir–hardwood forest. By manually juxtaposing the identified combinations of habitat conditions until all species were included, we derived a configuration of habitats to accommodate the requirements of all species that use different habitat conditions for breeding and feeding.

The evaluation of the effects of the structural composition of habitats on wildlife species richness focused on the presence of three habitat components commonly recognized as being heavily influenced by intensive timber management: dead-and-down wood, snags, and hardwoods (trees and shrubs). We queried the WHR model to determine which wildlife species use each of the special habitat components as breeding or feeding substrates in each habitat condition.

Results

Analysis of habitat diversity

Habitat size

Cumulative numbers of wildlife species increased as a function of greater area (Fig. 22.2). The resulting species-area curves show potential increases in species numbers in relation to increases in patch size for each habitat condition. The inflection points indicate minimal habitat areas that accommodate at least one breeding pair of most species. Minimal habitat areas varied by seral stage and were larger for both pole-medium tree (16–20 ha, 40–50 acres) and mature–old-growth tree (20–24 ha, 50–60 acres) stages than for grass-forb (3–6 ha, 7–15 acres) and shrub-seedling-sapling (5–10 ha, 12–25 acres) stages. In all stages except grass-forb, species-area curves increased slightly beyond 100 ha (250 acres) because of the addition of a few species of predators with large home ranges.

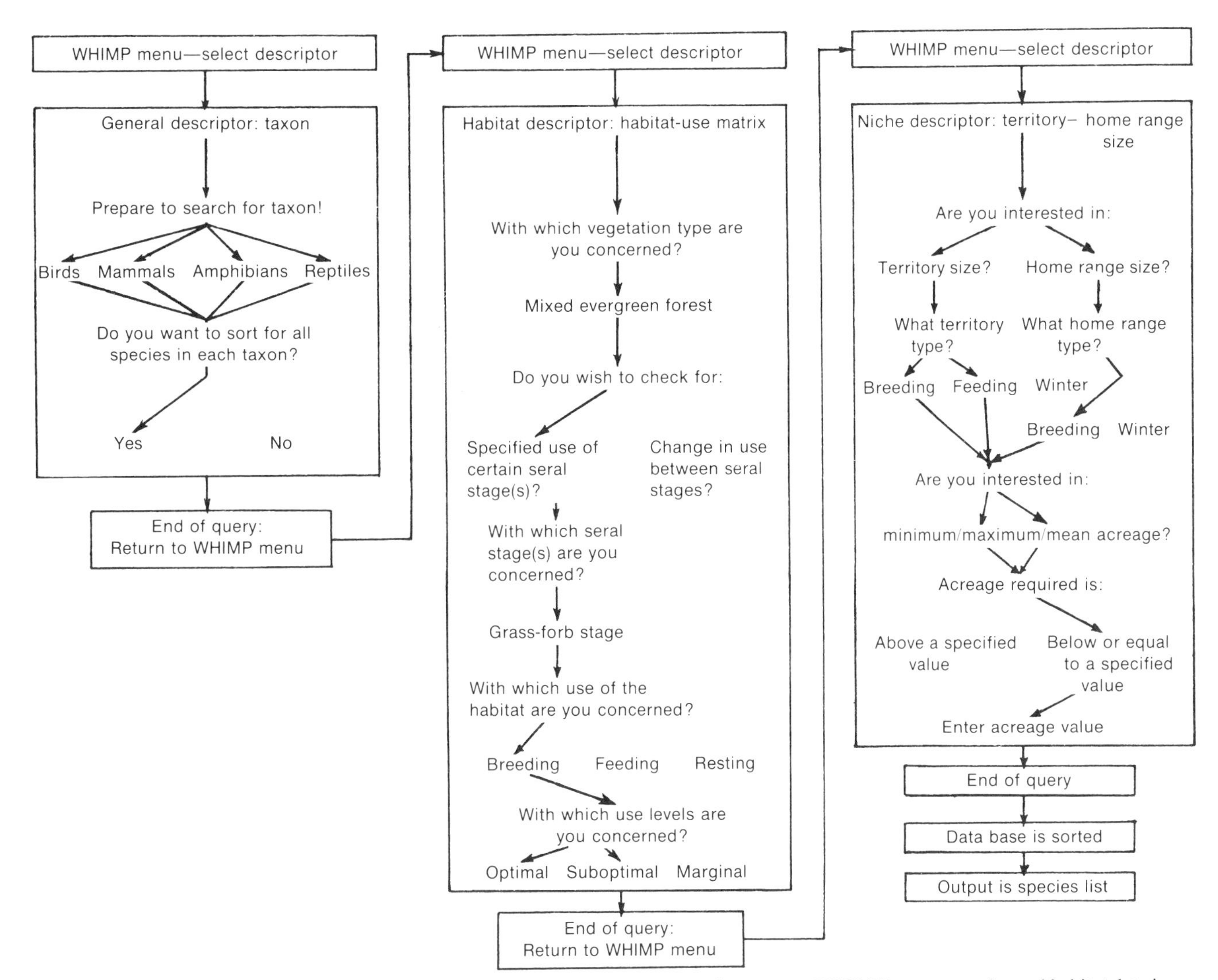

Figure 22.1. Logic flow of a query used with the Wildlife Habitat Inventory Matrix Program (WHIMP) to sort species and habitat data in the Wildlife Habitat Relationships (WHR) model.

Configuration of habitat patches

We identified 16 species of wildlife (6 bird and 10 mammal species) with multihabitat requirements for breeding and feeding (Table 22.1). Although the list represents a small percentage (5%, n = 16 of 304) of the total number of wildlife species occurring on the SRNF, it includes key game species and nongame species whose life histories are poorly understood (e.g., several species of bats).

Our configuration of habitat patches (Fig. 22.3) illustrates one pattern of habitat conditions to accommodate the needs of species with multihabitat requirements for breeding and feeding. In general, species richness increased with increased edge contrast, such as closed-canopy mature–old-growth stage adjacent to shrubfields or open-canopy forested stages. A high percentage (63%, n = 10 of 16) of species requiring two habitats used closed-canopy forested stages for breeding and early seral or open-canopy forested stages for foraging.

Special habitat components

The presence of habitat components positively influenced wildlife species richness in all habitat conditions. The potential increase in wildlife species richness because of the presence of special habitat components (dead-and-down woody material, snags, and hardwood trees and shrubs) varied by

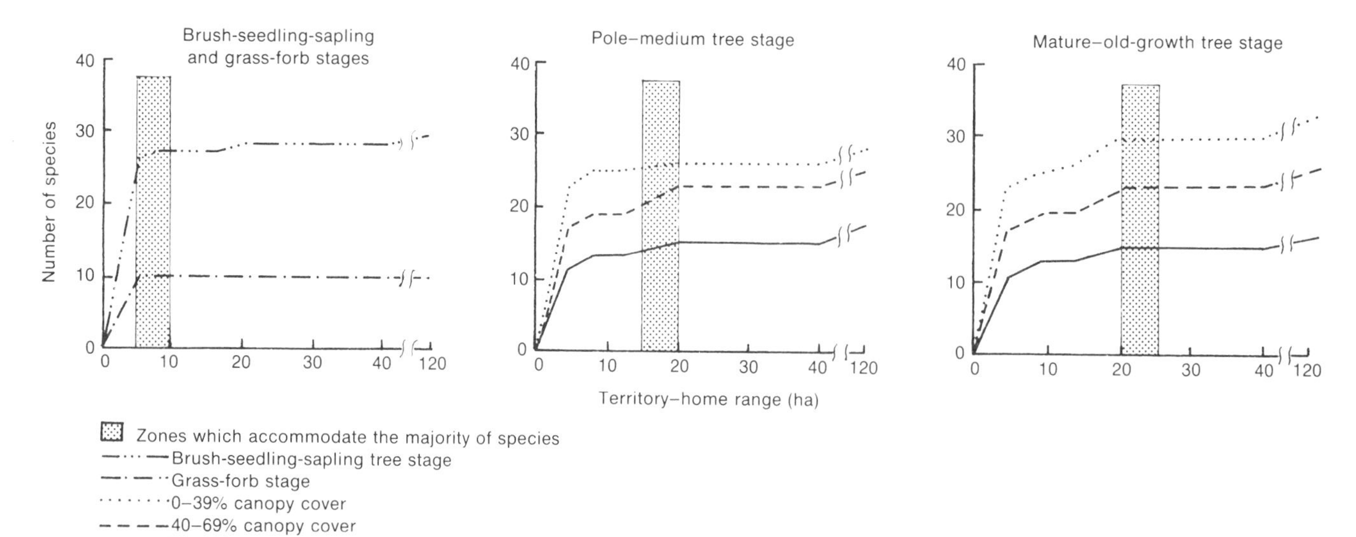

Figure 22.2. Cumulative wildlife species richness by seral and structural stage of mixed Douglas-fir–hardwood forest as a function of territory–home range size.

seral stage, and ranged from 2% to 33% (Fig. 22.4). Dead-and-down woody material generally provided breeding and feeding substrates for a greater number of species in each habitat, followed by snags and hardwoods in decreasing levels of importance. We were unable to assess the degree of species' dependence on habitat components because the information is not available in the WHR model. We expect that species with rigid habitat requirements would be most heavily influenced by the presence or absence of a particular special habitat component. Other species with the flexibility to use alternative breeding or feeding substrates would not be affected as much by the absence of one or all three habitat components.

Discussion

Collectively, natural forces influence the composition of wildlife communities. On the SRNF, fire, topographic diversity, geologic activity, and soil type have produced a mosaic of habitats which differ in age, structure, configuration, and patch size. Our generalized habitat-management strategy would maintain a semblance of the natural, unmanaged

Table 22.1. Species in Douglas-fir–hardwood forest with multihabitat requirements for breeding and feeding. (This list was derived by querying a wildlife-habitat relationships [WHR] model of the North Coast–Cascades Zone of California.)

Species	Habitat conditions for breeding and foraging requirements[a]
Red-tailed hawk (*Buteo jamaicensis*)	2/3B, 2/4B, 3A/4B
American kestrel (*Falco sparverius*)	2/3B, 2/4A, 2/4B, 2/4C, 3A/4B, 3A/4C
Blue grouse (*Dendragapus obscurus*)	4A/4C
Band-tailed pigeon (*Columba fasciata*)	2/4C, 4C/riparian
Long-eared owl (*Asio otus*)	2/3B, 2/4A
Common raven (*Corvus corax*)	3C/riparian, 4C/riparian
Little brown bat (*Myotis lucifugus*)	2/4C, 3A/4C, 4A/4C
Yuma myotis (*Myotis yumanensis*)	2/4C, 3A/4C, 4A/4C
Long-legged myotis (*Myotis volans*)	3A/4B, 3A/4C, 4A/4C
Silver-haired bat (*Lasionycteris noctivagans*)	3A/4B, 3A/4C, 3C/4C, 4A/4C
Big brown bat (*Eptesicus fuscus*)	3A/4B, 3A/4C, 3C/4C, 4A/4C
Pallid bat (*Antrozous pallidus*)	3A/4C, 4A/4C
Black bear (*Ursus americanus*)	2/4C, 3A/4C, 3B/3C, 3C/riparian
Raccoon (*Procyon lotor*)	2/4A, 2/4B, 2/4C
Elk (*Cervus elaphus*)	2/3A, 2/3B, 2/4B
Mule deer (*Odocoileus hemionus*)	3B/4A, 3B/4B

[a] See Figure 22.4 for description of habitat codes (seral stages and canopy cover classes).

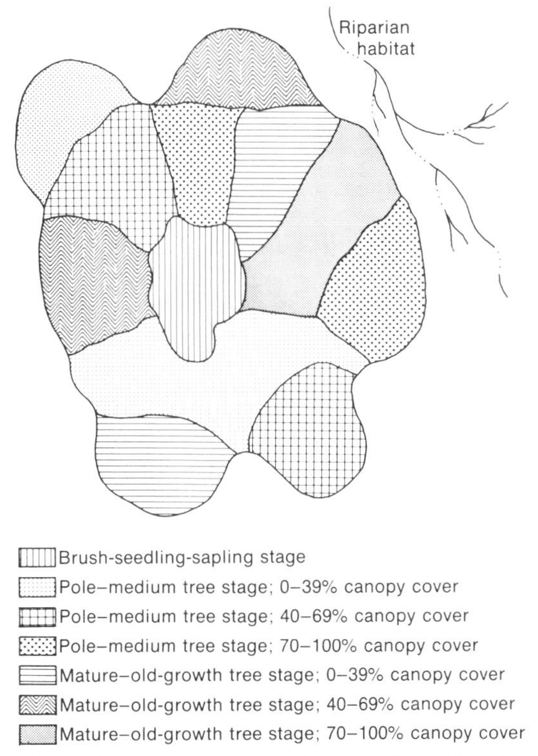

Figure 22.3. Mosaic pattern for horizontal habitat diversity. The configuration (juxtaposition) and composition of habitat patches would meet the needs of wildlife using multiple patches of habitat in a mixed Douglas-fir–hardwood forest.

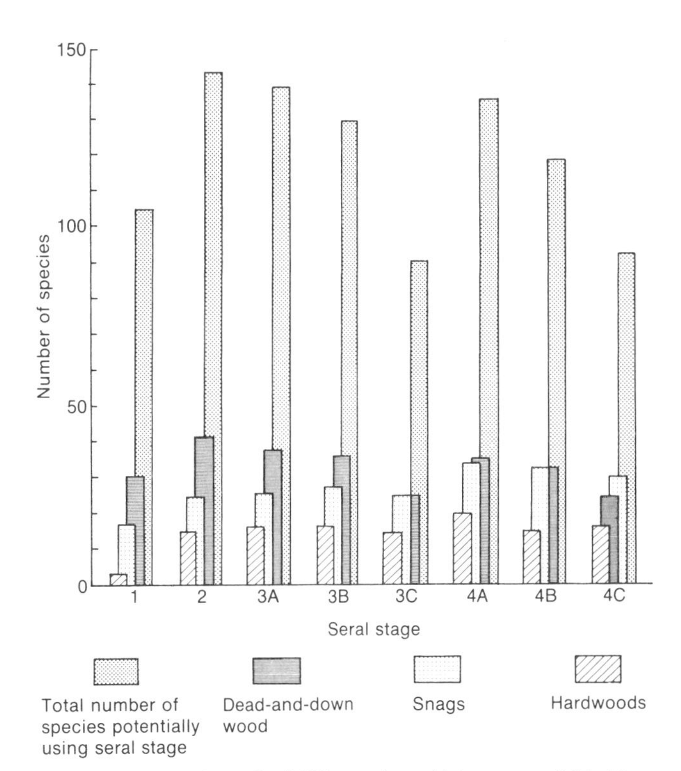

Figure 22.4. Number of wildlife species which use special habitat components (snags, hardwoods, or dead-and-down wood) for breeding or feeding at optimal or suboptimal use levels (hardwoods analyzed for breeding only) by seral or structural stage of a mixed Douglas-fir–hardwood forest (seral stages: 1 = grass-forb, 2 = brush-seedling-sapling, 3 = pole–medium tree, and 4 = mature–old-growth tree; cover classes: A = 0–39%, B = 40–69%, and C ≥ 70% [Marcot 1979]).

habitat characteristics specific to the Douglas-fir–hardwood forest on the SRNF.

Although our estimates of minimal habitat area of forested stands (16–24 ha, 40–60 acres) were slightly lower than those reported by Galli et al. (1976) in eastern deciduous forests 24–34 ha, 60–84 acres) and the theoretical estimate in eastern Oregon of Thomas et al. (1979) (30–40 ha, 75–100 acres), any similarities would probably be fortuitous, owing to differences in data, study area, or technique of estimation.

Our analyses may underestimate minimum patch size needs. Habitat patch sizes greater than those in our study may be required to ensure that species have a high enough probability of finding the correct proportions of their habitat needs in a typical habitat stage. However, the alternative of not setting management criteria for minimum patch size could lead to habitat homogeneity (decreasing alpha diversity) under the guise of increased edge effect (increasing beta diversity). Increased habitat fragmentation could produce such small and intermixed habitat patches that they approach a sameness (Thomas et al. 1979).

Although wildlife species richness apparently would benefit by combining the results on special habitat components, patch size, and configuration of patches, a lack of qualitative and quantitative data (e.g., wildlife response to patch size, their ability to use alternate habitats or combinations of habitats, and their reliance on special habitat components) in the WHR model precluded detailed levels of sensitivity analyses. Additional research is needed on travel corridors, dispersal requirements, and population responses to different configurations of habitat patches (e.g., Wiens 1976). For example, the WHR model failed to identify 11 additional species of wildlife suspected to have multihabitat requirements for breeding and feeding: northern goshawk (*Accipiter gentilis*), ruffed grouse (*Bonasa umbellus*), California quail (*Callipepla californica*), mountain quail (*Oreortyx pictus*), great horned owl (*Bubo virginianus*), olive-sided flycatcher (*Contopus borealis*), dark-eyed junco (*Junco hyemalis*), white-crowned sparrow (*Zonotrichia leucophrys*), fringe-tailed bat (*Myotis thysanodes*), California myotis (*M. californicus*), and red bat (*Lasiurus borealis*). Raphael and Marcot (Chapter 21) are now validating the WHR data base. Their results will increase our understanding of wildlife-habitat interactions specific to the North Coast–Cascades Zone of the WHR model and will be incorporated into the model to reassess our findings and to refine our management strategy for habitat diversity. We expect that our estimates of size and configuration of habitat patches will be apt to vary by geographic area and vegetation type. Factors aside from the quality and structure of habitats—such as social structure of populations, competition, and predation—also influence habitat use by wildlife. Again, the WHR model lacks this kind of information.

A scientifically based general habitat-management strategy is highly important to a wildlife-management program on a national forest. It (1) provides habitat requirements for most species; and (2) allows land managers to focus their attention on species- or habitat-specific management concerns. The latter would include species with special habitat requirements or of special management concern (e.g., game, threatened, or endangered species) or that require an intensity of management unlikely to occur under general habitat management.

The types and patterns of habitat stages we suggest can be translated to timber-harvest scheduling to help develop interdisciplinary goals for optimal layout patterns (e.g., see Salwasser and Tappeiner 1981; Mealey et al. 1982). The timing and type of habitat modification and specific objectives of wildlife management should be considered both spatially and temporally. The suggested size and configuration of habitat patches could be modified to meet habitat needs of species that require larger patch sizes or specific habitat configurations. However, the primary objective could still be to increase species richness.

The results of our analyses were used on the SRNF to develop management standards for the retention of special habitat components through time and guidelines for kinds, sizes, distributions, and amounts of habitat patches through space. Forest biologists currently use these standards and guidelines in conjunction with tabular and mapped data on habitat distribution to design forestwide and site-specific projects, and to assess the impacts of such projects on wildlife. Biologists also use the standards and guidelines in the management of species of special interest (e.g., spotted owls, *Strix occidentalis*—Solis and Toth, pers. obs.).

However, the application of our results required a translation of vegetation descriptors of the Timber Stand Classification System on the SRNF into habitat codes which were compatible with descriptions of habitat conditions in the WHR model. We used a modified version of the Wildland Resource Information System (WRIS; Russell et al. 1975; R. Zane, pers. comm.) to store the habitat codes for all mapped habitats (36,000 WRIS mapping units) on the forest. The data base also included 47 other resource attributes, thus allowing us to query and map wildlife-habitat conditions, alone or with other resource data.

Acknowledgments

We thank Jake Rice, Bill Laudenslayer, C. J. Ralph, and Patricia Manley for helpful reviews and comments on the draft of this manuscript.

23

Use of Expert Systems in Wildlife-Habitat Modeling

BRUCE G. MARCOT

Abstract.—The next phase of modeling wildlife-habitat relationships may involve use of expert systems technology. An expert system is a computer program that reasons like a human expert to solve such problems as diagnosis and classification. A wildlife-habitat expert system may predict response of wildlife species to habitat conditions and changes and may use probabilistic structures of program control and computation. Two demonstration programs (BRUSH and GUILD) illustrate such predictions by using "data-driven" reasoning, program explanations of its reasoning process, and capabilities for suggesting habitat-management prescriptions. Existing computer programs that help create and maintain expert systems may reduce model development time by an order of magnitude. Questions of utility, audience, cost, peer review, and validation require careful scrutiny. Other factors may include deciding which human experts to model the reasoning process after and selecting appropriate problems for expert system formalism.

Predicting the response of wildlife species to habitat conditions (vegetation type and successional stage) and changes in conditions is the goal of many wildlife managers and land-use planners. Currently, data storage and retrieval systems (e.g., Patton 1978; Marcot 1980; Verner 1980) and models that index habitat capability, such as those of the U.S. Fish and Wildlife Service (Schamberger et al. 1982) and USDA Forest Service (Hurley et al. 1982), are being developed and used.

A next generation of predictive modeling may extend these approaches by using computers to integrate information from field studies, the literature, and expert opinion. Computer software engineering that would aid in predicting the response of species to habitat conditions is known as expert systems engineering. An expert system is a computer-based consultation program consisting of facts and expert knowledge to help classify, diagnose, or plan (Duda and Shortliffe 1983). The use of currently available data storage and retrieval systems and habitat-capability models requires the user to ask all pertinent questions and develop lines of reasoning, whereas with expert systems the computer conducts much of the querying and reasoning by using built-in rules. In this chapter I review expert systems technology, present two demonstration models, discuss expert performance and the use of expert systems to build knowledge bases, and present guidelines and cautions for the possible next generation of wildlife-habitat models.

Expert systems technology: A brief review

An expert system is a computer program that uses facts and "if-then" choices or rules to solve a problem in technology or management. Specifically, expert systems may be used in problems of classification, such as classifying chematographic profiles or habitat conditions, or diagnosis, such as diagnosing patients' symptoms or habitat suitability for wildlife species. Hundreds of such rules may be combined into what is termed "rule networks." The expert system transcends traditional data storage and retrieval programs in its ability to keep track of its own reasoning process, to handle uncertainty and rules of thumb in computations, and to revise its own data base and logic structure from experience. Expert systems aimed at practical application often reach performance levels comparable to those of a human expert in some specialized problem domains (Nau 1983).

Expert system programs have been used to assist in the diagnosis of medical symptoms (MYCIN; Shortliffe 1976), in the interpretation of mass spectroscopy (DENDRAL; Lindsay et al. 1980), and in the design of computer hardware configuration (R1; McDermott 1982). A consulting system for mineral exploration, PROSPECTOR (Duda and Shortliffe 1983), has successfully located new mineral deposits. Duda and Shortliffe (1983), Hayes-Roth et al. (1983), and Nau (1983) have reviewed other system applications. None of the existing applications, however, has addressed problems in ecology or wildlife management.

An expert system consists of two integrated parts: a knowledge base and a logic control structure, sometimes referred to as an "inference engine" (Brachman et al. 1983). A knowledge base is a coded list of fundamental facts and a set of rules for using the facts under different contexts. Facts may be represented as relations, such as "Natal roosts of hoary bats = Dense tree foliage." Rules may be represented as if-then syllogisms, such as shown in Figure 23.1. (Note the use of a probability statement in line 3850, rule R11.)

The inference engine is a set of controls consisting of general problem-solving knowledge (Buchanan et al. 1983). An inference engine is essentially the logic structure of program execution. One example of a high-level control rule may be to execute a particular subset of facts and rules that pertain to deducing species' use of deciduous foliage in a forest

BRUCE G. MARCOT: Cooperative Wildlife Research Unit, Department of Fisheries and Wildlife, Oregon State University, Corvallis, Oregon 97331. *Present address*: USDA Forest Service, Fish and Wildlife Staff, 319 SW Pine Street, Box 3623, Portland, Oregon 97208

```
3630 REM
3640 REM   ****** RULES FOR DEDUCTIVE INFERENCE ******
3650 REM
3660 DATA "BRUSHFIELD"
3670 DATA "R1", "IF", "HAS <20% WOODY VEGETATION COVER", "THEN", "IS GRASS STAGE"
3680 DATA "R2", "IF", "CONTAINS >5% COVER BULL THISTLE", "THEN", "IS GRASS STAGE"
3690 DATA "R3", "IF", "HAS MOST SHRUBS <2 M TALL", "THEN", "IS EARLY STAGE"
3700 DATA "R4", "IF", "HAS SHRUBS >2 M TALL", "HAS >30% LITTER COVER", "THEN"
3710 DATA    "IS LATE STAGE"
3720 DATA "R5", "IF", "IS GRASS STAGE", "CONTAINS >5% COVER Festuca OR Elymus"
3730 DATA    "THEN", "HAS LESSER GOLDFINCH"
3740 DATA "R6", "IF", "IS GRASS STAGE", "CONTAINS SNAGS =>3 M TALL", "THEN"
3750 DATA    "HAS WESTERN BLUEBIRDS"
3760 DATA "R7", "IF", "IS EARLY STAGE", "IS ADJACENT TO MATURE STANDS"
3770 DATA    "CONTAINS LARGE DOWN LOGS", "IS <1 KM FROM OPEN WATER", "THEN"
3780 DATA    "HAS MOUNTAIN QUAIL"
3790 DATA "R8", "IF", "IS EARLY STAGE", "HAS DENSE, DECIDUOUS BRUSH", "THEN"
3800 DATA    "HAS POTENTIAL FOR WRENS"
3810 DATA "R9", "IF", "HAS POTENTIAL FOR WRENS", "IS >45% SLOPE", "THEN"
3820 DATA    "HAS WRENTITS"
3830 DATA "R10", "IF", "HAS POTENTIAL FOR WRENS", "IS <45% SLOPE"
3840 DATA    "HAS DENSE BRUSH", "THEN", "HAS BEWICK'S WRENS", "HAS WRENTITS"
3850 DATA "R11", "IF", "IS LATE STAGE", "CONTAINS ALDER IN RIPARIAN STRIPS", "THEN"
3860 DATA    "HAS WILSON'S WARBLERS", "HAS MACGILLIVRAY'S WARBLERS (P < 0.01)"
3870 DATA "R12", "IF", "IS LATE STAGE", "IS ADJACENT TO MATURE STAND"
3880 DATA    "CONTAINS SNAGS >4 M TALL", "THEN", "HAS DUSKY FLYCATCHERS"
3890 DATA "STOP"
```

Figure 23.1. A computer-generated listing of rules used in program BRUSH.

canopy. Inference engines are employed to quickly trim a myriad of possible solutions to only a few, which then are filtered by prompting the user for more specific information. Inference control strategies include evidence-to-hypothesis reasoning (forward-chaining), hypothesis-to-evidence reasoning (backward-chaining), or some combination. Technical reviews of inference control structures may be found in Winston (1977) and Stefik et al. (1983).

Some expert systems represent their control strategy in terms of conditional states or probabilities. Probabilistic control structures vary widely in expert systems. The reason for using uncertainty measures is to increase reliability by combining evidence. A probabilistic approach may prove useful for predicting wildlife responses to habitat conditions by computing probabilities that certain species are present, given that specific habitat conditions have been observed.

The construction of an expert system (Hayes-Roth et al. 1983; Weiss and Kulikowski 1984) begins with a dialogue between the knowledge engineer (a computer programmer) and the human expert. First, the problem is clearly identified. For example, a problem statement may be to diagnose habitat conditions of a particular vegetation type and successional stage in a particular geographic area for the purpose of deducing the presence of species of birds. Second, characteristics of the problem, such as species-habitat interactions, are represented and coded as concepts, facts, and decision rules. Facts may include a classification system of habitats that could assist in predicting species presence. The dialogue becomes critical at this stage, because human experts often apply rules of inference and rules of thumb that are articulated and codified only after careful discussion with the knowledge engineer. Other methods of obtaining expertise may include surveys of a priori professional judgments, such as those gathered by the Delphi technique (Zuboy 1981). The Delphi technique is very powerful and, when properly used, can give a high degree of reliability. At this stage, whether the original problem was defined too vaguely, broadly, or incompletely or whether the problem itself cannot be represented well in this framework will become clear. Usually one then returns to step 1 and refines or redefines the problem. In the third step, the system of rules and facts is tested to verify that they are encoded adequately (Buchanan et al. 1983). Finally, peer review and field validation are used for determining whether the fundamental facts and rules are incomplete or fallacious.

An expert system approach to wildlife-habitat modeling: Two examples

In general, an expert system that predicts wildlife response to habitat conditions should (1) identify species which may occur together under general habitat conditions, such as forest cover types and stages of development; (2) evaluate the response of a species or a set of species to changes in habitat conditions; (3) suggest which habitat attributes would best predict species' patterns of abundance; (4) allow the user to offer information as well as prompt the user for specific information; (5) give a rationale for hypotheses or conclusions reached; (6) be designed to be updated with new facts and rules; and (7) prescribe habitat conditions and recommend methods for creating these conditions to maintain or enhance particular species.

An example of a narrowly defined problem domain for use in wildlife management is predicting bird species' presence in a brushfield habitat following clearcut timber harvesting in the Coast Range of northwestern California. Two demonstration programs, in which I have encoded my own knowl-

```
4030 REM
4040 REM  ****** HYPOTHESES FOR DEDUCTIVE INFERENCE ******
4050 REM
4060 DATA "HAS LESSER GOLDFINCH"
4070 DATA "HAS WESTERN BLUEBIRDS"
4080 DATA "HAS WRENTITS"
4090 DATA "HAS BEWICK'S WRENS"
4100 DATA "HAS MOUNTAIN QUAIL"
4110 DATA "HAS WILSON'S WARBLERS"
4120 DATA "HAS MACGILLIVRAY'S WARBLERS"
4130 DATA "HAS DUSKY FLYCATCHERS"
4140 DATA "STOP"
4150 END
```

Figure 23.2. A computer-generated listing of contentions regarding presence of eight bird species in brushfield habitats, taken from the demonstration expert system BRUSH written in BASIC.

edge and control rules, will serve to highlight some of the features outlined above.

A rule-based demonstration program BRUSH was written in the BASIC programming language on an IBM Personal Computer and was modeled after the examples in Winston (1977) and Duda and Gaschnig (1981). I supplied data on habitat conditions in brushfields of Douglas-fir (*Pseudotsuga menziesii*) resulting from clearcutting. Through a series of 12 rules of species-habitat relationships (Fig. 23.1), BRUSH deduces the suitability of the site for a variety of bird species. BRUSH's contentions that establish suitability of habitat conditions for eight bird species are presented in Figure 23.2.

A sample run of BRUSH (Fig. 23.3) demonstrates the forward-chaining or data-driven nature of the control structure and the ability of the program to trace and present its own lines of reasoning. A more advanced version of BRUSH may (1) trigger hypotheses from conditional probabilities; (2) incorporate additional rules and hypotheses; (3) allow the user to suggest solutions and to volunteer information; and (4) learn from previous query sessions which rules may be more likely to provide correct deductions under different combinations of responses.

BASIC is a poor language for rule-based deduction systems, although an earlier, full-scale wildlife-habitat retrieval model that used some elements of expert system programming successfully employed BASIC on the Tektronix 4050-series microcomputer (Marcot 1980). BASIC is not designed to manipulate symbols and names extensively, as would be necessary in an expert system. However, the programming language LISP is specifically designed for relating and comparing symbols and is commonly used in expert system programming. I wrote a second example program, GUILD, in LISP (dialect ALISP) on a CDC Cyber 170/720 mainframe computer to demonstrate some advantages of this symbol-based language.

GUILD is based on lists of items and their properties that allow search and retrieval of entities whose properties have user-specified values. For example, the entity "Species-name" is assigned a number of properties, including "Diet," "Foraging-Substrate," and "Habitat." "Habitat" itself is composed of further properties, specifying vegetation types and successional stages. The values of properties are qualita-

```
RUN
******** PROGRAM 'BRUSH' ********
      DEMONSTRATION EXPERT SYSTEM
          BASED ON RULE-HYPOTHESIS STRUCTURE
          AND FORWARD-CHAINING INFERENCE
This program uses   12   rules to establish one of the
following   8   hypotheses:
  BRUSHFIELD HAS LESSER GOLDFINCH
  BRUSHFIELD HAS WESTERN BLUEBIRDS
  BRUSHFIELD HAS WRENTITS
  BRUSHFIELD HAS BEWICK'S WRENS
  BRUSHFIELD HAS MOUNTAIN QUAIL
  BRUSHFIELD HAS WILSON'S WARBLERS
  BRUSHFIELD HAS MACGILLIVRAY'S WARBLERS
  BRUSHFIELD HAS DUSKY FLYCATCHERS

Respond with YES, NO, or WHY.

Is this true:   BRUSHFIELD HAS <20% WOODY VEGETATION COVER? NO
Is this true:   BRUSHFIELD CONTAINS >5% COVER BULL THISTLE? NO
Is this true:   BRUSHFIELD HAS MOST SHRUBS <2 M TALL? NO
Is this true:   BRUSHFIELD HAS SHRUBS >2 M TALL? YES
Is this true:   BRUSHFIELD HAS >30% LITTER COVER? YES

Rule R4 deduces BRUSHFIELD IS LATE STAGE
Is this true:   BRUSHFIELD CONTAINS ALDER IN RIPARIAN STRIPS? NO
Is this true:   BRUSHFIELD IS ADJACENT TO MATURE STAND? YES
Is this true:   BRUSHFIELD CONTAINS SNAGS > 4 M TALL? WHY
I am trying to use Rule R12
The inference structure has already deduced that:
BRUSHFIELD IS LATE STAGE
BRUSHFIELD IS ADJACENT TO MATURE STAND
IF:
BRUSHFIELD CONTAINS SNAGS >4 M TALL
THEN:
BRUSHFIELD HAS DUSKY FLYCATCHERS

Is this true: BRUSHFIELD CONTAINS SNAGS >4 M TALL? YES
Rule R12 deduces BRUSHFIELD HAS DUSKY FLYCATCHERS
I conclude that BRUSHFIELD HAS DUSKY FLYCATCHERS.
```

Figure 23.3. A sample run of BRUSH showing features of forward-chaining deduction. The program progresses through hypotheses by querying the user for pertinent information. The user's responses (underlined) may include "Why," from which the system discloses its reasoning process that led to the asking of a particular question.

tive codes, such as diet items and foraging substrates, or continuous variables, such as mean nest height and home range size.

A set of LISP functions can then be called to access and manipulate the data base. A brief dialogue with GUILD generated a partial list (Fig. 23.4) of breeding species representing the potential negative impact of a reduction of brush foliage volume in clearcuts 6–10 years old. Field-derived estimates of species densities before and after the brush reduction are given by the program, along with probabilities of species presence (field-derived estimates of percent occurrence of the species in sites having the specified foliage volume levels). Specifying the "mitigate" function triggered the system to output further information, including a set of prescriptions for brush management and species' expected densities resulting from the mitigation activities, estimated from field-derived regressions of species' densities on brush volumes. A more advanced version of GUILD may suggest

PROGRAM 'GUILD': A LISP-BASED QUERY SYSTEM

? (SETQ BEFORE-STAGE LATE-SHRUB AFTER-STAGE EARLY-SHRUB)
? (TELL-IMPACT)

	HABITAT STAGE				
	DENSITY (N/40 HA)			PROBABILITY OF OCCURRENCE	
	BEFORE	AFTER		BEFORE	AFTER
SPECIES	LATE-SHRUB	EARLY-SHRUB	PERCENT CHANGE	LATE-SHRUB	EARLY-SHRUB
BLACK-HEADED-GROSBEAK	28.8	5.5	—81	0.95	0.65
CALLIOPE-HUMMINGBIRD	29.8	0.0	—100	0.29	0.00
CHESTNUT-BACKED-CHICKADEE	2.7	2.3	—15	0.38	0.46
FOX-SPARROW	27.3	2.4	—91	0.33	0.31
HERMIT-THRUSH	8.5	2.4	—72	0.57	0.54

DO YOU WANT TO MITIGATE (Y/N)? Y

PRESCRIPTIONS

THE 5 SPECIES IN THIS LIST REPRESENT NEGATIVE IMPACTS ON BREEDING DENSITIES BY A CHANGE IN HABITAT STAGE FROM LATE-SHRUB TO EARLY-SHRUB. THE NEGATIVE IMPACT IS MOSTLY FROM REDUCTION OF SHRUB VOLUME AND COVER. TO MITIGATE THE NEGATIVE IMPACTS ON SPECIES DENSITIES, THE FOLLOWING MANAGEMENT ACTIONS MAY BE TAKEN:

1) RETAIN DECIDUOUS AND EVERGREEN SHRUB COVER ALONG ALL PERMANENT AND EPHEMERAL WATERCOURSES ON THE SITE;

2) RETAIN OR ENCOURAGE POCKETS OF LOCALLY DENSE SHRUB COVER WITHIN THE SITE, AVERAGING AT LEAST 2 M TALL AND 5 M ACROSS; SUCH POCKETS MAY BE SPATIALLY ARRANGED SO AS NOT TO SUBSTANTIALLY INTERFERE WITH REFORESTATION ACTIVITIES;

3) TOTAL SHRUB FOLIAGE VOLUME SHOULD NOT AVERAGE LESS THAN 10,000 CU. M. PER HA.

EXPECTED SPECIES DENSITIES WITH MITIGATION ACTIVITIES

SPECIES	DENSITY WITH MITIGATION PROCEDURES	PERCENT IMPROVEMENT OVER NO MITIGATION
BLACK-HEADED-GROSBEAK	16.5	38
CALLIOPE-HUMMINGBIRD	14.0	47
CHESTNUT-BACKED-CHICKADEE	2.5	7
FOX-SPARROW	14.1	43
HERMIT-THRUSH	5.3	34

SRU 0.310 UNITS.

RUN COMPLETE.

Figure 23.4. A sample run of GUILD showing features of LISP program implementation. User responses are underlined. Note that program response to the "mitigation" option triggered an output of possible habitat prescriptions and expected species' densities.

to the user other habitat features that help predict and influence the species' abundances.

Expert performance

What constitutes expert performance and lends credibility to professional advice? We choose among experts, according to Simon (1977), by "forcing the experts to disclose how they reached their conclusions, what reasoning they employed, [and] what evidence they relied upon." Disclosing the reasoning process lends credibility. Credibility also depends on the expert's actual experience in the field, his or her contribution to the primary literature, and his or her record of validated predictions. Expertise extends beyond familiarity with existing literature and, especially in modeling, involves the ability to distinguish between realistic and unrealistic assumptions.

An explanation facility is an important facet of an expert

Table 23.1 Knowledge-engineering system

System	Problem domain	Inference structure	Features
AGE	General	Forward-, backward-chaining; blackboard[a]	Flexible in knowledge representation and processing
EMYCIN	Deduction, diagnosis	Backward-chaining	Employs certainty factors
EXPERT	Classification	Rules ordered by user	Hypotheses expressed with uncertainty values
HEARSAY–III	General	Blackboard[a]	Supports incremental construction, testing; relational data base
KAS	Deduction, diagnosis	Forward- and backward-chaining	Chooses promising rules via heuristic evaluation function
OPS5	General	User-defined	Flexible in representation schemes
RLL	General	Agenda (flexible) priority system	Library of various control structures
ROSIE	General	Rules ordered by user	English-like syntax

[a]A "blackboard" is a central control medium, used for representing partial solutions and pending program executions.

system because it enables the program to describe its line of reasoning, why it is requesting certain pieces of information, and how it reached a particular conclusion (e.g., Clancey 1983). Such disclosure also helps the system accept new lines of reasoning (new rules or facts) and grow with its use (Winston 1982). A disclosure of reasoning was demonstrated above with BRUSH. The expert system may function better than a human expert because it can easily expose for review its chain of reasoning and inferences, allowing the user to carefully assess its credibility. However, just as with the human expert, output and advice from an expert system should be viewed critically. The system is no better than the data, relations, and reasoning processes it contains.

Quality control of the knowledge base of a wildlife-habitat expert system should include field testing of model predictions and peer review of the adequacy and accuracy of the facts, reasoning process, and controls used in the system. The goal would be to show explicitly, under specified field conditions or ecological contexts, how well or how poorly a system performed. Validation should also include a test of the system's utility, i.e., applicability in an actual management and decision-making environment. Criteria of model validation, which may also be useful for judging "expert" contributions to such a system, were reviewed by Marcot et al. (1983).

Knowledge-engineering systems

Several expert-system-building tools have been constructed that may help reduce development time by an order of magnitude (Table 23.1) (Barstow et al. 1983; see also van Melle 1981). Using such tools allows programmers to compile a knowledge base and to develop inference structures without programming in general-purpose languages such as BASIC, LISP, PROLOG, and FORTRAN. Three of the knowledge-engineering systems—EMYCIN, EXPERT, and KAS—are designed for specific problem domains; the others are general-purpose systems and allow for a greater variety of inference (control) structures, but may sacrifice some ease of use. EMYCIN (van Melle 1979), KAS (Duda et al. 1981), and OPS5 (Forgy 1981) are all well suited to the problem of diagnosing habitat conditions and inferring species' responses. However, knowledge-engineering systems for use on personal computers are coming of age (e.g., Konopasek and Jayaraman 1984).

Questions and cautions

The effort required to produce a full-scale wildlife-habitat expert system is likely to be measured in years of working time (Duda and Shortliffe 1983). Although decision-support systems are becoming increasingly common (Wagner 1982), careful considerations of the cost, need, and utility of such systems seem warranted. Who are the intended users and what are their specific information needs? What specific areas of habitat management could fit into and benefit from an expert system approach? How should a wildlife-habitat expert system be updated and validated? Which human experts should the reasoning processes be modeled after?

Wildlife biologists and resource planners may be the first audience which uses such systems for assessing project impacts and planning alternatives. Other specialists may later integrate their information needs. Predicting the response of wildlife species to habitat conditions and prescribing management activities for mitigation are two functions that can help biologists and planners.

Validation must be an integral part of an expert system. Many ecological problems of habitat management may be ill-suited to the fact- and rule-based structures of expert systems. For example, problems of habitat fragmentation and species' interactions are poorly understood and would be unsound candidates for expert system formalism. Three critical stages in developing a full-scale system are (1) adequately specifying the problem and surveying expert knowledge in a particular problem area; (2) adequately encoding the knowledge into facts and rules of inference and deduction; and (3) validating the system with new field data to determine whether the facts and rules have been represented fully and correctly. Failure to attend to each of these stages would probably result in the building of models in which little confidence could be placed. The expert system should not be used to completely supplant essential field work, such as basic research, population monitoring, wildlife inventory, or reconnaissance for project impact assessment.

The knowledge base must be evaluated for quality, correctness, and completeness. Evaluation also reveals how well an expert system may be expected to perform, given missing or false information. Evaluation by domain experts, such as avian ecologists, would help determine the accuracy of the knowledge base and any advice or conclusions the system provides; evaluation by users, such as habitat managers, would help determine the utility of the system (Gaschnig et al. 1983). Characteristics of expert systems to be evaluated include quality of the system's decisions and advice, correctness of the reasoning techniques used, the nature of the interactions with the human user, the system's efficiency in using facts and rules, and the system's cost-effectiveness. Although no expert system of wildlife-habitat relationships has been formally evaluated, several other types of expert systems have been (Gaschnig et al. 1983).

The relationship between errors in or incompleteness of the knowledge base and errors in the output of an expert system is variable, depending on the level of the rules in the logic structure and the frequency with which the rules are called. In some cases, erroneous or missing rules may accentuate errors in the output. Sensitivity analysis of model output to changes or additions of the knowledge base would help quantify the relationship and thus the need for corrective actions. For example, sensitivity analysis of the MYCIN program revealed that the certainty factors used in the program to weight different responses influenced the output less that did the semantic and structural context of the rules per se (Gaschnig et al. 1983).

Error rates of fully evaluated systems, such as MYCIN, R1, or PROSPECTOR, are generally low as long as the systems are used within appropriate problem domains. Such evaluations may serve to show the usefulness and evolutionary development of a wildlife-habitat expert system (e.g., see Buchanan and Shortliffe 1983). Our current knowledge of wildlife-habitat relationships requires much additional development and testing, and a wildlife-habitat expert system cannot be expected to perform any better than our own knowledge allows. The greatest benefit of such an expert system, however, would be in distributing existing expertise in narrowly defined problem domains (such as response of songbirds to clearcutting Douglas-fir forest) to users that require but lack such expertise.

It is my opinion that an expert systems approach that incorporates field-monitoring information, discloses its reasoning process, and helps prescribe habitat conditions to suit particular species may be a valuable tool for habitat managers and decision makers, if adequately validated and applied to appropriate problem domains. However, risks of applying untested systems may be high if pertinent facts and reasoning processes are developed in isolation from extensive peer review and field validation.

Acknowledgments

I thank Andrew Carey, William Landenslayer, Vivian Marcot, C. John Ralph, and Jake Rice for their reviews of earlier versions, and Kelly Laberee for typing the manuscript. A. Carey suggested, among other things, the need for a validation procedure that necessarily integrates population monitoring. This work was conducted under the auspices of the Oregon Cooperative Wildlife Research Unit, with Oregon State University, Oregon Department of Fish and Wildlife, U.S. Fish and Wildlife Service, and the Wildlife Management Institute cooperating. The USDA Forest Service, Pacific Southwest Region, funded the effort. Oregon State University Agricultural Experiment Station Technical Paper 7253.

Summary: Development, Testing, and Application of Wildlife-Habitat Models— The Manager's Viewpoint

JANET F. HURLEY

The chapters presented in Part I demonstrate the impressive energy devoted to the growing field of wildlife-habitat modeling. Clearly, the development, application, and testing of wildlife models is a major concern in wildlife management today.

These chapters have described efforts in model development and testing that cover several regions of the country and that have been conducted by state and federal agencies as well as university and private personnel. They have also included models that focus a great deal of information on a single species, and others that encompass the entire vertebrate community within a region. Because of the wealth of information presented here, I regretfully will not be able to mention each chapter. Instead, I will discuss several major themes emerging from this effort.

Goals of wildlife-habitat modeling: What do managers want?

The managers most likely to use species-habitat models in daily work have biological training and are responsible for evaluating land-management activities that affect wildlife habitat. These managers require a decision-making tool that displays the consequences of habitat change resulting from management activities. For example, see the goal statements of Salwasser et al. (1980), Sheppard et al. (1982), and the Fish and Wildlife Service (1980a).

A system, therefore, must be based on habitat identification that is useful to managers, and preferably integrated with classification systems developed for timber, range, or other resource uses. Second, the system should yield information about the response of wildlife species to habitat changes at a reasonable level of accuracy. This concerns us with validation, a recurring theme throughout the chapters. Finally, managers must have reasonable access to the information, along with guidelines for use of the system and interpretation of the results.

Two levels of wildlife-habitat models

Community-level and single-species models form two broad categories of the several modeling approaches presented. Community-level models are represented by the Wildlife Habitat Relationships (WHR) efforts of the USDA Forest Service. Another type of community model, derived from statistical analysis of empirical data, is presented by Rice et al. (Chapter 13). Single-species models are exemplified by the Habitat Evaluation Procedures (HEP) spearheaded by the Fish and Wildlife Service. The emphasis of the Forest Service on community models may stem from the agency's mandate, under the National Forest Management Act, to maintain viable populations of all species of native vertebrates. Yet single-species models are conceptually easier to understand and apply. Perhaps for this reason, the Fish and Wildlife Service uses detailed models for representative species to approximate the community response to habitat changes. Indeed, some Forest Service planning applications are being answered by single-species models developed as an adjunct to WHR systems (Hurley et al. 1982). Despite the difference in starting points, the ultimate goals of both types of models are the maintenance and management of desired wildlife communities.

Some distinctions may be drawn between these two levels of models. First, single-species models may be relatively easier to validate. This is particularly true with models built from several years' research on one species in one region. For example, Johnson et al. (Chapter 5) achieved impressively accurate model predictions with extensive data on one species. In contrast, field verification of community-level models is far more time-consuming and costly. Because the quality of information is uneven, the community model may be less accurate in total performance. The largest verification effort of a WHR model (Raphael and Marcot, Chapter 21) showed that the overall accuracy of the model was low, although parts of the model performed well.

Community-level models are usually built by cooperative efforts, involving agency support and several years of preparation, review, and testing. For managers at the local level, this means that a product becomes available for use with little investment beyond the time needed to learn how to use and interpret the system. Ideally, single-species models are also built in this fashion. However, the pressing needs of coordination often force managers to build their own models for specific projects. The HEP documents provide guidelines for model building at the local level, but these local models are often quickly put together and are less likely to be validated than are agency-backed cooperative efforts. Doering and Armijo (see Chapter 58 in Part V) show that considerable time must be invested in building the model before it

JANET F. HURLEY: USDA Forest Service, Star Route, Box 1295, Sonora, California 95370 can be applied—time that is likely to be prohibitive to managers facing tight deadlines and restricted budgets.

If the aim of these information systems is the management of desired wildlife communities, then community-level models provide the most direct approach for doing so. However, this requires a great deal of synthesis and interpretation on the part of managers. Toth et al. (Chapter 22) show how the responses of individual species within the community can be aggregated into a few relatively simple guidelines for management of a community-supporting habitat. The HEP process works in the other direction—managers must carefully select evaluation species and then develop guidelines for managing their habitat to achieve a program that will maintain desired habitat for this species and the community they represent. The key here is the careful selection of evaluation species. Fry et al. (Chapter 17) fused the two types of models, extracting information contained in the WHR model and using it to select evaluation species for which HEP models would be built. This approach is rational and replicable, and I think it holds promise for reconciling the needs of wildlife communities with the practical simplicity of dealing with selected species.

The concern for validation

Most authors in this section emphasize validation of wildlife models as a requirement for acceptability to management. Here, validation means measurement of the degree of correspondence between model predictions and observed species' abundance in relation to habitat. Validation is thus limited to the concept of "veracity" (see Salwasser, Chapter 59 in Part VI). Because other components of management needs, such as utility, are overlooked in this framework, the conclusions derived from these validation efforts should be viewed as one part of the evaluation of a model's application to management.

Few investigators set criteria for acceptable model performance (i.e., percent accuracy of model predictions as compared to field observations), instead preferring to use commonly accepted statistical measures of significance. In my view, this can lead to setting unrealistically high thresholds when applied outside of normal statistical tests. For example, some investigators have proposed that a species be observed at all times to meet the model definition of "optimum" habitat (Dedon et al., Chapter 19), or that similarity indices of "1" be achieved between model predictions and field observations (Block et al., Chapter 18). These stringent thresholds may result in model failure by definition. Because of such confounding factors as stochastic effects on species' distributions (Haila, Chapter 45 in Part IV), variations in detectability of species, and sampling error, perfect performance in the real world is an unrealistic expectation. Investigators may thus reject a potentially useful management tool by setting narrower tolerances than managers need.

To determine how accurate models should be, we need to know how stochastic effects and sampling error cause model evaluations to deviate from theoretical potentials. We want models that perform better than predictions made by chance. However, we should recognize that managers are quite comfortable with accuracy levels of 75–80% for total model output, because the planning applications to which models are put do not require greater precision. This issue warrants further discussion among wildlife professionals so that we can agree on verification standards that make acceptable trade-offs between costs, effort, and information yielded.

Appropriate measures and tests of veracity

Field testing of model predictions is an essential step in determining the veracity of models. A major problem centers on the difficulty of identifying appropriate field measures of a species' response to the quality of its habitat. Several authors, including Hamel et al. (Chapter 20), Lancia et al. (Chapter 11), Rice et al. (Chapter 13), and Schamberger and O'Neil (Chapter 1) have determined that absolute and relative measures of abundance give misleading impressions of model failure. Van Horne (1983) demonstrated that density may give misleading indications of habitat quality if demographic studies of the species are lacking. The reasons for this are several, including the cyclical nature of some wildlife populations, the many factors of population dynamics that are not accounted for in habitat models, and the difficulty of measuring certain populations in the field. Model performance varies with the geographic scale of application (Hamel et al., Chapter 20) and improves with the inclusion of multiple years of data (Rice et al. 1984). Some alternate measures have been suggested, such as frequency or consistency of habitat use by a species. Some measures of community similarity also show promise for exploring the performance of models compared to real-world observations (Raphael and Marcot, Chapter 21). This important issue deserves further attention by both researchers and managers.

Several techniques for verification studies have been presented. Two types of sampling efforts are species-centered plots and random block sampling. The species-centered method collects data efficiently but may be limited to single-species models or species-poor habitats. Random sampling of habitats requires extensive field work but yields information on a variety of species for community-level models. Larson and Bock (Chapter 7) and Raphael and Marcot (Chapter 21) provide careful discussions of these two techniques.

The wave of the future

Marcot's chapter on expert systems (Chapter 23) extends community-level models into a new realm of accessibility. Expert systems codify the available knowledge of species-habitat relationships into an interactive computer program with which managers can evaluate the effects of habitat changes. I believe the greatest value of the expert system lies in its ability to disclose the reasoning behind predictions of species-habitat relationships. A statement of reason for a predicted change gives the manager information that can be related to the current problem. In effect, it allows managers to agree or disagree with the model predictions on the basis of correspondence between model assumptions and the situation at hand. This, I believe, increases the intuitive acceptance of models in the same way that verification indicates that a model may be scientifically sound. In the advance of expert systems, I encourage designers of conventional models to explore ways to increase disclosure of reasoning.

In summary, the development, application, and testing of wildlife-habitat models will be a thriving effort for some years. The challenge, particularly with the advent of expert systems, will be the assembly and codification of a vast array of information. As models are developed and tested, we will be forced to investigate species, habitats, and relationships that have received little attention to date. This will increase our shared knowledge base, to the benefit of greater professionalism of wildlife managers and better management of wildlife and their habitats.

Summary: Development, Testing, and Application of Wildlife-Habitat Models— The Researcher's Viewpoint

DAVID E. CHALK

The recent surge in the development of wildlife-habitat-relationships models was a predictable course for wildlife management. In fact, it is surprising that modeling took so long to develop in this field, considering the widespread use of models in other disciplines. Ideally, given the title of this volume and the title of Part I, I expected to find a reasonable balance among chapters explaining the history and methods of model development, the various ways in which models have been applied by managers, and tests of their validity. In a broad sense, this expectation was satisfied, but I am left with the sense that the development of models is outstripping their application and that empirical testing of them is lagging far behind.

The chapters in Part I represent a remarkable effort on the part of the research community to respond to needs of resource planners and managers by providing them with an array of tools to assist them in their decision making. With the development and use of models, it is obvious that we are leaping into a new and exciting era of wildlife management. The first chapters in this section generally deal with single-species models (e.g., Hammill and Moran, Chapter 3; Latka and Yahnke, Chapter 4; Johnson et al., Chapter 5). Middle chapters described ways to improve model development (e.g., Mosher et al., Chapter 6; James and Lockerd, Chapter 9) and measure species' habitat needs (e.g., Larson and Bock, Chapter 7; Brush and Stiles, Chapter 10). Later chapters discuss sources of flaws in models or the modeling process (e.g., Stauffer and Best, Chapter 12; Rice et al., Chapter 13; Laymon and Barrett, Chapter 14; Laymon and Reid, Chapter 15). The final chapters deal with guild concepts (Short and Williamson, Chapter 16; Fry et al., Chapter 17; Block et al., Chapter 18) and with multiple-species models (Dedon et al., Chapter 19; Hamel et al., Chapter 20; Raphael and Marcot, Chapter 21; Toth et al., Chapter 22). The last chapter (Marcot, Chapter 23) takes us into the future.

Schamberger and O'Neil (Chapter 1) begin by pointing out various problems associated with model development. They make a clear point that habitat models are not population models, a subject that surfaces in several of the chapters. They also plead for consistency in standards, objectives, and definitions. These problems appear to be consistent whether we deal with single-species models or assemblages of several species.

Blenden et al. (Chapter 2) correctly conclude that to evaluate the overall performance of a model, one must evaluate its objectives and assumptions as input into the model itself. This illustrates the importance of selecting variables within the model that are relevant, as stressed by Stauffer and Best (Chapter 12). This approach would probably answer questions regarding overall model performance, especially when it is apparent that a model failed. It appears that we may need better measures of the habitat requirements of animal species, or at least better ways to measure them. Some promising directions are suggested in the chapters by Larson and Bock (Chapter 7), Brush and Stiles (Chapter 10), Smith and Connors (Chapter 8), and James and Lockerd (Chapter 9).

Several chapters in this section showed both favorable and unfavorable results when testing model predictions with actual field data. This should not be surprising when the model is based on scanty literature or on the subjective professional judgment of specialists, or both. Indeed, most existing models have been developed in that way. But it is especially disconcerting when models based on extensive field work also fail to give consistently good predictions of a species' abundance or distribution. In this category, Stauffer and Best (Chapter 12) concluded that their models for bird species of hardwood forests in floodplains and adjacent uplands were reasonably reliable only when based on rather large data sets. They also found that models developed from data for one habitat type (bottomland or upland hardwood forests) did not predict well for the other. Reporting results from a long-term study of bird populations along the Colorado River, Rice et al. (Chapter 13) found species during the fourth year in habitats in which they had not been detected during the previous 3 years. These examples support the need for continued research, not only to improve existing models empirically, but also to give us better insights into the modeling process itself. Only in this way will we be able to achieve the level of accuracy needed to apply models in a land-management context.

Laymon and Barrett (Chapter 14) also experienced poor predictive results from their HSI models for spotted owls, marten, and Douglas' squirrels, "even though they were based on what was believed to be good information." The geographic scale at which these models could be applied was suggested as one problem (a similar conclusion emerges from the study by Hamel et al., Chapter 20), and Laymon and Barrett have also identified predation as a possible source of

DAVID E. CHALK: USDA Soil Conservation Service, West National Technical Center, 511 N.W. Broadway, Portland, Oregon 97209 error (this topic is considered further in Part III). As one solution, they suggest lowering the suitability rating based on assessments of habitat quality when predators are likely to limit a population. I disagree that the suitability rating of the habitat should be lowered, but the population variation should be explained by the presence of high numbers of predators. A habitat-suitability rating is a measure of the capability of the vegetation (its structure and composition) to provide the needs of a species. Predation is an important but separate issue.

The potential for applying the guild concept in wildlife management is apparently still debated, as seen in the contrasting viewpoints of Short and Williamson (Chapter 16) and Fry et al. (Chapter 17), on the one hand, and those of Block et al. (Chapter 18) on the other. I agree with Block et al. that existing evidence does not support the use of an indicator species to be representative of other species with which it is grouped into a guild. But I also believe the guild concept has other potential benefits for wildlife management that need to be explored.

Chapters by Hamel et al. (Chapter 20), Raphael and Marcot (Chapter 21), Rice et al. (Chapter 13), and Lancia et al. (Chapter 11) provide good examples of what I believe wildlife managers really need. Their models have clear objectives, they identify the assumptions and possible errors in the models, and they explain the variation of their results. They also recommend both the temporal and spatial scales at which these models would be most appropriately applied.

Managers are understandably interested in the accuracy of the predictions from models they use in their planning. Hurley, in the preceding Summary, states that the limits defined by Hamel et al. (Chapter 20) and Mosher et al. (Chapter 6) of 75–80% are acceptable. This seems to me to be a reasonable expectation, but accuracy is only one side of the coin. The other is precision. In my opinion, a model that can give a reasonably accurate prediction of some mean value is still suspect if the limits to variation about that mean value are large. Researchers need to report measures of both the accuracy and the precision of their models.

I think it is particularly fitting that this part concludes with a chapter that examines a possible future direction in the development of wildlife-habitat relationships models—expert systems. Marcot (Chapter 23) has clearly defined a challenge for us—to look at such probabilistic approaches through the next generation of wildlife-habitat models. He used two demonstration programs to show the possibilities of data-driven reasoning, which "may reduce model development time by an order of magnitude." Marcot is cautious about the potential for using expert systems in this arena, but the possibilities for cost savings and objective model development demand serious attention.

I sense considerable reluctance among researchers, for a variety of reasons, to release their models for application in management contexts. For example, model assumptions often are not met, uncertainty hampers the linking of models, input data are often flawed, and models are commonly applicable only in a restricted geographic area. Researchers must not get so involved with the complexity of their modeling efforts or with their desire to create perfect models that they lose sight of the reality of decision making. We must heed the admonition of Thomas in his introductory remarks in this volume that model builders cannot hide from the realities of resource management. The trucks are rolling now, friends, and managers are making decisions. If an imperfect model helps to make those decisions even a little wiser, it should be used. And I believe that professionally trained wildlife biologists, using state-of-the-art information and technologies, even though imperfect, will be better advocates for wildlife resources than decision makers with other resource interests. The researcher's responsibility is to identify the limits of a model, but not necessarily to decide when the model is ready to use. The mere fact that a model is published indicates that it can be used, with due caution as identified by the author.

Managers have responsibilities, as well. Model outputs can be useful guides to assist in making wise decisions. Managers must not expect the models to make decisions for them or to provide them with a perfect version of a real-world system. They need to put forth the effort to understand the models they use—their assumptions and limitations, as well as their potential applications.

In the long run, if modeling technology is ever to attain its full potential in wildlife management, researchers and managers must work together much more closely in the future than they have in the past. Only then will researchers fully understand management's needs, so they can develop appropriate models that can be readily applied in a planning context. And only then will managers develop a sense of ownership in the models they use, so they will be willing to make the effort to understand them better. A recent modeling experience in Southeast Alaska, discussed by McNamee et al. (Chapter 56 in Part V), is exemplary in this regard.

II *Biometric Approaches to Modeling*

Introduction: Biometric Approaches to Modeling

FRED A. STORMER and DOUGLAS H. JOHNSON

Every science, with the possible exception of genetics, has had a descriptive beginning followed by the study and quantification of process. Ecology, as well as its applied derivative, wildlife management, is founded in natural history. From this descriptive base, it is perhaps inevitable that a point of scientific development is reached when a symposium on modeling habitat relationships of terrestrial vertebrates is timely. Legislation, such as the National Environmental Policy Act and the National Forest Management Act, has provided an impetus toward defining quantitatively the outcome of land management on terrestrial vertebrates, while increased availability of high-speed computers and the development of statistical procedures have provided advanced tools for accomplishing this task. In fact, biometric tools available to represent, test, and process models have outstripped our capability to formulate and measure relationships in ecological systems.

Mathematical models can be viewed as quantitative abstractions of the essential parts of real-world situations. Because statistics quantitatively express uncertainty about real-world phenomena, ecological modeling must incorporate statistical procedures. As such, the use of statistics in biological models is not new. Francis Galton (1822–1911), the father of biostatistics, studied the application of regression models to biological measurements a century ago (Sokal and Rohlf 1973).

The tie between statistics and biological models goes beyond representation of the outcome of variable real-world events. Statistical induction goes hand-in-hand with model deduction toward understanding. Models represent what we know or believe we know about the real world. They complement reality in a two-way flow of information. Models suggest questions to ask of nature; nature provides answers that further shape our models. In this light, model building can be viewed conceptually as a deductive-inductive process as described by Box et al. (1978:1–15). Based on certain assumptions, information derived from literature, personal experience, and expert opinion, one contrives a model depicting species-habitat relationships. At this stage the model comprises one or more hypotheses, which lead by deduction to certain necessary consequences. Consequences of the model are tested against actual data, and when consequences and data fail to agree, a second model is contrived. A second iteration in the model-development process is thus initiated. Model building is complete when deduced consequences agree with real-world data. Like theories, however, models are always tentative and never absolutely proven.

Viewed another way, model building involves (1) knowledge of subject matter to formulate models and their consequences; and (2) a strategy to efficiently and effectively test whether or not deduced consequences agree with real-world data. In this case, a proper strategy involves the prudent use of experimental design and statistical analysis. However, statistics is more than a collection of tests and techniques; it is the embodiment of an attitude toward scientific inquiry. It demands objectivity, rigor, and honest skepticism—attributes that should also be features of a good modeling effort.

The rapid proliferation of biological models in the last few years has been striking. Indeed, habitat suitability index models for more than 50 species have been developed by the U.S. Fish and Wildlife Service. Nevertheless, the frequent lack of adequate testing of models before application and the lack of evaluation during operational use are troublesome. In essence, we have proceeded only through the initial hypothesis-formation phase of model building, but have not yet carried the deductive-inductive process to its logical and necessary conclusion.

Rigorous application of biometric procedures comes into play in the examination of assumptions upon which models are based. Quantitative modeling serves a valuable function by requiring that assumptions be stated explicitly, which in turn allows them to be isolated and tested. Application of any model involves extrapolation in time, area, or condition; i.e., the model will be applied in the future, whereas data were collected in the past, perhaps in a different area or under different conditions. Extrapolation is always risky, but the danger is more acute if fundamental statistical and biological assumptions of the model are not satisfied. Some statistical assumptions are purely operational in nature, addressing such topics as model linearity, effects of interaction among explanatory variables, model sensitivity, and adequate prediction of the outcome of habitat changes within acceptable limits of error. Other statistical assumptions include those relating to normality, randomness, and independence of observations. Biological assumptions may relate to some of the theoretical underpinnings of wildlife-habitat relationships and include assumptions made in model development about such concepts as carrying capacity, population growth, and interspecific interactions. Also, we need to understand the biological reasons for quantitative relationships to avoid confusion of correlation with causation and to

FRED A. STORMER: USDA Forest Service, Forest Environment Research, P.O. Box 2417, Washington, D.C. 20013

DOUGLAS H. JOHNSON: U.S. Fish and Wildlife Service, Northern Prairie Wildlife Research Center, P.O. Box 1747, Jamestown, North Dakota 58401

know when and how to apply, or not to apply, any given model.

Even a cursory review of models being developed to aid our understanding of habitat relationships of terrestrial vertebrates gives the strong impression that development and application are moving faster than research can move to provide the necessary foundation for model building. The economic facts of life are such that the need for research information is greater than the supply, and we frequently must develop models based on tenuous information. Nevertheless, there sometimes is a tendency among some model developers, planners, and managers to dismiss easily the need for a rigorous scientific foundation for habitat-relationships models on the presumption that models built on the "best information" can be applied with impunity, whether the "best information" is actually reliable or not. Managers require information at least as good as that needed by researchers, because they must predict actual events and their outcomes. The cost of mistakes by planners and managers greatly exceeds that of researchers. Even nonengineers know that the height to which a structure can be built is in direct proportion to the strength of its foundation, and that structures built on weak foundations will inevitably collapse.

Part II of this volume, dealing with the wedding of statistics and biology, then, should serve not only as a watermark for the state of the knowledge, but also as a focus for emphasizing the necessity of a rigorous application of statistical design, procedures, and attitudes in the development and application of modeling habitat relationships. Success of this effort could be judged by how much of the substantive body of these chapters is eventually incorporated into general practice. Nevertheless, we are dealing with a field that is very much in its youth. Perhaps a better measure of success is how rapidly we grow and, hence, how rapidly the information presented here becomes dated.

24

Improving Vertebrate-Habitat Regression Models

KEVIN J. GUTZWILLER and STANLEY H. ANDERSON

Abstract.—The regression approach to modeling animal-habitat relations receives frequent use by wildlife scientists. Underlying the valid use of regression is the assumption that the model is appropriate for the data. We describe how to find and eliminate three problems that invalidate this assumption: nonconstant error variance, nonindependence of error terms, and outliers. We also demonstrate that by resolving these problems, researchers can obtain more accurate and precise estimators for regression coefficients. Such model improvements are valuable in developing realistic vertebrate-habitat regression equations.

Ecologists expect individuals, populations, and communities to depend on various attributes of the environment. Regression, with its underlying assumption of a dependency between a response variable and one or more explanatory variables, is well suited to explore and test such expectations. Indeed, it is regression's intuitive appeal and broad applicability that have led to its frequent use by the wildlife profession. Relations identified by regression analyses may help form wildlife-management plans, support or cast doubt on ecological theory, and generate new hypotheses. But the value of regression relations for these purposes depends, in part, on their accuracy. If wildlife management and ecology are to advance through such endeavors, it is important to assess the validity of models.

A first step toward validating any form of regression equation is to determine whether statistical problems associated with model estimation exist. If such problems are left unremedied, a model with low precision and poor descriptive and predictive abilities may result; if they are corrected, confidence in the model is strengthened and fresh insight into relations among variables may materialize. Our objectives are to show how to detect nonconstant error variance, nonindependence of error terms, and outliers, and to outline techniques for eliminating or reducing their effects.

With published examples, we demonstrate how these three statistical problems may influence model interpretation. We did not find examples in which the effects of these departures were recognized in the interpretation of animal-habitat models. Instead, we use examples related to business that illustrate principles equally pertinent to all regression relations, but we recast these examples to reflect processes and variables involved in wildlife-habitat modeling.

Studies of wildlife-habitat relations often involve multiple-regression equations, both linear and nonlinear. The techniques we describe are applicable to all such models (Neter and Wasserman 1974:111), but our discussion and examples focus on the simple linear model:

$$Y_i = \beta_0 + \beta_1 X_i + \epsilon_i. \quad (1)$$

Y_i is the ith value of the dependent variable, β_0 and β_1 are parameters to be estimated, X_i is the ith observation of the explanatory variable, and ϵ_i is the random error term for observation i (Neter and Wasserman 1974:30). A residual (e_i) is the observed value of the dependent variable (Y_i) minus its predicted value ($\hat{Y}_i$). A standardized residual is a residual divided by $\sqrt{MSE}$, where MSE is the error mean square for the model in question. An error term is the difference between the observed value (Y_i) and the theoretical expected value $[E(Y_i)]$ (Neter and Wasserman 1974:97–98).

Nonconstant error variance

Ordinary least squares (OLS) estimators of regression coefficients are based on the assumption that the error terms have equal variances at all levels of the explanatory variable (X). When this is not true, the estimators are still unbiased, but the variance of the estimators is larger than would otherwise be the case (Neter and Wasserman 1974:131). Consequently, tests of significance for the coefficients are less sensitive, and confidence intervals for β_0 and β_1 are larger (Chatterjee and Price 1977:44). In short, the estimated effect on Y of a unit increase in X is less precise.

Researchers can detect nonconstant error variance by plotting the residuals against the values of the explanatory variable or the predicted values ($\hat{Y}$). If the error variance is constant, most residuals will be contained in a horizontal band of approximately uniform width around the zero line. If the variance of the error terms changes, however, a trapezoidal pattern, with the spread of the residuals increasing or decreasing with X or $\hat{Y}$, will likely result (Neter and Wasserman 1974:103).

Statisticians recommend using a weighted least squares (WLS) fit to eliminate unequal error-term variance so that minimum-variance estimators for the coefficients can be obtained. When nonconstant error variance is a problem in biological applications, the standard deviation of the residuals often increases as the value of the explanatory variable increases (Chatterjee and Price 1977:47). If this is the case, one can use the OLS procedure on transformed variables to

KEVIN J. GUTZWILLER and STANLEY H. ANDERSON: Wyoming Cooperative Fishery and Wildlife Research Unit, Department of Zoology and Physiology, University of Wyoming, Laramie, Wyoming 82071

get a WLS fit. The transformed variables for the simple, linear model are

$$Y' = Y/X \quad (2)$$

and

$$X' = 1/X. \quad (3)$$

The equation for the model using transformed data is

$$Y/X = \beta_1' + \beta_0'/X, \quad (4)$$

and that for the model based on the original variables is

$$Y = \beta_0' + \beta_1'X. \quad (5)$$

One fits equation (4) using OLS and then substitutes β_0' and β_1' into model (5). The residuals from (5) should not show evidence of unequal error variance.

The effects of using the above procedure are demonstrated in the following example adapted from Chatterjee and Price (1977:44–49). To study the relation between snake density (Y) and distance to hibernaculum (X) for 27 desert communities, researchers fitted Y vs X using OLS; the estimates were 0.115 (SE = 0.011) for the slope (β_1) and 14.448 (SE = 9.562) for the intercept (β_0). A plot of the residuals vs X (Fig. 24.1*a*) suggested nonconstant error variance, so they fitted a WLS equation using the transformations described above. The estimates became 0.121 (SE = 0.009) for the slope and 3.803 (SE = 4.570) for the intercept, and the plot of the residuals vs $1/X$ (independent variable in the transformed model) showed no strong evidence of unequal error variance (Fig. 24.1*b*). These differences in the coefficients and the reduction in the standard error of the estimates resulting from the WLS fit demonstrate that model precision and interpretation may depend significantly on whether the assumption of constant error variance has been met.

A WLS fit can also be used to eliminate nonconstant error variance in multiple regression models. If nonconstant error variance arises from more than one explanatory variable, weights are estimated from empirical, intuitive, or theoretical information, and the coefficients are obtained through iterative estimation (e.g., Chatterjee and Price 1977:107–115).

Transformations of just the Y data are also used to remedy nonconstant error variance. If the Y values are counts, statisticians recommend the square root transformation; if they are percentages or proportions, analysts apply the arc sine transformation; and if the variance of Y increases with X, modelers use the logarithmic transformation (Zar 1974:220–221). These and other transformations (Montgomery and Peck 1982:88–89) are applicable to both simple and multiple regression models.

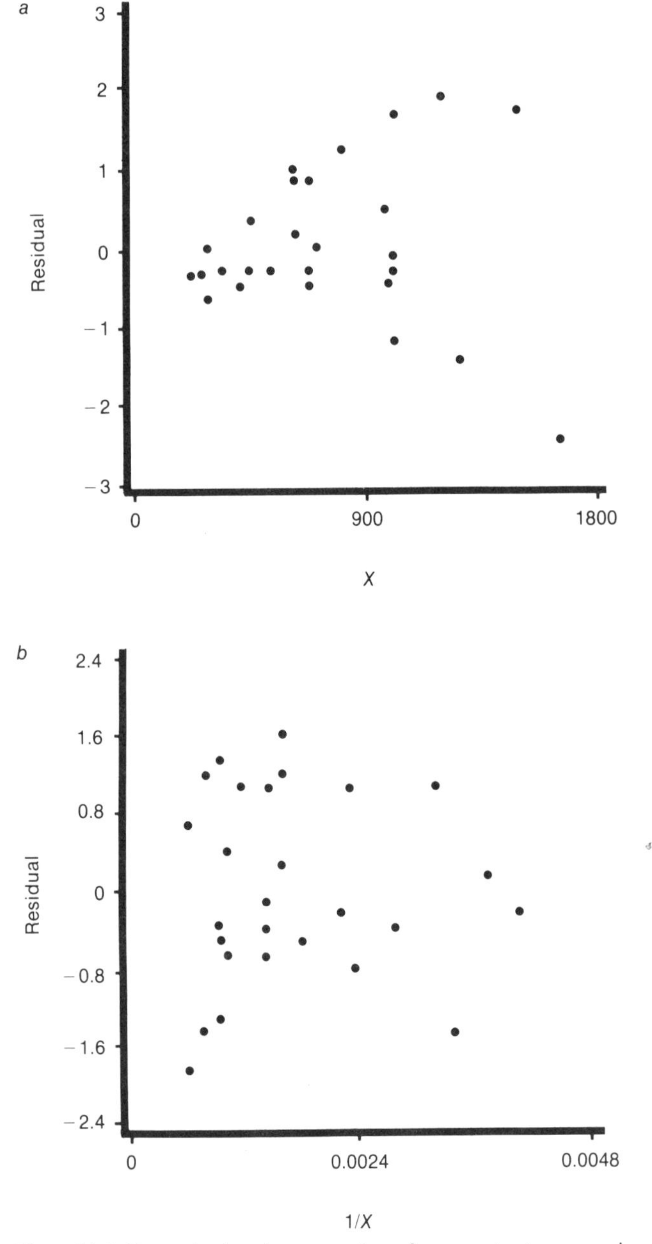

Figure 24.1. Example showing correction of nonconstant error variance using the weighted least squares procedure: (*a*) standardized residuals vs X; (*b*) standardized residuals vs $1/X$. (Modified and reprinted, with permission, from S. Chatterjee and B. Price. 1977. Regression Analysis by Example. John Wiley and Sons, Inc., New York.)

Nonindependence of error terms

In regression one assumes the error terms are not correlated, but if only moderate departure from this assumption exists, valid results are still obtainable (Freund and Minton 1979:92). When error terms are correlated and the observations constitute a natural sequence in time or space (autocorrelation), regression coefficients do not have minimum variance, estimates for MSE and the standard errors of the coefficients (as computed by OLS) may be significant underestimates, and confidence intervals and tests involving t- and F-statistics are not valid (Neter and Wasserman 1974:352).

Two common methods exist to detect autocorrelation in simple or multiple regression models. One is to plot the residuals against time. For positive autocorrelation, a group of positive residuals is followed by a group of negative residuals, and so on (Chatterjee and Price 1977:126); a rising-then-falling trend is also possible (Montgomery and Peck 1982:351). Negative autocorrelation is indicated by the lack of residual groups and a rapid alternation of signs for successive residuals (Montgomery and Peck 1982:67). A second method is to compute the Durbin-Watson statistic (d), and compare it to appropriate critical values (Chatterjee and Price 1977:125–127).

A significant Durbin-Watson d-value or a residual plot indicating temporal or spatial dependency may arise because of real autocorrelation or because of apparent autocorrelation due to missing explanatory variables (Montgomery and Peck 1982:353–354). Chatterjee and Price (1977:128–130) demonstrated how to remove autocorrelation resulting from an actual time dependency. The following example, based on Montgomery and Peck (1982:351–355), shows how to resolve the problem of correlation when there is an apparent time dependency arising from a missing X variable. Using data collected during a 20-year period, ecologists regressed small-mammal density (Y) against shrub cover (X) in an area. Checking for autocorrelation, they found a significant d-statistic and a marked rising-then-falling trend in a plot of the residuals vs time (Fig. 24.2*a*). They suspected winter snow depth to be the missing regressor (because it was correlated with time and small-mammal survival), so they included it in the analysis and reestimated the model. After this step, neither the Durbin-Watson d nor the residual plot (Fig. 24.2*b*) showed evidence of autocorrelation. Thus, the ecologists corrected the apparent time dependency and avoided the problems listed earlier by including an important explanatory variable in their model.

Outliers

Outliers are observations that are far from typical of most of the data set. In OLS regression, "a fitted line is pulled disproportionately toward an outlying observation because the sum of the squared deviations is minimized" (Neter and Wasserman 1974:107). Extreme points can, therefore, significantly affect model estimation. The analyst's first reaction may be to eliminate outliers from the analysis. But atypical points can be very informative because they may arise from events of special interest. "A safe rule frequently suggested is to discard an outlier only if there is direct evidence that it represents an error in recording, a miscalculation, a malfunctioning of equipment, or a similar type of circumstance" (Neter and Wasserman 1974:107).

The following example, adapted from Chatterjee and Price (1977:19–27), demonstrates how to detect outliers, and it illustrates the possible effects of outliers on model estimation and interpretation. This example involves the elimination of four extreme points; the cause of these was not identified, but ordinarily one would follow the above guidelines for deleting atypical points. Using data from 30 different regions, biologists regressed the number of salamander species (Y) against an index of wetland diversity (X). A plot of Y vs X indicated a linear relationship, but potential outliers were also shown. So they assessed the adequacy of

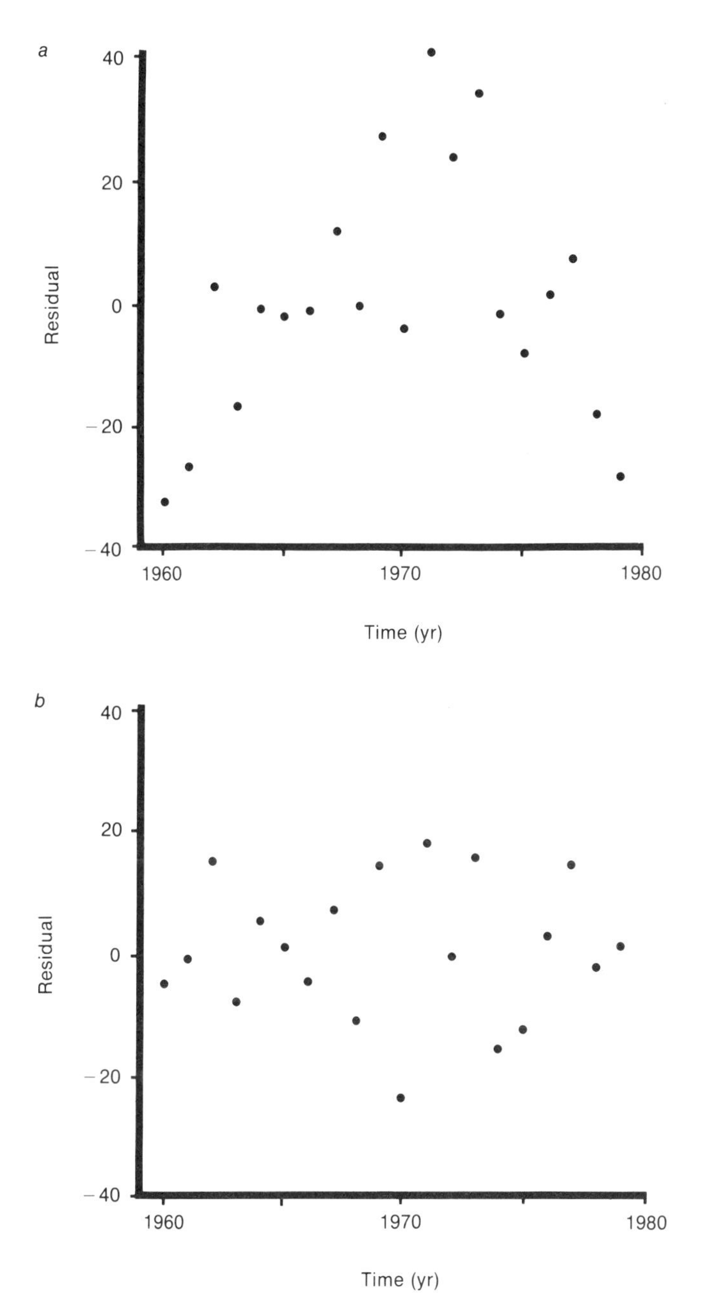

Figure 24.2. Example illustrating correction of correlated error terms: (*a*) residuals vs time for uncorrected model; (*b*) residuals vs time for corrected model. (Adapted and reprinted, with permission, from D. C. Montgomery and E. A. Peck. 1982. Introduction to Linear Regression Analysis. John Wiley and Sons, Inc., New York.)

their model by examining several residual plots. Graphs of the standardized residuals vs X (Fig. 24.3*a*) and the standardized residuals vs the predicted values (Fig. 24.3*b*) again showed four extreme points. In addition, the remaining 26 observations formed a sloping pattern instead of the expected horizontal band. The results confirmed the presence of four outliers and that the linear relation the biologists assumed was not appropriate. To check their work further,

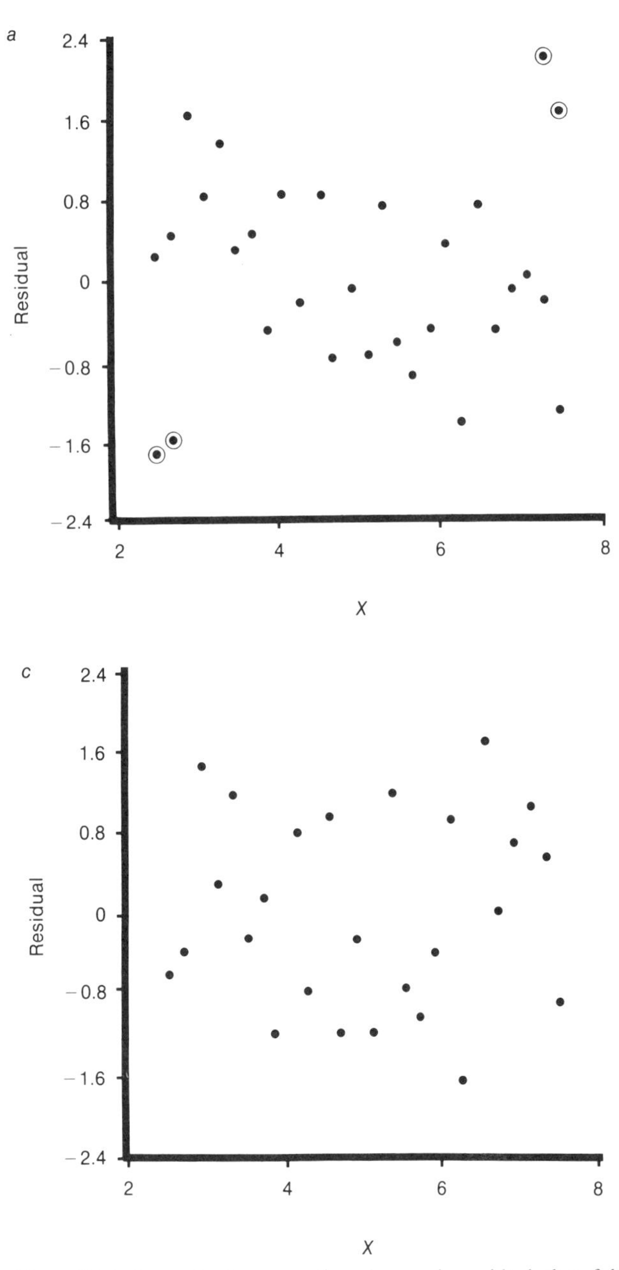

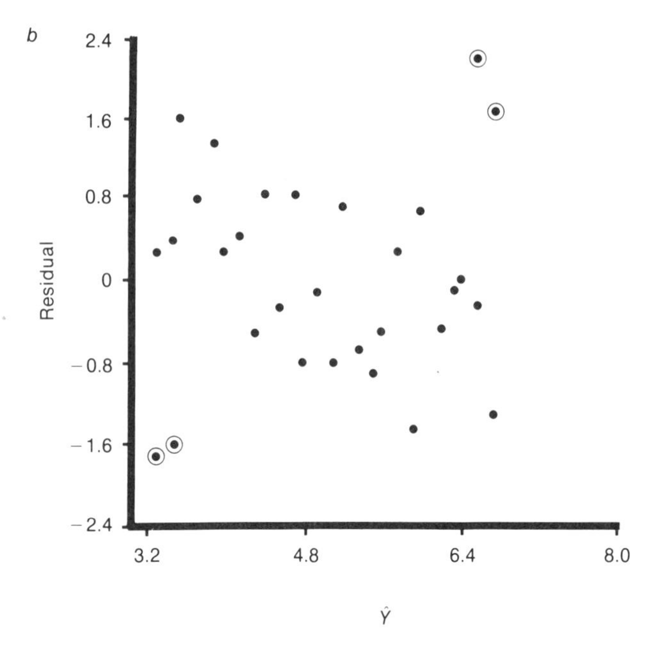

Figure 24.3. Example demonstrating changes in residual plots following elimination of outliers (circled points): (*a*) standardized residuals vs *X;* (*b*) standardized residuals plotted against predicted values; (*c*) standardized residuals vs *X*, after removal of outliers. (Modified from S. Chatterjee and B. Price. 1977. Regression Analysis by Example. John Wiley and Sons, Inc., New York, with permission.)

they dropped the four extreme observations from the analysis and reestimated the model. With the reduced data set ($n = 26$), the plot of the standardized residuals vs X (Fig. 24.3*c*) showed no evidence of the above departures. Various regression summary statistics also changed after the four points were deleted; R^2 decreased from 0.396 to 0.161, and b_1 (coefficient for X) decreased from 0.665 ($SE = 0.155$, $t = 4.287$) to 0.260 ($SE = 0.121$, $t = 2.147$). Considering the magnitude of change here, it is clear that just a few outliers can have a large effect on model estimation and, therefore, model interpretation.

Sometimes atypical points cannot be positively identified as errors and eliminated, and they are therefore included in the analysis. But if such points are still suspect, the analyst can use robust regression to place less emphasis on them (relative to the remaining data) and thus reduce their disproportionate influence on model estimation. Montgomery and Peck (1982:364–381) discussed the underlying theory of robust regression and provided two examples for understanding its application.

Acknowledgments

We thank D. A. Anderson, R. S. Cochran, D. M. Finch, L. L. McDonald, and A. L. Ward for reviewing this manuscript; P. Gutzwiller and H. Kulas-Adler for preparing the figures; and M. Standard and C. Steadman for typing several drafts.

25

Assumptions and Design of Regression Experiments: The Importance of Lack-of-Fit Testing

JOHN M. MARZLUFF

Abstract.—I present an experimental design tailored for regression analyses of avian-habitat relations but applicable to all correlation-based multivariate procedures. Prior knowledge of wildlife response is required; therefore, this design will be most useful for testing hypothesized wildlife-habitat relationships. Pure error is calculated so that fit of the proposed population model can be tested and maximum R^2 determined. Maximum R^2 indicates how much of a wildlife response can be explained, because it measures the degree to which organisms respond to factors other than measured habitat variables. Measurement of organism response should be replicated at endpoints and midpoints of the habitat gradient in order to test biologically realistic curvilinear models. This is possible if habitat measurements are assigned to discrete categories. Category width should reflect the execution precision of management recommendations. Replications should be spatial and temporal and within biological time periods (e.g., territory establishment, nesting, fledging). Site comparisons are limited to 21 geographic localities, and I suggest that no more than three explanatory variables be investigated at once. Unambiguous recommendations concerning wildlife response to key habitat components will result; however, these are limited in generality by the variability in explanatory variables not investigated and the geographic range covered.

Statistical appraisal of wildlife-habitat relationships depends upon fulfillment of test assumptions. When assumptions are violated, the probability of observing a given value of a test statistic cannot be appraised by conventional methods. Carefully designed experiments provide data meeting test assumptions. However, many investigators of wildlife-habitat relationships proceed without adequate experimental design and then subject the data collected to multivariate procedures, hoping that assumptions will be approximated or that the consequences of their violation will not affect management recommendations. Rigorous model validation (Johnson 1981a) and objective testing of assumptions are needed to appraise the seriousness of such violations. This chapter was prompted by my concern that wildlife researchers are developing increasingly complex models (e.g., Shugart 1981) before adequately validating current ones and without incorporating statistically based experimental designs to guide data collection.

A basic assumption of multivariate procedures is that the proposed model follows the form (e.g., linear, quadratic) of the population response. Although Shelford's (1913) ecological rule of tolerance relates individual fitness, and possibly population density, to habitat gradients in a nonlinear fashion, models of nonlinear responses are rare in wildlife-habitat studies (Johnson 1981a, 1981b) or are selected over linear response models without reference to the population response (e.g., Meents et al. 1983).

My objective is to relate lack-of-fit testing to studies of wildlife habitat. This test objectively appraises the congruence of modeled and population responses and is a powerful tool to judge between competing linear and nonlinear models similar in other ways (e.g., R^2) (Sugihara 1981). I propose an experimental design that measures pure error, thus allowing lack of fit to be tested. Regression models are specifically considered because regression as a multivariate technique has been only partly developed for wildlife studies (Johnson 1981a). This chapter, together with those by Gutzwiller and Anderson (Chapter 24), Rotenberry (Chapter 31), and Rice et al. (Chapter 13), is intended to fill this void.

The lack-of-fit testing procedure

Lack-of-fit testing is a standard part of all thorough expositions on regression (e.g., Box et al. 1978; Weisberg 1980; Draper and Smith 1981), but it is not a part of standard statistical packages such as SPSS, MINITAB, or BMDP. Moreover, it has not been used in studies of wildlife habitat, to my knowledge, even when data suitable for testing were available (e.g., Meents et al. 1983). The underlying assumption of least-squares estimation is that lack of fit does not occur. Violation of this assumption precludes valid probability estimation for F-tests of regression significance and t-values used for computing confidence limits. Residual analyses, particularly examination of e_i (observed minus predicted Y values) vs $\hat{Y}_i$ (Y value predicted by the regression equation) plots, may suggest that proposed models suffer from lack of fit (Gutzwiller and Anderson, Chapter 24), although residual analyses are somewhat subjective. Lack-of-fit testing constitutes an internal form of model evaluation because adherence to assumptions is objectively appraised.

JOHN M. MARZLUFF: Department of Biological Sciences, Box 5640, Northern Arizona University, Flagstaff, Arizona 86011

TEST DEVELOPMENT

Mathematical development of the lack-of-fit procedure includes several parameters and constants. I have followed the notation in Draper and Smith (1981:33–43) and list computational formulas for these in Table 25.1.

Lack-of-fit testing compares the variation about the regression line, $\sigma^2_{y \cdot x}$, with random variation in the Y population, σ^2. Note that $\sigma^2_{y \cdot x}$ and σ^2 are population constants; $\sigma^2_{y \cdot x}$ is the variation of the population response about the regression line, and σ^2 represents the portion of this variation due to the variance in the normal distribution of error in the population response. If the proposed model has the same form as the true model, then $\sigma^2_{y \cdot x} \approx \sigma^2$ and the expected value of the residual mean square (s^2, Table 25.1) is σ^2; hence, s^2 is an appropriate quantity to relate to the amount of variation explained by the covariation of X and Y in the assignment of overall regression significance. If the postulated model suffers from lack of fit with the true model, $\sigma^2 < \sigma^2_{y \cdot x}$, because systematic bias error in $\sigma^2_{y \cdot x}$, due to lack of fit, is added to random error (σ^2). For example, when a postulated linear model is fitted to data from an inherently quadratic relationship, the residuals are likely to differ systematically in relation to the magnitude of X, being negative at very low and at high X values, but positive at intermediate values.

Testing for lack of fit requires an estimate of σ^2 to compare with s^2, the estimate of $\sigma_{y \cdot x}$. Prior study of the system may provide such an estimate (Weisberg 1980), but this is not typical in wildlife studies. Alternatively, replicate measures of Y at identical levels of X can provide an independent estimate of σ^2, termed pure error.

Total residual sum of squares (SSE) is divided into two components: pure error and lack of fit. Pure error sum of squares (SS_{pe}) is calculated by summing, for each level of X having repeats, the squared deviation of each Y from the average Y at that X level (Table 25.1). SS_{pe} is associated with n_e = (number of replications − number of X-levels replicated) degrees of freedom. Pure error mean square, $S_e^2 = SS_{pe}/n_e$, is an estimate of σ^2 regardless of whether the model being fit is correct or not. Sum of squares due to lack of fit, $SS_L = SSE - SS_{pe}$, is associated with $n_L = df_{SSE} - n_e$ degrees of freedom. Mean square due to lack of fit, $MS_L = SS_L/n_L$, estimates σ^2 if the proposed and population responses are of the same form, but it overestimates σ^2 if lack of fit occurs.

The F-test for lack of fit is performed by comparing the ratio MS_L/S_e^2 with the $100(1 - \alpha)\%$ point of an F-distribution having n_L and n_e degrees of freedom. If this F-value is significant, regression significance is not meaningful because the proposed model suffers from lack of fit at the $100(1 - \alpha)\%$ level. If it is not significant, no evidence for lack of fit between proposed and true models is evident and both S_e^2 and MS_L can be pooled into s^2 as an estimate of σ^2. Overall significance of regression can then be tested as usual with $F = MSR/s^2$. Whether the lack-of-fit test is significant or not, residuals should still be analyzed for detection of outliers, violation of assumptions about normality and variance equality, and possible additional independent variables to be included in the model (Gutzwiller and Anderson, Chapter 24).

Table 25.1. Computational formulas of parameters cited in text

Parameter	Computational formula	Source[a]
s^2[b]	$\sum_{i=1}^{n} (Y_i - \hat{Y}_i)^2/(n - m)$	21
SSR[b]	$\sum_{i=1}^{n} (\hat{Y}_i - \bar{Y})^2$	21
SST[b]	$\sum_{i=1}^{n} (Y_i - \bar{Y})^2$	21
SSE[c]	$\sum_{j=1}^{m} \sum_{u=1}^{n_j} (Y_{ju} - \hat{Y}_j)^2$	37
SS_{pe}^2[c]	$\sum_{j=1}^{m} \sum_{u=1}^{n_j} (Y_{ju} - \bar{Y}_j)^2$	37

[a] Page number in Draper and Smith (1981).

[b] i indexes the n individual response measurements (Y_i). $\hat{Y}_i$ are the responses predicted by the regression model, and m is the number of predictor variables plus the intercept included in the model.

[c] j indexes each of the m levels of X for which repeat measures are available, and u indexes the n_j individual repeat Y measures at each X_j.

A HYPOTHETICAL EXAMPLE

Suppose we are interested in modeling the relationship between warbler abundance and percent cover, from 0% to 100%, at canopy and shrub layers. Warblers were counted at 11 combinations of canopy and shrub cover (Table 25.2). Six replicate plots of canopy and shrub cover were surveyed at five canopy and shrub configurations. The sum of individual pure error sums of squares for all replicated cover values gave $SS_{pe} = 35.33$, with $(6 - 1)(5) = 25$ df; $S_e^2 = 35.33/25 = 1.41$. I initially entertained the linear model $\hat{Y} = b_0 + b_1X_1 + b_2X_2$ (Table 25.3), which accounted for 34% of the variation in warbler abundance ($F_{2,33} = 8.49$, $P < 0.01$). Here $SS_L = 284.46 - 35.33 = 249.13$, with $df_L = 33 - 25 = 8$, and $MS_L = 249.13/8 = 31.14$. The ratio of these two variance estimates gave $F = 22.09$ ($P \ll 0.01$, $F_{8,25}$). The proposed model suffered from lack of fit, and regression significance ($F_{2,33} = 8.49$) could not be validly determined.

Table 25.2. Pure error calculations for hypothetical warblers counted at various levels of canopy and shrub cover

Percent Canopy	Percent Shrub	No. warblers (Y)	$\bar{Y}$	$\Sigma (Y - \bar{Y})^2$	df
0	0	0,2,0,0,0,0	0.33	3.33	5
0	100	0,0,2,0,1,0	0.50	3.50	5
100	0	4,7,4,7,6,7	5.83	10.83	5
100	100	3,6,4,6,5,5	4.83	6.83	5
50	50	10,8,7,10,9,11	9.17	10.83	5
50	10	2			
10	50	8			
25	25	3			
75	75	8			
50	75	5			
75	50	9			

Note: $SS_{pe} = 35.33$, $df_{pe} = 25$.

Table 25.3. Components of variation for hypothetical warbler data fitted to a general linear model

Source	*df*	*SS*	*MS*	*F*
Regression	2	146.29	73.15	8.49
Residual	33	284.46	8.62	
Lack-of-fit	8	249.13	31.14	22.09
Pure error	25	35.33	1.41	
Total	35	430.75		

Residual analyses suggested that warblers responded quadratically to cover in this example, being most abundant at intermediate shrub and canopy densities. A quadratic model of the form $\hat{Y} = b_0 + b_1X_1 + b_2X_2 + b_3X_1^2 + b_4X_2^2$ fit the data much closer ($R^2 = 87\%$, $F_{4,31} = 54.85$). Pure error calculations were unchanged; however, *SSE* was reduced from 284.46 to 53.41. The $SS_L = 53.41 - 35.33 = 18.08$, with $31 - 25 = 6$ *df*; $MS_L = 3.01$, which was closer to S_e^2. The *F*-test for lack of fit was not significant ($F_{6,25} = 2.14$, $P > 0.05$). Therefore, the proposed model provided a better representation of these data. With this assumption fulfilled, both MS_L and S_e^2 were pooled, giving $s^2 = 1.72$. Significance of the regression could now be validly appraised by comparing $F = 54.75$ to $F_{\alpha,3,32}$. The result was significant at $\alpha = 0.01$.

Significant models with high R^2 that suffer from lack of fit misrepresent the population response. Management recommendations based on such models may be incorrect. In the above example, if high levels of warblers were desired, a manager using the "significant" linear model would propose management steps to increase shrub and canopy cover; however, because the population was actually responding quadratically to cover, warbler abundance would not increase. The correct management recommendation, suggested by the model without lack of fit, would be to create intermediate levels of canopy and shrub cover.

Replication

Independent replicate measures of *Y* measure random variation in *Y*. If independent replicates are not obtained, S_e^2 will underestimate σ^2 and the lack-of-fit test will incorrectly indicate nonexistent lack of fit (Draper and Smith 1981). In the multivariate case, each explanatory variable must be equal, or nearly so, on replicate plots. "Similar" is a subjective criterion of equality, one which may differ among disciplines. For wildlife studies, I suggest that all replicates of a habitat variable be within ±5% of that variable's mean value. For example, if old-growth plots have an average canopy cover of 80%, then all old-growth replicates should have between 75% and 85% canopy cover. This is an initial recommendation based on the range of precision timber managers can be expected to obtain when executing a prescribed treatment (B. Grecko, pers. comm.). Alternatively, plot similarity may be based on biological considerations, because species may respond differently to vegetation measures within 10% of one another. From a management perspective, however, models incorporating this variation as pure error will more realistically represent wildlife responses after prescribed treatments are executed.

Genuine replicates should be subject to all error sources affecting previous counts (Box et al. 1978), which means that replicate counting is done under similar environmental conditions (e.g., time of day, precipitation regime) and biological states (i.e., point in species' annual cycle). All replicate plot parameters need not be equal. Model generality is increased if attributes other than explanatory ones (e.g., aspect, altitude, and soil characters of the site) are allowed to vary freely (i.e., plot selection is random with respect to them). Variation among replicates increases pure error, but such models represent wildlife response to habitat regardless of site location.

How does one sample so that replicates are statistically valid? Counting two similar, geographically isolated study sites during biologically equal time periods (e.g., nest building) produces true replicates. Surveying the same study site along different survey routes or by randomly choosing a starting point on the same route (Gates 1981) produces pseudoreplicates (Hurlbert 1984). Counts such as these may be highly dependent upon one another and serve only to measure intrasite variability. Counts of the same route in rapid succession, or by multiple observers, provide virtually simultaneous readings of the response variable and are therefore pseudoreplicates which may conservatively assess intrasite variation. Whether samples are true replicates or pseudoreplicates, pure error is not attributable to vegetation differences. The most detailed management recommendations and their precise implementation will suffer from this error.

Maximum R^2 and management

The concept of maximum attainable R^2 allows managers to appraise potential effectiveness of management models. With replicate measures of *Y*, R^2 cannot equal 100% because of random variation in *Y*. Maximum attainable $R^2 = (SST - SS_{pe})/(SST)$, where $SST = SSR + SSE$ (Table 25.1). The R^2_{max} represents response variation that can be accounted for by changes in the explanatory variables. The model R^2 divided by R^2_{max} gives the percentage of the explainable variation accounted for, which may be a more realistic summary of model adequacy. The hypothetical relationship (Table 25.2), for example, had $R^2_{max} = (430.75 - 35.33)/(430.75) = 0.918$. At best, a manager might expect to account for 92% of the variation in warbler abundance. The selected model accounted for $0.876/0.918 = 95\%$ of this explainable variation.

Experimental design

Background

Frequently wildlife researchers view multivariate techniques, especially regression, as "cure-all" procedures to be used when data are collected haphazardly with respect to underlying habitat variation. Alternatively, studies should be designed that will produce data suitable for the preferred test. Well-designed studies may enable researchers to induce independence among the explanatory variables so that multicollinearity does not obscure effects of single variables (Hinchen 1970). Inability to assess independently the importance of individual habitat variables is perhaps the greatest limitation to interpretation and implementation of many habitat models.

The initial step in designing studies is to specify goals. Two main goals in wildlife-habitat studies are understanding how habitat changes affect wildlife and predicting the responses of wildlife to these changes. Understanding may increase predictive ability, but prediction is not entirely dependent upon understanding. The predictive equation may not incorporate biologically relevant predictors and may produce biased results, especially beyond the range of sampled data and sampling conditions, even though it may be a valid representation of what was observed. R^2 is not affected by violation of assumptions, although its significance is, and it may be appropriately used to indicate variation in the response accounted for by explanatory variables.

Wildlife-habitat studies designed to produce predictive models provide hypotheses of wildlife response and may suffice as short-term management tools; however, if long-term use of the model is a goal, then understanding hypothesized wildlife-habitat relationships is highly desirable. Studies designed for understanding may take longer to produce highly predictive equations, but these equations will reflect true biological relationships and hence may produce robust recommendations applicable over a wider range of habitats. Below, I develop a design that I believe addresses the problem of understanding wildlife-habitat relationships. Although it is most suitable for studies of avian habitat requirements, it is generally applicable to other taxa.

Proposed sampling design

The importance of testing for lack of fit and the usefulness of pure error to appraise management potential dictate that studies include replicate measures. The obvious consequence of replicating measures at some X-levels is that fewer levels can be studied with a fixed effort. Current recommendations emphasize measurement of many different levels along a vegetation gradient. Johnson (1981b) recommends sampling at evenly spaced intervals along a gradient. Verner (1981:546) stated that "sampling many sites a few times is better than sampling a few sites many times." Such designs produce results applicable over a wide geographic range because many sites are surveyed; however, replicates during biological time periods may not be possible, which precludes lack-of-fit testing. Time spent surveying each plot also may be so short that species use of the habitat (e.g., for foraging or nesting) may not be determined. Therefore, habitat quality may not be accurately assessed (Van Horne 1983).

Quadratic relationships are the most likely alternative to linear ones in wildlife-habitat studies (Westman 1980; Johnson 1981a, 1981b). Draper and Smith (1981) illustrated level allocation for testing linear vs quadratic relationships when a total of 14 surveys are to be conducted. Designs allocating each survey to a different X-level, or splitting surveys equally among maximum and minimum X-levels, are quickly rejected, because lack-of-fit testing in either case is not possible. Replicating seven sites two times each (similar to recommendations by Johnson 1981a and Verner 1981) should be avoided because seven levels are not needed to test for quadratic fit and the standard deviation of b_1 is high with respect to sampling variance about Y. Strictly speaking, only $n + 1$ levels of X are needed to distinguish between models of degree n and $n - 1$ (e.g., three levels to distinguish linear from quadratic) (Draper and Smith 1981). Designs with three or four X-levels and the majority of replicates taken at the endpoints are recommended by weighing trade-offs in (1) the variability of the b_1 estimate, which decreases with a decreasing number of sites; and (2) balancing lack-of-fit and pure error degrees of freedom (df_L increases and df_{pe} decreases with the number of replicated samples).

Sampling a few levels of the explanatory variables many times is better than sampling many levels a few times because higher-order relationships can be tested against linear ones. Biologically, more information, such as nest location and perhaps nesting success, is obtained with such a sampling regime because plots are visited repeatedly throughout the season (Conner and Dickson 1980; Wiens 1981a). A balanced design of replicated endpoints may be preferred to variance proportional replication (see Gates 1981), because independence of explanatory variables is assured and unweighted ANOVA, ANCOVA, and other tests can be performed on the data set. I suggest selecting nonreplicate plots near habitat configurations with high response variation to assess this variation adequately. These plots also increase the number of X levels that can be investigated, so that third- and higher-order relationships can be tested. Plot selection from a wide geographic range, incorporating as pure error any variation in nonexplanatory variables, increases model generality. My recommendations pertain to selection among polynomial response models. If other nonlinear functions are to be tested (e.g., step functions or sigmoidal functions) other designs may be more appropriate.

As an illustration of level allocation, consider the design of plots varying in shrub and canopy cover presented in Table 25.2. Replicates were taken at each of the four endpoints in the array of canopy (c) versus shrub (s) cover (0%c/0%s, 0%c/100%s, 100%c/0%s, 100%c/100%s) and at the midpoint (50%c/50%s). This balanced design induces independence (in the sense of a 0 correlation) of the two explanatory variables. Six other single samples were taken at various intermediate combinations of shrub and canopy cover and were concentrated somewhat at the middle to upper cover levels, because warbler abundance was more variable there. The design could be readily expanded to incorporate more explanatory variables (Box et al. 1978); however, the number of points needing replication equals $2^n + 1$, where n is the number of explanatory variables. The design I have introduced here and expand upon below is a response surface design, which would allow for fitting of polynomial regression equations (see Rotenberry, Chapter 31).

This experimental design may not be attainable if, for example, plots of 100%c/100%s do not exist. In such cases, the maximum and minimum values found in nature determine endpoints of the study design. Mixture designs (see Snee 1971) address this problem and may be useful in studies in which some endpoints do not exist. Additionally, 0% or 100% endpoints of a gradient may contain nonrepresentative populations of the study organism. This would probably not be determined in studies of wide geographic scope, in which plots are briefly sampled. Management recommendations incorporating nonrepresentative habitat use may be misleading. If such relationships are determined, selection of endpoints nearer to the gradient mean, perhaps with ± 1 *SD* of the mean, should be selected for replication. The important consideration is that conclusions can be applied only to the joint range of explanatory variables investigated.

How should sampling be replicated?

Replication can be either temporal or spatial (Gates 1981); I recommend a combination of both. Relative numbers of each should be based on their respective costs and inherent variation (Cochran 1963). Temporal replication of the response at a given site is usually less expensive, but it may conservatively estimate pure error relative to spatially isolated sites. Spatial replicates are true replicates (Hurlbert 1984), but sites exactly replicating all explanatory variables may be difficult to locate if variation is viewed as continuous. However, if vegetation is assigned to discrete categories, with increments set according to the precision of treatment execution (precise execution requires precise replication), then exact replicates are easier to obtain. Given this difficulty and a limited budget, supplementing with temporal replicates may be desirable. Counts from temporal replicates should be averaged for each geographic site before inclusion in the regression analysis. This would alleviate pseudoreplication (Hurlbert 1984). Residual degrees of freedom would then be correct to test regression significance, and temporally averaged counts from geographically isolated "similar" sites could test for lack of fit. Seasonal trends within the same site may be of interest, but the temporal replicates I refer to are within biological time periods (see below). Averaging these does not greatly reduce information on seasonal influences but does give a better summary of activity within one season than a single measurement would. Temporal and spatial replicates should be used without averaging in the computation of total pure error needed to calculate R^2_{max}. Pure error comes from two sources, sampling error within a plot and random variation among true replicates. Temporal replicates estimate sampling error, whereas spatial replicates estimate random error in the response. Because both errors reduce the predictive ability of a model, I recommend including both in R^2_{max}.

Survey results are affected by the biological period (i.e., territory establishment, incubation, etc.) at which they are recorded (see Slagsvold 1977). Replicates at each X level should be within each of these periods. Models for each period individually or an overall model blocking biological periods could then be constructed (see Meents et al. 1983).

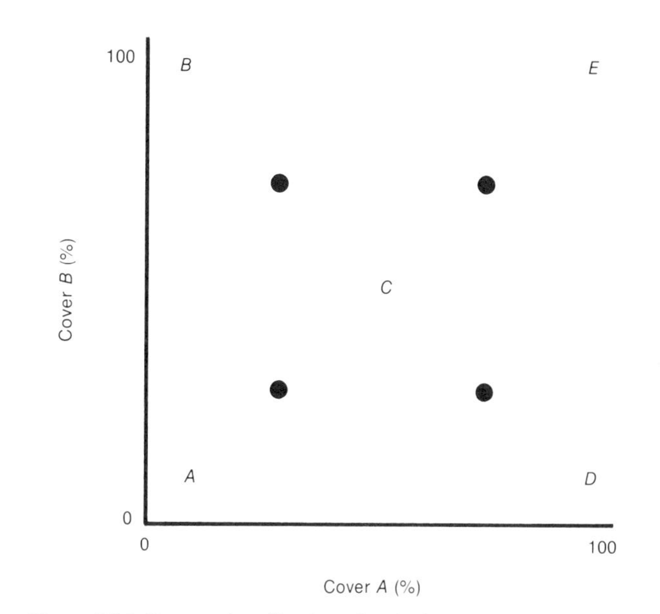

Figure 25.1. Proposed replication of endpoints (A,B,D,E) and midpoint (c) if two vegetation variables (Cover A and Cover B) are investigated. Solid circles are nonreplicated intermediate points.

Preliminary counting may be necessary to determine replicate intervals.

If replications must be completed within 2-week periods to ensure biological consistency (Lyon and Marzluff 1985) and counting is done twice a day for 6 days a week, 24 counts are available to allocate among replicate and nonreplicate sites. In a study investigating two explanatory variables, each of the five configurations to be replicated could be counted four times, leaving four nonreplicated intermediate configurations to be selected (Fig. 25.1). I suggest that each replicated configuration have eight survey plots, four in each of two spatially isolated sites (Fig. 25.2) (see Gates 1981 for plot

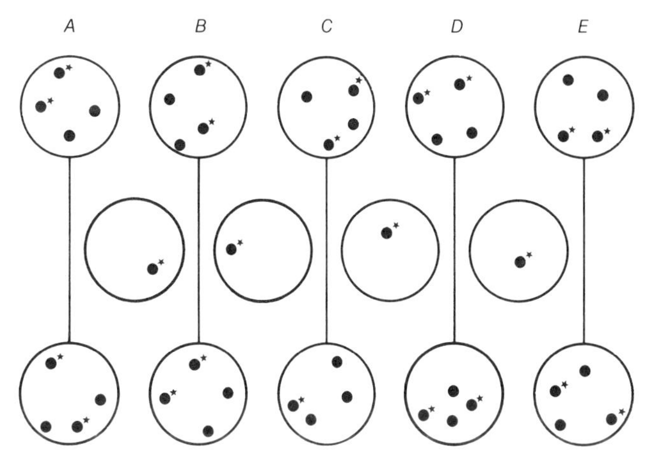

Figure 25.2. Plot location of spatial and temporal replicates. Large circles represent geographic sites encompassing survey plots (solid circles). Lines connect spatially isolated replicates. Starred plots exemplify those to be censused in one biological time period.

layout). Initially, two plots from each of the two spatial replicates would be randomly selected for counting animals (Fig. 25.2). The remaining two plots in each spatial replicate should be surveyed during the following 2-week period. This study design would allow for 14 geographic study sites (2 replicates × 5 sites + 4 nonreplicated intermediate sites) with 44 study plots (4 plots/replicate × 2 replicates/site × 5 sites + 4 nonreplicated sites) distributed among them.

If three counts per day are possible (see Marzluff and Lyon 1983), then 36 could be done in 2 weeks. I suggest modifying the above scheme in three ways: (1) let each replicate include three geographic sites if they are available, each containing four plots; (2) survey each replicate level six times per period, again by randomly selecting plots per geographic site; and (3) distribute six nonreplicate plots among intermediate vegetation levels. This design allows 21 geographic study sites with 66 plots distributed among them.

Constraints of experimental design

Estimating pure error reduces model applicability, because geographic sites are limited to 21 (3 counts/day); however, because plots are surveyed more often (12 times during a 12-week breeding season), biological use of plots would be better known. Choice of a counting method is certainly area-specific (Verner 1985), but if transects were used, they could be of sufficient length (1000 m) (Conner and Dickson 1980) and visited often enough to obtain good estimates of the population (Shields 1979; Conner and Dickson 1980).

Replication also limits studies to investigation of no more than three explanatory variables, because replication of the 17 ($2^4 + 1$) X-levels required to investigate four variables would consume all survey time. This recommendation certainly goes against the flow of current uses of multivariate analyses in wildlife studies, which follow Green (1971) in claiming that species' niches are n-dimensional (Hutchinson 1957), so the number of explanatory variables assessed should approach n (but see Johnson 1981a). Because vegetation variables are not orthogonal, interpretation of significant correlations between vegetation and wildlife abundance is difficult when numerous explanatory variables are investigated simultaneously. Wildlife response may be better represented by including numerous habitat variables in a model, but recommendations to practicing managers are ambiguous. Results from a designed experiment are necessarily more clearcut because multicollinearity between explanatory variables is reduced (Hinchen 1970). Recommendations, although limited to a few key vegetation variables and applicable only to a reduced geographic range, are therefore not ambiguous. Successful application of these recommendations will depend upon whether the key variables causally influence wildlife or just co-vary with causal factors.

Implementation of this study design is contingent upon selection of key habitat components. This may necessitate 1 or 2 years of prior study in which many habitat variables and their covariable influence on wildlife are investigated (T. A. Larson and J. Verner, pers. comm.). Reliance on literature or on initial reconnaissance to determine habitat components used for foraging (Holmes 1981) or nesting may reduce time searching for key components. I also suggest that variables meaningful to timber managers, such as basal area, be considered in wildlife models (see Marzluff and Lyon 1983). I do not advocate ignoring other habitat variables; they can be measured. But site selection and model construction should be based on two or three key ones.

For effective implementation of study design the response variable should also be limited. Instead of trying to account for variation in all species of wildlife in one study, several studies, each focusing on a few ecologically similar species, would be more effective. Sites could be chosen to cover accurately the range of vegetation used by these species, and replicate intervals would be sufficient for all species.

Conclusions

The experimental design I have detailed will produce robust, predictive models reflecting our understanding of wildlife responses to a few key habitat components. These models are not confounded by multicollinearity of explanatory variables and match the form of the population response. Such models are limited in generality in proportion to the geographic region surveyed, the variability in uninvestigated explanatory variables, and the study duration. Inclusion of nonreplicate sites in intermediate habitat configurations and use of temporal pseudoreplication increases model generality. Influential components of the habitat, the variation in wildlife response along these components, and acceptable limits of dissimilarity among spatial replicates must be identified prior to use of this design. This may take a year or two of site reconnaissance and wildlife sampling, which should be viewed only as a beginning. Initial management recommendations should be updated with results from studies such as I propose, and these results should ideally be tested with manipulation experiments so that causal relationships between habitat and wildlife can be determined.

The type of designed, experimental approach I advocate is nothing new, but it is still not a common part of wildlife research. Current studies attempt to relate many habitat components to an entire wildlife community simultaneously, without adhering to sound experimental design. Ambiguous correlations may result and may become the basis for management recommendations. My proposals have been nicely summarized by Johnson (1981a): "Where we cannot design experiments freely, select optimal levels of variables at will, and replicate as often as we desire, we must be reserved about our findings." I believe a flexible approach to the design I have presented will allow adequate replication at appropriate levels and will enable us to be less reserved about our findings.

Acknowledgments

I thank Jeff Brawn, Graydon Bell, Douglas Johnson, Jack Lyon, Bruce Marcot, Brian Maurer, Barry Noon, Fred Stormer, Rich Turek, and Jerry Verner for suggesting appropriate improvements in the development of these ideas. I also thank Colleen Sanford for proofreading the manuscript and Debbie Meier for drafting the figures.

26

Multivariate Models of Songbird Habitat in New England Forests

DAVID E. CAPEN, JAMES W. FENWICK, DOUGLAS B. INKLEY, and ALLEN C. BOYNTON

Abstract.—Three studies led to the development of species-habitat models for songbirds in a northern hardwood forest. Each study was designed to discriminate between used and unused study plots, but plots were assigned to groups based on different criteria: bird use, territory site, or nest site. Models which predicted used and unused potential habitat were developed by discriminant function analysis and logistic regression, and then evaluated by a jackknife procedure, cross-validation, and classification of independent data. Models for three species of birds are presented to illustrate unacceptable results of classifying indpendent data. Reasons for failure of these models include inappropriate design, multicollinearity of habitat variables, and inadequate sampling of habitat diversity. Models for a fourth species, the red-eyed vireo (*Vireo olivaceus*), illustrate some inconsistency, but two of four models show reasonable rates of classification.

The demand for predictive models that relate single species or groups of species to measurable components of their habitats has escalated in recent years, largely as a result of habitat mitigation programs (Schamberger and Krohn 1982) and legislative directives for managing wildlife indicator species (Verner 1983). Such predictive models depend on quantitative information about the habitat requirements of selected wildlife species.

The recent interest in quantifying components of a species' habitat has largely been based on the theoretical perception of an animal's niche as a multidimensional space (Hutchinson 1957). One portion of the niche, the habitat niche, can be described by measuring appropriate environmental variables and using multivariate statistical analyses to reduce the number of variables to those which describe important dimensions of the niche (Green 1971). The classic paper by James (1971), in which she described the "niche-gestalts" of breeding birds, stimulated much research on the topic of bird-habitat relationships, but many subsequent studies dealt mostly with niche theory and community dynamics, and were not designed to address management needs. Shugart (1981) described the synthesis of scientific research which has progressed from niche theory to the practical objective of predicting the distribution and abundance of animals in space and time.

A useful application of multivariate statistics in developing predictive habitat models is to employ discriminant function analysis (DFA) to distinguish between "used" and "unused" areas of potential habitat. This experimental design has been used in numerous studies (e.g., Martinka 1972; Anderson and Shugart 1974; Conner and Adkisson 1976; Smith et al. 1981a; Whitmore 1981; Conner et al. 1983; Rice et al. 1983a). We report on three studies designed to develop species-habitat models for songbirds (Passeriformes) during the breeding season.

The purposes of this paper are (1) to review study designs and the statistical development of predictive single-species habitat models; (2) to demonstrate validation procedures for the models; and (3) to critique the field and analytical procedures used in constructing multivariate species-habitat models.

Methods

Data were collected from 1978 to 1982 in southern Vermont on the Grafton Forest Resources Project study area in the eastern foothills of the Green Mountains (43° 11′ lat, 72°37′ long). The study area was 1200 ha in size and ranged from 244 to 549 m in elevation. Second-growth deciduous trees characterized the forest, but patches of conifers were present on some sites. Detailed descriptions of the study sites can be found in Boynton (1979) and Inkley (1980).

Design of field studies

The three study designs differed most importantly in the manner in which habitat plots were classified as used or unused. In one study (Inkley 1980), plots were classified by noting the repeated occurrence of singing male birds; these

DAVID E. CAPEN: Wildlife and Fisheries Biology Program, University of Vermont, Burlington, Vermont 05405

JAMES W. FENWICK: Wildlife and Fisheries Biology Program, University of Vermont, Burlington, Vermont 05405. *Present address:* Department of Statistics, University of Wyoming, Laramie, Wyoming 82071

DOUGLAS B. INKLEY: Wildlife and Fisheries Biology Program, University of Vermont, Burlington, Vermont 05405. *Present address:* Wyoming Cooperative Research Unit, University of Wyoming, Laramie, Wyoming 82071

ALLEN C. BOYNTON: Wildlife and Fisheries Biology Program, University of Vermont, Burlington, Vermont 05405. *Present address:* North Carolina Wildlife Resource Commission, Cullowhee, North Carolina 28723

Table 26.1. Habitat variables used in species-habitat models for songbirds (red-eyed vireos, American redstarts, ovenbirds, and Blackburnian warblers)

Variable	Variable description	Unit
BATOTL	Total basal area of trees and shrubs	m^2/ha
BATOD	Total deciduous basal area (including shrubs)	m^2/ha
BACSL	Basal area of coniferous trees >4 cm dbh	m^2/ha
BADSAW	Basal area of deciduous trees >28 cm dbh	m^2/ha
BACSAW	Basal area of coniferous trees >23 cm dbh	m^2/ha
BADPLP	Basal area of deciduous trees 4–28 cm dbh	m^2/ha
BACPLP	Basal area of coniferous trees 4–23 cm dbh	m^2/ha
BACLC	Basal area of coniferous trees 10–15 m high	m^2/ha
BASNAG	Basal area of trees >4 cm dbh, <50 % live	m^2/ha
DDOM	Mean basal area of deciduous trees	m^2/ha
MEAND	Mean dbh of deciduous trees	cm
NOCPLP	Number of coniferous trees 4–23 cm dbh	stems/ha
NOCSAW	Number of coniferous trees >23 cm dbh	stems/ha
NODPLP	Number of deciduous trees 4–28 cm dbh	stems/ha
SHRSTM	Number of shrub stems >1.4 m high, <4 cm dbh	stems/ha
NOCSHR	Number of coniferous shrub stems >1.4 m high, <4 cm dbh	stems/ha
TOCSHR	Total number of coniferous stems (including shrubs)	stems/ha
CANHT	Canopy height	m
LTTR	Ground covered by litter	%
LIVE	Ground covered by herbs	%

were labeled bird-use plots. A second study (Boynton 1979) determined the boundaries of each species' territory, then chose plots which were categorized as either territory or nonterritory. Finally, nest searches resulted in the location of 475 nests of 25 species of birds; habitat variables were then measured on nest plots. Table 26.1 is a partial list of habitat variables from all three studies, including only those used in models presented in this chapter.

Bird-use plots

Thirty-five 0.08-ha bird-use plots were located randomly throughout the 1200-ha study area. Use of these plots by singing birds was recorded during five 1-hour periods on each plot—a design similar to that used by Anderson and Shugart (1974). Seventeen habitat variables were measured on each plot. Variables were similar to those described by James and Shugart (1970) but were modified somewhat to conform with standard forest-inventory measurements.

Territory plots

Territory plots were established by gridding a 13-ha square study area into 324 sample plots measuring 20 × 20 m and mapping territories of breeding birds (Robbins 1970) throughout this study area. Twenty habitat variables were measured on each 0.04-ha plot and were most similar to those described by Noon (1981a).

Nest plots

Nests were located throughout the 1200 ha and were discovered without a search-image bias because nest locations of all species on the study area were sought. Circular plots of 0.06 ha were established, with the nest centered in each plot. Nineteen habitat variables were measured on each plot.

Development and assessment of models

Each of the three study designs allowed plots to be assigned to one of two categories, used (present) or unused (absent), for each bird species. Assignment of bird-use plots was straightforward; where use by a certain species was observed, plots were assumed to represent habitat used by that species, and plots where use was not observed represented areas where the species was absent. All plots in the study area that was territory-mapped were assigned to groups in a similar fashion based on whether a plot was within or outside of the territory boundaries for a selected species. Nest plots obviously represented used habitat for the bird species which nested on the plot; the sample of unused plots was obtained from nest-centered plots where other species nested. It was assumed, then, that a plot where one species nested represented a plot where another species did not nest and that interspecific competition did not preclude nesting. We do not know, however, whether these assumptions were valid.

The distribution of measurements for each habitat variable was evaluated for normality by examining skewness, kurtosis, and the correlation with a probability plot. The lower limits for normality were 1.0, 3.0, and 0.95 for the coefficients of these three measures. Only about one-third of the variables displayed a normal distribution; another third were marginally normal; and the remainder deviated greatly from normality. The high proportion of univariate nonnormality or marginal normality indicated that the different data sets used in developing our models rarely, if ever, satisfied the assumption of multivariate normality.

The initial step in comparing habitat characteristics between the two groups—i.e., used-habitat and unused-habitat plots—was to reduce the number of variables used in model development for each species. We screened variables by making univariate comparisons between group means and eliminating those variables that did not show differences ($P > 0.10$). For nest-plot data, however, variables were sometimes retained for model development based on our knowledge of the preferred habitat of a given species, even if the univariate test was not significant; variables were screened further for multicollinearity.

Two-group DFA was the principal method used to develop predictive equations. Stepwise procedures (Nie et al. 1975) selected a subset of variables from those which had been screened previously, to create a function which distinguished between the two groups for each species. A less commonly used multivariate technique, logistic regression (LR), was also employed to develop predictive functions which gave probabilities of individual plots being classified into each of the two groups (Dixon and Brown 1979). Cutpoints for LR were selected to balance the correct classification rates between present and absent plots. Only the nest-plot data were analyzed by LR in a study to compare results of DFA classification with those of LR (Fenwick 1983).

Performances of the discriminant models were evaluated by using a jackknife procedure (Lachenbruch 1975), a cross-validation process (Lachenbruch and Mickey 1968), and by classifying independent data. The jackknife procedure classified each sample, using the discriminant equation derived from all samples except the one currently being classified. We set prior probabilities proportional to group sizes. Cross-validation was used to test the predictive capabilities of models developed from nest-centered-plot data and to evaluate the stability of variables in these models. The cross-validation technique took random samples of 75% of the data and used this sample as a "training set" to create a function. The remaining 25%, the "validation set," was then classified using this function. Prior probabilities of group membership were set at 0.5 in these analyses. This random division of data was repeated 10 times, which allowed the generation of means and standard deviations for each variable in the model. The coefficient of variability was calculated to assess the stability of each variable. Means of the training-set functions were then compared with full-model parameters to provide insight into the reliability of variables as predictors.

The best validation method is to use independent data to test models, which we were able to do for bird-use and territory models by using data from one study to test models from the other. We also gathered limited independent data in 1984, on another study area, to test models from nest-plot data.

Table 26.2. Parameters and classification results of three songbird habitat models

Discriminating variable[a]	Unstandardized coefficient[b]	Classification results (% correct)[c]		
		Group	Jackknife	Independent
OVENBIRD: BIRD-USE MODEL (23 used, 12 unused)				
BADSAW	0.1331	Used	87	45
SHRSTM	0.0537	Unused	42	48
constant	−2.6451		$P < 0.05$	$P > 0.98$
AMERICAN REDSTART: TERRITORY MODEL (63 used, 103 unused)				
NOCS	−0.2772	Used	94	44
CANHT	0.1187	Unused	98	75
BATOTL	−0.0266		$P < 0.01$	$P > 0.25$
TOCSHR	0.0004			
LITTR	−0.0214			
BACLC	−0.1494			
constant	1.0035			
BLACKBURNIAN WARBLER: BIRD-USE MODEL (8 used, 27 unused)				
BACPLP	−1.1818	Used	100	0
NOCPLP	0.3074	Unused	100	96
BADPLP	0.0519		$P < 0.01$	$P > 0.98$
constant	−2.8060			

[a] See Table 26.1 for variable descriptions.
[b] Coefficients of the predictive models; these coefficients do not indicate the relative weight of each variable.
[c] *P*-values shown reflect probability of observing classifications this accurate or better based on a chance model (Morrison 1969).

Results and discussion

Songbird-habitat models

Ovenbird (Seiurus aurocapillus)

The ovenbird model was developed from habitat measurements on bird-use plots. Five habitat variables differed between 23 ovenbird-present plots and 12 ovenbird-absent plots, and two of these variables were selected by the stepwise discriminant analysis process (Table 26.2).

The model characterizes used habitat as a mature deciduous forest with ample regeneration in the understory. The jackknife classification was much higher for used than for unused plots, but classification rates of independent plots did not differ from a chance model and were not acceptable (Table 26.2).

The difference in rates of classification between plots used to construct the model for the ovenbird and independent plots emphasizes the importance of validating models on different sites. Even though the variables in the ovenbird model could reasonably be associated with ovenbird habitat, densities of shrub stems were not different between plots assigned to the two groups on the validation area. Also, ovenbirds were found on plots dominated by conifers on the validation area. Thus neither of the two variables in the model was a good predictor of independent data. The weakness of this model was that it was generated on a study area which did not include an adequate portion of the continuum of habitat used by ovenbirds, especially mature conifer cover. Also, sample sizes were inadequate, especially for plots representing unused habitat.

American redstart (Setophaga ruticilla)

Territory and nonterritory plots were used to develop the model for the American redstart. Univariate screening identified 27 variables as significantly different between groups, and stepwise DFA retained six of these for the model (Table 26.2). This model seemed to describe adequately the habitat of this species as a stand with a high canopy and few coniferous stems in the understory and midcanopy strata. Results of jackknife classification were quite encouraging, but classification rates of independent bird-use plots were poor (Table 26.2).

This model is probably better than the classification of independent data indicated. The chief failing of the function was that it did not predict well in conifer stands. Coniferous plots on the study area where data for model development were gathered had a coniferous understory and were not used by American redstarts. In the validation area, however, several plots in mature conifer stands with a deciduous understory were used by this species. This model probably could be refined to adequately predict habitat used by redstarts in either area. The importance of developing models from data representing a wide range of used habitat and of validating with independent data is obvious.

Blackburnian warbler (Dendroica fusca)

A predictive model for Blackburnian warbler habitat was composed of three variables measured on bird-use plots (Table 26.2). We observed this species regularly in forest stands that contained conifers, although this warbler tolerates a moderate component of deciduous trees in the canopy. The predictive equation is not compatible with such a perception of the habitat of this bird, despite a 100% correct classification of plots by the jackknife procedure (Table 26.2).

The problem with the Blackburnian warbler model was a high correlation ($r = 0.90$) between two variables (NOCPLP and BACPLP), which led to a positive rather than a negative sign on the third variable, basal area of deciduous pulpwood-sized trees (BADPLP). This is not an uncommon occurrence when stepwise procedures are used to formulate a model and the objective is to discriminate as much as possible between groups. This situation should have been avoided by eliminating one of the redundant variables before DFA. Nevertheless, models containing highly correlated variables can result in impressive rates of classification of data used in model development. Again, the classification of independent data (Table 26.2) emphasized the weakness of the Blackburnian warbler model.

Red-eyed vireo (Vireo olivaceus)

We developed four separate models for the red-eyed vireo—DFA models from data collected from each of the three study designs and a LR model from nest-plot data (Table 26.3). The territory model was developed from plots in ten red-eyed vireo territories. Seventeen habitat variables were significantly different between territory plots and nonterritory plots, but only two were selected for the model. This simple model contains a positive coefficient for basal area of deciduous trees and a negative term for understory coniferous shrubs. Only slightly more than half of the plots were classified correctly by the jackknife method. The development and assessment of this model may have been complicated because territories for this species were difficult to delineate; male red-eyed vireos tend to stay in one portion of their territory more than other species do. Hence, some plots were probably assigned incorrectly to the nonterritory group.

The red-eyed vireo model developed for bird-use plots was comprised of three variables, two of which were intercorrelated (BACSAW and NOCSAW), making the model difficult to interpret. The classification rates of this model were erratic. It is noteworthy that 95% of the independent territory plots, but only 20% of the nonterritory plots, were classified correctly (Table 26.3). These classification results further suggest that some plots were assigned incorrectly to the nonterritory group, as suggested above.

The model generated by DFA from nest-centered plots was similar to the territory-plot model; the function described a deciduous stand with a dense subcanopy lacking coniferous stems (Table 26.3). The four-variable model was developed from only five variables which differed between groups. Correct classification rates by jackknife and cross-validation methods were almost identical. Three nest plots and eight territory plots were measured in 1984, and all were classified correctly with this model. The same four variables were components of the LR model, which classified plots using the cross-validation method almost as well as the DFA model. In both the DFA and LR models, means of coefficients derived in the training sets were almost identical to the coefficients of the full models, indicating the appropriateness of variables used in the models and their stability. Coefficients of variation for the variable coefficients ranged from 0.095 to 0.261 in the DFA model and from 0.202 to 0.284 in the LR model. These were much lower than comparable statistics for models of nine other species analyzed in a similar manner (Fenwick 1983).

Table 26.3. Parameters and classification results of four habitat models for the red-eyed vireo

		Classification results (% correct)[c]			
Discriminating variable[a]	Unstandardized coefficient[b]	Group	Jackknife	Cross-validation	Independent
	TERRITORY MODEL (91 used, 116 unused)				
BATOD	0.0986	Used	56	—	88
NOCSHR	−0.0020	Unused	65	—	64
constant	−1.6626		$P < 0.01$		
	BIRD-USE MODEL (29 used, 6 unused)				
BACSAW	−2.0070	Used	86	—	95
NOCSAW	0.5147	Unused	50	—	20
NODPLP	0.0010		$P > 0.12$		$P < 0.01$
constant	−2.2817				
	NEST-PLOT MODEL (DFA) (40 used, 435 unused)				
LITTR	0.0281	Used	72	74	100[d]
BASNAG	−0.0005	Unused	60	62	—
NOCSHR	−0.0240		$P < 0.01$	$P < 0.01$	
DDOM	−0.0016				
constant	−0.4090				
	NEST-PLOT MODEL (LR) (40 used, 435 unused)				
LITTR	0.0244	Used	—	68	—
BASNAG	−0.0006	Unused	—	73	—
NOCSHR	−0.0736			$P < 0.01$	
DDOM	−0.0021				
constant	−3.5100				

[a] See Table 26.1 for variable descriptions.
[b] Coefficients of the predictive models; these coefficients do not indicate the relative weight of each variable.
[c] *P*-values shown reflect probability of observing classification this accurate or better based on a chance model (Morrison 1969).
[d] A small sample of 11 plots.

Design of field studies

A problem basic to the design of use vs nonuse studies of habitat is obtaining the assurance that plots classified as nonuse are not suitable habitat. Although it is usually the objective of habitat models to predict suitable habitat rather than areas unsuitable as habitat, the proper identification of unsuitable habitat is critical to the development of the models. Many investigators routinely assume that breeding habitats for songbirds are saturated. Even though this assumption is usually not stated, it is implicit in use/nonuse study designs. There is much evidence that this assumption of equilibrium and community saturation is not valid, though it continues to be a subject of lively debate (Wiens 1983).

Another concern, and one illustrated by one of our study designs, is the lack of independence of habitat samples when a number of sample plots are located in a single territory. Because the effect of this problem is difficult to evaluate, such designs should be avoided. For this reason, nest-centered plots may be preferred; they also probably represent habitat-selection patterns better than territory plots or use plots (Collins 1981).

A dilemma which presented itself in several of our modeling efforts was whether or not to eliminate outlying habitat plots before data were screened and models developed. Some randomly located plots in radically different habitat types (e.g., clearcuts for forest bird species) were eliminated before deriving models. This allowed the derivation of more habitat-sensitive models that more closely met the statistical assumption of homogeneity in the variance-covariance matrices of the two groups.

Statistical considerations

Sample sizes of the two groups in use vs nonuse designs are practically always disparate, as illustrated in the models discussed in this chapter and elsewhere (e.g., Conner et al. 1983; Rice et al. 1983a). Statistical problems associated with this disparity are discussed in detail by Williams (1983). Predictably, the species-present group will be more homogeneous and almost always smaller than the larger, heterogeneous species-absent group. This further complicates the already difficult task of identifying and sampling the nonuse habitat. Hence, equality in the variance-covariance matrices is practically impossible to achieve. Fortunately, the objective of formulating predictive functions is affected less by this problem than investigations that emphasize interpretations of the predicting variables are. Useful functions can be derived and used, but the stability of coefficients must be assessed by employing cross-validation or similar iterations and by validating models with independent data.

Unequal group sizes in data sets can also create difficulties in interpreting classification tables (Morrison 1969). Authors of discriminant habitat models for wildlife species should recognize the problem of establishing prior probabilities of group membership. Likewise, tests comparing classification results to chance models should be reported. As illustrated by models in this chapter, it is common to achieve high rates of correct classification but to discover that this could have happened by chance because group sample sizes were disparate.

Most investigators now recognize the importance of obtaining an adequate sample of independent variables in relation to the number of observations of the dependent variable. Ignoring this can be a serious weakness in the derivation of predictive models using multivariate methods (Williams 1983); see also Magnusson's (1983) critique of Hale et al. (1982). Small sample sizes probably contributed to the unstable classification rates in our models developed from bird-use plots.

We favor a varied, manipulative approach for assessing the utility of discriminant models for predicting suitable or unsuitable habitat. Obviously, the best approach is to validate models with independent data which vary in time, space, and along habitat dimensions. Prior to field validation, other statistical procedures can refine models. Logistic regression, which may indeed be a favorable alternative to discriminant analysis for wildlife habitat model development, can be used as a check against discriminant analysis by comparing models obtained by both methods and examining the stability of variables. We used both classification procedures with nest-plot data for 10 bird species and concluded that logistic regression did not seem to be a better technique for developing habitat models, but that it was useful in testing stability of variables. Logistic regression is considered superior for data sets involving mixtures of discrete and continuous variables (Efron 1975; Press and Wilson 1978), and many habitat variables are best suited for discrete measures.

The potential problems in deriving single-species habitat models from multivariate analyses are numerous: field studies are time-consuming; adequate sample sizes are hard to obtain; many habitat parameters are difficult to estimate within normally acceptable margins of error; statistical assumptions are practically impossible to meet; and biological components of a species' niche other than habitat dimensions usually are not estimated. Yet the statistical approaches demonstrated by this chapter seem to be the most reasonable means for describing quantitatively the multidimensional space in which an animal lives. Some multivariate models of wildlife habitat do work. However, they must be based on sound field studies where group membership is determined confidently and where appropriate habitat variables are measured.

Acknowledgments

We thank the directors of the Windham Foundation, Grafton, Vermont, for their support of field work from 1976 to 1982. Financial support was also received from Cooperative Forest Research Funds administered by the School of Natural Resources, University of Vermont. For review of this manuscript, we thank R. J. Henke, D. H. Hirth, M. L. Hunter, Jr., D. H. Johnson, M. L. Morrison, B. R. Noon, and F. A. Stormer.

27

The Use of Multivariate Statistics for Developing Habitat Suitability Index Models

LEONARD A. BRENNAN, WILLIAM M. BLOCK, and R. J. GUTIÉRREZ

Abstract.—We develop Habitat Suitability Index (HSI) models based on linear and nonlinear multivariate statistical analyses. The combination of discriminant function analysis (DFA) and all possible subsets regression (APSR) was used to develop a linear HSI model, and stepwise logistic regression (SLR) was used to develop a nonlinear HSI model. As examples, we describe the development of mountain quail (*Oreortyx pictus*) HSI models based on a two-group analysis of used (mountain quail present) and available (randomly located) habitat data. We used the linear and nonlinear methods to derive classification functions, and we compared the classification power of both functions. We also compared the error terms, confidence regions, and goodness-of-fit statistics associated with each model. The nonlinear SLR fit our data better than the linear DFA–APSR model and provided better group separation as well. The output of the linear and nonlinear models was tested with habitat data from different areas of mountain quail use. We conclude that HSI models can be developed with either linear or nonlinear regression techniques and that the structure of HSI models should be tested with long-term population data for validity.

The U.S. Fish and Wildlife Service Habitat Evaluation Procedures (HEP) require Habitat Suitability Index (HSI) models that are used to evaluate habitat quality for selected vertebrate species (Fish and Wildlife Service 1980a, 1980b, 1981a). In HEP analyses, habitat suitability is defined as the potential of an area to support a particular evaluation species and is represented by an index ranging from 0 (completely unsuitable) to 1 (optimal) (Fish and Wildlife Service 1981a). The suitability (sensu an HSI model) and size of an area are combined to calculate the habitat units that are available to an evaluation species.

Thus, HEP analyses are only as good as the HSI models used therein. Most HSI models have been developed on the basis of qualitative accounts and general statements about a species' habitat preferences, and verified by an authority on the evaluation species rather than by an analysis of empirical data (Lancia et al. 1982; Cole and Smith 1983).

Our objective is to evaluate the use of linear and nonlinear multivariate statistics in the development and testing of HSI models for mountain quail (*Oreortyx pictus*). We address the assumptions of these models and compare the merits and limitations of each statistical method.

LEONARD A. BRENNAN: Department of Wildlife, Humboldt State University, Arcata, California 95521. *Present address:* P.O. Box 1144, Laytonville, California 95454

WILLIAM M. BLOCK: Department of Wildlife, Humboldt State University, Arcata, California 95521. *Present address:* Department of Forestry and Resource Management, 145 Mulford Hall, University of California, Berkeley, California 94720

R. J. GUTIÉRREZ: Department of Wildlife, Humboldt State University, Arcata, California 95521

Methods

Study areas

Four areas in northern California (Modoc Plateau, northern Sierra Nevada, Klamath Mountains, and Coast Range Mountains of northern California) were chosen to sample the habitat use of mountain quail in a variety of conditions. Each study area contained two discrete sites, one for model development and the other for model tests. At each study area, development and test sites were at least 10 km apart but were within the same seral stage and vegetation type of the Küchler (1977) classification system. Each site was about 500 ha in size; elevations ranged from 1000 m at the Coast Range area to 2100 m at the northern Sierra Nevada area.

The Modoc area was dominated by pine–western juniper (*Pinus* spp.–*Juniperus occidentalis*) forest, shrubsteppe vegetation, and basalt lava reefs. Both the Sierra Nevada and the Klamath areas were dominated by stands of mixed conifers and mixed species of brush. The Coast Range site was dominated by mixed-conifer and broad-leaved forest, mixed species of brush, and oak woodland. Descriptions of the vegetation, soils, and geology of each study area can be found in Barbour and Major (1977). Brennan (1984) described the species composition of the vegetative cover types at each study area.

Model development

We measured 15 habitat variables on organism-centered (used) and randomly located (available) 0.02-ha (15-m diameter) plots. Organism-centered plots were obtained by walking transects and using the location of the first quail seen in a covey as the center of a plot. Thus, it is possible our estimate of habitat use may be biased in favor of the habitat structure used by the most conspicuous individuals rather than the entire population. An estimate of available habitat structure was obtained from a random sample of plots stratified by cover type. We acknowledge that the available habitat sample possibly contained an unknown amount of suitable habitat. It was impossible, however, to map the spatial limitation of mountain quail territories to obtain a used/unused habitat contrast. See Brennan (1984) for details of the sampling methods.

We first performed a one-way analysis of variance (ANOVA) to determine which variables had significant ($P < 0.01$) between-group differences. Second, we calculated all product-moment correlation coefficients between pairwise combinations of the significant variables and eliminated one of a pair if $r \geq 0.4$. The variable with the lowest between-group significance was eliminated from further analysis. On the basis of these criteria, five habitat variables (distance to water, distance to escape cover, minimum shrub height, maximum shrub height, and percent shrub canopy cover) were included as predictor variables in the models (see Results). These five variables provided better classification success than any other combination of the original 15 variables. The addition of variables that did not have significant ($P < 0.01$) between-group differences in the ANOVA or that had r-values ≥ 0.4 with any of the five predictor variables did not improve the classification success and therefore they were not included. Third, we used the linear and nonlinear statistical analyses to rank the variables in the order of their ability to improve the fit of the data to each model, and to derive classification functions.

Linear multiple regression and discriminant function analysis (DFA) are mathematically similar in the two-group case, and the regression coefficients are proportional to the discriminant function coefficients when the dependent variable assumes values of 0 and 1 (Tatsuoka 1971). Therefore, the linear HSI model was developed with the all possible subsets regression (APSR) program (Frane 1981) and the stepwise discriminant analysis program (using the jackknife method, Jennrich and Sampson 1981). Our linear HSI model was based on the general linear model:

$$Y = a + B_1 X_{1i} + \ldots B_m X_{mi}, \quad (1)$$

where Y = a predicted value; a = a constant; B = the regression or discriminant coefficients; and X = the predictor variables in the model. APSR analysis was used to obtain standard errors of the regression coefficients. DFA was used to derive a classification function and examine between-group separation. The posterior probability of correct classification in the used-habitat (group 1) category given a vector of habitat measurements $P(1|x)$ in the linear model was calculated by

$$P(1|x) = \frac{1}{1 + \frac{q_2}{q_1} e^{-x'k + \frac{t_1 + t_2}{2}}}, \quad (2)$$

where q_1 = the prior probability that the habitat is from group 1 (used); $q_2 = 1 - q_1$; x = a vector of mean-corrected habitat measurements—the X variables in equation (1); k = a vector of unstandardized discriminant function coefficients; t_1 and t_2 = the mean discriminant scores for the respective groups (Green 1978). Prior probabilities of group membership were based on the relative sample size of each group.

The nonlinear HSI model was developed with the stepwise logistic regression (SLR) program (Engelman 1981), based on the general model:

$$P(1|x) = \frac{e^{a + B_1 X_{1i} + \ldots B_m X_{mi}}}{1 + e^{a + B_1 X_{1i} + \ldots B_m X_{mi}}}, \quad (3)$$

where $P(1|x)$ = the estimated probability that the area represents mountain quail habitat given a vector, x, of habitat measurements; a = a constant; B = the regression coefficients; and X = the predictor variables in the model. Cox (1970) provides details on logistic regression computation methods.

The relationship of each predictor variable to the dependent variable in both models was evaluated by performing separate linear and logistic regression analyses to plot suitability-index curves and their associated 95% confidence regions.

EVALUATION OF MODEL OUTPUT

At each of the four model test sites, we measured habitat data (the five variables included in the HSI models) from areas of known mountain quail use. In an analysis of habitat selection by mountain quail at the four study areas, Brennan (1984) observed that mountain quail used all vegetated cover types in proportion to their availability. We therefore did not attempt to use model output to assess the suitability of unused areas because we were unable to be completely sure that the habitats were, in fact, totally unused. Twenty habitat samples were randomly located at each test site in areas where mountain quail were observed, and the classification functions—equations (1) and (3)—were used to test whether the models associated the habitat data with the correct group.

Results

The five predictor variables entered both the linear and nonlinear analyses in similar order, the only difference being the position of percent shrub canopy cover (Table 27.1). The addition of each variable (except percent shrub cover) significantly improved the fit of the nonlinear model (chi-square ≥ 6.2, $df = 1$, $P \leq 0.01$, Improvement test, Engelman 1981). The inclusion of all five variables significantly improved the fit of the linear model (Wilks' lambda ≥ 0.78, $df = 1, 5$; $P < 0.001$, Bartlett's V-test). The ratios of the regression coefficients to their associated error terms were similar between models (Table 27.1). The average classification was better for the four study sites using the nonlinear model (Table 27.2). With both models, the greatest number of classification errors were in the available-habitat group (Table 27.2). The direction of the relationships of each variable in both models appeared similar, as did the

Table 27.1. Order of variable selection, regression coefficients, and error terms in the linear and nonlinear HSI models for mountain quail

Variable	Selection order		Linear multiple regression			Nonlinear logistic regression		
	Linear model	Nonlinear model	Coefficient[a]	*SE*	Ratio[b]	Coefficient	*SE*	Ratio[b]
Distance to water	1	1	−0.0007	0.0001	−5.05	−0.0050	0.001	−4.86
Distance to cover	3	2	−0.0167	0.0064	−2.61	−0.2590	0.081	−3.12
Minimum shrub height	4	3	0.1250	0.0464	2.70	1.9400	0.766	2.53
Maximum shrub height	5	4	0.0440	0.0214	2.09	0.2880	0.137	2.10
Percent shrub cover	2	5	0.0020	0.0010	2.47	0.0070	0.006	1.01
(Constant)[c]			0.5600	0.0710	7.87	0.5500	0.442	1.25

[a] Multiply linear coefficients by 5.05 to obtain unstandardized discriminant function coefficients.
[b] Coefficient/*SE*.
[c] Not a habitat variable but a term in each regression model. The constant in the linear discriminant function = 0.45.

confidence regions around the bivariate functions that described these relationships (Fig. 27.1, Table 27.3). A test of linearity ($F = 16.0$, $df = 5, 284$; $P < 0.001$) indicated that the data were not linear, while a goodness-of-fit test in the SLR analysis (chi-square = 285.8, $df = 282$, $P = 0.43$, Engelman 1981) indicated that the nonlinear logistic regression model fit the data well.

In our analysis of model output, both models correctly associated the test-site data with the used-habitat group (Table 27.4). Model output from two test sites consisted of probability coefficients that were lower than expected, especially because the data were collected from areas of known quail use. The probability of correct association with the used habitat at the Modoc and Sierra test sites was just slightly better than random when based on a probability cutpoint of 0.5. If, however, one interprets the probability coefficients from the tests of model output (Table 27.4) as suitability values which follow the 0–1 HSI scale, then both models predicted that the habitats at the Modoc and Sierra test sites were average in quality, while habitats at the Klamath and Coast Range sites were above average in quality. Until information is available on any possible differences in the survivorship of mountain quail in the habitats at these test sites, we cannot completely assess the accuracy of our test results.

Discussion

Our results suggested that HSI models can be developed and assessed by use of either linear or nonlinear multivariate statistical techniques, specifically a combination of discriminant function analysis (DFA) and all possible subsets regression (APSR), or stepwise logistic regression (SLR). The role of each variable in the models can be identified and graphed in bivariate space (e.g., Fig. 27.1) much like the suitability-index graphs that accompany traditional HSI models, thus

Table 27.2. Summary of classification success of available and used mountain quail habitat groups by discriminant function analysis and stepwise logistic regression

Site/Group	No. plots (0.02 ha)	Discriminant function (linear model)			Logistic regression (nonlinear model)		
		Available	Used	Percent correct[a]	Available	Used	Percent correct[a]
Modoc Plateau							
Available	25	15	10		23	2	
Used	50	4	46	81	5	45	91
Northern Sierra Nevada							
Available	25	13	12		17	8	
Used	50	6	44	76	6	44	81
Klamath Mountains							
Available	25	10	15		13	12	
Used	40	2	38	74	3	37	77
Northern Coast Range							
Available	25	9	16		9	16	
Used	50	5	45	72	0	50	79
All sites							
Available	100	37	63		59	41	
Used	190	14	176	73	19	171	79

[a] Values given are the average percentages of all samples classified in the correct group. Prior probabilities of correct classification were based on the relative sample sizes of each group.

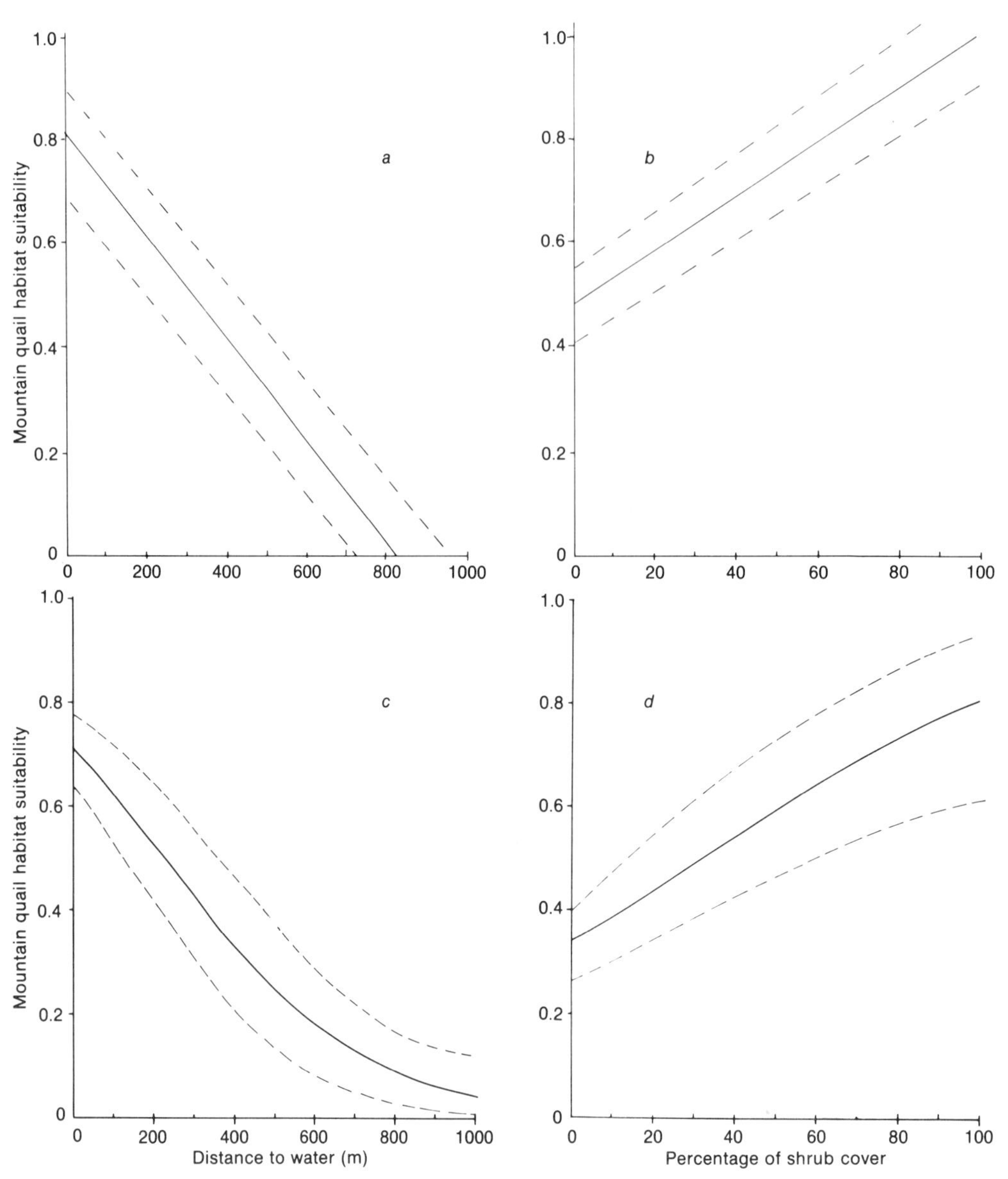

Figure 27.1. Bivariate relationships of two predictor variables in the mountain quail HSI models. Linear graphs (*a* and *b*) are based on equation (1); nonlinear graphs (*c* and *d*) are based on equation (3). Areas within the dashed lines represent 95% confidence regions.

Table 27.3. Regression coefficients and error terms in the equations which represent the direction of the relationships of each predictor variable

	Linear HSI Model[a]				Nonlinear HSI Model[b]			
Variable	Constant	*SE*	Coefficient	*SE*	Constant	*SE*	Coefficient	*SE*
Distance to water[c]	0.810	0.036	−0.0008	0.0001	0.913	0.230	−0.004	0.001
Distance to cover	0.700	0.030	−0.0320	0.0060	0.470	0.160	−0.220	0.070
Minimum shrub height	0.630	0.030	0.1000	0.0500	−0.140	0.180	1.490	0.700
Maximum shrub height	0.450	0.060	0.0880	0.0210	−0.590	0.290	0.330	0.120
Percent shrub cover[c]	0.460	0.050	0.0040	0.0010	−0.670	0.260	0.021	0.006

[a]Based on equation (1) (see text).
[b]Based on equation (3) (see text).
[c]Graphed in relation to the range of values for each variable in Figure 27.1.

Table 27.4 Evaluation of model output with habitat data from four northern California sites

	Discriminant function (linear model)		
Test site[a]	Discriminant score	Posterior probability[b]	Logistic regression[c] (nonlinear model)
Modoc Plateau	0.45	0.51	0.54
Northern Sierra Nevada	0.59	0.53	0.55
Klamath Mountains	1.13	0.66	0.74
Northern Coast Range	1.51	0.74	0.85

[a] Tests at each site based on 20 samples of the five HSI variables measured on 15-m-diameter circular plots.

[b] Probability that the discriminant score classifies the area in the correct habitat group.

[c] Probability that the test site represents mountain quail habitat.

allowing the investigator to isolate the relationship of each variable and check for ecological meaning. Confidence regions associated with each function illustrate the variation in the relationship of each predictor variable to the dependent variable. The linear combination of a vector of predictor variables can be used to derive either linear or nonlinear classification functions, which can then be used to calculate the probability that an area represents a suitable habitat structure for a species. Because probability coefficients and HSIs are both scaled within a range of 0 to 1, it should be possible to incorporate HSI models based on multivariate methods into HEP analyses. This statement is made, however, with the assumption that aspects of model validity have been addressed and that model output is known to be correlated with some ultimate measure of habitat suitability such as fitness.

Assumptions and model validity

The assumptions that the data were sampled from multivariate normal populations which have equal covariation underlie both the linear and the nonlinear multivariate methods described above. With observational data in ecology these assumptions are almost never met; furthermore "separate and combined effects of these violations on canonical analysis are almost totally unknown" (Williams 1981:68). We know of no accepted method to test directly the hypothesis that our data were sampled from multivariate normal populations. It is meaningful, however, to infer from a series of univariate tests for skewness and kurtosis that the data approximate multivariate normal distributions (B. R. Noon, pers. comm.); therefore, we tested this assumption indirectly. Inspection of the frequency distributions obtained from our data indicated that only one variable (maximum shrub height) did not have a skewed or kurtotic distribution, allowing us to infer that our data departed from multivariate normality. Also, our data did not have equal within-group covariation (Box's $M = 57.2$, $df = 1$, $P < 0.001$). Violation of these assumptions was justified because the large ratio of samples to variables (290:5) probably reduced the likelihood of obtaining "unstable" (sensu Williams 1981, 1983) classification results. Alteration of the sample sizes of the data sets used in model development by random selection of different subsets did not change the classification success of either model.

A central assumption of HSI models is that model output is correlated with population fitness. We thus acknowledge that our HSI models rest on the untested assumption of a correlation between the suitability coefficients obtained from model output and ultimate habitat suitability. Tests of this assumption with long-term population data are required before any HSI models based on multivariate statistics can be considered valid.

Another important assumption of HSI models based on multivariate statistics is that as population density increases, habitats of decreasing suitability will be used (cf. O'Connor, Chapter 34). To minimize this confounding effect, we obtained our habitat-use samples from mountain quail populations with densities that ranged from 4 to 25 birds per 40 ha (Brennan 1984).

In addition to testing the major assumptions of a model, a complete validation test requires that numerous aspects of the model be evaluated. For example, Marcot et al. (1983) summarized 23 criteria from the literature that should be considered when the validity of wildlife-habitat models is tested. Model utility, however, can be assessed before validation tests are complete. Our evaluation of model output, for example, indicated a potential negative bias in the suitability coefficients from areas of known quail use. This potential bias may be the result of basing model development on a used-available contrast in which nearly all of the organism-centered, but only half of the randomly located, habitat samples were correctly classified (Table 27.2). Thus, our estimate of the structure of available habitat appeared to contain some potentially suitable habitat. Alternatively, the negative bias observed in our tests may be a result of stochastic variation in habitat structure. The impact of this bias on management decisions regarding mountain quail habitat would be that the manager using our models might obtain a conservative estimate of habitat suitability.

Merits and limitations of each method

With our data, the SLR provided a slightly better group separation and a significantly better fit than the DFA–APSR method. This merit, and the potential of being more robust than DFA–APSR to deviations from multivariate normality and equal covariation (Efron 1975; Press and Wilson 1978), led us to decide that the HSI model based on SLR was superior. Halperin et al. (1971) and Press and Wilson (1978) also observed that SLR provided better between-group separation and classification success than did DFA. Further, the nonlinear tendencies in our data were probably distorted by the linear analysis (cf. Meents et al. 1983). Goldstein (1977) and Johnson (1981a) argued that because many ecological phenomena have an inherent nonlinear nature, nonlinear statistics should often provide more realistic results than linear statistics.

SLR has at least two limitations from the HSI modeler's perspective. First, software with SLR routines is not as widely available as DFA and multiple regression programs.

SLR programs are available in the BMDP series and the Statistical Analysis System (SAS) supplemental library (Helwig and Council 1979; Draper and Smith 1981), and as a separate program developed by the USDA Forest Service (Hamilton 1974). DFA and multiple regression programs are available in all major software packages. A second potential limitation of SLR is that it requires at least three times more computation time than DFA does. This can be a potentially serious problem if budget allocations for computer time are limited.

Both the SLR and DFA–APSR methods have several common merits. First, after initial model development, the calculations required to classify habitats in unknown areas can be accomplished with any standard calculator equipped with a natural antilogarithm function. Thus, use of these models, once developed, does not require access to a computer. Second, it is possible to examine residuals, outliers, and other diagnostics with both methods, though the particular approach to such analysis is different for SLR than for multiple regression (Landwehr et al. 1984). Third, both models can accommodate two-group contrasts obtained from different types of sampling methods. Because the available group in a used-available contrast can potentially contain unknown amounts of suitable habitat, a used/unused contrast might be preferable if the investigator can be confident that certain habitats are in fact unused. HSI models for many passerine birds, for example, could be developed by mapping territories and contrasting habitat samples from within territories with nearby unused areas. HSI models for aquatic species such as amphibians and fish might be developed by contrasting characteristics of streams or watersheds that contain a species with those that do not. Habitat-use patterns of small mammals could be obtained by trapping in different habitats and HSI models developed by contrasting characteristics of used and unused areas.

Advantages of data-based HSI models

Development of data-based HSI models with regression techniques allows the investigator to evaluate aspects of model behavior and distortion in an objective, quantitative manner. Real and potential biases can be assessed, thus providing the manager with a realistic measure of model utility before validation tests are completed. Such rigorous, objective evaluation is simply not possible when HSI models are based on general, qualitative assertions about a species' habitat preferences. Although the cost of developing HSI models using the data-based approach is clearly greater than developing them from literature reviews and existing general information, the information gained from the quantitative approach can be of much value. Empirical data, for example, can be used to test the qualitative natural history accounts on which many current HSI models are based. Such data can be used in site-specific mitigation studies and impact assessments. Accurate, quantitative data can be of great value when litigation is involved with management issues. In addition to management applications, data used to develop HSI models can also be used to explore basic theoretical questions of habitat selection in terrestrial vertebrates. Thus, the value of a well-designed HSI model development project should justify the cost.

Acknowledgments

G. M. Allen, C. W. Barrows, D. H. Johnson, B. G. Marcot, B. A. Maurer, K. E. Mayer, B. R. Noon, M. G. Raphael, and F. A. Stormer reviewed earlier drafts of this chapter and offered many helpful suggestions. This research was supported in part by the California Department of Forestry Forest Resources Assessment Program and the International Quail Foundation. Computer services were provided by the Humboldt State University Computer Center.

28

A Keyword Census Method for Modeling Relationships Between Birds and Their Habitats

MARC C. LIVERMAN

Abstract.—A simple method for quantifying habitat physiognomy and vegetation that is compatible with point-count census techniques and useful in arid, mountainous terrain is presented. Field-test observations were made at 120 clustered sample points in southeastern Oregon over a 20-day period. Ordination and cluster analysis were used to analyze results in terms of habitat gradients with corresponding bird species' distributions and to classify sites in an ecological series. Some basic assumptions, advantages, and disadvantages of this method are discussed.

Each modeling process is guided by a theoretical framework that determines the possibilities for observation and by objectives that set an appropriate level of model generalization. The theory underlying models presented in this chapter states that the bird community occurring in any given area results from the direct and interactive effects of relatively few habitat factors. These factors include the area's climate, topography, vegetation, and avifauna.

The interpretation of data concerning any one factor is dependent on its interactions with other factors, a relationship that will probably differ with locality and geographic region. Moreover, the relative importance of each factor within the same area will be influenced by time (cyclic and progressive), biotic interactions, such as predation and competition, and by the rate and intensity of fire, grazing, and/or other forms of disturbance.

In areas having steep or mountainous terrain, topography and vegetation interact strongly to determine habitat characteristics. Topography acts as a relatively long-term template which mediates the distribution of temperature, moisture, soil formation, and disturbance. Processes such as surface erosion, debris slides, and frost heaving can influence both the rate and the direction of primary and secondary succession. Topography also determines the geographic distribution of unique geomorphic habitats such as caves, cliffs, and talus. Vegetation influences topography by regulating the movement and storage of soil and sediment on slopes and in small- and intermediate-sized streams. It also acts as the primary substrate for life-sustaining activities of birds, such as cover, rest, feeding, and reproduction. The mobility of birds allows them some independence from the habitat conditions encountered at a given site, but it is still habitat in a broader sense that comprises the proximate or ultimate factors to which a species must respond (Rotenberry 1981).

My objective was to develop a method of surveying topography, vegetation, and bird-species' distributions that would be useful in mountainous terrain, sample each factor at a similar level of intensity, and lead to an explanation of the ecological structure of local bird-habitat relationships in terms of both underlying habitat gradients and a classification of habitat types.

MARC C. LIVERMAN: Department of Forestry and Resource Management, 145 Mulford Hall, University of California, Berkeley, California 94720

Study area

The Alvord Basin occupies approximately 1600 km^2 of Harney County in southeastern Oregon. Elevations range from 1200 to 1400 m, and the highest surrounding point is over 2300 m. The climate is cold desert with an average annual temperature of 9.4°C and 16.8 cm of annual precipitation (Johnsgard 1963). Vegetation is dominated by shrubs, including *Artemisia* spp., *Atriplex* spp., *Chrysothamnus* spp., *Tetradymia* spp., and scattered annual and perennial grasses (Cronquist et al. 1972:114; Franklin and Dyrness 1973:234).

Methods

Bird and habitat measures

Twelve sampling sites were selected to represent the range of habitat conditions present throughout the study area. These sites correspond to points along a transect oriented approximately northeast by southwest, and extending the length of the basin. Ten sampling stations were randomly selected along a second transect established at each of the sampling sites. Stations were separated by at least 200 m along the transect so that the chance of observing the same bird from more than one station was minimized.

I recorded the presence or absence of bird species at each station during two 4-min periods. Both 4-min censuses were taken during a 24-hour span, one near sunrise, the other near sunset. Weather conditions were clear and calm. The radius of observation was determined by the distance to the farthest identifiable bird, excluding soaring species.

Following each bird census, I also recorded the presence or absence of topographic features and woody and/or emergent plant species within the same radius of observation used for birds. Terms (''keywords'') describing topographic features and including soil and moisture conditions were selected from an initial training checklist. Terms on this checklist were gleaned from published descriptions of Great Basin bird habitats similar to the Alvord Basin (Grinnell and Miller 1944; Hubbard 1970; Hayward et al. 1976; Marks et al. 1980; Monson and Phillips 1981). Precise keyword definitions were obtained from the Dictionary of Geological Terms (American Geological Institute 1976). No time limit was specified for this aspect of data recording. Each plant species and landform characteristic attributed to a sample station was confirmed or deleted following the final bird census of that station.

I conducted 240 censuses allocated to 12 sites, which yielded approximately 16 hours of timed bird observation and 60 hours of habitat observation. All censuses were completed between 22 June and 11 July 1982.

I calculated the frequency of every site attribute (i.e., bird species, plant species, and topographic features) at each sampling site as the proportion of station censuses where that attribute was present. All frequency values were compiled into a matrix of attributes (rows) by sites (columns).

Analysis

I used reciprocal averaging (RA) ordinations (Hill 1973, 1974), detrended correspondence analysis (DCA) ordinations (Hill 1979a), and TWINSPAN cluster analysis (Hill 1979b) to analyze relationships between sampling sites in terms of each of the three sampling criteria. In RA, sites and their attributes are ordinated simultaneously to maximize correlation between their scores (Pielou 1984:176). DCA is initiated by an ordinary RA ordination and applies a series of adjustments to flatten the "arch" often associated with ordinations of nonlinear data and to rescale the ends of each ordination to compensate for scale contraction induced by the RA process.

TWINSPAN is a form of divisive, hierarchical classification and, like DCA, is initiated by RA ordination. The centroid of the preliminary ordination is used to split the data into two groups. Each of the resulting groups is then reordinated and the process is repeated through as many iterations as necessary to completely subdivide the collection. TWINSPAN also includes a number of refinements designed to emphasize indicator species at each step, and thereby define classes that are as "natural" as possible.

Assumptions

Models presented in this chapter depend on the following assumptions: (1) observations are unbiased and independent; (2) differences in bird species' distributions are a result of underlying habitat factors; and (3) analytical methods maximizing the association between sampling sites and their attributes are ecologically informative.

The first assumption lists conditions of sampling protocol that must be maintained so that limited observations can be usefully generalized to explain a system of wider interest. These conditions apply both to the selection of sampling sites and the observation of site characteristics. Bias can arise from nonrandom sampling, observer error, differences in detectability between species, physical site differences, and effects related to the time and season of sampling. It should be clear that these conditions are interrelated and may require trade-offs for their relative satisfaction.

The second assumption is well established in the ornithological literature (see Lack 1937; Hildén 1965), but should not be taken to imply a simple resource-defined equilibrium between birds and environment (Wiens 1977). Factors such as behavior (e.g., time lags, territoriality, and responses to events at distant locations) can be unrelated to the intrinsic characteristics of a given site but may nonetheless influence its occupancy by birds.

The final assumption states that examining the association between sites and their attributes is ecologically informative. This is intuitively true but unfortunately leads to a logical circularity underlying many studies of wildlife-habitat relationships. Such studies assume that an animal is present because of habitat, then measure habitat to determine why the animal is present. Although there are compelling reasons to believe that animals are adapted through natural selection to live in some habitats and not in others, this circularity is not broken until a recognition of patterns between species and habitats is used to develop predictions that are subsequently tested experimentally. Measuring a statistical correlation between animals and habitat is not equivalent to deduction, and only deduction can be strictly true or untrue. Other forms of reasoning, including statistical inference, can only be relatively strong or weak.

Results

Values describing the occurrence of topographic features, plant species, and bird species at each sampling site were compiled into separate matrices (e.g., Table 28.1, summarizing the occurrence of topographic features). These raw matrices convey few patterns or relationships in themselves but provide the basis for subsequent ordination and cluster analyses.

Comparison of preliminary ordination results indicated that one site was compositionally extreme with respect to plant and bird species and tended to compress all other sites into a relatively small section of the first ordination axis. The unusual site was located in a crested wheatgrass (*Agropyron cristatum*) planting. It differed from the other sites primarily in the dominance of a single plant species and the occurrence of only one species of bird, the horned lark (*Eremophila alpestris*). Deletion of this site from later calculations improved the ordination results by spreading the remaining 11 sites more evenly over the primary axis.

The next stage of analysis required comparing the relative positions of site features along each ordination axis. Groups of features associated with the ends of each axis were examined for systematic differences that could be interpreted in terms of underlying habitat gradients. Note that this is a heuristic process (reification—Marriott 1974:9; Pimentel 1979:7) and not a strictly statistical one. The data are not unambiguous for this process, and scientists may differ in their interpretation of them. Relationships developed in this

Table 28.1. Occurrence of topographic features by sampling site. (Scores range from 0 to 10.)

Feature[a]	Sampling site 1	2	3	4	5	6	7	8	9	10	11	12
Alkali				9			2					
Bajada	10			10	3	10		3				
Basin floor			10				10		10		10	
Building					2					2	3	
Canyon		10			8							
Cliff	3	3			8			1				
Dune							5		9			
Escarpment					1	10						
Fence			2							4	5	2
Flat			10				10		8	8	10	10
Gully	2					7		4				
Hill					2					2		
Lake			5									
Outcrop	10	10			8	10		7				
Ravine			2	3	4							
Road	3	6			10					1	3	
Saturated soil			4		1		7		6	5	2	
Scree	2	2			7	10		7				
Spring		1	2								1	
Standing water		1	4						4	1		
Stony	8	10		10	10	10		10				3
Stream		10	2		10			3		2	5	
Talus	2	10		9	7							
Terrace	2			10		5						

[a] Definitions follow American Geological Institute (1976).

way must be considered as hypothetical until experimentally proven.

I found that RA provided three interpretable axes for the topographic and plant species matrices and two for the bird species matrix. Only one DCA axis could be interpreted for each of the same data sets. I conclude that in this case, the "corrections" applied by DCA to adjust RA scores actually caused a loss of ecologically useful information. Consequently, the analysis that follows was developed using only RA results.

Each RA axis has an associated eigenvalue that is equivalent to the coefficient of determination (R^2) between the sampling sites and their attribute scores along that axis (Pielou 1984:183). If one sets the sum of all eigenvalues provided in the solution to a particular RA ordination problem equal to 1, the relative contribution of each axis to the problem's solution can be easily compared. A graph of the percent eigenvalue of each RA axis produced in this analysis illustrates how interpretable axes differ from noninterpretable ones in terms of this statistic (Fig. 28.1). Note that in each case, the value of subsequent axes declined rapidly but irregularly, only those axes with the highest percent eigenvalues for each data set were interpretable, and the proportionate difference between axes did not necessarily predict which axes would be interpretable and which would not.

Topographic features

The ranking of topographic features along the first three RA axes can be understood by focusing on features associated with the origin and terminus of each axis (Table 28.2).

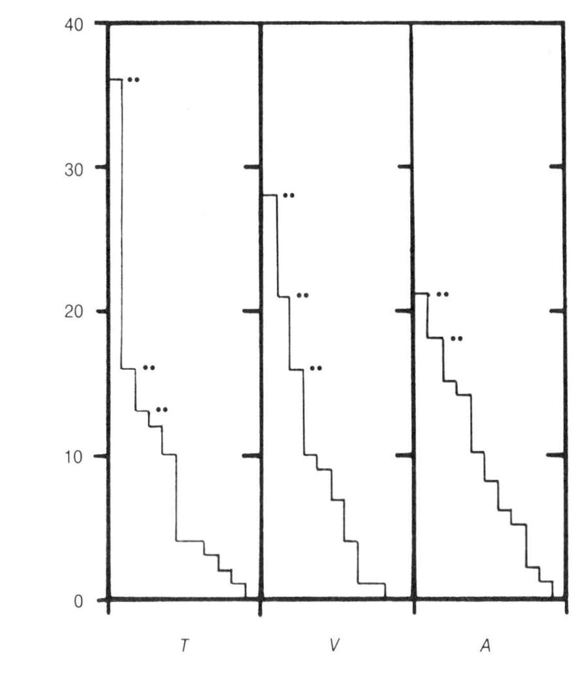

Figure 28.1. Percent eigenvalue associated with each RA axis by data set; T = topography, V = vegetation, and A = avifauna (dots indicate axes interpreted in this analysis; the sum of all eigenvalues provided for a given data set equals 100).

The position of features along the first axis are inferred to indicate a complex gradient of increasing relief and decreasing soil depth and moisture. In the Alvord Basin, these changes typically occur with increasing elevation.

Other important changes in geomorphological conditions taking place along this gradient include a shift from aggradational, relatively fine sediment environments toward those dominated by fractured bedrock. A corresponding change

Table 28.2. Positions of topographic features along RA axes (percent eigenvalue)

Axis	Origin Score	Origin Feature	Terminus Score	Terminus Feature
1	0.0	Dune	94.7	Ravine
(35.5)	5.0	Floor	96.1	Gully
	6.9	Saturated soil	96.8	Escarpment
	7.4	Flat	97.9	Bajada
	11.2	Lake	100.0	Terrace
2	0.0	Dune	64.9	Road
(15.7)	12.6	Terrace	72.5	Stream
	22.6	Floor	86.0	Fence
	23.0	Escarpment	89.9	Building
	26.4	Bajada	100.0	Hill
3	0.0	Lake	85.4	Saturated soil
(13.0)	12.2	Alkali	85.7	Terrace
	34.0	Spring	88.1	Building
	63.2	Standing water	92.0	Dune
	64.8	Canyon	100.0	Hill

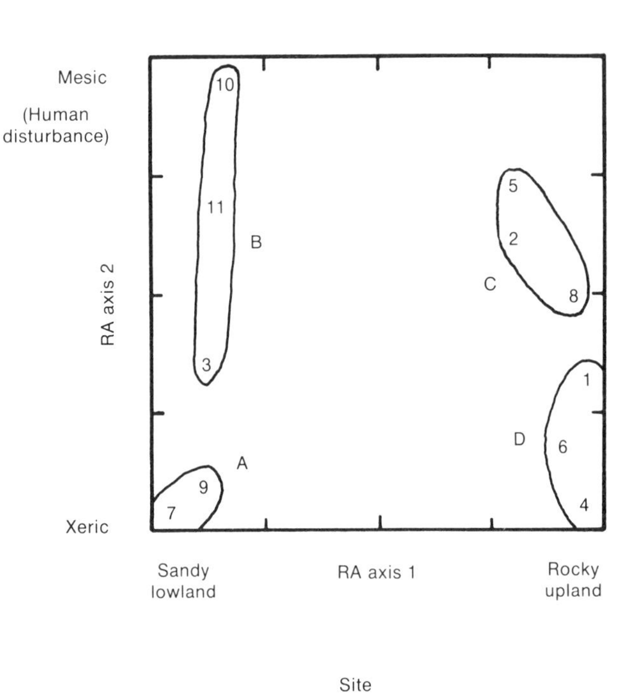

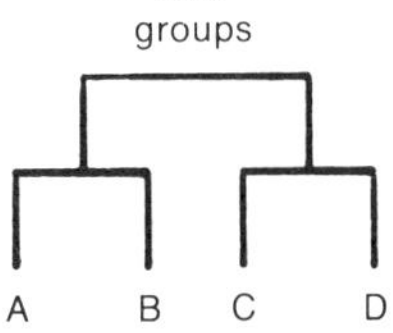

Figure 28.2*a*.

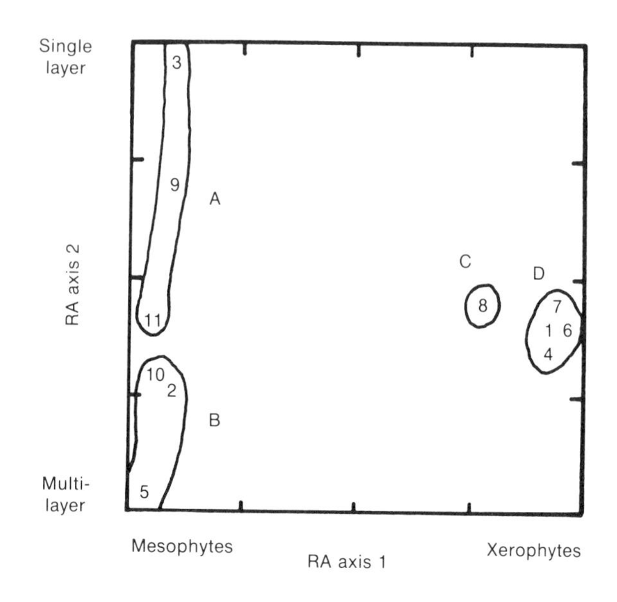

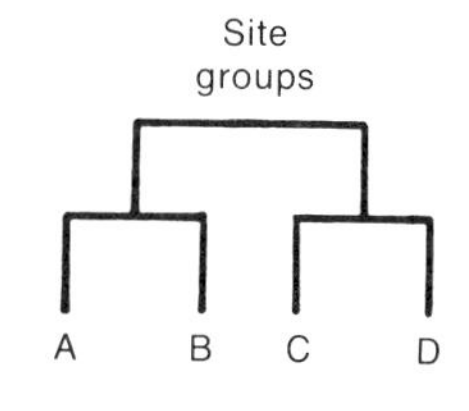

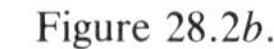
Figure 28.2*b*.

Figure 28.2*a–c*. RA ordinations of sampling sites by (*a*) topography, (*b*) vegetation, and (*c*) avifauna. Qualitative interpretations of each axis were obtained from corresponding species ordinations (see text). Dendrograms summarize TWINSPAN classification for each data set.

occurs in the balance of erosion processes, from sheetwash and sheetflood to lateral stream planation. Although the Alvord Basin's climate is extremely arid, its landforms are nonetheless sculpted primarily by water. The combination of sustained snowmelt from the summit of Steen's Mountain and rare but violent rainstorms feeding numerous, short, intermittent and discontinuous streams gives the area its distinctive topographic character.

Features associated with the terminus of the second axis differ from those at the origin in one important respect—they include terms describing features created by man. None of the sites included in this study was occupied by people during the study period, although several showed signs of former occupancy, including abandoned wooden structures and crumbling sod houses. There were also signs of continuing intermittent use for grazing, mining, or recreation. Such sites were usually not far from either a permanent stream or a spring.

Features associated with the origin of the third axis include "alkali," "lake," and to a lesser extent, "spring." These features are particularly characteristic of a single site occurring at a remote location on the basin floor. Nearly all drainage in the basin flows into a series of intermittent lakes (playas), but the lake at this site is unique for its large, permanent volume maintained by artesian springs. This site is arguably an outlier with respect to all other sites sampled, but it is included because of its great contribution to the basin's total bird diversity.

When all sites are grouped by TWINSPAN according to topographic features and then plotted in a scatter diagram defined by the first two ordination axes, a sharp separation between lowland groups (A and B) and upland groups (C and D) is apparent on the first (= X) axis (Fig. 28.2*a*). Moisture differences are less clear because the lowland sites in group A show only saturated ground as a moisture source, but upland sites in group C have a diverse combination of standing water, springs, saturated ground, and stream features. Conversely, lowland sites in group B are relatively moist, but upland sites in group D are almost completely dry. Sites with the highest values on the second (= Y) axis (i.e., 11, 10, 5, and 2) have the greatest evidence of human use and also tend to have more moisture available than other sites. If the third (= Z) axis were added, it would clearly distinguish site

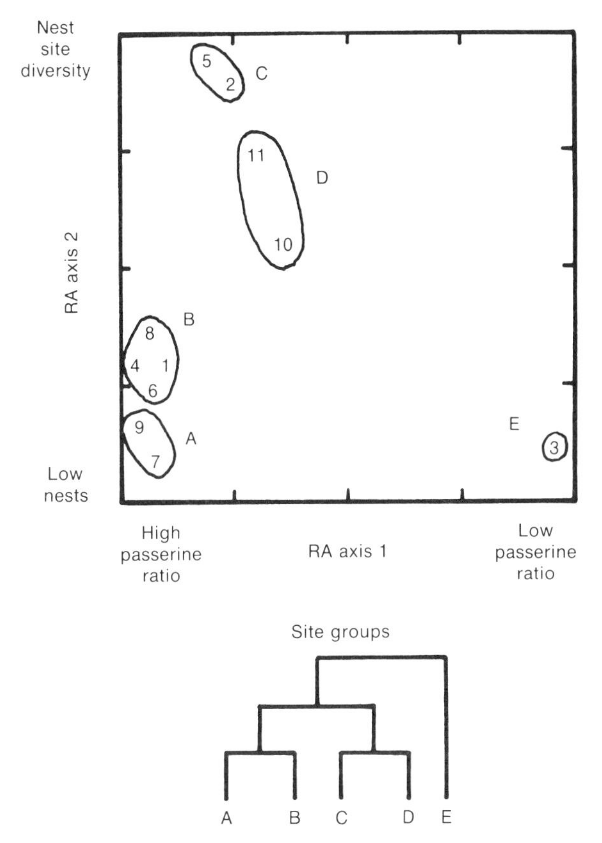

Figure 28.2*c*.

3 with its extensive system of artesian springs and alkali flats.

Vegetation

Plant species associated with the origin of the first ordination axis are a mixture of small- to medium-sized trees, including black cottonwood, boxelder, apple, and mesic shrubs (Table 28.3). Species associated with the terminus are small- to medium-sized shrubs that are also often sclerophyllous and spinose, such as spiny hopsage, and saltbush. This axis is interpreted as a gradient of plant physiognomy, dominated by changes in stand composition and structure that result from site conditions shifting toward increasing dryness.

Species associated with the origin of the second axis are very similar to the first but are now contrasted with a smaller group of emergent wetland species, including hardstem bulrush, cattail, and rushes. The simplification of vertical stand structure seen here is similar to the first axis but occurs in the opposite direction, toward increasing moisture.

Plant species positioned at the terminus of the third axis include boxelder, apple, Russian-olive, and black locust. These species are not native to the Alvord Basin yet are found in a number of areas, particularly sites associated with historic human occupancy. This axis is interpreted as a shift in the composition of plant species resulting from introductions by man.

If all sampling sites are grouped by TWINSPAN according to plant species and then plotted in a scatter diagram defined by the first two ordination axes, a strong division between groups A and B, and C and D is again seen (Fig. 28.2*b*). This time the contrast is attributable to plant physiognomic differences. Groups A and B represent the more mesic sites and show greater diversity of vertical structure than do groups C and D, which are dominated by xeric shrub species.

Site 8 is anomalous in that it forms a group by itself. This is explained by the presence of a discontinuous stream that emerges at a number of points around this site. Wherever the stream emerges, it supports mesic shrubs such as willow and Wood's rose (*Rosa woodsii*) and leads to a plant assemblage with intermediate structure.

Differences between groups A and B are also attributed to a gradient of decreasing vertical structure in the direction of increasingly moist sites. This trend culminates at site 3, an area characterized by extensive emergent vegetation.

If a third axis were added, it would distinguish sites with a large component of introduced vegetation, including sites 9, 10, and 11.

Avifauna

Bird species ranked at the beginning of the first axis include four species of sparrow: black-throated, lark, sage, and Brewer's sparrows (Table 28.4). Only two species listed here are nonpasserines: the common nighthawk and the chukar. Of those species ranked at the terminus of this axis, only two are sparrows: the savannah and song sparrows. All of the remaining species are nonpasserines. This axis is interpreted as a gradient of sparrow dominance and high passerine ratio changing toward an increasing proportion of nonpasserine species.

A comparison of the nest-site locations used by bird species along the first axis indicates that species associated with the axis origin prefer low nesting sites, ranging from a simple scrape on the ground to a low shrub location, but species associated with the terminus typically nest in cavities, high in tree branches, or in other specialized sites. This axis is interpreted as a gradient of nest-site preference, extending from ground to tree canopy, with a corresponding increase in dependence on vertical stand structure.

All sampling sites were once again classified by TWINSPAN, this time according to bird species, and plotted in a scatter diagram defined by the first two ordination axes, so that site relationships could be examined with respect to bird distributions (Fig. 28.2*c*). Group E consists only of site 3, the same location previously described as lowland with extensive artesian springs, alkali crusts, and predominately emergent vegetation. Birds at this site included a variety of shore and aquatic species not encountered elsewhere, such as the snowy plover, long-billed curlew, and cinnamon teal.

Both groups A and B were dominated by sparrows, particularly the black-throated, sage, and lark sparrows, but group

Table 28.3. Positions of plant species along RA axes (percent eigenvalue)

	Origin		Terminus	
Axis	Score	Feature	Score	Feature
1	0.0	Utah serviceberry (*Amelanchier utahensis*)	92.3	Little horsebrush (*Tetradymia glabrata*)
(27.9)	0.9	Black cottonwood (*Populus trichocarpa*)	95.4	Budsage (*Artemesia spinescens*)
	1.4	Boxelder (*Acer negundo*)	96.7	Winterfat (*Eurotia lanata*)
	1.4	Apple (*Malus sylvestris*)	99.1	Saltbush (*Atriplex canescens*)
	2.3	Golden currant (*Ribes aureum*)	100.0	Spiny hopsage (*Atriplex spinosa*)
	3.0	Chokecherry (*Prunus virginiana*)		
2	0.0	Utah serviceberry (*Amelanchier utahensis*)	62.5	Russian-olive (*Elaeagnus angustifolia*)
(20.9)	11.7	Golden currant (*Ribes aureum*)	64.8	Black locust (*Robinia pseudoacacia*)
	18.4	Chokecherry (*Prunus virginiana*)	78.0	Rush (*Juncus* spp.)
	19.0	Black cottonwood (*Populus trichocarpa*)	86.9	Cattail (*Typha latifolia*)
	24.9	Willow (*Salix* spp.)	100.0	Hardstem bulrush (*Scirpus acutus*)
3	0.0	Utah serviceberry (*Amelanchier utahensis*)	52.2	Sedge (*Carex* spp.)
(15.8)	4.5	Golden currant (*Ribes aureum*)	55.9	Black cottonwood (*Populus trichocarpa*)
	10.1	Quaking aspen (*Populus tremuloides*)	68.8	Black locust (*Robinia pseudoacacia*)
	19.2	Chokecherry (*Prunus virginiana*)	72.2	Russian-olive (*Elaeagnus angustifolia*)
	23.9	Hardstem bulrush (*Scirpus acutus*)	100.0	Boxelder (*Acer negundo*)
			100.0	Apple (*Malus sylvestris*)

Table 28.4. Positions of bird species along RA axes (percent eigenvalue)

	Origin		Terminus	
Axis	Score	Feature	Score	Feature
1	0.0	Black-throated sparrow (*Amphispiza bilineata*)	27.5	Song sparrow (*Melospiza melodia*)
(21.3)	2.6	Rock wren (*Salpinctes obsoletus*)	27.5	Great horned owl (*Bubo virginianus*)
	2.7	Lark sparrow (*Chondestes grammacus*)	30.8	California quail (*Callipepla californica*)
	2.9	Common nighthawk (*Chordeiles minor*)	41.3	Red-winged blackbird (*Agelaius phoeniceus*)
	3.8	Sage sparrow (*Amphispiza belli*)	47.4	Brewer's blackbird (*Euphagus cyanocephalus*)
	5.1	Sage thrasher (*Oreoscoptes montanus*)	56.6	Mallard (*Anas platyrhynchos*)
	7.2	Chukar (*Alectoris chukar*)	63.5	Killdeer (*Charadrius vociferus*)
	7.4	Brewer's sparrow (*Spizella breweri*)	100.0	Cinnamon teal (*Anas cyanoptera*)
			100.0	Snowy plover (*Charadrius alexandrinus*)
			100.0	American avocet (*Recurvirostra americana*)
			100.0	Long-billed curlew (*Numenius americanus*)
			100.0	Wilson's phalarope (*Phalaropus tricolor*
			100.0	Savannah sparrow (*Passerculus sandwichensis*)
			100.0	Yellow-headed blackbird (*Xanthocephalus xanthocephalus*)
2	0	Short-eared owl (*Asio flammeus*)	77.6	Northern oriole (*Icterus galbula*)
(17.9)	0	Common yellowthroat (*Geothlypis trichas*)	86.0	Ash-throated flycatcher (*Myiarchus cinerascens*)
	10.2	Brewer's sparrow (*Spizella breweri*)	89.7	MacGillivray's warbler (*Oporornis tolmiei*)
	12.0	Sage thrasher (*Oreoscoptes montanus*)	89.7	Yellow-breasted chat (*Icteria virens*)
	12.9	Cinnamon teal (*Anas cyanoptera*)	91.9	Northern flicker (*Colaptes auratus*)
	12.9	Snowy plover (*Charadrius alexandrinus*)	92.7	Barn swallow (*Hirundo rustica*)
	12.9	American avocet (*Recurvirostra americana*)	94.2	Lazuli bunting (*Passerina amoena*)
	12.9	Long-billed curlew (*Numenius americanus*)	94.7	Bushtit (*Psaltriparus minimus*)
	12.9	Wilson's phalarope (*Phalaropus tricolor*)	100.0	Western wood-pewee (*Contopus sordidulus*)
	12.9	Savannah sparrow (*Passerculus sandwichensis*)	100.0	House wren (*Troglodytes aedon*)
	12.9	Yellow-headed blackbird (*Xanthocephalus xanthocephalus*)	100.0	American dipper (*Cinclus mexicanus*)
	13.4	Red-winged blackbird (*Agelaius phoeniceus*)		
	17.7	Common nighthawk (*Chordeiles minor*)		
	20.3	Lark sparrow (*Chondestes grammacus*)		
	20.8	Sage sparrow (*Amphispiza belli*)		

B was distinguished by a larger number of Brewer's sparrows and the occurrence of sage thrashers.

Groups C and D consisted of sites with greater vertical plant diversity and consequently greater nest-site diversity. Group C included sites with greater relief and was typified by species such as the black-billed magpie (*Pica pica*), yellow-breasted chat, and lazuli bunting. Group D included lower, slightly drier, and more open sites where human introduction of plant species has made an important contribution to the vertical habitat structure. These sites were characterized by bird species such as the song sparrow, great horned owl, California quail, and Brewer's blackbird. A number of species were found at sites in both groups C and D, including the ash-throated flycatcher, northern flicker, northern oriole, and American robin (*Turdus migratorius*).

Discussion

This chapter focuses on conditions within a contiguous area of moderately large proportions. This scale reduces the confounding influence of gross geographic variation and may more closely approximate the operational scale of planned management activities than would either very large or very small study tracts.

The site characters chosen for analysis in this method included terms that describe the presence of physical conditions distinctive to the study area, as well as plant and bird species. The total number of characters selected was indeterminate until all observations were completed. This is contrary to popular practice, which limits the number of characters to a few with either management significance or alleged ecological importance, or with an arbitrarily set power of statistical determination. Each of the above ways to select characteristics is valid for a particular purpose, but the system presented here offers several unique advantages: it is flexible regarding unforeseen circumstances, maximizes the use of field observers' knowledge, adapts to rapid and efficient point counts, is easily updated, and provides a general-purpose data base that can be tailored for many wildlife species.

The data-analysis portion of this method applied RA to frequency indices having a relatively small range. Although this tends to compress variation, multivariate methods such as ordination appear to be little affected by the use of imprecise data (Green 1977; Gauch 1982a). A potentially more serious problem is that frequency tends to confound distribution with density (Mueller-Dombois and Ellenberg 1974:72). Further research may show that although frequency seems well suited for species-poor communities such as the Alvord Basin, it is an inappropriate parameter for species-rich communities or for species with such high density that frequency is always near 100%.

Statistical tests of the significance of ordination results are currently unavailable (Gauch 1982b:244). This is partly because the relevant statistical tests are poorly developed and partly because data concerning ecological communities rarely fulfill the rigorous assumptions of statistical tests. As a result, the interpretation of relationships identified using DCA must be regarded as tentative. Three things can be done to increase confidence in ordination until more definitive tests are developed: (1) several random subsets of data may be ordinated to see if basic patterns remain stable (Wilson 1981); (2) new samples not included in the original ordination may be used to see if relationships can be successfully predicted from ordination results (Gauch 1982b:172); and (3) results can be considered in terms of the known autecology of species involved and the known relationships of environmental variables. Samples for the second and most direct test may be randomly withheld from the original ordination or a portion of the study area may be randomly selected and reserved for later sampling and analysis.

What is most lacking in the example given in this chapter is a suggestion of how a temporal snapshot could be extended to include longer periods of time. One solution may be to view the data as having more than two dimensions, i.e., species × samples × times (Williams and Stephenson 1973; Gillard 1976; Austin 1977; Swain and Greig-Smith 1980). Unfortunately, DCA, RA, and other current ordination algorithms accept only two-way matrices as input. If ordination methods are to be used, three-way (or larger) matrices must be initially separated into simpler two-way matrices, and analyzed, and their results compared using an additional level of analysis.

It is impossible to specify a sampling design or to select variables that will be suitable for studying every wildlife-habitat combination of interest. Nevertheless, approaches that prove successful will probably emphasize completeness in representing a total natural system, will be flexible, will use efficient field methods, and will offer easily communicable results.

Acknowledgments

This study was partly supported by a grant from the Oregon Department of Fish and Wildlife Nongame Fund. I thank W. E. Grenfell, S. A. Laymon, C. D. McIntyre, E. F. Moore, B. R. Noon, C. J. Ralph, M. G. Raphael, and W. E. Waters for many helpful comments and suggestions regarding this chapter.

29

Discrete State-Space Models for Analysis and Management of Ecosystems

YOSEF COHEN

Abstract.—A linear, discrete, multivariable state-space model that can incorporate processes such as interspecific interactions and inputs (e.g., logging, hunting, and other potential habitat manipulations) to the ecosystem is developed. The hypotheses on which it is based include (1) the population size of a species depends on its past population size, on past and present population sizes of other species, and on inputs; and (2) the interactions postulated in (1) may operate with time lags. The model parameters are estimated with multivariable linear regression, and the z-transform technique is used to develop the state-space model. The model is applied to a system whose states are the population sizes of northern pintail (*Anas acuta*) and blue-winged teal (*A. discors*), and whose input is the number of ponds in the habitat. System characteristics such as stability and transient response are analyzed, and the system is found to be stable with the transient reaction of pintail and blue-winged teal populations to perturbation in the number of ponds similar.

Models that are designed to describe ecosystem dynamics, to offer optimal control policy, and to forecast species' populations must be dynamic; that is, they must include the projected behavior of species' populations over time. Such models should account for interspecific interactions if these are suspected to exist. When cast in an input/output form, such models should include the fact that, because of human intervention and natural processes, habitats are constantly changing. In addition, because large-scale ecosystem experiments are generally unfeasible, ecosystem models should allow simulation of possible perturbations.

The model

Multivariable state-space models (Csaki 1977) are potentially applicable to resource-management problems (Goh 1980) and have several desirable features: they consider systems as a whole; they allow consideration of inputs to, outputs from, and feedbacks within the system; and they provide a framework for the investigation of optimal control. By optimal control one usually refers to the process of maximizing an objective (output, or cost function), given the following data: (1) a model of the process; (2) the initial state and the target state of the system; and (3) the set of feasible controls.

The proposed model belongs to a class of general multivariable discrete state-space system models (Csaki 1977). At a particular discrete time instant k, the model considers M inputs denoted by $u(k)$, N different states $x(k)$, and S different outputs $y(k)$. The column vectors $u(k)$, $x(k)$, and $y(k)$ are of dimensions M, N, and S, respectively. In their most general form, these models describe the relationships between the inputs, state of the system, and the outputs as follows:

$$x(k+1) = f[x(k), u(k), k],$$
$$y(k) = g[x(k), u(k), k], \qquad (1)$$

where f and g are vector-valued functions and k is the discrete time interval, $k = 0, 1, 2, \ldots$. Because my main purpose is to introduce the state-space methodology and its potential applications, I consider only linear models with constant coefficients that constitute a subclass of those in equation (1). The use of linear models does not preclude processes such as population fluctuations and a limitation on population growth. This is especially true for linear difference equations of order higher than 1. Furthermore, evidence suggests that for some nonlinear models (e.g., the logistic), a linearized form gives qualitative results similar to those of the nonlinear model (Nisbet and Gurney 1982). As with any other model, when the mechanisms operating within the system are actually changing, such models fail.

Suppose that populations of S species in an ecosystem are censused at intervals k ($k = 0, 1, 2, \ldots$) where k may be in any time units. Furthermore, at each of these times, M inputs to the ecosystem are measured (e.g., rainfall, number of ponds, logging, hunting, pollutants, and any other habitat perturbation). Denote by $y_i(k)$ ($i = 1, 2, \ldots, S$) the population size of species i at time k, and by $u_j(k)$ the level of the jth input ($j = 1, 2, \ldots, M$) at time k. Usually, $y_i(k)$ depends on (1) present and past population sizes of other species within the ecosystem; (2) present and past inputs to the ecosystem; and (3) its own population size in the past. The effects of these on the species' population are mediated through demographic processes such as birth, death, emigration, and immigration rates. Note that demographic processes may be modeled in a similar manner; that is, one may define additional inputs or outputs, depending on the modeling purpose, as with these variables. In this case, the dimensions of y and u will increase accordingly.

Based on the modeled system and modeling purpose, the time units can be arbitrarily chosen. For example, let the

YOSEF COHEN: Department of Fisheries and Wildlife, 200 Hodson Hall, University of Minnesota, St. Paul, Minnesota 55108

time unit be in years. Then $y_i(k - l)$ denotes the population size of species i at year $k - l$, and a discrete, linear system model with S species, M inputs, and lags operating up to L years may be represented by:

$$\sum_{l=0}^{L} a_{iil} y_i(k - l) = \sum_{\substack{j=1 \\ i \neq j}}^{S} \sum_{l=0}^{L} a_{ijl} y_j(k - l)$$

$$+ \sum_{j=1}^{M} \sum_{l=0}^{L} b_{ijl} u_j(k - l) \quad i = 1, 2, \ldots, S \tag{2}$$

where $a_{ii0} = 1$. Based on equation (2), the discrete state-space model may be derived (details are available from the author). For example, consider a system of two species and one input, and denote by $y_1(k)$, $y_2(k)$, and $u(k)$ the population sizes of the two species and the input, respectively, at time k. Furthermore, assume that lags operate for one time unit only. The following linear model would then describe the dynamics of the system:

$$y_1(k) = a_{111} y_1(k - 1) + a_{120} y_2(k) + a_{121} y_2(k - 1) + b_{10} u(k) + b_{11} u(k - 1)$$

$$y_2(k) = a_{221} y_2(k - 1) + a_{210} y_1(k) + a_{211} y_1(k - 1) + b_{20} u(k) + b_{21} u(k - 1).$$

The parameters a_{ijl} describe the strength of the interaction between species i and species j for time lag l ($i, j = 1, 2$, $l = 0, 1$), and the parameters b_{il} describe the strength of the interaction between species i and the input u for time lag l. For a system with two species and one input, which includes lags of up to 2 years, the state-space model is of the form:

$$\begin{aligned} x(k + 1) &= Ax(k) + bu(k) \\ y(k) &= Cx(k) + du(k), \end{aligned} \tag{3}$$

where A is a 4×4 matrix, $x(k)$ is the state vector of dimension $N = S + L = 4$, b is a vector of dimension 4, C is a 2×4 matrix, and d is a vector of dimension 2. The matrices A and C, and the vectors b and d, do not have a readily apparent interpretation, and they are defined for mathematical convenience only. The derivation of model (3) from model (2) permits analytical solution and analysis of system characteristics such as stability and response to perturbations. Model (3) has the following solution:

$$x(k) = A^k x(0) + \sum_{i=0}^{k-1} A^{k-i-1} bu(i)$$

$$x(0) = x_0$$

$$y(k) = CA^k x(0) + \sum_{i=0}^{k-1} CA^{k-i-1} bu(i) + du(k).$$

Although model (3) is linear with constant coefficients, it includes lags of more than one time period (the model is of lth order). Therefore, the model can potentially mimic the behavior of natural populations—such as damped oscillations, overshoot or undershoot, when approaching steady state (May 1976b), and transient responses. The matrix A characterizes a closed-loop system; it incorporates the fact that a present output depends on present outputs other than the one considered, on the present inputs, and on the past history of all outputs and inputs, thereby allowing potential control of ecosystems through manipulations of $u(k)$ and $y(k)$, where appropriate.

The data

The U.S. Fish and Wildlife Service has been estimating the abundance of major waterfowl species in North America since 1955 (Martin et al. 1979). Data were collected over 49 strata (Fig. 1 in Martin et al. 1979) during the waterfowl breeding season, and the number of ponds was counted in strata 26–49. To estimate model parameters, I used the data for strata 26 through 49. Stratum number 30 was excluded from the data during the process of parameter estimation, and was later compared to the results of the model. I chose to exclude this stratum from the data for the purpose of model verification simply because it is situated near the center of the survey area and thus may represent a typical breeding habitat. Because the model uses time lags of up to 2 years, the data for 1955 and 1956 had to be excluded from analysis. Thus, the total number of breeding-population counts for each species in the data file was 621 cases (23 strata $\times$ 27 years). To isolate closely related species (in terms of their population fluctuations), I used data provided by the Fish and Wildlife Service (Trends in duck breeding populations, 1955–1983. Office of Migratory Bird Management. Administrative Report. 11 July 1983: Table 1).

Model development

To isolate a subsystem from a larger system, three questions need to be answered for the present data set: (1) Of the 10 waterfowl species, how many should be included in the model? (2) How many inputs should be considered? And (3) How many time lags should be included? To answer the first question, I used the abundance data for the 10 waterfowl species from 1955 through 1979 to construct the matrix of correlation coefficients. The absolute values of the correlation coefficients were subjected to complete linkage cluster analysis (Gnanadesikan 1977; Anonymous 1983). The changes in population sizes of northern pintail (*Anas acuta*) and blue-winged teal (*A. discors*) were the most closely related, so I included these two species in the model. To answer the question of how many inputs to include, I fitted the model with the data for the number of ponds. Because this resulted in a good fit (see below), I included one input only. Note that other variables could be included as inputs: e.g., harvesting effort, weather parameters, and number of predators in the area. Furthermore, other variables could be defined as outputs: e.g., birth rate, death rate, and number of ducks harvested.

To determine the number of lags to include in equation (2), I calculated the correlation coefficients between the abundances of these two species with lags of up to 10 years. The

r-values for lags of up to 2 years were high (e.g., rs between the abundance of the pintail and the lagging abundance of the blue-winged teal for 0, 1, and 2 years were 0.81, 0.45, and 0.15, respectively). For lags of more than 2 years the r-values remained generally small, even when the lagged abundances were switched (e.g., blue-winged teal vs lagged abundance of pintail). The correlations between each of these species and the number of ponds displayed a similar pattern, and therefore I chose $L = 2$.

Note that significant autocorrelations do not necessarily imply density-dependent population changes. To examine density dependence, a different test is needed (Vickery and Nudds 1984). Autocorrelations may be significant, without density dependence, when populations are regulated through other intermediary mechanisms. Other differences between the results here and those of Vickery and Nudds (1984) may be attributed to the following factors: (1) The test of density dependence was applied for 1-year lag only. It may be plausible that mechanisms of density dependence operate with a delay of more than 1 year. (2) The data set used by Vickery and Nudds (1984) was from the Lousana Waterfowl Study Area in Alberta and from Redvers Waterfowl Study Area in Saskatchewan. These data constitute a subset of the data used here.

Parameters a_{ijl} and b_{ijl} of equation (2) ($i = 1, 2; j = 1, 2;$ and $l = 0, 1, 2$) were estimated by multiple regression (Anonymous 1983). Based on the above considerations, the model is then:

$$\begin{aligned} y_1(k) - 0.4123y_1(k-1) - 0.2447y_1(k-2) = {} \\ 0.2925y_2(k) + 0.0649y_2(k-1) - 0.2661y_2(k-2) \\ + 0.2413u(k) + 0.0368u(k-1) - 0.0625u(k-2) \quad (5) \end{aligned}$$

for the blue-winged teal, and

$$\begin{aligned} y_2(k) - 0.04734y_2(k-1) - 0.1876y_2(k-2) = {} \\ 0.2290y_1(k) - 0.0687y_1(k-1) - 0.1702y_1(k-2) \\ + 0.4972u(k) - 0.2042u(k-1) - 0.0873u(k-2) \quad (6) \end{aligned}$$

for the pintail.

All the coefficients in equations (5) and (6) are statistically significant ($P < 0.05$), and they may be interpreted as reflecting the strength of interaction among the variables. When standardized, these coefficients reflect the number of standard deviations of change in one variable due to change in one standard deviation of another variable. For both equations (5) and (6) the intercepts were less than 5% of the mean population size and were therefore ignored. For the blue-winged teal [$y_1(k)$, equation (5)], r^2 was 0.65 and for the pintail [$y_2(k)$, equation (6)], r^2 was 0.84.

Equations (5) and (6) were used to construct the state-space model (3) with the following matrices (details are available from the author):

$$A = \begin{bmatrix} -0.2549 & 0.0748 & 0.2557 & 0.2557 \\ -0.0167 & 0.7253 & -0.0569 & -0.0569 \\ 0.0390 & 0.0390 & -0.2479 & 0.0089 \\ 0.0036 & 0.0036 & 0.0008 & 0.7311 \end{bmatrix} \quad b = \begin{bmatrix} 0.1904 \\ 0.0588 \\ 0.0159 \\ 0.0304 \end{bmatrix}$$

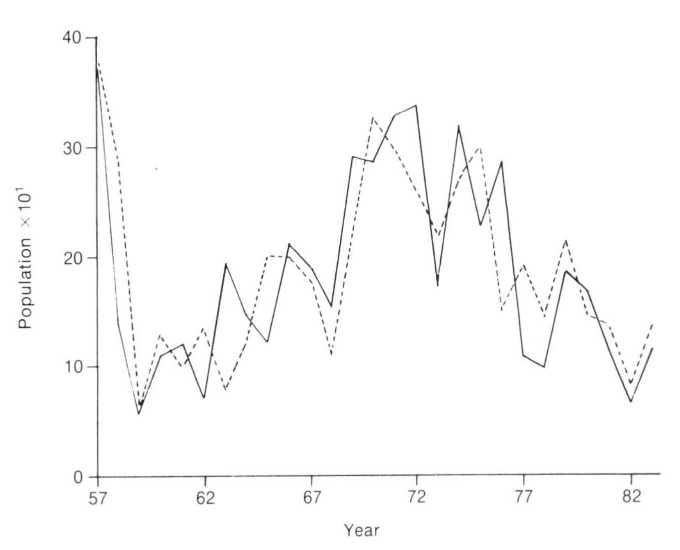

Figure 29.1. Counted (solid line) and model (broken line) population fluctuations of the pintail. Data are for transect number 30 (Martin et al. 1979).

$$C = \begin{bmatrix} 1.0718 & 1.0718 & 0.2454 & 0.2454 \\ 0.3136 & 0.3136 & 1.0718 & 1.0718 \end{bmatrix} \quad d = \begin{bmatrix} 0.3807 \\ 0.6085 \end{bmatrix}.$$

These matrices provide a starting point for analysis of system characteristics such as stability, transient response, and optimal control.

The model parameters were estimated from the data, excluding transect number 30. These parameters were then used to compare the model to the actual data from transect number 30 (Fig. 29.1). Apparently, the model fits the data well and thus lends credibility to the model as a descriptor of the system. Because my major motivation is to introduce state-space models, I did not compare the fit of this model to competing models.

Use of the model to analyze system characteristics

Model (3) mimics population interactions for blue-winged teal and pintail and their reaction to changes in the number of ponds over the years. Therefore, it may be used to address the topics of system stability and transient response of the species' populations to changing inputs. A word of caution is in order here. When the model is applied this way, an implicit assumption (of the model) is that the structure of the system does not change as inputs and states (population sizes) are varying. By invariable structure I mean that the matrices A and C and the vectors b and d do not change and the form of equations (5) and (6) remains unchanged. This is a plausible assumption when variables in the system do not fluctuate widely. This implicit assumption is made in the great majority of cases in which models are analyzed. When the system is prone to experimentation, or when the inputs vary widely, this assumption may be verified. In any case, the model is valid for the range of input fluctuation that was observed. If that range is believed to be wide enough to

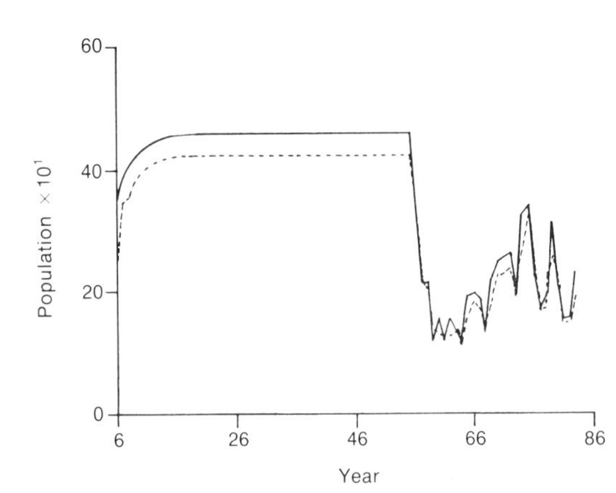

Figure 29.2. The response of the populations of pintail (solid line) and blue-winged teal (broken line) to a step input starting at 1906 and continuing to 1955 in transect number 30. Starting at 1955, the input (number of ponds) was set according to the data for transect number 30.

include most typical fluctuations of inputs, then model analysis is a useful exercise.

For system (3) to be stable, the eigenvalues of the matrix A must be inside the unit circle (Cadzow 1973). The four eigenvalues of A are: -0.35177, -0.14825, $0.72675 + i(0.0148)$, and $0.72675 - i(0.0148)$, which indicate that the system is stable. Because the last two eigenvalues are complex, and because the complex part is relatively small compared to the real part, the populations of these two species will oscillate slightly in response to perturbations.

If the input to the system resembles a step function—that is, it changes from a prolonged period of low level to a high level and remains high for some time (by virtue of, say, a number of very dry years followed by a number of very wet years)—the question is then how fast will the populations of these two species react to such a drastic change and how long will it take the populations to reach a steady state. To answer this question, one may consider a step input. In this case, the input is zero, and at some point in time it is increased to a certain level at which it remains.

Figure 29.2 demonstrates the following scenario: Up to 1906 (when $k = 0$ by definition), the input $u(0)$ and the states $x(0)$ were set to zero. In 1906 the input was set equal to the number of ponds in 1955 in transect number 30. From 1906 on, the input remained constant for the next 50 years. From 1955 through 1983 the input $u(k)$ was set equal to the observed value. The responses of the outputs $y(k)$ were plotted with this scenario. The first part of the plot (1906 through 1955) then simulates the combination of the reaction of the populations of these two species to a step input and to the interaction between their populations. The time it takes both species to reach a steady state (i.e., the transient response) is about 15 years for both species. The populations reach this level with some fluctuations, which are not evident in Figure 29.2 because of its scale, but which were evident from the numerical results and anticipated from the complex eigenvalues.

Starting at 1955, the actual input of the number of ponds was used, and the right-hand portion of Figure 29.2 (1956 through 1983) illustrates the population fluctuations of these two species. Although the time of the transient response of these two species is equal, the pintail reaches a higher population size than the blue-winged teal. Furthermore, the population of the pintail fluctuates more violently than that of the blue-winged teal (Fig. 29.2).

Discussion

The analysis outlined herein provides an important tool to resource managers, for often the question is not only what the reaction of an ecosystem to perturbations will be, but also how long it will take before the system reaches a steady state. Furthermore, it may be important to know, before perturbing the system, how the various species will reach the equilibrium state: will they overshoot the equilibrium and, if so, by how much? These system characteristics are available from matrix A, provided the model is reasonably accurate.

The technique of testing the system under step input also serves the following purpose. Since $x(0)$ is not generally known, the analytical solution of the model—equation (4)—can assume that $x(0) = 0$. By starting the solution back in time, the system reaches a steady state before the input starts to change (in 1955). Thus, the effect of the initial state of the system on the reaction of the system to fluctuating inputs is removed. For stable systems this effect will always be very small after a few time periods—as can readily be verified from equation (4).

The model can be criticized for a number of reasons: it is linear and therefore neglects nonlinear processes in ecosystems and populations; it includes lags for which I did not present biological justification; it overlooks demographic processes such as birth and death rates; and it is deterministic and thereby ignores stochastic processes.

Because the purpose is to devise a model that will mimic population fluctuations with regard to inputs and outputs, data of demographic processes are unnecessary. Furthermore, demographic processes, inputs, and interspecific interactions are all reflected in the population size of each species and include lags, and precisely these are the data that were used. When data are available, and when management intervention can potentially change these demographic processes directly, demographic variables can be easily incorporated in such models. This may be done by increasing the dimension of the input vector $u(k)$ accordingly.

When the design of a systems model is considered, system identification becomes very complicated, especially in situations where feedback mechanisms intervene in the process of parameter estimation. Nonlinearities should generally be introduced to biological models only when they are necessary and when simpler models fail. The second-order linear difference model mimics the data closely (Fig. 29.1) and therefore further complication was unnecessary. Furthermore, evidence suggests that when some nonlinear models (e.g., the logistic) are linearized (e.g., the Taylor series expansion), the dynamics of the linear models follow the behavior of nonlinear models closely (Nisbet and Gurney 1982).

The residuals from fitting equations (5) and (6) appeared to be distributed randomly with means of zero and variance v, where v is a vector of dimension 2. This result can be used to derive a model similar to (3) with one additional input vector $z(k)$ of length S, which represents a random shock. Thus, including stochasticity in the model is simple. Tiwari and Hobbie (1976) and Tiwari et al. (1978) proposed other methods that incorporate randomness in ecosystem models.

The model is relatively complex to apply. Nevertheless, the formulation lends itself to writing a general and easily applied computer program. Although the model resulted in a good fit to a different data set, the validity of the model is still at question. Adequacy may be judged by its ability to forecast changes in populations. This was not done because of the shortness of the time series (27 years). As data accumulate, the same data set may be used to reevaluate the parameters based on part of the data set, say 25 years, and the rest of the time series can then be used to validate the model.

The outlined input/output approach is general and may be applied to many different situations. For example, the time units may be expressed in days and populations may be followed for a year, where both habitat and weather variables may be recorded. The resulting model may then be used to simulate the effects of habitat perturbations on the system. According to equation (1), the approach outlined herein may be used with nonlinear models as well. The theory of such models has been developed (Goh 1980).

Since state-space models can be used in optimal-control processes, they are particularly useful in management decisions. For example, the following optimal-control problem can be solved. Given that (1) the ecosystem processes are described by the state-space models (1) or (3), with $u(k)$, $x(k)$ and $y(k)$, ($k = 0, 1, \ldots, T$), defined as the control sequence (inputs), the states of the system, and the outputs, respectively; (2) the initial conditions are $x(0)$; (3) $u(k)$ are bounded within a certain range of plausible management actions; and (4) a management objective, which is some function of $y(k)$, is to be achieved at the end of a planning period T, find the sequence $u(0), u(1), \ldots, u(T)$ that will maximize that objective over the period $0, \ldots, T$. A variant of this optimal-control problem seeks to minimize the time T required to bring the system from an initial state $x(0)$ to a final desired state $x(T)$. Other more basic questions may be addressed. (1) Is the system controllable; that is, can we use a feasible control sequence $u(k)$ to bring the system to a desired final state $x(T)$? (2) What is the set of final states that can be reached from the set of initial states for the whole range of plausible management actions? Further work with similar systems will show how useful such models are to wildlife managers.

Acknowledgments

This chapter is dedicated to the memory of Gad Boneh. Financial support was provided by University of Minnesota Graduate School (grant number 0350–4935–62) and by University of Minnesota Agricultural Experiment Station (grant number 0302–4841–32). Data were kindly provided by the U.S. Fish and Wildlife Service. The manuscript benefitted from reviews by Al Ek, George Spangler, and Larry Jacobson. Douglas Johnson and Fred Stormer, who refereed the manuscript, provided helpful remarks. The repeated reviews and suggestions by Mike Morrison greatly improved the manuscript.

Summary: Biometric Approaches to Modeling— The Researcher's Viewpoint

BARRY R. NOON

The incentive for modeling relationships between habitat and population parameters of wildlife species is based on the belief that animals respond to habitat in an adaptive fashion. That is, habitat selection is believed to be a specific behavior, exposed to the forces of natural selection, that has evolved to maximize lifetime reproductive output. According to the argument of Southwood (1977), habitat acts as a template for ecological strategies. From Southwood's model we assume that decisions about a habitat's suitability are based on proximal cues received from the habitat that are associated with probabilities of survival and reproductive success. Formalization of the theory of animal-habitat relationships in an evolutionary framework serves two main functions. First, it establishes an underlying theoretical basis and justification for attempting to model the relationships between wildlife and habitat. Second, it suggests that predictive statements about animal behavior can be made; that is, we can affect an animal's decisions by altering specific habitat components.

Predictive models are, in turn, valuable because they lend themselves readily to tests of their validity. For example, by specific manipulation of certain habitat components, we can examine whether an animal's decisions are altered according to the habitat features changed. If a species does not behave as predicted, we can measure whether a "wrong" decision has imposed a cost. Failure to detect a cost, or observing a disparity between observed and predicted responses, makes us skeptical of the model and leads to its further refinement or rejection.

Note that models of habitat selection, developed initially from an ecological and evolutionary perspective, are easily translated into the statistical framework of hypothesis statement and testing, with concomitant decisions on model acceptance or rejection. Thus, it seems logical that wildlife ecologists have incorporated biometric models into their attempts to understand wildlife-habitat relationships.

Discussion of chapters in Part II

A survey of wildlife-habitat literature reveals two facts: (1) the past decade has seen a large-scale trend to quantify habitat patterns in terms of multivariate, biometric models; and (2) most biometric approaches have been restricted to multiple regression (MR), discriminant function analysis (DFA), or principal components analysis (PCA). Multiple regression is usually framed in the context of explaining the variance of a dependent variable (e.g., breeding density) by considering its relationship with two or more independent (habitat) variables (e.g., shrub density, tree basal area). The resulting model is a linear function of the independent variables, each weighted by their partial degree of association with the dependent variable. Multiple regression is often viewed as having applications beyond the immediate curve-fitting procedure because it can be interpreted in terms of a model that predicts the value of the dependent variable conditioned on a new vector of independent variables. Its appeal to wildlife managers is obvious.

Discriminant function analysis (DFA) is often framed in the context of testing the null hypothesis of no difference in the position of two or more group centroids in some multivariable space. An equally popular use and interpretation of DFA is in the context of classification. The researcher is interested in determining the probable group membership of an observation given a vector of predictive variables. The two contexts of DFA are often combined; rejection of the null hypothesis of no difference serves as a justification to examine the classification results. Also, given a significant result, the linear canonical functions are interpreted in terms of those variables whose coefficients suggest a strong contribution to group separation. This is the step at which biological inferences are made (see Tatsuoka 1971; Williams 1981, 1983).

The principal components model is most often used for data exploration. The concepts of dependent and independent variables are not relevant; there is no attempt at statistical inference or tests of hypothesis; and the data have no group substructure. The researcher seeks one or more linear composites of the data that recapture most of its original variance, but in fewer dimensions. Dimension reduction is straightforward, as the principal components are ordered in terms of the amount of variance they explain. Principal components, as regression equations and canonical variates, can be given biological interpretations in terms of those variables with large associated coefficients.

A variant of PCA that is receiving increasing attention is reciprocal averaging (RA). Like PCA, RA is based on an eigenanalysis of the data matrix, but it differs from PCA in the way in which weights are assigned to the variables. In RA the observations and variables are ordinated simultaneously to maximize the correlation among them. The reader is referred to Pielou (1984:176–188).

All of the models previously discussed have inherent as-

BARRY R. NOON: Department of Wildlife, Humboldt State University, Arcata, California 95521

sumptions, both biological and statistical, that affect model validity and interpretation. The most stringent is the restriction to linear relationships among the variables. A less commonly acknowledged restriction is that in almost all cases any inferences that are made about biological processes are not in fact tested by any of the models. For the most part, these models are restricted to documenting patterns that may or may not have any biological significance (cf. James and McCulloch 1985).

Marzluff (Chapter 25) emphasizes the importance of testing for the appropriateness of a linear response in the multiple regression model. The test for linearity depends on multiple observations of the dependent variable at each combination of the independent variables. As the number of replicates required rapidly increases with the number of independent variables, this approach, he points out, is unfeasible for most initial, exploratory studies. An additional topic of experimental design discussed by Marzluff is the spatial and temporal allocation of samples. Specific combinations of independent variables (i.e., sample sites) must be replicated so as to allow tests for the aptness of a linear model, but not clumped so as to restrict biological inference. If we are interested in determining wildlife-habitat patterns, we must sample extensively along the habitat gradient. Allocation of samples is ultimately constrained by opposing extensive and intensive requirements and by the important requirement that replication be greatest at those habitat configurations where response variation is high.

Gutzwiller and Anderson (Chapter 24) also discuss the appropriateness of a linear regression model by examining problems detectable through an analysis of residuals. Plots of residuals on independent variables, predicted values, and so on, can often be interpreted unambiguously. In addition, probability plots of the residuals provide an easy way to inspect the normality assumption of the model. Of the problems discussed by the authors, I believe that nonconstant error variance is the most common in wildlife-habitat studies. To illustrate, if we assume population density to be the dependent variable, its behavior is often most variable near the ends of habitat gradients. Nonconstant error terms are particularly distressing because they greatly affect the precision of the parameter estimates. However, as Gutzwiller and Anderson explain, remedial methods are available through a weighted least-squares model. Despite the clarity of their presentation, Gutzwiller and Anderson have unfortunately restricted their discussion to simple linear regression, when multiple regression is the usual model for wildlife-habitat studies. The interpretation of residual plots and the implementation of remedial methods is not so clearcut when the model contains multiple independent variables.

Capen et al. (Chapter 26) contrast the performance of two-group DFA and logistic regression (LR), on the basis of their analysis of avian breeding habitat in New England. I believe the primary strength of their contribution is found in the outline they present of methods to validate DFA models. The authors also contrast a variety of available experimental designs for comparing used vs unused habitat groups. The strength of their comparison is the finding that the same problems, from both biological and statistical perspectives, are encountered regardless of the experimental design. An important point is that when planning used/unused (presence/absence) habitat comparisons, one must have a strong rationale for defining the sampling frame. One reason is that the size of the unused (or absent) group is totally arbitrary. This introduces an indeterminacy into the classification results, and the interpretation of the canonical functions, because of their dependency on prior probabilities of group membership.

In Chapter 26 Capen et al. discuss several frequently encountered problems in the application of DFA to observational studies. The most common is multicolinearity of the predictor variables. The main source of concern is that the estimation of the discriminant (regression) coefficients may become highly specific to the particular data set being analyzed. Further, addition or deletion of a few observations may substantially affect the magnitude (and sign) of the coefficients, and thus invalidate biological interpretation. Remedial methods for multicolinearity have received much attention from statisticians but, unfortunately, without any general agreement.

The chapter by Brennan et al. (Chapter 27) also contrasts DFA and LR models, but specifically in the context of developing Habitat Suitability Index models (HSI). Their empirically based approach to HSI model development should result in predictive models that regularly outperform those based on "expert opinion." I suspect that their approach to HSI model development will be generally adopted for future studies. In addition, the authors, in line with suggestions of Capen et al. (Chapter 26), tested their model with independent data. This essential step of independent model testing is often bypassed in wildlife-habitat studies. In the absence of independent verification, one cannot judge whether the model is applicable to situations beyond that used to construct the model. Developing wildlife-habitat models that are statistically significant says nothing about their relevance to data sets collected in other parts of a species' range, or at different times in the same location. Developing statistically significant discriminant or regression models is an easy task, particularly in presence/absence studies. Developing models that predict or classify well in different years or at different locations has proven to be a much more formidable undertaking.

Several authors in this volume have used LR in comparison with, or in lieu of, two-group DFA. Logistic regression leads to the same results as a two-group DFA when the data meet all the assumptions of the latter model. However, when these assumptions are violated, or the real model is a nonlinear one, LR seems to regularly outperform DFA, at least in terms of classification accuracy (Press and Wilson 1978; Brennan et al., Chapter 27). Care must be used when comparing the classification accuracy of DFA and LR, as the models differ in the estimation of posterior probabilities of group membership and LR classification is very sensitive to the probability cutpoint that is selected. LR may be used when the predictor variables are categorical or a mixture of categorical and continuous, even though the joint distribution of such variables will not be multivariate normal. This may be its key strength.

The experimental design used by both Capen et al. (Chapter 26) and Brennan et al. (Chapter 27) needs to be accompanied by a cautionary note because of the difficulties in presence/absence analyses (Johnson 1981a). We would like to be able to interpret the absent samples as simply representing unsuitable habitat. However, absence can derive from reasons other than lack of suitability.

Liverman (Chapter 28) estimated environmental attributes and bird species composition at specific locations along a steep environmental gradient. He then performed RA ordination analysis simultaneously on the site and their attributes, then separately on bird species by site and on site-data matrices. Separate RA analyses allow a visual comparison of the ordination patterns and a qualitative association of birds with environmental features. Liverman's application of the RA model represents a form of indirect gradient analysis. An important issue is whether the application of this statistical model has yielded insights into the relationships between birds and habitat features that would not have been apparent from a much simpler approach. It is often much easier to give biological interpretations to direct ordinations of species with environmental factors, particularly when the data represent extensive environmental variation.

The chapter by Cohen (Chapter 29) differs from the usual biometric approach to modeling wildlife-habitat relationships. Cohen's model is a linear, discrete, multivariate state-space model whose states are the population sizes of two species of waterfowl and whose inputs are the number of ponds available for breeding. Multiple regression is used to estimate the parameters for the model by modeling a species' abundance in year *i* (dependent variable) as a function of both its abundance and the abundance of other species in the 2 previous years, and the current and past number of ponds. The model was successful as judged by the fit between predicted annual population sizes and estimates based on data not used in model parameterization. For this approach to gain in popularity in the future, the mathematical competence of most wildlife ecologists would have to increase.

Cohen's approach in Chapter 29 also differs from previous efforts at modeling the population dynamics of waterfowl species. Previous studies have explicitly incorporated demographic parameters such as birth and death rates into their models (e.g., Anderson 1975b; Martin et al. 1979). This latter approach has the advantage of interpreting model behavior in terms of those biological processes determining the population pattern. To a large degree, waterfowl-management efforts of the U.S. Fish and Wildlife Service are directed toward influencing recruitment and mortality rates. In Cohen's model, the relative contributions of these two parameters are masked, but the model could easily be modified to include them (Y. Cohen, pers. comm.). The importance of partitioning out these two effects for effective management is easily illustrated. A population limited by recruitment, but not by mortality, may require additional breeding habitat. On the other hand, a population limited by mortality, but not by recruitment, may be helped by limiting hunting pressure or by managing for increased winter habitat.

Limitations of biometric models

At the outset we need to recognize that multivariate techniques are essentially correlational. That is, they do not necessarily yield insights into the true causal relationships that exist between animals and their habitats. It is my belief that many wildlife ecologists have often mistakenly confused biological with statistical significance. Multivariate models have generally not yielded great insights into the relationships between population processes and habitat structure. This failure may arise for several reasons, some of which are discussed below.

Many studies suffer from poor experimental design. That is, data have been collected in such a fashion that they are unable to meet the assumptions of the models used to analyze the data, or sample sizes do not yield the power needed to test the proposed hypotheses. A common fault of many experimental designs is the failure to recognize that relationships between habitat structure and the population biology of wildlife species are less precise when narrow ranges of habitat are studied. This occurs most often when predictive models are developed from, and applied to, local studies. Such models often perform poorly when they are applied outside the area of model development.

Researchers often attempt to use linear models to describe relationships between populations and habitats that are nonlinear, at least when examined over the entire length of the habitat gradient occupied by a species. Linear models may be close approximations over parts of the gradient, but they can lead to contrasting inferences about the relationship between population size and habitat when developed piecewise.

Most wildlife-habitat research has been based on observational studies and has used statistical models in an exploratory sense. Even though models such as DFA are framed in the context of statistical inference and hypothesis testing, wildlife ecologists have seldom used them in this confirmatory sense (Williams 1983). Rather, DFA, like PCA, is often used as a method of data reduction and for projecting sample units (PCA) or groups (DFA) into spaces of reduced dimensionality. Both techniques are useful for initial data exploration, and they may suggest specific hypotheses that can be tested in future confirmatory studies.

The distinction between observational and experimental studies is that, in the latter, treatments or factors are controlled by the researcher (James and McCulloch 1985). For successful, long-term management of most wildlife species, information on biological processes (i.e., birth and death rates) will be essential. Thus it is imperative that wildlife ecologists begin to conduct manipulative experiments, make comparisons across habitats that differ in specific ways, or design their studies in such a way that confounding covariates are controlled. The latter undertaking will require experimental designs with replication of treatment factors, as outlined by Marzluff (Chapter 25) and Hurlbert (1984).

Considerable indeterminacy is introduced into many multivariate models because of low observation-to-variable ratios. This occurs for a variety of reasons, the most obvious of which is the low precision in the estimate of the variance and covariance structure. Because analysis of the variance-covariance matrix is at the heart of most multivariate models, this indeterminacy is not surprising. A related issue is the number of samples necessary to estimate the habitat variables with specified precision and power to discriminate among contrasting hypotheses. Guidelines are available in the statistical literature (e.g., Stauffer 1982; Zar 1984). Recall, however, that the necessary sample sizes will need to be determined separately for each study, on the basis of its objectives and the magnitude of habitat variation.

The success of predictive models using empirical data depends on (1) the constancy of the system being studied; and (2) the extent to which the variables in the model reflect the true, underlying causal relationships being modeled (James and McCulloch 1985). Thus we should expect the predictive accuracy of wildlife-habitat models to be good only if they meet these requirements. An additional factor contributing to low accuracy may be that our multivariate models need to include variables other than vegetation. For example, it may prove valuable to consider past and present regional population levels for the target and other species (Cohen, Chapter 29); behavioral variables such as site fidelity; and migration potential, climatic influences, and other factors that may introduce time lags into a species' response to habitat change. If we accept that habitat is a necessary but not sufficient condition on which to base predictions of local population levels, then it is not surprising that even data-rich habitat models have predicted poorly in some instances.

Future directions

If our models, and the inferences that we make from them, are to improve, then we must be more concerned with the way we conduct our studies. James and McCulloch (1985) identified several discrete phases that collectively define the scientific method. Borrowing from their ideas, I suggest that we may be able to define discrete stages in the modeling process, which, if followed in sequential order, might lead to more accurate wildlife-habitat models. These are (1) an initial observation period in which the system is roughly defined and a set of questions posed; (2) an exploratory stage using statistical techniques (i.e., multivariate analysis) designed to find patterns in the data; (3) a period of empirical model building still largely based on data gathered from observational studies; (4) a period in which the predictions of empirical models are statistically tested with observational data, the model refined, retested, and so on; (5) a phase in which the results of the empirical model testing are synthesized into a theoretical model that makes statements (predictions) about the relationships between population processes and habitat components; and (6) the final stage at which the predictions of the theoretical model are tested by confirmatory statistical methods. This final stage lends itself most readily to experimental studies, or to creatively designed observational studies in which confounding covariates are controlled statistically. The theoretical model may be rejected at this time and the researcher forced to go back to an earlier stage and begin the process again. A quick survey of the wildlife-habitat literature suggests that most wildlife ecologists are at the third stage in the process, with some notable exceptions (e.g., Rice et al., Chapter 13 in Part I).

From a statistical perspective, I believe that we need to focus attention on the sources of error in parameter estimates. In our modeling attempts, we estimate two major components—wildlife populations and habitat variables. Population ecologists and statisticians have dealt in considerable detail with the problems of estimating animal demographic parameters (e.g., Otis et al. 1978; Burnham et al. 1980; Ralph and Scott 1981; Seber 1982), but the estimation problems of vegetation variables have been largely ignored. Statistical models have two major sources of error, sampling error and measurement error. Sampling error, which results from the fact that we use a sample to estimate the value of a population parameter, is familiar to most of us. Measurement error, on the other hand, is often ignored and assumed to be compensating (when averaged, compensating errors result in zero error). In the usual development of most statistical models, such as MR and DFA, the predictor variables are either assumed to be fixed or to be subject only to sampling error. However, when variables are measured with error, severe biases may be introduced into the coefficients.

Philosophical considerations

I believe at this time that additional insights into wildlife and their habitat dependencies will not derive from more elaborate statistical models. What is required are more carefully designed studies and greater attention to model assumptions. The most basic model assumptions reflect how the data were collected and the distributions of the variables. In addition, we must be able to write clearly about the results of our biometric modeling attempts in order to communicate our goals, methods, and results to wildlife managers.

I have long perceived a wide communication barrier between researchers and managers. It is time that we ask ourselves why this communication barrier exists and how we can overcome it. If we do not make this effort, we will end up talking to ourselves and continuing to feel frustrated that few of our research findings are incorporated into management decisions. How do we close this gap?

1. Researchers should take more responsibility for targeting the results of their research to a management audience. The burden is on the authors of research papers to clarify the methods and statistical models used. This will require those of us using these sophisticated statistical models to have more than a superficial understanding of the methods, their assumptions, and limitations.

   One way to increase the clarity of our methods and research goals is to give more consideration to experimental design before data collection begins. Often overly detailed and confusing statistical models are the consequence of a poor experimental design. The points made by Green (1979) should be reread by most of us on a regular basis. Remember that remedial methods are never a substitute for a good initial experimental design, and they are sure to add confusion to the presentation of our findings.

2. A second way to bridge the gap between researchers and managers is to minimize the apparent dichotomy. This is a task that should be addressed by educators as well as by the administrators of federal and state agencies. As an educator, I believe we can contribute to the solution by creative changes in curriculum and by a clearer integration of research and management skills in our courses.

The proceedings of this symposium are readily compared with those from the Vermont multivariate conference (Capen 1981). The comparison suggests to me that in the interim we have not advanced much in our understandings of wildlife population dynamics and their relationships to habitat variation. The cautionary notes expressed at the Vermont Symposium by D.H. Johnson, L.L. McDonald, and B.K. Williams are still applicable today—we are still making the same mistakes. I am not convinced that sophisticated biometric models have yielded great insights into the variables most closely associated with variation in a species' demographics. One thing I am convinced of, however, is that as our models become more complex, the gap between researchers and managers widens.

Summary: Biometric Approaches to Modeling—The Manager's Viewpoint

BRUCE G. MARCOT

Much of Part II has addressed problems and applications of specific biometric approaches to modeling wildlife-habitat relationships. Appropriately, the emphasis has been on multivariate techniques. Gutzwiller and Anderson (Chapter 24) and Marzluff (Chapter 25) have called for attention to the assumptions of regression analyses when designing studies and interpreting data. Specific approaches to developing and validating models have been presented by Brennan et al. (Chapter 27) and by Cohen (Chapter 29). However, several topics of special interest to the model-using manager remain to be addressed.

The manager requires information about biometric models at a higher degree of synthesis than the researcher does. Most chapters in this section are more appropriate to researchers' needs. Specifically, managers need well-documented models that can be used with existing habitat-inventory data. They also need readily intelligible information on how to use the models, including precise definitions of habitat types and other model inputs. In addition, a clear explanation of model output is needed, including discussion of correct and fallacious interpretations, and levels of confidence that can be placed in predictions. It follows, then, that the manager has the onus of weighing the levels of confidence associated with using a model in any management decision.

Four interrelated problem areas are of particular importance to enable the manager to use a biometric model successfully, and they may also suggest future directions for research. First, along with the degree of confidence that may be placed in model predictions, managers need to know how well a biometric model of wildlife-habitat relationships can be expected to perform. For example, only a moderate proportion of the variation in species' abundance is accounted for by habitat variables alone, typically averaging about 50%. In a management context, this means that half of the variation of species' abundances will be influenced by factors external to any manipulation of those habitat factors considered in the model. Managers need to know how well habitat-management activities can be expected to influence species' densities to help them decide whether they have sufficient control over densities to make a given habitat-management activity feasible.

BRUCE G. MARCOT: Cooperative Wildlife Research Unit, Department of Fisheries and Wildlife, Oregon State University, Corvallis, Oregon 97331. *Present address:* USDA Forest Service, Fish and Wildlife Staff, 319 SW Pine Street, Box 3623, Portland, Oregon 97208

How many, and which, additional nonhabitat variables should be included in field studies and resultant biometric models and management objectives? The answer to this may vary according to management goals and the need to attain some minimal level of management influence. Conversely, monitoring and validation studies can help describe the actual utility of predicting (and managing) species' abundance from habitat variables alone (Holling 1978; Salwasser et al. 1983). Here, Marzluff (Chapter 25) reviews the use of maximum R^2, which may be used to measure the degree to which nonhabitat factors influence the variation of species' numbers.

The second problem area involves different types of model failure. The habitat manager should be more concerned when a model predicts the presence of a species that is missing than when it predicts the absence of a species that is present. This translates into a Type II statistical error, the rate of which is estimated by the conditional probability of the model predicting a species' presence, given that the species is absent. Power measures how well the prediction may avoid this type of error. If biometric modelers supplied the levels of power associated with a model and its predictions (Toft and Shea 1983), the manager could better assess the usefulness of the models. None of the chapters in this section addressed statistical power.

The researcher can more easily rationalize when a species is predicted to be present but is not observed (e.g., considering sampling error) than when a species is predicted to be absent but is observed. Further, the implications of different sorts of model failure may vary according to different management perspectives, as suggested by J. Rice (pers. comm.). If the aim of management is mitigation, such as by preservation or acquisition of habitat, or enhancement of unsuitable habitat to a suitable state, then high costs in money and effort demand that the model not fail in predicting species' presence or positive responses. If, however, the aim is impact assessment, then errors in predicting species' presence or abundance may be more tolerable, and false predictions of species' absence or negative responses to habitat conditions may be of greater concern than in the case of mitigation. The different consequences of model failure should help aim efforts of model validation and refinement in different directions, depending on the management context within which the model is used.

The third problem is that managers need to know the accuracy, reliability, and level of detail of the habitat information that drives a model, such as that derived, for example, from wetland or forest inventories. This helps the manager assess how useful the model may be and at what scale it can be employed with any degree of confidence. Models founded on general habitat characteristics (e.g., Cohen's, Chapter 29) cannot be expected to perform well at local scales or for site-specific use. More generally, this translates into the difficult statistical problem of introducing error terms into the independent variables (Kendall and Stuart 1979:399–443). Future work in biometric modeling may address this problem.

Finally, given budgetary, workforce, and time constraints, the manager is concerned with the overall need for rigorous biometric approaches, including model development and testing. Good field biologists can fairly accurately predict a species' presence and general patterns of its numbers without the use of biometric models. Biometric approaches that are too abstruse or that fail to extend existing professional knowledge will have low utility for managers. A prime role of statistical testing of biometric models should be to help the manager understand the ecological contexts within which the model may be reliably used. Such contexts include ranges of habitats, seasons, and environmental conditions. The converse is also true; that is, the model should be initially devised from a sufficient range of habitat conditions along appropriate habitat gradients to which it will be applied.

Effective ranges of habitat gradients to be investigated may be set by management scales and constraints. For example, to assess the association between warbler abundance and hardwood tree density in relation to silvicultural manipulation of mixed-conifer–hardwood stands, the gradient may be defined as the proportion of the forest stand that occurs as hardwoods. Endpoints to the gradient may be the low and high proportions likely to be found in a particular vegetation type or on a particular district, forest, or land area classified as usable for commercial timber harvest (thus excluding noncommercial forest). Accordingly, the model should then be validated and used only within this specific range of habitat conditions, unless validation is designed to specifically test the generality of the model in other habitats.

The transfer of information from biometric modeler to manager involves defining new technologies and new roles for the researcher. The success of such a transfer or "mission-oriented" science, to borrow from Regier (1978), depends in part on how well each player understands the other's unique constellation of information needs.

III

When Habitats Fail as Predictors

Introduction: When Habitats Fail as Predictors

L. JEAN O'NEIL and ANDREW B. CAREY

Consistent, observable patterns of abundance of vertebrates occur in both natural and human-dominated landscapes. The general associations of many species of wildlife with particular plant communities or elements of the landscape (cliffs, riparian zones, barns, etc.) are well accepted and documented.

Human activities have become so pervasive as to leave few landscapes intact. Even ecological reserves are affected by changes on adjacent lands. Thus, there is increasing concern about the ability of land managers to maintain the natural diversity of wildlife and viable populations of indigenous species.

Effective conservation requires specific, accurate, predictive models of wildlife-habitat relationships. The question is: Can models based solely on habitat elements consistently and accurately estimate species' population responses (given well-defined objectives, experimental designs appropriate to the objectives, sampling plans that meet statistical and mathematical assumptions, methods that estimate accurately and without bias carefully chosen parameters of interest, wise analysis of data, and thoughtful building of models)? In many cases the answer is no. Why is this so? One or more of at least six explanations are possible. Hutchinson (1978) discussed many of these situations, and Carey (1984) has recently given a brief overview:

1. A species' population response (population size, birth rates, death rates, population structure) to its environment can be partitioned into two categories. One can be labeled "habitat"; it refers to the range of ecological conditions a species occupies. This range may be largely a function of the species' tolerance of physical conditions, life history requirements, and affinities for particular types of environments. A species may be limited altitudinally by its ability to adjust to climatic conditions and within an altitudinal belt by specific life requirements, such as deep soil for a burrow or tree cavities for nesting. The other category can be labeled "niche." This refers to the relationship of the species to the other members of an ecological community—prey, predators, competitors, parasites, and facilitators (species that alter the environment in a way that benefits the species, e.g., cavity excavators that improve an environment for occupancy by secondary cavity users). Thus, a particular community may be made suitable by the presence of prey, parasites that affect competitors or predators, and facilitators. The community could be made less suitable by predators, parasites, and an absence of facilitators. Introduction of species, range expansions and contractions, and other aspects of zoogeography can further complicate matters. Habitat descriptors may not adequately account for these intracommunity relationships. Indeed, general descriptors may reflect correlative factors only. If the correlative factors are only moderately related to causative factors, the predictive utility of the descriptors will be less than ideal.
2. Migratory species are under the influence of often dissimilar and geographically separated environments. Descriptors of the physical and vegetative portions of a unit within a landscape in one region cannot predict the influence of another region or a route of migration. Hence, the temporal and spatial scale of investigation becomes important.
3. Weather can be a major influence. Weather occurs in cycles measured in decades and in random patterns in the short term. Habitat descriptors usually cannot predict the influence of weather on a population or the interactions of weather, prey, predators, competitors, and parasites.
4. Some landscapes are in inherently harsh environments where species occur in low and variable abundance. Habitat descriptors cannot account for this kind of variability (although confidence intervals on a model could suggest a range of abundance from data collected over years).
5. Factors controlling populations vary among and within types of environments and from year to year. Interactions of limiting factors may be as important as the factors themselves. And certain species in certain environments may be density-dependent in population response and use of types of environments.
6. Most populations are subject to stochastic events that may be genetic, demographic, or environmental in nature (including natural catastrophes). Stochastic events cannot, by definition, be predicted by habitat descriptors. Indeed, a current ecological theory suggests that only metapopulations may persist, whereas local populations inevitably fluctuate and become extinct—only to have the vacated community recolonized by immigrants from other nearby local populations.

Recent literature has contributed some examples and clues for recognizing when habitat characteristics are not the dominant factors in determining population responses.

L. Jean O'Neil: Waterways Experiment Station, U.S. Army Corps of Engineers, P.O. Box 631, Vicksburg, Mississippi 39180

Andrew B. Carey: USDA Forest Service, Forestry Sciences Laboratory, 3625 93rd Avenue, SW, Olympia, Washington 98502

Bergerud et al. (1983) and O'Connor (Chapter 34) have demonstrated dominance of intracommunity factors. Hejl and Beedy (Chapter 35) found weather to be a determinant; their work and that of Rotenberry (Chapter 31) illustrate the importance of scale as well. Stochastic factors were hypothesized by Rotenberry.

In addition to direct demonstration of the role of nonhabitat elements, experimental manipulation of a critical feature of the habitat with no observed population response (S. Droege and B. Noon, pers. comm.) is a clear signal. Van Horne (1983) suggested certain characteristics of species that might cause the decoupling of population densities and habitat. Finally, if a "good" habitat model is applied under new conditions and fails to perform well, it may have fallen prey to other influences.

Does this gloomy list mean all is lost? No. But it does mean that the situation is complex and that simple models will not accomplish complex objectives. Long-term, complex, and costly studies will be needed to accomplish multifaceted objectives. And a large degree of uncertainty will remain to be both accepted and accounted for in management efforts.

30

Factors Confounding Evaluation of Bird-Habitat Relationships

LOUIS B. BEST and DEAN F. STAUFFER

Abstract.—We explore five factors that can confound attempts to model wildlife-habitat relationships: (1) nonlinear and nonmonotonic relationships; (2) incomplete sampling along habitat gradients; (3) variability in species' responses to habitat parameters; (4) coarseness in measuring habitat variables; and (5) sampling scale. Data were obtained for 28 study plots in Iowa in habitats ranging from hayfields to closed-canopy forest. Habitat Suitability Index (HSI) models were developed for eight bird species during the breeding season by using three habitat variables.

Nonmonotonic, wildlife-habitat relationships violate assumptions of linearity and may not be amenable to some statistical procedures. Incomplete sampling along habitat gradients can result in contradictory wildlife-habitat models, particularly if interpolation or extrapolation is used to compensate for missing data. Confidence limits for HSI curves depend upon how frequently species are encountered during counts; more abundant and more detectable species have the most reliable HSI curves. Wildlife-habitat relationships can be obscured when variables are measured too coarsely. HSI curves derived from microplot and macroplot data generally are similar, particularly for species observed most frequently during counts.

Delineating relationships between wildlife abundance and environmental features has become a major focus of concern among both research biologists and land-use managers. Advanced computer technology, allowing multivariate analysis and other sophisticated quantitative procedures, has facilitated these efforts. But models of wildlife-habitat relationships are often limited by our efforts or ability to collect reliable and representative empirical data. Measurement and sampling errors, although present, often are undetected or ignored and may result in models that are neither informative nor predictive.

Many statistical methods used in examining wildlife-habitat interactions assume linear relationships among variables. These methods include commonly encountered versions of principal component analysis, discriminant analysis, and linear regression (Meents et al. 1983). Additionally, the range of environmental conditions selected for study in developing wildlife-habitat models is often determined by constraints imposed upon the sample design (e.g., manpower, budget, and time limitations; availability and accessibility of study sites; and ease of measuring habitat parameters). Deciding what, where, or how much to sample can be somewhat arbitrary, and requires investigators to "perceive" the habitat requisites of the species under study. Consequently, certain habitat variables may be under- or overemphasized, or habitat gradients may be nonuniformly or incompletely represented. Also, some species may be sampled more intensively than others because of their greater abundance or detectability, consequently affecting variation in measured responses to habitat parameters.

Wildlife-habitat relationships can be obscured when the variables used to characterize habitat are measured too coarsely. Additionally, the scale at which habitat use is measured may be important. For some researchers, sample units have been entire study plots (macroplots), each comprising several hectares (e.g., Shugart and James 1973; Webb et al. 1977; Rotenberry and Wiens (1980); in other instances, usage was measured on microplots less than 0.1 ha in size (e.g., James 1971; Anderson and Shugart 1974; Whitmore 1975).

A major prerequisite in reducing measurement and sampling error is a greater cognizance of the factors that can confound attempts to model wildlife-habitat relationships. Our objective is to illustrate five such factors. Although we use data from a study of avian communities associated with riparian habitats in Iowa, the general principles discussed apply to other taxa and habitats as well. Confounding factors we considered are (1) nonlinear and nonmonotonic relationships; (2) incomplete sampling along habitat gradients; (3) variability in species' responses to habitat parameters; (4) coarseness in measuring habitat variables; and (5) sampling scale.

Methods

The data were obtained from 28 study plots in Guthrie County, Iowa, representing a gradient of habitats ranging from hayfields to closed-canopy deciduous forest (herbaceous, 8.9 ha; savannah, 5.5 ha; scrub, 3.8 ha; wooded edge, 6.8 ha; floodplain woodland, 28.4 ha; and upland woodland,

Louis B. Best: Department of Animal Ecology, Iowa State University, Ames, Iowa 50011

Dean F. Stauffer: Department of Fisheries and Wildlife Sciences, Virginia Polytechnic Institute and State University, Blacksburg, Virginia 24061

88.8 ha). Vegetation composition and physiographic features of the study area are detailed in Stauffer and Best (1980). Study plots consisted of parallel transects, marked at 25-m intervals and positioned 50 m apart. The length (350–500 m) and number (1–5) of transects per plot were determined by the extent of the particular habitat.

Breeding birds were censused on each plot by the spot-map method (Robbins 1970). Twelve early-morning counts were conducted approximately weekly on each plot from mid-April to mid-July by walking transect lines and recording all birds observed. Composite maps for individual species were used to estimate breeding-bird densities (number of pairs/40 ha) for the 15 largest study plots (4.2–12.2 ha), hereafter referred to as macroplots. Ideally, some of our macroplots should have been larger, but the riparian plant communities available for sampling were often limited in area. The number of observations of each bird species in the 25 × 50-m (0.125 ha) microplots (n = 1349), centered at each grid point along all transects, was also compiled from composite maps. We calculated a gross index of detectability (DI) for each bird species by dividing the total number of observations on all 15 macroplots by the estimated number of breeding pairs on the 15 plots. To illustrate confounding factors, we chose the eight woodland bird species observed most frequently (>450 times each) during all census counts.

Although a plethora of measurements has been used to characterize wildlife habitat (e.g., James 1971; Anderson and Shugart 1974; Stauffer and Best 1980; Rotenberry and Wiens 1980), we chose only three. In doing so, we are not suggesting that these are preferred habitat variables, nor that our procedures used to measure them are the most desirable, but rather that they illustrate factors that can confound attempts to model wildlife-habitat relationships. The confounding factors we discuss probably are inherent in other habitat measurements as well.

Habitat variables selected were tree, vertical vegetation, and snag (standing dead tree) density. Tree (dbh > 5 cm) density was determined for each grid point by the point-centered-quarter method (Cottam and Curtis 1956) and expressed as trees/ha. To measure vertical vegetation density, we recorded whether or not live vegetation was present within each of 10 height classes (0–0.5, >0.5–1.5, >1.5–3.0, >3.0–5.0, >5.0–7.0, >7.0–9.0, >9.0–12.0, >12.0–17.5, >17.5–22.5, and >22.5 m) at each grid point. Vegetation density within the shrub (<3 m), sapling (3–9 m), and tree (>9 m) layers was calculated as the percentage of height classes within each layer containing vegetation. Location and hardness were recorded for every snag within 15 m on either side of each transect line. Snag hardness was based upon the percentage of the original limbs still present (soft, <34%; intermediate, 34–66%; hard, >66%). Snag density, recorded as the number of snags per 25 × 30-m plot (centered at each grid point), was determined for soft, hard, and all snags (snags of intermediate hardness occurred too infrequently to be used). The sampling procedure within microplots is illustrated in Stauffer and Best (Chapter 12).

Values for the three habitat variables were pooled into classes in the microplot analyses to ensure adequate sample sizes; each class contained observations from at least 50 microplots. For tree density, classes (and their respective sample sizes) were 0 (162), 1–50 (70), 51–100 (115), 101–200 (199), 201–300 (160), 301–400 (118), 401–500 (77), 501–600 (52), 601–700 (57), 701–900 (75), 901–1100 (55), 1101–1300 (54), 1301–1600 (59), and >1600 (96) trees/ha. Class midpoints were used to construct HSI models (see below). Vertical vegetation density classes for the shrub and sapling layers were 0% (n = 191 and 540, respectively), 33% (533 and 335), 67% (484 and 303), and 100% (141 and 171); for the tree layer, classes were 0% (799), 25% (262), 50% (192), and 75–100% (96). Categories for soft snag density were 0 (866), 1 (257), 2 (113), 3 (53), and >3 (60) snags/plot; and for hard snags, 0 (957), 1 (203), 2 (77), 3–4 (55), and >4 (58); for all snags combined, they were 0 (656), 1 (265), 2 (151), 3 (100), 4 (61), 5–6 (57), and >6 (59).

Habitat Suitability Index (HSI) models (Fish and Wildlife Service 1981a) were developed for the representative bird species by using the three variables mentioned. For each species and variable, the number of bird observations per microplot was averaged for all microplots within each class of the variable and the average values were then standardized by setting the value of the class with the largest mean bird count at 1.0 and expressing all other values as a proportion of this. Data for the 15 macroplots were standardized in a similar manner, except that density estimates (pairs/40 ha) were used instead of observations per microplot. We fitted selected HSI curves using polynomial regression, with linear and quadratic terms, to determine whether or not curvilinear relationships were present. Analysis of covariance (ANCOVA) was used to test for heterogeneity of slopes among selected models (Freund and Littell 1981). As has been the practice with most habitat-suitability studies (see references cited herein), we assumed that the intensity of habitat use (frequency of observation or population density) adequately represents habitat suitability for the species studied. We constructed HSI models because they are currently being promoted by the U.S. Fish and Wildlife Service. We acknowledge that this modeling system may not necessarily be the best for our data base or for studies in general, but our aim in this chapter is to illustrate confounding factors and not to present definitive models of bird-habitat relationships.

Results and discussion

Nonlinear and nonmonotonic relationships

The relationships between bird habitat usage and habitat parameters can be linear, curvilinear but monotonic (sign of the slope remains unchanged), or nonmonotonic. For example, the HSI curves for the gray catbird (*Dumetella carolinensis*) relative to vegetation density in the shrub, sapling, and tree strata are approximately linear (Fig. 30.1, Table 30.1), as is the HSI curve of the downy woodpecker (*Picoides pubescens*) for hard snag density (Fig. 30.2, Table 30.1). Curvilinear relationships (either monotonic or nonmonotonic), however, are common in biological data. Although the blue jay (*Cyanocitta cristata*) HSI curve for vegetation density in the shrub stratum is linear, the curve for the

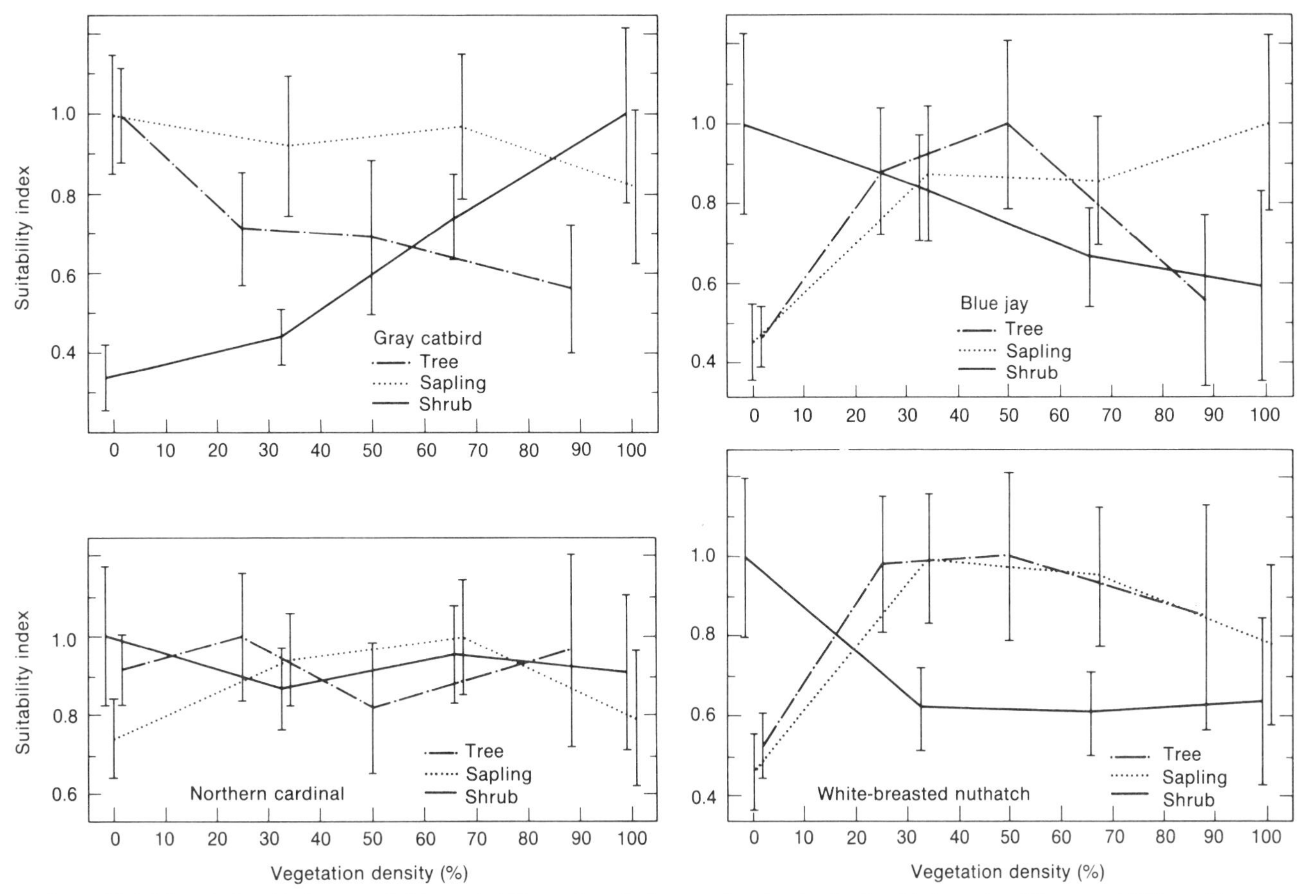

Figure 30.1. The relationship between habitat suitability and vertical vegetation density for four bird species. (Means and 95% confidence limits [±2 *SE*] are given for frequency classes in the three vertical strata.)

tree layer approximates a parabola (Fig. 30.1, Table 30.1). Suitability curves for vegetation density in all three strata are curvilinear for the white-breasted nuthatch (*Sitta carolinensis*). Habitat suitability for the eastern wood-pewee (*Contopus virens*) and northern cardinal (*Cardinalis cardinalis*) increases abruptly and then gradually decreases with greater tree density (Fig. 30.3); for the gray catbird and blue jay, suitability increases and then decreases abruptly, followed by a second increase and gradual decrease. The latter up-and-down pattern could result if habitat characteristics associated with various life history requisites (e.g., nesting sites, feeding sites, singing perches) differ, thus resulting in high suitability ratings for disjunct segments of a habitat gradient.

Although many correlational and regressional analyses and other commonly used statistical procedures are based upon assumptions of linearity, such relationships may be the exception rather than the rule. Nonmonotonic relationships pose the greatest problem in developing linear models to describe wildlife-habitat relationships. Meents et al. (1983) have presented a method to modify linear techniques in order to consider some basic curvilinear relationships.

Table 30.1. Significance of linear and quadratic terms in polynomial regression models of selected HSI curves. (Significance based on *F*-test with 1 and 1346 *df*.)

	Vertical vegetation density						Snag density			
	Shrub		Sapling		Tree		Soft		Hard	
Species	Linear	Quadratic	Linear	Quadratic	Linear	Quadratic	Linear	Quadratic	Linear	Quadratic
Gray catbird	0.001	0.190	0.320	0.820	0.001	0.280	—	—	—	—
Blue jay	0.002	0.670	0.001	0.042	0.001	0.001	—	—	—	—
White-breasted nuthatch	0.008	0.007	0.001	0.001	0.001	0.001	—	—	—	—
Northern cardinal	0.980	0.540	0.068	0.004	0.890	0.990	—	—	—	—
Downy woodpecker	—	—	—	—	—	—	0.005	0.023	0.001	0.53

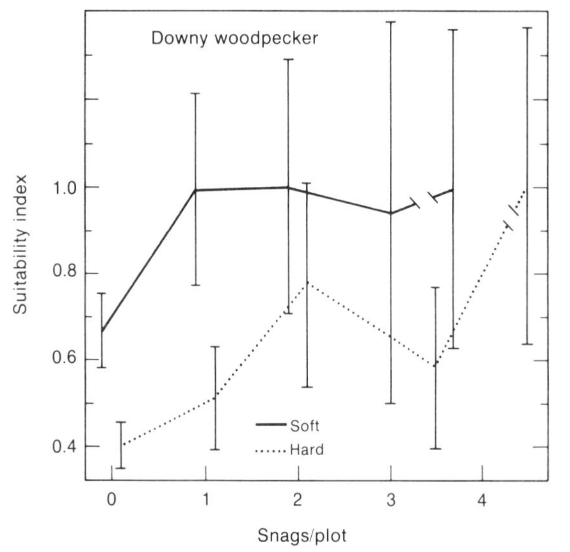

Figure 30.2. The relationship between habitat suitability and snag density for the downy woodpecker. (Means and 95% confidence limits are given for density classes of soft and hard snags.)

Incomplete sampling along habitat gradients

The consequences of incomplete or nonuniform sampling along habitat gradients can be demonstrated by regressing habitat suitability on tree density for selected portions of the tree-density continuum (Fig. 30.4). The relationship between habitat suitability and tree density for the house wren (*Troglodytes aedon*) is positive if the sample includes only densities below 600 trees/ha, but negative at tree densities above 600; if the entire tree-density gradient is included, there is no relationship. For the black-capped chickadee (*Parus atricapillus*), if tree densities less than 100 trees/ha are sampled, the relationship between habitat suitability and tree density is positive; if the samples are excluded, there is no relationship. Such shifts in pattern demonstrate the ambiguities that may be generated from incomplete sampling and attest to the drawbacks of extrapolating (or interpolating) from only a subset of the habitat gradient.

Problems with incomplete sampling are also illustrated by the HSI curves for house wrens and downy woodpeckers relative to snag density (Fig. 30.5). At low snag densities, habitat suitability for the wren is lower than that for the downy woodpecker (at snag density = 0, t = 5.08, df = 1310, $P < 0.001$; at snag density = 1, t = 2.36, df = 528, $P < 0.025$), but at high densities, it is similar for both species (t-tests, $P > 0.10$ at snag densities >1). The disparity between the two species at low snag densities may be because of differences in their territory sizes. Downy woodpecker territories are several times larger than those of house wrens (Evans and Conner 1979), and to assure that snag abundance is not below requisite levels, smaller territories may require greater snag density to compensate for their reduced area. If a study design did not include samples from areas with low snag density (<2 snags/plot), the conclusion could be drawn that snag density has little influence on habitat suitability or that the two species respond to snag abundance similarly. Wiens and Rotenberry (1981b) have illustrated how incomplete sampling along a shrubsteppe–tallgrass-prairie continuum can lead to different interpretations of the habitat relations of species.

Variability in species' responses to habitat parameters

Some bird species were encountered much more frequently during counts than others (Fig. 30.3), and these differences are reflected in the 95% confidence limits for the HSI curves. The less frequently a species was counted, the more variable its measured response to habitat parameters and, hence, the less reliable its HSI curves. Differences in confidence limits among species are not the result of variation in sample size per se, because the 1349 microplots were censused for all species; rather, differences result primarily from variation in relative abundances and detectability.

A species' relative abundance in samples may be influenced by many factors, including range of habitats sampled, home range or territory size, interactions with other species, location of the study site within the species' geographic range, and breeding success and adult mortality. The wider the range of habitats sampled, the more variable a species' relative abundance within the various habitat types. Individuals of species with larger home ranges or territories are more dispersed, and all else being equal, they would be seen less often. For example, the confidence limits for the house wren HSI curves are relatively narrow, owing largely to high population densities and smaller territories on our study plots; the reverse is true for the blue jay (Fig. 30.3). Interspecific competition or predator-prey interactions may exclude or substantially reduce the occurrence of species within portions of the range of habitats they potentially could occupy (Noon 1981b; Kotler 1984). When this occurs, wildlife-habitat-relationships models based solely on comparisons of abundance or usage estimates with various habitat measurements may misrepresent habitat suitability. Near the periphery of a species' geographic range, population levels may be low, even though many environmental features appear to be ideal (see Dow 1969); here, other factors (e.g., climate) may largely determine abundance patterns. Our representative species were well within their geographic ranges. Poor breeding success or high adult mortality can also cause low population densities even in the most suitable habitats. Whenever populations are below the carrying capacity of the environment (i.e., when intraspecific competition is low), not all available habitats will be used to the extent that they could, particularly those of marginal quality (see Noon et al. 1980; O'Connor, Chapter 34). Consequently, wildlife-habitat-relationships models developed from data collected from populations at or near saturation levels may differ substantially from those derived from populations below carrying capacity.

Bird species differ in detectability during counts (Best 1981; Emlen and DeJong 1981). The effect this had on the confidence limits of HSI curves was evaluated by considering the index of detectability (DI, Fig. 30.3) values for each species. The eastern wood-pewee had the lowest index of

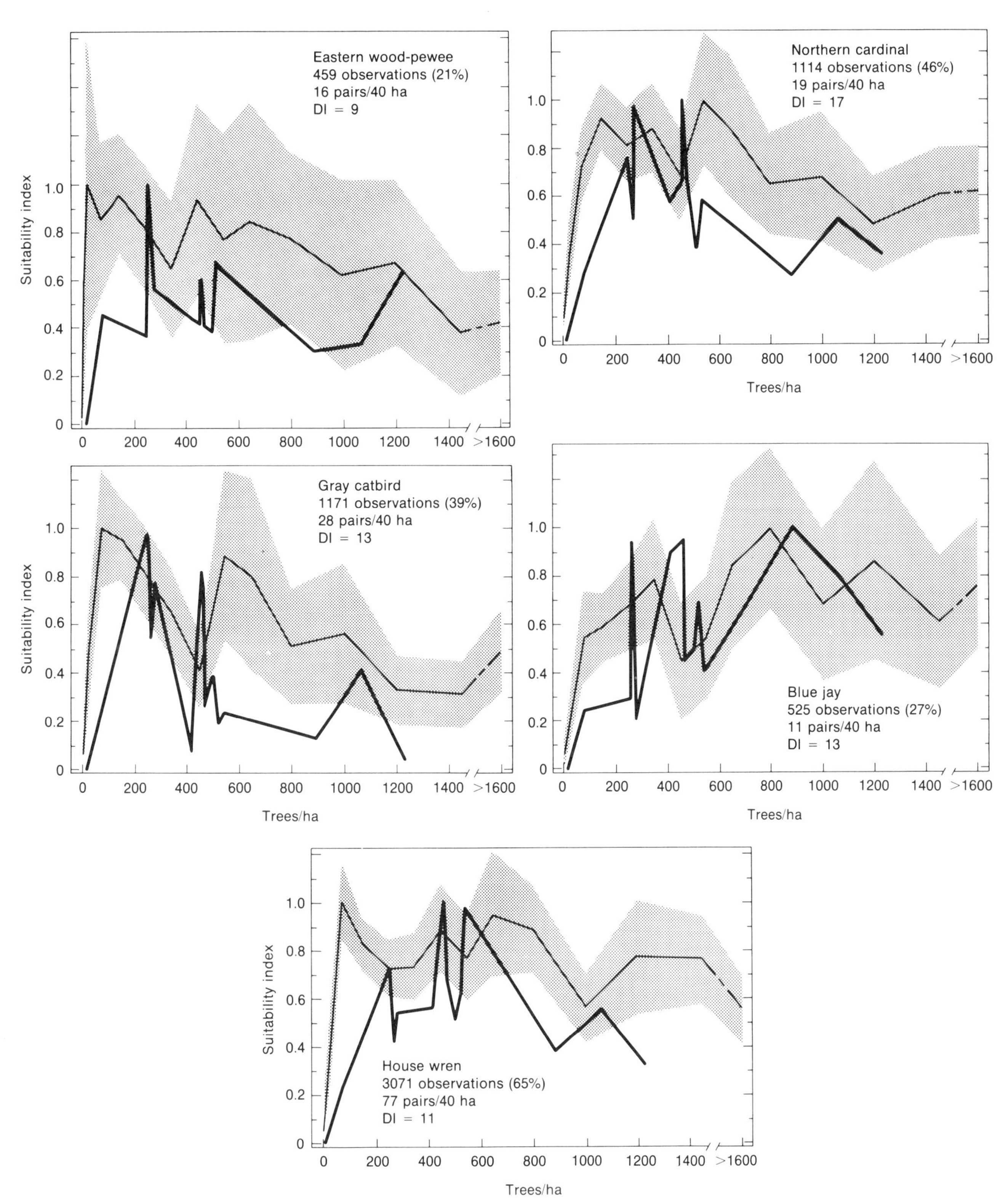

Figure 30.3. The relationship between habitat suitability and tree density for five representative woodland bird species. The narrow lines and shaded areas represent means and 95% confidence limits, respectively, derived from microplot data; the heavy lines were obtained from macroplot data. Also indicated are the total observations in 28 study plots, the percentage of microplots where birds were observed at least once, the average density estimate for 15 macroplots, and the detectability index (DI) value.

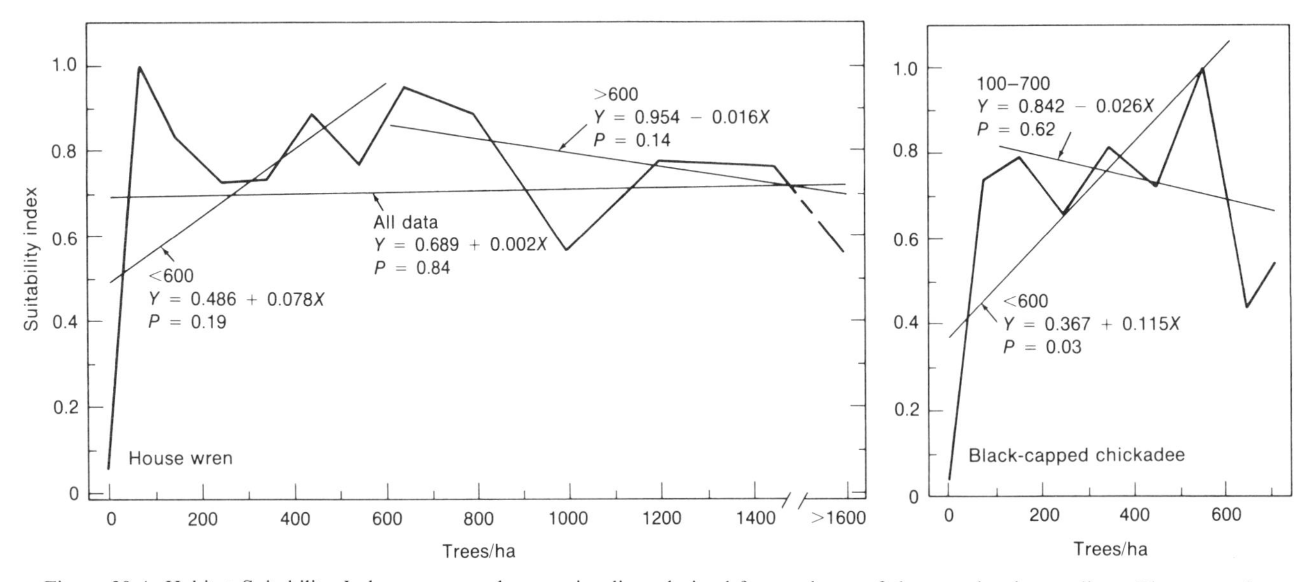

Figure 30.4. Habitat Suitability Index curves and regression lines derived from subsets of the tree-density gradient. The regression equations were determined by using mean values of habitat suitability for each class of tree density.

detectability, and this, coupled with its relatively low population densities, could account for the wide confidence limits around its HSI curve for tree density. The interplay between detectability and relative abundance is suggested in the HSI curves for the gray catbird and northern cardinal. The confidence limits for both species were similar, as were the total observations, but whereas cardinals were more detectable, catbird population levels were higher.

Some variation in HSI curves undoubtedly reflects differing importance of the habitat variables for the species studied. The less closely related habitat variables are to species' life history requisites (cover, food, etc.), the less well defined will be species' measured responses to such variables. Efforts to determine meaningful wildlife-habitat relationships involve sorting through potentially important variables and identifying those that best predict habitat use. The degree of specialization in habitat use also influences species' measured responses to habitat variables. Habitat generalists would tend to have less well defined HSI curves with wider confidence limits than their more specialized counterparts.

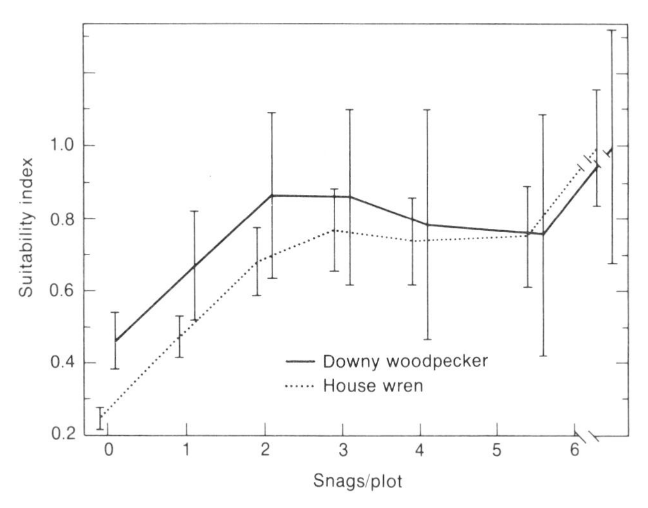

Figure 30.5. The relationship between habitat suitability and density of all snags for two birds species. (Means and 95% confidence limits are given for the snag-density classes.)

Coarseness in measuring habitat variables

When variables are measured too coarsely, some wildlife-habitat relationships may be obscured. For example, snag density could be based on all standing dead trees, or hard and soft snags could be distinguished. Hardness influences how easily snags can be excavated for nest sites and how long they may have been used as a nesting substrate; it also influences food availability for insect-probing birds. Distinguishing hard from soft snags may be unimportant for species that neither feed on nor nest in snags, but for cavity-nesting birds, such a distinction could be relevant. For the downy woodpecker, habitat suitability changes more with increases in hard snag density than with increases in soft snags, and maximum suitability is reached at lower densities of the latter (Fig. 30.2, Table 30.1). This suggests that downy woodpeckers may be affected more by changes in density of hard snags than soft snags.

Habitat-use patterns relative to vegetation density also illustrate the importance of coarseness in measuring habitat variables. For the northern cardinal, the HSI curves for vegetation density within the shrub, sapling, and tree layers are similar (Fig. 30.1; ANCOVA, $F = 1.30$, $df = 1, 4040$, $P = 0.27$), and combining all vertical strata into a single vegetation density measurement evidently would be acceptable. In contrast, habitat-use patterns of blue jays differ for all three strata ($F = 8.92$, $P < 0.001$). Some birds respond to changes in vegetation density in some but not all strata. For example, habitat use by gray catbirds evidently is influenced more by vegetation density in the shrub and tree strata than by that in the sapling layer, for which no significant relationship was found (Fig. 30.1, Table 30.1). In some instances, the habitat-suitability pattern for one stratum may be opposite to that for another. Compare, for example, the HSI curves for vegetation density in the shrub and tree strata for the gray catbird (ANCOVA, $F = 34.93$, $df = 1, 2692$, $P < 0.001$) and white-breasted nuthatch ($F = 24.30$, $P < 0.001$) and the HSI curves for the shrub and sapling layers for the blue jay ($F = 30.49$, $P < 0.001$) (Fig. 30.1). If layers with opposite patterns are not distinguished, the cancelling effect associated with combining them might lead to the erroneous conclusion that vegetation density did not influence habitat use.

Sampling scale

To evaluate the influence of sampling scale, we compared HSI curves for tree density derived from both microplot (<0.2 ha) and macroplot (>4 ha) data. Generally, HSI curves produced for each species from the two data sets were similar (Fig. 30.3), although curves tended to be more congruent for the species most frequently observed. Compare, for example, curves for the house wren, gray catbird, and northern cardinal with those for the eastern wood-pewee and blue jay. The peaks and troughs in curves derived from macroplot data often were displaced somewhat (along the tree-density axis) from those obtained from microplot data, but the use of plot averages to construct macroplot curves could account for the disparity.

There are advantages and disadvantages to both macroplot and microplot approaches. A single population-density (or relative abundance) estimate is derived for each macroplot, and this is then coupled with average values for all habitat variables measured on the plot. Sample size is determined by the number of macroplots in the research design (15 in our study). Full representation of a habitat gradient, with the necessary replication for statistical analyses, may require more study plots than time, budget, or manpower constraints will allow. Thus with macroplots, all portions of the habitat gradient are less likely to be represented. For example, tree-density values between 78 and 256 trees/ha are not represented by a macroplot sample (Fig. 30.3), and this may account for the notable disparity between microplot and macroplot HSI curves for this portion of the tree-density gradient. Also, by using average values to characterize habitat conditions on each plot, extremes of the habitat gradient will tend to be truncated. For example, compare the upper extremes of tree-density values for microplot and macroplot data in Figure 30.3.

Heterogeneity of the vegetative cover within macroplots also influences how accurately wildlife-habitat relationships can be quantified. The more heterogeneous the plant cover on plots, the less meaningful will be average values used to represent habitat conditions and bird abundances. To determine relative homogeneity of vegetation before finalizing study-plot selection may first require extensive plant sampling, a sequence that rarely is feasible. Using smaller macroplots would reduce plant heterogeneity and would also reduce the time required for bird counts and plant sampling, but small plots introduce greater error in estimating bird abundance (see Verner 1981). To alleviate this problem, investigators have recommended that plot sizes be maintained above certain minima (e.g., Robbins 1970).

Measuring bird usage within microplots can greatly increase the number of observational units for statistical analyses (1349 in our study). Such an approach should not be restricted to intensive sampling within a limited area; otherwise, major segments of habitat gradients may be missed. The size and number of microplots relative to territory or home range size will influence how accurately habitat can be characterized and usage documented. The more densely packed the microplots (i.e., the greater the proportion of habitat sampled), the more representative will be the sample. The rationale for positioning microplots within a habitat also will influence the results. If microplots are centered around singing perches (James 1971), foraging sites, or nesting sites, rather than being located randomly or systematically (Anderson and Shugart 1974), they will tend to characterize habitat features associated with particular aspects of a species' life history rather than with habitat use in general. Such an approach is not necessarily undesirable, as long as interpretations of habitat suitability are qualified.

Circular-plot census procedures (Reynolds et al. 1980) are useful when evaluating bird-habitat relationships on a microplot scale. Plots are small enough to eliminate most problems with plant heterogeneity. A stratified sampling design can be used to assure replicate and complete sampling along habitat gradients, and the plots can be separated spatially to assure independence of samples (a limitation in our study).

Concluding comments

A knowledge of confounding factors is important in any effort to describe wildlife-habitat relationships. Researchers who ignore such factors may find that their attempts to predict wildlife responses to habitat change are inconsistent with other studies or are not confirmed through subsequent validation in the field. We have considered only a subset of the factors that can thwart attempts to model wildlife-habitat relationships accurately; other chapters in this volume focus on other constraints.

Identifying confounding factors is not easy, but perhaps the greater challenge is to develop the means whereby the effects of such factors may be eliminated or at least substantially reduced. In the case of nonlinear wildlife-habitat relationships, researchers must be aware of such relationships and then avoid using linear statistics unless the data are appropriately transformed. With more complicated nonmonotonic relationships, the data may not be amenable to some statistical procedures. Incomplete sampling along habitat gradients is best diagnosed by constructing preliminary wildlife-habitat models and, if necessary, augmenting the empirical data base where critical gaps exist. Such a process is necessarily laborious and will be easiest to carry out for the common, well-studied species and most challenging to do for the rare, little-known species. This attests to the advantages of long-term studies with "robust" funding. Variability in species' responses to habitat parameters that is caused by sampling artifacts, such as differences in relative abundance or detectability, is difficult to remedy and may militate against constructing models for some uncommon or secretive species. The fineness required in measuring habitat variables ultimately will depend on research objectives, but it should be determined initially from intensive studies where parameters are quantified on a fine scale. Coarser sampling may be used subsequently if, through data analysis, it is determined that finer sampling is not warranted. There are advantages to sampling on either a macroplot scale or a microplot scale, but there seem to be more drawbacks to using the former.

Acknowledgments

The field work was funded by the U.S. Fish and Wildlife Service, Office of Biological Services; computer costs were furnished by the School of Forestry and Wildlife Sciences, Virginia Polytechnic Institute and State University. David F. Cox and Eric P. Smith provided advice on computer analyses; Andrew B. Carey, James J. Dinsmore, L. Jean O'Neil, Donald J. Orth, and Mark R. Ryan reviewed earlier drafts of the manuscript and made helpful comments.

31

Habitat Relationships of Shrubsteppe Birds: Even "Good" Models Cannot Predict the Future

JOHN T. ROTENBERRY

Abstract.—Response surface analysis is used to describe relationships among bird species' abundances and habitat attributes in semiarid shrubsteppe environments. Models based on data from 14 plots sampled for 3 years yielded excellent descriptions of the abundances of sage sparrow (*Amphispiza belli*) and Brewer's sparrow (*Spizella breweri*), accounting for up to 70% of the total variation in the densities of these species. The same models, however, failed to predict adequately the densities of sparrows on five of the original plots on which sampling continued for four more years. Additionally, the models failed to predict sparrow densities on new plots in similar habitats located 10–450 km from the original plots. Likewise, the models provided an inadequate description of the response of sparrows to an "experimental" manipulation that resulted in major changes in the physical structure and plant species composition of the habitat. The failure of these apparently "good" models reflects stochastic variation in population size independent of the details of habitat relationships and is probably an important feature of the biology of many migratory passerines.

Documenting relationships between vertebrates and physical or biotic features of their habitat has become a cornerstone of modern ecology. Such associations are significant to the wildlife manager concerned with preserving adequate numbers of individuals or species in an environment increasingly disrupted and fragmented by humans, and to theoretically inclined ecologists seeking clues to natural principles that govern the patterns of abundance of animals. The establishment of consistent, repeatable relationships between an animal and components of its habitat not only permits the prediction of species' abundance as a function of those components, but also defines the environmental setting in which the animal's adaptations and ecological strategies are expressed (Southwood 1977; Rotenberry 1981).

It is usually assumed that natural selection for some sort of optimal habitat response is relatively strong and continuous and that populations are generally in equilibrium with respect to the resources that the habitat provides. But both population and environment vary in time and space, and as variation increases, pressure favoring selection of optimal habitat may not always be intense. Thus, species may not be able to track shifting resource abundance, and population densities may become uncoupled from habitat parameters that otherwise might have influenced changes in population size (Wiens 1977).

Here I address the constancy and predictability of habitat associations of shrubsteppe passerine birds over time and space by seeking to determine if patterns that are detected over a large scale are generalizable and if they can be used to assess species' responses to major habitat alterations. If consistent associations fail to be present, such failure itself may yet provide insight to the processes operating on populations.

JOHN T. ROTENBERRY: Department of Biological Sciences, Bowling Green State University, Bowling Green, Ohio 43403

Methods

Response surface analysis

I used response surface analysis (Myers 1971) to describe the relationships among bird species' abundance and a variety of habitat attributes. In response surface analysis, a multidimensional nonlinear surface is fit to species abundances using multiple regression. Thus, if Y is density and X_i the ith habitat variable, the response surface is described by:

$$Y = b_o + \sum_i^h b_i X_i + \sum_i^h b_{ii} X_i^2 + \sum_i^h \sum_j^h b_{ij} X_i X_j, \text{ for } i < j,$$

where h is the number of habitat variables and bs are fitted coefficients. A surface regression model has several desirable attributes: (1) it considers both linear and nonlinear distribution of species' abundance along gradients (see Meents et al. 1983); (2) it includes two-way interaction terms, recognizing that species may respond differently to factors in combination than to factors assessed independently; and (3) it describes "contours" of species' density in habitat space in a manner equivalent to the concept of the Grinnellian niche (Maguire 1973; Rotenberry and Wiens 1981; James et al. 1984). Because of statistical problems associated with application of regression techniques to predictor variables sampled in a nonexperimental context, my emphasis here is less on the statistical properties of the response surface model and more on its descriptive attributes.

Samples, species, and habitat variables

During the breeding seasons of 1977–1979, vegetation features and bird abundances were sampled at 14 sites scattered throughout the Great Basin shrubsteppe of eastern Oregon and northern Nevada. Sites ranged from 5 to 150 km apart. Vegetation typically was dominated by big sagebrush (*Artemisia tridentata*), greasewood (*Sarcobatus vermiculatus*), or various species of saltbush (*Atriplex* spp.) or rabbitbrush (*Chrysothamnus* spp.).

Bird densities were estimated using 600-m strip censuses (Emlen 1971). Of the 14 species encountered during the sampling period, I report here the pattern observed for the two most abundant, sage sparrow (*Amphispiza belli*) and Brewer's sparrow (*Spizella breweri*). One hundred point samples of vegetation composition and structure taken throughout each census area permitted description of two classes of variables: (1) floristics, the percent coverage of each of nine shrub species; and (2) physiognomy, 20 measures designed to index the physical structure and spatial heterogeneity of the vegetation. Because of the results of previous work with these data (Wiens and Rotenberry 1981b), each class was analyzed separately. Within each class, rather than deal with intercorrelated raw variables, I used principal components analysis to extract the major patterns of covariation represented by the vegetation data. Those components with eigenvalues greater than 1 were readily interpretable in an ecological context (three physiognomic, four floristic—described in Table 31.1), and form a set of new, uncorrelated habitat variables on which the response surfaces are generated. More detailed descriptions of localities, methods, and principal components interpretation appear elsewhere (Wiens and Rotenberry 1981b).

The 42 original observations (14 sites × 3 years) are used to construct the response surface models to which all subsequent samples are compared. Five of the original 14 were selected as validation sites, and sampling was continued on them from 1980 through 1983. These 20 observations (5 sites × 4 years) allow me to address the question of temporal consistency of habitat association. Four new plots were sampled to assess the presence of geographic variation in habitat association. These geographic sites consisted of two (Owyhee B and Jack Creek) located within the same general area as the original plots (neither was more than 20 km from the nearest original plot) and two (ALE–Sage and ALE–IBP) located in similar sagebrush-dominated habitat 450 km north in the lower Columbia Basin of Washington. Finally, one of our original plots, Guano Valley, was aerially sprayed with the herbicide 2,4–D to eradicate sagebrush and was reseeded to non-native grasses during winter 1979–1980. Monitoring of birds and habitat variables continued through 1983 to assess the degree to which species responded to this major habitat change.

Table 31.1. Habitat variables (principal components) against which species' densities are regressed

Number	Ecological description
	PHYSIOGNOMIC COMPONENT
I	Increasing horizontal heterogeneity
II	Grassland to shrubland gradient
III	Flat basin to rocky slope gradient
	FLORISTIC COMPONENT
I	Sagebrush to small spinescent shrubs gradient
II	Sagebrush to large spinescent shrubs gradient
III	Increasing rabbitbrush
IV	Rabbitbrush to sagebrush gradient

Note: Each component is a linear combination of original, measured habitat variables (20 physiognomic and nine floristic). Each set accounts for more than 70% of the total variation in the original data sets. (Modified from Wiens and Rotenberry [1981b].)

Statistical methods

I derived two response surfaces for each bird species, using the 42 original observations, one based on the three physiognomic components, the other on the four floristic components. These surfaces define the basic bird-habitat associations, which, if consistent in time and space, should predict species' abundance at any other site based on its vegetation structure and composition. This assumes, of course, that the original models provide a satisfactory description of the bird's habitat relationships. A "satisfactory" description was determined by the magnitude of R^2 and its associated statistical significance level. Each site subsequently sampled (validation, geographic, or experimental) was scored on the principal components based on its measured values of the original 29 variables. The resulting scores were then used to generate an expected density for each species based on the original response surface models. Expected densities were compared to those actually observed in three ways: (1) for the validation sites, a cross-validation correlation (correlation between predicted and observed values) was calculated and tested for significance; (2) because cross-validation correlations have only 2 *df* for combined geographic plots, I treated each plot separately by calculating the standard error of each predicted density (Zar 1974:266) and used Student's *t*-test to derive the probability that the observed density was sampled from the same population as the predicted density; and (3) because statistical tests were of very low power for the small sample of Guano Valley, I simply describe and comment on the difference in magnitude between predicted and observed densities. All analyses were performed using SAS–82 (SAS Institute 1982b).

Results

Clear patterns of habitat associations emerged from fitting response surfaces (Table 31.2). Over 70% of the variation in abundances of sage sparrows is described by the floristic response surface, with all three components associated with sagebrush (Table 31.1) contributing significantly. On the other hand, densities of these birds appear to vary randomly with respect to physiognomy, at least over the scale represented here. Physiognomy is useful, however, in describing distribution of Brewer's sparrows. They reach their greatest

Table 31.2. Results of response surface regressions of original plots (n = 42) (see Table 31.1 for ecological description of components)

Type of component	R^2 (%)	Significant components	Shape of significant relationships
SAGE SPARROW			
Physiognomic	35.5	—	—
Floristic	70.4***	I*II***IV*	Linear***
BREWER'S SPARROW			
Physiognomic	46.6**	III**	Linear **
Floristic	67.7***	I**	Linear*, quadratic**, interaction**

*$P < 0.05$.
**$P < 0.01$.
***$P < 0.001$.

Table 31.3. Individual probabilities that observed density of a species on a plot was sampled from the population estimated by the response surface

Type of component	Great Basin		Columbia Basin	
	Jack Creek	Owyhee B	ALE-IBP	SLE-Sage
SAGE SPARROW				
Floristic	*	**	—	—
BREWER'S SPARROW				
Physiognomic	—	*	—	—
Floristic	—	***	***	—

Note: Statistical significance indicates poor fit to expectation.
*$P < 0.05$.
**$P < 0.01$.
***$P < 0.001$.
—$P > 0.05$.

abundance in broad, flat valleys characteristic of the northern Great Basin (component III). Their relationship to shrubs is considerably more complex, involving linear, quadratic, and interaction terms, but is nonetheless highly significant. Thus, three basic patterns appear reasonably strong (i.e., apparently nonrandom) and will form the basic habitat-response surface against which observed densities will be compared.

Comparisons of the five validation sites between the 1977–1979 calibration period and the 1980–1983 validation period showed no statistically detectable change in either bird abundances or habitat attributes averaged over each space-time block. Sage sparrow densities averaged 100 and 101 birds/km^2 in consecutive periods (n = 15, 20), while Brewer's averaged 215 and 203 birds/km^2. Despite relative constancy in abundances, there was a complete lack of agreement between predicted and observed densities for sage sparrows for the floristic model, which yielded a cross-validation correlation of only $-$ 0.07 (18 df, $P > 0.75$). This poor fit was not the result of some geographic peculiarity of the validation sites, because the cross-validation correlation between predicted and observed during the calibration period was 0.80 (13 df), which compares favorably to an overall r = 0.84 (40 df) for the entire original set. By contrast, Brewer's sparrows were reasonably predictable from both physiognomic and floristic models (cross-validation r = 0.52 and 0.53, $P <$ 0.05 for both), although the R^2 values (27% and 28%) are somewhat low.

Extension of the original models through space (Table 31.3) proved as disappointing as their extension through time. Worse yet, no consistent pattern of failure seemed obvious. Agreement was reasonably close (at least, not significantly bad) at one of the Columbia Basin sites and uniformly bad at one of the Great Basin sites. Abundances of sage sparrows were, somewhat surprisingly, adequately predictable from their response to shrub species when extrapolated 450 km north, but not in the region where that response was determined.

Although sample sizes are small, the lack of correspondence between observed and expected densities based on the nature of the habitat manipulation (which profoundly affected the physical structure and shrub-species composition of the plot [J. T. Rotenberry and J. A. Wiens, pers. obs.]) is plainly evident, especially for Brewer's sparrows (Fig. 31.1). Neither of these species appeared to respond to the rather massive environmental perturbation in a manner at all consistent with (i.e., predictable from) their previous habitat associations.

Discussion

The habitat relationships observed in this analysis (Table 31.2) are consistent with previous, more conventional analyses (Wiens and Rotenberry 1981b) and a decade of experience with these species in the field. Sage sparrow abundances are highly correlated with increasing coverage of sagebrush but largely independent of vegetation structure, whereas Brewer's sparrows are rarely found in areas characterized by short, spinescent shrubbery. Yet these relationships fail to predict consistently the response of these birds to habitat projected through time, space, or via perturbation. As none of the sites sampled lay outside the limits of the habitat domain described by the 42 original observations, these results do not appear confounded by excessive extrapolation.

Instead, I believe they reflect what appears to be an essential feature of shrubsteppe bird biology: a stochastic variation in population size independent of the details of habitat and other biotic relationships. Such variation in these species has been termed the "checkerboard" effect, where checkerboard variation is present in time as well as space (Rotenberry and Wiens 1980; Wiens 1981b). When averaged over space-time blocks, species appear to be in a regional equilibrium that displays no significant differences in their densities between periods. However, this apparent "equilibrium" is an artifact of measurement scale, as densities on any single plot may be extremely variable from year to year.

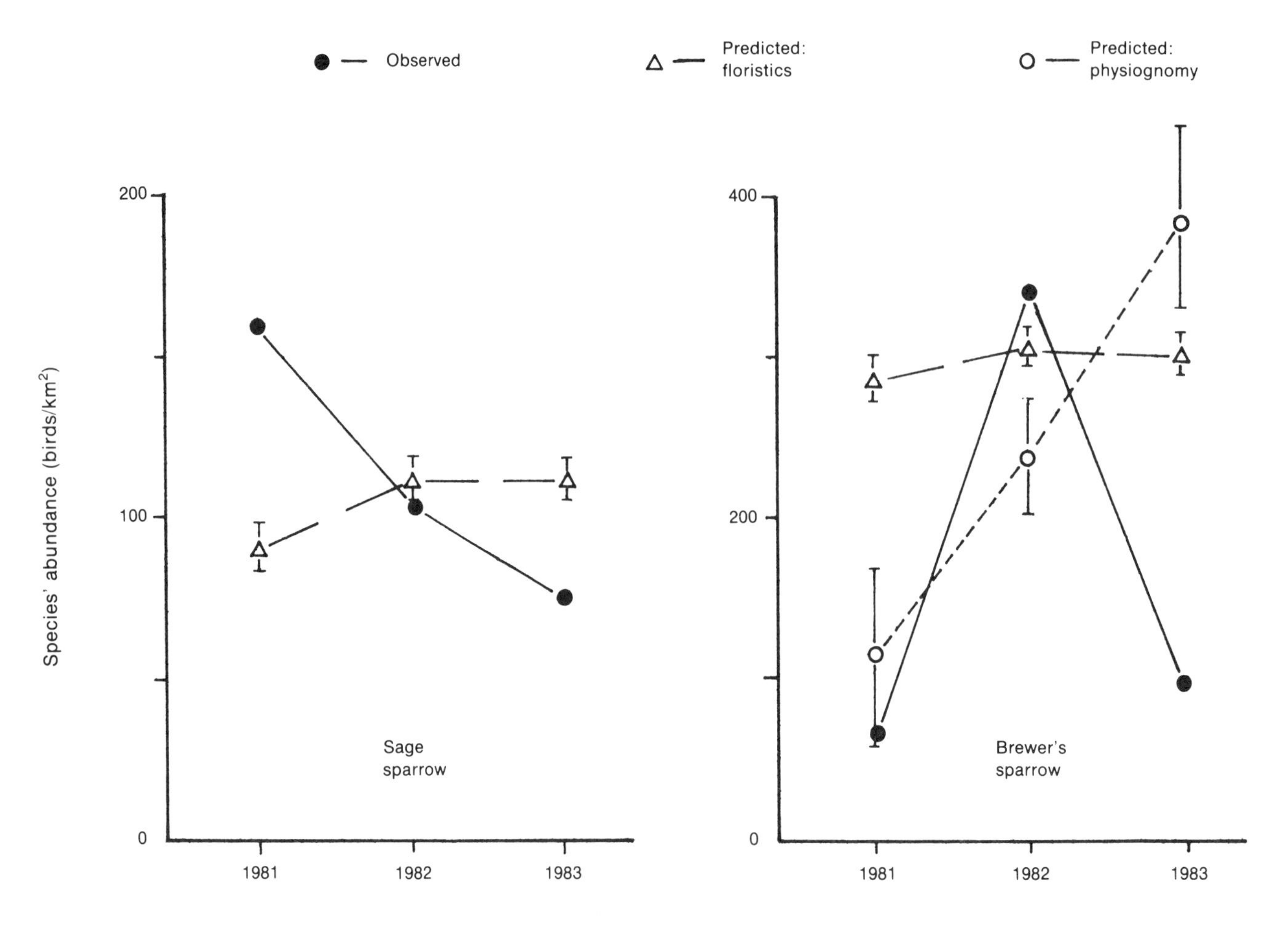

Figure 31.1. Observed and predicted (±2 SE) values of sage and Brewer's sparrows at Guano Valley after habitat manipulation.

Although physical and floristic features of undisturbed shrubsteppe environments may change little over a period of several years (or probably decades for slow-growing perennial shrubs), annual bird abundances may change by orders of magnitude (Rotenberry and Wiens 1978; Wiens and Rotenberry 1981b; and pers. obs.).

Thus, although there appears to be a basic component to habitat association in these birds (e.g., sage sparrows are rarely found in areas in the Great Basin lacking sagebrush), once this coarse habitat preference is expressed, fluctuation in density becomes largely uncoupled from the details of habitat variation. It seems likely, too, that such a pattern is not simply a quirk of shrubsteppe habitats but probably occurs in any species that is migratory, and for which a large component of mortality may be expressed in areas unrelated to where it breeds (Fretwell 1972; May 1981). If population dynamics on the breeding grounds include a large component of unrelated, apparently stochastic variation (regardless of how deterministic the birds' nonbreeding season relationships may be), what we are probably observing with seemingly good regression or correlation models is a form of overfitting—developing a rather large R^2 but only at the expense of capitalizing on chance variation. Including such "noise" in a model can only erode our ability to predict abundance through time and space, an erosion that becomes even more pronounced the greater the variation present.

The relatively coarse habitat associations that do emerge probably represent "niche-gestalt" (James 1971), a response by a particular species to a basic configuration of the environment, and are therefore likely to be highly consistent within a species. Such consistencies are likely to be evident only over a relatively large scale and are largely insensitive to local density variation through time. Thus, when bird-habitat relationships are considered over a large and somewhat heterogeneous scale, predictable associations should recur; such patterns may be submerged, however, when examined at a local level (see Wiens and Rotenberry 1981b).

Detection of meaningful patterns is further confounded because analytical techniques differ in their sensitivity to scale and stochastic population variation. Whereas regression or correlation techniques include estimates of a species' abundance as an integral feature of data collection, discrimi-

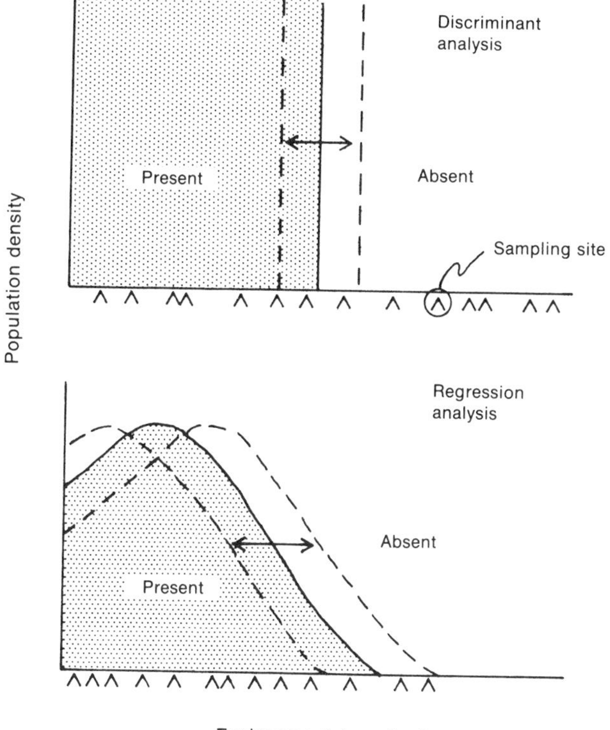

Figure 31.2. Relationship of different statistical techniques to sampling scale and sensitivity to stochastic variation in population density—*above*: discriminant function or canonical variate analysis; *below*: regression or correlation analysis. Dashed lines and arrows represent annual variation in density independent of habitat features.

nant analysis is usually concerned merely with the presence or absence of a species. The latter, therefore, must contrast habitat occupied with habitat not occupied, and thus of necessity involves sampling over a larger environmental scale (Fig. 31.2). The former are more concerned with documenting variation in abundance where the species is already present, and clearly unsuitable habitat is rarely sampled (Fig. 31.2). The disparity in sensitivity to variation is evident; substantial shifts in the pattern of habitat occupancy that alter the predictability of abundance from regression or correlation have little effect on the power of a discriminant function to predict presence/absence. It seems clear that studies conducted over a large scale, and that involve heterogeneous habitat types, will profit from a combined statistical approach, such as that advocated by Rice and his colleagues (Rice et al., Chapter 13).

Consistent prediction of habitat associations still does not necessarily imply cause and effect as far as population change is concerned, because it is possible that mechanisms of population dynamics are not affected by habitat variables that are good predictors. Although having a reliable predictor may suffice in some management situations, biological understanding is nonetheless limited.

These observations lead to the conclusion that while there are features of the habitat to which birds may respond, such responses are likely to be evident (i.e., detectable and predictable) only over a relatively large scale. At increasingly local scales, however, population numbers may reflect influences geographically and temporally external to the locale, leading to a much greater variation in apparent habitat selection (see Collins 1983a). Although this reduces our ability to predict accurately the numbers of individuals and weakens our confidence in the predictive value of theoretical models that rely on populations in equilibria, it nonetheless remains an essential feature of the biology of species.

Acknowledgments

John Wiens shared the pleasures of collecting these data, as well as the frustrations of analyzing them. I thank him for a decade of intellectual stimulation and leadership. Brian Maurer, Scott Collins, and Kim Smith offered constructive criticism of early drafts of this work, and the final version was significantly improved by the efforts of Andy Carey, Mike Morrison, and Jean O'Neil. This research was supported in part by NSF grants DEB–8017445 and BSR–8307583 to John Wiens.

32

The Dilemma of Plots or Years: A Case for Long-Term Studies

WILLIAM S. GAUD, RUSSELL P. BALDA, and JEFFREY D. BRAWN

Abstract.—We used the spot-map method to estimate the densities of five species of breeding birds over 10 years at eight sites (for a total of 33 site-years of observations) in a ponderosa pine (*Pinus ponderosa*) forest subjected to different cutting practices. Our objective was to choose from several alternative experimental designs the best approach for collecting data for predicting species' densities. We analyzed the data in subsets of observations to develop predictive equations based on (1) single-year, multiple-plot models; (2) multiple-year, single-plot models; and (3) multiple-year, multiple-plot models. Stepwise regression was used to select the best predictive equation for each bird density. We examined these models for consistency in the predictive variables chosen and for the amount of variability in species' density explained by the models. We also compared the models with models resulting from the entire 33 site-years. The single-year, multiple-plot approach produced a reliable result for Grace's warbler (*Dendroica graciae*) but a misleading result for the Steller's jay (*Cyanocitta stelleri*). The multiple-year, single-plot models were site specific for some species' densities and general for others. The economy of time and resources inherent in these two approaches needs to be balanced against the necessity for a more reliable prediction from multiple-year, multiple-plot models that can be used in management. We suggest that a sampling design adequate for predicting the densities of most ponderosa pine forest birds consists of 5 years of observation at a minimum of three different sites.

Densities of birds vary from place to place and fluctuate from year to year. The density of a species is affected by differences in the quality and quantity of vegetation, weather, season, food, interspecific interactions, and use of resources (Karr 1980). Bird populations in western montane ponderosa pine (*Pinus ponderosa*) forests show greater year-to-year variation than bird populations in eastern mixed deciduous forests (Holmes et al. 1979).

The best design for explaining population fluctuations in densities of birds of the ponderosa pine forest may be long-term studies over a variety of sites (Green 1979; Rice et al. 1981; Wiens 1981a). This approach, however, is not error free (Johnson 1981a), and costs can be considerable (Gates 1981).

We have been studying the effects of logging practices on bird populations in north-central Arizona since 1973. Details of the study plots and early results have appeared previously (Balda 1975a; Szaro and Balda 1979; Cunningham et al. 1980; Balda et al. 1983).

We contrast three approaches to predicting population densities of five species of birds that breed in the ponderosa pine forest. We hope that this analysis will contribute to the effective and economical design of future field studies.

WILLIAM S. GAUD, RUSSELL P. BALDA, and JEFFREY D. BRAWN: Department of Biological Sciences, Box 5640, Northern Arizona University, Flagstaff, Arizona 86011

Methods

STUDY AREAS AND FIELD METHODS

Vegetation (Table 32.1) and breeding-bird density data were collected on eight plots near Flagstaff, Coconino County, Arizona. All plots were in ponderosa pine forest but, because of silvicultural treatments, they had different total foliage volumes (Fig. 32.1). For example, site "PN" had not been logged and consisted of mature ponderosa pine mixed with dense thickets of pole or sapling pine stands. In contrast, site 17 had been heavily cut in 1966 and was relatively open with few understory trees. Ponderosa pine and Gambel oak (*Quercus gambelii*) were the only tree species on all plots.

Plots were sampled using the point-centered-quarter method (Cottam and Curtis 1956). Standard measures of frequency, density, and dominance were derived. Foliage volume of pine and oak, as well as snag densities, were estimated (Szaro and Balda 1979).

Breeding-bird densities were estimated using the spot-map method (Kendeigh 1944). Eight to 10 census visits were made on each plot at equal intervals from mid-May to late June (Robbins 1970). In addition, we made supplemental visits to the sites to verify that pairs counted were actually breeding.

We compiled a weather picture for each plot-year (Table 32.1). Weather factors relevant to the ecology of the bird species were identified (Szaro and Balda 1979; Balda et al. 1983). For example, weather data were divided into dispersal

Table 32.1. Climatic and vegetation factors (and their mnemonic codes) used in the analysis of breeding bird densities

Mnemonic code	Habitat factor
FVONE	Foliage volume 0–2 m above the ground
FVTWO	Foliage volume 0–4 m above the ground
FVTOT	Total foliage volume
SAP	Sapling volume
TRNK	Trunk volume
OAK	Oak foliage volume
IVPON	Importance value of ponderosa pine
IVOAK	Importance value of oak
SNAG	Snag density
TREES	Absolute density of trees
RATIO	Proportion of oak to total foliage volume
P1	Total precipitation, 25 Mar–2 Jun
P2	Snow, 25 Mar–2 Jun
P3	Total precipitation, 1 Jan–21 Apr
P4	Total precipitation, previous 8 Oct–2 Dec
DD1	Degree days, 1 Jan–25 Mar
DD2	Degree days, 26 Mar–2 Jun
DDT	Degree days, 1 Jan–2 Jun
DISPL	Dispersal mean low temperature
DISPH	Dispersal mean high temperature
DISPS	Dispersal precipitation as snow
DISPR	Dispersal precipitation as rain
DISPT	Dispersal total precipitation
WINL	Winter mean low temperature
WINH	Winter mean high temperature
WINS	Winter precipitation as snow
WINR	Winter precipitation as rain
WINT	Winter total precipitation
NESL	Nesting mean low temperature
NESH	Nesting mean high temperature
NESS	Nesting precipitation as snow
NESR	Nesting precipitation as rain
NEST	Nesting total precipitation

(previous fall), winter, and nesting (spring) periods of the birds. In addition, temperature and precipitation variables were constructed to represent divisions of fall, winter, and spring, which seemed ecologically relevant.

Analysis

Bird species selected for analysis were the western wood-pewee (*Contopus sordidulus*), Steller's jay (*Cyanocitta stelleri*), solitary vireo (*Vireo solitarius*), Grace's warbler (*Dendroica graciae*), and the dark-eyed junco (*Junco hyemalis*). We chose these species to represent a mixture of guilds (Szaro and Balda 1979; Monson and Phillips 1981) and residency status.

Densities for the five species of birds were compiled by site and year and analyzed using the SPSS (Hull and Nie 1981) and BMDP (Dixon 1983) statistical packages. Stepwise multiple regression analysis (both forward and backward stepping) was used to choose single variables and combinations of variables to explain variability in bird species' density. Criteria for model selection were R^2 and the probability level of its associated F-value. Residuals were analyzed to detect deviations from linearity. No transformations of data were employed. Interaction terms were not used in the general linear model. The data were analyzed in subsets of ob-

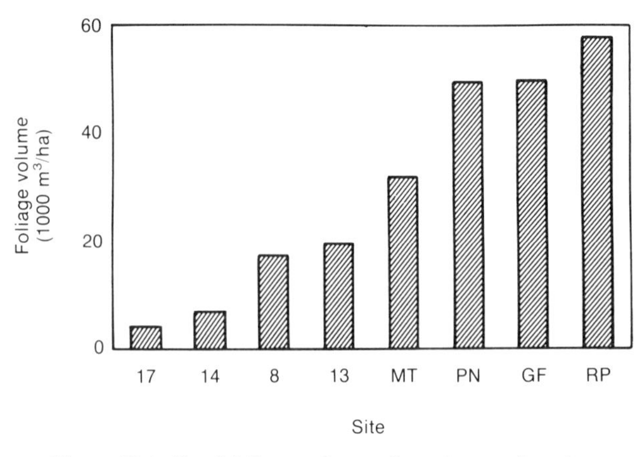

Figure 32.1. Total foliage volume of ponderosa pine sites.

servations that resulted in predictive equations comprising single-year, multiple-plot models (approach I: sites 8, 13, 14, 17, and PN); multiple-year, single-plot models (approach II: sites 8, 13, and 17 separately); and multiple-year, multiple-plot models (approach III: sites 8, 13, and 17 in IIIa and all plots in IIIb).

In approach I we constructed predictive models of breeding-bird densities, using vegetation characteristics from five plots in 1975 as independent variables. This, then, is a single-year, multiple-plot model that ignores weather-related factors. In approach II we compiled data over a 10-year period for three plots that differed vegetatively. We constructed one model for each plot. Approach II, therefore, assessed the influence of weather only. We measured changes in snag densities and assumed that other vegetation characteristics changed insignificantly in comparison to weather fluctuations during the study period. In approach IIIa we combined data, by 2-year increments, from three plots over the 10 years. Year-to-year weather patterns, as well as vegetation characteristics, were incorporated into the model at each species' density. In approach IIIb, we provided models using data for all sites censused for all years. All other available data (2 years on three sites and 4 years on one site) were added to those data used in approach IIIa to construct the data set for approach IIIb.

Results

Dark-eyed junco

The dark-eyed junco is a permanent resident of the ponderosa pine forest and belongs to the ground-foraging and ground-nesting guilds (Szaro and Balda 1979; Monson and Phillips 1981). Approach I identified foliage volume in the first 4 m above the ground (Table 32.2) as the variable most highly correlated with junco density. In approach II weather was not significantly related to density on two of the three sites. At the most open site (site 17), both temperature and precipitation during the nesting period significantly predicted junco density. Foliage volume close to the ground appeared as a consistent predictor of junco density when yearly obser-

Table 32.2. Predictors of breeding bird densities of five species in the period 1973–1982

No.	Dark-eyed junco		Solitary vireo		Grace's warbler		Steller's jay		Western wood-pewee		
Cases	Variable	R^2	Variable	R^2	Variable	R^2	Variable	R^2	Variable	R^2	$E(R^2)$[a]
					APPROACH I						
5	FVTWO	0.96	TREES	−0.31*	FVONE	0.67*	TREES	0.88	FVTWO	0.49*	0.25
					APPROACH II						
7 (Site 8)	DISPH	0.42*	WINR	0.78	WINR	0.73	P4	0.45*	P3	0.79	0.17
					P4	0.88	DDT	0.94			
7 (Site 13)	DISPH	0.33*	NESS	0.92	P2	−0.34*	WINR	−0.56	(Absent on this site)		
			P2	−0.99	P1	0.91	P2	−0.99			
7 (Site 17)	NEST	0.60	WINS	−0.60	DISPT	0.46*	NESS	0.34*	NESR	−0.61	0.17
	DD2	0.94	NEST	0.94			P1	−0.92	P4	0.92	0.33
					APPROACH IIIa						
5	FVONE	0.75*	TRNK	−0.77	FVONE	0.91	P2	−0.95	SAP	0.95	0.25
9	FVTWO	0.65	RATIO	0.77	FVONE	0.91	P4	−0.89	NESS	0.66	0.12
	DD2	0.85	NESR	−0.94	WINH	0.95			DDT	0.89	0.25
15	FVTWO	0.26	WINR	0.30	FVONE	0.58	P2	−0.34	IVPON	0.53	0.07
	DD2	0.48	IVPON	0.51	WINR	0.71	DISPR	0.63			
21	DISPH	0.24	IVPON	0.30	FVONE	0.65	NEST	0.38	IVPON	0.61	0.05
	WINL	−0.41	DISPL	0.48	WINR	0.76	P1	−0.05			
					APPROACH IIIb						
33	IVOAK	−0.45	TRNK	−0.25	FVTOT	0.11*	DISPH	0.17	TRNK	−0.22	0.03
	TRNK	−0.61	P1	0.26	FVTWO	−0.26	P3	−0.30	FVTWO	0.65	0.06
	DISPH	0.67	P2	−0.47	FVONE	0.57	WINT	0.43	IVOAK	−0.76	0.09

Note: A stepwise selection procedure picked variables from the list in Table 32.1. The R^2 value next to the first variable is that for the single-variable model. The R^2 value next to the second variable is that for the model with both variables present. Where a second variable does not occur in approach II and approach III with more than five cases, only one independent variable in the two-factor model was significant. The influence of the predictor variable as positive or negative is indicated by a positive (assumed) or negative sign preceding the R^2 value.

[a] The R^2 value expected by chance.

* $P > 0.05$.

vations were accumulated on the three sites (approach IIIa, Table 32.2). When this variable, identified as best in approach I, was followed through approach III as more data were used to build the model, the R^2-value decreased considerably (Table 32.3). Considering all variables as cases increased from five in approach I to 33 in approach IIIb, the R^2-value declined and the difference between it and the expected R^2 narrowed (Crocker 1972). The percentage of variation in density explained by the model decreased considerably and the factors identified as the best determinants of junco density changed to trunk volume and importance value of oak.

Solitary vireo

The solitary vireo, a summer resident of the ponderosa pine forest, nests and feeds in the foliage (Szaro and Balda 1979; Monson and Phillips 1981). No significant relationship between density and vegetation was found (approach I, Table 32.2). Tree density was correlated with vireo density in both the five and nine case models of approach IIIa (Table 32.3).

Approach II identified winter precipitation on site 8, precipitation in the nesting period on site 13, and a combination of winter precipitation and temperature during nesting on site 17 as significant predictors of vireo density. As yearly observations were accumulated (approach IIIa), both vegetation characteristics and weather were important predictors of density. Precipitation was consistently important, but vegetation factors changed as data were added. Approach IIIb identified trunk volume and precipitation before and during the breeding season as important variables.

Grace's warbler

Grace's warbler is a summer resident and nests and feeds in pine foliage (Szaro and Balda 1979; Monson and Phillips 1981). Data of this species displayed the most consistent relationship of density to environmental characteristics (Table 32.2). Foliage volume close to the ground explained most of the variability in warbler density in approaches I, IIIa, and IIIb (Table 32.2). In approach II, where only weather factors were considered, precipitation rather than temperature consistently and significantly predicted bird density. The particular precipitation variable selected, however, was site-specific.

Table 32.3. R^2 values of variables picked by approach I

No. cases	Dark-eyed junco FVTWO	Solitary vireo TREES	Grace's warbler FVONE	Steller's jay TREES	Western wood-pewee FVTWO
			APPROACH I		
5	0.96	0.31	0.67	0.88	0.49
			APPROACH IIIa		
5	0.69	0.59	0.91	0.00	0.11
9	0.65	0.65	0.91	0.03	0.02
15	0.26	0.00	0.58	0.04	0.22
21	0.20	0.05	0.65	0.03	0.30
			APPROACH IIIb		
33	0.13	0.07	0.05	0.04	0.03

Note: The R^2 values of each species and their respective variables are followed through the other approaches, except approach II, which did not include vegetation factors.

The foliage volume in the first 2 m was consistently correlated with warbler density through approach IIIa. Its correlation coefficient remained well above its expected value as cases increased. It failed, however, to be correlated directly with warbler density in approach IIIb (Table 32.3) and entered the model (Table 32.2) only because of its partial correlation.

Steller's jay

Steller's jays are omnivorous permanent residents that build open nests in pine (Szaro and Balda 1979; Monson and Phillips 1981). Approach I identified tree density as a highly significant predictor of bird density (Table 32.2). The other approaches did not identify vegetation characteristics as determinants of Steller's jay density. The Steller's jay was clearly different from the other species in the very small role vegetation played in explaining variability in its density. Weather factors, primarily precipitation, were subsequently chosen as predictors of jay density.

Western wood-pewee

The pewee is a summer resident, builds open nests, and is an aerial feeder (Szaro and Balda 1979; Monson and Phillips 1981). Foliage volume up to 4 m was identified as the most important variable predicting density in approach I, but it was not significant (Table 32.2). Weather variables identified in approach II were total precipitation during the winter and early spring period at site 8. For site 17, rain during the nesting period and precipitation during the fall were identified. The importance value of pine emerged in the 15 and 21 case models as a significant variable in approach IIIa. Pine, however, was not identified as important in approach IIIb.

The models for this species were characterized by a change in their predictor variables as the approach varied and cases were added. The fluctuation in the correlation of pewee density with FVTWO (Table 32.3) was additional evidence of the instability of the models for this species.

Discussion

Our contrasting approaches strongly indicate that models constructed for one or a few years (short-term studies) are not necessarily dependable as the basis for management guidelines. This conclusion is consistent with that of Wiens (1981a), although it was arrived at in a different manner. Variables identified in approach I were not consistently present in models built from larger data bases. For example, the Steller's jay and solitary vireo models contained a variable (absolute density of trees) that never appeared in models generated with other approaches. In contrast, models for Grace's warbler were relatively consistent among approaches in that foliage volume up to 2 m appeared in all models where vegetation variables were considered. Because of its lack of significance in approach I, however, this variable might have been discarded as a predictor for subsequent study. Short-term studies may thus suffice for some species but may be totally inadequate for other species.

Predictions and management policies based on short-term studies can be misleading because of among-year density variation on the same plots (Rice et al. 1981; Marzluff, Chapter 25). Managers have no control over this variation, which arises from the vagaries of weather or from other factors affecting the population dynamics of the birds. The solitary vireo and the western wood-pewee are examples of species whose predictor variables changed from site to site (in approach II) or from year to year (in approach IIIa) on the same sites.

Weather factors differed in their importance from site to site. It was clear from approach II that precipitation played an important role at all sites, whereas variations in temperature were of secondary importance. On each site, changes in weather were effective in explaining changes in species' densities (especially for the junco, the vireo, and the pewee). It was noteworthy, however, that the unique character of the vegetation on each site altered the importance of the various weather factors by site and by species (for example, compare the vireo on sites 8, 13, and 17). Each site's habitat structure seemed to determine whether precipitation or temperature was more important and in what season these effects were strongest (fall, winter, or spring). These effects suggest interaction between vegetation and years (weather).

No factors universally predicted the densities of breeding birds. Each bird species responded to its environment in an individual way. This is not surprising from an ecological point of view because by definition each species is unique. Nevertheless, from a management perspective, one would hope to identify a realistic set of habitat variables that would reliably determine the population density of the species of interest. This is not the case for ponderosa pine birds. One of the vegetation variables (trunk volume) comes close, but only in approach IIIb, which includes all cases available to us (10 years over eight sites).

Management considerations

The prediction of the density of forest birds (and, hence, effective management) depends on three factors. First, relevant variables should be measured (Rice et al. 1981). This chapter indicates that such variables may differ from species to species and even from plot to plot. However, important niche characteristics may be measured indirectly, and multivariate techniques, such as principal components analysis, may be used to reduce the number of factors examined (Green 1979; Johnson 1981a).

Second, enough habitats should be sampled to provide coverage of environmental extremes where the species live. Our analysis showed that, in general, studies restricted to a single habitat gave predictive models that were unique to each habitat. This result implies a different management strategy for each habitat.

Third, a span of years long enough to cover the natural variation in population density that occurs at a single site should be sampled. It is important to avoid drawing conclusions from either an extreme high or an extreme low in population density. Models for three of the five bird species considered contained a vegetative-site characteristic as a relatively stable density predictor after 5 years of measurement (approach IIIa: 15 cases, Table 32.2).

Each of the three approaches used in this analysis has advantages and disadvantages (Table 32.4). Approaches I and II contain serious drawbacks in that each misses observing an important source of population variation—yearly fluctuations and different site influences, respectively. Approach III is the most costly from an economic standpoint, but we feel that this cost is more than outweighed by the improvement in ecological information obtained.

Table 32.4. Advantages and disadvantages of the sampling approaches used in the present study

Advantages	Disadvantages
APPROACH I	
Short-term.	Misses year-to-year population variation.
Covers a wide variety of habitats.	Models have limited application.
APPROACH II	
Makes inferences about year-to-year population variation.	Limited to a single site.
Simple logistics.	Long term.
	Manager has no control over factors.
APPROACH III	
Makes inferences about year-site interactions.	Requires most resources and time.
Covers population fluctuation over a range of habitats under varying conditions.	Not always easy to interpret.

Our analysis suggests that 5 years of observation on a minimum of three sites representing a range of habitats should be sufficient to predict population densities of birds in the ponderosa pine forest. Models, such as ours, that significantly explain variability in species' densities are a good start in developing management approaches. One should keep in mind that these models perform better than R^2 would indicate when pure error is removed (Marzluff, Chapter 25). Replication at several sites is important (as in approach III) for assessing lack of fit and pure error (Draper and Smith 1981; Marzluff) and for establishing confidence intervals of species' densities. Our simultaneous measurement of year-to-year weather fluctuations made it possible to explain some of the intrasite variability observed, which otherwise would be included in the error terms.

Acknowledgments

We are grateful to John Marzluff, Gregory Goodwin, William Boecklen, Robert Szaro, Michael Morrison, and Andrew Carey for critically reviewing an earlier draft of the manuscript. We also acknowledge the financial support of the USDA Forest Service, the Federal Timber Purchasers' Association, the National Forest Products Association, and Organized Research at Northern Arizona University.

33

Factors Confounding Predictions of Bird Abundance from Habitat Data

BARBARA DIEHL

Abstract.—Bird-community dynamics was studied for 21 years in a heterogeneous and changing habitat. Spot mapping was used to estimate the relative abundance of birds. The total abundance of nesting passerine birds increased with time in response to increasing complexity of the habitat (shrub and tree invasion), suggesting that habitat structure allowed reliable predictions of total numbers of breeding pairs. However, a regular, annual pattern of alternating increasing and decreasing numbers of breeding passerines was superimposed on the long-term increasing trend. This could not be explained by changes in habitat structure. At the population level, departures from expectations based on habitat structure were large for some species. This was interpreted to be a result of changing intensities of intra- and interspecific competition as a direct result of nest predation. Four- and 5-year trends in numbers of certain species that ran counter to the 21-year trend, and counter to habitat predictions, argued strongly for long-term studies to recognize real patterns of population dynamics over a gradient of habitat changes.

The development of predictive models for population or community dynamics requires an understanding of the mechanisms that drive natural systems. In addition to habitat conditions, it may be important to know the role of biotic interactions, such as competition, cooperation, or predation, in shaping ecological systems. My 21-year observations of a bird community in a heterogeneous and changing habitat show that mean community density can be predicted from the structure of the habitat, but this is not necessarily the case for densities of individual species within the community. One of the most important reasons may be that individual species differ in their competitive abilities, and these species-specific differences are additionally modified by habitat conditions (Diehl 1985). As a result, no simple and identical process should be expected for different species in the same habitat, or for the same species in different or varying habitats. This chapter attempts to show the relative importance of biotic interactions, especially competition, and habitat conditions to the population dynamics of selected community members.

Study area and methods

The study was done on a shrubby meadow located 15 km west of Warsaw, Poland, in a 60-ha habitat island surrounded by forest. The closest similar habitat is almost 1 km away. The study area is a mosaic of dry and wet patches with unevenly scattered clumps of woody vegetation. These are remnants of a hardwood forest cleared early in this century. Mowing and cattle grazing prevented forest succession until the early 1960s, when a nature reserve was established. Young regrowth first appeared close to the existing clumps of trees. These were almost exclusively birches (*Betula verrucosa*) on dry patches and alders (*Alnus glutinosa*) on wet patches of the habitat. Moderately wet patches were spotted with willows (*Salix* spp.), a vigorous regrowth of which took place in the late 1970s. Now, after more than 20 years of succession, most birch and alder thickets are in the pole stage. Wooded patches are still interspersed with open spaces of different sizes. This extremely diversified habitat structure is reflected in a high diversity of birds in terms of species composition and population density.

Birds have been censused on a 44.1-ha plot from 1964 onwards, with a 3-year gap in 1966–1968. The mapping method (Kendeigh 1944; Enemar 1959) was used to estimate the number of breeding pairs. In most years the number of visits ranged from 25 to 35. They were rather evenly spaced from April through July, or August in some years.

BARBARA DIEHL: Wysockiego 22 m 82, 03 388 Warsaw, P.O. Box 163, Poland

Results

GENERAL CHARACTERISTICS OF THE COMMUNITY

The bird community consisted of 35–40 species in the first years of the study and up to 50 species recently (Fig. 33.1). Most species were small passerines, and the majority of those were migratory, territorial, open-nesting birds that feed on invertebrate animals. Only two species were purely granivorous. The number of passerine species was relatively constant ($\bar{x}$ = 27.4; 2 *SE* = 1.3) during the first 11 years of the study, but in 1978 the number jumped to a significantly higher plateau ($\bar{x}$ = 36.6; 2 *SE* = 1.4), where it has remained since (Fig. 33.1). The total number of breeding pairs (passerines and nonpasserines combined) increased from about 300 (70 pairs/10 ha) in the first few years to 800 (180 pairs/10 ha) in the last 2 years.

Wet habitats were dominated by the sedge warbler (*Ac-*

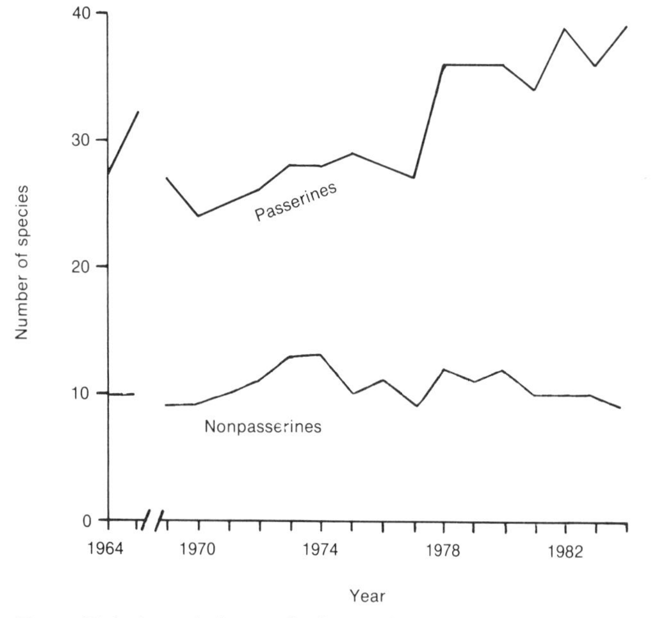

Figure 33.1. Annual changes in the numbers of species nesting on the study area.

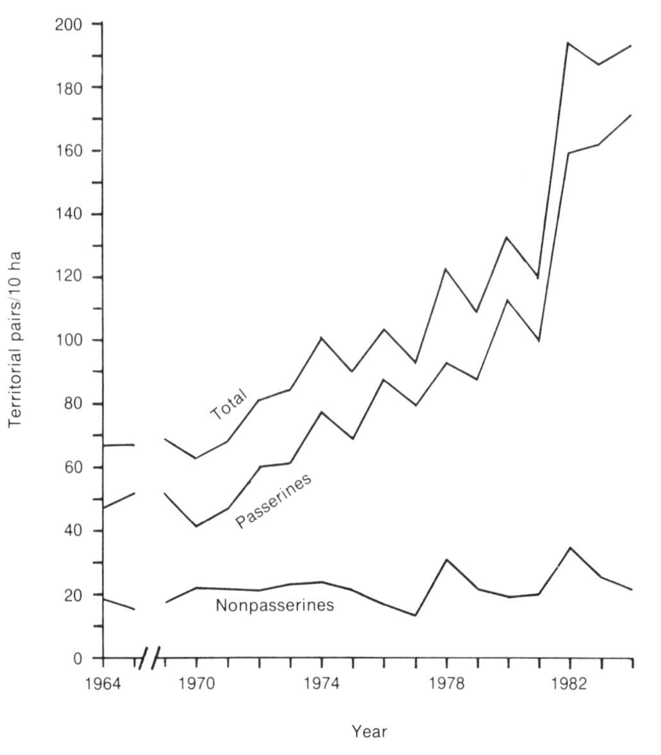

Figure 33.2. Annual changes in the number of territorial pairs (nonpasserines, passerines, and total) on the study area (adapted from Diehl 1985).

rocephalus schoenobaenus) and reed bunting (*Emberiza schoeniculus*). Dry habitats for most of the study period were dominated by the red-backed shrike (*Lanius collurio*) and barred warbler (*Sylvia nisoria*). Then the garden warbler (*S. borin*), icterine warbler (*Hippolais icterina*), and whitethroat (*S. communis*) became most abundant.

During the 21-year study period, some species disappeared and some new species appeared, but most species were present throughout the study period; only their numbers changed significantly.

Community dynamics

The pattern of change in density of nonpasserine birds largely differed from that of passerines (Fig. 33.2). The density of nonpasserines was fairly stable throughout the study, with an overall mean of 22.2 pairs/10 ha (SD = 4.41). The density of passerines was stable only for the first 8 years. Thereafter, passerines showed a steady long-term increase in density with small but regular declines in numbers, or at least in population growth rate, superimposed on the long-term trend (Fig. 33.2).

The sawtooth pattern of regular increases and decreases in alternating years, superimposed on the otherwise steady upward trend in numbers of breeding passerines on the study area, was notable. It continued for 11 years and could even be traced for 13 years when a comparison was made of year-to-year changes in population growth rates, not just changes in numbers, of all species of breeding passerines combined. This long series of alternate decreases and increases in growth rate was probably not a result of random factors ($P <$ 0.05, one-sample runs test; Siegel 1956). This does not imply that random factors such as weather in winter, losses during migration, or sampling error were not involved. But their influence, if present, was not sufficient to obliterate the effects of other, nonrandom factors.

Population changes

At the population level the situation was more complicated. The densities of some species in the community tracked changes in habitat structure, gradually increasing with development of woody vegetation. The icterine warbler was the best example of such a species, as its numbers increased steadily throughout succession (Fig. 33.3). Furthermore, numbers of some species typical of earlier successional stages were also accurately predicted by habitat changes. Their numbers exhibited bell-shaped curves such as that of the barred warbler (Fig. 33.4).

Most species in the community, however, showed large deviations in their population numbers from those expected on the basis of changes in habitat structure alone. The sedge warbler and the reed bunting were good examples of this group of species (Fig. 33.5). In general, the population of sedge warblers increased with time, but its pattern of growth deviated from that predicted from uniform changes in habitat structure. Numbers increased in two big jumps, each preceded by a period of almost stable numbers.

Similarly, the population curve for the reed bunting departed from predictions based on habitat structure. Instead, it tended to be bimodal, while changes in habitat structure over the 21-year study period moved steadily toward in-

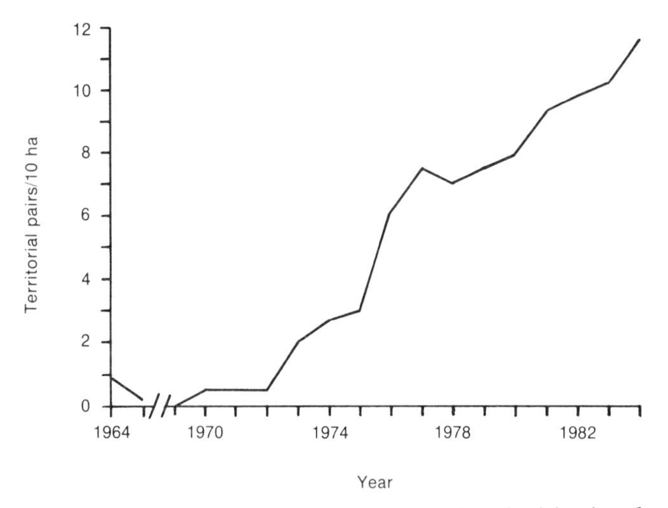

Figure 33.3. Annual changes in the number of territorial pairs of icterine warblers on the study area.

creasing complexity. Indeed, at least during early years of the study, trends in the population of reed buntings were opposite to those of the sedge warbler (Fig. 33.5).

The population growth curves for European greenfinches (*Carduelis chloris*) and linnets (*Acanthis cannabina*) (Fig. 33.6) also contradicted expectations based on habitat requirements alone. Both of these species nest in shrubs, so their numbers should have increased with increasing shrub volume. Greenfinches nested in higher vegetation than linnets did, so their numbers should have increased for a longer time than those of linnets. This was not the case in this community, however. The populations of both species increased by the mid-1970s, fitting expectations based on habitat changes. Thereafter, however, the population of linnets continued to grow, while that of greenfinches declined (Fig. 33.6).

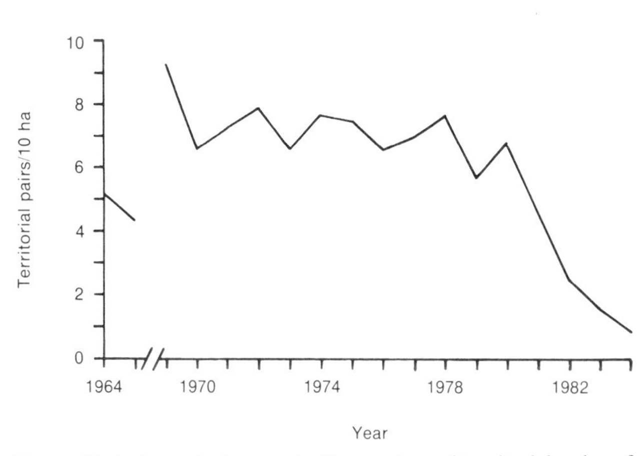

Figure 33.4. Annual changes in the number of territorial pairs of barred warblers on the study area.

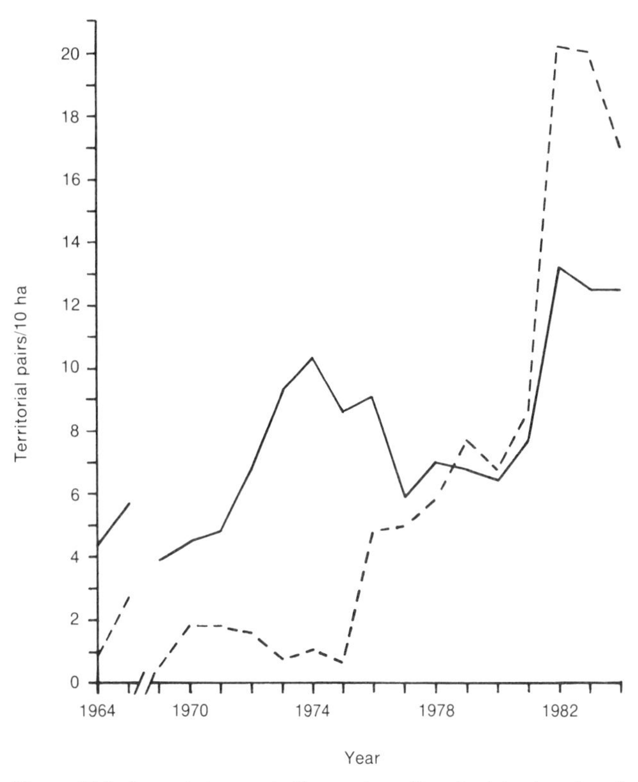

Figure 33.5. Annual changes in the number of territorial pairs of reed buntings (solid line) and sedge warblers (dashed line) on the study area.

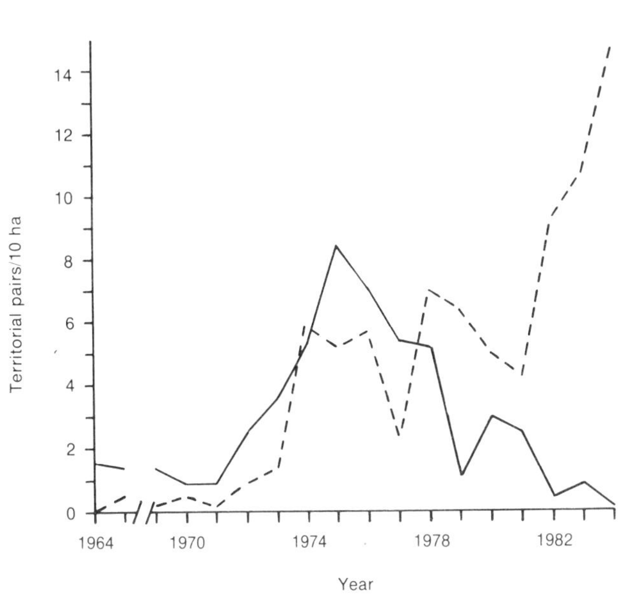

Figure 33.6. Annual changes in the number of territorial pairs of greenfinches (solid line) and linnets (dashed line) on the study area (adapted from Diehl 1985).

Discussion

Populations of individual species

Knowledge of the habitat relationships of certain species, in relation to plant succession on the study area, clearly allowed prediction of changing population densities over time, as for the icterine warbler. This species foraged and nested in shrubs, so one would predict that its density would increase with an increasing volume of shrubby vegetation. Based on habitat structure alone, one could predict that this upward trend in its population would continue until the habitat structure reached an optimum for this species. Then fluctuations about a steady state would be expected, followed eventually by a declining trend with further growth of trees and an associated decline in shrubby vegetation toward the climax forest community. As a result, the population growth of the icterine warbler should follow a bell-shaped curve over the complete series of time-related changes in habitat conditions suitable for the species.

The bell-shaped population curve of the barred warbler is one example of such a relationship found during my study. This species is adapted to earlier successional stages than the icterine warbler, so its optimum habitat was replaced by advancing development of the plant community toward denser shrubs and pole-sized trees. It was clear, however, that changing numbers of this species could also be reliably predicted from changes in habitat structure.

Populations of many species, however, showed marked changes that could not be predicted from habitat changes, as with the sedge warbler and reed bunting. Both species used shrubs for nesting and singing, although reed buntings also nested in clumps of sedges. Consequently their numbers should have increased with increasing shrub cover. Competition between them may have affected population growth of both of these species (Diehl 1985).

Similarly, a more detailed analysis of the unexpected decline in numbers of greenfinches that began in 1975 seemed to show that competition with icterine warblers was involved (Diehl 1985). It is also worth noting that the linnet population fluctuated widely about the upward trend, suggesting that it too was affected by factors other than changes in habitat structure.

Community patterns

Differences in the general patterns of population change shown by passerines and nonpasserines can probably best be explained by differences in their habitat requirements. Most nonpasserines were species such as mallards (*Anas platyrhynchos*) and common snipes (*Gallinago gallinago*), for which woody vegetation was of little importance. Most passerines, on the other hand, were species that rested, foraged, and nested in shrubs and trees. The development of woody vegetation in early successional stages could thus account for an upward trend in their numbers, as has been found in many other studies dealing with succession in land-bird communities (e.g., Johnston and Odum 1956; Blondel 1979).

The 2-year cycling pattern superimposed on the gradual upward trend in the total number of breeding passerines could not, however, have resulted from successional changes in habitat structure. Plant succession progressed steadily toward increasing structural complexity during the 21 years of the study. A detailed analysis of the pattern of change in passerine numbers suggested that it might be generated by yearly changes in the intensity of intra- and interspecific competition in response to changes in the rate of nest predation (Diehl 1985). Comparisons of yearly increases or decreases in numbers of individual species, using this hypothesis, gave consistent results according to predictions based on (1) vulnerability to nest predation; and (2) relative, interspecific social dominance of species.

As discussed in detail elsewhere (Diehl 1985), individual species differed in their competitive abilities, perhaps accounting for major differences in the patterns of their population changes in changing habitats. Superior competitors tended to track habitat changes very closely, so changes in their numbers could be reliably predicted from knowledge of habitat changes. This was not the case, however, for inferior competitors. Their population dynamics were largely determined by interactions with other community members, so habitat changes would have failed totally as predictors of their numbers. It is thus important to know the dominance status of the various species in a community to increase the reliability of predictions made from habitat changes. And this should be estimated with caution, because the competitive ability of each individual species may be relative in nature. It may be higher or lower, depending on competitive abilities of all other community members, so it may vary from one habitat to another.

Because nest predation may have influenced the intensity of intra- and interspecific competition (Diehl 1985), knowledge of a species' relative vulnerability to nest predation may be needed to increase the reliability of species-habitat predictions. In addition, studies by others (e.g., Hejl and Beedy, Chapter 35) have shown that factors such as extreme weather can substantially affect population dynamics and account for departures from habitat-based predictions of population dynamics. In view of the facts that so many factors can disturb simple species-habitat relationships, and that some species may be extremely vulnerable to these disturbances, thorough study seems to be necessary to obtain reliable baseline data for wildlife management.

A need for long-term studies

Had my study been confined to the years between 1970 and 1975, results would have suggested the erroneous conclusion that the growth of shrubby vegetation deteriorates habitat conditions for sedge warblers. No habitat variable, moreover, would have allowed prediction of the sudden increase in their numbers in 1976. Similarly, if their population had been studied only between 1976 and 1981, nothing would have suggested the next jump in numbers in 1982. This clearly shows that short-term, species-habitat correlations cannot be used as a reliable basis to predict numbers of at least some bird species and emphasizes the need for long-term studies to obtain a complete pattern of population dynamics under changing habitat conditions. Similarly, if one relied on short periods of time as a basis for drawing conclusions about habitat relationships of reed buntings, those conclusions could be misleading. The population of reed buntings increased from 1969 to 1973 and then steadily decreased for 6 years. Thus the conclusion might be that habitat structure reached an optimum for this species in the early 1970s. This was certainly not the case, as indicated by the subsequent growth of the reed bunting population to an even higher density than that of 1973. Factors other than changes in the habitat must have been of greater importance to the dynamics of this species' population, making unreliable any predictions of its numbers from habitat changes, especially short-term ones. Similar conclusions follow from examination of density changes of linnets during the course of my study.

Acknowledgments

Professor Kazimierz Tarwid, Assistant Professor Leszek Grüm, Dr. Waclaw Malinowski, and Dr. Janusz Uchmanski discussed an early draft of this chapter with me. Drs. Andrew Carey, Jean O'Neil, and Beatrice Van Horne offered helpful comments on the manuscript. Dr. Jared Verner was instrumental in early revisions of this chapter and in completing the final version. The sponsors of the Wildlife 2000 Symposium generously provided funds for my travel and stay at the symposium. To all I express my sincere appreciation.

34

Dynamical Aspects of Avian Habitat Use

RAYMOND J. O'CONNOR

Abstract.—I used long-term (1962–1980) bird population data from Britain to examine the validity of a habitat-dynamics model based on the Brown-Fretwell-Lucas model of territoriality. As a detailed example, I relate habitat use by the mistle thrush (*Turdus viscivorus*) population in Britain to a fourfold increase in numbers following the decline induced by the severe winter of 1962–1963. In the early stages of the increase new habitats were colonized, while in the later stages selective "filling-in" of habitats occurred. Average clutch size fell with population density, but only egg survival decreased systematically over the less preferred habitats. Because chicks in the resulting smaller broods were then more successful than at lower densities, these habitat-related changes could not regulate the population on their own. However, less frequently used habitats were also habitats characterized by late laying and (by inference) fewer broods, and this factor or density-dependent adult or juvenile mortality could then provide the extra regulatory power.

Empirical data show that cross-sectional analyses of habitat use may yield different results at different stages of population growth. I examined population stability as a measure of habitat preference and found it works in some but not all cases.

Many statistical models of habitat use by birds are based on cross-sectional studies of habitat replicated in space but not in time. These studies typically assume that local densities reflect the suitability of habitat features and that statistically linear correlations between avian densities and habitat then identify the habitat elements of most importance to birds. A growing body of evidence, however, indicates that modeling bird-habitat relationships as simple linear or (less restrictively) monotonic relationships predicting density is unsafe. Confounding factors include (1) nonlinearities in the response curve of bird densities to habitat characteristics (Osborne 1982; Rands 1982; Meents et al. 1983; O'Connor 1984); (2) breeding densities held below carrying capacity because limiting factors operate in the nonbreeding season (Cawthorne and Marchant 1980; Van Horne 1983); (3) the influence of site dominance in selecting for site fidelity (Fretwell 1980; O'Connor 1985); and (4) territorial behavior forcing subdominant individuals into suboptimal territory in which their breeding success is reduced or nonexistent (Brown 1969; Fretwell and Lucas 1970; O'Connor 1982). In the present chapter, I focus on this last source of nonlinear relationships between bird numbers and habitat characteristics because density-dependent migration between habitats of different quality can result in changes in correlations between density and habitat variables (O'Connor 1980).

A model of habitat dynamics

That there might be a limit to the dynamic range of the one-to-one relationship between density and habitat characteristics was originally developed by Brown (1969). He pointed out that sustained population increase would eventually bring the number of territorial birds in an area to some carrying capacity set by habitat area divided by minimal acceptable territory size. In populations larger than this carrying capacity, some individuals would be unable to acquire territories in which to breed. Fretwell and Lucas (1970) further developed this model to consider (1) the possibility of a hierarchy of habitat preferences with the best habitat used first, followed by the next best, and so on; and (2) the possibility that habitat quality (that is, the evolutionary fitness achieved by a typical member of the population when breeding in that habitat) might be density-dependent, decreasing with the density of birds breeding in the habitat. In this latter situation, a bird without territory could be equally well off (in fitness terms) by breeding in an uncrowded habitat of low intrinsic quality as by settling in some alternative habitat intrinsically preferred but already crowded.

These dynamic concepts of avian habitat allow a number of predictions of relevance to habitat modeling. First, the idea of a hierarchy of habitat preferences predicts that a greater variety of habitats should come into use as population density increases and the reverse should occur during decreases. Second, reproductive success within any habitat should be higher the earlier in the use-sequence it comes (unless adult or juvenile survivorship is the source of fitness benefit). The corollary of this is that average reproductive success in the population should decrease with population density and habitat diversity because more inferior habitats are in use at high population levels. This happens irrespective of whether colonization of inferior habitats occurs through exclusion by social dominance (the "despotic" distribution of Fretwell-Lucas) or through density-dependent equalization of fitness (their "ideal free distribution"). Finally, if density-dependence is severe enough, population

RAYMOND J. O'CONNOR: British Trust for Ornithology, Beech Grove, Tring, Hertfordshire, HP23 5NR, England

regulation to some equilibrium density is possible (Lack 1966; O'Connor and Fuller 1985).

Materials and methods

I analyzed habitat use and breeding biology of various species in relation to their population level. I chose species for which adequate population and breeding data were available. Population levels were taken from the Common Birds Census (CBC) of the British Trust for Ornithology (BTO) for 1962–1980 (Taylor 1965; Bailey 1967). Habitat maps for each CBC plot, prepared by the observer and updated periodically, provided materials for detailed analysis of habitat requirements (e.g., Williamson 1969; Morgan and O'Connor 1980). Data on breeding performance (fledgling production and its components of clutch size, egg success, and nestling success) were obtained from the Nest Records Scheme of the BTO for the years 1962–1980 (Mayer-Gross 1970). Nest habitats were classified (on the basis of notes supplied by ecologically untrained observers) only to broad habitat categories—e.g., "broad-leaved woodland," "coniferous woodland," or "mixed woodland"—but this classification was done in standard fashion by practiced technical assistants coding cards for computer analysis. An index of the diversity of nest habitats used was calculated for each year's data, using the Shannon-Weaver formula (May 1976a).

For some analyses, the relative frequency of habitat representation in the pooled 1962–1980 sample was used as an index of habitat preference within the Fretwell-Lucas model. Because relative frequency is the product of reporting rate for the habitat times its abundance, there is a risk that less preferred habitats are over-represented. Against this, the estimate was averaged over 19 years and nearly fivefold variation in densities, so that over-representation of a habitat in high-density years would be diluted by its absence in years of low density. Nest success was estimated throughout, using the methods of Mayfield (1961, 1975) to avoid sources of bias associated with the finding of nests partway through the nest cycle.

To examine variation in habitat-density correlations at different levels of population density, data were extracted for eight CBC farmland plots studied in each of the 18 years from 1962 to 1980. Within each year, the correlations were calculated between avian density and hedge abundance and between avian density and woodland abundance. These calculations were across the eight farms for each of six species and were then examined in relation to the logarithm of the farmland CBC index of the species.

Nonlinear trends were tested throughout for statistical significance, using Spearman's rank correlation coefficient (r_s), thus avoiding arbitrary choice of linearizing functions. Logarithmic transformations were used with population data, however, because of the natural multiplicative nature of population growth. Unspecified correlations were otherwise calculated as Pearson product-moment correlations (r).

Results

Population dynamics of the mistle thrush

Density-dependence of habitat use and breeding

On farmland plots, the CBC index (I) for the mistle thrush increased from its 1963 value of 32 (a low value following a 76% drop after the exceptionally severe winter of 1962–1963) to 113 in 1967, fluctuating thereafter (Fig. 34.1). Over the period 1965–1980, a similar increase in woodland (first indexed as a separate habitat for mistle thrush in 1965) was positively correlated with that on farmland (r = 0.504, P < 0.05). Subsequent analyses were therefore based on the farmland index alone, in effect using it as a single overall measure of population level. During the first 5 years of the population increase (after which the CBC index had largely leveled out), the number of new habitats (n) in use varied according to the relationship: $n = 6.88 \log_e(I) - 7.42$ (r =

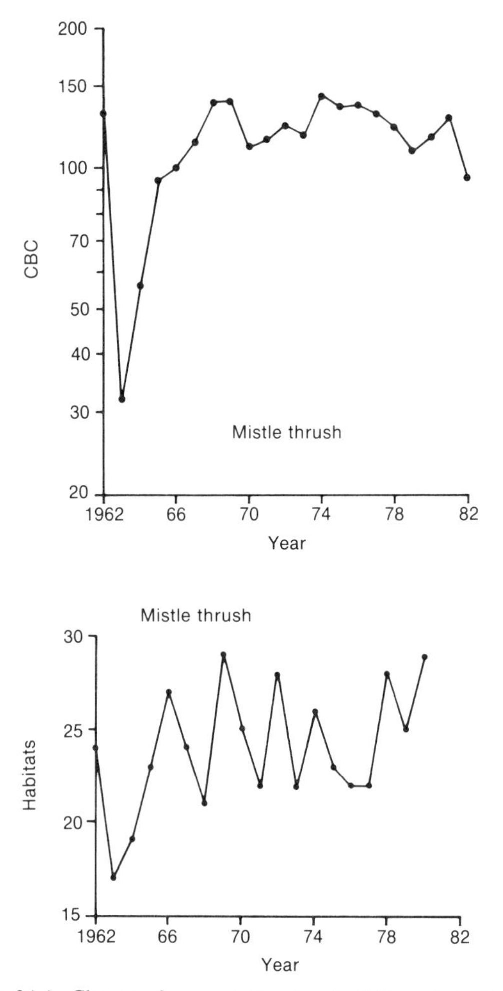

Figure 34.1. Changes in population level of the mistle thrush in Britain (as logarithm of Common Birds Census index) between 1962 and 1982 (*above*) and the numbers of habitat categories in use for nesting for the same years (*below*).

0.906, $P < 0.05$). Thereafter, however, the number of habitats in use in any given year fluctuated irregularly about the 24–25 level, thus tending to level out at the same time as did the population density on farmland and woodland. Although no net increase took place in the number of habitats in use at higher CBC levels, net nesting habitat diversity increased. Overall, nest habitat diversity (NHD) and Common Birds Census index correlated weakly (r_s = 0.47, $P < 0.055$), but in detail (results not presented here) the relationship was curvilinear, steepening at higher densities.

According to the Fretwell-Lucas model, net fitness should be correlated with density-dependent patterns of habitat use. The clutch size (C) of the mistle thrush averaged across all habitats in any given year, in fact, decreased in a density-dependent and nonlinear fashion as predicted: $C = 4.88 - 0.48 \log_e(I)$ ($r = -0.542$, $P < 0.05$). During 1963–1967, when new habitats were coming into use, clutch size decreased according to: $C = 5.38 - 0.77 \log_e(I)$ ($r = -0.950$, $P < 0.02$). After 1967, the density dependence was not apparent: $C = 2.35 + 0.74 \log_e(I)$ ($r = 0.282$, n.s.). Hence, clutch size was larger at low densities than at medium or high densities but was otherwise not greatly affected by density.

Egg mortality increased as the population density of mistle thrush rose ($r_s = 0.500$, $P < 0.05$), showing that greater per capita loss of eggs took place at high densities. Nestling mortality, on the other hand, actually decreased at high density ($r_s = -0.620$, $P < 0.01$). Nest habitat diversity paralleled the CBC index in relation to egg and chick mortality (r_s = 0.500 and 0.536, respectively; $P < 0.05$ for both). Egg mortality was positively correlated with clutch size (r_s = 0.501, $P < 0.05$) and with chick mortality ($r_s = 0.500$, $P < 0.05$), but chick mortality was independent of clutch size (r_s = 0.079, n.s.).

Habitat dependence of reproductive success

The 10 most common habitats recorded in the mistle thrush nest record cards (each with 40 or more cards) were gardens (19.6%), agricultural land (9.7%), broad-leaved woodland (7.8%), pasture (5.3%), unspecified woodland (4.5%), mixed broad-leaved–coniferous woodland (4.4%), roadside and other verges and margins (3.7%), suburbia (3.1%), orchards (2.6%), and parkland (2.4%). Clutch size differed significantly among these 10 habitats ($X^2 = 30.09$, df = 18, $P < 0.05$). Clutch size did not differ between the two agricultural habitats ($X^2 = 2.87$, $df = 4$, n.s.) nor between the two suburban habitats ($X^2 = 0.91$, $df = 2$, n.s.) but did differ across the various woodland classes ($X^2 = 17.36$, df = 4, $P < 0.01$); these three "pooled" classifications also differed from each other in clutch size.

Egg mortality was positively correlated with frequency of habitat use, with daily mortality increasing in a nonlinear manner from 4.0% in gardens to 8.3% in parkland (r_s = 0.624, $P = 0.055$). Chick survival was essentially independent of habitat sequence ($r_s = 0.127$, n.s.).

Population regulation by habitat use

The results so far presented indicate that reproductive performance fell as the population spread into new habitats. Are these effects strong enough to regulate the population to somewhere near its observed equilibrium (Fig. 34.1)? I constructed a simple deterministic model with which to evaluate their strength. In the model each pair of birds produced a clutch of C eggs which experienced daily mortality m for an incubation period of I days; those eggs that hatched then experienced daily mortality M through a nestling period of N days; the surviving young fledged and joined the adults to share an annual survival rate S before the population bred again the following year. Thus, if the population size in year n is P_n pairs, the following year's population is given by

$$P_{n+1} = [P_n + P_n C(1 - m)^I (1 - M)^N/2]S.$$

By setting C, m, and M to reflect the observed patterns in reproductive parameters, I could test whether the trends apparent with variation in population density were strong enough to regulate the population to an equilibrium value, i.e., one to which the population density tends to return following perturbation (Lack 1966). The tests were done in two steps: (1) by substituting the observed values of C, m, and M for each year from 1963 to 1980; and (2) by computing each year's value of C, m, and M from their respective regression equations on population density, substituting for each year the model density achieved at the start of that breeding season. In the equation above, an incubation period of 14 days, a nesting period of 14 days, and a fixed (density-independent) estimate of adult survival of 0.492 (as in O'Connor 1981) were first used; survival rate was also varied over the range 0.30–0.90 as a check against a biased survival estimate. As all combinations of parameters failed to bring the model to equilibrium, the detailed results of these simulations are not presented here, but some examples may be given to illustrate these failures. Using the empirical clutch size and egg and nesting mortalities, adult survival of 0.492 gave population estimates for the years 1963–1980 ranging from 0.2 to 0.6 times the observed values; a survival of 0.70 gave 0.6–69.6 times the observed levels; and 0.30 gave population extinction. Using density-dependent equations to estimate C, m, and M gave estimates 0.4–2.7 times and 0.8–363.0 times the true levels for S = 0.492 and 0.70, respectively. In both cases, the population increased steeply by 1980. Adult survival of 0.30 again led to extinction by 1980. Only by making adult and juvenile mortality density-dependent, or by making the number of broods reared per year density-dependent, could a regulated equilibrium be achieved. The density-dependent functions used were arbitrarily chosen in the absence of empirical data, so presentation of detailed results is unwarranted. The important point is that the density-dependencies of the reproductive parameters already discussed were inadequate to regulate the population over a wide range of assumptions as to density-independent mortality. Hence, the equilibration of mistle thrush numbers in Britain could be due to habitat specificity of breeding success only if the number of broods reared per season is habitat sensitive. This may be the case, because within the 10 most frequently used habitats, laying was later in the less frequently used habitats ($r_s = 0.706$, $P < 0.025$), with modal laying dates ranging from 24 March to 11 April. This span is about half the normal nest cycle of 40 days, so that late-laying birds would have difficulty in producing a second brood.

Density-related changes in habitat correlations

One consequence of Fretwell-Lucas patterns of habitat dynamics is that density-habitat correlations may alter with population density. I computed annual correlations between the amount of hedgerow present and the density of European wrens on each of eight farms censused annually from 1962 through 1980. I then plotted the resulting correlations against the national CBC index as a measure of global wren density (Fig. 34.2). With the small sample size, none of the individual correlations is statistically significant, but there is a statistically significant trend from positive correlations at low populations to negative correlations at high populations (r = 0.652, $P < 0.01$). To test whether the positive correlations at low densities were likely to be a chance outcome of small samples, I checked the 1965 data on a further 57 farms for which hedgerow data were available. Overall, on 65 farms (57 plus the eight long-term plots), wrens were statistically more numerous where hedge densities were greatest ($r = 0.51$, $P < 0.05$). The change in sign with wren density shown in Figure 34.2 thus suggests that wrens initially used farms with much hedgerow but thereafter settled in farms with little hedgerow as their population increased.

How prevalent are the risks of density-habitat correlations altering with density in this way? Table 34.1 summarizes an analysis of this point in relation to hedgerow and to woodland habitat for six species (including the wrens of Figure 34.2) and shows that seven of the 12 correlations varied significantly with density.

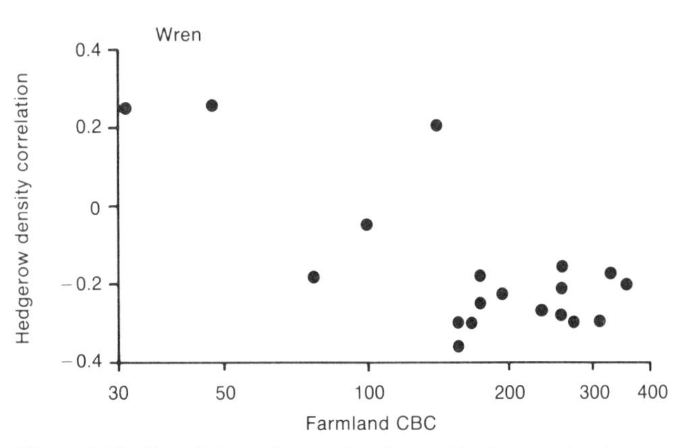

Figure 34.2. Correlation of wren density and hedgerow density on eight census plots in each year 1962–1980 in relation to the national population index for wrens in Britain (farmland Common Birds Census index). The correlation coefficient is −0.652, $P < 0.01$.

Table 34.1. Dependence of annual habitat-density correlations on the prevailing population density of six species in Britain 1962–1980

Species	Habitat element	
	Hedgerow	Woodland
Wren (*Troglodytes troglodytes*)[a]	−0.714***	0.482*
Dunnock (*Prunella modularis*)	−0.652**	−0.530*
European blackbird (*Turdus merula*)	−0.727***	0.025
Mistle thrush (*Turdus viscivorus*)	0.572*	−0.473*
Magpie (*Pica pica*)	−0.037	−0.148
Chaffinch (*Frigilla coelebs*)	−0.369	0.142

[a] See Figure 34.2 for detail of the wren-hedgerow case.
*$P < 0.05$.
**$P < 0.01$.
***$P < 0.001$.

Population stability as a habitat indicator

Pearson (1980) suggested that variation in population density can itself be used as an indicator of habitat quality. If less preferred habitats are filled to saturation only in years of high density, while preferred habitats are saturated in all but the lowest density years, population density will fluctuate more in the former (Kluyver and Tinbergen 1953; Krebs 1971). Pearson suggested that surveys in areas of optimal habitat should, therefore, yield more stable densities than those in less preferable areas. He showed for the European blackbird and the European robin (*Erithacus rubecula*) that optimal foliage profiles of the type described by James (1971) could be identified by this means. I tested this idea in relation to temporal variations in density (Fig. 34.3). Here the coefficient of variation of population density of the dunnock is plotted against hedgerow density on the long-term census farms already mentioned. From Pearson's argument, the most stable populations and lowest coefficients of variation should occur on the farms with optimal hedgerow density, here apparently around 5–12 m/ha. This conclusion is compared against a "conventional" analysis of density vs habitat based on census data from 65 farms for 1965, when dunnock populations were still low. Here the highest densities and, by implication, optimal hedgerow densities were at 7–9 m/ha of hedgerow—in the range suggested by the temporal analysis.

Similar analysis of temporal variation in population densities was undertaken on the other five species of Table 34.1, examining woodland and hedgerow densities as habitat variables. Only in the case of the blackbird, in relation to hedgerow density, was there any suggestion of an optimum of the type indicated by Figure 34.3. For the magpie, temporal population stability was negatively correlated with hedgerow density ($r_s = -0.744$, $P < 0.05$), reaching zero at hedge densities of around 2–3 m/ha. Mistle thrushes had a positive trend with farm woodland ($r_s = 0.74$, $P < 0.05$), with most stability apparent on farms with less than about 2% of their area under woodland. Other results were rather variable, with none showing clear habitat trends of interest.

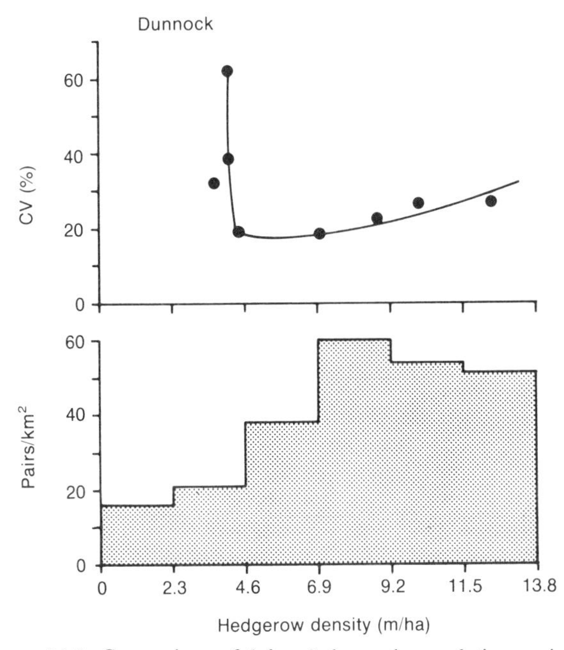

Figure 34.3. Comparison of (*above*) dunnock population variance (coefficient of variation in density over 1962–1980) as a function of hedgerow density on various farms and (*below*) median population density on farms of various hedgerow density classes. Density classes here were originally determined using English rather than metric units.

Discussion

Avian habitat use is dynamic and nonlinear; simplistic use of habitat-density correlates, therefore, may give misleading results. Both the Brown (1969) and Fretwell and Lucas (1970) models provide conceptual insight into the habitat dynamics observed in the wild and appear, on the basis of the evidence discussed here, to help in predicting the kinds of consequences that will result from habitat-density correlates. These models, however, do not permit quantitative predictions, and they do not take into account the importance of site fidelity in modifying birds' responses to habitat vacancies (Fretwell 1968; O'Connor 1985). In addition, we lack knowledge of the scale of habitat classification within which such dynamics operate, a point already emphasized in relation to habitat correlations by other studies (Anderson 1981; Wiens 1981b; O'Connor and Fuller 1985). Whether between-habitat dynamics would be quite so pronounced in relation to microhabitat features (cf. Anderson 1981; Holmes 1981; Wiens 1981b), as shown here for crude habitat variables (Fig. 34.2), is a moot point.

The Fretwell-Lucas model shows that nonlinearities in habitat-density relationships are most likely whenever species are subject to density-dependent depression of reproductive success within different habitats, but with the rate of depression being a function of each habitat. Where these conditions do not prevail, habitat correlations may be adequate indicators of avian habitat requirements, either directly (when they are measured for unsaturated populations) or after taking into account population stability, as suggested by Pearson (1980) (cf. Fig. 34.3). Where density relationships do vary among habitats, however, habitat correlations are likely to be valid only at low (unsaturated habitat) densities. Otherwise, the more densely populated areas identified in a spatial survey may merely be population "sinks," accommodating breeding birds that produce very few recruits to future generations (cf. Pienkowski and Evans 1982; Van Horne 1983).

The suggestion of Pearson (1980) that temporal stability in particular habitats would be a useful measure of habitat preferences by birds takes advantage of the dynamical aspects of habitat take-up. Figure 34.3 showed that population stability could serve as a measure of habitat preference in this way, but the broader analysis of Table 34.1 suggests that Pearson's technique is not of universal application across species. Further work on this elegant idea seems warranted, since such a stability measure could provide a substitute for the habitat correlations lost in the ebb and flow of populations between habitats.

Empirical studies of avian habitat dynamics are too few to permit extensive generalizations as to their significance for practical habitat management. As shown here for mistle thrushes, even quite marked density-dependencies associated with a broadening of habitat range may be inadequate to actually regulate the population to an equilibrium level; density-dependence in adult survivorship, or in the number of breeding attempts per season, may instead be the regulating factors. Nevertheless, these habitat dynamical processes can lead to habitat distributions that greatly alter one's view of the significance of particular habitat elements for birds (Fig. 34.2). At present, therefore, it behooves us to be aware of dynamical aspects of avian habitat use and to consider the possible impact of bird distributions governed by "despotic" and by "ideal free" processes in formulating habitat models. The relative importance of these two types of behavior will vary from species to species. For some species, despotic processes will take the population through a series of habitats of relatively fixed carrying capacity with little density-dependent reduction in fitness and with habitat diversity increasing with population level. For other species, a narrower range of habitats may be in use, but with considerably greater scope for density compression and associated reduction in fitness (O'Connor and Fuller 1985). Van Horne (1983) examined how populations within a single species might vary and considered which characteristics of populations and habitats might contribute to nonlinear habitat dynamical relationships of these types. She concluded that among populations with pronounced social interactions, a range of habitats is appropriate, and territory size is essentially incompressible within each habitat. Density-dependent fitness is more likely in less socially interactive populations or in habitats that are poorly predictable as to resource availability. The development of such insights, and their extension to interspecific consideration, may contribute to understanding the limits of applicability of currently available models of habitat needs of birds.

Acknowledgments

I thank Alan Eardley, Elizabeth McHugh, and Philip Whittington for data input of nest record cards. Detailed review of a draft of the manuscript was provided by Colin Bibby, Andrew Carey, Robert Fuller, Beatrice Van Horne, and Patrick Osborne, to whom my grateful thanks go. I thank Elizabeth Murray for artwork and Dorothy Rushton for typing support. The work reported here was conducted as part of a contract (HF3/03/192) from the Nature Conservancy Council to the British Trust for Ornithology. Preparation of the computer databanks of nest record data was funded by the Natural Environment Research Council under grant GR3/4357. This support is gratefully acknowledged.

35

Weather-Induced Variation in the Abundance of Birds

SALLIE J. HEJL and EDWARD C. BEEDY

Abstract.—We examined four breeding-bird species of true fir forests of the Sierra Nevada to assess the influence of weather on the abundances of two permanent residents, one short-distance migrant, and one long-distance migrant. Counts of all four species were lower in the summer of 1983, following an unusually harsh winter. These four species—mountain chickadee (*Parus gambeli*), red-breasted nuthatch (*Sitta canadensis*), yellow-rumped warbler (*Dendroica coronata*), and western tanager (*Piranga ludoviciana*)—are common breeders in true fir forests, yet we found significantly different abundances for three of the species between the 2 years. Current wildlife-habitat-relationships programs use matrices that are not based on field data and fail to reflect true variability in bird numbers. We suggest using 95% confidence intervals from presence/absence data in order to incorporate these two factors. In an example of this method, values ranged from 91–100% presence for mountain chickadees and western tanagers to 100% presence for yellow-rumped warblers.

Wildlife-habitat-relationships programs give land managers qualitative information on the predicted responses of wildlife species to land-management alternatives in forests and rangelands (Verner and Boss 1980). The California Wildlife-Habitat Relationships Programs for the North Cascades Zone (Marcot 1979), Northeast Interior Zone (Airola 1980), and the Sierra Nevada (Verner and Boss 1980) employ species-habitat association matrices that classify habitats by vegetation type and desirability to wildlife. Individual experts rated habitat quality for each wildlife species, based on a qualitative judgment of the breeding density or frequency of use for feeding, resting, and nesting cover—not on actual field data. Although the data are entered into subjective habitat-utilization categories (optimal, suitable, and marginal), completed species-habitat matrices imply a detailed knowledge of status and habitat preference.

Many researchers have found year-to-year variation in bird abundance (e.g., Rotenberry and Wiens 1978; Szaro and Balda 1979; Alatalo 1981; Wiens 1981a; Balda et al. 1983; Grzybowski 1983; Hall 1984; Gaud et al., Chapter 32; but see Winternitz 1976). Several of these authors (Wiens 1981a; Grzybowski 1983; Balda et al. 1983; Gaud et al., Chapter 32) attributed changes in bird abundance to annual differences in weather patterns. Beedy (1982) and Raphael and White (1984) found numbers of permanent resident species, but not migrants, to be depressed following a severe winter in the Sierra Nevada. These temporal variations in bird numbers will affect the construction and interpretation of a wildlife-habitat model.

Our objective was to compare the relative abundance of four breeding birds in true fir forests of the Sierra Nevada of California during years with extremely different weather patterns. Our null hypothesis is that the abundance of birds does not differ between years. We also suggest one method of constructing a wildlife-habitat model which reflects temporal variation in bird species' presence within a particular habitat.

Sallie J. Hejl: Department of Biological Sciences, Box 5640, Northern Arizona University, Flagstaff, Arizona 86011, and USDA Forest Service, Forestry Sciences Laboratory, 2081 E. Sierra Avenue, Fresno, California 93710

Edward C. Beedy: 320 W. 14th Street, Davis, California 95616

Methods

Site selection

Fifty-one study sites in stands of mature or older California red (*Abies magnifica*) and white (*A. concolor*) fir forest were randomly selected from 161 potential sites located in Yosemite National Park, Sierra National Forest, and Sequoia National Park. Site selection met the following criteria: (1) each site was located in a stand of trees that was homogeneous in terms of age structure and canopy cover; (2) stands were selected to obtain a wide distribution of canopy covers and age structures; (3) one sampling unit was located in each stand; (4) stands were at least 10 ha in size with dimensions accommodating a rectangle of at least 200×400 m; (5) most sites were at least 800 m apart, and when markedly different habitat intervened, they were no less than 400 m apart (this criterion increased the likelihood of independent counts at all sites); and (6) sites were selected in clusters of two or three that were sufficiently close to permit completion of bird counts in all between 06:00 and 11:00.

Weather data

Precipitation and snowfall data were obtained from Pacific Gas and Electric Company. Their climatological station was located at Wishon Dam, Fresno County, California, near the geographic center of our study sites.

Bird observations

At each site, the sampling unit consisted of three points connected by a transect. The first point was located by walking in a randomly chosen direction from a point on the forest edge, located 100 m along the stand on a trail or road, 100–200 m into the stand. The second and third points were located along a line in the same direction at 100-m intervals. A point was used if it was located at least 100 m from any discontinuity in the stand (meadow, edge, new vegetation type) or a previous point. All points were permanently marked with metal tags on the trunks of nearby trees.

The order of site sampling was balanced for diurnal and seasonal effects. All sites were sampled six times, three times each during the breeding season (23 May to 31 July) of 1983 and 1984. As nearly as possible, each site was counted in early (06:00 to 08:00), middle (08:00 to 09:30), and late (09:30 to 11:00) morning. Two observers counted in each year. One observer counted on all 51 sites twice; the other observer counted on all of them once. Birds were detected by visual or audio observations during three 8-min point counts and two 100-m transects. Abundance figures were calculated from all adult females and males detected.

Here we examined the temporal variations of four common breeders in true fir forests. These included two permanent residents (mountain chickadee [*Parus gambeli*] and red-breasted nuthatch [*Sitta canadensis*]), a short-distance migrant (yellow-rumped warbler [*Dendroica coronata*]), and a long-distance migrant (western tanager [*Piranga ludoviciana*]). We selected these four species because we had sufficient data to compare densities statistically between years.

Paired *t*-tests were used to compare species' abundances between years (Nie et al. 1975:267–275). We examined the normality of the data with histograms and chi-square goodness-of-fit tests. In the one case of violation of normality, matched-pairs Wilcoxon signed rank tests were used to compare the species' abundances between years (Conover 1971:206–215). Significance level was $P < 0.05$.

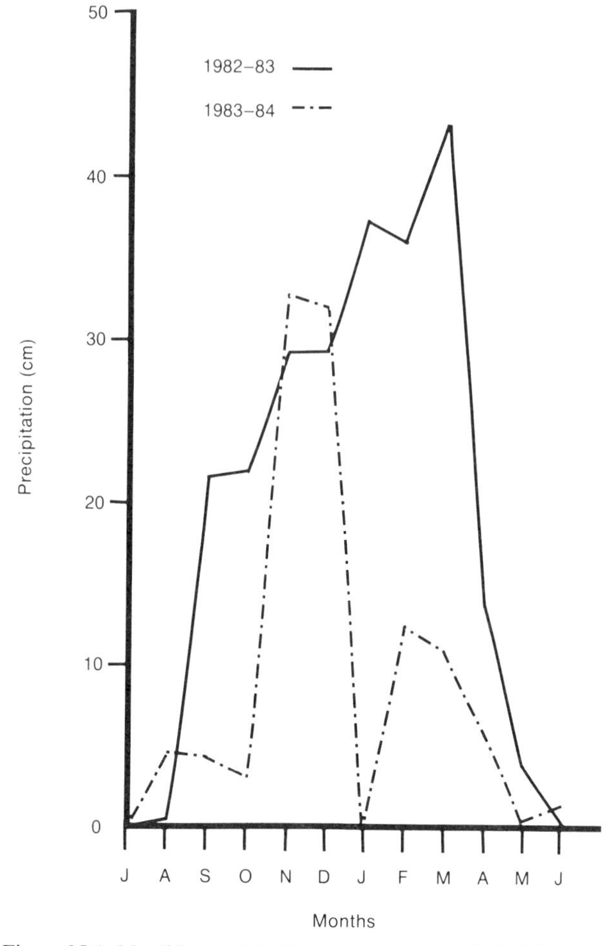

Figure 35.1. Monthly precipitation patterns (as total rainfall equivalents) from July 1982 to June 1983 and from July 1983 to June 1984.

Results

Precipitation

Dramatically different precipitation patterns occurred between 1983 and 1984 (Fig. 35.1). The summer of 1983 followed a winter with a record high snowfall (10,509 mm). In May 1983, most study sites were under 3–6 m of snow (Hejl, pers. obs.). A site in Yosemite National Park still had patches of snow up to 1.5 m deep on 31 July 1983. Less snow fell during the winter of 1983–1984 (2786 mm). When present on the study sites, snow was found in patches from 0.3 to 1.5 m deep in late May 1984, and all was melted by mid-June of that year.

Bird numbers

The abundances of three of the species (mountain chickadee, red-breasted nuthatch, and yellow-rumped warbler) were significantly lower in summer 1983 than in summer 1984 (Table 35.1). The paired *t*-test indicated significant differences for western tanager abundance too. However, the data for the western tanager were not normally distributed. The matched-pairs Wilcoxon signed rank test gave nonsignificant differences ($P = 0.10$) for western tanager abundance between the 2 years. Yet, the trend seen with this species is the same as with the others.

Residency status did not coincide with the magnitude of differences in total mean abundance between years (Table 35.1). The two migrant species had the greatest (yellow-rumped warbler) and the least (western tanager) differences between years, and differences in abundance of the permanent residents were intermediate. Average changes in abundance were estimated for each of the species and are indicated by the 95% confidence intervals for the true differences of the means between years. For example, the average change in abundance for the yellow-rumped warbler for the central and southern Sierra Nevada between 1983 and 1984 was at least 1.82 birds/site with a maximum of 2.93 birds/site.

Discussion

Our results agree with those of Beedy (1982) and Raphael and White (1984), who found that relative abundances of resident species varied dramatically between years in Sierra Nevada coniferous forests. A severe 2-year drought oc-

Table 35.1. Paired two-tailed *t*-test on average abundances for 51 study sites

Species	1983 $\bar{x}$	1983 *SD*	1984 $\bar{x}$	1984 *SD*	*P*	Differences	Confidence intervals
Mountain chickadee	1.74	1.26	2.52	1.37	0.000	0.77	0.36, 1.19
Red-breasted nuthatch	1.46	0.64	2.79	0.83	0.000	1.33	1.11, 1.55
Yellow-rumped warbler	6.27	1.34	8.64	1.71	0.000	2.37	1.82, 2.93
Western tanager	1.74	1.08	2.10	1.06	0.009	0.36	0.10, 0.63

Note: Values displayed are means, standard deviations of the means, true differences between means (1984 − 1983), and 95% confidence intervals of the true differences between means.

curred throughout most of California in 1976 and 1977, but the winter of 1977–1978 was among the wettest on record. Beedy noted pronounced decreases in relative abundances of permanent residents, such as woodpeckers and mountain chickadees, during the breeding season of 1978, following the severe winter of 1977–1978. Numbers of long-distance migrants like vireos, warblers, and tanagers, however, did not differ significantly, or predictably, between 1977 and 1978. Raphael and White found that differences in precipitation explained between 64% and 83% of yearly variation for all cavity-nesting birds (mostly residents). No correlation was found between precipitation and numbers of open-nesting birds (mostly migrants). Therefore, both of these studies (Beedy 1982; Raphael and White 1984) suggested that overwintering mortality might be the variable most affected by weather. Beedy and Raphael and White found that only permanent residents decreased significantly in numbers; long-distance migrants did not. The one long-distance migrant we examined did not decrease significantly in numbers, whereas the two permanent residents and the short-distance migrant did.

Graber and Graber (1979, 1983a) noted that emigration, local movement, and mortality are the most likely explanations for abrupt changes in bird population numbers. Although they found declines in bird populations in both mild and severe winters, the rate of the decline was greater in the severe winter. It is likely that between-year differences for permanent residents in our study were caused by overwinter mortality during the harsh winter of 1982–1983. This appeared to be a regional decline, because the pattern was noted in study sites throughout the true fir forests of the central and southern Sierra Nevada. However, migrant numbers could have been smaller in 1983 than in 1984 for two reasons. The migrants could have been affected by harsh weather in their wintering areas, or they could have stayed downslope instead of moving up into normally occupied breeding sites where snow lasted most of the summer, thus explaining the lower counts in 1983.

The results of the present study and those discussed above illustrate a fundamental problem facing the researcher attempting to interpret species' abundance data from different years: that of determining the relationship between relative abundances of birds and habitat associations. Local abundances vary, and the magnitude of these differences may or may not depend on the quality of the habitat. Järvinen and Väisänen (1979) noted that population fluctuations in optimal habitats tend to be less pronounced than those in suboptimal habitats. Both long-term and short-term population trends need to be considered when assessing species' habitat needs. Biotic and abiotic factors such as weather fluctuations, varying resource bases, philopatry, and competition may alter a species' numbers, even where appropriate habitat exists (Van Horne 1983).

Because of high temporal fluctuations of the populations discussed here, researchers using current wildlife-habitat matrices may fail to detect even common species in seemingly optimal habitats. Apparently, terrestrial bird communities are inherently variable, and assigning them to absolute categories (e.g., optimal, suitable, or marginal) on the basis of estimates of breeding densities will inevitably decrease the accuracy of these matrices. An alternate approach would be to construct confidence intervals from relative frequency data (number of sites in which the species was recorded per total number of sites) to indicate the probability of a particular species' presence in a given habitat. We suggest using the minimum and maximum values obtained in different years for the resulting wildlife-habitat matrices. We constructed such a table for the four species under study for 1983 and 1984 (Table 35.2). These matrices would recognize the uncertainty of species-habitat associations. After we summarize vegetation data from our sites, we plan to create more detailed matrices, which will predict bird numbers in finer categories of habitats than those now employed in wild-

Table 35.2. Confidence intervals of the presence of each species in true fir forests

Species/Year	Estimated CI	Suggested CI, for wildlife-habitat guidelines
Mountain chickadee		
1983	91–100%	91–100%
1984	100%	
Red-breasted nuthatch		
1983	94–100%	94–100%
1984	100%	
Yellow-rumped warbler		
1983	100%	100%
1984	100%	
Western tanager		
1983	91–100%	91–100%
1984	94–100%	

Note: 95% confidence intervals were constructed around the estimated percentages of species' presence within the 51 sites for each year.

life-habitat-relationships programs. These matrices will simulate habitat manipulation and will help wildlife managers make better-informed decisions.

Throughout this chapter we have discussed the variation created in relative abundance of bird species between 2 years with extremely different weather patterns. When we examine a coarser measure of occurrence, that of presence/absence in an area, we still find variability between years in common breeding species. We did not have large enough data sets to examine appropriately other, rare species also recorded during our counts in true fir forests, such as hermit warbler (*Dendroica occidentalis*), fox sparrow (*Passerella iliaca*), purple finch (*Carpodacus purpureus*), and evening grosbeak (*Coccothraustes vespertinus*). We expect even greater variation in both presence/absence and relative abundance data with rare species than with common breeders. We plan to analyze the occurrence of other birds as we obtain more data.

Extremes in weather and other factors (see Van Horne 1983; O'Connor, Chapter 34), therefore, can exert a strong influence on bird numbers, which, as a result, do not necessarily reflect a bird's choice of a habitat. Large, weather-induced variations in population densities reviewed here suggest that few, if any, of these habitats are saturated with birds. We suggest that one should not subjectively assign absolute categories (optimal, marginal, and suitable) to models of species-habitat associations. Rather, confidence intervals of a species' presence in a given habitat or other quantitative measures should be generated from actual field data to indicate the probability of a particular species' presence in a given habitat.

Acknowledgments

This study was funded by the USDA Forest Service. Russell P. Balda and Jared Verner assisted with the study design. Penny Allen, George Banuelos, Dawn Breese, Carol DiGiorgio, Lee Elliott, Nancy Gooch, Ellen Hammond, Zev Labinger, Kathryn Purcell, and Dan Taylor assisted with field work. James Baldwin, Graydon W. Bell, David Sharpnack, and Richard Turek gave statistical advice. Graydon W. Bell, Jeffrey D. Brawn, William S. Gaud, Stephen L. Granholm, John M. Marzluff, Susan Mopper, Peter W. Price, Susan Sanders, Terry A. Vaughan, and Thomas G. Whitham offered valuable comments on the manuscript. To all we express our sincere thanks.

36

Assessing Habitat Quality for Birds Nesting in Fragmented Tallgrass Prairies

RICHARD G. JOHNSON and STANLEY A. TEMPLE

Abstract.—We evaluated whether or not birds in tallgrass prairie fragments were most likely to nest in the habitat type that provided high-quality habitat. We defined habitat types according to the size of the fragment, its proximity to a forest edge, and the number of growing seasons since the vegetation was last burned. We defined habitat quality in terms of the productivity of nests in each habitat type. We developed regression models to identify the habitat characteristics correlated with nest productivity and nest occurrence for five species of tallgrass prairie birds. Nest productivity and, hence, habitat quality for all species were highest in the habitat type containing areas located far from a forest edge with one growing season since the vegetation was last burned. In none of the five species was the probability of nest occurrence greatest in the habitat type we identified as high quality. Management actions based solely on nest occurrences would have favored habitat types with relatively low rates of nest productivity and could have impaired the ability of these species to maintain stable populations.

More than 99% of the presettlement tallgrass prairie in Minnesota has been converted to agriculture and other uses. The areas of native prairie that remain usually occur as small, isolated fragments. Because of this severe reduction and fragmentation, appropriate protection and management of the remaining fragments for conservation purposes is imperative. We studied breeding birds on tallgrass prairie fragments in western Minnesota to determine which habitat characteristics provide high-quality habitat for these species, many of which are relatively dependent on tallgrass prairie habitat (Niemi 1982).

Although habitat quality for nesting birds is measured most reliably by the productivity of nests (i.e., the average number of young fledged from each nest), most studies have assumed that the density of nesting birds can be used as an accurate indicator of habitat quality. Nonetheless, Wiens and Rotenberry (1981a) and Van Horne (1983) identified several high-density situations in which annual survival rates or productivity would be too low for maintenance of a stable population. In these situations, high population densities would not, therefore, be an indicator of habitat quality.

We used the productivity of nests as our primary indicator of habitat quality. We developed models to estimate the two components of nest productivity (i.e., rate of nesting success and the number of young fledged from successful nests) in response to the three characteristics that defined our habitat types (i.e., fragment size, distance to forest edge, and the number of growing seasons since the vegetation was last burned).

For comparison, we also developed models to estimate the probability that a nest of a given species would occur in each habitat type. Finally, we compared our estimates of productivity in different habitat types with previously published estimates of annual adult and juvenile survival rates. In this chapter we discuss whether nest productivity in a particular habitat type is high enough to compensate for annual mortality and, thus, whether a population will be increasing, decreasing, or stable.

Study sites and methods

Field work was done in western Minnesota during the nesting seasons of 1983 and 1984. We searched for nests on eight isolated fragments of native tallgrass prairie located in Becker, Chippewa, Clay, Douglas, Lac Qui Parle, and Pope counties. These study sites, which are managed for conservation purposes, contained relatively undisturbed, dry to mesic, tallgrass prairie vegetation, interspersed with shelterbelts of cottonwood (*Populus deltoides*) and clumps of invading trees such as quaking aspen (*P. tremuloides*) and willows (*Salix* spp.). These fragments are intensively managed by prescribed burning, but only one-half or less of each fragment is usually burned during any one year.

The size of each fragment was categorized as either large (130–486 ha) or small (16–32 ha). Fragments contained prairie vegetation both far from (beyond 45 m) and near to (within 45 m) a forest edge. Three categories were established to codify the number of growing seasons since the vegetation had last been burned: one season, two or three seasons, and four or more.

We used a factorial experimental design (Box et al. 1978)

RICHARD G. JOHNSON: Department of Wildlife Ecology, 1630 Linden Drive, University of Wisconsin–Madison, Madison, Wisconsin 53706. *Present address:* The Nature Conservancy, 1313 Fifth Street S.E., Minneapolis, Minnesota 55414

STANLEY A. TEMPLE: Department of Wildlife Ecology, 1630 Linden Drive, University of Wisconsin–Madison, Madison, Wisconsin 53706

to identify our habitat types. There are 12 possible combinations ($2 \times 2 \times 3$) of size, distance from forest edge, and growing seasons since burn. Fragments were divided along section lines into 63 sample plots of approximately 16.2 ha; three of these sample plots were further divided along firebreaks. Each of the sample plots that we searched was divided into a near unit, consisting of all areas within 45 m of a forest edge, and a far unit, consisting of all areas beyond 45 m of a forest edge. Each near unit was often a combination of several discrete areas, whereas far units were usually contiguous.

We searched 31 sample plots in 1983 (31 far units and 20 near units) and 32 sample plots in 1984 (32 far units and 22 near units). We did not search for nests on sample plots that were burned during the year of the study, nor did we sample plots that contained a high proportion of either exotic vegetation, wet meadow vegetation, or shrubs. On aerial photographs of all sample plots that we searched for nests, we measured the total area of prairie vegetation both far from a forest edge and near to a forest edge. The extent of prairie vegetation in sample plots ranged between 2 and 17.6 ha (median = 14.5 ha).

We searched sample plots for nests between one and six times (median = 3 searches), using a 25- to 30-m rope drag. Each nest was marked with a uniquely numbered flag placed 5 m north of the nest. Nests were usually revisited at 3- or 4-day intervals until the nesting attempt ended. Once all nesting attempts were completed, we returned to each nest to record habitat characteristics.

We analyzed data for the five most common ground-nesting passerines encountered. We located 135 clay-colored sparrow (*Spizella pallida*) nests, 46 savannah sparrow (*Passerculus sandwichensis*) nests, 46 grasshopper sparrow (*Ammodramus savannarum*) nests, 48 bobolink (*Dolichonyx oryzivorus*) nests, and 76 western meadowlark (*Sturnella neglecta*) nests. Nests of these species constituted 351 (48%) of the 728 total nests we found.

Toward the end of a nestling period, it is difficult to determine whether the nestlings have fledged or have been preyed upon. In our study, a nest was considered to have been preyed upon if there were obvious signs that predators were present (e.g., partially consumed young or a drastically disturbed nest cup). If these signs were absent on a revisit and young were not observed near the nest site, we used a subjective evaluation of nestling development made during the previous visit to determine whether the nestlings were sufficiently developed to have fledged.

We used stepwise logistic regression (Engelman 1983) to identify habitat characteristics correlated with nest occurrence and nesting success. We used stepwise multiple regression to identify habitat characteristics correlated with the number of young that fledged from successful nests. In this model, we used dummy variables (Draper and Smith 1981) to code all explanatory variables. In addition to the habitat variables, we also evaluated the effect of several nuisance variables (i.e., species, year of study, total area searched, and number of searches) whenever appropriate.

At each step in the selection procedure for the logistic regression models, the habitat and nuisance variables (and any interactions) were evaluated for entry into or deletion from the current model. Variables with a P-value less than 0.10 were considered for entry into the model. A variable previously entered into the model was removed if its P-value subsequently exceeded 0.15. We considered interactions only if the corresponding main-effect variables were already in the model. The deviance chi-square statistic was used to evaluate the fit of the current model against the (saturated) model that predicts each observation exactly. A large P-value indicated that the explanatory variables in the current model provided an adequate fit to the data. See Kleinbaum et al. (1982), McCullagh and Nelder (1983), or Fox (1984) for a detailed description of logistic regression.

We used nest records from all five species in the nesting success model. Twenty-nine of these 351 nests were eliminated from the analysis because we could not determine their outcomes. The success or failure of each nest (the response variable) was considered an independent observation. Successful nests (n = 115) fledged at least one of their own young, whereas failed nests either were abandoned (n = 19), preyed upon (n = 184), or fledged only brown-headed cowbird (*Molothrus ater*) young (n = 4). We used nest records from the 115 successful nests to estimate the number of young fledged per successful nest. We assumed that a successful nest fledged the number of young present on our visit prior to finding the nesting attempt completed.

In the nest-occurrence models, we analyzed records for each species separately. The presence or absence of a nest in each near or far unit of each sample plot (the response variable) was considered an independent observation. At successful near or far units we found one or more nests of a species, whereas at unsuccessful units no nests were found.

We used the methods of Henny et al. (1970) to calculate the productivity of nests and to determine the annual survival rate of adults and juveniles that would be necessary to maintain a stable population. Because these birds produce only a single brood per nesting season, productivity was defined as the product of nesting success and the average number of young fledged from successful nests. Assuming that fledgling females nest in the following year, the adult and juvenile survival rates (S_a and S_j) necessary to maintain a stable population were calculated as

$$S_a = 0.5[2 - (\text{productivity} \times S_j)].$$

The calculated survival rates necessary for a stable population may overestimate the actual values because to some extent renesting probably occurred.

Results

Nest-productivity model

The rate of nesting success was significantly higher for nests located far from a forest edge and for nests located in recently burned vegetation (Tables 36.1 and 36.2). Because there were no significant interactions between species effects and habitat variables, all species responded similarly to

Table 36.1. Summary of significance levels for the explanatory variables used in each model

Explanatory variable	*P* for nesting-success model	*P* for nest-occurrence model: Clay-colored sparrow	Savannah sparrow	Grasshopper sparrow	Bobolink	Western meadowlark
Size of fragment	ns	0.001	ns	0.033	ns	0.004
Distance to edge	0.067	0.004	ns	0.005	ns	0.054
Years since burn	0.001	ns	ns	0.017	0.016	ns
Species	ns					
Year of study	ns	0.027	ns	ns	ns	ns
Search intensity		0.085	ns	ns	ns	ns
Hectares in unit		ns	ns	ns	0.001	0.004

ns = not significant, and blank indicates variable was not appropriate to the model.

the habitat variables. This model fit our data fairly well (deviance chi-square = 0.707, $df = 3$, $P = 0.872$).

The number of young fledged from a successful nest was significantly lower for savannah sparrow nests and for nests parasitized by cowbirds. The average number of young that fledged from unparasitized savannah sparrow nests was 3.2, whereas 1.3 young fledged from parasitized nests. An average of 3.7 and 2.3 young fledged, respectively, from unparasitized and parasitized nests of species other than savannah sparrows.

In summary, each species achieved its highest rate of nest productivity in the habitat type located far from a forest edge with one growing season since the vegetation was last burned. Habitat types, in decreasing order of nest productivity, are as follows (1) near to edges with one growing season; (2) far from edges with two or three growing seasons; (3) near to edges with two or three growing seasons; (4) far from edges with four or more growing seasons; and (5) near to edges with four or more growing seasons.

Nest-occurrence models

The probability of occurrence of a clay-colored sparrow nest was significantly higher for sample plots on small fragments and for units in sample plots located near to a forest edge. The probability of nest occurrence was also affected by search intensity and year of study (Tables 36.1 and 36.3). This model fit the data fairly well (deviance chi-square = 8.996, $df = 13$, $P = 0.773$).

The probability of occurrence of a savannah sparrow nest was independent of all habitat and nuisance variables. A trend for the probability of nest occurrence to be higher for sample plots on small fragments was not significant.

Table 36.2. Observed and expected probabilities of a nest fledging young, for each combination of distance to forest edge and growing seasons since burn

Distance to edge	Years since burn	No. nests	Probabilities of a nest fledging young[a]: Observed	Predicted ± *SD*
Far	1	59	0.559	0.537 ± 0.052
Far	2 or 3	78	0.385	0.386 ± 0.035
Far	≥ 4	68	0.235	0.254 ± 0.041
Near	1	46	0.391	0.422 ± 0.057
Near	2 or 3	38	0.290	0.283 ± 0.043
Near	≥ 4	33	0.212	0.176 ± 0.041

[a] Predicted probability of fledging young calculated from logistic regression model:

$$P(Y = 1|X) = \exp X/1 + \exp X,$$

in which $X = 0.760 - 0.462\,A - 0.613\,B$; $A = 1$ for near to a forest edge and 0 for far from a forest edge; $B = 1$ for one growing season, 2 for two or three growing seasons, and 3 for four or more growing seasons since a burn.

Table 36.3. Observed and expected probabilities of a clay-colored sparrow nest occurring in units on sample plots that varied in size of fragment, distance to forest edge, intensity of searches, and year of study

No. searches	Year of study	Size of fragment	Distance to edge	No. sample units	Probabilities of a nest occurring in sample unit[a]: Observed	Predicted ± *SD*
1	1983	Large	Far	5	0.000	0.033 ± 0.029
1	1983	Large	Near	3	0.333	0.119 ± 0.091
1	1983	Small	Far	0	—	— —
1	1983	Small	Near	0	—	— —
1	1984	Large	Far	7	0.143	0.091 ± 0.069
1	1984	Large	Near	3	0.000	0.280 ± 0.166
1	1984	Small	Far	0	—	— —
1	1984	Small	Near	0	—	— —
2 or 3	1983	Large	Far	12	0.083	0.139 ± 0.066
2 or 3	1983	Large	Near	4	0.250	0.386 ± 0.133
2 or 3	1983	Small	Far	0	—	— —
2 or 3	1983	Small	Near	0	—	— —
2 or 3	1984	Large	Far	10	0.400	0.318 ± 0.100
2 or 3	1984	Large	Near	7	0.714	0.645 ± 0.119
2 or 3	1984	Small	Far	3	0.667	0.752 ± 0.131
2 or 3	1984	Small	Near	2	1.000	0.922 ± 0.059
≥ 4	1983	Large	Far	9	0.111	0.158 ± 0.069
≥ 4	1983	Large	Near	9	0.556	0.422 ± 0.117
≥ 4	1983	Small	Far	5	0.600	0.550 ± 0.151
≥ 4	1983	Small	Near	4	0.750	0.826 ± 0.096
≥ 4	1984	Large	Far	8	0.375	0.352 ± 0.119
≥ 4	1984	Large	Near	6	0.500	0.679 ± 0.116
≥ 4	1984	Small	Far	4	0.750	0.779 ± 0.108
≥ 4	1984	Small	Near	4	1.000	0.932 ± 0.045

[a] Predicted probability of nest occurrence was calculated from the logistic regression model from Table 36.2, in which $X = -2.289 + A + 1.062B + 1.874C + 1.360D$; $A = -1.077$ for units searched once, 0.463 for units searched two or three times, and 0.614 for units searched four or more times; $B = 1$ for nests located in 1984 and 0 for nests located in 1983; $C = 1$ for small fragments and 0 for large fragments; and $D = 1$ for near to a forest edge and 0 for far from a forest edge.

Table 36.4. Observed and expected probabilities of a grasshopper sparrow nest occurring in units on sample plots that varied in size of fragment, distance to forest edge, and growing seasons since burn

Size of fragment	Distance to edge	Years since burn	No. sample units	Probabilities of a nest occurring in sample unit[a]	
				Observed	Predicted ± *SD*
Large	Far	1	16	0.188	0.204 ± 0.081
Large	Far	2 or 3	16	0.438	0.353 ± 0.070
Large	Far	≥ 4	19	0.474	0.537 ± 0.099
Large	Near	1	10	0.100	0.050 ± 0.034
Large	Near	2 or 3	9	0.000	0.100 ± 0.051
Large	Near	≥ 4	13	0.231	0.191 ± 0.089
Small	Far	1	5	0.000	0.038 ± 0.040
Small	Far	2 or 3	5	0.000	0.077 ± 0.075
Small	Far	≥ 4	2	0.500	0.151 ± 0.141
Small	Near	1	5	0.000	0.008 ± 0.010
Small	Near	2 or 3	5	0.000	0.017 ± 0.019
Small	Near	≥ 4	0	—	— —

[a]Predicted probability of nest occurrence was calculated from the logistic regression model from Table 36.2, in which $X = -2.155 - 1.876A - 1.593B + 0.755C$; $A = 1$ for small fragments and 0 for large fragments; $B = 1$ for near to a forest edge and 0 for far from a forest edge; and $C = 1$ for one growing season, 2 for two or three growing seasons, and 3 for four or more growing seasons since a burn.

Table 36.5. Observed and expected probabilities of a bobolink nest occurring in units on sample plots that varied in growing seasons since burn and hectares searched

Hectares in unit	Years since burn	No. sample units	Probabilities of a nest occurring in sample unit[a]	
			Observed	Predicted ± *SD*
1–2	1	7	0.000	0.001 ± 0.003
1–2	2 or 3	5	0.000	0.000 ± 0.001
1–2	≥ 4	10	0.000	0.000 ± 0.001
3–7	1	13	0.154	0.298 ± 0.101
3–7	2 or 3	14	0.286	0.161 ± 0.060
3–7	≥ 4	11	0.091	0.079 ± 0.047
≥ 8	1	16	0.563	0.522 ± 0.106
≥ 8	2 or 3	16	0.375	0.330 ± 0.074
≥ 8	≥ 4	13	0.077	0.182 ± 0.079

[a]Predicted probability of nest occurrence was calculated from the logistic regression model from Table 36.2, in which $X = -3.393 + A - 0.797B$; $A = -7.615$ for 1–2 ha, 3.335 for 3–7 ha, and 4.280 for ≥ 8 ha; and $B = 1$ for one growing season, 2 for two or three growing seasons, and 3 for four or more growing seasons since a burn.

The probability of occurrence of a grasshopper sparrow nest was significantly higher for sample plots on large fragments, for units in sample plots located far from a forest edge, and for sample plots with four or more growing seasons since the vegetation was last burned (Tables 36.1 and 36.4). This model fit our data fairly well (deviance chi-square = 6.036, $df = 7$, $P = 0.536$).

The probability of occurrence of a bobolink nest was significantly higher for sample plots with one growing season since the vegetation was last burned. The probability of nest occurrence was also affected by the total area searched within a sample plot (Tables 36.1 and 36.5). This model again fit our data fairly well (deviance chi-square = 4.284, $df = 6$, $P = 0.638$).

The probability of occurrence of a western meadowlark nest was significantly higher for sample plots located on large fragments and for units in sample plots located near to a forest edge. The probability of nest occurrence was also affected by the total area searched within a sample plot (Tables 36.1 and 36.6). This model also fit our data well (deviance chi-square = 3.253, $df = 4$, $P = 0.516$).

Discussion

All of the species we studied achieved their highest nest productivity in areas far from a forest edge with one growing season since the vegetation was last burned. Thus, we consider this habitat type to be high-quality habitat for these species. In none of the species we studied was the probability of nest occurrence highest in the habitat type we defined as high quality. The probability of occurrence of a clay-colored sparrow nest or a western meadowlark nest was higher for units in sample plots located near to a forest edge rather than far from it. Although the probability of occurrence of a grasshopper sparrow nest was higher for units in sample plots located far from a forest edge, nest occurrence was also higher for sample plots with four or more growing seasons, rather than one growing season, since the vegetation was last burned. Although the probability of occurrence of a bobolink nest was higher for plots with one growing season since the vegetation was last burned, nest occurrence was independent of distance to a forest edge. Because of these inconsistencies, nest occurrence (and perhaps other

Table 36.6. Observed and expected probabilities of a western meadowlark nest occurring in units on sample plots that varied in size of fragment, distance to forest edge, and hectares searched

Hectares in unit	Size of fragment	Distance to edge	No. sample units	Probabilities of a nest occurring in sample unit[a]	
				Observed	Predicted ± *SD*
1–2	Large	Far	2	0.000	0.088 ± 0.070
1–2	Large	Near	17	0.294	0.272 ± 0.103
1–2	Small	Far	0	—	— —
1–2	Small	Near	3	0.000	0.065 ± 0.049
3–7	Large	Far	9	0.444	0.353 ± 0.141
3–7	Large	Near	15	0.667	0.679 ± 0.108
3–7	Small	Far	7	0.000	0.093 ± 0.065
3–7	Small	Near	7	0.286	0.284 ± 0.130
≥ 8	Large	Far	40	0.600	0.621 ± 0.075
≥ 8	Large	Near	5	0.400	0.234 ± 0.119
≥ 8	Small	Far	0	—	— —
≥ 8	Small	Near	0	—	— —

[a]Predicted probability of nest occurrence was calculated from the logistic regression model from Table 36.2, in which $X = -0.818 + A - 1.677B + 1.357C$; $A = -1.522$ for 1–2 ha, 0.212 for 3–7 ha, and 1.310 for ≥ 8 ha; $B = 1$ for small fragments and 0 for large fragments; and $C = 1$ for near to a forest edge and 0 for far from a forest edge.

Table 36.7. Savannah sparrow productivity and annual survival rates necessary for a stable population in different habitat types

Distance to edge	Years since burn	Annual productivity and survivorship with 100% brood parasitism		Annual productivity and survivorship with no brood parasitism	
		Productivity	Survivorship[a]	Productivity	Survivorship[a]
Far	1	0.69	0.74	1.72	0.54
Far	2 or 3	0.50	0.80	1.23	0.62
Far	≥ 4	0.33	0.86	0.81	0.71
Near	1	0.54	0.79	1.35	0.60
Near	2 or 3	0.37	0.85	0.91	0.69
Near	≥ 4	0.23	0.90	0.56	0.78

[a]Assumes equal juvenile and adult survival rates.

indicators of abundance) cannot substitute for nest productivity in identifying high-quality habitat.

Without information on the actual age-specific mortality rates for these species, it is not possible to know for certain if our high-quality habitat actually supported stable populations. However, annual survival rates for most adult temperate-zone passerines appear to be approximately 40–60%, and juvenile survival is usually lower than adult survival (Ricklefs 1973). If the species we examined followed this general pattern (adult survival ≤ 60% and juvenile survival ≤ 30%), then nest productivity was too low to balance annual mortality for all species in all of the habitat types we studied.

Even if we assumed that juvenile survival equaled adult survival, nest productivity probably was still too low in some habitat types to maintain stable populations. For example, savannah sparrows had the lowest overall productivity of the five species we studied. The habitat type with the highest savannah sparrow nest productivity required an average annual survival rate of 54% without brood parasitism, and 74% if brood parasitism occurred in all the nests. The habitat type with the lowest savannah sparrow nest productivity required an average annual survival rate of 78% without brood parasitism, and 90% if brood parasitism occurred in all the nests (Table 36.7).

If management designed to provide high-quality habitat for these species was based on the occurrence of nests rather than on nest productivity, the resultant decisions would lead to relatively low rates of nest productivity. Whether this lowered productivity would cause the populations of these birds to decline would depend on the actual (but unknown) age-specific annual survival rates for each species. However, if the annual survival rates for these birds are typical of temperate-zone passerines, then many of the habitat types we studied are acting as population sinks, but few are population sources (sensu Wiens and Rotenberry 1981a). On the basis of this research, appropriate management of the remaining fragments of tallgrass prairie for nesting birds appears to entail the provision of prairie fragments that are devoid of forest edges and are frequently burned.

Acknowledgments

Financial assistance was provided by W. Dayton, the Minnesota Department of Natural Resources Nongame Wildlife Program, and the Minnesota chapter of The Nature Conservancy. We received assistance from the staff of both the Department of Natural Resources and The Nature Conservancy, especially G. Barvels, B. Dohlman, M. Heitlinger, M. Kohring, L. Pfannmuller, and J. Weigel. Field assistance was provided by B. Argue, J. Barzen, B. Bell, K. Berigan, C. Bradle, W. Cossette, B. Christensen, M. Dillman, A. Groteluesehen, G. Johnson, N. Johnson, A. Lettenberger, P. McIntyre, C. Miller, M. Minchak, L. Mueller, J. Rode, M. Walker, L. Wright, and A. Zusi. G. Cottam, R. McCabe, and M. Samuel reviewed the manuscript. This chapter is dedicated to G. Johnson.

37

Limitations of Existing Food-Habit Studies in Modeling Wildlife-Habitat Relationships

ERNEST A. GLUESING and DONNA M. FIELD

Abstract.—Forest management changes vegetation structure and plant species diversity. These changes affect wildlife by altering the availability of foods and/or the nutritional or energy content of the diet. Changes in availability, nutrition, and energy of foods affect growth rates of animal populations. These premises were used to assess the existing literature to see to what degree predictions could be made about the effects of forest-management practices on population growth rates of bobcats (*Felis rufus*), northern bobwhites (*Colinus virginianus*), wild turkeys (*Meleagris gallopavo*), and beavers (*Castor canadensis*). A bibliography of 2421 references was compiled. These references showed that general habitat requirements and general population characteristics for each species were well documented as a result of numerous studies on food habits, reproduction, and mortality. However, few studies were designed to look at cause-and-effect relationships. Food-habit studies were primarily listings of the contents of an individual's crop, gizzard, stomach, or feces, and the majority were conducted during the harvest season, not during the entire year. Few studies related food eaten to food available or to nutritional quality, or considered the relationships among diet, energetics, and reproductive output. Given the existing data, predictions about how forest management changes availability, nutritional quality, and energy content of foods, and how these changes affect population dynamics of any of the four species, could only be made in very general terms.

Models for evaluating actual or potential carrying capacity of different habitats often rely on published data to relate habitat variables such as plant foliage diversity and the diversity and density of food-producing plants to an animal's requirements for growth, survival, and reproduction (Fish and Wildlife Service 1981a; Schamberger et al. 1982). These models are not always able to evaluate carrying capacity accurately or to predict abundance successfully. Some of the possible reasons for this lack of success, such as not incorporating the effects of competition or predation, have been discussed by Flather and Hoekstra (1985). In this chapter, we explore a further possible reason: limitations of existing food-habit studies in modeling wildlife-habitat relationships.

The proximate and ultimate factors that regulate populations are poorly understood; however, numerous studies (reviewed by Watson 1970) support a hypothesis that food is a proximate factor and can limit the size and growth rates of populations. Therefore, as a means of assessing current knowledge about the relationships among food, wildlife, and wildlife habitat, we reviewed the literature for four species within the following conceptual framework: (1) forest management changes plant species composition and structural diversity, and this in turn alters the food resource directly for herbivores and indirectly for predators because of particular relationships between plants and animals (see MacArthur and MacArthur 1961; Gentry et al. 1968; Balda 1975b; Golley et al. 1975); (2) every species has a genetically determined minimum nutrient level needed for reproduction, growth, and survival, and a genetically determined ability to convert food resources into energy used for reproduction, growth, and survival; (3) the net energy gain to an individual, i.e., energy that can be partitioned among reproduction, growth, and survival, is the difference between energy expended to obtain food and the caloric content of the food (Hamilton and Watt 1970; Schoener 1971); and (4) if survival consumes most of the available energy, little is left for reproduction, and reproductive output is adjusted accordingly (Selye 1950; Strecker and Emlen 1953; Ferns 1980). The species we chose were a top forest carnivore, the bobcat (*Felis rufus*); two important forest game species, the northern bobwhite (*Colinus virginianus*) and the wild turkey (*Meleagris gallopavo*); and an important furbearer and forest economic species, the beaver (*Castor canadensis*).

Methods

Articles containing information on food habits, reproduction and growth, home range, and habitat requirements were used as the primary sources for assessing knowledge about the biological relationships between each of the four species and its forest habitat. To keep within our conceptual framework that not all foods are equally nutritious or avail-

ERNEST A. GLUESING: Department of Wildlife and Fisheries, P.O. Drawer LW, Mississippi State University, Mississippi State, Mississippi 39762

DONNA M. FIELD: Department of Wildlife and Fisheries, P.O. Drawer LW, Mississippi State University, Mississippi State, Mississippi 39762. *Present address:* Department of Plant Science/Entomology, University of Wyoming, Laramie, Wyoming 82071

able, and as a means of determining why habitat-evaluation models built from published data might not always produce the desired results, we used the following criteria to assess the literature: (1) food-habit studies should contain data not only on what foods are eaten, but also on the relationship between foods eaten and foods available, so that predictions could be made for other habitats in which the availability of specific food items differs; and (2) food-habit studies should contain data on the ability of different foods to meet a species' requirements for growth, reproduction, and survival.

Results

We compiled a bibliography of 603 references on turkeys, 962 on northern bobwhites, 234 on bobcats, and 622 on beavers, of which 107, 276, 34, and 67, respectively, were on food habits, growth, and reproduction.

Important prey items of bobcats, when ranked as a percentage of all diet components, include cottontails (*Sylvilagus* spp.), cotton rats (*Sigmodon hispidus*), white-tailed deer (*Odocoileus virginianus*), voles (*Microtus spp.*), white-footed mice (*Peromyscus* spp.), and gray squirrels (*Sciurus carolinensis*). The importance of these species, however, varied between studies (localities) and between seasons. Presumably this observed variation was, at least in part, a function of prey availability, although only Rollings (1945), Beasom and Moore (1977), and Jones and Smith (1979) studied prey selection by bobcats in relation to varying prey densities.

Landers and Johnson (1976) reviewed the literature on the food habits of northern bobwhites and developed a list of more than 650 plant species that were found in the birds' crops and gizzards. They ranked 45 foods as important because they were repeatedly eaten by northern bobwhites and another 33 that might be important because they contributed as much as 1% of the foods eaten in at least one of the 27 study areas. The abundance of plants, and the extreme variation among study areas, is also likely to be the result of availability and preference, yet we know of only five studies (Robel and Slade 1965; Brunswig and Johnson 1972; Hurst 1972; Robel et al. 1974; Sweeney et al. 1981) that examined both the frequency of foods in bobwhite diets and their availability in the environment.

Knowledge of food habits of turkeys was similar to that of bobwhites. Our review showed that at least 248 plant species were eaten by turkeys. Foods eaten varied substantially by season and study area, but only 26 plants constituted 1% or more of the diet, and only nine constituted 5% or more. Food-habit studies of turkeys, like those of bobwhites, used percent occurrence or volume in the crop, gizzard, or droppings to determine preference, not availability or nutrition.

Food-habit studies for bobwhites and turkeys showed a seasonality related to hunting. Of a random sample of 73 northern bobwhite food-habit studies, 86% were done during the fall and early winter hunting season. Similarly, 71% of the turkey food-habit studies were done during late winter and spring, corresponding to the spring turkey season.

Beavers eat a variety of vegetation ranging from grasses to woody plants. Plants eaten vary geographically, seasonally, and among colonies in the same area (Aleksiuk 1970; Jenkins 1975, 1979; Svendsen 1980; Shipes et al. 1980). Beaver studies occasionally compared plants eaten to plants available; however, except for the work of Jenkins (1976, 1979, 1980), no statistical analyses of plant preferences were made. Selection preferences shown by beavers appear to be related to nutrition and/or taste (Jenkins 1979), tree trunk size (Aldous 1938; Hall 1960; Nixon and Ely 1969; Jenkins 1975), distance from water (Hiner 1938), and the interaction between trunk size and distance from water (Jenkins 1980).

Except for bobwhites, studies of nutrition constituted less than 10% of the food-habit studies for each species. We did not find any study of bobcats that correlated the relationships between growth, reproduction, and survival with diet, nutrition, and energetics or defined them mathematically.

Embryo counts (Huey 1956; Rutherford 1964) and growth rates (Rutherford 1964) of beavers vary among habitats, and these differences are thought to be due in part to differences in available nutrients (Rutherford 1964). However, we did not find any study that mathematically linked data on available nutrition to data on reproduction, growth, or survival.

Many nutritional requirements for growth, survival, and reproduction are known for bobwhites (Nestler 1940, 1946, 1949; Nestler et al. 1942, 1948; Nestler, Bailey, Llewellyn, and Rensberger 1944; Nestler, Bailey, Rensberger, and Benner 1944; DeWitt et al. 1949; Scott et al. 1958; Wilson et al. 1972; Andrews et al. 1973; Serafin 1974) and to a lesser extent for turkeys (Marsden and Martin 1949; Dunkelgod 1961; National Research Council 1971). The nutritional value of some foods eaten by bobwhites and turkeys is also known (Nestler et al. 1949a; Billingsley and Arner 1970; Beck and Beck 1955; Robel et al. 1974; Robel, Bisset, Clement, Dayton, and Morgan 1979; Robel, Bisset, Dayton, and Kemp 1979), and energy requirements for bobwhites were recently studied by Robel (1972), Robel et al. (1974), Robel, Bisset, Clement, Dayton, and Morgan (1979), Robel, Bisset, Dayton, and Kemp (1979), Case (1972), and Case and Robel (1974).

Gluesing and Field (1982) used published data on nutrition and energetics to estimate how well important natural foods of bobwhites and turkeys met the daily minimum nutritional requirements of those animals. Importance was defined in the traditional manner as a high percentage of crop contents, such as percent occurrence, volume, or weight. Our analysis showed that of 24 foods studied, only half provided 80% or more of winter-existence requirements. None of the foods, at least in seed form, provided minimum requirements for egg laying. Although too few data were available to make similar calculations for turkeys, an analysis of protein requirements vs protein available suggested that of 15 plants studied, only four provided minimum requirements for growth or reproduction.

Discussion

Specifically, population growth is a function of the number of individuals in the breeding population and the number of their offspring that survive to become part of the breeding population. The number of individuals in the breeding population, the number of offspring they have, and the survival of these offspring is, in part, a function of each individual's condition. An individual's condition is a function primarily of diet, age, weather, and habitat. Forest management, by altering both the composition and structural diversity of vegetation, can affect the diet as well as the other components of an animal's habitat. Food is an essential component of an animal's habitat because the energy and nutrients it provides are used to meet different seasonal and physiological needs of the animal (Moen 1973). Thus, knowledge not only of what foods are eaten, but also of their availability, selection, and nutrient value for growth, survival, and reproduction is important for understanding and predicting the effects of forest management on an animal population.

The importance of every food item we reviewed varied across habitat types, and by seasons and years within the same habitat type, when importance was calculated as a percent occurrence or volume of the crop, gizzard, or stomach contents. Because of the lack of data about diet selection vs availability, we were unable to calculate regression equations that would have specifically, or at least more specifically, predicted diet contents in an unstudied area where the mix of available diet items differed from that of previously studied areas.

Minimum amounts of protein, minerals, and vitamins are needed for growth, maintenance, and reproduction. Our knowledge of how well available native foods meet these requirements is extremely meagre. For example, mortality as a result of deficiencies in Vitamin A occurs in wild northern bobwhite populations (Nestler and Bailey 1943; Nestler et al. 1949b; Lehman 1953), but at present Vitamin A content is known for only 17 foods eaten by bobwhites. Of these 17 foods, 11 meet minimum requirements for growth and maintenance and only five meet minimum breeding requirements (Gluesing and Field 1982). The content of Vitamin B complex, deficiencies of which can retard growth and feather development and cause leg paralysis, making young individuals more susceptible to predation, is not known for any of the native plants we examined. This general lack of nutritional knowledge, as suggested by our review, is significant and could be one of the reasons why habitat-evaluation models do not always work. The resolution of habitat-evaluation models developed from existing information may not be sensitive enough to detect subtleties of nutrition and energetics that affect population dynamics in a way that consistently produces the desired results. For example, a habitat-evaluation model could show that a particular habitat is good because it contains an ample amount of an animal's requirements, including foods known to be eaten by that animal. However, the habitat could in fact be poor because the nutrient value, rather than the abundance, of the foods in the habitat limits the population. The habitat may, moreover, be missing one or more food items that might provide essential nutrients lacking in available foods.

Land managers who make decisions on land use employ whatever means are available. Habitat-evaluation models are being used with increasing frequency as tools to assist the decision-making process. Our review showed that much general knowledge exists about activities, movements, food habits, and habitat selection for bobcats, beavers, turkeys, and northern bobwhites, but that little specific information was available about cause-and-effect relationships, particularly as these relationships pertain to growth, survival, and reproduction. If land managers need to make specific statements about species' densities or the carrying capacity of a particular habitat and use a habitat-evaluation model developed largely from the literature, they may encounter reliability problems. Reliable specific statements about density, carrying capacity, and maximum sustainable yield require a more exact knowledge of how habitat affects population growth than currently exists. Under the present circumstances, we could ascertain only in a very general way what factors, and in what combination and what order, limited the populations of the four species under review here.

Acknowledgments

This chapter is a contribution of the Mississippi Agricultural and Forestry Experiment Station, Mississippi State, Mississippi. The study was funded by the Mississippi Agricultural and Forestry Experiment Station, Project No. MIS–0608. We thank Dale Arner, Ed Hill, and Frank Miller for their helpful comments and suggestions on earlier versions of the manuscript.

Summary: When Habitats Fail as Predictors—The Manager's Viewpoint

EDWARD F. TOTH and JOHN W. BAGLIEN

As Noss (1983) pointed out in his synthesis of current concepts of wildlife-habitat diversity, researchers and managers must continually interact when addressing the issues, choices, and consequences of land-management decisions. The development and use of wildlife-habitat models play a very important role in the process. The basic premise is that these models, developed largely through research, will increase the ability of managers to judge the effects of habitat-management decisions on particular wildlife species and on the wildlife community as a whole. We believe this premise should not be accepted without due caution.

In many cases, managers have a difficult time determining what models, if any, to use in the decision-making process. Managers do not, as a general rule, possess the technical skills displayed in the many chapters in this volume to evaluate the quality of models available. Nor do we have a clear picture of the limitations that even reliable models have under different conditions in which physical and biological factors vary from those tested. This lack of understanding could very easily lead to the use of faulty models or to the misuse of solids ones. Subsequent failure of models to meet our expectations can lead to a basic distrust of the modeling concept.

Part III has provided an excellent forum to begin examining the less appealing aspects of models that rely solely on habitat variables, specifically, the reasons why they might not work. We hope that the various chapters are taken as a self-critique of the modeling concept to be used in a constructive manner, both to strengthen model development and to avoid the pitfall of overrating or overextending the use of models. The premise that a wildlife-habitat model is a useful management tool is true only when the model is technically sound, well tested, and properly applied.

The chapters in this part have approached the problem in a constructive manner. They have given us a better understanding not only of failure of models due to flaws in design or analysis, but also of factors external to the models that can have an overriding influence. In the first instance, Best and Stauffer (Chapter 30) conclude that supporting studies often fall short in terms of design, duration, geographic scale, the range of habitat variations studied, and appropriate analytical methods applied. They have done an excellent job of showing the level of detail necessary to ensure the validity of study results, as well as the pitfalls of incomplete sampling or data analysis. In light of their findings, we believe the research community, through the peer review process, should play a more active role in ferreting out models with these types of flaws before they reach management.

The dilemma of time as a management constraint was brought out in the chapters by Diehl (Chapter 33) and by Gaud et al. (Chapter 32), who have clearly shown a need for long-term studies. Unfortunately, time and dollar limitations do not at present allow managers to acheive high levels of study detail or to engage in long-term studies. Management decisions are usually based on very generalized data. The yearly demands of assessing the impacts on wildlife of proposed timber sales scattered in 5- to 75-ha patches on a 120,000-ha ranger district of a national forest preclude site-specific study. A wildlife biologist on a district is fortunate to have the opportunity to supplement aerial photo interpretation with a 4-hour walk through a stand proposed for harvest!

The chapters of Diehl and of Gaud et al. also raise another important issue related to management: How can one design a forest-monitoring program for selected wildlife populations to indicate accurately trends in the habitat and wildlife community? These chapters clearly show that for the results to be meaningful, a much more intensive monitoring program of populations is needed than managers currently anticipate. As monitoring programs become an additional part of a forest biologist's workload, the dependence on wildlife-habitat models as tools to predict the impacts of habitat alteration will increase. Faced with this situation, we need to know the level of confidence we can place in them.

The results of the various chapters in this part challenge the wisdom of using a limited number of indicator species to monitor overall, long-term trends in a wildlife community. This strengthens the position of Verner (1983) in favor of an integrated monitoring program involving specific species, guilds, and habitat trends.

The chapters by Rotenberry (Chapter 31) and by Hejl and Beedy (Chapter 35) show that models can have serious limitations because of stochastic variations in populations and climatic influences. However, we believe that Hejl and Beedy's study was weakened by linking the results with the use of species-habitat matrices. Even biological factors such as interspecific competition, intraspecific competition, and predation can have a major effect on wildlife populations,

EDWARD F. TOTH: USDA Forest Service, Six Rivers National Forest, 507 F Street, Eureka, California 95501

JOHN W. BAGLIEN: USDA Forest Service, Wallowa-Whitman National Forest, P.O. Box 907, Baker, California 97814

apart from influences of the habitat as suggested by Diehl in Chapter 33. These results should caution managers that the provision of suitable habitat for wildlife does not necessarily mean that population objectives will be achieved. Nor does the loss of a population of some featured species necessarily mean that efforts to manage its habitat have failed or that continuation of those efforts is unwarranted. On the other hand, findings by Johnson and Temple (Chapter 36) point out the need for periodic reevaluation of habitat-management programs to ensure that their emphasis has not been misdirected because of faulty assumptions.

Several of the chapters in this part appear to be validation studies to develop data bases for wildlife-habitat relationships. As will be discussed in depth in Parts IV and V, managers have a pressing need for models that can predict responses of wildlife to perturbations in natural systems over time—responses to type, amount, size, and juxtaposition of community types and successional stages. O'Connor (Chapter 34) tested the Brown-Fretwell-Lucas models, showing changes in the patterns of habitat use by birds through time. His chapter focuses on a major problem facing managers who must deal with maintenance of population viability for species whose abundances will significantly decline as a direct result of human activities, as with the spotted owl (*Strix occidentalis*) (Carrier et al. 1985). As O'Connor's chapter points out, there is a need to accurately identify and manage esential core habitat for species of management interest, particularly if viability is at risk. Because of the current level of precision of models as tools for prediction, heavy reliance on them for making viability or habitat-management decisions appears to be unwise. Direct monitoring of populations and habitats is still the best choice.

This leads us to venture a comment about research priorities. It should not be a surprise to anyone that large variations in numbers occur among species that are habitat generalists and have high reproductive and mortality rates. At what point do we stop trying to account for the last increment of variation in habitat used by a species known to breed successfully in a variety of habitats and to exhibit extreme population fluctuations? This thought also extends to some of the more popular habitat specialists as well. It does not, however, discount the work by those such as Raphael and Marcot (Chapter 21 in Part I); such studies are still crucially needed to build more accurate, overall models of species-habitat relationships. This is particular apparent when we consider Gluesing and Field's (Chapter 37) point that we have such an incomplete picture of the cause-and-effect relationships between habitat modifications and wildlife energetics.

Researchers and managers must continually reevaluate programs to identify and implement high-priority research items. Realizing that many researchers feel that their studies need not necessarily be in response to management priorities, we should not always expect to find a consensus about the priorities given to research projects. Also, because research is usually a long-term venture, managers must temper their expectations with patience in their search for answers. Once a research project is started, it is extremely important that it produce a final package, including complete analysis of the data generated. Both management and research are sometimes guilty of losing interest at the most critical point, data analysis. With the budget constraints we all face, it is important to choose projects wisely, carry them out thoroughly, and integrate them with other past and ongoing research projects to the extent feasible.

This part clearly indicates the need for constant, stringent assessment of wildlife-habitat models to detect technical weaknesses before the models are passed on to management. In addition, model developers must provide managers with criteria for determining model applications and limitations. Managers, after all, must know when to be cautious in the choice, use, and analysis of results.

On the basis of the studies presented in Part III, we conclude that managers should use wildlife-habitat models in the decision-making process, but they must be cautious not to overestimate the accuracy of those models. All indications are that models are and will become increasingly important in that process, and it is to be hoped, for the better. To improve the effectiveness of models, the researcher-manager team must continually reevaluate the standards for model design and application. Managers should continue to participate in the selection of research priorities, parameters to be measured, and determination of the level of acceptable accuracy. Research studies should be of sufficient duration and scope, as well as being thoroughly analyzed, to provide meaningful information for model development. Resulting models must be sufficiently generalized or translated into understandable terms so that managers can apply the information and analytical processes as they become available. Managers, for their part, must make a concentrated effort to keep informed about the current state-of-the-art knowledge in wildlife research and modeling to make the job of technology transfer easier. Part III of this volume is a positive step in that direction.

Summary: When Habitats Fail as Predictors— The Researcher's Viewpoint

BEATRICE VAN HORNE

A common premise underlying the chapters in this volume is that studies of animal-habitat relationships should permit managers to predict the effects of habitat alteration on a species or group of species. Yet contributors disagree about the primary obstacles to achieving predictability. Thus, they have identified a range of "considerations," but no consensus exists regarding the direction that future studies, and associated habitat models, should take. One reason for this lack of consensus is that the two very different ways of increasing levels of predictability appear to be contradictory, and each has vastly different implications for how studies of species-habitat relationships should be done. On the one hand, one can improve predictability by conducting more studies over more sites and over longer periods of time, that is, by making studies more extensive. Such extensive studies are generally based on correlating species' densities with habitat features. On the other hand, one can increase predictability by looking very closely at processes that occur in a relatively small area, that is, by making studies more intensive. Intensive studies generally involve investigation of processes that influence survival and productivity, with the assumption that if these processes are understood, the effects of habitat alteration can be predicted.

What problems does an investigator face in producing strong and predictable correlations in an extensive study? According to results presented by Gaud et al. (Chapter 32) and by Diehl (Chapter 33), the study must be done over several years and must include the extremes of habitat occupied. O'Connor's (Chapter 34) observation that species-habitat relationships can change with species' density bolsters this argument, particularly in light of Hejl and Beedy's (Chapter 35) demonstration of yearly density variation resulting from weather patterns. In addition, Gaud et al. point out that the variables relevant to the species must be measured with the correct degree of precision or specificity. Obviously, the difficulty of doing this will increase as an increasing number of species are included in the study.

Best and Stauffer (Chapter 30) and O'Connor (Chapter 34) argue that one should pay attention to nonlinear relationships—something we all seem to agree with but generally prefer to ignore because of statistical problems. Best and Stauffer also emphasize the importance of sampling not only the extremes of habitat gradients, but also an adequate number of points along those habitat gradients. Finally, they identify the closely related problem of scale, illustrating that correlations between species and habitat variables can change markedly, depending on both the range of the gradient over which such correlations are measured and how finely the gradient is divided (how sampling plots are lumped).

Rotenberry's (Chapter 31) use of response surfaces to describe species-habitat relationships shows how a close fitting of species-habitat relationships, one that describes and allows for the nonlinearities in these relationships that we know exist, can lead to a loss in predictability because too much attention is paid to "noise." In a nonlinear system, noise can cause much larger changes in the pattern than in a linear system. Perhaps this is another reason why researchers so often rely entirely on describing linear relationships; assuming linearity is one way (although perhaps not the best way) to ignore such biological noise.

What, then, can one conclude about these extensive studies? One must sample often and over a wide area to swamp out the noise in the system, and, as a consequence, one can describe habitat relationships only in a very general way. Massive correlations are required; specificity leads to a loss in predictability. Such a notion, however, is really contrary to the basic tenets of the scientific method. Ideally, a researcher should develop a hypothesis about processes governing species-habitat relationships and then test that hypothesis rigorously. If the hypothesis is falsified, it must be reformulated or abandoned. In extensive studies, one normally assumes that such processes will become evident in the general pattern of species-habitat correlations.

In intensive studies, clear hypotheses regarding the importance of processes regulating species-habitat relationships can be tested with strong inference or experimental work. Diehl's (Chapter 33) work represents careful data collection over a long period of time, and the stage is set for further intensive work to demonstrate the mechanism of competition.

This very basic difference in outlook represented by those who conduct intensive or extensive studies—i.e., by those who would achieve predictability by going in completely opposite directions—is in part responsible for a split among community ecologists. One group, employing the extensive approach, looks for evidence of competition or other processes, using broad-scale biogeographic patterns. The other group sees no evidence for a process such as competition unless the mechanism by which it occurs can be identified and tested experimentally—an intensive approach. Community ecology as a discipline is moving toward more experi-

BEATRICE VAN HORNE: Department of Biology, University of New Mexico, Albuquerque, New Mexico 87131

mental studies and studies incorporating population biology. This reflects an increasing dissatisfaction with the more extensive approach. At the same time, urgency and impatience among managers, which have been caused by reduced budgets and critical environmental problems, have led to an increasing emphasis on quicker, more general, and more pattern-oriented studies—the type I have labeled "extensive." Thus management- and research-oriented investigators are moving in opposite directions. This divergence is often accompanied by mutual disregard of each other's approaches, thinking, and findings. More cross-fertilization between these approaches is needed.

I believe that some of the problems inherent in extensive studies are insoluble without a certain amount of intensive study, no matter how many plots or years are sampled. To achieve predictability, researchers must somehow identify which habitats among the range available are of highest quality. Habitat "quality" should best be measured by the reproduction and survival rates of a species in a given habitat, relative to other habitats (see Van Horne 1983). High-quality habitats are those in which a species survives and reproduces relatively well. Most extensive studies fail to measure survival or reproduction, using instead some measure of abundance as an indicator of habitat quality. Abundance, however, may not always be a good measure of habitat quality (Van Horne 1983). The disassociation of density and habitat quality occurs for several reasons. First, territory packing can lead to spatial saturation in a series of habitats that differ in quality. Animals on these territories may differ greatly in survival and reproductive rates. Second, during years when species' density is high, animals may achieve very high densities in low-quality habitats where survival and production are low because offspring have been forced into these habitats by intraspecific competition. I have observed this phenomenon in my own work on small mammals in Southeast Alaska (Van Horne 1982), and it is shown very clearly by O'Connor's (Chapter 34) synthesis of data for birds in England. Such a phenomenon should be observable in all territorial species from time to time. The intensive work represented by O'Connor's measures of clutch size, egg success, and chick success is a good example of how information on survival- and productivity-related processes occurring in species-habitat relationships can greatly strengthen the interpretation of more extensive information represented by surveys relying on abundance. In another demonstration that density (of nests, in this case) may not be coupled with survival and productivity, Johnson and Temple (Chapter 36) show that productivity of bird species in the tallgrass prairie habitats is not correlated with nest density. Habitats with high nest density may have low productivity associated with poor nestling survival rates.

Thus, intensive studies can shed light on the nature of the relationship between density and habitat quality in a manner that is extremely useful in the interpretation of extensive work. Similarly, some of the other problems with extensive studies, such as those raised by Gaud et al. (Chapter 32) and by Best and Stauffer (Chapter 30), can best be addressed through intensive work. Best and Stauffer have provided few guidelines for deciding what is adequate sampling. Perhaps it is only by using intensive studies to identify key processes that influence survival and productivity in different habitats that useful extrapolation and interpolation of existing data on habitat occupancy can be done. Hejl and Beedy's contribution (Chapter 35) shows that the influence of weather on density may vary greatly among years. I think it is only by knowing how this influence occurs that we will know how and when to include weather effects in habitat models.

Several contributors have emphasized the importance of identifying the relevant variables in extensive studies. It is not good form to "go fishing" for relevant variables. The more variables one measures, the more difficult it is to find strong relationships, because one loses degrees of freedom, and researchers are constantly faced with reducing variable sets. It is more straightforward to identify a priori those habitat variables that are likely to be biologically relevant. By establishing which habitat variables are biologically relevant, an intensive study could provide a sound foundation for selecting the variables to be measured in an extensive study.

Intensive studies thus have the potential to provide an understanding of the processes governing species-habitat relationships that is not available from survey data. A habitat-density correlation based on survey data only reveals a pattern and says nothing about what is causing the pattern. This is usually inferred, but such inferences may be risky, especially when they are founded on inadequate, sketchy, or incorrect knowledge of the biology of the species. Intensive studies are necessary to determine why a given correlation pattern occurs, or why it may not accurately reflect the requirements of the species.

The essence of useful management guidelines is predictability: the ability to make long-range estimates of the effects of habitat manipulations on a species. Without some understanding of the processes mediating observed patterns of habitat occupancy, predictability will remain low and the data describing habitat relationships will remain site-specific and be applicable to describing habitat relationships at only a rather gross scale. On the other hand, extensive studies provide a necessary context for the intensive work. Thus I recommend that we increase the breadth of habitats and the duration of extensive studies, and that these studies be explicitly coordinated with intensive studies that include not just density-habitat correlations, but also independent investigations of processes influencing survival and productivity, such as those associated with food habits and mating patterns. We should eliminate those "middle ground" studies that do not include a large enough range of habitat types over a long enough time frame to be useful as extensive studies, or that do not investigate processes behind the observed patterns in sufficient depth to qualify as useful intensive studies.

IV

Predicting Effects of Habitat Patchiness and Fragmentation

Introduction: Predicting Effects of Habitat Patchiness and Fragmentation

STANLEY A. TEMPLE and BRUCE A. WILCOX

As we approach the year 2000, habitat fragmentation looms on the horizon as perhaps the single most significant challenge to the development of models applicable to wildlife management, if not ultimately to the survival of wildlife altogether. Although habitat fragmentation traditionally has not been a consideration in the study of wildlife-habitat requirements, neither has habitat fragmentation occurred on the enormous scale witnessed at present. Given this unfortunate trend, it is difficult to imagine how any model of wildlife-habitat relationships could long have much practical value outside this context.

Habitat fragmentation is the universal mode of habitat disturbance accompanying extensive land use characteristic of a growing society. As housing, industry, agriculture, roads, freeways, pipelines, and all the accoutrements of modern civilization spread outward from metropolitan centers, continuous tracts of natural landscape become increasingly fragmented by a network of developed land. Although habitat is destroyed outright in the process of fragmentation, it is not just the loss of habitat per se that is at issue. Rather, it is an insidious process that ensues in the remaining habitat, exhibited as a syndrome of effects ranging from the extinction of "area sensitive" large mammals (Picton 1979) to the decline of "edge sensitive" birds (Ambuel and Temple 1982; Brittingham and Temple 1983; Wilcove and Whitcomb 1983). Thus even wildlife in "protected" areas—national parks, national forests, wildlife refuges, and the like—is not immune from the effects of habitat fragmentation. Protected areas are becoming ecological islands in a sea of development and, as such, are subject to a wide range of threats to their integrity (see Wilcox 1980a; Diamond and May 1982). Furthermore, areas protected or managed for wildlife are subject to internal fragmentation, because of habitat conversion associated with timber harvesting, petroleum and mineral development, and development to support recreation. That the dominant feature of the future natural landscape will be its fragmented nature is inescapable.

The essence of this problem, from the standpoint of modeling wildlife-habitat relationships, is that to have practical utility models must do more than just reflect the immediate habitat requirements of a species (or a guild or community). Models must also consider the size, shape, proximity, and spatial arrangement of fragments of natural landscape containing other essential habitat components. Although these criteria are only partially known for a handful of species, the survival of many species—perhaps most species—ultimately depends on the extent to which such criteria can be determined, and can thus provide the basis for wildlife-management programs.

The impetus for concern about the effects of fragmentation comes largely from studies of true islands and from natural or human-created fragments of continental habitat. Island biogeographers have long known that islands characteristically have biotas that are depauperate in comparison to those of similar mainland areas. The essentially universal tendency for smaller islands to have fewer species (the "area effect") and the frequent tendency for more isolated islands to have fewer species (the "distance effect") undoubtedly prompted MacArthur and Wilson (1967) to state in the opening pages of *The Theory of Island Biogeography* that the "same principles apply, and will apply to an accelerating extent in the future, to formerly continuous natural habitats now being broken up by the encroachment of civilization." Although some of the details of their theory have been the subject of dispute, this consideration should not detract from the value of the island paradigm as a useful conceptual framework for developing practical strategies for mitigating the effects of habitat fragmentation (see Temple 1981; Harris 1984).

Some of the most relevant implications of island biogeographic theory to conservation are often overlooked in the midst of debate about the validity of the equilibrium theory (e.g., Simberloff 1976; Simberloff and Abele 1982). The underlying tenets of island biogeographic theory predict the equilibrium condition as only one possible outcome of immigration-extinction dynamics. The loss and insularization of habitat that accompany fragmentation produce a non-equilibrium, supersaturated condition in which extinction rates exceed immigration rates. The ensuing "relaxation" results in a steady decline of species diversity that, judging from studies of land-bridge islands (Diamond 1972; Terborgh 1975; Wilcox 1978), involves the loss of numerous species over a protracted period of time.

The relaxation process has been followed for several decades on at least one artificially created island preserve, Barro Colorado, in Lake Gatun, Panama. This hilltop of lowland tropical forest, marooned by the flooding of Lake Gatun in 1914, has lost since its isolation more than 30% of its original avifauna of approximately 235 bird species (Willis 1974; Karr 1982). In another ongoing study, tracts of forest in Brazil were censused immediately before fragmentation and again afterwards. In less than 4 years of such moni-

STANLEY A. TEMPLE: Department of Wildlife Ecology, 1630 Linden Drive, University of Wisconsin–Madison, Madison, Wisconsin 53706

BRUCE A. WILCOX: Center for Conservation Biology, Department of Biological Sciences, Stanford University, Stanford, California 94305

toring, early signs of ecosystem decay are evident (Lovejoy et al. 1984).

Many of the same effects are apparent from the growing research on existing temperate ecosystem fragments (Burgess and Sharpe 1981). Yet fragmentation studies have been more documentary than prescriptive, more basic than applied. If this research has revealed anything of practical value so far, it is that little time can be wasted in correcting this condition. We need to have models to predict the consequences of habitat fragmentation for various taxa and communities. From the standpoint of the land manager, the size, shape, and dispersion of tracts of habitat required to minimize the effects of fragmentation can often be relatively straightforward to control. The proper guidance is lacking, however, and it is up to researchers to provide it.

38

Effects of Forest Fragmentation on Vertebrates in Douglas-fir Forests

KENNETH V. ROSENBERG and MARTIN G. RAPHAEL

Abstract.—In this study, most terrestrial vertebrates were inventoried in Douglas-fir (*Pseudotsuga menziesii*) forests in northwestern California during 1981–1983. Using relative abundance or frequency data from 46 forest stands and 136 study plots within those stands, we tested for associations among vertebrate species and seven measures of forest fragmentation, including stand area, insularity, and proximity to adjacent clearcuts or pure hardwood patches. Fragmentation was assessed on three hierarchical scales: plot, stand, and 1000-ha block surrounding each stand. Stands ranged from complete islands (100% insularity) to continuous forest, and half the plots contained some clearcut edge.

Bird and amphibian species richness increased significantly in more fragmented stands and in plots containing more edge. Twenty bird species were detected relatively frequently ($\geq$10%) along plot edges during variable circular-plot counts, and another 20 species were detected rarely on edges ($\geq$1%). Partial correlation analyses and analyses of variance among five classes of stand area and insularity indicated that relatively few species exhibited negative responses to forest fragmentation. Those showing greatest sensitivity included fisher (*Martes pennanti*), gray fox (*Urocyon cinereoargenteus*), spotted owl (*Strix occidentalis*), and pileated woodpecker (*Dryocopus pileatus*), with ringtail (*Bassariscus astutus*), northern flying squirrel (*Glaucomys sabrinus*), sharp-shinned hawk (*Accipiter striatus*), blue grouse (*Dendragapus obscurus*), and Pacific giant salamander (*Dicamptodon ensatus*) also of potential concern.

Recentness of fragmentation in Douglas-fir forests precludes making definite conclusions about species' tolerances, and we will probably see further changes as the break-up of this habitat continues. However, we suggest the incorporation of conservative minimum stand-size characteristics (i.e., >20 ha) into a current working definition of old-growth forest.

Perhaps the most critical problem facing forest wildlife, worldwide, is the systematic shrinking and fragmentation of their habitat. Using the analogy of forest patches as biogeographic islands, a growing body of theory has been developed to predict the diversity, colonization, and extinction of species in such island situations (MacArthur and Wilson 1963, 1967; Simberloff 1974; Strong 1979). The effects of newly created edges or ecotones on species composition are superimposed on questions of patch size and configuration. Whereas these edge effects are well documented in many forest systems, an empirical data base that links island biogeographic theory to applied ecosystem conservation is only in its infancy.

KENNETH V. ROSENBERG: USDA Forest Service, Redwood Sciences Laboratory, 1700 Bayview Drive, Arcata, California 95521. *Present address:* Museum of Zoology, Louisiana State University, Baton Rouge, Louisiana 70803

MARTIN G. RAPHAEL: Department of Forestry and Resource Management, 145 Mulford Hall, University of California, Berkeley, California 94720. *Present address:* USDA Forest Service, Rocky Mountain Forest and Range Experiment Station, 222 South 22nd Street, Laramie, Wyoming 82070

Perhaps the best examples of faunal studies in fragmented forests come from the deciduous forests of eastern North America where the composition, population trends, and life history correlates of bird species have been examined extensively on local (Forman et al. 1976; Galli et al. 1976; Whitcomb et al. 1977) and regional scales (Whitcomb et al. 1981). An understanding of avifaunal response to rapidly diminishing tropical forests is also beginning to emerge (Lovejoy 1975; Terborgh 1975; Morton 1978; Keast and Morton 1980). However, studies of other taxa and in other types of forests (e.g., coniferous) have been generally lacking.

In the Pacific Northwest, the rate of depletion and fragmentation of commercially harvestable Douglas-fir (*Pseudotsuga menziesii*) forests rivals that in any other forest type on earth. In roughly 30 years, nearly half of the mature and old-growth Douglas-fir forests in northwestern California (about 400,000 ha) have been clearcut, with an even larger proportion cut over a longer time period in western Oregon and Washington. In most areas logging has resulted in the rapid conversion of continuous forest to a mosaic of forest islands, corridors, and early-successional brushfields. Where forest and clearcut abut, the edge is abrupt and without ecotonal development.

Concomitant with increased timber harvesting on public

land, a preliminary inventory of most terrestrial vertebrates in the forests of northern California has been completed (Raphael and Barrett 1981; Raphael 1984). In this chapter we investigate distribution and abundance patterns of species in relation to forest patch size, insularity, and edge characteristics. Our specific objectives are to (1) present a scheme for measuring forest fragmentation in relation to vertebrate populations; (2) discuss the interaction of man-induced fragmentation with natural environmental gradients such as elevation, plant succession, and moisture; and (3) identify vertebrate species most sensitive to the fragmentation of Douglas-fir forests.

Methods

Study area

We selected 46 forest stands ranging in size from 5 to >300 ha, all within the Six Rivers, Klamath, and Shasta-Trinity national forests in northwestern California. A stand was defined as a continuous area of similarly classified forest (age, stocking level) as delineated by USDA Forest Service timber-type maps (1975–1978). All stands selected were dominated by Douglas-fir with varying amounts of tanoak (*Lithocarpus densiflorus*), Pacific madrone (*Arbutus menziesii*), and canyon live oak (*Quercus chrysolepis*). Nearly all study stands were bordered in part by recent clearcuts, forming abrupt, high-contrast edges. Thus, our study area was a web of interconnected forest patches (see Fig. 38.1); only four stands were true islands, and none was isolated from nearby forest.

Within each stand, we randomly located from one to five circular 10-ha plots (180-m radius), whose centers were the focus of all vegetation and vertebrate sampling (Fig. 38.1). These 136 sample plots varied in elevation from 427 to 1220 m, averaging 838 m. Because of changes in topography within stands, the plots differed in their slope exposure and ranged in estimated age from 55 to 370 years. Associated with these gradients of stand age and elevation were changes in moisture regime, understory plant composition, and forest floor structure (e.g., downed logs). In general, most clearcutting has been on ridges where old-growth forests predominate; thus, many fragmentation measures were correlated with stand age. At present, most of the extensive and unfragmented stands are in younger forests with a large component of noncommercial hardwoods, especially oaks.

Fragmentation measures

We measured forest fragmentation on three scales after locating each stand and plot on timber-type maps and aerial photographs (1980 series). In several instances, these measurements were modified to reflect very recent clearcuts. From each plot center we determined the distance to the nearest clearcut and pure hardwood edge, as well as the length of each edge that was within each 10-ha circle. For each study stand, we defined insularity as the percentage of its perimeter that was clearcut edge and measured the total contiguous area of adjacent land that was cleared or pure hardwood.

To account for the possible influence of forest-fragmentation patterns in a larger area surrounding our study stands, we centered a 1000-ha square around each stand (Fig. 38.1). Although vertebrate populations were not sampled throughout these 1000-ha blocks, this area approximated the home range size of the spotted owl and fisher, two species of concern because of their association with old-growth timber. Using a standard map-area grid, we determined the percentage of each 1000-ha square that was either clearcut (designated as "D1" on Forest Service maps) or old-growth forest (D4, D5, D6), and the total linear distance of forest/clearcut edge.

Vertebrate inventory

Our vertebrate sampling methods have been described in detail by Raphael and Barrett (1984), Raphael and Rosenberg (1983), and Raphael (1984). Data from three general techniques were used in our fragmentation analysis. First, we counted birds and diurnal squirrels using variable circular-plot (VCP) counts. We conducted 12 such 10-min counts at each plot center each spring (May–June) in 1981–1983. For each animal detected along the edge of a stand, we recorded the edge type (e.g., clearcut, hardwood) as determined with a field map and compass. Second, we employed an array of 10 pitfall traps at each sample plot to capture small mammals, reptiles, and amphibians. These traps were run nearly continually and were checked at varying intervals (usually 1–3 weeks) throughout the 3-year period. Third, we used sooted aluminum track stations to determine presence of carnivores and other mammals. Each plot was sampled during a 10-day period in August–September each year. In addition, nocturnal surveys were conducted during spring 1981 and 1982 to detect presence of owls in each stand.

Analytical design

Because we found many vertebrate species to be strongly associated with gradients of stand age and elevation, and because most fragmentation measures were also correlated with these gradients, our analyses were designed to separate the effects of natural and man-induced factors. We therefore used partial correlation analysis to measure the association of species with fragmentation variables, simultaneously controlling for the effects of stand age and elevation. We used relative abundance data (number/VCP count or number/1000 pitfall trap-nights) for these analyses, except when only presence data were available (owls, large mammals).

For comparisons of plot-level measures, relative abundance or presence data for each plot were considered separately ($n = 136$). For stand or 1000-ha block-level measures, vertebrate data were averaged for all plots within each stand ($n = 46$).

Finally, to assess possible limits to the distribution of species, we compared relative abundances or frequencies of occurrence among five classes each of stand size and insularity. We used a one-way analysis of variance to test the

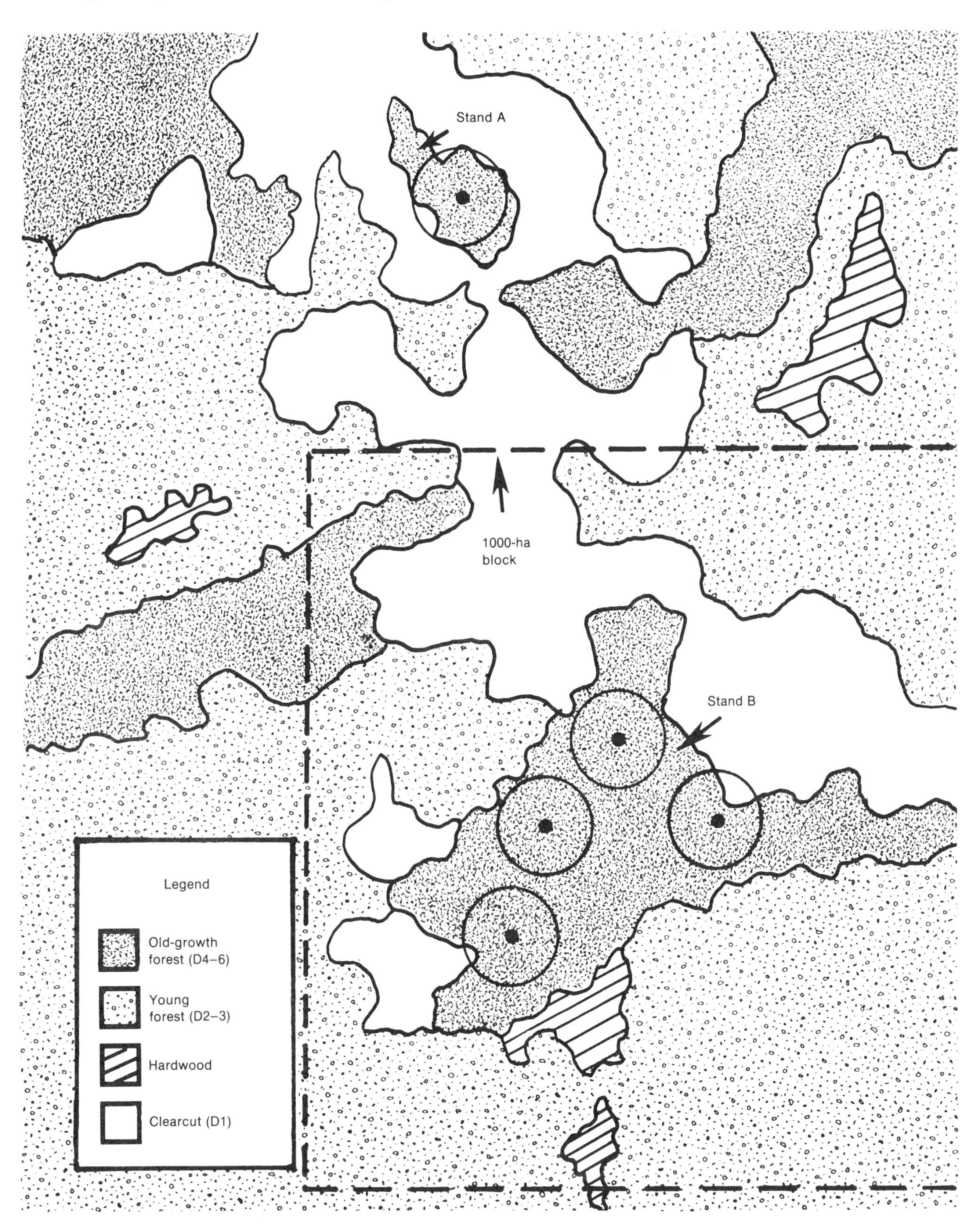

Figure 38.1. One of the study area configurations in fragmented forest of northwestern California. Circles represent 180-m radius plots within study stands. Fragmentation was measured in relation to study plots, stands, and within 1000-ha blocks (see text).

significance of increases or decreases in abundance among the stand classes. Visual inspection of actual distributions helped to suggest tolerance limits and identified additional species of potential concern. A crude tolerance index was computed as the ratio of a species' abundance or frequency in the smallest or most insular class to that in the largest or least insular class (Whitcomb et al. 1981:148).

Results

Forest configuration

Stands ranged from complete islands (100% insularity) to continuous forest, and half of the plots included some clearcut edge within the circular boundary (Table 38.1). In general, 1000-ha blocks of land with many clearcuts also contained much old-growth forest and much edge. Thus many of these measures were correlated.

Stands of pure hardwoods were relatively infrequent and rarely contributed much to the amount of edge present. Furthermore, these hardwood-edge measures did not necessarily represent the influence of hardwoods on the vertebrate community, as many Douglas-fir stands contained much oak or madrone or both in their understory.

The vertebrate community

Of 101 species of forest birds detected on spring counts, 44 occurred frequently enough (>10% of plots) to be included in correlation analyses. In addition, of five owl species recorded on nocturnal surveys, only spotted owl occurrence is considered here. About 32 mammal species occurred on the plots, but 14 were rare and were not included here. In addition, we found seven salamanders (three were rare), two anurans, and five lizards; none of the 10 snake species found was recorded on enough plots to allow statistical comparisons.

Table 38.1. Measures of fragmentation of Douglas-fir forests in northwestern California

Fragmentation measure	n	Mean	Minimum	Maximum
Distance to clearcut (m)	136	356	30	2232
Distance to hardwoods (m)	136	794	40	2115
Length of clearcut edge (m)	136	178	0	1045
Length of hardwood edge (m)	136	19	0	385
Insularity (%)	46	39	0	100
Area of adjacent clearcut (ha)	46	53	0	200
Area of adjacent hardwoods (ha)	46	22	0	257
Clearcut in 1000-ha block (%)	46	18	0	44
Old-growth in 1000-ha block (%)	46	39	0	81
Total edge in 1000-ha block (km)	46	18	0	39

Bird species richness in each plot or stand was positively correlated with the proximity and extent of adjacent clearcuts and was higher in more fragmented stands (Table 38.2). Among the other animal groups, number of amphibian species was also higher in plots with nearby clearcut edges.

Associations with edges

Of about 70,000 detections of birds and diurnal squirrels on VCP counts, 4% were on the edges of plots. Twenty species of birds were detected along edges at least 10% of the time (Table 38.3). Another 20 species were very rarely detected on an edge (≤1%). The olive-sided flycatcher was the only common species that was detected more often on an edge than in the forest interior. Because edges were usually >100 m from the counting points, these numbers uniformly underrepresented the frequency of occurrence of most species along the edge. However, because pairs of species with similar detectabilities (e.g., house wren vs winter wren, dusky vs Hammond's flycatcher, white-breasted vs red-breasted nuthatch) exhibited opposite tendencies, we believe these percentages constitute a valid index to the relative associations of species to edges.

Among the three squirrels, western gray squirrels were never found along edges. Douglas' squirrels and Allen's chipmunks were rarely detected there (1–2%); however, we found chipmunks to be common within brushy clearcuts.

With the effects of stand age and elevation controlled, partial correlation analysis revealed a pattern of responses to fragmentation among Douglas-fir vertebrates (Table 38.4). In plot-level comparisons, relative abundance of 11 bird species, as well as Allen's chipmunk, deer mouse, and ensatina, increased with proximity and length of a clearcut edge. Eight other bird species showed an opposite pattern of abundance, as did the frequency of occurrence of ringtail, fisher, and striped skunk.

When whole stands were considered, abundance patterns among birds were essentially similar to those described

Table 38.2. Correlation of measures of species richness and fragmentation in Douglas-fir forests

	Plot (n = 136)		Stand (n = 46)			1000-ha block (n = 46)	
Species group	Distance to clearcut	Length of edge	Area	Insular index	Adjacent hardwood	Percent clearcut	Total edge
Birds	− − −	+ + +	ns	+ + +	ns	+ + +	+ + +
Mammals	ns	ns	ns	ns	ns	+	+
Amphibians	− − −	+ + +	ns	+	ns	+	+ + +
Reptiles	ns	ns	ns	ns	ns	−	ns

Note: Partial correlation analysis; ns = not significant; + or − = 0.10 > P > 0.05; + + or − − = 0.05 > P > 0.01; + + + or − − − = 0.01 > P.

Table 38.3. Bird and squirrel species detected on edges of Douglas-fir forest plots during variable circular-plot counts

	Number detected	Percent on edge
Species associated with edges		
Fox sparrow (*Passerella iliaca*)	36	75
Olive-sided flycatcher (*Contopus borealis*)	402	52
Rufous-sided towhee (*Pipilo erythrophthalmus*)	191	28
Mountain quail (*Oreortyx pictus*)	224	27
Western wood-pewee (*Contopus sordidulus*)	284	26
House wren (*Troglodytes aedon*)	71	23
Wrentit (*Chamaea fasciata*)	130	22
MacGillivray's warbler (*Oporornis tolmiei*)	429	20
White-breasted nuthatch (*Sitta carolinensis*)	31	19
Chipping sparrow (*Spizella passerina*)	26	19
Northern flicker (*Colaptes auratus*)	429	18
Song sparrow (*Melospiza melodia*)	101	18
Red-tailed hawk (*Buteo jamaicensis*)	24	17
Lazuli bunting (*Passerina amoena*)	25	16
Acorn woodpecker (*Melanerpes formicivorus*)	329	15
Northern pygmy-owl (*Glaucidium gnoma*)	44	14
Dusky flycatcher (*Empidonax oberholseri*)	57	14
Pileated woodpecker (*Dryocopus pileatus*)	279	13
Steller's jay (*Cyanocitta stelleri*)	2,683	12
Nashville warbler (*Vermivora ruficapilla*)	2,794	12
Species avoiding edges		
Red-breasted nuthatch (*Sitta canadensis*)	2,278	1
Winter wren (*Troglodytes troglodytes*)	476	1
Yellow-rumped warbler (*Dendroica coronata*)	685	1
Pine siskin (*Carduelis pinus*)	551	1
Douglas' squirrel (*Tamiasciurus douglasii*)	1,174	1
Hammond's flycatcher (*Empidonax hammondii*)	654	<1
Western flycatcher (*Empidonax difficilis*)	6,235	<1
Chestnut-backed chickadee (*Parus rufescens*)	2,279	<1
Brown creeper (*Certhia americana*)	1,706	<1
Golden-crowned kinglet (*Regulus satrapa*)	1,365	<1
Hermit warbler (*Dendroica occidentalis*)	15,077	<1
Sharp-shinned hawk (*Accipiter striatus*)	15	0
Cooper's hawk (*Accipiter cooperii*)	19	0
Blue grouse (*Dendragapus obscurus*)	88	0
Ruffed grouse (*Bonasa umbellus*)	14	0
Band-tailed pigeon (*Columbia fasciata*)	25	0
Downy woodpecker (*Picoides pubescens*)	22	0
Townsend's warbler[a] (*Dendroica townsendi*)	110	0
Red crossbill (*Loxia curvirostra*)	14	0
Western gray squirrel (*Sciurus griseus*)	169	0

[a]Detected as a spring migrant on the study sites.

above, with about 17 species significantly more numerous in more insular stands, or in stands surrounded by 1000-ha blocks containing more clearcuts and more edge. Only five species were more abundant in continuous forest stands. Of those, only the winter wren was also associated with older forests. A few forest-interior species (e.g., western flycatcher, brown creeper, golden-crowned kinglet) attained higher densities in more insular stands or in stands within highly fragmented 1000-ha blocks, even though they were seldom detected along edges.

Most medium- and large-sized mammals occurred less frequently in more insular stands or in 1000-ha blocks that were more fragmented. In particular, we found the presence of fishers to be more highly correlated with stand insularity ($r = -0.44$, $P < 0.001$) than with any other habitat measure. Among the amphibians, reptiles, and small mammals, only the deer mouse exhibited a consistent stand-level response, reaching higher abundance in stands that were more fragmented, and in areas with more clearcuts and more edge.

Species attaining higher abundance or frequency in stands with adjacent hardwoods were generally those associated with the hardwood component within Douglas-fir forests, especially the acorn woodpecker, black-headed grosbeak, dusky-footed woodrat, gray fox, and pinyon mouse.

Associations with Area

Correlations with stand area were generally weak; however, all but one were positive (Table 38.4). Abundances of winter wrens and acorn woodpeckers were the most highly associated with large stands. In addition, counts of most large, widely dispersed bird species, such as spotted owl, pileated woodpecker, sharp-shinned hawk, and ruffed grouse, were at least weakly correlated with this measure ($P < 0.1$). Occurrences of fisher, gray fox, dusky-footed woodrat, and Pacific giant salamander were also positively associated with stand area.

Tolerances to Fragmentation

To assess tolerance, we compared mean abundance or frequency of all species among five classes of stand area and insularity. Species that differed significantly among these classes (Table 38.5) were much the same as those associated with area and insularity in the correlation analysis (Table 38.4).

As before, no species decreased significantly in abundance or frequency with stand area, and 13 increased. Eight species decreased significantly in more insular stands. In addition, several species were either absent, or occurred much less frequently, in the smallest or most insular stands, or both. Even though their associations were not statistically significant, the rarity of these species in highly fragmented stands is of potential concern.

The abundances of acorn and pileated woodpeckers and winter wrens fell sharply in stands <20 ha (Fig. 38.2). In addition, northern flying squirrels and ringtails occurred in only one stand <20 ha; sharp-shinned hawks and ruffed grouse were found in only one stand <50 ha, and both were absent from the smallest stand class. Western gray squirrels decreased steadily in both abundance and frequency as stand size decreased, and they were virtually absent from stands <11 ha. Fishers also decreased sharply in frequency of occurrence in stands <100 ha. Although we found spotted owls less frequently in stands <20 ha, this trend was not statistically significant.

Many of the same species were also rarely detected in highly insular stands (Fig. 38.3), although most of those stands were also small. Blue and ruffed grouse, sharp-shinned hawks, and gray foxes were all most numerous or frequent in stands without clearcut edges. Other species appeared to tolerate as much as 50% or 75% of the stand perimeter being clearcut.

Table 38.4. Associations (partial correlations, controlling for stand age and elevation) between vertebrate species and fragmentation measures in Douglas-fir forests

	Plot (n = 136)		Stand (n = 46)			1000-ha block (n = 46)	
	Distance to clearcut	Length of edge	Area	Insular index	Adjacent hardwood	Percent clearcut	Total edge
Amphibians							
Pacific giant salamander (*Dicamptodon ensatus*)	ns	ns	++	ns	ns	ns	ns
Rough-skinned newt (*Taricha granulosa*)	ns	+	ns	ns	ns	ns	++
Ensatina (*Ensatina escholtzi*)	−−	ns	ns	+	ns	ns	ns
Del Norte salamander (*Plethodon elongatus*)	−	ns	ns	ns	ns	ns	ns
Western toad (*Bufo boreas*)	−	ns	ns	ns	ns	ns	ns
Pacific tree frog (*Hyla regilla*)	ns	ns	ns	ns	ns	ns	++
Reptiles							
Western fence lizard (*Sceloporus occidentalis*)	ns	ns	ns	ns	ns	ns	ns
Sagebrush lizard (*Sceloporus graciosus*)	+++	ns	ns	ns	ns	ns	ns
Western skink (*Eumeces skiltonianus*)	ns	ns	ns	ns	ns	ns	−
Southern alligator lizard (*Gerrhonotus multicarinatus*)	ns	ns	ns	ns	−−	−	ns
Northern alligator lizard (*Gerrhonotus coeruleus*)	ns	ns	ns	ns	ns	ns	ns
Birds							
Sharp-shinned hawk (*Accipiter striatus*)	ns	ns	++	ns	+++	−−	ns
Blue grouse (*Dendragapus obscurus*)	+++	ns	ns	ns	ns	−−−	−−−
Ruffed grouse (*Bonasa umbellus*)	ns	ns	+	ns	+++	ns	ns
Mountain quail (*Oreortyx pictus*)	−	+	ns	ns	ns	ns	+
Spotted owl (*Strix occidentalis*)[a]	ns	ns	+	ns	++	ns	ns
Northern pygmy-owl (*Glaucidium gnoma*)	−−	++	ns	+++	ns	ns	+++
Acorn woodpecker (*Melanerpes formicivorus*)	ns	ns	+++	ns	+++	ns	ns
Red-breasted sapsucker (*Sphyrapicus ruber*)	ns	ns	+	ns	++	ns	ns
Downy woodpecker (*Picoides pubescens*)	++	ns	ns	ns	ns	ns	−−
Hairy woodpecker (*Picoides villosus*)	ns	ns	ns	ns	ns	ns	ns
Northern flicker (*Colaptes auratus*)	ns	ns	ns	ns	ns	ns	ns
Pileated woodpecker (*Dryocopus pileatus*)	ns	ns	+	ns	++	ns	ns
Olive-sided flycatcher (*Contopus borealis*)	−−	+++	ns	+++	ns	ns	+++
Western wood-pewee (*Contopus sordidulus*)	−−	+++	ns	++	ns	ns	++
Hammond's flycatcher (*Empidonax hammondii*)	ns	ns	ns	ns	ns	ns	++
Western flycatcher (*Empidonax difficilis*)	−−−	+++	ns	+++	+	+++	+++
Steller's jay (*Cyanocitta stelleri*)	++	ns	ns	−−	ns	ns	ns
Common raven (*Corvus corax*)	ns	ns	ns	ns	+++	+	ns
Chestnut-backed chickadee (*Parus rufescens*)	−	ns	++	ns	ns	ns	ns
Red-breasted nuthatch (*Sitta canadensis*)	ns	ns	ns	ns	−−−	ns	−
White-breasted nuthatch (*Sitta carolinensis*)	+++	ns	+++	ns	+	+++	ns
Brown creeper (*Certhia americana*)	ns	ns	ns	ns	ns	+++	+
House wren (*Troglodytes aedon*)	−	+++	ns	+++	ns	+++	+++
Winter wren (*Troglodytes troglodytes*)	ns	−	+++	−−	ns	ns	ns
Golden-crowned kinglet (*Regulus satrapa*)	ns	ns	ns	−−−	+++	ns	ns
Townsend's solitaire (*Myadestes townsendi*)	+++	−−	ns	−−−	ns	−−−	−−−
Hermit thrush (*Catharus guttatus*)	ns	ns	ns	ns	++	ns	ns
American robin (*Turdus migratorius*)	+++	ns	+	−−	+	−−−	ns
Solitary vireo (*Vireo solitarius*)	ns	ns	ns	ns	ns	ns	+
Hutton's vireo (*Vireo huttoni*)	ns	ns	ns	ns	ns	ns	ns
Warbling vireo (*Vireo gilvus*)	−−	++	ns	++	++	ns	++
Orange-crowned warbler (*Vermivora celata*)	ns	ns	++	ns	ns	ns	ns
Nashville warbler (*Vermivora ruficapilla*)	+++	ns	ns	ns	ns	ns	ns
Yellow-rumped warbler (*Dendroica coronata*)	ns	ns	ns	ns	ns	ns	ns
Black-throated gray warbler (*Dendroica nigrescens*)	ns	ns	++	ns	ns	ns	ns
Hermit warbler (*Dendroica occidentalis*)	ns	ns	ns	ns	ns	+	ns
MacGillivray's warbler (*Oporornis tolmiei*)	−−−	ns	−	++	ns	++	ns
Wilson's warbler (*Wilsonia pusilla*)	−−−	++	ns	+++	ns	+++	+++
Western tanager (*Piranga ludoviciana*)	++	−	ns	ns	++	−−−	−−
Rufous-sided towhee (*Pipilo erythrophthalmus*)	−−−	ns	ns	++	ns	+++	++
Dark-eyed junco (*Junco hyemalis*)	ns	ns	ns	ns	ns	ns	ns
Brown-headed cowbird (*Molothrus ater*)	ns	ns	ns	++	ns	ns	++
Purple finch (*Carpodacus purpureus*)	−−	++	ns	+++	ns	+	++
Pine siskin (*Carduelis pinus*)	−−	ns	ns	ns	ns	++	ns
Black-headed grosbeak (*Pheucticus melanocephalus*)	+++	−−	++	−−	+++	−−−	−−−

Table 38.4. *(Continued)*

	Plot (n = 136)		Stand (n = 46)			1000-ha block (n = 46)	
	Distance to clearcut	Length of edge	Area	Insular index	Adjacent hardwood	Percent clearcut	Total edge
Mammals							
Pacific shrew (*Sorex pacificus*)	–	ns	ns	ns	ns	ns	ns
Trowbridge's shrew (*Sorex trowbridgii*)	ns	ns	ns	ns	ns	ns	ns
Shrew-mole (*Neurotrichus gibbsii*)	ns	–	ns	ns	ns	ns	ns
Coast mole (*Scapanus orarius*)	ns	ns	ns	ns	ns	ns	ns
Allen's chipmunk (*Tamias senex*)	– –	ns	+	ns	ns	ns	ns
Western gray squirrel (*Sciurus griseus*)	+ +	ns	ns	ns	+ +	– –	– –
Douglas' squirrel (*Tamiasciurus douglasii*)	ns	ns	ns	ns	ns	ns	ns
Northern flying squirrel (*Glaucomys sabrinus*)[a]	ns	ns	ns	–	ns	ns	ns
Deer mouse (*Peromyscus maniculatus*)	– –	+ + +	ns	+ +	ns	+ + +	+ + +
Pinyon mouse (*Peromyscus truei*)	ns	ns	ns	ns	+ + +	ns	– –
Western red-backed vole (*Clethrionomys californicus*)	ns	ns	ns	ns	ns	ns	ns
Dusky-footed woodrat (*Neotoma fuscipes*)[a]	ns	+	ns	+ + +	ns	+ + +	ns
Gray fox (*Urocyon cinereoargenteus*)[a]	ns	–	+	– – –	+ + +	– – –	– – –
Black bear (*Ursus americanus*)[a]	ns	ns	ns	– – –	ns	ns	ns
Ringtail (*Bassariscus astutus*)[a]	+ + +	ns	ns	ns	ns	ns	– –
Fisher (*Martes pennanti*)[a]	+ +	– –	+	– – –	ns	– – –	– – –
Western spotted skunk (*Spilogale gracilis*)[a]	ns	ns	ns	· –	ns	ns	ns
Striped skunk (*Mephitis mephitis*)[a]	+ + +	– –	ns	– –	ns	ns	ns

Note: Significance levels are as given in Table 38.2.
[a]Only frequency data used.

Table 38.5. Species most sensitive to fragmentation of Douglas-fir forests

	Significant differences[a] among		Tolerance ratio[b] to stands	
	Area classes	Insularity classes	<10 ha	>75% insular
Amphibians				
Pacific giant salamander	+	ns	0	0.29
Western toad	+ +	ns	0	0.49
Reptiles				
Western fence lizard	+	– – –	0.39	0.01
Sagebrush lizard	+	ns	0.20	0
Birds				
Sharp-shinned hawk	ns	ns	0	0
Blue grouse	ns	ns	0.48	0
Ruffed grouse	ns	ns	0	0
Spotted owl[c]	ns	–	0.52	0.73
Acorn woodpecker	+ + +	ns	0.12	0.84
Downy woodpecker	ns	ns	0	0
Pileated woodpecker	+ + +	– – –	0.33	0.50
Common raven	+ + +	ns	0.33	0.63
Winter wren	+	ns	0.24	0.60
Townsend's solitaire	ns	– – –	1.0	0.28
Mammals				
Western gray squirrel	+ + +	ns	0.23	0.36
Northern flying squirrel[c]	+ + +	ns	0	0.29
Gray fox[c]	ns	– – –	0.52	0.16
Ringtail[c]	+	–	0.31	0.36
Fisher[c]	+ +	–	0.26	0.20
Striped skunk[c]	ns	ns	0	0

[a]One-way analysis of variance; significance levels are as given in Table 38.2.
[b]Ratio of abundance or frequency in smallest or most insular stands to that in largest or least insular stand class (Whitcomb et al. 1981).
[c]Only frequency data used.

Discussion

As mature forests are converted to early-successional brushfields on a large scale in the Pacific Northwest, we are witnessing regionwide changes in wildlife species composition. Overall populations of forest species are decreasing, whereas several species favoring open habitats are expanding their ranges into the Douglas-fir forest zone (Raphael, pers. obs.). Within the remaining forest patchwork, a few species with large home-range requirements are showing signs of avoiding edges and withdrawing from highly fragmented regions. However, the majority of the species we studied exhibited no negative responses either to reduced patch size or to increased length of adjacent clearcut edge.

These results contrast in several ways with studies in eastern deciduous forests. Whereas Whitcomb et al. (1981) found bird species typical of forest interior to be largely replaced by ecotonal species in small woodlots (<14 ha), we found the composition of birds, small mammals, reptiles, and amphibians to be very similar in all classes of forest patch size and insularity. Increased bird species richness in more fragmented Douglas-fir stands resulted largely from the addition of early-successional and chaparral species that have colonized recent clearcuts. These species occurred frequently along edges, but penetrated the forest only where a shrubby hardwood understory was prevalent. Furthermore, whereas neotropical migrants were more adversely affected than permanent residents in eastern forests, only two of 18 migratory species showed negative responses to fragmentation measures in our study. However, nine of the 26 residents tested were associated with unfragmented forest interiors.

A notable difference between Douglas-fir forests subject to clearcutting and other fragmented habitats was the lack of

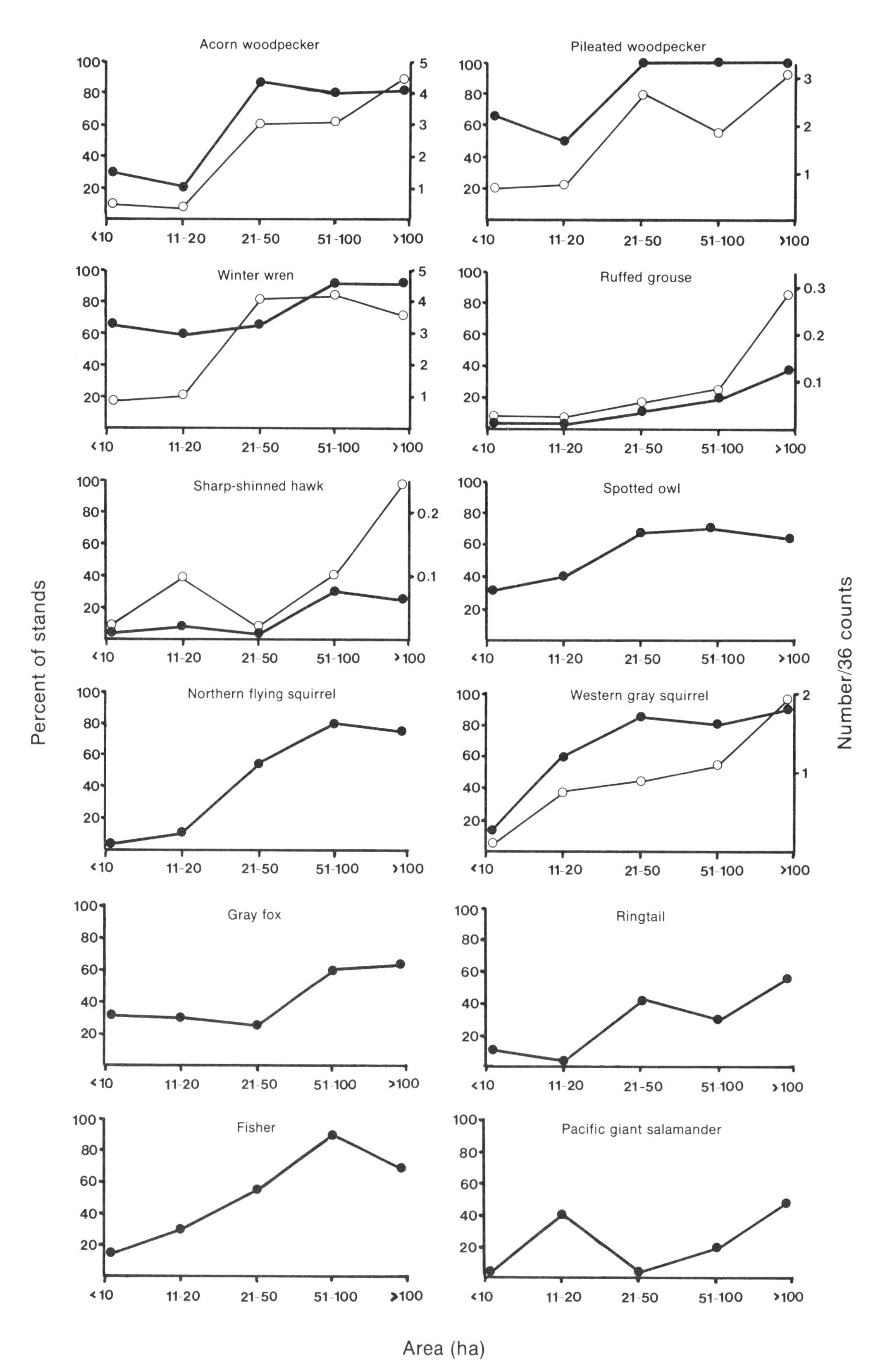

Figure 38.2 Distribution of 12 species among five classes of stand area. Solid circles represent frequency of occurrence; open circles indicate relative abundance, when available. We sampled 6 stands of ≤10 ha, 10 of 11–20 ha, 9 of 21–50 ha, 10 of 51–100 ha, and 11 of >100 ha.

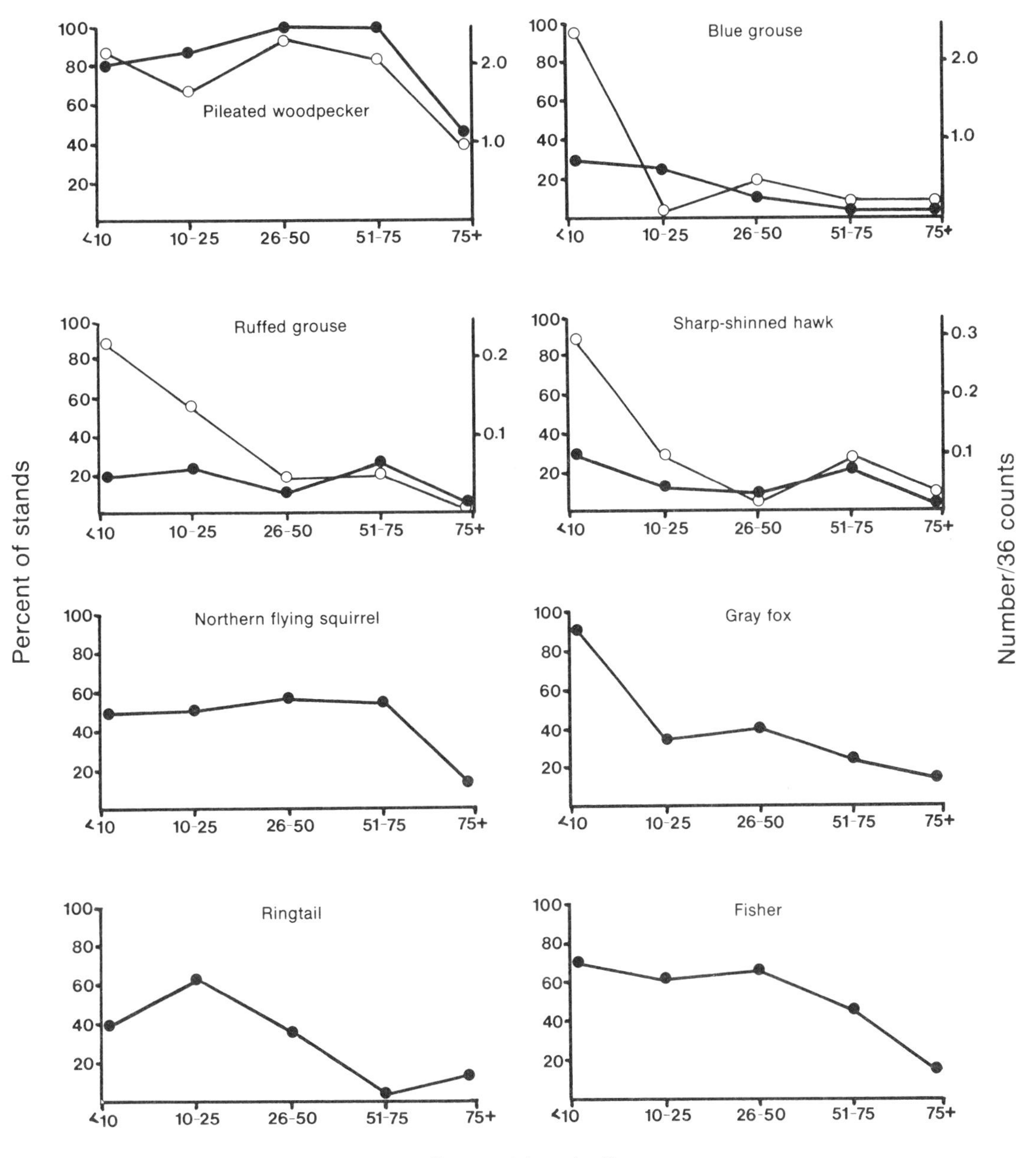

Figure 38.3. Distribution of eight species among five classes of stand insularity. Solid circles represent frequency of occurrence; open circles indicate relative abundance. We sampled 10 stands with <10% insularity, 8 with 10–25% insularity, 12 with 26–50% insularity, 9 with 51–75% insularity, and 7 with >75% insularity.

a well-developed ecotone. Therefore, little interaction was evident between forest and clearcut species, and only a very few species (e.g., olive-sided flycatcher) occurred along this abrupt edge more often than within each adjacent habitat. Even small forest patches that retained such microhabitat components as downed logs or large snags seemed to support forest-interior species associated with those elements, even within a few meters of an unvegetated clearing.

Among the species suspected of being most sensitive to forest fragmentation in our study, only the fisher and spotted owl were also associated with old-growth forests. The pileated woodpecker, another species of regional interest, showed only moderate intolerance of very small forest islands and was numerous in young forests containing many hardwoods and few commercially sought large firs. This species remains common in fragmented eastern deciduous forests (Whitcomb et al. 1981).

Rare or low-density species may have been missed in our smaller stands because of reduced sampling effort (number of plots); however, several factors suggested that the observed trends were real. First, species exhibited similar patterns in both stand- and plot-level tests, where sampling effort was equal among plots. Second, with a greater proportion of the area of a small stand being sampled, the chance of detecting rare animals there may have been equal to or even greater than in a large stand containing more study plots. Third, even at our lowest level of sampling intensity, large resident species (e.g., spotted owl) should have been detected in 3 years of study.

If our results appear to differ markedly from those of eastern U.S. forests, we must emphasize that the fragmentation of Douglas-fir forests is relatively recent in comparison with forests of other regions. In our study areas, it is likely that much of the clearcutting has occurred within the lifetimes of many individual animals we have sampled, or at least within very few generations for most species. Therefore, it is almost certain that we are not able to discern the true long-term responses of these species to the break-up of their habitat, and we will probably see further changes as an equilibrium of forest and early-succesional patches is reached. We also cannot yet evaluate the effects of forest island shape or isolation.

Whitcomb et al. (1981) found that some forest-interior birds increased in abundance initially in small stands after the destruction of adjacent habitat; those species were later greatly reduced in numbers or eliminated from those areas. Presumably, a "packing" of individuals into the remaining available habitat may lead to increased territorial aggression, higher nest predation, increased competition for resources, and subsequent low reproductive success. We may be witnessing just such a response in the early stages of the fragmentation process in the forests we studied.

In spite of the tentative nature of our findings, we can offer some conservative prescriptions to forest managers in the Pacific Northwest. Most important, our results suggest minimum characteristics of stands that are to be included in forest totals to meet planning guidelines. Specifically stands <20 ha tend to lack the full complement of vertebrate species and should be excluded as viable stands. For isolated stands (>50% insularity), this minimum size should be increased to 50 ha. Furthermore, a working definition of an old-growth stand in northwestern California should be modified to include these size criteria; that is, very small stands should not be tallied as old-growth, regardless of their age or stocking characteristics.

We have learned much in recent years about vertebrate-habitat relationships in Douglas-fir forests. Because the data that have been gathered so far are primarily inventory in nature, we have little understanding of how individual animals or local populations respond to clearcutting, or of the differential reproductive fitness and dispersal capabilities of species that appear to be persisting in small islands of forest. This chapter represents the infancy of forest-fragmentation research in the Pacific Northwest, and it is hoped that it will suggest directions for study in the near future. The facts that few species have shown visible negative responses to date, and that apparently no species has been lost from the community as a whole, suggest that it is not too late to set in motion a management plan that will ensure the permanent integrity of the Douglas-fir forest ecosystem.

Acknowledgments

This study was supported by the Pacific Southwest Region and the Pacific Southwest Forest and Range Experiment Station of the USDA Forest Service, and by the University of California, Agricultural Experiment Station project 3501 MS. We thank those who assisted with field work and with data analysis, particularly J. A. Brack and C. A. Taylor. This chapter benefitted from the comments of B. R. Noon, F. B. Samson, M. L. Shaffer, J. Verner, and B. A. Wilcox.

39

Avian Demography in Mosaic Landscapes: Modeling Paradigm and Preliminary Results

DEAN L. URBAN and HERMAN H. SHUGART, JR.

Abstract.—We pursued a mechanistic explanation for the local extinction of some bird species from small and/or isolated patches in a habitat mosaic. A model was developed to simulate demographic processes of natality, dispersal, and mortality as these might occur in a mosaic of discrete habitat patches. These processes were adjusted to reflect life history attributes that seem to confer sensitivity to habitat fragmentation. We used the model to explore the effects of restricted vagility and reduced natality resulting from biotic interactions for a hypothetical bird species. Restricted vagility resulted in local extinctions in small, isolated patches, but this effect was buffered by an available pool of surplus nonbreeders. Lower natality reduced this surplus so as to increase the frequency of local extinction and allowed restricted vagility to play a more important role in affecting the species' distribution. Moderate levels of restricted vagility and reduced natality were synergistic in effect. In general, small patches in isolation tended to lose their populations permanently, large patches rarely suffered extinctions, but small and intermediate patches adjacent to other patches suffered extinctions and were recolonized repeatedly. The basic modeling approach can be expanded to include greater mechanistic detail, but implementation has been limited by a lack of appropriate data. Our preliminary results suggest which data are necessary to pursue a mechanistic, patch-scale understanding of the demography of species at the landscape scale.

A central concern of landscape ecology is to account for the distribution and local abundance of species inhabiting a mosaic of habitat patches. In particular, strong motivation exists to explain the local extinctions of some species from small and/or isolated habitat patches. Species sensitive to habitat fragmentation can be categorized according to life history attributes that seem to confer sensitivity. Among birds, such traits reflect migratory status, vagility, habitat specialization, and nesting behavior (Lynch and Whitcomb 1978; Whitcomb et al. 1981).

Much of our current understanding of avian sensitivity to habitat fragmentation is inferential, based as it is on observed distributions of birds in woodlots of various sizes (Forman et al. 1976; Galli et al. 1976; Lynch and Whitcomb 1978; Whitcomb et al. 1981; Ambuel and Temple 1983). In this chapter, we use life history attributes as a basis for pursuing a more mechanistic explanation of the distribution of species in habitat mosaics.

For a population to maintain itself in a mosaic, (1) potential breeding habitat must exist; (2) the patches must be accessible to the birds; and (3) net reproduction must be positive. Habitat availability is mechanistically explained by the dynamics of the habitats themselves: succession, disturbance regimes, land-use changes, and so on. We ignore these for now and instead make two assumptions about habitats: (1) given a mosaic, potential breeding habitats can be so classified for a species of interest; and (2) habitat dynamics are so slow compared to avian population dynamics that the habitats can be treated as being constant.

The mechanistic explanations underlying habitat accessibility and net reproduction reflect life history attributes of the bird species: dispersal capabilities, natality, and mortality. Pursuing this, we develop a model that simulates the demographic processes of natality, dispersal, and mortality as these might occur in a mosaic of discrete habitat patches. Our intent is to develop a modeling paradigm that can be applied with comparable reliability to a large number of species in a variety of landscapes.

We use this approach to examine the demography of a hypothetical bird species subject to two mechanisms affecting its distribution in a habitat mosaic.

1. Limited dispersal ability, whether due to physical or behavioral constraints, leads to restricted vagility: a bird tends to remain in or very near its natal habitat patch. Restricted vagility reduces the chance that an isolated habitat that suffers local extinction will be recolonized, thus contributing to the observed "isolation effect" in fragmented habitats.

2. Brood parasitism, nest depredation, predation, and competition are biotic interactions that lead to decreased

Dean L. Urban: Environmental Sciences Division, Building 1505, Oak Ridge National Laboratory, P.O. Box X, Oak Ridge, Tennessee 37831. *Present address:* Graduate Program in Ecology, University of Tennessee, Knoxville, Tennessee 37916

Herman H. Shugart, Jr.: Environmental Sciences Division, Building 1505, Oak Ridge National Laboratory, P.O. Box X, Oak Ridge, Tennessee 37831. *Present address:* Department of Environmental Sciences, University of Virginia, Charlottesville, Virginia 22903

natality and/or increased mortality in some habitats. Gates and Gysel (1978) have demonstrated such effects in forest edges. More recently, brood parasitism (Brittingham and Temple 1983) and nest depredation (Temple 1983) have been quantified as these occur near forest edges. Because small woodlots have proportionately more edge than large woodlots (Forman and Godron 1981), we would expect biotic interactions to be more pronounced in small habitats, thus contributing to the "area effect" observed for many species in habitat mosaics.

Our preliminary examination of species' demography in habitat mosaics is exploratory rather than definitive. Our purpose is to discover the nature of the effects of restricted vagility and biotic interactions and the possible range in magnitude of these effects. Our results illustrate the utility of the modeling paradigm, suggest modifications necessary for a more comprehensive model, and indicate the empirical data needed to implement such modifications.

Methods

Model development

The model simulates spatial and temporal dynamics of a species inhabiting a mosaic of discrete habitat patches. Each patch has a carrying capacity determined by its area and the species' territory size, relative to which two subpopulations are defined. Birds are "breeders" until carrying capacity is reached; birds in excess of this are nonbreeding "floaters." A fractional territory is allowed to support a breeder with a probability equal to its fractional size. This approach reflects natural variability in territory size, and while the implementation is rather cavalier it does not seem to affect model performance qualitatively. For computational ease, the model tracks males only, so a breeding male represents a territorial pair.

In the course of a simulation year, (1) natality is computed for breeders; (2) birds "migrate south" and experience stochastic overwinter mortality (which is the same for both subpopulations); (3) habitat patches are recolonized during "spring migration," in which birds disperse probabilistically among patches; and, optionally, (4) floaters are allowed to redisperse in search of available territories. After dispersal, breeding and floating subpopulations are redefined and the cycle begins again.

At the heart of the model are the probabilities for dispersal among patches, reflecting the spatial complexity of the habitat mosaic. We have viewed avian dispersal as a process roughly analogous to electrical flow. Flow of electrons through a conductor decreases with the length and resistance of the conductor and increases with the voltage of the power source. Similarly, the flow of emigrants from one habitat patch to another decreases with interpatch distance and the resistance (navigability) of the intervening matrix, and increases with the carrying capacity of the donor patch (reflecting the probable availability of emigrants).

In application, distance can be a resistance-weighted measure, where elements in the intervening matrix have empirically derived resistance values. Total interpatch distance would be the sum of resistance-weighted distances for all intervening elements. This approach allows for dispersal corridors (Forman and Godron 1981; Forman 1983) and habitat bridges (Whitcomb et al. 1981) while absorbing these effects into the distance measure for computational ease. This is especially amenable to grid-based information-management systems that include algorithms for finding weighted distances (e.g., the Map Analysis Package, Tomlin 1980).

Actual dispersal probabilities are determined by emigrant flow rates and the site fidelity of subpopulations. Breeders have high site fidelity; floaters, low fidelity. The initial dispersal matrix for each subpopulation has site fidelity as its diagonal and flow rates as off-diagonal elements. Because each row of the matrix must sum to 1.0, the elements are adjusted accordingly:

$$P_{jik} = \begin{cases} (F_{ji}/S)\,(1 - \gamma_k), & S > 1.0 - \gamma_k, \\ F_{ji}, & S \le 1.0 - \gamma_k, \end{cases}$$
$$P_{iik} = \begin{cases} \gamma_k, & S \ge 1.0 - \gamma_k, \\ 1.0 - S, & S < 1.0 - \gamma_k, \end{cases} \quad (1)$$

where P_{jik} is the probability of a bird of subpopulation k dispersing from patch j to i; P_{iik} is adjusted site fidelity; F_{ji} is relative flow from j to i; S is the sum of all F_{ji} over $j \neq i$; and γ_k is initial site fidelity for subpopulation k. A biological translation of these equations is this: for a patch adjacent to many other patches ($S > 1.0 - \gamma_k$) dispersal probabilities are relativized so that birds are apportioned according to the relative proximity of the nearby patches; for a patch in isolation ($S < 1.0 - \gamma_k$), site fidelity is adjusted so that birds are retained as nonbreeders. This can effectively reshape the dispersal-distance function to reflect the spatial distribution of patches.

In simulation, birds surviving the winter return to the same patch they occupied the previous year, and subpopulations are redefined. This allows floaters (including juveniles) to fill vacant territories in their home patch before dispersing elsewhere. The "homing" approach allows use of edge-to-edge interpatch distance (which is computationally convenient) without denying juveniles the chance to disperse within very large patches. In spring migration, the expected number of emigrants from all patches j destined for patch i is

$$E_i = \sum_{j=1}^{m} P_{ji,1} N_{j,1} + \sum_{j=1}^{m} P_{ji,2} N_{j,2}, \quad (2)$$

where m is the number of patches, and $N_{j,1}$ and $N_{j,2}$ are the number of breeders and floaters, respectively, in patch j. During subsequent redispersals of floaters, only the latter half of equation (2) applies. As implemented, individual birds move according to random numbers, guided by the dispersal probability matrix.

Implementation

The distance effect was formulated as a negative exponential function based on banding data for song sparrows (*Melospiza melodia*) (Nice 1937). Relative avian flow from patch j to i was defined:

$$F_{ji} = \exp(-bD_{ji}), \qquad (3)$$

where D_{ji} is edge-to-edge interpatch distance, and b is a fitted constant.

Realized vagility reflects the distance that a species will move in a single dispersal episode (its range) and the number of times it will move before setting (its mobility). We indexed range as the distance at which the distance function tailed to a value of 0.01 (at which point it was truncated to 0). Mobility was varied by specifying a number of postmigration redispersal episodes for floaters.

Biotic interaction effects were implemented as a reduction in realized natality only. There were no effects on adult mortality, although we assumed that only half of a typical clutch survived until the first winter. Natality was reduced as a $\log_2$-linear function of patch area, with natality in the largest patches equal to potential fertility. The $\log_2$-linear function was arbitrary but was computationally convenient and ensured that biotic effects were most pronounced in small patches. Our intent here was to determine whether these effects were sufficiently important to the model's performance to warrant more explicit attention in future model applications.

Species' parameters of fertility, mortality, territory size, and site fidelity were estimated to represent a typical neotropical migrant (Table 39.1). Some parameters were difficult to know precisely (e.g., the number of dispersal episodes). Others were estimated from data collected in contiguous habitat (e.g., the shape of the distance function), and it is not clear whether these parameters are strictly appropriate to discrete habitat patches. We manipulated these parameters to infer their importance to the model's behavior.

Simulations

We generated 33 random landscapes of circular habitat patches in homogeneous nonhabitat. The number of patches varied between 10 and 65 within a 100-km^2 landscape. Patch sizes ranged from 1 to 128 ha on a $\log_2$ scale. Discrete patch sizes allowed categorical analysis of population response. The size distribution was subjectively skewed so that most patches were small (1–8 ha), and very large patches were rare. Patch locations were determined by grid coordinates drawn from uniformly distributed random numbers.

We ran triplicate simulations for four levels of vagility, four levels of biotic interaction (including a control), and a combination of moderate levels of each mechanism (Table 39.2). Each model run simulated 50 years of population dynamics. Populations were allowed to equilibrate for 20 years, then their status was assessed for the final 30 years of simulation. Each run was initialized with patch populations at carrying capacity.

Table 39.1. Implementation of life-history attributes

Attribute	Implementation
Territory size	4 ha. A constant.
Site fidelity	Breeders: 0.9, floaters: 0.1; may be adjusted for isolated patches.
Mortality	Mean survivorship = 0.66667; *SD* = 0.1; stochastic each year; invoked individually per bird.
Natality	Potential fertility 1♂/♂/year; decreased as function of area to simulate biotic interaction.
Dispersal range	Distance to tail of negative-exponential function; varied experimentally.
Mobility	Number of postmigation redispersals of floaters; varied experimentally.

Table 39.2. Combinations of mechanisms examined in simulations

Mechanism/Level	Range	Mobility	Realized natality[a]
Vagility			
1	750 m	0	1.00
2	1500 m	0	1.00
3	750 m	3	1.00
4	1500 m	3	1.00
Biotic interaction			
1	1500 m	3	0.25
2	1500 m	3	0.50
3	1500 m	3	0.75
4	1500 m	3	1.00
Combination	1500 m	0	0.50

[a]Proportion of potential fertility realized in patches. Value is intercept of $\log_2$-linear function of area, with natality in 128-ha patches equal to 1.00.

Analysis

We summarized population response as persistence, defined as the mean number of breeders supported by a patch, divided by its carrying capacity. Note that "50% persistent" for a very large patch loosely translates as "half full"; for a very small patch (near territory size for the species), it means "full half the time." For each patch, we related persistence to patch area and to a measure of isolation derived from equation (2). We defined "area-weighted connectedness" for patch i as

$$C_i = \sum_{\substack{j=1 \\ j \neq i}}^{m} P_{ji,2} K_j, \qquad (4)$$

where K_j is the carrying capacity of patch j.

After preliminary inspection of results, we illustrated the relationships between persistence and connectedness for patch sizes 1–64 by fitting curvilinear regression lines to model output. The regression model was of the form:

$$\hat{P} = b_1 - b_2\exp(-b_3C), \qquad (5)$$

where $\hat{P}$ is predicted persistence, C is area-weighted connectedness, and other parameters are fitted (SAS Institute 1982b). Small sample sizes degraded regressions for patch size 128; we do not include these, but their behavior was similar to 64-ha patches. The regressions were not meant to be rigorously inferential, but rather were used for purposes of graphic presentation of model results.

Results

The curvilinear relationship between persistence and connectedness was most pronounced for small patches. Connectedness was less important for patches sufficiently large to be self-sustaining. This general relationship held for all combinations of mechanisms and parameters that we examined.

The effects of four levels of vagility were not substantially different from one another (Fig. 39.1). It seemed that a moderate dispersal range and a small number of redispersal episodes were sufficient to maintain all but the most isolated patches at carrying capacity. This result in part reflected a large floating subpopulation on most landscapes. These surplus floaters effectively buffered patches from local extinction.

Populations were quite responsive to biotic interaction when realized natality was reduced so as to eliminate most of the floating subpopulation. Without this buffering capacity, patch populations were more susceptible to stochastic extinctions. The consequent effect on persistence was most pronounced in small patches (Fig. 39.2)

Removing the buffering floaters in a mosaic allowed restricted vagility to play a more important role in affecting species' distribution among patches. Moderate levels of vagility and biotic interaction were interactive in effect. Biotic interaction increased the frequency of local extinction, and restricted vagility decreased the frequency of recolonization. The net result was more severe than either mechanism by itself (Fig. 39.3).

The interplay between local extinction and recolonization was quite dynamic in most habitat mosaics. In general, small patches in isolation tended to lose birds permanently. Large patches rarely suffered extinction, but small and intermediate patches connected to other patches suffered extinctions and were recolonized repeatedly (Table 39.3).

Population dynamics in a mosaic can be examined at the scale of the individual patch or at the scale of the entire mosaic. At the patch scale, population dynamics were typically quite lively (Fig. 39.4*a*). The mosaic metapopulation (the sum of the population in all the patches) was usually less so (Fig. 39.4*b*). This was primarily because the floating subpopulation absorbed most of the variation in abundance, and secondarily because the patches' population dynamics were not in synchrony.

Although the mosaic's metapopulation was merely the sum of the populations in its component patches, metapopulation dynamics were not always as straightforward as simple addition implies. The demographic processes were stochastic, and because patches were interactive, this stochasticity was compounded across the mosaic. With small populations, this resulted in a wide variety of behaviors. At one extreme, mosaics of as many as 20 small patches occasionally suffered

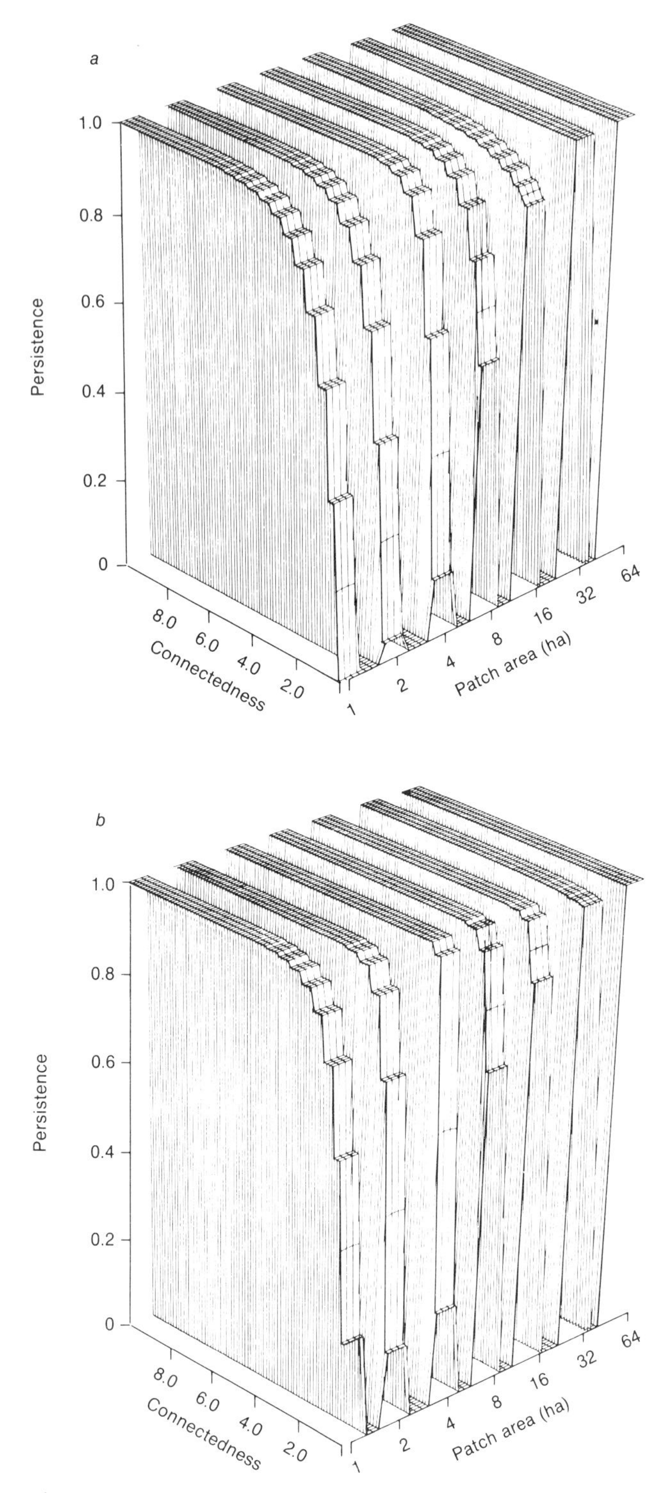

Figure 39.1. Effects of restricted vagility on the relationship between patch connectedness and population persistence for seven patch sizes; (*a*) corresponds to vagility simulation 1 in Table 39.2, and (*b*) to vagility simulation 4.

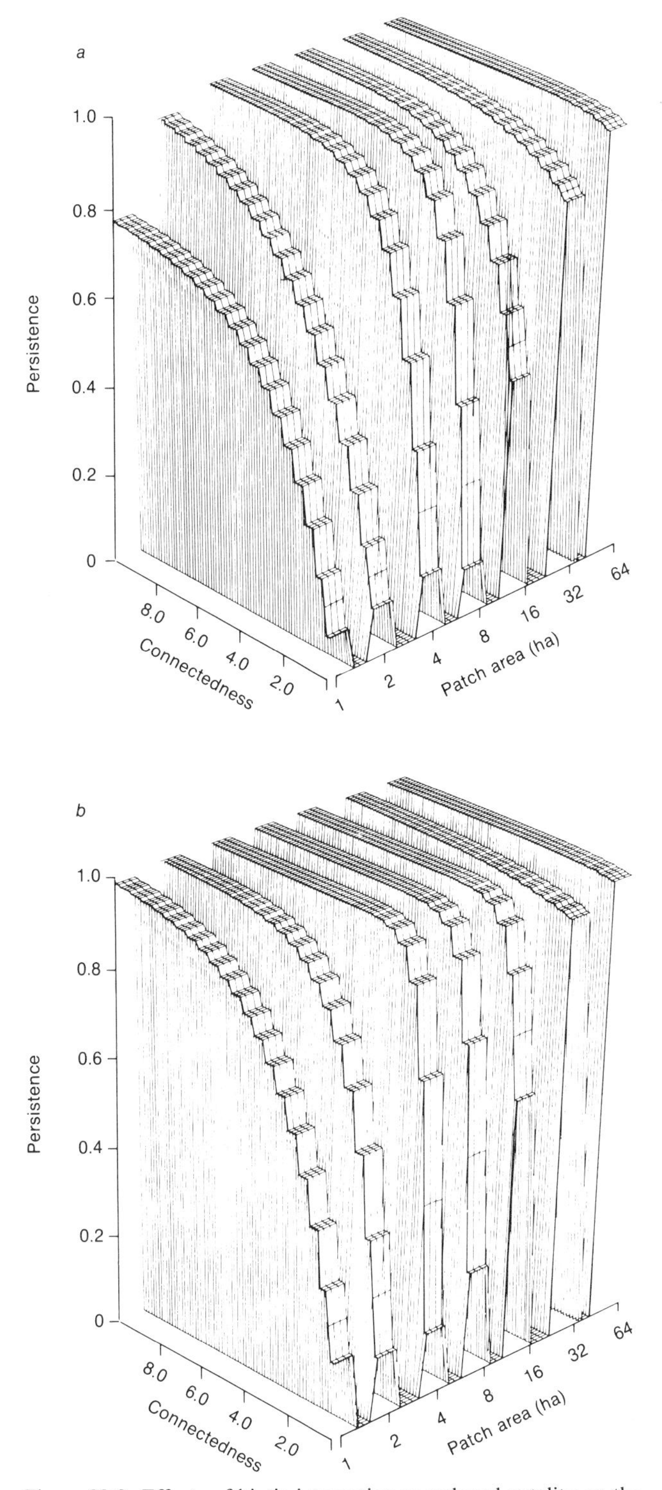

Figure 39.2. Effects of biotic interaction as reduced natality on the relationship between patch connectedness and population persistence for seven patch sizes; (*a*) corresponds to biotic interaction simulation 1 in Table 39.2, and (*b*) to simulation 2.

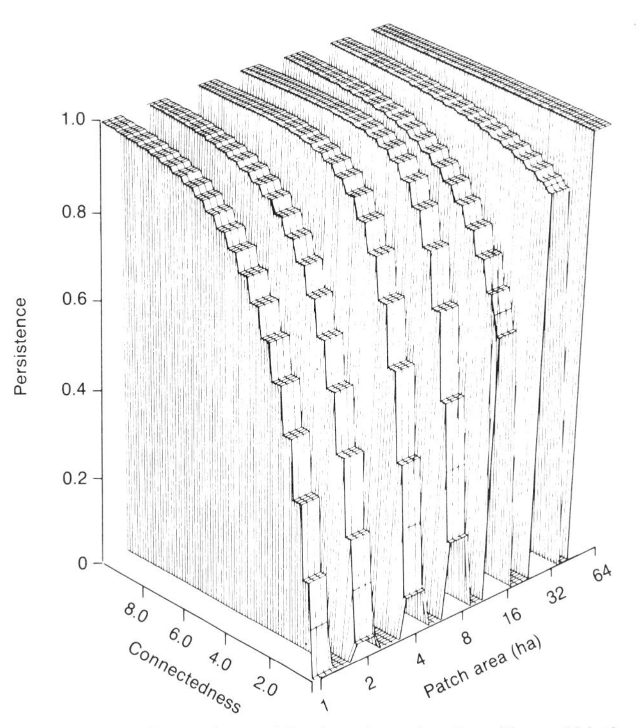

Figure 39.3. Effects of a combination of restricted vagility and biotic interaction on the relationship between patch connectedness and population persistence for seven patch sizes, corresponding to combination simulation in Table 39.2.

global extinctions. At the other extreme, some landscapes initially lost populations in a few patches, after which the remaining patches exhibited 100% persistence. In intermediate cases (e.g., Fig. 39.4*b*), mosaic metapopulations were 80–90% persistent, even though as many as 50% of the patches had no breeders in any given year. Our understanding of how a mosaic landscape's behavior translates from the behaviors of its component patches is still incomplete.

Discussion

Our preliminary results seem to justify this basic approach to modeling demography in habitat mosaics. The relationships of area and isolation to population persistence are intuitively reasonable and in accord with available empirical evidence. Patch dynamics in our simulations are consistent with recent suggestions that frequent local extinctions and recolonizations are typical of habitat mosaics (e.g., Middleton and Merriam 1981, 1983).

Our model results suggest that a relatively severe reduction in realized natality is required to effect a decline in the total landscape population. Recent evidence (Gates and Gysel 1978; Brittingham and Temple 1983; Temple 1983) indicates that brood parasitism and nest depredation may be sufficiently severe to produce such an effect in a mosaic of small forest patches.

From our results, it seems extremely unlikely that avian distribution in habitat mosaics is limited by species' vagility

Table 39.3. Dynamics of extinction of species on a selected landscape

Patch size (ha)	1	2	4	8	16	32	64	128	Total
Number of patches	7	7	14	7	5	2	1	1	44
Vagility 2[a]									
Extinctions	22	13	16	0	0	0	0	0	51
Recolonizations	21	11	12	0	0	0	0	0	44
Biotic interaction 2[a]									
Extinctions	27	21	25	2	2	0	0	0	77
Recolonizations	25	19	22	2	2	0	0	0	70
Combination									
Extinctions	36	14	23	6	3	0	0	0	82
Recolonizations	30	11	19	4	2	0	0	0	66

Note: Total extinction or recolonization events over final 30 years of a 50-year simulation.
[a]Simulation level from Table 39.2.

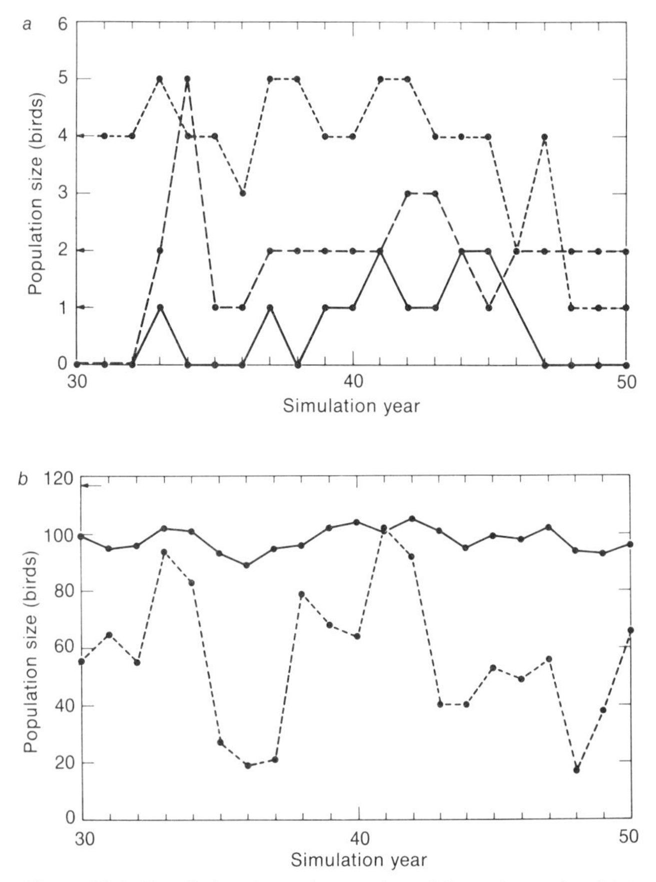

Figure 39.4. Population dynamics as viewed from the scale of (*a*) patches and (*b*) the entire mosaic. (*a*) Dynamics of total populations in three patches with carrying capacities for one, two, and four breeders (solid, dashed, and dotted lines, respectively). (*b*) Dynamics of breeders (solid line) and floaters (dotted line) for a mosaic with carrying capacity for 117 birds (arrow on ordinate).

in any but the most exceptional cases. The dispersal range that we used, based on song sparrow banding data, was sufficient to colonize most patches. Similar data collected for black-capped chickadees (*Parus atricapillus*) (Weise and Meyer 1979) indicate a dispersal range an order of magnitude greater—surely sufficient to colonize even more isolated patches.

The basic approach of adjusting demographic processes of natality, dispersal, and mortality to reflect life history attributes seems especially justified if the model is to be applied to situations other than those under which the model was parameterized. Most management-oriented applications that assess the effects of hypothetical land-use changes would fit this definition. In assessing a species' response to a novel situation, it is important that the modeled species experience the new conditions in the same way as a real animal would. An approach based on life history attributes permits this. In comparison, models based on inferred relationships (e.g., minimum area requirements or incidence functions) constrain the species to behave as noted in the baseline data set.

The mechanistic, life history attributes approach is expandable beyond the rather simplistic version presented here, but the approach has been limited by a lack of appropriate data. Although the life histories of most bird species are well documented (anecdotally and empirically), very few data on dispersal have been collected in habitat mosaics. To verify the shape of the dispersal-distance function, species' range, and mobility would require intensive telemetric or banding studies conducted in entire constellations of habitat patches. It is important that appropriate data be made generally accessible as they become available. Meanwhile, usable data may already exist (perhaps in obscure places) that might be made available as their importance becomes known.

To understand population dynamics in a habitat mosaic, we must think and work at three levels (Fig. 39.5). Landscape pattern (the abundance and distribution of patches) is a higher-level context that constrains the demography of individual birds. These constrained demographics generate, from below, the dynamics of the populations in patches. The populations of the patches collectively generate the dynamics of the mosaic metapopulation. The metapopulation is, in turn, a constraint on bird demography (empty patches are a different context than patches full of birds are).

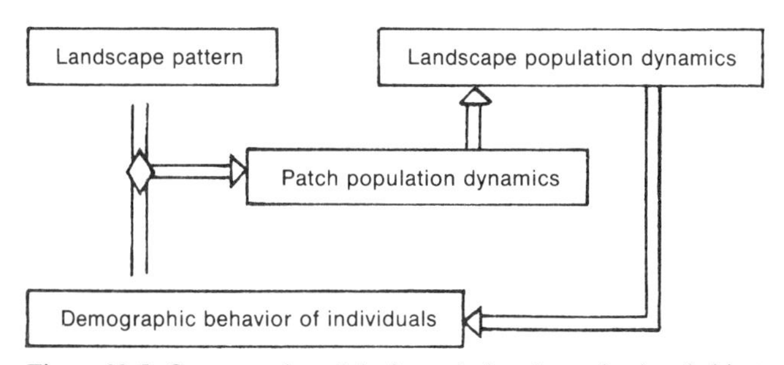

Figure 39.5. Conceptual model of population dynamics in a habitat mosaic as a multileveled system.

This approach illustrates a conceptualization of demographics in mosaic landscapes that is motivated by a hierarchical perspective (Allen and Starr 1982). This conceptualization offers an alternative to approaches couched in island-biogeographic theory, but dictates a change in our modeling technique and analytic methods. This chapter represents our preliminary efforts in this direction.

Acknowledgments

This manuscript was reviewed by R. V. O'Neill, D. L. DeAngelis, and E. Beals. The electrical flow paradigm stemmed from discussions with G. Sugihara. P. Lesslie wrote the graphics program. This research was supported by the National Science Foundation's Ecosystem Studies Program under Interagency Agreement No. (BSR–8315185) with the U.S. Department of Energy, under Contract No. DE–AC05–84OR21400 with Martin Marietta Energy Systems, Inc. Publication No. 2474, Environmental Sciences Division, Oak Ridge National Laboratory.

40

Generation of Species-Area Curves by a Model of Animal-Habitat Dynamics

STEVEN W. SEAGLE

Abstract.—A first-order Markov model was developed to simulate the effect of landscape dynamics on species richness for landscapes of various sizes. Colonization of these dynamic landscapes was allowed by 150 hypothetical species, on the basis of availability of appropriate habitat. Replicate model runs resulted in a species-area curve with an average maximum of 38 species and a slope of 0.31. If colonizing species were allowed to compete, the average maximum number of species dropped to 17, but the slope of the curve changed little (0.29). Habitat diversity and constancy increased with increasing landscape area. This increasing constancy raised the equilibrium number of species and thus determined the slope of the species-area curve. Increasing the disturbance frequency, with no competitive effects, lowered the slope of the curve to 0.22; decreasing the disturbance frequency increased the slope to 0.34. All of these slopes fall within empirically suggested ranges and thus indicate that succession and disturbance of animal habitats may be primary factors in controlling species-area relationships.

A landscape may be viewed as a mosaic of ecosystem components, in which each patch of the mosaic consists of a vegetation cover type. The mosaic depends on vegetation succession and periodic disturbance to develop and maintain a repetitious pattern characteristic of the landscape (Forman and Godron 1981). Frequency and intensity of the disturbance regime will thus play a major role in patch composition and the pattern of a landscape. This interplay of succession and disturbance may be referred to as landscape dynamics.

Within terrestrial landscapes, vertebrate animal species are distributed at various spatial scales as a result of characteristics of individual species (e.g., vagility) or distribution of the vegetation cover types that constitute the habitat of the species. Such varied distributions of vertebrate species on mosaic landscapes make them subjects for biogeographic studies, especially studies of habitat fragmentation (Whitcomb et al. 1981). Similar landscapes of differing area often display an increase in the number of species with increasing landscape area. Connor and McCoy (1979) enumerated various hypotheses proposed to explain this species-area relationship: (1) passive sampling, in which larger areas simply serve as larger targets for randomly dispersing individuals; (2) the area–per se hypothesis, in which area inherently controls number of species; and (3) the habitat-diversity hypothesis, which suggests that greater habitat diversity of larger areas allows existence of more species. Despite various attempts to correlate area, habitat diversity, and other environmental variables with animal species richness (e.g., Johnson and Simberloff 1974; Dueser and Brown 1980), a suitable explanation for the species-area relationship does not exist.

Interspecific interactions, especially competition, are often invoked to account for the distribution of species at local microhabitat scales (Seagle 1983), along elevational gradients (Hairston 1980), and at large geographic scales (MacArthur 1972). Competition also figured strongly in the development of island biogeographic theory and concepts of species richness (MacArthur and Wilson 1967). Ecological theory, then, suggests that intense competition may influence spatial distributions of species on landscapes. The effects of competition on community-level attributes such as species-area relationships, however, are unknown.

In this chapter, I present a theoretical model that incorporates the concepts of vertebrate habitat use and landscape dynamics. Specifically, I consider the effects of landscape dynamics, disturbance frequency, and competition on habitat diversity and on the species-area relationship.

Methods

This model is intended to simulate internal habitat variability and colonization of landscape islands. The model structure is based on three primary assumptions: (1) vegetation succession and disturbance form a mosaic of vegetation cover types; (2) colonization by animal species occurs; and (3) animal species require certain vegetation types as habitat. Colonization is a process recognized by empirical observation, and modern wildlife management has based much of its success on manipulation of appropriate habitat availability. Vegetation succession and disturbance are well-documented processes (White 1979). The characterization and dynamics of landscape mosaics, as well as the effects of a dynamic mosaic on animal abundances and distribution, are, by comparison, poorly understood.

STEVEN W. SEAGLE: Department of Forestry, North Carolina State University, Raleigh, North Carolina 27695

The landscape-dynamics regime described here is based on the succession and disturbance of a Tasmanian wet sclerophyll rainforest, which has been thoroughly described elsewhere (Noble and Slatyer 1980). Noble and Slatyer recognized 10 vegetation cover types in this forest and estimated the successional time between types, as well as the frequency of fire occurrence. Using the reciprocal of successional times and fire frequency as transition probabilities between cover types, I formulated a 10 × 10 habitat transition matrix. The matrix describes the probability of a patch of one cover type becoming a patch of another cover type in any year and allows simulation of the landscape as a first-order Markov process. Thus, during a model run, each patch of vegetation was considered for succession and disturbance each year.

Each model run consisted of 300 simulated years. A vegetation patch that did not undergo succession, and was not disturbed, could reach an age of 300 years. This is an unlikely event in nature or in simulation but necessitated record keeping of patch distribution in a 10-vegetation-cover-type × 300-year matrix. Each patch of vegetation was thus classed by both cover type and number of years remaining within a cover type. When the cover type of a patch changed, because of either succession or disturbance, its age returned to 1 to indicate the length of time the patch had been in its current cover type.

Each model run was initiated with the same distribution of patches within the cover type–age matrix. This distribution was determined during model development as the final distribution of patches at the end of 300 simulated years. Each patch was considered to be of the same size, so that the area of each dynamic landscape was controlled by the number of vegetation patches. In this study, landscapes consisted of 200, 800, 2000, 8000, or 20,000 patches, to give a total of five landscape sizes. The physical positions of vegetation patches were not tracked. In this respect the model is nonspatial and does not examine responses of animal species to structural phenomena such as patch dispersion and corridors. Such responses may influence the distribution and abundance of vertebrate populations but have been omitted here to focus on patterns of species richness resulting from succession and disturbance.

Population growth, colonization, and extinction

Detailed life history data are seldom available for all the vertebrate species of any particular region. This problem is particularly pertinent when a large pool of species is needed for sampling within a model. Therefore, the species pool used in this model consisted of 150 simulated species. Each characteristic of each species was generated using a normal distribution with fixed parameters. The characteristics consisted of an intrinsic rate of increase ($\bar{x}$ = 3.0, SD = 0.05), a territory size measured in number of vegetation patches ($\bar{x}$ = 20.0, SD = 5.0), a range of vegetation cover types used ($\bar{x}$ = 2.0, SD = 0.5), a range of vegetation ages used ($\bar{x}$ = 15.0, SD = 2.0), and an index of competitive ability ($\bar{x}$ = 5.0, SD = 1.5). The intrinsic rate of increase was used in population-growth equations, and the index of competitive ability served to calculate competition coefficients for pairs of species. These coefficients depend on both the competitive-ability index and the habitat overlap of pairs of species. The ranges of vegetation cover types and ages defined a species' niche and were located randomly within the cover type–age matrix. Each niche was thus represented as a rectangular portion of this matrix and could overlap with other species' niches. The number of vegetation patches found within a species' niche space, along with the territory size needed for individuals of that species, determined the carrying capacity of the landscape for the species. Because the landscape represents a dynamic mosaic of patches, carrying capacities varied through time.

Each simulation began with no animal species occupying the landscape. During each year of a simulation, potential colonists were chosen randomly from the pool of 150 species. Completely random selection of potential colonists gave each species the same probability of immigration. This assumption was biologically unrealistic but made colonization a function solely of habitat availability on the landscape. Nonrandom selection would result in lower species richness. The number of potential colonists was chosen from a normal distribution ($\bar{x}$ = 12.0, SD = 2.0). This mean number of potential colonists represented 8% of the species pool and, during model development, seemed to be the lowest value capable of sustaining colonists on the smallest dynamic landscape. If the carrying capacity of the landscape was equal to or greater than twice the territory size of a potential colonizing species, colonization was allowed and a propagule of two established to represent the minimal breeding population of most vertebrate species.

Once it has colonized, a species may undergo population growth. Because the dynamics of the landscape were modeled in discrete time intervals, population growth was modeled using the logistic growth model as a difference equation:

$$N_{i,t+1} = N_{it} + r_i N_{it}\,[1 - N_{it}/K_i]. \qquad (1)$$

In this equation, N_i is the population size, r_i is the intrinsic rate of increase, K_i is the carrying capacity of the landscape for species i, and the subscript t denotes time in years. As noted previously, carrying capacity may vary and was calculated yearly as

$$K_i = \frac{\sum_{x=\text{MINHAB}}^{\text{MAXHAB}} \sum_{y=\text{MINAGE}}^{\text{MAXAGE}} (P_{xy})}{T_i}. \qquad (2)$$

Here, the carrying capacity of species i is represented by K_i and the species' territory size by T_i. The subscripts on the summation symbols denote the position of the species' niche within the cover type–age matrix. Thus, MINHAB denotes the earliest successional stage and MAXHAB the latest successional stage used by species i. Likewise, MINAGE and MAXAGE symbolize the minimum and maximum vegetation ages used by the species. In those model runs in which interspecific competition was allowed, the equation

$$N_{i,t+1} = N_{it} + r_i N_{it}\left[1 - \frac{N_{it}}{K_i} - \frac{\sum_{j=1}^{n} (\alpha_j N_j)}{K_i}\right] \qquad (3)$$

was used. Here, N_j is the population size of species j, α_j is the product of the proportion of patches that species j shares with species i and the ratio of their competition indices, and n is the number of species established on the landscape.

I considered probability of extinction for species on the landscape as a function of population size, using the expression

$$P = \frac{1}{2^{N-1}}. \quad (4)$$

If the probability P, which is based on the population size N, was greater than a random probability generated each simulated year, extinction of that species occurred. This calculated probability decreased rapidly as population size increased, and therefore extinction was generally reflective of small population sizes. Such small population sizes can result from low landscape carrying capacity (as in the case of specialized or ephemeral habitats) or interspecific competition and low carrying capacity, depending on the population-growth equation used.

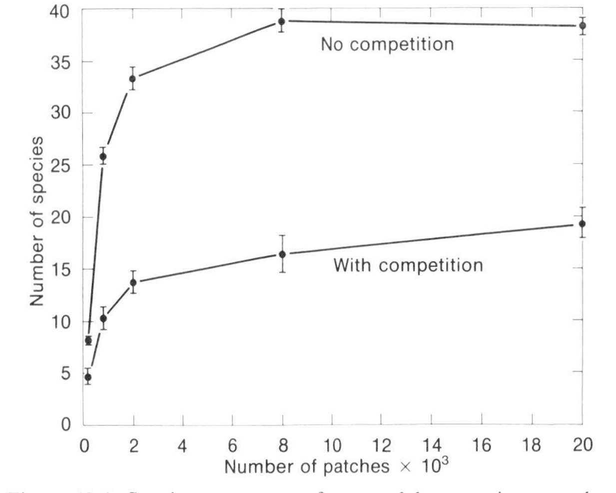

Figure 40.1. Species-area curves from model runs using a random species pool both with and without competition among colonizing species. The mean ±1 *SD* is represented for five landscape areas.

Model runs

Species-area relationships and the effect of landscape dynamics on habitat diversity were explored, using the model. Habitat diversity was calculated, using the Shannon-Weaver Index (Pielou 1975), with each element of the cover type–age matrix representing an individual observation. Results of this exercise should reveal the effect of the interaction between landscape dynamics and landscape size on habitat diversity through time. Ten model runs were made at each of the five landscape sizes to explore the effect of landscape dynamics on the mean and variability of species richness. In addition, 10 other model runs were made allowing competition between colonizing species. Comparison of these results with those of the model with no competition allowed examination of the additional effect of competition on species richness. Finally, I examined the effect of disturbance frequency on species-area relations by establishing two new transition matrices. In one of the new matrices, disturbance frequency was doubled; in the other, disturbance frequency was half that of the original matrix. Note that only those elements of the matrix controlling disturbance frequency were altered; successional rates remained the same. Because of small standard deviations in equilibrium species number, only five model runs were executed at each landscape size for this exploration. Only model runs without competition were made with these new transition matrices. For comparison with empirical species-area studies, I calculated slopes of the simulated species-area curves by log transformation of both species number and area and by linear regression of these variables.

Results

The dynamic landscapes of increasing size resulted in increasing species richness and plateauing species-area curves. With 150 species and no interspecific competition, the species-area curve rose rapidly and plateaued at approximately 38 species (Fig. 40.1). This increase in the number of species with increasing area occurred even though the number of potential colonists varied around a mean of 12 for all landscape sizes. In contrast, when these species were allowed to compete, the curve plateaued at about 17 species (Fig. 40.1). These results corresponded to 25.3% and 11.3%, respectively, of the species pool. Although the equilibrium number of species for these two curves varied, the slopes of the curves were similar. Without competition the slope was 0.31 (r = 0.86), and with competitive interactions among the species the slope was 0.29 (r = 0.94). Colonizing success, and thus species richness, was dependent on habitat availability.

Landscape dynamics and landscape area interacted to affect the diversity and constancy of available habitat (Fig. 40.2). The average diversity for 200, 2000, and 20,000 patch landscapes did not differ greatly (Fig. 40.2); however, the constancy of diversity varied dramatically over the 300 simulated years, because landscape dynamics interacted with landscape island area.

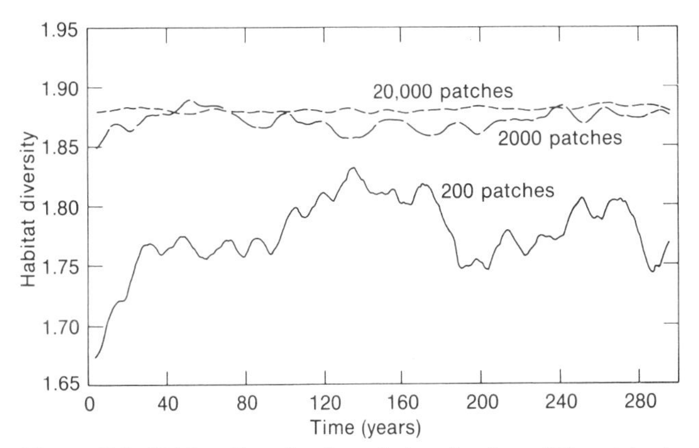

Figure 40.2. Habitat diversity through time for three different landscape-island sizes.

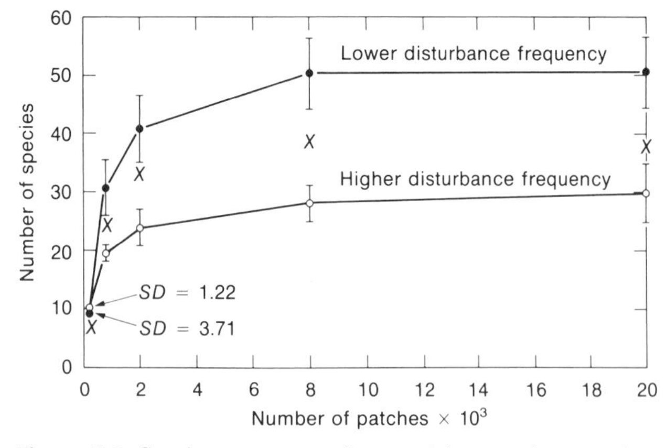

Figure 40.3. Species-area curves from model runs using a random species pool with low and high disturbance frequencies. Mean equilibrium number of species ±1 *SD* are plotted for each landscape area. Mean number of species for the moderate disturbance frequency of Figure 40.1 are indicated by an *X*.

Empirical slope values are known to vary considerably. Doubling the frequency of disturbance over that of the original transition matrix resulted in a slope of 0.22 ($r = 0.89$) for a curve that plateaued around 29 species, or 19.3% of the random species pool (Fig. 40.3). In contrast, when disturbance frequency was reduced by half from the original transition matrix, the curve plateaued at about 51 species, 34.0% of the species pool, and had a slope of 0.34 ($r = 0.81$). These curves bracketed the no-competition curve (Figs. 40.1 and 40.3) and displayed the increasing gradients of species richness and species-area slopes from high to low frequency disturbance landscapes. Although gradual response to disturbance by species-area characteristics was apparent, all three curves converged at smaller landscape areas. In fact, with a landscape composed of 200 patches, the order of mean species richness observed at all larger areas for the three disturbance regimes broke down.

Discussion

The observation that interspecific competition decreased the equilibrium number of species on a landscape is not a surprising result. However, the species-area curves resulting from model runs with and without competition had very similar slopes, which fall within the empirical range of 0.20–0.35 suggested by MacArthur and Wilson (1967). These results indicate that the shape of the species-area curve may be controlled by habitat availability resulting from the interaction of landscape dynamics and landscape area. The imposition of a single succession-disturbance regime on different landscape areas resulted in different patterns of vegetation or habitat diversity and constancy of diversity through time (Fig. 40.2). In terms of animal habitat use, this result reflects different patterns of habitat diversity. Vegetation mosaics are thus affected differentially by area, and small landscapes may be viewed as effectively nonequilibrium (Shugart and West 1981). Because of this condition, small landscapes support populations of animal species subject to higher extinction probabilities, either directly through habitat loss or through low population levels due to minimal habitat availability. The higher chances of extinction ultimately translate to lower species richness. In the case of 200-patch landscapes in Figure 40.3, the landscape is so small that even a regime of low-frequency disturbance creates nonequilibrium conditions, and the order of increasing number of species with decreasing disturbance frequency found at all other landscape areas disappears. Increasing landscape area, while maintaining the same succession-disturbance regime, shifts the landscape mosaic toward an equilibrium state (Fig. 40.2) and decreases the extinction of animal species. This interpretation is illustrated by the plateauing species-area curves of Figures 40.1 and 40.3. Alteration of the succession-disturbance regime, however, resulted in an increased range of species-area curve slopes. Frequency of habitat disturbance may thus be a factor in the variation of species-area slopes.

Of the hypotheses suggested by Connor and McCoy (1979) to explain species-area relationships, my model results are pertinent to two: the area–per se hypothesis and the habitat-diversity hypothesis. Listing these hypotheses separately is somewhat misleading, because habitat diversity is generally a positive correlate of area (Forman and Godron 1981). However, the dynamic landscapes of this model do not produce radically different habitat diversities. Constancy of diversity (Fig. 40.2) is the key characteristic distinguishing different landscape areas. The results suggest that elements of both the area–per se and habitat-diversity hypotheses contribute to the development of species-area relationships through interaction between area and landscape dynamics. Because of the necessary sampling and time considerations, it is not surprising that this interaction has been overlooked in empirical island biogeographic studies, even though changes in habitats through time have been suggested as important biogeographic considerations (Lynch and Johnson 1974; Willis 1974; Karr 1982).

The size and design of nature reserves is a viable area for application of landscape-dynamics concepts. Pickett and Thompson (1978) referred to the smallest area capable of maintaining all ecosystem components in the face of a natural disturbance regime as a "minimal dynamic area." They also suggested that this would be the smallest area necessary to approximate a viable nature reserve. The species-area curves presented here exemplify their concept. Small landscapes that approach disequilibria would be inherently poor nature reserves. Larger landscape areas are able to buffer habitat disruption and thus maintain a higher equilibrium number of species. Only careful analysis of the disturbance frequency for a given ecosystem can reveal its minimal dynamic area and answer the question of nature reserve size and design (Diamond 1975a; Simberloff and Abele 1976; Abele and Connor 1979; Higgs 1981).

Within the framework of this model, the role of interspecific competition in structuring the general shape of the species-area curve is minimal (Fig. 40.1). Although the concept of competition still thrives in ecological theory, evidence indicates that its importance in structuring communities has been overstated (Connor and Simberloff 1979; Strong et al. 1979). The results of this model suggest that competition plays little role in determining at least one community-level characteristic, the slope of the species-area curve. Yet, competition is capable of increasing extinction rates and lowering the equilibrium number of species. The importance of considering competition in such endeavors as developing nature reserves thus becomes a question of the pervasiveness of competition. Ideally, a single functional nature reserve should maintain a complete complement of animal species and, therefore, also maintain any competitive interactions among the species. Preservation of the entire assemblage under such a condition is best addressed by consideration of individual species' characteristics (such as dispersal ability) and the spatial configuration of habitats within a landscape. Spatial habitat considerations are influenced by landscape dynamics as described here but are not addressed explicitly by the model.

The model I have presented here is a simplification of natural ecosystem properties and is based on important concepts of landscape dynamics. I find it gratifying that such an abstraction agrees with empirical observations such as MacArthur and Wilson's (1967) range of slope values. However, other reasons for this characteristic range of slope values have been proposed (May 1975; Connor and McCoy 1979; Sugihara 1980), as have causes of variation within this range (Preston 1960; MacArthur and Wilson 1967; Schoener 1976). Obviously, the ideas presented here are in need of empirical validation; this will be no simple task, because the time and space scales implicit in such studies are much more amenable to computer modeling than to field experimentation. These concepts, though, are the type that must be addressed in future studies of vertebrate ecology on landscapes.

Acknowledgments

I thank H. H. Shugart for his aid in the conceptualization of this model. T. M. Smith, A. J. Hansen, and D. L. DeAngelis commented on earlier drafts of this manuscript. M. L. Tharp provided computer programming assistance. This research was sponsored by the National Science Foundation's Ecosystem Studies Program under Interagency Agreement No. BSR–8315185 with the U.S. Department of Energy, under Contract No. DE–AC05–84OR21400 with Martin Marietta Energy Systems, Inc. Publication No. 2473, Environmental Sciences Division, Oak Ridge National Laboratory, Oak Ridge, Tennessee 37831.

41

Butterfly Diversity in Natural Habitat Fragments: A Test of the Validity of Vertebrate-Based Management

DENNIS D. MURPHY and BRUCE A. WILCOX

Abstract.—The validity of relying solely on vertebrates as indicators of biological diversity is considered by examining butterfly diversity in natural insular habitats. Data on three geographic scales—insular montane ranges, within-range riparian habitat, and riparian 1-ha plots—are analyzed and compared with similar data for mammals and birds. The relatively small effects of habitat area and isolation per se on butterfly diversity suggest that vertebrates provide an adequate protective umbrella for invertebrates at most levels. However, the existence of special habitat requirements for butterflies indicates that consideration must be given to the protection and management of such habitat to provide for overall biological diversity.

Despite a large and growing literature relevant to habitat fragmentation, little attention has been paid to invertebrates. This is particularly distressing for two reasons. First, virtually no basis exists for judging whether wildlife-management programs that traditionally have been designed to protect vertebrates will also protect invertebrates. Second, making the first reason even more compelling, there are ethical and practical reasons (Council on Environmental Quality 1980; Ehrlich and Mooney 1983; Myers 1983), as well as a legal mandate (Kirby 1984), to protect the full range of biotic diversity on federal lands.

Given the number of invertebrates endangered as a consequence of habitat loss (Wells et al. 1983), the threat of habitat fragmentation to the invertebrate component of biotic diversity certainly warrants consideration. Indeed, the known extent of endangered invertebrates undoubtedly represents a mere fraction of those actually endangered or recently extinguished (Pyle et al. 1981). However, from a practical standpoint, monitoring even a minor component of an invertebrate fauna to ensure its persistence is difficult. It is therefore imperative to ask, Will managing for vertebrate diversity, particularly in the context of habitat fragmentation, assure protection for the invertebrate component of biotic diversity as well?

An obvious approach to this problem is to determine whether vertebrate diversity or its habitat determinants correspond to invertebrate diversity in a set of sample plots, forest tracts, or some other conveniently delineated areas. One might suppose, for example, that because of the virtual universality of the species-area relationship, variation in diversity of taxa among a set of such sample areas varying in size ought to be generally concordant. Nonetheless, to our knowledge no studies have examined the concordance of vertebrate and invertebrate diversity in a common suite of sample areas, habitat patches, or true islands. Nor have any general comparative studies of habitat requirements been done.

Here we attempt to address this problem by comparing butterflies with vertebrates on three geographic scales, using natural habitat fragments as analogues of man-made habitat fragments. Butterflies provide a convenient, representative invertebrate taxon for study, as they are taxonomically well described, ecologically well studied, and amenable to visual survey. Butterflies, like most phytophagous insects, are constrained to a relatively narrow range of food plants as larvae; consequently, their habitat requirements could be quite distinct from those of vertebrates.

Methods

Mountain ranges of the Great Basin, representing islands of boreal habitat isolated by arid scrub, were selected as large-scale habitat fragments (Table 41.1). Once more or less continuous throughout the Great Basin, boreal habitat became fragmented and restricted to elevations approximately >2300 m by the warm-dry post-Pleistocene climate. We compiled butterfly species lists for 18 ranges in central and eastern Nevada and western Utah from survey data that we collected as part of a larger study, with additional records from museum collections and from field notes of several private collectors. Data on resident boreal mammal and bird species and biogeographic variables are from Brown (1971, 1978). Size, highest elevation, and distance of each range from the nearest "mainland" (either the Rocky Mountains or the Sierra Nevada) have been readjusted slightly, on the basis of our own measurements. In addition, we use a habitat-diversity index developed specifically for birds in these ranges by Johnson (1975). This index is based on the number of coniferous tree species and habitat types in a mountain range.

On a smaller geographic scale, similar data (excluding mammals) were obtained from linear islands of riparian habitat in 12 canyons on the eastern slope of the Toiyabe Range in central Nevada. Species lists for butterflies, birds, and vascular plants were collected during the summers of

DENNIS D. MURPHY and BRUCE A. WILCOX: Center for Conservation Biology, Department of Biological Sciences, Stanford University, Stanford, California 94305

Table 41.1. List of Great Basin mountain ranges with biogeographic variables

Range	Area (km^2)	Elevation (m)	Distance (km)	Habitat	Birds	Mammals	Butterflies
White-Inyo	1911	4341	16	11	8	11	79
Panamint	122	3367	84	3	5	3	37
Sheep	140	3021	138	5	5	3	40
Spring	324	3633	201	7	6	6	90
Desatoya	215	2991	134	2	4	7	54
Diamond	412	3235	306	—	—	4	54
Roberts Creek	135	3089	348	—	—	4	65
Ruby	943	3471	278	9	6	12	74
Toiyabe-Shoshone	1772	3593	177	7	6	13	96
Toquima-Monitor	3051	3642	183	—	—	10	77
S. Pequop–Spruce	127	3128	251	4	4	4	63
Schell Creek–Egan	2642	3622	183	—	—	8	86
Snake	1080	3982	143	14	9	10	84
Grant–Quinn Canyon	389	3444	222	9	5	5	66
White Pine	679	3410	241	—	—	7	59
Deep Creek	578	3688	167	11	7	8	50
Oquirrh	212	3239	31	9	6	6	63
Stansbury	145	3362	63	8	6	6	74

Note: Area is that above 2300 m in each range. Distance is that to the closest mainland source of colonists, either the Rocky Mountains or Sierra Nevada. The habitat-diversity score and species number for birds and mammals are from Brown (1978), adapted in part from Johnson (1975).

1983 and 1984. Canyon area, elevation, and other physical measurements of the canyons were taken from U.S. Geological Survey 15-min series topographic maps.

Additional data were gathered from 17 individual 1-ha plots selected as representative samples of a variety of riparian habitat types in a subset of the canyons. Bird species lists for each plot were obtained by equal effort surveys during three visits: early June, late June, and early July in 1984. Butterfly lists were assembled from repetitive surveys taken during single visits to each plot in early July. All plots were visited within a 6-day period to minimize the effect of phenological differences on sampling.

Results

Montane-habitat islands

Concordance in number of species was found to exist between butterflies and vertebrates on the scale of entire mountain ranges (Table 41.2); the number of butterfly species was significantly correlated with the number of both mammal and bird species (Table 41.2).

The differences between butterflies and the vertebrate taxa extended to, and were partly explained by, their relationships with biogeographic variables. First, the correlation between the number of butterfly species and area was weak,

Table 41.2. Correlation matrices with physical variables and numbers of species in the Great Basin ranges

	Area	Elevation	Distance	Habitat	Birds	Mammals	Butterflies
Area		0.605 **	−0.085 ns	0.524 *	0.637 *	0.690 **	0.592 **
Elevation	0.737 **		−0.394 ns	0.759 **	0.871 **	0.665 **	0.561 **
Distance	−0.022 ns	−0.379 ns		−0.101 ns	−0.039 ns	−0.138 ns	−0.020 ns
Habitat	0.622 *	0.722 **	−0.176 ns		0.897 **	0.542 *	0.452 ns
Birds	0.700 **	0.869 **	−0.367 ns	0.851 **		0.591 *	0.496 *
Mammals	0.823 **	0.676 **	−0.151 ns	0.538 *	0.610 *		0.705 **
Butterflies	0.614 **	0.550 **	−0.036 ns	0.512 **	0.486 *	0.740 **	

Note: Horizontal headings are untransformed correlations; vertical headings are log-transformed correlations.
ns = not significant.
$^{*}P < 0.05$.
$^{**}P < 0.01$.

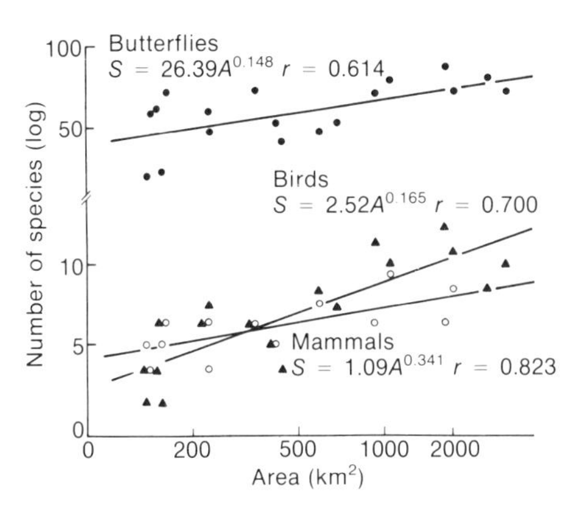

Figure 41.1. Species-area curves (where $S = cA^z$) on a log-log scale for boreal mammals, birds, and butterflies in the Great Basin ranges.

compared with that of mammals and birds (Fig. 41.1, Table 41.3). The discrepancy between taxa was even more marked in the correlation of the number of species with elevation. However, butterflies did share with mammals and birds the lack of a significant correlation with distance from mainland sources of colonists. The weakest concordance in diversity found among the taxa was that between butterflies and birds, which is consistent with a relatively weak correlation found between the number of butterfly species and the habitat-diversity index. This in turn can be explained by differences in the habitat requirements of the two taxa (see Discussion). The birds and butterflies had statistically indistinguishable z-values, and both differed substantially from that of mammals (Fig. 41.1, Table 41.3).

Riparian-habitat islands

On the scale of riparian canyons, the concordance between the numbers of butterfly and bird species improved dramatically ($r = 0.786$, $P < 0.001$). The relative strength of the correlations with environmental variables on this scale for butterflies paralleled that for birds (Table 41.4). Both butterflies and birds were most highly correlated with overall plant diversity and herbaceous plant diversity. Note that in neither taxon was the number of species significantly correlated with the numbers of tree and shrub species.

Again, as with the insular montane faunas, the apparent effect of area was relatively small: for birds $z = 0.421$, for butterflies $z = 0.072$ (Table 41.5, Fig. 41.2). Despite a significant but weak correlation with area, the z-value for butterflies was not significantly different from zero.

Riparian hectare plots

Among the hectare plots within riparian canyons, no significant concordance existed between the numbers of butterfly and bird species ($T = -0.047$, $P = 0.417$, Kendall's rank test). The number of butterfly species was positively correlated with that of plants overall ($T = 0.481$, $P < 0.02$), as well as with herbaceous plants in particular ($T = 0.458$, $P < 0.02$). However, the number of butterfly species was not correlated with that of trees and shrubs ($T = 0.079$, $P = 0.359$). The significances of the correlations for birds were virtually reversed from those for butterflies with the same variables (birds vs overall plant diversity: $T = 0.313$, $P = 0.078$; herbs: $T = 0.339$, $P = 0.063$; and trees and shrubs: $T = 0.432$, $P < 0.05$).

Table 41.3. Parameter values for log-linear and log-log models of the species (S)–area (A) relationships for boreal mammals, birds, and butterflies in the Great Basin ranges

				S vs log(A)		log(S) vs log(A)*			
	$\bar{x}$	SD	n	r	P	r	P	Z	S_o
Mammals	7.05	3.09	18	0.824	<0.0005	0.823	<0.0005	0.341	1.09
Birds	5.92	1.44	13	0.702	0.004	0.700	0.004	0.165	2.52
Butterflies	66.50	17.07	18	0.624	0.003	0.614	0.003	0.150	29.42

Table 41.4. Correlation matrices for physical and biological variables, including species richness of butterflies, birds, and plants for riparian canyons

	Area (ha)	Elevation (m) Highest	Elevation (m) Lowest	Rise (m)	Total plants	Trees and shrubs	Herbs	Grasses	Butterflies	Birds
Area		0.272	−0.474	0.367	0.834**	0.735**	0.787**	0.822**	0.540*	0.742**
Highest	0.123		0.193	0.984**	0.145	0.481	0.015	0.324	0.152	−0.157
Lowest	−0.579*	0.191		0.017	−0.569*	−0.271	−0.583*	−0.477	−0.482	−0.690**
Rise	0.177	0.988**	0.056		0.342	0.566	0.219	0.485	0.242	−0.036
Total plants	0.777**	0.052	−0.640*	0.257		0.857**	0.983**	0.933**	0.764**	0.823**
Trees/shrubs	0.579*	0.476	−0.241	0.543*	0.772*		0.767**	0.921**	0.472	0.484
Herbs	0.734**	−0.093	−0.686*	0.130	0.980**	0.647*		0.858**	0.822**	0.860
Grasses	0.790**	0.264	−0.471	0.408	0.931**	0.877**	0.851**		0.575*	0.693
Butterflies	0.486**	0.166	−0.497	0.251	0.714*	0.456	0.735**	0.569*		0.786**
Birds	0.799**	−0.194	−0.711**	−0.105	0.793**	0.380	0.817**	0.697**	0.753**	

Note: Horizontal headings are linear transformations; vertical headings are log-transformed variables.
* $P < 0.05$.
** $P < 0.01$.

Table 41.5. Physical and biological variables for riparian canyons

	Area (ha)	Length (km)	Elevation (m) Highest	Elevation (m) Lowest	Total plants	Trees	Shrubs	Herbs	Grasses	Butterflies	Birds
Broad	1465.4	6.28	3463	1951	104	8	13	62	18	39	16
Wall	1758.4	8.21	2689	1926	—	—	—	—	—	39	24
Pablo	3356.0	10.06	3412	1951	76	2	10	49	11	41	22
Ophir	1040.1	6.41	3266	1896	97	4	11	65	12	48	25
Summit	919.4	4.88	3180	1935	116	6	10	80	16	51	25
Cove	609.1	4.20	3463	1926	65	4	11	38	9	—	10
Timblin	472.6	3.78	3286	1981	80	4	12	50	11	35	8
Kingston	6170.2	14.61	3497	1926	195	6	23	132	25	56	34
Shoshone	531.1	4.44	3497	2024	52	3	13	25	10	39	10
Sante Fe	636.5	4.44	3497	2024	74	3	15	41	12	42	11
North Twin	4005.3	9.21	3475	1920	134	7	18	75	23	44	19
South Twin	4424.8	12.25	3588	1902	134	7	15	80	21	44	28

Note: Canyon area is that of the entire drainage. Canyon length is from measurable headwaters. Elevation refers to highest and lowest points within the drainage area. Species lists include any occurrence.

In addition, two potentially key determinants of habitat suitability for butterflies, the availability of nectar (see Murphy et al. 1983, 1984) and of standing water (Arms et al. 1974; Adler and Pearson 1982), were tested for their effect on species number. Evidence was found for the importance of both. Plant species diversity and nectar availability were highly correlated ($T = 0.769$, $P < 0.0005$), as were butterfly diversity and nectar ($T = 0.657$, $P < 0.002$). The number of bird species, on the other hand, was not significantly correlated with nectar abundance ($T = 0.123$, $P = 0.238$). Although all but two of the riparian lots were located on watercourses that had running water at the time of the study, only about half also had standing water in the form of seeps or pools, allowing puddling behavior in butterflies. Plots with standing water averaged nine more butterfly species than those having the same plant diversity but lacking standing water (Fig. 41.3). However, no such relationship existed between birds and standing water.

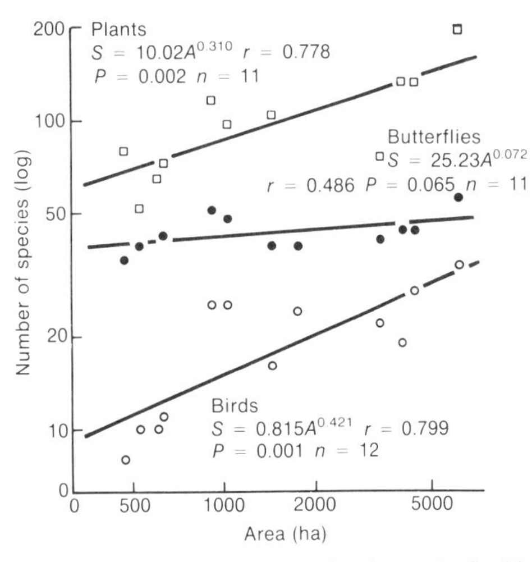

Figure 41.2. Species-area curves on a log-log scale for birds, butterflies, and plants in riparian canyons.

Discussion

The foregoing analysis is relevant to two practical management issues emerging from the currently unprecedented rate of conversion of natural habitat. One is recognition that management efforts must be considered in the context of habitat fragmentation on virtually every geographic level (Harris 1984; Wilcox 1984; Wilcox and Murphy 1985). The other is that management efforts must address the concept of "wild-

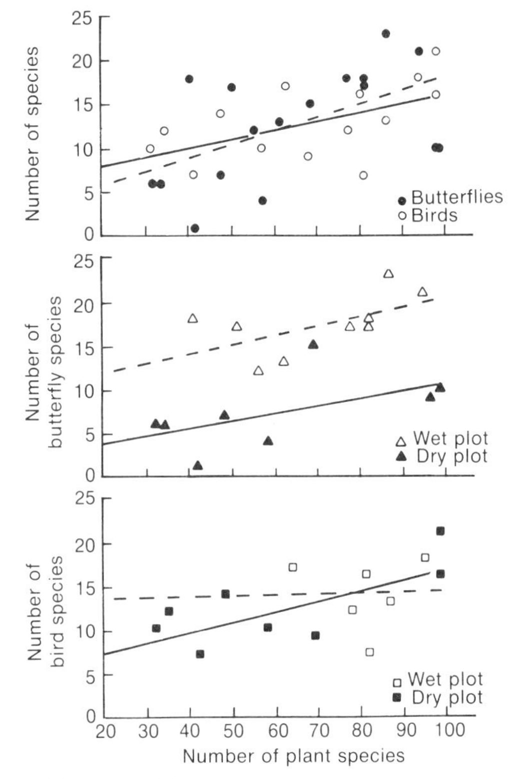

Figure 41.3. Number of bird and butterfly species plotted against the number of plant species in "wet" and "dry" riparian sample hectare plots.

life" in its broadest sense—as the full range of biotic diversity, including vertebrates and invertebrates. Here, by examining the relationships of species diversity to area, isolation, and various characteristics of natural habitat fragments, we draw inferences concerning the effects of habitat fragmentation on biotic diversity. Also, we assess the efficacy of relying solely on vertebrate-oriented management to fulfill the legal requirements to provide for diversity on federal lands.

The potentially negative impacts of habitat fragmentation on diversity are the result of two accompanying factors: decreased habitat area and increased habitat insularity (Wilcox 1980a). The problem of decreased habitat area can be further broken down into the loss of habitat diversity (e.g., types of habitat) and simply the loss in amount of habitat (e.g., extent of area). Both area and insularization (isolation) effects are well documented by island biogeographic studies (see Diamond and May 1982; Diamond 1984a, 1984b), although the effects vary in intensity among taxa.

The slope of the species-area curve, or the z-value (from the relation $S = cA^z$, in which S is the number of species and A is area), and its interpretation have been the subject of frequent discussion in the literature on island biogeography since MacArthur and Wilson (1967). Although the validity of attaching biological significance to the z-value has been disputed (Connor and McCoy 1979), overwhelming evidence shows that variation among observed values is, to a significant degree, the result of differences in immigration and extinction rates among other factors (see MacArthur and Wilson 1967; Schoener 1976; Diamond and Mayr 1976; Wilcox 1980a, 1980b; Martin 1981a; and Wright 1981). Differences in z-values among taxa thus reflect characteristic differences such as those that confer colonizing ability (e.g., flight) and extinction susceptibility (e.g., low population density).

This is illustrated, for example, by the Great Basin boreal mammal and bird faunas (Brown 1971, 1978). The natural fragmentation of boreal habitats precipitated, as Brown (1978) has argued, a dramatic decline in mammalian diversity and at least a moderate decline in avian diversity. Because interrange dispersal is virtually nonexistent for small boreal mammals, these mammal faunas are considered to be nonequilibrium. The difference in z-value (of the species-area relationship) exhibited by birds most certainly can be ascribed to their dispersal advantage (Brown 1978). The similarly low z-value for butterflies probably also follows from the dispersal advantage conferred by flight. Flight, even in very small organisms, confers a certain resistance against habitat insularization not shared by small nonvolant mammals.

The importance of interfragment dispersal in maintaining diversity, however, depends on the effect that fragment size has on the frequency of extinction. The creation of habitat fragments of sizes that affect vertebrates may be of little consequence to invertebrates. This is supported again by the very low z-values for butterflies on both large, boreal-habitat islands and smaller riparian islands, suggesting that the creation of moderate-to-large (about 10^2–10^5-ha) habitat fragments has a relatively small effect on invertebrates. However, two important caveats must be added. First, the butterfly faunas of the Great Basin ranges are already depauperate compared to their probable "mainland" sources of colonization (the Sierra Nevada and the Rocky Mountains). Indeed, certain butterfly genera found in the source ranges are missing from the Great Basin ranges. Most notably, the arctic-alpine component of the boreal fauna is completely lacking from the central Great Basin, although a few species penetrate its fringes. (This is not surprising because the arctic-alpine habitat is the most fragmented as a consequence of post-Pleistocene climatic change.)

Second, the degree of fragmentation, that is, the amount of habitat lost and the distance between the natural fragments, is small compared with that which often occurs as a result of habitat conversion in densely populated regions. We estimate that the boreal-habitat fragments in this study constitute about 30% of the total surrounding land area. Although riparian-habitat fragments are relatively much smaller, most of the fragments are isolated from others by less than a kilometer. Thus, so long as fragmentation is not more extensive than this, these results suggest that managing for the effects of habitat fragmentation for vertebrates may well provide simultaneously for invertebrates.

This conclusion does not address the entire suite of effects associated with habitat fragmentation. Only the overall habitat loss and isolation of fragments are considered and not the habitat disturbance or modification that typically accompanies human-created fragmentation. Our results, in fact, also indicate that the latter effects may render vertebrate-directed management ineffective in providing for invertebrate diversity in some instances. Correlations from analysis using the rangewide bird-habitat diversity score indicates that the habitat determinants of vertebrate and invertebrate diversity differ substantially. This is further supported by our riparian canyon and hectare plot data. Analysis of riparian fragments, the small "islands," indicates that although factors such as plant diversity are important to both groups, other habitat components important to butterflies (standing water and nectar sources) are unimportant to birds. This may explain, at least in part, the discordance between the two taxa on this scale.

Dobkin and Wilcox (Chapter 42) discuss in detail the habitat relations of birds in some of the same canyons. These relations include significant correlations of species diversity with the number of tree species, as well as the occurrence and abundance of certain birds with the abundance of particular tree species. While habitat requirements for avian diversity are difficult to characterize because of their complexity, those for butterflies appear equally elusive because of their subtlety. And, although birds exhibit habitat-specific characteristics, their specificity, particularly in regard to food resources, does not compare in extent to that of butterflies. The habitat requirements for a butterfly species include several key components: (1) adequate numbers of a single or very few host-plant species for oviposition; (2) nectar-source plants; and (3) more cryptic resources, ranging from mutualistic dependencies (for instance, obligatory early-stage associations with ants) to pools of standing water (providing critical minerals).

This study has revealed that less frequently studied habitat components, such as nectar or standing water, may be extremely important determinants of habitat suitability. Indeed, the significant positive correlation between butterfly and plant species diversities masks the presence of two distinct "habitat types"—plots with and without standing water. Such subtle but critical resource requirements must exist for many, if not most, organisms (see Karr and Freemark 1983).

Habitat conversion or modification affecting such components would thus negatively affect diversity. Indeed, it would be possible to disturb habitat in such a way that bird species diversity would be unaffected while butterfly diversity might be severely disturbed, or vice versa. Given the nature of these components, however, fragment size may not be as critical for invertebrates, a thesis supported by the nearly zero z-value found for butterflies in riparian canyons. So long as critical habitat components remain intact, fragment size may have little effect on invertebrates because of the relatively small area requirements for invertebrate populations. Nevertheless, studies of the population structure and dynamics of invertebrates suggest that fragmentation on a scale imperceptible to many vertebrate species can have a serious impact on some invertebrates (Wilcox and Murphy 1985). Many, if not most, invertebrates exist as groups of local populations (a metapopulation), supported by disjunct patches of suitable habitat, which may be subject to local extinction and recolonization. Where both a sufficient number of patches and the capacity for interpatch dispersal exist, a metapopulation can persist. Yet fragmentation can disrupt this immigration-extinction balance through (1) the loss of habitat patches, and thus potential sources of recolonization; and (2) increased insularization of remaining patches, increasing the likelihood of extinction. The latter follows in theory from the reduction of interpatch dispersal and gene flow necessary to provide genetic variability in local populations (Schonewald-Cox et al. 1983) and demographic recruitment for declining populations (Brown and Kodric-Brown 1977).

The efficacy of vertebrate-based management in providing for overall biotic diversity depends upon geographic scale. Geographically extensive fragmentation, although a threat to large vertebrates, may pose little threat to invertebrate diversity. Management of a regional landscape for vertebrate diversity, therefore, will probably satisfy the requirements for invertebrate diversity. On the other hand, fragmentation that is locally intensive may have the opposite effect. Local, selective losses of certain habitat components may cause declines in invertebrates without affecting vertebrates. Thus, to provide for wildlife diversity as specified by many recent legislative and regulatory actions, a broader perspective will be required than traditionally has been taken in wildlife management.

Acknowledgments

Our field team included David Dobkin, Eric Fajer, Jack Fisher, Jennifer Holmes, Steve Kramer, Marian Menninger, Frank Smith, and many others. Critical butterfly data were provided by George Austin and P. J. Savage. We thank Mary Beck for adroitly handling the data analysis, and Paul Ehrlich, Marian Menninger, Julie Armstrong, Carol Boggs, and Richard Holm for reviewing the manuscript. Funding was provided by NSF DAR 8022413, The Nature Conservancy, World Wildlife Fund, Alfred Heller, the Janss Foundation, and the Koret Foundation.

42

Analysis of Natural Forest Fragments: Riparian Birds in the Toiyabe Mountains, Nevada

DAVID S. DOBKIN and BRUCE A. WILCOX

Abstract.—Avian communities were examined in 20 riparian forest fragments of Nevada's Toiyabe Range to assess the relationships of biogeographic factors and habitat components to species richness and community composition. Visual and acoustic surveys were used to sample (1) the avifauna of selected plots in each riparian vegetation type within a fragment; and (2) the avifauna of the entire fragment. The resulting surveys were analyzed for patterns associated with physical and biotic characteristics of the canyon habitats. The species-area relationships for riparian bird species exhibited relatively high z-values, indicative of an insular system, although we found evidence that habitat diversity as well as habitat area affected avian diversity. Although the effect of area appeared to be largely a consequence of its high concordance with habitat diversity, the distribution of many species was affected by reduced area, independent of habitat diversity. In these avian communities, effects of habitat fragmentation might be largely mitigated by managing for habitat diversity and not habitat area per se, although large, continuous habitat fragments are essential for some key bird species.

Studies of habitat "islands" have become increasingly common in the literature of applied island biogeography (e.g., Moore and Hooper 1975; Galli et al. 1976; Higgs and Usher 1980; Whitcomb et al. 1981; Lynch and Whigham 1984). Indeed, for a discipline that began with the study of the geographic distributions of taxa on island archipelagoes (Wallace 1869; MacArthur and Wilson 1967), relatively few reminders of this remain in the contemporary ecological literature. Given that today's continental landscape is becoming predominantly archipelago-like as the result of mankind's accelerating population growth and development, the application of fundamental principles of island biogeography to continental habitat islands may provide us with a tool for mitigating the impacts of habitat fragmentation.

Two important questions are frequently posed regarding the application of island biogeographic theory to noninsular habitat islands: (1) To what extent do habitat islands exhibit the properties of true islands? (2) How do those properties translate into the distribution of species and the composition of communities? We examined these questions by studying the avifauna of an archipelago of habitat islands: fragments of riparian forest confined to canyons along mountain slopes in the central Great Basin. This system has two qualities not shared by most systems that have been studied to date. First, these habitats provide a "natural experiment" in forest fragmentation as a result of post-Pleistocene climatic changes that caused mesic habitats to recede from valley floors. Second, the habitats surrounding the riparian areas are also natural rather than a result of human activities.

David S. Dobkin and Bruce A. Wilcox: Center for Conservation Biology, Department of Biological Sciences, Stanford University, Stanford, California 94305

Study areas and methods

The avifauna of riparian habitats in 20 canyons of central Nevada's Toiyabe Range (approximately 39°N 117°W) was examined from May through July 1983. Within each canyon, sample plots were established in each major vegetation type; from one to four plots were situated in each canyon, depending on canyon size (Table 42.1) and number of habitat types. Elevations ranged from 1896 to 3588 m. The major tree species constituting the riparian habitats were six species of willow (*Salix* spp.), three species of aspen and cottonwoods (*Populus* spp.), and one of birch (*Betula* sp.). Riparian habitats seldom exceeded 30–50 m in width and generally were quite open, thus providing excellent visibility for observers detecting birds.

Within each plot, breeding birds were sampled by using a modified line-transect method (Emlen 1971, 1977). On two consecutive mornings, transects measuring 150 m in length were surveyed by walking both sides of the riparian habitat for a total of 30 min. For each species that was detected in sample plots on both slopes, its relative abundance was compared between slopes with Mann-Whitney U-tests of sample-plot data (Table 42.2). Relative abundance also was estimated qualitatively from sample-plot data combined with foot surveys conducted over the entire length of each canyon. These combined estimates could not be treated statistically because of biases introduced in the collection of foot-survey data. Bird species diversity (BSD) for each canyon avifauna was calculated by using the Shannon information formula for the sample-plot data (Pielou 1969). Species richness (total number of species) was based on the combined results of plot data and foot surveys.

Canyon avifaunas were analyzed for their relationships with physical characteristics (area, length, maximum and

Table 42.1. Physical and biotic attributes of the 17 canyons in central Nevada's Toiyabe Range for which complete bird and plant data were available

Canyon	Area (ha)	Length (km)	Elevation (m)		Plant species			Bird species[a]			BSD[b]
			Highest	Lowest	Total	Tree	Shrub	1	2	3	
East slope											
Timblin	473	3.8	3,286	1,981	80	4	12	6	5	8	1.89
Shoshone	531	4.4	3,497	2,024	52	3	13	3	6	7	1.89
Cove	609	4.2	3,463	1,926	65	4	11	6	5	6	1.95
Summit	919	4.9	3,180	1,935	116	6	10	10	6	11	2.73
Broad	1,465	6.3	3,463	1,951	104	8	13	8	9	9	2.40
Pablo	3,356	10.1	3,412	1,951	76	2	10	9	7	13	2.73
North Twin	4,005	9.2	3,475	1,920	134	7	18	13	9	14	2.58
South Twin	4,424	12.2	3,588	1,902	134	7	15	10	11	11	2.60
Kingston	6,170	14.6	3,497	1,926	195	6	23	16	14	15	3.06
West slope											
Brook	457	3.8	3,103	2,073	80	4	13	5	7	11	2.15
Reeds	656	3.5	2,968	2,024	53	1	13	5	5	12	1.63
Crane	1,011	7.7	3,266	2,201	122	7	19	12	10	15	2.89
Stewart	1,499	8.5	3,353	2,316	155	6	19	15	13	16	3.01
Washington	1,593	9.7	3,345	2,097	138	5	15	13	9	10	2.75
Clear	1,839	7.2	3,162	2,256	118	6	15	13	12	9	2.74
San Juan	2,734	8.4	3,126	2,097	127	4	16	12	10	6	2.39
Reese	14,234	22.3	3,588	2,121	174	4	19	16	10	12	3.11

[a]Number of bird species belonging to each of three groups: 1—riparian species; 2—intermediate species; 3—nonriparian species.
[b]Bird species diversity (Shannon Index).

Table 42.2. Distribution and relative abundance of bird species in riparian canyons of the Toiyabe Range in central Nevada

	East-slope canyons		West-slope canyons		Relative abundance[a]	
	<1000 ha (n = 5)	>1000 ha (n = 6)	<1000 ha (n = 2)	>1000 ha (n = 6)	East slope	West slope
Riparian species						
Northern harrier (*Circus cyaneus*)	—	1	—	—	x	
Spotted sandpiper (*Actitis macularia*)	—	—	—	1		x
Broad-tailed hummingbird (*Selasphorus platycercus*)	5	6	2	6	ns	
Belted kingfisher (*Ceryle alcyon*)	—	1	—	1	ns	
Yellow-bellied sapsucker (*Sphyrapicus varius*)	—	3	—	6		*
Downy woodpecker (*Picoides pubescens*)	—	1	—	3	ns	
Hairy woodpecker (*Picoides villosus*)	1	2	1	4	ns	
Western wood-pewee (*Contopus sordidulus*)	1	—	—	4		o
Western flycatcher (*Empidonax difficilis*)	3	5	2	5	ns	
House wren (*Troglodytes aedon*)	2	6	1	6		*
American dipper (*Cinclus mexicanus*)	—	4	—	2	o	
Swainson's thrush (*Catharus ustulatus*)	—	2	—	5	ns	
Hermit thrush (*Catharus guttatus*)	—	3	—	5		o
American robin (*Turdus migratorius*)	2	5	1	6		*
Warbling vireo (*Vireo gilvus*)	5	6	1	6	ns	
Yellow warbler (*Dendroica petechia*)	—	1	—	4		*
MacGillivray's warbler (*Oporornis tolmiei*)	3	6	2	6	ns	
Lazuli bunting (*Passerina amoena*)	4	5	—	1	o	
Fox sparrow (*Passerella iliaca*)	2	5	—	6	ns	
Song sparrow (*Melospiza melodia*)	—	3	—	4	ns	
Intermediate species						
Cooper's hawk (*Accipiter cooperii*)	2	4	1	3	o	
Northern goshawk (*Accipiter gentilis*)	—	2	—	4		o
Chukar (*Alectoris chukar*)	2	—	—	—	x	
Northern flicker (*Colaptes auratus*)	1	5	—	6	ns	
Empidonax flycatcher (*Empidonax* spp.)	4	[illegible]	—	6	ns	
Violet-green swallow (*Tachycineta thalassina*)	1	6	2	6	ns	
Black-billed magpie (*Pica pica*)	—	2	—	—	x	
Mountain chickadee (*Parus gambeli*)	3	5	2	6		*
Common bushtit (*Psaltriparus minimus*)	4	6	2	6	ns	
White-breasted nuthatch (*Sitta carolinensis*)	—	1	—	4		o
Brown creeper (*Certhia americana*)	—	2	—	2	—	—
Orange-crowned warbler (*Vermivora celata*)	—	1	—	—	x	

Table 42.2 (*Continued*)

	East-slope canyons		West-slope canyons		Relative abundance[a]	
	<1000 ha (n = 5)	>1000 ha (n = 6)	<1000 ha (n = 2)	>1000 ha (n = 6)	East slope	West slope
Yellow-rumped warbler (*Dendroica coronata auduboni*)	1	4	—	6	ns	
Western tanager (*Piranga ludoviciana*)	4	4	2	4	ns	
Black-headed grosbeak (*Pheucticus melanocephalus*)	—	3	1	1	o	
Brewer's blackbird (*Euphagus cyanocephalus*)	—	1	—	1	ns	
Brown-headed cowbird (*Molothrus ater*)	—	2	—	3	ns	
Cassin's finch (*Carpodacus cassinii*)	3	5	2	6	ns	
Nonriparian species						
Sharp-shinned hawk (*Accipiter striatus*)	1	1	—	2	ns	
Red-tailed hawk (*Buteo jamaicensis*)	—	2	2	4	ns	
Golden eagle (*Aquila chrysaetos*)	1	—	2	1		o
American kestrel (*Falco sparverius*)	2	2	2	5	ns	
Prairie falcon (*Falco mexicanus*)	1	2	1	1	o	
Blue grouse (*Dendragapus obscurus*)	1	2	—	—	x	
Band-tailed pigeon (*Columba fasciata*)	—	1	—	—	x	
Mourning dove (*Zenaida macroura*)	1	1	2	2		o
White-throated swift (*Aeronautes saxatalis*)	2	6	1	—	o	
Olive-sided flycatcher (*Contopus borealis*)	1	1	—	—	x	
Scrub jay (*Aphelocoma coerulescens*)	3	2	—	3	o	
Pinyon jay (*Gymnorhinus cyanocephalus*)	1	5	1	5	ns	
Clark's nutcracker (*Nucifraga columbiana*)	3	6	2	6	ns	
Rock wren (*Salpinctes obsoletus*)	4	5	—	3	ns	
Canyon wren (*Catherpes mexicanus*)	1	1	—	—	x	
Mountain bluebird (*Sialia currucoides*)	—	2	—	5		o
Townsend's solitaire (*Myadestes townsendi*)	1	2	—	1		o
Sage thrasher (*Oreoscoptes montanus*)	—	—	—	1		x
Virginia's warbler (*Vermivora virginiae*)	—	1	—	—	x	
Black-throated gray warbler (*Dendroica nigrescens*)	4	5	2	3	ns	
Green-tailed towhee (*Pipilo chlorurus*)	2	4	2	4	ns	
Rufous-sided towhee (*Pipilo erythrophthalmus*)	5	6	2	4	*	
Chipping sparrow (*Spizella passerina*)	2	6	2	4	ns	
Brewer's sparrow (*Spizella breweri*)	—	1	—	3		o
White-crowned sparrow (*Zonotrichia leucophrys*)	—	—	—	1		x
Dark-eyed junco (*Junco hyemalis*)	—	3	1	5		o
Pine siskin (*Carduelis pinus*)	—	2	1	5	ns	

Note: The number of canyons in which each species was detected is indicated for small and large canyons of the east and west slopes. Comparisons of relative abundances between slopes used sample-plot data analyzed by Mann-Whitney U-tests. One small, dry canyon which contained no true riparian habitat is omitted from this analysis.

[a] * = significantly greater abundance in sample plots of the indicated slope ($P < 0.05$); ns = no significant difference in abundance between slopes; x = detected on one slope only; o = detected in sample plots of one slope only.

minimum elevations) and with biotic characteristics (total vascular plant species richness and woody [trees plus shrubs] plant species richness) of the canyon habitats. (Plant data were collected by other members of our research group and are based on collections made by foot survey over the entire length of each canyon.) To disentangle the extensive intercorrelations among variables, higher-order partial correlations were carried out to control for effects of physical variables (area, length, and elevation) and for effects of variables assumed to be related to habitat diversity and quality (total vascular plant species richness and total woody plant species richness). Partial correlation analyses using log-transformed data were performed with standard SPSS routines; nonparametric analyses were derived from Siegel (1956). Classic island biogeographic analyses of species-area relationships were assessed with the logarithmic model $S = cA^z$ (Preston 1962). Correlation analyses were based on the 17 canyons for which we had complete bird and plant data (Table 42.1).

Results

As determined by their presence, condition (based on mist-net samples), and behavior (carrying nest materials, feeding young, etc.), 65 bird species were considered as breeding in the 20 surveyed canyons of the Toiyabe Range. On the basis of our observations of habitat use and of the species' ecologies and behaviors (and information derived from the literature), species were classified into three groups (Tables 42.1 and 42.2): (1) species dependent entirely upon riparian habitat (riparian—20 species); (2) species that make extensive use of riparian habitat but are not entirely dependent upon it (intermediate—18 species); and (3) species that rarely use or entirely avoid riparian habitat (nonriparian—27 species).

Species' distribution patterns

Of the 65 breeding species, all but 12 were detected on both the east and west slopes (Table 42.2); nine species were found only on the east slope, and three species were found only on the west slope. Each of these 12 species was detected in only a single canyon, and in each case very few individuals were detected (a single singing male, a pair, or one family group). A ubiquitous "core group" of nine species occurred in nearly every canyon on both slopes: broad-tailed hummingbird, western flycatcher, Clark's nutcracker, mountain chickadee, common bushtit, warbling vireo, MacGillivray's warbler, Cassin's finch, and rufous-sided towhee.

On the basis of sample-plot data, six species exhibited significant ($P < 0.05$) differences in relative abundance between slopes (Table 42.2). Yellow-bellied sapsuckers, mountain chickadees, house wrens, American robins, and yellow warblers were more abundant on the west slope; only rufous-sided towhees were more abundant on the east slope.

On the basis of the combined plot and survey data (see Methods), many additional species appeared to differ in relative abundance between slopes. The northern goshawk, red-tailed hawk, American kestrel, mourning dove, downy woodpecker, western wood-pewee, violet-green swallow, white-breasted nuthatch, hermit thrush, Swainson's thrush, mountain bluebird, pine siskin, and dark-eyed junco appeared to be more abundant on the west slope. In addition to the rufous-sided towhee, only the white-throated swift, rock wren, and lazuli bunting displayed greater abundances on the east slope.

Canyon and plant attributes—Comparisons between east and west slopes

East-slope canyons are steeper and tend to be somewhat shorter than west-slope canyons of comparable area (Table 42.1). In addition, because the valley floor to the west of the Toiyabe Range was higher than the valley to the east, canyons on the east slope began at significantly lower elevations than canyons on the west slope ($P < 0.001$, Mann-Whitney *U*-test). Similarly, east-slope canyons achieved significantly higher maximum elevations than west-slope canyons ($P = 0.05$; Table 42.1).

Species richness of vascular plants was not significantly different ($P > 0.10$) between slopes, although west-slope canyons tended to be richer (Table 42.1). Woody plant species diversity, however, was significantly greater ($P = 0.025$) for the west-slope canyons, because of the presence of a greater number of shrub species. Birch was more abundant on the east slope and was missing entirely from several canyons on the west side. Aspen was more abundant and occurred in more extensive stands on the west slope. The only extensive stands of large aspen on the east slope occurred in the three largest canyons: Kingston, North Twin, and South Twin.

No statistically significant differences were found in bird species richness, species diversity, or the number of species in any of the three bird categories between slopes. West-slope canyons, however, tended to have higher numbers of riparian and intermediate species and tended to be richer in total species number (Table 42.1). Among the east-slope canyons, only the three largest harbored bird and plant communities that were comparable in species richness to medium and large canyons on the west slope.

Canyon and plant attributes—Comparisons within slopes

Simple correlation analysis (zero-order partials) revealed many statistically significant relationships among variables, not all of which were congruent between slopes (Table 42.3). For example, total avian species richness, species richness for each group, and bird species diversity were highly significantly correlated with total plant species richness on both slopes. In contrast, woody plant species richness was highly correlated with riparian, intermediate, and total bird species richness on the west slope, but it was well correlated ($P < 0.01$) only with intermediate bird species richness on the east slope. Although length and area were highly correlated ($P < 0.001$) on both slopes, consistent correlations between length or area and bird species richness were only found for the east slope.

Higher-order partial correlation analyses indicated strong associations between avian community attributes and both area and habitat parameters (Table 42.4). Both amount and diversity of habitat were strongly associated with total bird species richness on the east slope, but habitat diversity appeared to be more strongly associated with bird species richness on the west slope. Habitat diversity, rather than amount, was more strongly associated with riparian species richness on both slopes. For intermediate bird species, strong associations with both amount and diversity (especially woody plant species richness) of habitat were seen for east-slope canyons, but west-slope canyons showed a strong association only with habitat diversity. Nonriparian birds showed no significant second-order partial correlations. Avian species diversity was strongly associated with features of habitat diversity on the west slope, but showed no statistically significant relationships on the east slope.

Species-area relationships

Nearly half of the 65 breeding species exhibited area-related distributions and rarely or never occurred in canyons of less than 1000 ha (Table 42.2). Such distributions were found for 75% (15 species) of the riparian species, 50% (nine species) of the intermediate species, and only 25% (seven species) of nonriparian species.

Species-area relationships for the total avifauna (Fig. 42.1*a*) showed a significant correlation only for the east slope (Table 42.3), with a relatively high z-value of 0.32 (west-slope z-value $= 0.13$). Combining the three categories of birds, however, masked important patterns that were revealed by separately analyzing the riparian, intermediate, and nonriparian components of the avifaunas (Fig. 42.1*b* and *c*). All three categories of birds exhibited statistically significant relationships on the east slope (z-values $= 0.40$, 0.32, and 0.28 for riparian, intermediate, and nonriparian, respectively). Although the combined west-slope species-area relationship was not significant, that for riparian birds alone was significant (Table 42.3), with a z-value of 0.33 (0.14 and -0.05 for intermediate and nonriparian, respectively).

Table 42.3. Pearson product-moment correlation analyses of avian, plant, and physical parameters for east- and west-slope canyons of the Toiyabe Range. (All data were log-transformed.)

	East-slope canyons						West-slope canyons					
	Total plants	Woody plants	Length	Area	High	Low	Total plants	Woody plants	Length	Area	High	Low
All birds	0.92***	0.66*	0.90***	0.93***	0.20	−0.67*	0.87**	0.93***	0.74*	0.54	0.79*	0.83**
Riparian birds	0.92***	0.57	0.75*	0.82**	0.02	−0.81**	0.94***	0.85**	0.87**	0.74*	0.78*	0.65
Intermediate birds	0.83**	0.83**	0.90***	0.89***	0.57	−0.52	0.88**	0.88**	0.63	0.49	0.60	0.86**
Nonriparian birds	0.79**	0.45	0.83**	0.88**	0.05	−0.50	0.02	0.31	0.00	−0.18	0.32	0.37
Bird species diversity	0.82**	0.45	0.83**	0.86**	0.05	−0.63	0.96***	0.93***	0.79*	0.57	0.84**	0.68
Total plants			0.73*	0.77*	0.11	−0.71*			0.89**	0.73*	0.87**	0.58
Woody plants			0.57	0.57	0.44	−0.42			0.69	0.47	0.73*	0.78*
Length				0.99***	0.54	−0.59				0.93***	0.92***	0.30
Area					0.48	−0.66					0.76*	0.14
High						−0.19						0.38

*$P < 0.05$.
**$P < 0.01$.
***$P < 0.001$.

Discussion

Differences between bird communities and between the physical and vegetational characteristics of the east- and west-slope canyons were largely attributable to the steeper, narrower nature of the east-slope riparian habitats and to the more extensive stands of older aspen with relatively well-developed herbaceous understories in the west-slope canyons. Species associated with these latter habitats included the yellow-bellied sapsucker and downy woodpecker, both of which were found primarily on the west slope. The sapsucker and, to a lesser extent, the downy woodpecker are especially important components of these riparian communities, because they are primary cavity-nesters that provide nest sites for secondary cavity-nesters. These include the violet-green swallow, mountain chickadee, white-breasted nuthatch, house wren, and mountain bluebird, all of which were more abundant in west-slope riparian habitats. Similarly, the hermit thrush, Swainson's thrush, and dark-eyed junco were associated with the herbaceous understories of aspen stands. In contrast, species that were more

Table 42.4. Second-order partial correlation analyses of avian, plant, and physical parameters for canyons of the Toiyabe Range, statistically controlling for area vs habitat effects

	Length and area				Total plants and woody plants			
	East-slope canyons		West-slope canyons		East-slope canyons		West-slope canyons	
	Total plants	Woody plants	Total plants	Woody plants	Length	Area	Length	Area
All birds	0.87**	0.49	0.49	0.79	0.88**	0.92**	0.18	0.17
Riparian birds	0.81*	0.31	0.70	0.64	0.38	0.51	0.28	0.37
Intermediate birds	0.65	0.91**	0.91**	0.80*	0.94**	0.90**	−0.62	−0.29
Nonriparian birds	0.32	−0.10	−0.34	0.27	0.69	0.72	0.26	0.02
Bird species diversity	0.44	−0.08	0.91**	0.85*	0.70	0.69	−0.32	−0.43

* $P < 0.05$.
** $P < 0.01$.

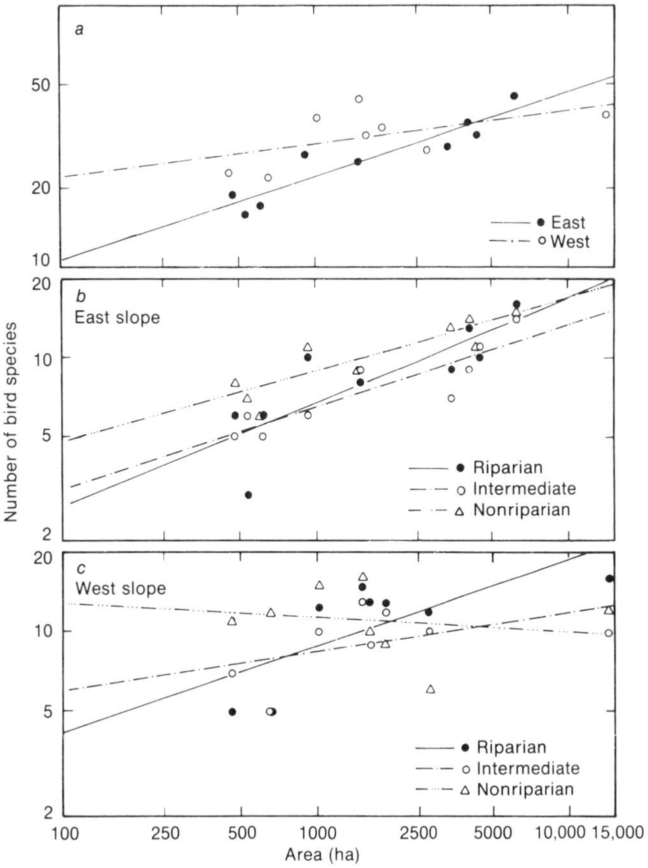

Figure 42.1. (*a*) Species-area relationships for avifaunas of riparian canyons in central Nevada's Toiyabe Range. East-slope avifauna had a *z*-value = 0.32; west-slope avifauna had a *z*-value = 0.13. (*b*) Species-area relationships for east-slope avifaunas divided into three components based on each species' degree of ecological dependence on riparian habitat. Fully dependent riparian species show a *z*-value = 0.40; intermediate species' *z* = 0.32; nonriparian species' *z* = 0.28. (*c*) Species-area relationship as in (*b*) for west-slope avifaunas. Riparian species' *z* = 0.33; intermediate species' *z* = 0.14; nonriparian species' *z* = −0.05.

abundant on the east slope are not riparian species and are associated strongly with habitat attributes of narrower, steeper canyons (white-throated swift, rock wren, and rufous-sided towhee). It is not clear why the lazuli bunting was more abundant on the east slope.

Although birch occurred in steeper, narrower portions of the canyons on both slopes, it was more abundant and stands were more extensive on the east slope because canyons were significantly steeper there. Bird communities associated with birch-dominated stands were impoverished, both in density and in number of species. In contrast, mature stands of aspen contained a relatively diverse avian community, as described above. Although canyon physiography may largely account for the distribution of birch, an important contributor to the scarcity of large, older aspen stands on the east slope is the history of more extensive mining and logging operations there, compared to the west slope. The field journals of Jean Linsdale (housed in the Museum of Vertebrate Zoology, University of California, Berkeley) refer repeatedly to such activities in east-slope canyons. For example, an entry of 29 May 1930 noted that aspen trees in Ophir Canyon had been cut for mining operations, hence only young trees were present. Linsdale also noted the presence of extensive logging roads in the Wisconsin Creek drainage. That such anthropogenic activities have had significant impact on the structure and composition of bird communities in the Toiyabe is corroborated by the complete absence of yellow-bellied sapsuckers and downy woodpeckers from the Toiyabe prior to 1940 (and probably for a decade or more later) (Linsdale 1936, 1938). Linsdale (1938) also noted the relative rarity of hairy woodpeckers, which were most common in stands of aspen found in the uppermost reaches of a few canyons.

The impact of cattle grazing on these riparian habitats was difficult to assess because of the absence of grazing records for individual canyons. Canyons with signs of recent sustained use by cattle contained plant and animal communities that were structurally and taxonomically impoverished and were excluded from further study. The most floristically diverse canyons appeared to have had little if any recent cattle grazing, as judged by the absence of bovine fecal material.

Three-fourths of the riparian species and one-half of the intermediate species showed definite area-related distributions and did not occur in canyons smaller than 1000 ha. Uncommon species (e.g., northern harrier, spotted sandpiper, and white-crowned sparrow) were especially likely to occur in only the largest canyons. The belted kingfisher was a particularly good example of a species with a distribution that was dependent on the complex interrelationship between area and habitat variables. It occurred only in the largest canyon of each slope: Kingston on the east and Reese on the west. Only these two canyons contain bodies of water large enough to provide the vertical sandbank walls where nest burrows can be placed and persistent enough to provide reliable sites to forage for fish.

These riparian habitats are highly linear, regardless of total area. Although it has been shown that forest-interior birds of eastern North America show pronounced edge effects, illustrating the importance of "core area" rather than total forest area (Temple, Chapter 43), riparian habitats are inherently narrow, necessitating a lowered sensitivity to "edge" on the part of species that use it. Nevertheless, below some threshold width, riparian habitats will begin to lose species (Stauffer and Best 1980).

Interpretation of the slope of the species-area relationship has received considerable attention (Connor and McCoy 1979; Gould 1979; Sugihara 1981; Martin 1981a; McGuinness 1984). As a matter of convenience, we have referred to the slope as "*z*" from the power relation $S = cA^z$, although the exponential relation can provide an equally good fit (McGuinness 1984). Although Connor and McCoy (1979) argue that no basis exists for attributing any biological significance to the *z*-value, others have shown that differences among archipelagoes can be explained by environmental diversity, size of the species pool, isolation, and geographic scale (area) (see Martin 1981a). Also, differences in *z*-values among taxa can be explained by characteristic immigration and extinction rates (Diamond and Mayr 1976; Schoener 1976; Wilcox 1980a, 1980b; Wright 1981).

Regardless of the underlying explanations for the *z*-value, it is a potentially useful empirical tool for testing sensitivity to fragmentation effects. In theory, *z*-values should increase along with sensitivity to area or isolation, or both. Comparison of the insular boreal butterfly, bird, and mammal faunas of the Great Basin ranges (Brown 1978; Murphy and Wilcox, Chapter 41) provides a good illustration. The dramatically higher *z*-value for small boreal mammals, which follows from their relatively low vagility, demonstrates a high degree of sensitivity to fragmentation.

Similarly, the high *z*-values found here for riparian birds suggest increased sensitivity to fragmentation for those species dependent upon riparian habitat, relative to intermediate and nonriparian species. This also confirms the view that these riparian fragments tend to represent islands for some species more than for others. Overall, the results suggest an effect of reduced habitat area on bird species that are at least partly dependent on these highly linear forest habitats. However, the statistical effect of area appears to be largely a consequence of its high concordance with habitat diversity. Therefore, at least in these avian communities, effects of habitat fragmentation might be largely mitigated by managing for habitat diversity and not habitat area per se, although large, continuous habitat fragments (especially of aspen communities) are essential for some key bird species.

Acknowledgments

We thank Jennifer Holmes and David Fortna for help in collecting the bird data, Frank Smith and Jaynie Belnap for collecting plant data, and Charles Metzler and Mary Beck for computer programming and assistance. Paul Ehrlich, Jennifer Holmes, Michael Morrison, and Jared Verner provided helpful comments on the manuscript. Funding was provided by NSF grant DAR 8022413, Alfred Heller, the Janss Foundation, the Koret Foundation of San Francisco, and the World Wildlife Fund–U.S. Special thanks to the Tonapah Ranger District of the USDA Forest Service for logistical support.

43

Predicting Impacts of Habitat Fragmentation on Forest Birds: A Comparison of Two Models

STANLEY A. TEMPLE

Abstract.—Most models used to predict responses of bird species to fragmentation of deciduous forest habitat in eastern North America have focused on total forest area as the main predictor of each bird species' presence and abundance in a fragment. I present a new model that gives more accurate predictions of the presence and abundance of those bird species sensitive to fragmentation; it relies on core area (the area of forest more than 100 m from an edge) instead of total area as the major habitat variable. The core-area model is more accurate because the total-area model does not correctly predict the presence and abundance of birds in forest fragments that are large in total area but, because of their shapes, have little core area. Several ecological explanations can be given as reasons why forest birds respond to core area rather than to total area of fragments. Management recommendations derived from these two models diverge in important ways and illustrate some of the pitfalls in making management recommendations on the basis of wildlife-habitat models.

In the extensively fragmented habitat of the eastern deciduous forest of North America, the total area of a forest fragment is a major habitat variable commonly used to predict the presence and abundance of forest bird species in a fragment (Forman et al. 1976; Galli et al. 1976; Whitcomb 1977; Robbins 1979; Goldstein et al. 1981; Whitcomb et al. 1981; Ambuel and Temple 1983; Lynch and Whigham 1984). The relationship between total area of a fragment and bird populations is so consistent that the resulting predictive model has been used to develop specific management recommendations for forest birds (e.g., Robbins 1979; Whitcomb et al. 1981).

In this chapter I show that the total-area model is not the most accurate predictor of bird populations in a fragment and, furthermore, that some of the management recommendations derived from the total-area model may not correctly address the specific ecological problems encountered by many forest birds in a fragmented habitat. An alternative model that uses the core area of a fragment (the area more than 100 m from an edge) is compared with the total-area model.

Methods

Forty-nine forest fragments in the south-central Wisconsin counties of Columbia, Dane, Dodge, Fond du Lac, Green, Green Lake, Jefferson, Rock, and Sauk were selected as study sites. These study sites shared several characteristics: all contained mature upland-forest types characteristic of the region and were surrounded by open nonforest habitat; none had standing or running water, recent cutting or grazing, buildings, conifer plantations, or steep topography. I used a planimeter to measure total areas of forest fragments on aerial photographs; I measured core areas in a similar fashion, but I measured only the area of forest more than 100 m from edges at which forest adjoined nonforest habitat. I used a 100-m buffer zone around the core area because previous studies have shown that, at distances of over 100 m from an edge, many of the negative impacts associated with edge habitat have ameliorated (Gates and Gysel 1978; Ranney et al. 1981; Brittingham and Temple 1983; Temple 1983). The ranges of total areas and core areas included in the study areas are shown in Figure 43.1.

Breeding birds (except raptors, crows, gallinaceous birds, and shorebirds) on each study area were counted, using a line-transect method, as described in detail by Ambuel and Temple (1982). Results yielded density estimates that were used to calculate the estimated number of individuals of each species in each forest fragment.

Bird species sensitive to fragmentation were identified by comparing their frequencies of occurrence in forest fragments with core areas of 0–10, 11–100, and >100 ha. If significant differences in frequencies of occurrence of a species were detected, that species was considered to be sensitive to fragmentation.

For each bird species that was sensitive to fragmentation, I calculated linear regressions in which the estimated number of individuals of each species in a fragment was a dependent variable, and either total area or core area was an independent variable. All variables were log-transformed before analysis. For each of the resulting regressions, I calculated a coefficient of determination (r^2), which revealed the percentage of total variation in number of individuals of each species that was accounted for or explained by the regression model (Zar 1984). The coefficients of determination for the total-area and core-area models were then compared to see which model accounted for the greatest proportion of variation in bird numbers.

STANLEY A. TEMPLE: Department of Wildlife Ecology, 1630 Linden Drive, University of Wisconsin–Madison, Madison, Wisconsin 53706

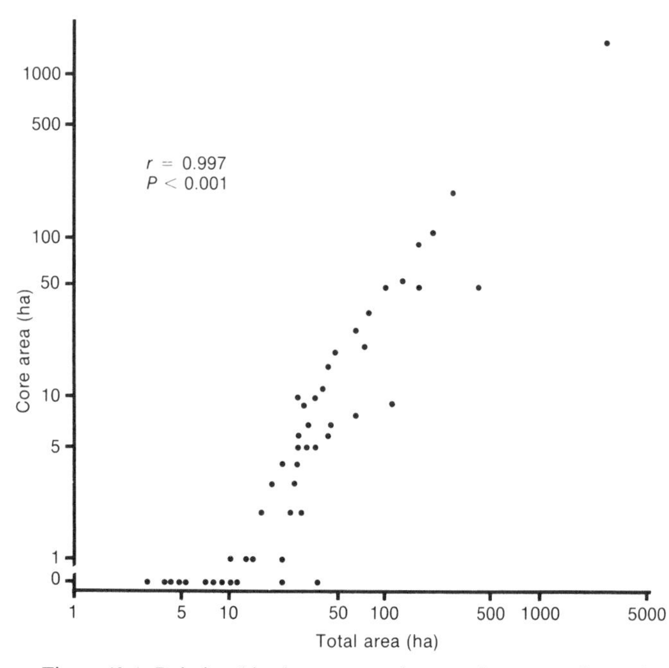

Figure 43.1. Relationships between total area and core area for each of the 49 forest fragments used to develop regression models.

To further test predictions generated by the core-area and total-area models, I selected 10 additional forest fragments, not used to develop the regression models, that met the criteria used for selecting study areas and that had been censused for birds as part of another study (Tilghman and Rusch 1981). For each of the species that proved to be sensitive to fragmentation, I used two-tailed paired-sample t-tests (Zar 1984) to compare the population numbers predicted by the total-area and core-area models with actual density estimates from the 10 fragments.

Results

Sixteen of the 43 species encountered on the study areas showed a distributional pattern of occurring less frequently, if at all, in smaller-sized fragments (Table 43.1); these were the species sensitive to fragmentation. The remaining species occurred at similar frequencies in forest fragments of all sizes.

Linear regressions of the estimated number of individuals of each species sensitive to fragmentation against either total area or core area of fragments were always significant. Both the total-area and core-area models were, therefore, fairly good predictors of the presence and abundance of birds, but the two models were not equally accurate. Comparisons of the coefficients of determination for the two linear regression models revealed consistent differences in the accuracy with which they predicted the number of individuals of each bird species in forest fragments. The core-area model consistently accounted for more of the variation in population estimates of species sensitive to fragmentation than the total-area model (Table 43.2). Inspection of the data revealed that this difference was primarily the result of the total-area model's erroneous predictions of population sizes of birds in fragments that were relatively large in total area but, because of their shapes, had little core habitat. For example, Figure 43.2 contrasts the occurrence of birds sensitive to fragmentation in two forest fragments that were similar in total area but, because of their geometric configurations, had markedly different core areas.

A further indication of the validity of the core-area model came from comparisons of the population estimates predicted by the two models and the observed population sizes for 10 forest fragments that had not been used in developing the models. The predicted population sizes from the core-area model were significantly different ($t > 2.26$, $P < 0.05$) from observed population estimates for two of the 16 species. On the other hand, predictions from the total-area model were significantly different from observed population estimates in nine of the 16 species.

Discussion

Most investigators have found significant associations between the presence and abundance of birds sensitive to fragmentation and the total areas of forest fragments. As a result, total area has been a useful predictor of the bird species likely to be found in a fragment. The strong association has also led to the formulation of specific management recommendations for forests birds. These recommendations place great emphasis on total area of forest fragments as one of the major habitat variables affecting the distribution of these birds in fragmented habitat. In this chapter I, too, have demonstrated a significant association between bird populations and total area of fragments, but I have also discovered an even stronger association between bird population and the core area of fragments.

Total area and core area were significantly correlated (Fig 43.1), and the relationship was particularly strong among fragments that had a relatively compact shape. The relationship was weaker, however, when I included fragments whose shapes result in a low ratio of core area to total area, because of the high proportion of edge habitat in such fragments.

Recommendations that stress total area per se as the principal habitat variable of concern for managing forest birds may passively permit practices that are actually detrimental to birds sensitive to fragmentation—the very group that the recommendations are often supposed to favor. As my results have suggested, management practices that preserved large areas of forest, but permitted those areas to have elongated rather than compact shapes, indented rather than entire unbroken perimeters, or inclusions of open habitat within the fragment rather than a solid forest stand, would not benefit forest birds sensitive to fragmentation. The results of this study suggest that it would be far better for managers concerned with forest birds to maximize core areas of forest fragments rather than area per se.

Table 43.1 Percentages of forest fragments in each of three size-classes that supported breeding populations of bird species

Species[a]	Percent occurrence in fragments of indicated size		
	0–10 ha (n = 10)	11–100 ha (n = 31)	>100 ha (n = 8)
Red-headed woodpecker (*Melanerpes erythrocephalus*)	90	100	100
Downy woodpecker (*Picoides pubescens*)	80	94	100
Hairy woodpecker (*Picoides villosus*)*	50	42	100
Northern flicker (*Colaptes auratus*)	100	100	75
Pileated woodpecker (*Dryocopus pileatus*)*	0	0	75
Eastern wood-pewee (*Contopus virens*)	90	87	100
Acadian flycatcher (*Empidonax virescens*)*	0	0	75
Least flycatcher (*Empidonax minimus*)*	10	6	75
Great crested flycatcher (*Myiarchus crinitus*)	100	100	100
Eastern kingbird (*Tyrannus tyrannus*)	10	30	25
Blue jay (*Cyanocitta cristata*)	100	100	100
Black-capped chickadee (*Parus atricapillus*)	90	94	100
Tufted titmouse (*Parus bicolor*)*	10	45	75
White-breasted nuthatch (*Sitta carolinensis*)	100	94	100
House wren (*Troglodytes aedon*)	100	100	88
Blue-gray gnatcatcher (*Polioptila caerulea*)*	0	16	75
Veery (*Catharus fuscescens*)*	20	32	88
Wood thrush (*Hylocichia mustelina*)*	40	65	100
American robin (*Turdus migratorius*)	100	100	100
Gray catbird (*Dumetella carolinensis*)	100	100	100
Brown thrasher (*Toxostoma rufum*)	40	42	50
European starling (*Sturnus vulgaris*)	80	48	75
Yellow-throated vireo (*Vireo flavifrons*)*	20	94	100
Red-eyed vireo (*Vireo olivaceus*)	90	94	100
Chestnut-sided warbler (*Dendroica pensylvanica*)*	0	6	88
Cerulean warbler (*Dendroica cerulea*)*	0	16	100
American redstart (*Setophaga ruticilla*)*	0	0	75
Ovenbird (*Seiurus aurocapillus*)*	10	39	100
Mourning warbler (*Oporornis philadelphia*)*	0	6	50
Common yellowthroat (*Geothlypis trichas*)	100	65	88
Hooded warbler (*Wilsonia citrina*)*	0	0	88
Scarlet tanager (*Piranga olivacea*)*	20	100	100
Northern cardinal (*Cardinalis cardinalis*)	90	100	100
Rose-breasted grosbeak (*Pheuticus ludovicianus*)	80	100	100
Indigo bunting (*Passerina cyanea*)	100	100	100
Rufous-sided towhee (*Pipilo erythrophthalmus*)	20	32	75
Song sparrow (*Melospiza melodia*)	50	48	50
Red-winged blackbird (*Agelaius phoenicius*)	100	100	100
Common grackle (*Quiscalus quiscula*)	100	100	100
Brown-headed cowbird (*Molothrus ater*)	100	100	100
Northern oriole (*Icterus galbula*)	70	87	75
American goldfinch (*Carduelis tristis*)	20	19	25
House sparrow (*Passer domesticus*)	40	23	50

[a]Species marked with an asterisk are considered sensitive to fragmentation because the original data showed significant differences (chi-square contingency table analyses, $P < 0.05$) in frequency of occurrence between size-classes of fragments.

There seems to be a fundamental ecological explanation for these conclusions. Because of the detrimental impacts of nest predation, brood parasitism, and competition that are associated with edge habitats (Gates and Gysel 1978; Ambuel and Temple 1983; Brittingham and Temple 1983; Temple 1983), large forest fragments may favor forest birds, not so much because of their total areas, but because they often have secure core areas within them. As my results have shown, forest birds do not occur regularly in fragments that are large in total area but lack a secure core area.

This discussion highlights one of the pitfalls that attends the modeling of wildlife-habitat relationships with the goal of making specific recommendations for habitat management. On occasions, such as the one under discussion, two or more strongly intercorrelated habitat variables may appear to be important predictors of a species' response, but they lead to divergent management recommendations that could have contradictory impacts on wildlife. Only by understanding the ecological processes underlying the patterns used to generate the models can managers decide which models and resulting recommendations are most likely to have the desired impact on target populations.

Table 43.2 The percentages of total variation (r^2) in population estimates for bird species in fragmented forests that are explained by fitted regressions from either the total-area model or core-area model

Species	Coefficient of determination (r^2)[a]	
	Total-area model	Core-area model
Hairy woodpecker	0.615	0.718
Pileated woodpecker	0.892	0.994
Acadian flycatcher	0.949	0.984
Least flycatcher	0.928	0.986
Tufted titmouse	0.499	0.647
Blue-gray gnatcatcher	0.829	0.987
Veery	0.814	0.985
Wood thrush	0.698	0.766
Yellow-throated vireo	0.927	0.944
Chestnut-sided warbler	0.968	0.982
Cerulean warbler	0.947	0.992
American redstart	0.508	0.820
Ovenbird	0.948	0.994
Mourning warbler	0.949	0.982
Hooded warbler	0.897	0.992
Scarlet tanager	0.929	0.974

[a]Values were calculated as r^2 (regression sum of squares/total sum of squares) for those regression equations that were shown by analysis of variance to be significant ($P < 0.05$).

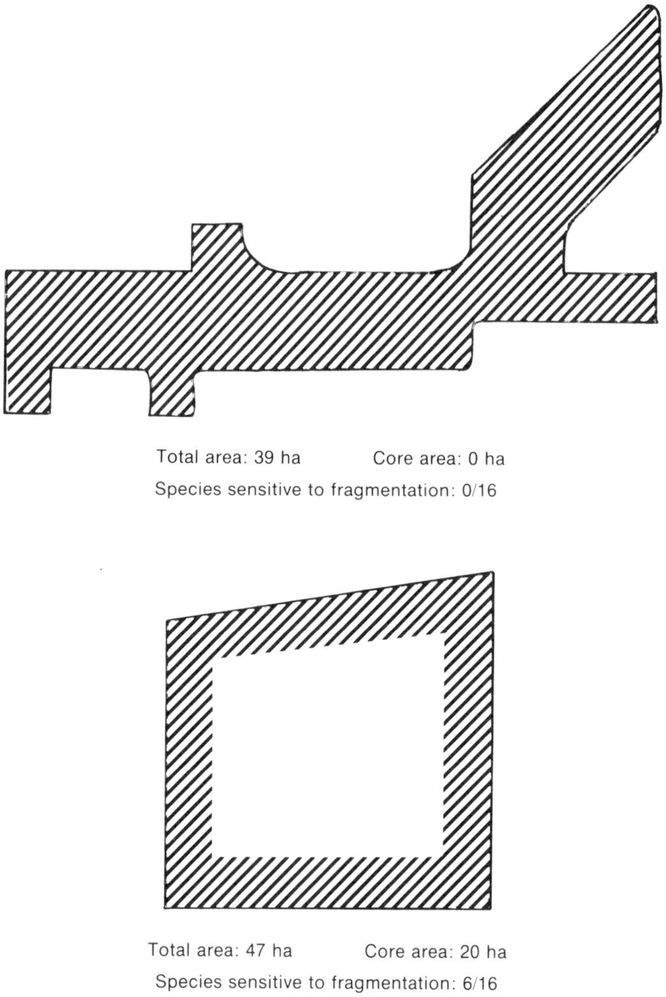

Acknowledgments

This research was supported by the Agricultural Experiment Station, College of Agricultural and Life Sciences, University of Wisconsin–Madison. Count data were collected by B. H. Ambuel and N. G. Tilghman. B. H. Ambuel, J. J. Hickey, and M. C. Brittingham reviewed the manuscript.

Total area: 39 ha Core area: 0 ha
Species sensitive to fragmentation: 0/16

Total area: 47 ha Core area: 20 ha
Species sensitive to fragmentation: 6/16

Figure 43.2. A comparison of the occurrence of birds sensitive to fragmentation in two forest fragments with similar total areas but markedly different core areas.

44

Effects of Forest Fragmentation on New- and Old-World Bird Communities: Empirical Observations and Theoretical Implications

CHARLES H. McLELLAN, ANDREW P. DOBSON, DAVID S. WILCOVE, and JAMES F. LYNCH

Abstract.—The effects of forest fragmentation on the bird communities of England and the eastern United States are considered, using complementary sets of empirical data. Nomograms for both countries reveal that a series of small reserves will contain more species than a single large reserve of the same total area, but that the large reserves are needed to preserve a number of area-sensitive species. A simulation model is then presented to illustrate the key effects of fragmentation on the species pool of an originally contiguous habitat. The model suggests that extinctions of species are initially low, but increase rapidly once a critical percentage of the original habitat has been destroyed. This percentage depends crucially upon both the territory sizes and dispersal abilities of the species pool under consideration. We then discuss how the optimum conservation strategy for preserving woodland birds will depend upon the number of area-sensitive species, the slope of the species-area curves, and the extent to which the habitat has already been fragmented. We conclude by discussing the work's more general implications for conservation policymakers on a variety of different geographic, taxonomic, and administrative scales.

Habitat fragmentation, the process whereby contiguous tracts are reduced to numerous smaller tracts isolated from one another by development of some kind, poses two major conservation problems: (1) What is the effect of fragmentation on the pool of species originally present in the intact habitat? (2) Given that many habitats are already fragmented, what are the guidelines available for determining the optimal size distribution and spatial configuration of fragments that are to be designated as nature reserves?

The majority of previous studies on this topic have concentrated on answering the second of these questions (e.g., Diamond 1975a; Wilson and Willis 1975; Simberloff and Abele 1976, 1982; Higgs and Usher 1980). We will continue this tradition by concentrating particularly on the question of whether it is better (in terms of the total number of species present) to concentrate a given total area of habitat in a few large reserves or to subdivide it among numerous small ones. We will then address the first question by developing a simulation model, based on the MacArthur and Wilson (1967) equilibrium model, which calculates the proportion of the original species pool expected to remain in a region as the habitat in the region becomes progressively fragmented.

The initial empirical analysis in Section I uses two extensive sets of data on breeding woodland birds, one from England and the other from the eastern United States. Woodland in both regions is now considerably fragmented. However, marked differences exist in the time-scales concerned. English deciduous woodland has been fragmented for over 2000 years (Rackham 1976), whereas the eastern deciduous forest of the United States was largely intact as recently as 300 years ago (Terborgh 1975). Present patterns of species richness are discussed in the light of these differences and the results of the modeling exercise described in Section II.

Although the empirical analysis concentrates on the effects of fragmentation on the bird community, we conclude by suggesting some more general rules that should apply to conservation strategies designed to protect all animal and plant species. These rules suggest that the controversy over reserve area strategies may be best resolved by considering variation in key biogeographic parameters with the geographic scale of conservation decision making.

Section I: Reserve-area strategies

Methods

Data for England were taken from the Register of Ornithological Sites. This is the result of an extensive investigation into the bird species composition of a variety of habitats organized by the British Trust for Ornithology (see Fuller 1982). Most of the 197 woods selected were located in central and southern England. The woods were all either deciduous or mixed deciduous-coniferous, with a tendency for coniferous stands to be more prevalent in the larger size-classes. Woods ranged in size from 1.5 to 1117 ha. Only discrete

Charles H. McLellan: Centre for Environmental Technology, Imperial College, London, SW7 2BB, England

Andrew P. Dobson and David S. Wilcove: Department of Biology, Princeton University, Princeton, New Jersey 08544

James F. Lynch: Smithsonian Environmental Research Center, P.O. Box 28, Edgewater, Maryland 21037

woods separated from similar habitat by at least 50 m of open land were used. Breeding-species lists were compiled by volunteer observers, largely during the period 1972–1976. Most of these woodlots were censused repeatedly throughout the breeding season for a period of 2–3 years. Analysis was restricted to a pool of 54 species that regularly breed in woodlands and also do most of their feeding there. It should, however, be noted that none of these species is entirely restricted to forest habitat.

The study area in the United States was Maryland, where the data were collected by Lynch and Whigham (1984). The 270 woods censused were broadly similar to the British ones: they were discrete patches, mostly deciduous, and ranged in size from 2 to 1136 ha. They were all separated from another wood, in this instance, by a minimum of 10 m of open land, though more than 98% of the woods were separated by at least 100 m. Breeding-species lists were compiled by means of point surveys, forest tracts < 50 ha being censused at one point, tracts 50–100 ha at two points, and tracts > 100 ha at three. Sixty-two species were included in the analysis, of which 45 were classified as forest-dwelling species. The latter group is largely restricted to forest habitat. Further details of the data collection are given in Lynch and Whigham (1984).

The species lists for the British woodlots are probably close to complete, given the extensive coverage of each site. This is less likely to be the case for the American woodlots, where only 60 min of field time were spent at any one survey point. Although more point surveys were made in the larger (>50 ha) tracts, species lists from these tracts are probably less complete than species lists from the smaller ones. However, despite the shortcomings inherent in data collected in this way, we believe the similarities between the two data sets far outweigh the differences between them. They are without question the most extensive data sets available on woodland birds for the two continents and, provided their limitations are clearly understood, their use in a comparative study of this sort can only be profitable.

Both data sets were stored as arrays of presence/absence records for each species in each wood, along with the area of the wood. Relationships between breeding-bird species number (S) and woodland area (A) were investigated. Two functions, a log-transformed power function $\ln S = \ln k + z(\ln A)$, and a log-transformed exponential function $S = a + b(\ln A)$, were fitted to each data set, using standard linear regression techniques (Sokal and Rohlf 1969).

We compared the relative merits of concentrating a given total area of woodland in a few large reserves or subdividing it among many small ones, over a wide range of area, by constructing nomograms. To achieve this, the woods in each data base were divided into six area-classes (0–10, 11–20, 21–40, 41–80, 81–160, and >160 ha). We first determined the mean number of species in a single wood for each area-class. Next, we calculated the mean number of species in 30 combinations of two randomly chosen woods from each area-class. Each species was counted as present only once so as to obtain the total number of species present in the combination. We continued this procedure for three, four, and so on,

Table 44.1. Species-area regressions for the numbers of species of birds expected in woods of different sizes in England and in eastern North America

Country	Function	Slope	Intercept	r	r^2
Britain	$S = a + b\ln A$	3.395	19.78	0.646	0.418
USA	$S = a + b\ln A$	1.533	15.85	0.406	0.165
Britain	$\ln S = \ln k + z\ln A$	0.105	3.066	0.628	0.395
USA	$\ln S = \ln k + z\ln A$	0.067	2.797	0.370	0.137

Note: All of the fitted regression lines were significant at the $P < 0.001$ level.

up to the number of woods in the area-class. Equal species contours were then fitted to the resultant array of numbers. By superimposing contours of equal total area, the nomograms can be used to get a direct estimate of the reserve area strategy that will maximize breeding-bird species richness. This is done by following the appropriate area contour until it intersects the maximal species contour.

Results

The species-area correlations between bird species number and woodland area were considerably higher in England than in the United States. In both countries, the exponential function $S = a + b\ln A$ provided slightly the better fit to the data (Table 44.1). The main feature of both species-area relationships was the large amount of variation in S that was unaccounted for by dependence on A (58.2% and 83.5% in England and America, respectively). Both log-log slopes (commonly denoted by z) were low compared to the 0.15–0.35 range commonly encountered in insular studies (see May 1975; Connor and McCoy 1979). There was a significant positive correlation between woodland area and the proportion of forest-dwelling species breeding in Maryland woods ($r = 0.248$, $n = 270$, $P < 0.001$; arcsine-transformed). In the case of the American woodlots, the low slope may also stem in part from the incomplete surveys from the larger forests.

Nomograms

Both nomograms (Figs. 44.1 and 44.2) show that one to a few large woods contain fewer species than the same total area subdivided among up to 20 small woods, at least over the area ranges encompassed by the data. In England a single 200-ha wood will, on average, hold about 60% of the species pool, while five 40-ha woods or 20 10-ha woods will contain over 80% of the pool (Fig. 44.1). In Maryland (Fig. 44.2) a 200-ha wood will hold less than 50% of the regional species pool, whereas 10 20-ha woods will contain about 70%. The increase in species richness with subdivision is matched by a decrease in the variance in the number of species we would expect the woodland to contain (Fig. 44.3). As all the area-classes contain approximately equal numbers of woods, this reflects a real change in variation rather than an artifact of sampling size. However, although subdivision is seen to initially give an increase in species number, the curve soon levels off at high levels of subdivision and may

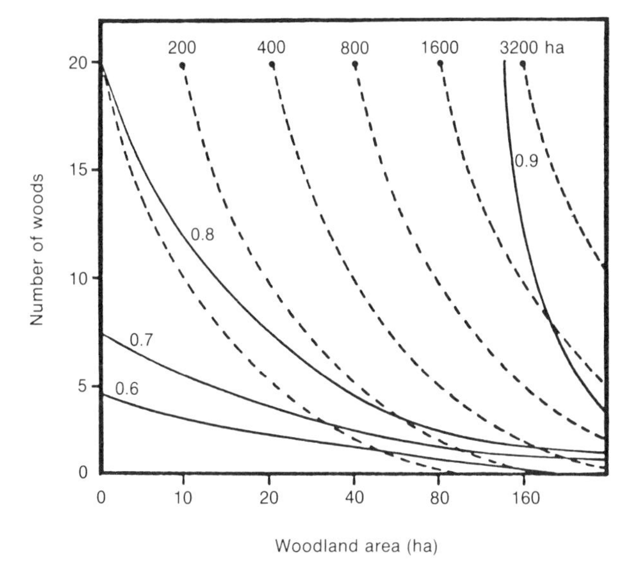

Figure 44.1. Nomogram of numbers of bird species expected in combinations of woodland fragments in southern and central England. The broken lines are contours of equal total area; the solid lines are contours of proportions of the total species pool.

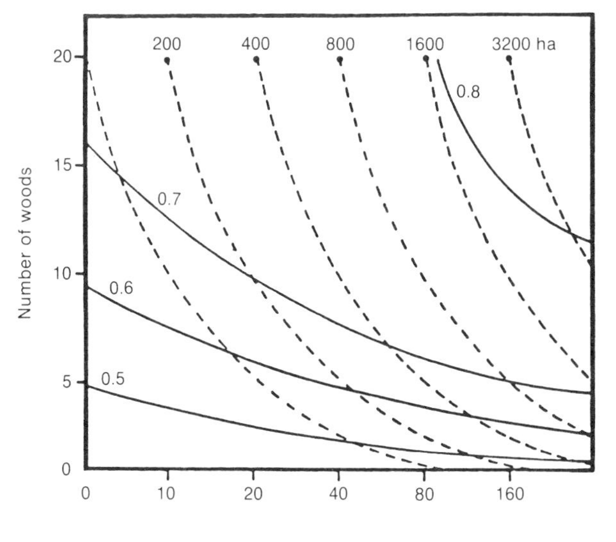

Figure 44.2. Nomogram of numbers of bird species expected in combinations of woodland fragments in Maryland, USA. The broken lines are contours of equal total area; the solid lines are contours of proportions of the total species pool.

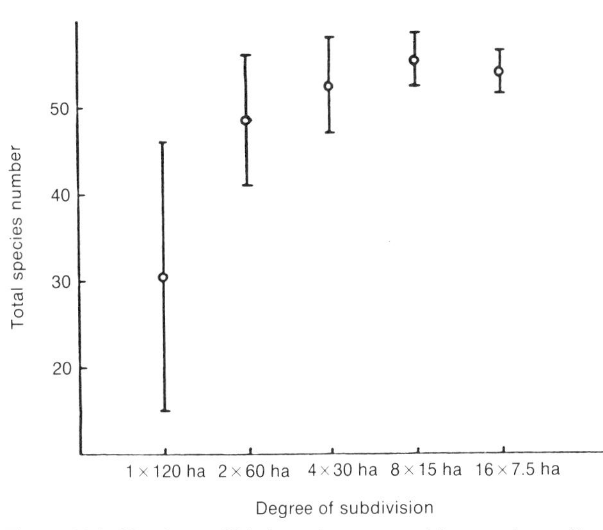

Figure 44.3. Numbers of bird species expected in a total woodland area of 120 ha at several different levels of subdivision. The data are for the British woods; the confidence intervals are 2 *SE*.

even turn over once the mean size of the fragments approaches the mean territory size of many of the species (Galli et al. 1976). Such an effect would be more apparent if our samples had included more very small woods (e.g., <5 ha).

Two features of the nomograms deserve special attention. First, neither nomogram depicts a strategy whereby combinations of similar-sized woods can capture 100% of the species pool. Second, the maximal percentage of the pool that can be captured in this way is greater in Britain (approximately 90%) than in the United States (approximately 80%).

Discussion

Whether one reserve or two or more adding up to the same total area hold more species depends empirically upon the interaction between the slope of the species-area relationship and the similarity in species composition among the small reserves (Simberloff and Abele 1976; Abele and Connor 1979; Higgs and Usher 1980). When the slope (z) is steep, a single reserve is likely to be favored, unless the smaller reserves are sufficiently dissimilar in species composition to outweigh the area effect, in which case subdivision will be optimal. Both z-values reported here are low (<0.2), so it is not surprising that area subdivision emerges in our nomogram analysis as the strategy that maximizes the number of species preserved.

A major difference between the nomograms for Britain and the United States is the maximum proportion of the species pool that can be captured in combinations of similar-sized woods. In Britain increasing subdivision of resources can allow a comparatively modest total area to contain over 90% of the total species pool. By contrast in the United States, we are unable to obtain more than 75% of the species pool by subdivision. The prime reason for this involves differences in the species pools of the two countries. About 30% of the American species pool is composed of edge species that breed in forest clearings or other habitats besides undisturbed forest interior. Clearly, conservation efforts should be directed preferentially to the forest-dwelling species which make up the remaining 70% of the species pool. Several of these forest-interior species are confined to large (>100 ha) patches. In Britain the picture is somewhat different because few, if any, of the "woodland" species are confined to undisturbed woodland interior. This difference is probably due largely to the histories of fragmentation in the two countries. The larger members of the original forest-interior guild in Britain, such as the goshawk (*Accipiter gentilis*), went extinct long ago, while other species, such as the robin (*Erithacus rubecula*), have been able to adapt to the new range of habitats in the fragmented landscape.

Although the data have shortcomings, we believe that the methods we have used are of a priori interest to conservationists, who should perhaps try applying the above techniques to their own data. Although the American set of data might be improved by more intensive sampling, it is unlikely that an operation on the scale of the BTO sites registry could be mounted. It therefore seems sensible to initiate conservation efforts based on the data available now, while understanding their shortcomings, rather than to spend valuable time and funds in increasing sampling effort in order to produce more rigorous guidelines at a later time, when the actual conservation situation could be much worse.

Section II: Habitat fragmentation

Methods

In this section, we describe the results of a computer simulation designed to mimic the effects of fragmentation on species pools with different minimum area requirements and dispersal abilities. The fragmentation of our hypothetical habitat is based on the broad patterns revealed in studies by Moore (1962) and Jones (1973) on heathland and chalk grassland, respectively, in Dorset (England) and by Sharpe et al. (1981) on woodland in Wisconsin (USA). For simplicity we have shown the total area of habitat decreasing linearly over time (Fig. 44.4); in reality, the rate of destruction varies (usually increasing) with time. The total number of fragments tends to increase exponentially with time, reflecting an increasingly skewed distribution toward a large number of very small fragments. We have accordingly assigned a roughly lognormal shape to the distribution of fragment size at each of the stages of fragmentation (Fig. 44.5).

The assumptions of the model used to calculate the total number of species remaining in the distribution of fragments at successive stages of exploitation are described in Appendix I. We have illustrated the results of the simulation for two different hypothetical species pools (Fig. 44.6). In one, the species have comparatively modest area requirements and are able to migrate readily between habitats. In the other case, the species are less vagile and require much larger areas of land to sustain viable breeding populations.

Results

The numbers of species remaining, expressed as percentages of the original species pool P, are depicted for our two

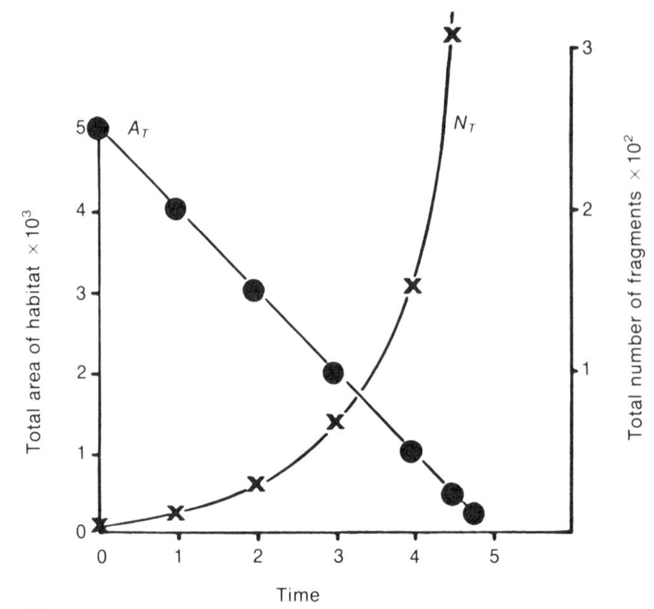

Figure 44.4. Data used in the simulation study of the effects of fragmentation: the circles represent the total area remaining (A_T) at seven different stages of fragmentation; the crosses represent the number of fragments (N_T), or individual woodlots, at each successive stage.

species pools for seven sequential stages of fragmentation (Fig. 44.7). As the habitat is reduced to 5% of its original extent and fragmented from five extremely large tracts to over 450 medium-sized to very small ones, the pattern of species loss in the two species pools is interestingly different. At first, when there is a large amount of habitat and many large fragments, no species are lost from either pool. When about 50% of the original habitat remains, the increasingly skewed distribution of fragment area, and increasing average level of isolation, render the landscape incapable of supporting several species from the pool with larger area requirements and restricted vagility. Thereafter, species are lost from this pool at a fairly steady rate, until only 30% of the original pool remains when 95% of the original habitat has been destroyed. The second species pool, with its less extensive area requirements and greater vagility, is much more resilient to fragmentation. Species begin to go extinct only when about 20% of the original habitat remains. However, it appears that once this critical total area is reached, the subsequent collapse may be both severe and rapid, since the habitat is much more highly fragmented by this stage. The effect of fragmentation per se, independent of reduction in the total area of habitat, is further illustrated by the dotted line in Figure 44.7. This is the proportion of the first species pool that would be present in the same total area of habitat when minimally fragmented (e.g., the total area in each frequency-class is concentrated into just one wood). It is evident that many fewer extinctions occur with this alternative.

Discussion

Habitat fragmentation is an insidious process which, because of its gradual, piecemeal nature and its occurrence

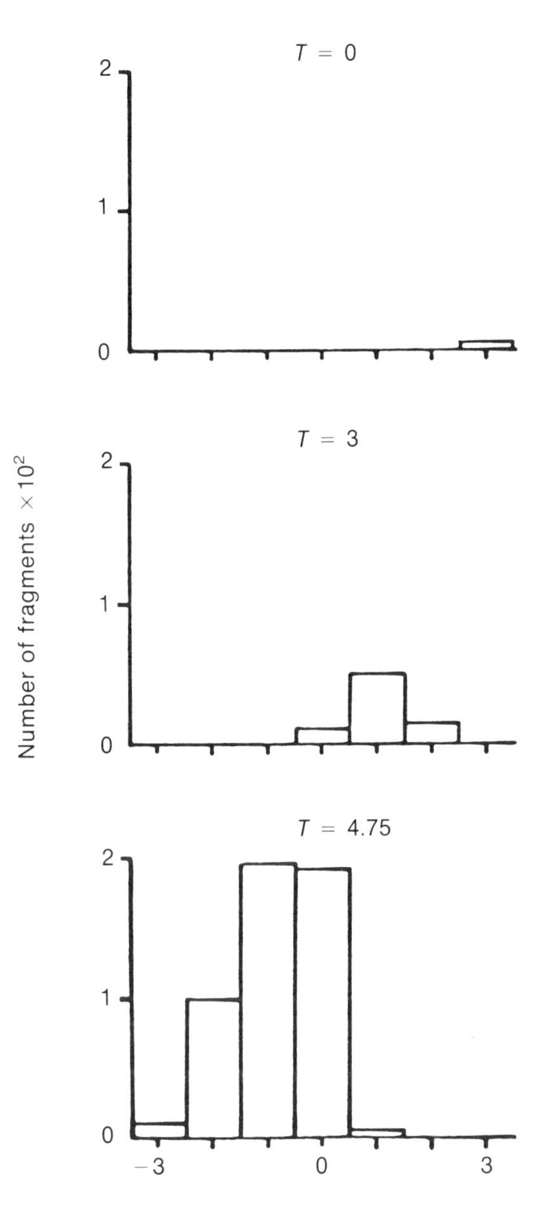

Figure 44.5. The frequency distributions of fragment sizes at various stages of the fragmentation sequence. The top diagram corresponds to the original, largely contiguous, forest; the two lower diagrams to different, smaller total areas with increasing numbers of fragments.

over large geographic areas, is likely to be difficult for land-use managers to monitor and control. Unfortunately fragmentation is almost certainly the ultimate cause, if not always the proximate cause, of many reported declines and extinctions of valued species of wildlife (Terborgh and Winter 1980). For example, Wilcove (1985) has shown that nest predation rates in small (<15 ha) forest fragments in Maryland are extremely high by comparison with rates in large (>500 ha) tracts. Differential rates of nest predation are clearly an important cause of the well-documented reductions in abundance and losses of migratory forest-interior songbirds from small fragments of eastern deciduous forest in the United States (see Whitcomb et al. 1981). Yet it is clearly the fragmentation of the original contiguous forest into numerous small, isolated tracts that has allowed this and other proximate causes to come into operation.

Thus, although we do not believe that the oversimplified equilibrium model we have employed here is an entirely adequate description of these processes, it may serve as a useful first approximation. The most crucial effect it highlights is the increased rate of local extinction in small fragments. If species within a given pool differ markedly in their propensity to local extinction, then the pattern of species loss both locally and regionally will be highly nonrandom. Those species with the highest extinction rates and lowest colonization rates will disappear first. Terborgh (1974) called these "extinction-prone" species. In general, species confined to habitats found only in the larger fragments (interior species) will be more extinction-prone than species which can utilize habitats also found in the surroundings (edge species). Among birds, two categories of interior species are particularly extinction-prone (Terborgh and Winter 1980): (1) large, often predatory, species with extensive territory requirements; and (2) species that specialize on habitats which are patchily distributed within and between fragments. Since the occurrence and abundance of suitable habitat may not always be a simple function of fragment area, extensions of the models employed here will have to be developed which explicitly include the effects of habitat heterogeneity and the habitat requirements of species.

Despite its limitations, our model emphasizes several points which we feel land-use managers should be aware of. First, different taxa in the same habitat will respond differently to fragmentation, depending on their relative extinction and colonization rates. For example, among the vertebrates, the nonflying mammals (with their relatively poor dispersal ability) are more susceptible to fragmentation than the birds considered above. The amphibians and reptiles (with high population densities) may be particularly resilient to fragmentation. The bats have low population densities but good dispersal ability, and are therefore likely to be intermediate in their susceptibility to fragmentation (Wilcox 1980a). Second, it appears that fragmentation per se is a bad thing, particularly when the total area of habitat remaining is small. Below about 50% of the original area of habitat, a large fragment of area A_T contained more species from our susceptible pool at equilibrium than did the same area when fragmented and distributed roughly lognormally (see Fig. 44.7). Although our model may slightly overstate the case against fragmentation because of its omission of the effects of habitat heterogeneity, we feel that the general message is clear: in regions where a large amount of habitat (>50%) still remains, subsequent fragmentation should be minimized; where reduction and fragmentation are considerably advanced, special priority should be given to maintaining the integrity of the remaining large fragments.

Finally, it should also be borne in mind that if the timescale over which reduction and fragmentation have occurred is short by comparison with that of species' population dy-

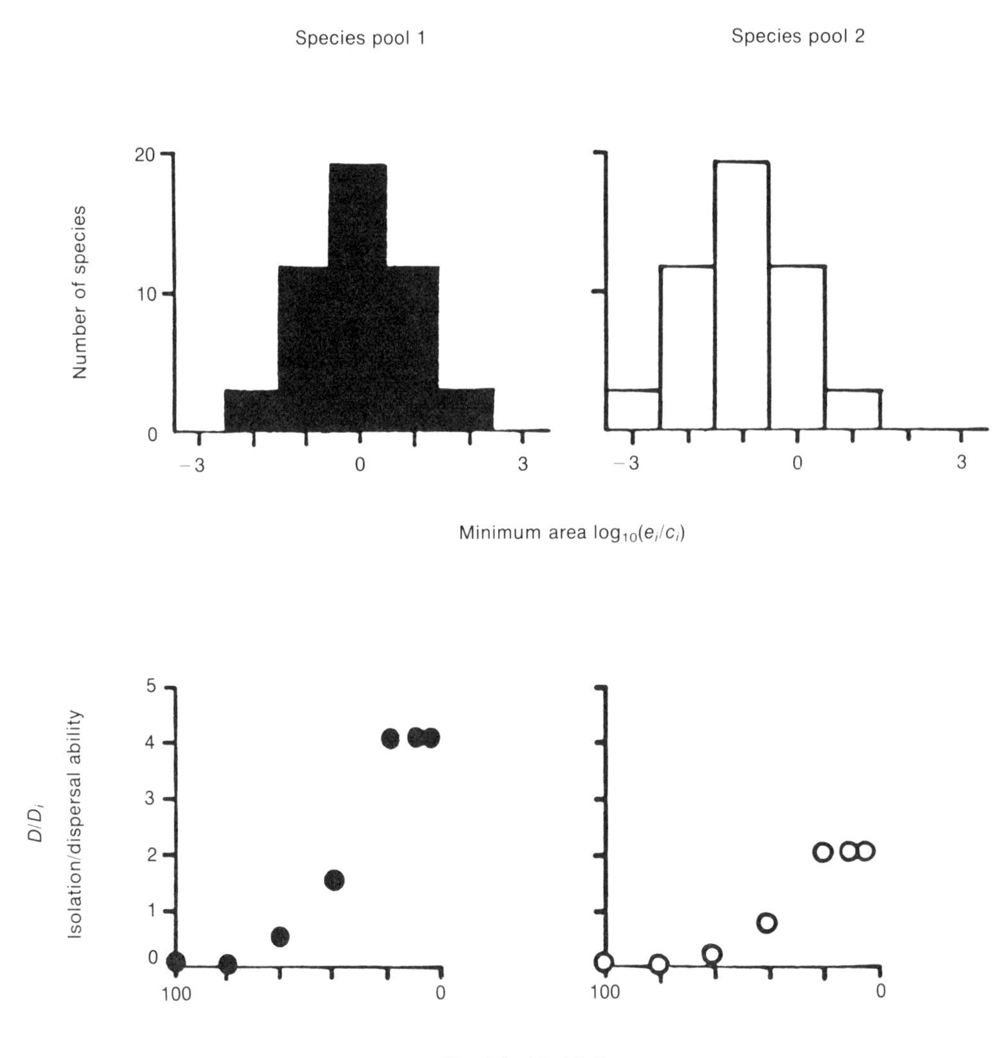

Figure 44.6. The area requirements and dispersal abilities of the two hypothetical species pools in the fragmentation simulation. The species pool on the right has comparatively modest area requirements (histogram above) and is able to disperse between habitats comparatively easily, even at high levels of fragmentation (point graph below). The species pool depicted on the left requires much larger areas and is less vagile. For a formal definition of the indices of area requirement and dispersal, see Appendix I.

namics, then it is likely that the attainment of equilibrium will lag some way behind the process of habitat destruction.

Conclusions

We believe that much of the debate about reserve-area strategies can be clarified by considering how the slope of the species-area relationship and similarity in species composition are likely to vary with the degree to which a habitat has already been fragmented and the geographic scale of conservation decision making. Species-area slopes (z-values) are rarely constant when a wide enough range of areas is considered, but, rather, they tend to be steep (>0.35) at very small areas, intermediate (0.15–0.35) at intermediate areas, and shallow (<0.15) at very large areas (Connor and McCoy 1979; Martin 1981a). Data from a number of studies on woodland bird communities confirm this relationship (Fig. 44.8). This steepening at small areas is predicted both by the equilibrium theory (see Schoener 1976) and by the statistical theory of species-area relationships (see May 1975); the shallowing at large areas is also predicted by the equilibrium theory as a consequence of the exhaustion of the regional species pool.

This variation in the slope of the species-area relationship

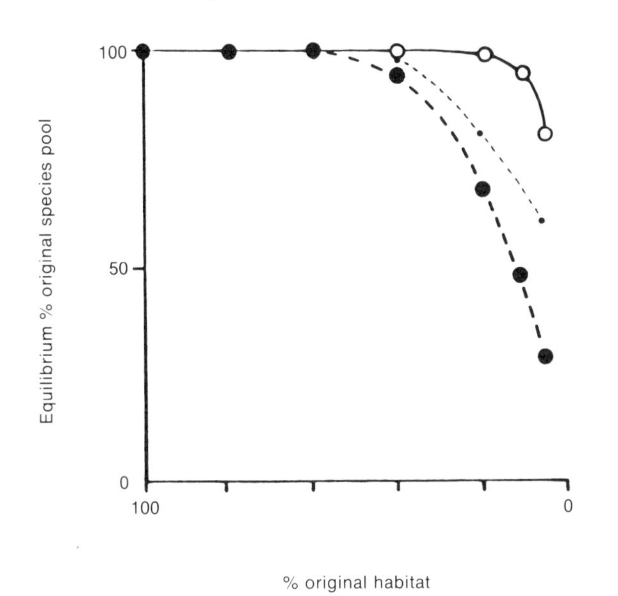

Figure 44.7. The numbers of species remaining in each species pool as fragmentation proceeds. Large, solid circles show the pool of species with larger area requirements and low vagility; open circles show the species pool with less stringent area requirements; the small, solid dots depict the proportion of the first species pool that would be present when the habitat is minimally fragmented.

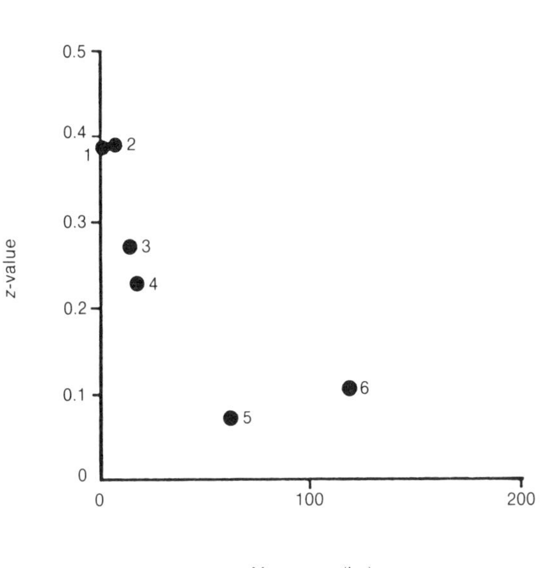

Figure 44.8. The slopes of fitted species-area curves from a number of studies of woodland birds, plotted against the mean size of the woods included in the study sample. The studies used were (1) Martin 1981b; (2) Galli et al. 1976; (3) Moore and Hooper 1975; (4) Woolhouse 1983; (5) present study USA data; and (6) present study UK data.

is particularly pertinent when we consider that different conservation organizations operate on different geographic scales and with different-sized budgets. For example, on a local scale, a farmer may be willing to set aside 10 ha of woodland; a county-based organization may have funds for 100 ha; a regional one, 1000 ha; and a national one, 10,000 ha. The variation in species-area slope suggests that decisions as to how to apportion the total area available among reserves will vary with geographic scale. Consideration of the likely (but as yet uninvestigated) trends in the other crucial parameter, similarity in species composition, confirms this view. On a farm, small woods should share many species because of their physical proximity, their probable similarity in habitat, and because the effective size of the species pool for such small woods may be less than the regional pool as a result of minimum-area effects. With increasing scale and size, the similarity of small reserves is likely to decrease, largely because of physical separation and habitat differences. At the national scale, although "small" reserves will be thousands of hectares in extent, similarity will still be low because they will generally be located in different biogeographic regions and, therefore, be colonized by different species pools.

These general trends in species-area slope and similarity have a simple consequence for reserve-area strategies. When the total area of habitat available for conservation is small, it will usually be best to go for single large reserves. As the total available area increases (with increasing geographic scale) several (or many) small reserves will usually constitute the best strategy. The optimal level of subdivision will depend on the relative advantages of capturing more habitat variation (by subdivision) and maintaining area-sensitive species (by concentration).

The net result of all conservation organizations applying such reserve strategies should result in (1) a fair number of small reserves, which serve local needs and contain a variety of habitats attractive to area-insensitive species; and (2) a small number of very large reserves which are designed to maintain populations of extinction-prone species with large area requirements. In between, there should be a spread of large to medium-sized reserves that are designed to maintain the bulk of the regional species pool (cf. the nomogram analysis above). This is clearly an intermediate strategy between a single enormous reserve and huge numbers of small reserves, both of which we believe are unrealistic and likely to be suboptimal. The frequency distribution of woodland areas preserved is midway along the spectrum that runs from pristine unbroken forest to complete urbanization. Given that further development will continually move the fragment-size distribution toward smaller, more isolated woods, we believe that special priority must be given to securing large reserves in different biogeographic regions. In our view, this remains the primary responsibility of national conservation organizations.

Acknowledgments

We thank Rob Fuller of the British Trust for Ornithology for access to data and for comments on the manuscript. Most of the Register of Ornithological Sites data were collected under contract to the U.K. Nature Conservancy Council. We thank all those who surveyed sites as part of this exercise. McLellan would like to thank the U.K.'s Departments of the Environment and Health and Social Security for financial support.

Appendix I.

This appendix describes the assumptions underlying the simulation model used to mimic the effects of fragmentating an originally contiguous habitat. The fate of the pool of P species in the original habitat is modeled, using a species-by-species formulation of the equilibrium model (see Gilpin and Diamond 1981). The basic variable is the probability J_i that a given species i ($i = 1 \ldots P$) occurs as a breeding population in a fragment. This probability, termed the "incidence" of species i (Diamond 1975b), increases with fragment area (A) as a consequence of increasing population size (and therefore decreased chance of stochastic extinction). Incidence decreases with increasing isolation (D) as a result of a decreasing frequency of colonization. On any given fragment, the equilibrium incidence level (J_i^*) is determined by the balance between the (area-dependent) extinction rate and the (isolation-dependent) colonization rate for the species concerned. Values of J_i^* plotted against A for a given D (or vice versa) are known as "incidence functions" (Diamond 1975b).

The incidence function model used here was first developed by Levins and Culver (1971) and has the general form:

$$dJ_i/dt = a_i J_i (1 - J_i) - b_i J_i , \quad (1)$$

where dJ_i/dt is the rate of change of incidence of species i; $a_i J_i$ is the probability per unit time that an unoccupied fragment (of which the proportion is $1 - J_i$) will be colonized; and b_i is the probability per unit time that a population in an occupied fragment will become extinct. This model differs from the one described by Gilpin and Diamond (1981) by having the colonization rate ($a_i J_i$) a function of J_i rather than a constant. We believe this is a more appropriate model for fragmented systems, where colonists are unlikely to be available in constant supply from a large "mainland" source area, as is often the case for islands in the sea. At equilibrium, when net rates of colonization and extinction are equal, equation (1) has the solution:

$$J_i^* = 1 - (b_i/a_i). \quad (2)$$

Note that no positive equilibrium incidence exists unless the colonization rate a_i exceeds the extinction rate b_i.

To generate incidence functions, we must specify the dependence of the colonization rate on fragment isolation (D) and the extinction rate on fragment area (A). We have used an exponential model for $a_i = f(D)$,

$$a_i = c_i \exp(-D/D_i), \quad (3)$$

which is appropriate for organisms that disperse in a constant direction and suffer a constant risk of death per unit distance (Gilpin and Diamond 1976). The species-specific parameters c_i and D_i are a colonization coefficient and the "mean dispersal distance" of migrants, respectively (note: ($1/D_i$) is the death rate per unit distance of migrants). If extinctions are simply the result of stochastic fluctuations in population size, and population size is directly proportional to area, then a simple reciprocal model suffices for $b_i = f(A)$ (Gilpin and Diamond 1976):

$$b_i = e_i/A, \quad (4)$$

in which e_i is a species-specific extinction coefficient. Substituting equations (2) and (3) into equation (1) and rearranging gives the expression for the incidence function:

$$J_i^* (A,D) = 1 - [e_i/c_i \exp(D/D_i)/A]. \quad (5)$$

For nonisolated fragments ($D/D_i \simeq 0$), equation (5) describes a hyperbolic function in which the parameter e_i/c_i (the ratio of the species-specific extinction and colonization coefficients) corresponds to the fragment area at which J_i^* is zero. As fragments become more isolated ($D/D_i > 0$), the actual "minimum area" [$= e_i/c_i \exp(D/D_i)$] increases as extinctions in small fragments fail to be reversed by recolonizations.

One may expect e_i/c_i to be distributed lognormally within the species pool on theoretical statistical grounds (Gilpin and Armstrong 1981), and there is some empirical evidence that this is the case (Gilpin and Diamond 1981). We have accordingly defined two species pools with lognormal e_i/c_i distributions; these have the same variance ($\simeq 1.0$), but differ by an order of magnitude in their mean area requirements (see Fig. 44.6). The second parameter needed to define the incidence function is D/D_i, the ratio of the distance to a source of colonists and the mean dispersal distance (equation [4]). Sharpe et al. (1981) have shown that the average nearest-neighbor distance between fragments increases until about 20% of the original unfragmented habitat remains, and we use this parameter as our measure of isolation (D). We do not attempt to divide fragments into isolation classes, but assume that all fragments have the average value of D at any given stage of fragmentation. We similarly assume D_i to be the same for all species in the species pool, whereas it is likely to vary and be correlated with e_i/c_i. To explore the effects of fragmentation on two extreme types of species pool, we assign a low D_i-value (poor dispersal ability) to the pool with extensive area requirements (high average e_i/c_i) and a high value to the pool with small minimum areas (see Fig. 44.6).

The result we seek is the quantity $S_T^*(A_T,N_T)$, the total number of species present at equilibrium in a landscape containing a total area A_T of habitat divided among a total of N_T fragments. We shall calculate $S_T^*(A_T,N_T)$ for each of the seven stages of fragmentation depicted in Figure 44.4. At any given stage, the N_T fragments are distributed among seven logarithmic area-classes (Fig. 44.5), such that n_x fragments occur in the xth area-class. That is, $N_T = \Sigma_{x=1}^{7} n_x$ and $A_T = \Sigma_{x=1}^{7} a_x n_x$, where a_x is the midpoint of the xth area-class. $S_T^*(A_T,N_T)$ is then calculated as follows. Since species colonize fragments independently of one another, the mean number of species in a fragment of given area A and isolation

D at equilibrium $S_T^*(A,D)$ is simply the sum of the individual incidence values $J_i^*(A,D)$ over the species pool P:

$$S_T^*(A,D) = \sum_{i=1}^{P} J_i^*(A,D). \quad (6)$$

The total equilibrium number of species in n fragments each of area A and isolation D, $S_T^*(n,A,D)$, is the sum of the probabilities that each species is in at least one of the n fragments:

$$S_T^*(n,A,D) = \sum_{i=1}^{P} \{1 - [1 - J_i^*(A,D)]^n\}. \quad (7)$$

Finally, the total equilibrium species number $S_T^*(N_T,A_T)$ in N_T fragments distributed over a total area A_T as $\Sigma a_x n_x$ is the sum of the probabilities that each species is in at least one of the n_x fragments in the xth area-class, each of area a_x (and isolation D), given that it has not already been counted in any of the preceding area-classes:

$$S_T^*(N_T,A_T) = \sum_{i=1}^{P} \sum_{x=1}^{7} \left[\prod_{j=1}^{(x-1)} [1 - J_i^*(a_x,D)]^n \right] \quad (8)$$

$$\{1 - [1 - J_i^*(a_x,D)]^n x\}.$$

45

North European Land Birds in Forest Fragments: Evidence for Area Effects?

YRJÖ HAILA

Abstract.—I elaborate a sampling model for the colonization of habitat islands by North European land birds. The model is based on the hypothesis that different species colonize islands independently of each other and in numbers that are compatible with their regional abundances and habitat requirements. Data from several forested archipelagoes in northern Europe agree with this view. The propensity of different species to colonize is determined by the availability of suitable habitats and, in some cases, by specific autecological mechanisms; no area effects are discernible. Forest fragmentation per se is not likely to be important for forest birds in northern Europe. The challenge for conservation is to identify the indirect consequences of fragmentation, such as habitat changes and edge effect, that are important for different species. The total area of habitats important for birds is of greater concern than the spatial configuration of the habitats.

The equilibrium hypothesis of island biogeography (MacArthur and Wilson 1967) has dominated studies on bird-community structure in northern forest fragments (Moore and Hooper 1975; Forman et al. 1976; Galli et al. 1976; Howe and Jones 1977; MacClintock et al. 1977; Whitcomb 1977; Whitcomb et al. 1977; Nilsson 1978; Svensson 1978; Robbins 1979; Butcher et al. 1981; Whitcomb et al. 1981). In northern areas, however, most bird species are migratory, and nonmigratory species usually disperse widely on a regional scale during the nonbreeding season. Consequently, the application of the equilibrium hypothesis for northern birds is problematic on both ecological and methodological grounds. Because of efficient dispersal of the birds, populations of species residing in nonisolated woodlots are not dynamically independent units but parts of the regional populations in surrounding areas. The proper frame of research is the whole archipelago of habitat fragments, and often no real "mainland" exists in such systems.

Methodologically, the main flaw introduced by the equilibrium hypothesis has been an almost exclusive restriction to qualitative presence/absence data. However, some of the main variables postulated by the hypothesis, such as immigration and extinction rates, are functions of population numbers. Moreover, the probability of presence in a small area, whether a habitat fragment or part of a mainland, is a function of abundance as well (Haila and Järvinen 1981; Williamson 1981; Simberloff and Abele 1982; Haila et al. 1983; Järvinen and Haila 1984).

In this chapter I elaborate an alternative view of the colonization of forest fragments by birds. This view is based on data from northern Europe, stating that communities in small woodlots may often be regarded as random samples from surrounding areas (Haila 1983a; Haila and Järvinen 1983; Järvinen and Haila 1984). I review evidence from bird surveys in nonisolated, forested archipelagoes, as well as studies from the literature.

YRJÖ HAILA: Department of Zoology, University of Helsinki, P. Rautatiekatu 13, SF-00100 Helsinki 10, Finland

Sampling colonization model

In nonisolated woodlots of boreal and northern temperate forests, seasonal turnover of birds is almost total, as most of the birds leave the woodlots in winter. The community of breeding birds is formed anew each year as a result of successful colonizations. A sampling model of colonization is based on the hypothesis that different species settle into woodlots independently of each other. Populations in the surrounding areas form a universe from which each woodlot draws its sample (Preston 1948; Haila 1983a). The hypothesis emphasizes the inherent stochasticity of the colonization process. Abundance of each species in the universe determines an expected number of pairs that would settle into a particular forest patch, but the actual population size is a random variable. The habitat characteristics of the patches constrain the sampling process; a habitat specialist is probably absent from a patch where its requirements are not met.

The simplest way to formalize a process of random sampling is by the terms in the Poisson series (see Fisher et al. 1943; Engen 1977):

$$Pr(n_i = r|X) = \frac{X_i^r}{r!}e^{-X_i},$$

where n_i is the number of pairs of species i, and X is the vector of expected abundances ($r = 0,1,2, \ldots$). According to the formula, population sizes of different species in the sample (i.e., in the community of a habitat patch) are random variables that vary independently of each other, and the expected mean population size of each species (X_i) is included in the vector X. A positive relationship between patch size and species number results by two not mutually excluding mechanisms: (1) species number is expected to increase with sample size, which is greater in large than in small patches; and (2) habitats in a large patch are probably more heterogeneous than those in a small one and, consequently, more habitat specialists are included in the sample.

The model predicts that population sizes of different species in single woodlots concur with expectations derived from regional abundances of the species and habitat composition of the woodlots. However, because of stochasticity of single colonization events, the realized breeding community in any single woodlot would be variable from year to year, even though population numbers on the regional scale would remain stable.

Evidence from island surveys

Survey data from two Finnish archipelagoes, the Åland Islands (60° N; 44 islands, size range 0.5–582 ha) (Haila et al. 1983) and the archipelago of Lake Inari (69° N; 41 islands, 0.5–885 ha) (Haila 1983b) agree with the sampling colonization model; the conclusion is summarized in Table 45.1 and the original survey data presented in Haila et al. (1983:Appendix) and Haila (1983b:Appendix). Both study areas comprise forested islands that are 10–15 km from the mainland (Main Åland, 970 km^2, and the mainland surrounding Lake Inari, respectively). More than 90% of the total estimated populations of both study areas comprise species that occur on the islands about as frequently as can be expected from survey data from the mainland and data on habitat structure of the islands. Specialized habitat or food requirements, or both, and, in some cases, interspecific interactions between particular species pairs, seem to explain the colonization

Table 45.1. Colonization success of land birds in two Finnish archipelagoes

Colonization success	Åland Islands	Lake Inari
Observed in expected numbers		
Abundant species	46	14
Scarce species	45	17
Total	91	31
More abundant than expected because of		
Mosaiclike island habitats	13	. . .
Abundant food	1	. . .
Obscure reasons	1	1
Total	15	1
Less abundant than expected because of		
Habitat scarcity	3	8
Nest predation (?)	2	. . .
Obscure reasons	2	1
Total	7	9
Absent because of		
Lack of habitat	5	14
Competition (?)	1	. . .
Wintering problems (?)	1	. . .
Total	7	14

Note: Survey results from islands were compared with expected population numbers derived from mainland data; species number in each colonization category is given. For detailed analysis and original data, see Haila (1983a), Haila and Järvinen (1983), and Haila et al. (1983) for Åland; and Haila (1983b) for Inari.

patterns of the rest of the species (see Haila and Järvinen 1983; Haila et al. 1983; Haila 1983b for details).

Two patterns observed on the Åland Islands emphasize the stochastic nature of the colonization process (Haila 1983a). First, Coleman (1981) presented a model of random placement, stating that, within an archipelago, the probability that a breeding pair settles into a particular island can be estimated by the proportion of the area of that island from the total area of the archipelago. Data from islands with similar habitat composition in Åland agree with the model of random placement. Second, on islands with similar habitat composition in Åland, species number increases with increasing sample size at an equal rate, irrespective of whether sample size is increased in space by surveying several islands in one year or in time by surveying one island in several years (Haila 1983a; see also Williamson 1983).

Another set of data originates from central Sweden (59°30′ N) where Ahlén and Nilsson (1982) surveyed birds in the forested parts of 93 islands with a size range of 0.6–776 ha. No reference surveys were conducted on the nearby mainlands, but Ahlén and Nilsson (1982) compared the occurrence of 49 forest breeding birds in the archipelago with data on their estimated densities in comparable habitats in other parts of southern Sweden. For 34 species, the minimum island area agreed with territory size deduced from mainland densities. Five of the species were rarities. Ten species were less frequent in the archipelago than expected. Eight of these were attributed by Ahlén and Nilsson (1982) to scarcity of suitable habitats on the islands, and to the rarity of the remaining two species they gave specific autecological explanations. In conclusion, the only effect of insularity revealed by the data of Ahlén and Nilsson (1982) was the absence of species from islands that were smaller than the territory size of the birds.

Nilsson (1977, 1978) published survey results from 39 small islands (size range 0.03–1.32 ha) in another Swedish lake (about 56°30′ N) and concluded that some species avoid these small islands. However, habitats on very small islands differed significantly from those on the mainland (as shown by the habitat measurements of Nilsson [1977]), and, furthermore, only the most abundant species can be realistically expected to reside in wooded patches of 1.3 ha or less in size. Less abundant species would most probably be absent from equal-sized mainland patches.

In agreement with the sampling model, habitat availability seems to be the most important determinant of the colonization of forested archipelagoes in northwestern Europe by land birds. The conclusion is further supported by older, mainly faunistic data from several archipelagoes along the Finnish coast (Sundström 1927; Välikangas 1937; Bergman 1939) as well as by data from long-term surveys on the Krunnit Islands off the Bothnian coast (65°25′ N) (Helle and Helle 1979, and pers. comm.), and by semiquantitative data from the islands of Lake Onega, Soviet Karelia (62° N) (Hohlova 1977). These data indicate that the species lists and population numbers of the breeding communities in these archipelagoes agree with the habitat composition of the islands.

Exceptions to this rule are known from Finnish islands,

such as the virtual absence of the yellow wagtail (*Motacilla flava*) as a breeding bird from the Åland archipelago (Palmgren 1935) and the scarcity of the great spotted woodpecker (*Dendrocopos major*) in the outer archipelago (see Haila and Järvinen 1983). Reasons for the exceptions are mostly unknown, but it is prohibitively difficult to derive a common explanation for them from an equilibrium view of island bird communities.

Domain of the sampling model

Validity of the sampling model is restricted in conditions (1) where isolation of the islands is important; or (2) where the habitat structure of the islands is different from that on the mainland, bringing forth specialization of island populations (e.g., on small islands north of Britain [Lack 1942] and on the Faroe Islands [Bengtson and Bloch 1983]). These factors are not important in the Finnish archipelagoes, which are close to the mainland, but they constrain the biogeographic realm of the model.

Also, the Poisson formalization implies that different colonization events are stochastically independent. For rare species, this is a reasonable assumption (Haila and Järvinen 1983; Järvinen and Haila 1984). For abundant species, however, the assumption is unrealistic. Because of limited availability of space for establishing territories, the probability that a new male acquires territory in a woodlot is a function of the number of males already present. Site tenacity of breeding birds counteracts random sampling as well, but first-year breeders in northwestern Europe usually show little tendency to return to areas where they were born, as emphasized for the pied flycatcher (*Ficedula hypoleuca*) by von Haartman (1960) and Järvinen (1983), the willow warbler (*Phylloscopus trochilus*) by Tiainen (1983), the scarlet rosenfinch (*Carpodacus erythrinus*) by Stjernberg (1979), and the chaffinch (*Fringilla coelebs*) by Dolnik (1982). The same pattern is evident even for sedentary forest tits such as the willow tit (*Parus montanus*) in the vicinity of Oulu, northern Finland (about 65° N) (M. Ojanen, pers. comm.).

Fidelity for a given nest site is presumably strongest in species that are very specialized in their habitat or their nest-site requirements, or both. An example is provided by the population of the raven (*Corvus corax*) on Skokholm island off the coast of Wales. This species has been remarkably stable at one breeding pair over a period of 47 years (Williamson 1983).

However, despite these factors, a varying degree of stochasticity remains (see Danilov 1980; Rice 1981; Rice et al. 1983b; Wiens 1983), and the basic approach of the sampling colonization model remains valid as well, although the Poisson approximation is unrealistic.

Implications for forest fragmentation

The sampling colonization model predicts that fragmentation per se would not influence population numbers (if the obvious case of fragments smaller than single territories is ruled out). Availability of different habitats, relative to the requirements of birds, is the decisive characteristic of small woodlots and determines the probability that a particular species will establish a territory there. Likewise, the model implies that the shape of the woodlots and the existence of ecological corridors are likely to be important only in so far as they influence specific habitat characteristics that are important for particular bird species. Habitat quality as a characteristic of small woodlots can be defined only by the requirements of the birds. Thus, for example, edge effect is relatively more important in elongated than in compact woodlots because of a greater proportion of edge zones in their area.

The concept of habitat fragmentation needs clarification. A clear distinction must be made between (1) changing spatial structure (fragmentation); and (2) decreasing total area (destruction) of forests, as potential factors affecting bird population sizes.

May (1981) constructed a model of population changes following forest fragmentation that predicts reduced population numbers and even local extinction as an outcome. However, May (1981) did not draw the necessary distinction between fragmentation and destruction of forests. Decreasing total area of suitable breeding habitats is followed by a decrease in overall population numbers, and this is likely to influence immigration rates between areas as well, especially if the remaining patches of suitable habitats are far from each other. For fragmentation per se, in contrast, this consequence is not necessary. An obvious alternative would be that the colonization pressure per unit of area would stay constant. Thus, an observed population decrease in an isolated forest patch is not necessarily a consequence of fragmentation per se but may reflect an overall population decline that is caused by a reduction in the total area of suitable habitats.

Data from North European archipelagoes agree with the prediction of the sampling model; it is, in fact, difficult to find evidence for negative effects of forest fragmentation (for a discussion of the Åland data, see Haila and Hanski 1984). These data refer to islands surrounded by water, but as pointed out by Haila and Hanski (1984), the comparison is conservative for the habitat-fragmentation issue, because real islands are likely to differ more from continuous habitat tracts on the nearby mainland than forest fragments on the mainland do.

In the formerly continuous coniferous forests of northern Finland, the situation seems to be different, however. Data from long-term bird surveys indicate that the populations of such resident taiga species as the Siberian tit (*Parus cinctus*) and the Siberian jay (*Perisoreus infaustus*) have declined relatively more than the area of primeval taiga forests has decreased. Järvinen (1980) attributed this decline to the fragmentation of the northern Finnish taiga due to forestry; lowered winter survival of these resident habitat specialists in fragmented forests is a good candidate for an explanatory mechanism.

Minimum area requirements

Ahlén and Nilsson (1982) used data on average densities to infer minimum area requirements of forest birds. This is problematic, however. An average density of, say, one pair per km^2 definitely does not mean that a pair would need 1 km^2 of space for successful breeding. Minimum area requirements should be defined by the actual territory size and habitat use of each species.

For example, Ahlén and Nilsson (1982:174) concluded that the red-breasted flycatcher (*Ficedula parva*) cannot tolerate fragmentation produced by modern forestry. Their data, however, included one probable breeding record on an island of 14.0 ha; as the species is scarce in Sweden (with about 100 observations per year in the early 1970s [Tjernberg 1984]), more records could hardly be expected. Furthermore, Tjernberg (1984) summarized data on the territory locations of the red-breasted flycatcher in eastern Sweden on the basis of 50 records. He concluded that "the areal requirements of the red-breasted flycatcher seem to be quite small. Breedings and probable breedings occurred in several stands with areas of 0.5–2 ha" (Tjernberg 1984:281). It is true that the red-breasted flycatcher is mostly found breeding in inner parts of large forests (see von Haartman et al. 1963–1972), but this reflects the fact that its favored habitats, stands of old trees, are mostly situated inside contiguous forests; it is thus a consequence of its habitat requirements, not its area requirements.

Similarly, other scarce migratory forest birds of northwestern Europe are occasionally recorded breeding on small islands with suitable habitats. For instance, the list of breeding records from a 31-ha forest stand on the Krunnit Islands includes the greenish warbler (*Phylloscopus trochiloides*) and arctic warbler (*P. borealis*), both of which belong to the faunal element of the Palearctic taiga and are rare in Finland (Helle and Helle 1979).

These cases underline the necessity of using data on regional densities as a reference for island distributions. For rare species, in particular, mere distribution data easily lead to wrong conclusions (Haila and Järvinen 1983; Järvinen and Haila 1984).

Minimum area requirement is an elusive concept in areas where isolation is not important; it is not definable at the population level. If colonists move widely in search of suitable breeding localities, it does not make sense to search for a minimum area where the subpopulation breeding there would be self-sustaining. The relevant problem is, What is the minimum area of suitable habitat that a pair of a particular species needs for successful breeding? The species would probably be absent from fragments below that size.

However, birds may be able to use several small areas as their territory or hunting area. This is true in North European archipelagoes; examples include the goshawk (*Accipiter gentilis*), honey buzzard (*Pernis apivorus*), and black woodpecker (*Dryocopus martius*) (Ahlén and Nilsson 1982; Haila et al. 1983). Such species tolerate fragmentation better than could be expected from their territory sizes, provided the fragments are sufficiently close to each other.

Area-dependent differences

Environmental characteristics of a woodlot depend on its size. Edge effect is one of the main factors causing such differences; small woodlots consist exclusively of edges (see Hooper 1971; Williamson 1975). Several quantitative studies have emphasized the significance of edge effect for forest birds in northwestern Europe (e.g., Hogstad 1967; Haila et al. 1980; Helle and Helle 1982; Hansson 1983; Vickholm 1983; Helle 1984). Vickholm (1983) concluded that total breeding densities in southern Finland were about 25% higher at forest edges than in interior parts of forests and that almost all species either favored or avoided edges.

These differences in the avian occupancy of forest edges vs forest interiors are usually attributed to autecological responses of the species to differences in habitat structure. Another factor of potential importance is the population pressure from surrounding habitats on woodlots; the pressure would probably be relatively higher in small woodlots than in large ones (Butcher et al. 1981; Ambuel and Temple 1983). Its importance in northern Europe has not been estimated, however.

Because of edge effect, the area that can be used by forest-interior species in woodlots is smaller than their total area. Increasing edge effect in fragmented forests thus means a reduction of the total area of forest-interior habitats. This certainly is relevant for conservation but should be separated from the consequences of habitat fragmentation per se.

Another possible source of area-dependent differences is demonstrated by the population fluctuations of the fieldfare (*Turdus pilaris*) in the Åland archipelago. The population plummeted between 1978 and 1979, probably because of the exceptionally harsh winter in central Europe. The decline was almost total on small islands but only about 40% on Main Åland (Haila and Järvinen 1981). We conjectured that small islands may be suboptimal for the species. The island population has steadily increased since the decline, but survey data from 1983 still indicate lower population numbers than in 1978.

The fieldfare seems to exhibit a case of source-sink population structure (Wiens and Rotenberry 1981a). That is, small islands may be suboptimal sinks that constantly need new colonists from a source area on the mainland. Similarly, Nilsson and Ebenman (1981) observed greater population fluctuations of the willow warbler on small islands than on the mainland in a lake in southern Sweden, and they concluded that the islands may be suboptimal for the species. Ultimately, the reproductive success of the subpopulations of any given species in different parts of a habitat archipelago should be known for an assessment of insularity effects.

Conclusions

It is hard to find evidence for area effects on the avifauna of northwestern European forests. This conclusion agrees with that of Helliwell (1976) from Britain, but it contrasts with a prevalent view that neotropical migrants breeding in interior parts of North American forests are excluded from forest fragments by area effects (Robbins 1979:198; Whitcomb et al. 1981:126). A detailed comparison of the evidence is, however, beyond the scope of this chapter.

I emphasize that forest fragmentation is an urgent issue for conservation of the bird fauna, but the specific ecological mechanisms through which its influence is mediated on different bird species need attention, rather than focusing only on woodlot size. An understanding of the detailed mechanisms is necessary for efficient conservation. Fragmentation is likely to cause changes in the composition of bird communities, but habitat differences, including edge effects and specific autecological mechanisms, are good candidates for explaining these differences.

The sampling view of colonization of forest fragments by birds proposed in this chapter implies that the difference between spatial population structure in contiguous forests vs fragmented forests may not be particularly great. The breeding population of a scarce species is also distributed patchily in uniform forest areas, where it consists of individual pairs located far from each other. For the protection of such species, the crucial problem is the preservation of suitable habitats in as vast areas as possible to assure that the overall population would remain self-sustaining. The size of individual preserves is an important matter as far as it influences the quality of habitats; small fragments comprise only edges and are excluded as preserves of species of forest-interior habitats. It is, however, unrealistic to demand that every preserve should have a self-supporting population of a rare species. The decisive question is the stability of the overall population, and for this purpose small areas of high-quality habitats can be extremely valuable.

Acknowledgments

O. Järvinen, R. A. Väisänen, and K. Vepsäläinen commented on earlier drafts of the manuscript. Financial support from the Emil Aaltonen Foundation and the Finnish Academy is gratefully acknowledged.

46

Prediction of Bird-Community Metrics in Urban Woodlots

ARNIE GOTFRYD and ROGER I. C. HANSELL

Abstract.—Birds were censused and the habitat sampled for 19 patches of deciduous forest in the vicinity of Toronto, Canada. Linear regression was used to find subsets of 21 habitat variables for precisely predicting the species richness, abundance, and density (pairs/ha) of breeding birds. The length of plot edge accounted for 82% of species richness and 96% of abundance variation. Edge length was a better predictor than woodlot area. The best model for species richness had three variables and accounted for 87% of the variation. The model proved robust to annual bird census fluctuations. Breeding-bird density was best predicted by the quality of habitat in the neighborhood surrounding the woodlot. The uniqueness of the urban context requires a modified approach to songbird preservation. The function and optimal structure of large regional preserves are different from those of smaller urban natural areas.

Patches of forest may be viewed as habitat islands within a sea of farmland or urban sprawl. This analogy was used by MacArthur and Wilson (1967), who developed a biogeographic theory to explain species impoverishment on oceanic islands. Principles of island biogeography have since been conceptually incorporated into the design of nature preserves at both regional (Diamond 1976; Terborgh 1976; Balser et al. 1981; Cole 1981; Bostrom and Nilsson 1983) and local (Davis and Glick 1978; Goldstein et al. 1981, 1983) scales. Simberloff and Abele (1976, 1982) have criticized some of these applications as being unjustifiable on both theoretical and empirical grounds. Two further problems with applying island biogeography to urban woodlot birds are (1) the high mobility of the colonizing species pool; and (2) the fact that the urban "sea," unlike a real ocean barrier, is densely inhabited by birds.

The urban avifauna is characterized by a great abundance of individuals distributed among a very few species (Nuorteva 1971; Beissinger and Osborne 1982). An increase in street and yard vegetation structure leads, however, to the presence of more species of breeding birds (Woolfenden and Rohwer 1969; Savard 1978). A suburban woodlot commonly has two to three times more species than the surrounding neighborhood (Gotfryd and Smith 1980, and pers. obs., Gotfryd). This is analogous to rural avifauna, where the presence of fence-row vegetation increased bird species richness of Iowa farmland from 10 species to 62 (Best 1983).

In this chapter, we empirically modeled the avifauna of urban woodlots, using habitat variables. In this way, we identified the primary habitat determinants and developed a management approach based on actual data.

ARNIE GOTFRYD and ROGER I. C. HANSELL: Department of Zoology, University of Toronto, Toronto, Ontario, Canada M5S 1A1

Methods

Field procedures

We chose for study 19 woodlots that varied in size from 0.8 to 26 ha and were situated in the suburban neighborhoods of four adjacent cities in southern Ontario. The following characteristics constituted the criteria for inclusion: discrete patch of forest; >400 m from nearest woods; regular shape; size >0.2 ha; mostly residential context; topographically flat; no stream or pond nearby; average canopy height >12 m; canopy >80% deciduous; essentially continuous canopy; internally consistent habitat; natural understory present; and leaf litter and ground cover present. Every woodlot met most of the selection criteria, and 17 of them satisfied 11 or more of the 13 criteria.

Breeding birds were censused in each of the 19 woodlots, using the standard spot-mapping technique, with transects 40 m apart (Robbins 1970). Censuses were done between 06:00 and 09:30, from 24 May to 7 July in 1978, 1979, 1980, and 1981. Sampling intensity varied among years, with 1980 censuses based on eight visits per woodlot and 3–6 man-hours of observation per ha of habitat, while the other censuses were usually based on only from three to six visits and an average of 15% fewer hours of observation per ha.

Vegetation was sampled during summer 1980 according to the standard technique developed by James and Shugart (1970). Six to 10 circular 0.04-ha sites were randomly chosen in each woodlot and were centered at least 20 m from the forest edge. These samples formed the basis of the data on vegetation cover, density, and structure (Table 46.1). Field data were collected by four skilled and trained naturalists. Perimeter, area, and neighborhood-habitat data were taken from aerial photographs and maps at 1:2400 scale. The vegetation structure in the surrounding neighborhood was summarized by the Surrounding Habitat Quality Index (SHQI), calculated as percent vegetation cover within 100 m of the woodlot, summed over three growth forms—trees, shrubs,

Table 46.1. Habitat variables used in analyses

Variable		
Name	Type	Details
A	Area	Measured from 1:2400 scale maps
LOGA	Area	Logarithm to base 10 of A values
ROOTA	Area	Square root of A values
P	Perimeter	Length of plot edge as measured on maps
LOGP	Perimeter	Logarithm to base 10 of *P*-values
ROOTP	Perimeter	Square root of *P*-values
GC	Ground cover	% cover of vegetation <0.7 m in height
CC	Canopy cover	% cover of vegetation >7 m in height
CHAV	Canopy height	Average of locally maximum tree heights
CHCV	Canopy relief	Standard deviation of canopy heights/CHAV
SDEN	Shrub density	Number/ha of woody stems <7.6 cm in diameter
SHQI	External vegetation	Surrounding Habitat Quality Index
BATOT	Productivity	Basal area/ha of all woody plants
BAA	Small trees	Basal area/ha of trees 7.6–15.2 cm in diameter
BA1	Understory	Basal area/ha of shrubs + BAA
BA2	Medium trees	Basal area/ha of trees 15.2–38.1 cm diameter
BA3	Overstory	Basal area of trees >38.1 cm dbh
STEV	Structural diversity	− Standard deviation of BA1, BA2, BA3
STDI	Structural diversity	(BATOT)(BA1)/(BATOT − BA1)
STDI2	Relative understory	BA1/(BATOT − BA1)
STDI3	Under:overstory	BA1/BA3

and herbs. The habitat and avifauna of each woodlot are described in detail elsewhere (Gotfryd 1984).

Analytical methods

Our intent was to develop a habitat model for songbird management in urban woodlots. Linear combinations of the 21 habitat variables were found that would best predict the abundance, density, and species richness of breeding birds in the study woodlots. By "best" prediction, we mean one that maximizes variation explained with a minimal number of biologically distinct variables. The motive for limiting model size was to effect a trade-off of variation explained for biological and statistical robustness.

Each of the three bird variables was separately regressed on all combinations of seven or fewer habitat variables, using a computer procedure whose output included only the names of the variables used and the R^2 value for each multiple regression model (SAS RSQUARE procedure, Helwig and Council 1979). The set of predictors was then reduced to 15 variables by (1) keeping only the edge and area variables with the highest individual and combined R^2 values (LOGP and A); and (2) eliminating redundant productivity variables (BAA and STD13). Then the all-subsets regression was rerun with the reduced set of variables. Within each model size the 10 or more regressions having the highest R^2 values were more carefully scrutinized after the biologically redundant variables had been systematically removed. This scrutiny involved (1) running partial F-tests to assess the significance of each variable's contribution to the regression; (2) plotting and analyzing residual errors to assess the statistical validity of the model; (3) checking the robustness of the model after removal of the largest (26 ha) woodlot from the regression; and (4) when more than one n-variable model met the above criteria, choosing the one with the most terms shared with the best $(n - 1)$-variable model. These steps resulted in the selection of six models having various numbers of terms.

The analytic procedure described above was done independently on two bird data sets from different years. One data set was based on 1980 bird data ($n = 19$ woodlots) and the other ($n = 16$ woodlots) had mainly 1979 data but also included two bird censuses from 1981. The mixing of census years in the second set was tolerated in order to maximize n of woodlots for the regression. The two data sets were not lumped for analysis for three reasons. First, there were inconsistencies in sampling intensity among years, a problem that affects the size and reliability of bird density estimates (J. Verner, pers. comm.). Second, lumping temporal repeats of data from the same woodlots would have greatly increased pure error, thereby complicating regression analysis. Finally, the separate analyses allowed us to test the temporal repeatability of our results and thereby give an indication of the biological robustness of the model. When the repeat analyses yielded the same results (i.e., same variables in the best n-term regression model), then the model with the highest $R_a{}^2$ value for 1980 data was chosen as the best habitat model for a given bird variable. In this final analytic step, $R_a{}^2$, a modification of R^2, was chosen as the selection criterion because it compensates for degree-of-freedom effects, thereby facilitating comparison of models with different numbers of terms (Draper and Smith 1981).

The empirical relationship between perimeter and area of the study woodlots was compared graphically to curves generated by regular shapes having the following perimeter-to-area ratios: circle, $2r:r^2$; square, $4l:l^2$; rectangle, $8l:3l^2$; and Nabisco (right isosceles) triangle, $(2 + 2^{1/2})b:b^2/2$.

Results

The 19 woodlots contained over 400 pairs of breeding birds belonging to 36 species (Table 46.2). On average, each woodlot had 21 pairs (range = 6–112, SD = 24.8), 13 species (range = 4–25, SD = 5.4) and a nesting density of 7.7 pairs/ha (range = 4–13, SD = 2.5).

The length of the woodlot's edge (perimeter) and the woodlot area were the only habitat variables which had significant ($P < 0.05$) simple correlations with either bird abundance or species richness. Bivariate regression analyses showed edge to account for up to 96% of the variation in bird abundance (r^2) and 67% of the variation in species richness (Table 46.3). When a logarithmic transformation was applied to the edge values, the explained variation in number of species increased to 82%. The other important variable, area, was comparably effective for predicting abundance when square root transformed. Woodlot area was considerably less powerful than edge, however, as a predictor of species richness, regardless of how area values were transformed.

Addition of other habitat variables into the models was done to see whether the accuracy of bird-community predictions could be improved. No combination of up to four habitat variables was able to account for even 1% more variation in abundance (R^2) than that explained by edge alone. Edge was therefore the only term in the best abundance model.

Breeding-bird density was best predicted by the quality of habitat in the neighborhood surrounding a woodlot. This variable, however, accounted for only 20% of density variation among woodlots (Table 46.3). Adding up to six habitat terms to the bird-density model boosted R^2 to 40% but barely affected R_a^2. In all cases, the SHQI was the only statistically significant variable (F-test, $P < 0.05$) in the set of 21 predictor variables. Adding six terms to the density model boosted the explanatory power up to 40%, but in all cases SHQI was the only statistically significant variable in the predictor battery ($P < 0.05$). Also, for any given number (>2) of variables in the density model, a wide variety of combinations met with comparable success.

Table 46.2. Summary of 1980 bird censuses in 19 woodlots

Species	Frequency (% of woodlots)	Abundance (total count of breeding pairs)	Density (pairs/km^2)
Common starling (*Sturnus vulgaris*)	100	97.0	156
Common grackle (*Quiscalus quiscula*)	95	33.0	53
American robin (*Turdus migratorius*)	89	34.5	55
House sparrow (*Passer domesticus*)	84	—	—
Blue jay (*Cyanocitta cristata*)	79	23.0	37
American crow (*Corvus brachyrhynchos*)	79	21.0	34
Brown-headed cowbird (*Molothrus ater*)	63	15.5	25
American goldfinch (*Spinus tristis*)	63	14.0	23
Eastern wood-pewee (*Contopus virens*)	63	12.0	19
Northern cardinal (*Cardinalis cardinalis*)	58	17.5	28
Northern flicker (*Colaptes auratus*)	58	16.5	27
Northern oriole (*Icterus galbula*)	53	21.0	34
Black-capped chickadee (*Parus atricapillus*)	53	15.5	25
Red-winged blackbird (*Agelaius phoeniceus*)	53	11.0	18
Song sparrow (*Melospiza melodia*)	37	16.0	26
Great crested flycatcher (*Myiarchus crinitus*)	37	12.5	20
Gray catbird (*Dumetella carolinensis*)	32	14.0	22
Cedar waxwing (*Bombycilla cedrorum*)	32	11.5	18
Downy woodpecker (*Picoides pubescens*)	32	7.0	11
Mourning dove (*Zenaida macroura*)	26	3.5	6
White-breasted nuthatch (*Sitta carolinensis*)	26	4.5	7
Red-eyed vireo (*Vireo olivaceus*)	16	6.0	10
Rose-breasted grosbeak (*Pheucticus ludovicianus*)	16	1.0	—
Indigo bunting (*Passerina cyanea*)	11	2.5	—
Wood thrush (*Hylocichla mustelina*)	11	2.0	—
Eastern kingbird (*Tyrannus tyrannus*)	5	3.0	—
Yellowthroat (*Geothlypis trichas*)	5	2.0	—
House wren (*Troglodytes aedon*)	5	1.5	—
Canada warbler (*Wilsonia canadensis*)	5	1.0	—
Barn swallow (*Hirundo rustica*)	5	—	—
Chipping sparrow (*Spizella passerina*)	5	—	—

Note: Species are listed in order of frequency. Abundance and density estimates include territories half in the woodlots, but exclude regular visitors whose nests and primary food sources are outside of the woodlots (e.g., house sparrow). The following species bred in census years other than 1980: red-tailed hawk (*Buteo jamaicensis*), great horned owl (*Bubo virginianus*), red-headed woodpecker (*Melanerpes erythrocephalus*), tree swallow (*Tachycineta bicolor*), and black-throated blue warbler (*Dendroica caerulescens*).

Table 46.3. Variance in bird data explained by the most powerful habitat variables

	Bird metrics		
Habitat variable	Species richness	Abundance	Density
Edge	0.67	0.96	0.04
Log(edge)	0.82	0.84	0.00
Root(edge)	0.77	0.92	0.00
Area	0.42	0.88	0.09
Log(area)	0.73	0.90	0.04
Root(area)	0.58	0.95	0.06
Neighborhood[a]	0.11	0.00	0.20

Note: Data are r^2 values for all habitat variables having at least one $P < 0.05$ correlation with a bird metric.

[a] This variable measures three layers of vegetation cover in neighborhood surrounding a woodlot.

Addition of terms to the species richness model improved predictions in a more consistent manner. The same subset of seven habitat variables (out of 21) explained ($R_a^2 =$) 87% of the 1979 variation, and 93% of the 1980 variation in bird species richness (Table 46.4). Also, the best one-, two-, and three-variable predictor models for both years had identical terms. The three strong variables (logarithm of perimeter, structural diversity, and ground cover) were also found in all of the best models containing four or more terms. The three-variable richness model was judged to be most robust, both biologically and statistically.

At least one structure variable occurred in every maximally efficient model having two or more variables (Table 46.4). The most effective structure variable (STEV) measured the evenness of vegetation distribution among three size-classes of woody plants. This variable's contribution to all the three- to six-variable bird species richness models was statistically significant (partial F-test, $P < 0.002$, 1979 data, $n = 16$; $P < 0.03$, 1980 data, $n = 19$). The other important structure variable (STDI) combined evenness and productivity in one measure of diversity. This variable's contribution was significant only for the 1979 four- and five-variable richness models ($P < 0.03$). Oddly, in all our models, the habitat-structure variables were negatively related to bird species richness.

The amount of ground layer vegetation (GC) was significantly and positively related to bird species richness after controlling for the effects of other variables ($P < 0.01$, 1979 data; $P < 0.03$, 1980 data).

Discussion

Avian response to woodlot geometry

The shape of a forest fragment has a considerable influence on bird abundance and species richness over and above the effect of plot size. This fact deserves emphasis because studies of wildlife in habitat patches usually ignore the perimeter of the plot and concentrate on area (e.g., Oelke 1966; Moore and Hooper 1975).

In order to understand the ecological effects of plot perimeter, we shall first consider its idealized relationship to the more familiar habitat factor, plot area. In general, larger plots have longer perimeters. For any regular shape, arithmetic increases in perimeter are associated with geometric increases in area. In particular, perimeter is directly proportional to the square root of the area. Real forest patches,

Table 46.4. Best habitat models of bird species richness

Number of variables in model	Variation explained R_a^2	Habitat variables: LOGP(+)	STEV(−)	GC(+)	A(−)	SDEN(−)	STDI(−)	CHAV(−)
		1980						
1	82%	***						
2	85	***	ns					
3	87	***	*	ns				
4	91	***	**	*	*			
5	92	***	**	**	*	ns		
6	93	***	**	*	*		ns	ns
		1979						
1	64%	***						
2	72	***	*					
3	80	***	**	*				
4	88	***	**	**			*	
5	86	***	**	**	ns		*	
6	87	***	**	ns		ns	ns	ns

Note: The most powerful one- to six-variable models are tabulated for each year separately with the significance of each variable in the model shown (partial F-tests). Models based on different census years are tabulated consecutively.

ns = not significant.

* $P < 0.05$.

** $P < 0.01$.

*** $P < 0.001$.

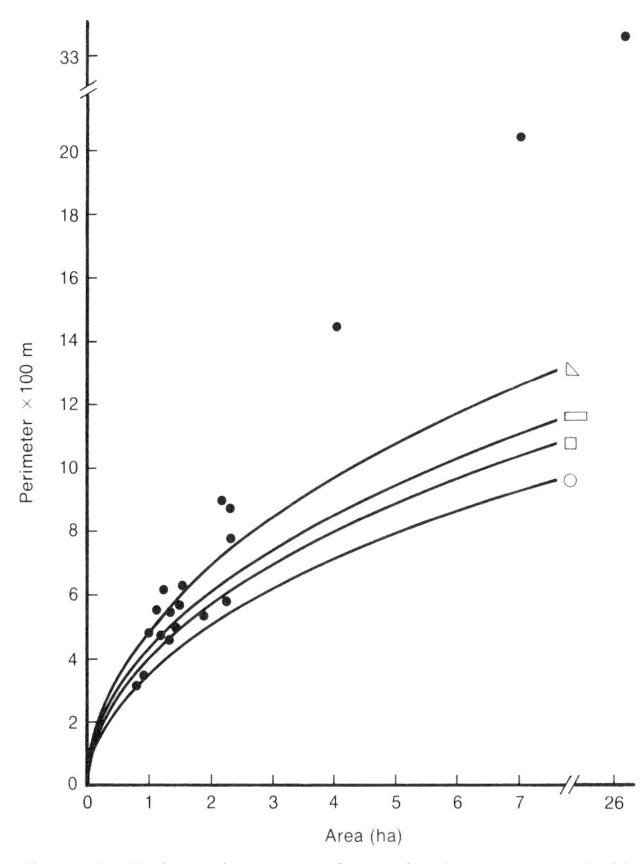

Figure 46.1. Perimeter/area curves for regular shapes compared with study woodlots (solid dots). Each curve was generated by the shape drawn on it. The 26-ha study woodlot is off-scale.

however, are varied in shape, and hence the perimeter variable gains some independence from area (Fig. 46.1).

A comparison of the four *P-A* curves for regular shapes shows that the perimeters become more divergent as area increases. Our study woodlots depart dramatically from these ideal curves in two ways. First, the slope is much steeper, which suggests that our sites are quite irregularly shaped. This effect is especially pronounced for woodlots larger than 2 ha. In fact, site maps and aerial photographs reveal that the overall woodlot shapes are not so unusual, but the additive effect of small edge undulations is the main cause of the steepness of the *P-A* slope. The second difference between our data and the ideal curves is, of course, that actual data have residual scatter. For our woodlots $r^2 = 90\%$, which means that 10% of the variation in perimeter is not accounted for by area. This apparently minute amount of variation is very sensitively traced by bird species richness, which is predicted better by perimeter than by area.

In contrast to our findings, Martin (1981b) stated that area is more important than edge in determining both species number and total abundance of birds. However, since his study sites were narrow shelterbelts, his area measurements were actually his best estimates of edge habitat. In contrast, his edge measurements (length + width) did not account for shape variation within or among plots. The point here is only that the importance of edge is not undermined by findings such as Martin's or by the relative paucity of literature pertaining to the ecological significance of the perimeter and shape of habitat islands.

The ecology of edge

The perimeter variable measures the amount of edge habitat, habitat which qualitatively and quantitatively differs from forest interior. The border of a wooded area supports a greater diversity and density of both plants and animals than does the forest interior (Kendeigh 1946; Balda 1975b; Gates and Gysel 1978; Helle and Helle 1982; Kroodsma 1982; Hansson 1983). The edge condition is attractive to birds for several reasons: (1) the great majority of passerine bird species in our region prefer to nest close to the ground (Preston and Norris 1947), and forest edge has relatively high cover in this layer; (2) edge vegetation supports a greater proportion of utilized food resources (see McDiarmid et al. 1977); (3) some species use trees for nesting and open areas for foraging and are thus often observed at forest edge (e.g., American robin [*Turdus migratorius*]); and (4) many species tend to use an edge location for singing (e.g., brown thrasher [*Toxostoma rufum*]).

Most of the bird species that colonize woodlots, and especially urban woodlots, have been ecologically classified as edge species (Whitcomb et al. 1981). Some species (e.g., gray catbird [*Dumetella carolinensis*]) nest only in vegetation within a few meters of an edge (Nickell 1965). This kind of colonization pattern results in the area of woodlot interior being more sparsely populated, especially because those birds classified as forest species do not often breed in woodlots in the size range considered in this study, even where apparently ample habitat exists (Lack 1976).

Management implications

Urban planners, homeowners, and other land managers are sometimes interested in managing land so as to enhance its natural beauty. Songbirds are an integral component of the natural landscape, even within the urban context. Managing for bird species richness has the dual benefit of increasing a desirable wildlife component and promoting the overall aesthetic appearance of the vegetational community.

Several of our results are relevant to the practical achievement of such a management goal. To maximize the number of breeding species of songbirds in urban deciduous woodlands, we suggest the following: (1) maximize the area of woodland, which should be effective from 1 to about 7 ha, after which the "law of diminishing returns" may become dominant; (2) for any woodlot size in this range, shape is an important factor, with circles and squares being least effective and irregular shapes working best because of their long perimeters; (3) the limiting condition to suggestion (2) is that minimum territory widths of birds in the colonizing species pool should form the lower bound of woodlot width; (4) for any given size and shape, undulating edges (perceptible at 5–10-m grain) are more effective than straight ones; and (5) concerted effort should be put into sustaining a dense layer of forb vegetation within 1 m of the ground throughout the interior of the woodlot.

Acknowledgments

This work was supported by grants from the National Science and Engineering Research Council of Canada. Thanks are extended to Jim Rising and Michael Gates, who critically reviewed the manuscript. We also thank the Creator for designing nature so that its patterns are interesting and almost tractable.

Summary: Predicting Effects of Habitat Patchiness and Fragmentation—The Researcher's Viewpoint

JARED VERNER

Natural landscapes are typically composed of discontinuities in their physical and biological dimensions, e.g., their slope, aspect, relief, and type of substrate. The result is a mosaic of habitats differing in size, shape, and the structure and composition of their vegetation. Wildlife managers often refer to this pattern in terms of the patchiness of habitats. If undisturbed long enough, such landscapes tend to reach a stage in which units of the mosaic retain fairly stable local plant communities. Whittaker (1953) labeled this the "climax pattern." But disturbances in the form of fires, storms, landslides, frost heaving, erosion, and so on (White 1979) are commonplace in nature. Such disturbances normally reset the affected area to an earlier stage in its progression toward the climax community and tend to reduce the average size of patches in the mosaic. Because so many species of terrestrial vertebrates are adapted to breed successfully in such disturbed habitats, we might safely infer that natural disturbance has been a frequent and widespread phenomenon in geologic history.

Human-induced disturbance of natural landscapes is a recent phenomenon, at least on a geologic time-scale, and it is generally like natural disturbance in setting the affected area back to an earlier stage in its progression toward the climax community. Unlike natural disturbance, however, it often affects enormous areas and increases mean patch size. And larger patches often possess few of the attributes required for successful reproduction by terrestrial vertebrates endemic to the area. This is especially true of disturbances associated with agriculture, which usually result in almost total loss of diversity in plant species composition. Often only fragments of the original plant communities remain as small, isolated islands that are unable to support viable breeding populations of many or most of the species well adapted to such habitats.

We still have an opportunity to study fragmentation as a process leading to remnant islands of formerly extensive habitats, because the process is relatively new in forests of the western United States, where extensive areas of reasonably pristine habitats still occur. In most of the central and eastern United States, fragmentation is well advanced, and in Europe it is still further advanced. Ideally, the various chapters in this part should give us insights into the effects of the process (fragmentation) and its by products (fragments of formerly extensive habitats) on populations of terrestrial vertebrates. And we should find examples of ways to model these effects and incorporate the findings into tools useful for resource managers.

An overview of the chapters, however, reveals an unbalanced coverage of the process of fragmentation and its byproducts. Only one chapter (Rosenberg and Raphael, Chapter 38) gives empirical results from an area in early stages of fragmentation. Chapters by Seagle (Chapter 40) and by Urban and Shugart (Chapter 39) present models for landscape mosaics, but the remainder of the chapters report results of studies in habitats already markedly fragmented. Furthermore, the studies were heavily biased toward birds. Rosenberg and Raphael studied amphibians, reptiles, birds, and mammals. Murphy and Wilcox (Chapter 41) studied the effects of fragmentation on butterflies, which they compare here to their results on birds and mammals in the same fragments. Seagle modeled for 150 hypothetical species, without specifying which vertebrate classes could have been represented. The seven remaining chapters present empirical results only for birds. But these biases are not restricted to the studies in this volume; they mirror the broader field of research on fragmentation and its effects on terrestrial vertebrates. Most of the literature in this field deals with birds, and most researchers have concentrated on the end products of fragmentation—the habitat islands. Such research is inherently easier than that needed to understand early stages in the process. Moreover, it is a logical extension of the classic work by MacArthur and Wilson (1967) on island biogeographic theory.

The messages to researchers from this overview are clear. Greater emphasis must be given to (1) studies of amphibians, reptiles, and mammals in relation to fragmentation; and (2) studies of the process of fragmentation as opposed to studies of the effects of its end products, the habitat islands. In Chapter 38 Rosenberg and Raphael have shown some ways in which this can be done, but I believe this field is ripe for innovation of techniques for studying the effects of patch size and juxtaposition on the abundance and distribution of vertebrate species. In western coniferous forests, for example, habitat fragments of a certain type are typically connected by corridors of similar habitat that probably permit most species to move freely among fragments. Such an abundance of corridors may have been a major reason why Rosenberg and Raphael found so few effects of patch size or distance to nearest clearcut on the relative abundances of the species they studied. And the interpretation of results in western forests must account for the fact that early stages of fragmentation have tended to produce two-dimensional is-

JARED VERNER: USDA Forest Service, Forestry Sciences Laboratory, 2081 East Sierra Avenue, Fresno, California 93710

lands (clearcuts) in three-dimensional seas (forests), but in the eastern states and in Europe the later stages of fragmentation have resulted in three-dimensional islands (forest fragments) in two-dimensional seas (e.g., agricultural land).

The key controversy about the mechanism accounting for lower species richness in small fragments is clearly evident in this part of the volume. The dominant view (e.g., Diamond 1975a; Robbins, 1979; Whitcomb et al. 1981) is that of MacArthur and Wilson (1967): the composition of animal communities on islands results from a dynamic equilibrium between immigration and extinction. These interact with distance from the nearest mainland source of colonizing animals and their relative dispersal capabilities to give smaller faunas on smaller and more distant islands. The competing hypothesis (e.g., Haila and Järvinen 1981; Williamson 1981; Simberloff and Abele 1982) is that "different species colonize islands independently of each other and in numbers that are compatible with their regional abundance and habitat requirements" (Haila, Chapter 45). Larger fragments are more likely to have a greater variety of habitat types and more edges, so they provide suitable habitats for more species. On this point, probably all would agree. Obviously more research is needed to sort out the reasons for the observed species-area relationships, but whatever the mechanism, the important result is that smaller habitat fragments have fewer animal species than larger ones. Land and resource managers must incorporate this fact into their planning.

A second controversy, and one more directly related to conservation planning, involves the species-area relationship discussed above. Should we preserve many small habitat fragments or a few large ones? According to Simberloff and Abele (1982, 1984), many small ones are preferable, because they can support more species than one to a few large reserves equal in total area to many small ones. The modeling studies reported here by McLellan et al. (Chapter 44) and by Seagle (Chapter 40) seem to support this conclusion. But maintaining the maximum number of species in relation to the area of habitat preserved is not the only appropriate goal for conservation. Species richness, per se, can be a deceiving measure of the success of a nature reserve; one must know which species are conserved as well. For example, the substitution of European starlings (*Sturnus vulgaris*) for northern flickers (*Colaptes auratus*) or western bluebirds (*Sialia mexicanus*) would preserve species richness, but I doubt whether many conservationists would be content with the substitution. Conversely, many have argued that the prudent conservation strategy is to provide very large reserves, because certain species are absent from small ones and because only large reserves can assure a stable ecosystem (e.g., Willis 1974; Diamond 1975a; Pickett and Thompson 1978; Abele and Connor 1979; Robbins 1979). The modeling reported in this part by Seagle suggests that this difference may depend in part on the frequency of disturbance in the fragments under study. Here is another important area for further research.

Because some species do not persist in small fragments (Willis 1974; Robbins 1979; Whitcomb et al. 1981; Ambuel and Temple 1983; Lynch and Whigham 1984), it seems obvious that appropriately large reserves must be maintained for those species, as recommended in several chapters in this part (e.g., Dobkin and Wilcox, Chapter 42; Temple, Chapter 43; McLellan et al., Chapter 44; Seagle, Chapter 40). Researchers must identify all such species and determine their minimum area requirements as soon as possible, and it seems that the core-area method described here by Temple may be the appropriate measure of minimum area for most of these species. None of this, of course, is to say that we should not continue to establish as many habitat reserves as possible, regardless of their size (see McLellan et al.; Seagle).

On the basis of results of his core-area analysis, Temple (Chapter 43) concludes that relatively smooth edges of habitat fragments are preferred to irregularly shaped or deeply incised edges. This may appear to conflict with conclusions of Gotfryd and Hansell (Chapter 46), who found total edge to be the best predictor of bird species richness in the urban woodlots they investigated. As pointed out above, however, achieving high species richness is not necessarily the wisest goal for conservation. Temple's results apply to problems of maintaining selected species that have specific requirements for large core areas in habitat fragments; species richness is not the issue here, nor should it be. In fact, few of the species detected in stands studied by Gotfryd and Hansell are in the group that most students of the area requirements of birds consider to be dependent on preserves with large core areas.

Another major question is whether or not it is even possible to develop suitable models for evaluating the full effects on wildlife of changes in habitat patchiness and fragmentation. The various combinations of different landscape patterns promise to overwhelm even the most sophisticated computer and modeling technologies. Island biogeographic theory should not be expected to apply to such situations; after all, it was not developed for them. But the studies by McLellan et al. (Chapter 44), Seagle (Chapter 40), and Urban and Shugart (Chapter 39), in particular, seem to show some ways that models can help. And their results are generally consistent with biological intuition. Chapters by Seagle and by Urban and Shugart point to a very great need for studies of the basic ecology and demography of most wildlife species. In any case, this appears to be a promising area for future research that will ultimately need to be integrated with models of species-habitat relationships; that is, the spatial aspects of habitat change (patchiness and fragmentation) should not be evaluated apart from the temporal aspects (succession).

Finally, researchers in this field need to give more attention to the adequacy of their methods in relation to specific objectives of their studies. Although I believe that results of most published studies on the effects of fragmentation on wildlife correctly reflect real patterns, I am nonetheless concerned that systematic biases inherent in the methods used by some researchers may have distorted the true picture. We know, for example, that spot mapping, as used to estimate densities of birds, is inherently biased by plot size. The method tends to overestimate bird densities on small plots (Marchant 1981; Verner 1981), so it may not be the best method to use for comparing the abundances of birds in habitat fragments of different sizes. Similarly, methods that give more effort to sampling in large fragments than in small ones introduce a bias from effort alone, which should result in detection of more species in the large fragments. This compounds the experiment, with the effects of both more effort and large size tending to give longer species lists. Researchers seldom acknowledge this problem, assuming instead that fragment size alone explains the length of species lists. Perhaps the reported differences in species richness with fragment size are not really so great as the literature suggests, at least in some instances. Whatever methods are used, I believe researchers should show that they gave sufficient effort to all fragments to result in a complete or nearly complete list of the target species. This could be accomplished by examining species accumulation curves in relation to sampling effort, and perhaps using some arbitrarily chosen rate of accumulation of new species to determine when sufficient effort has been given to fragments of different sizes (e.g., Verner and Ritter 1985). Continued use of methods known to give biased measures of biological populations in fragments of different sizes borders on the cavalier. At worst, it will result in wrong answers that lead to mismanagement; at best, it will slow progress toward a true understanding of this important phenomenon.

Acknowledgments

I thank Michael L. Morrison, Dean L. Urban, and Bruce A. Wilcox for their insightful comments.

Summary: Predicting Effects of Habitat Patchiness and Fragmentation—The Manager's Viewpoint

WILLIAM F. LAUDENSLAYER, JR.

Habitat fragmentation is the process whereby areas of homogeneous habitat are broken into a mosaic of smaller, dissimilar habitat patches. Fragmentation may occur as a consequence of natural disturbance or of management activity.

Natural landscapes, as Verner points out in the preceding summary, are mosaics of generally small vegetation patches characterized by differences in vegetation structure and composition. Natural patches also typically support relatively diverse plant assemblages of different ages and quantities of components, such as snags and dead-and-down materials. Vegetation succession in these patches has the potential to develop to "climax" conditions, which persist until the plant community is set back to an earlier successional stage by random instances of disturbance (e.g., fires, landslides, insect kills).

Land-management practices may result in planned fragmentation of the landscape into a mosaic of patches that often are larger than most natural patches, that may be converted to different vegetation compositions or structures, that have low species diversity, that are often monocultures, that are more even-aged, that do not retain natural quantities of components, and that are not retained to climax. This shift from natural to managed landscapes has generally resulted in a larger proportion of the landscape being dominated by relatively large patches of more even-aged plant communities. In forested lands, these generally are of a relatively young age, whereas in shrublands, patches often are old and decadent. Such changes in landscape have influenced communities of terrestrial vertebrates. For example, species that require large tracts of relatively unbroken forested lands have declined in number and distribution, whereas the reverse has occurred with species adapted to mixes of vegetation structure and composition with associated edges or large expanses of decadent shrubs.

Regardless of the approach selected to manage the landscape (i.e., allowing nature to take its course or adopting a plan to manage for some specific objective) assemblages of wildlife species will continue to persist. The critical questions for managers are, How many of each species should there be? And, Where will those animals reside? The public, through passage of various laws and other ways of expressing its desires (e.g., purchase of hunting licenses, founding and funding wildlife-advocacy groups), has given land managers broadly formulated answers to these questions. Laws such as the Endangered Species Act, the National Forest Management Act, and the Federal Land Policy and Management Act and their guiding regulations mandate that all species must be retained on public lands. In addition, production of animals for recreational and commercial pursuits has clearly been demanded by the public. To complicate the management situation further, the public also demands from the land other benefits that may conflict with each other and with wildlife goals—wood products, red meat, minerals, "clean" watersheds, and areas for recreation—snowmobiles, off-highway vehicles, skiing, camping, and hiking.

The challenge for managers, then, is to produce wildlife, various commodities, recreational opportunities, water, and other benefits, several of which cannot be provided simultaneously from the same parcel of land, in quantities to meet public demand. Trading off some benefits for others, even some wildlife species for others, is an inescapable consequence of decision making. Unfortunately, the public has not given managers clear directions to produce certain benefits at the expense of others, so they must continue to choose between producing spotted owls (*Strix occidentalis*) or mule deer (*Odocoileus hemionus*) or, perhaps, some commodity or suite of benefits from a unit of land.

Obviously, the only practical way to produce a desired set of wildlife is to manage habitat for these species. To effectively manage the land to produce desired numbers of wildlife, managers need information about species' requirements, including the effects of landscape fragmentation on wildlife. Further, managers need guidelines to assist them in using habitat fragmentation as a tool to produce a desired suite of wildlife.

As a manager, I had hoped that this part of the volume would include presentations on habitat fragmentation in two broad categories: (1) the impacts of fragmentation of large pieces of land on faunal elements; and (2) information that would be applicable to small-scale management activities (e.g., timber sales). Knowledge of the impacts of fragmentation on wildlife is essential for evaluating the impacts of virtually any management activity. In California, there are about 880 species of amphibians, reptiles, birds, and mammals (Laudenslayer and Grenfell 1983), not to mention the myriad species of invertebrates. Managers need to know how fragmentation will affect all of these species to ensure that they are retained as viable components of the system. And, we must have fragmentation theory translated into a

WILLIAM F. LAUDENSLAYER, JR.: USDA Forest Service, Pacific Southwest Region, R-5, Tahoe National Forest, Highway 49 and Coyote Street, Nevada City, California 95959

form that can be readily applied in the development and implementation of projects that benefit wildlife. Without such information, wildlife biologists will, again, be restricted to reacting to projects proposed by other functions and to giving best-guess recommendations.

Chapters in this part give much information about the impacts of fragmentation on wildlife. Unfortunately, of seven such studies, five were restricted to the effects of fragmentation on birds. This phenomenon probably should be expected, because the majority of terrestrial vertebrate species in many localities are birds—in California, for example, 62% of all terrestrial vertebrate species are birds—and the field of avian habitat relationships appears to be more advanced than that for other taxa. Two studies, however, treat taxa in addition to birds. Murphy and Wilcox (Chapter 41) examined effects of fragmentation on butterflies and birds, and Rosenberg and Raphael (Chapter 38) looked at fragmentation relative to all classes of terrestrial vertebrates.

All of the chapters on birds are concerned with forested habitats, including western riparian (Dobkin and Wilcox, Chapter 42), and urban woodlots (Gotfryd and Hansell, Chapter 46), as well as those of Europe (Haila, Chapter 45; McLellan et al., Chapter 44) and eastern North America (McLellan et al.; Temple, Chapter 43). This distribution points out the need to look at other systems in addition to forests, including grasslands, shrublands, woodlands, and deserts.

The chapters by Seagle (Chapter 40) and by Urban and Shugart (Chapter 39) present models that help to explain the effects of habitat fragmentation and may have application for predicting consequences of management actions. Seagle presents a model "to simulate the effects of landscapes of various sizes." His model allows colonization of landscapes by up to 150 hypothetical species, depending on habitat characteristics exhibited by the landscapes, and then simulates the effects of different patches of habitat on the species that colonize the land. Urban and Shugart present a model "to simulate demographic processes of natality, dispersal, and mortality as these might occur in a mosaic of habitat patches." Both of these approaches may be useful for simulating the cumulative effects of management actions on a land unit.

Only one study, that by Temple (Chapter 43), gives information that may be directly applicable to small-scale management questions. Temple discusses the concept of the core area of a patch in contrast to the entire area of the patch. He presents information suggesting that the core area explains differences in avifaunas more accurately than total patch area does. This concept will be exceedingly useful to managers, as it should provide more accurate predictions of the effects of management activities on wildlife and a rough guide to estimating the effective size of a patch.

Where do we go from here? Researchers can address several concerns about future habitat fragmentation. For example: (1) When is a habitat island not a habitat island? (2) When must we provide ecological reserves, as opposed to managed landscapes? (3) What guidelines can managers use when designing a habitat mosaic that will produce wildlife as well as other benefits? (4) How can managers appropriately evaluate the impacts of fragmentation on wildlife and other resources?

In my opinion, question (1) above is a very major concern. Much of the theoretical basis of habitat fragmentation is predicated on MacArthur and Wilson's (1967) theory of island biogeography. This theory is based on the idea that islands are surrounded by areas uninhabitable by species which colonize the islands. This assumption is applicable to terrestrial species on oceanic islands, but is a patch of old-growth forest an island if it is surrounded by pole-sized timber? Or do the old-growth and pole-sized timber areas together constitute an island? The answer, of course, depends on the species of interest. Much more research is needed to assist managers in making correct decisions about such questions.

Much of the conservation literature (see Soulé and Wilcox 1980; Frankel and Soulé 1981) advocates the retention of ecological reserves—large tracts of lands that are relatively homogeneous in plant composition and structure—to ensure the perpetuation of a desired set of biological attributes. Although this approach is necessary and will help conserve biotic elements, I question whether enough lands can be put aside to effectively meet all conservation objectives. Because the vast majority of lands in most areas will be under some form of management, and because even reserves may require management to ensure retention of the characteristics they are intended to preserve, a reserve system will not necessarily free us from the need for techniques that will produce wildlife as a consequence of resource management. Thus, research is needed to develop guidelines for large- and small-scale management projects.

Land-management activities are based on guidelines designed to meet certain objectives. Guidelines often have been designed to produce certain kinds of wildlife with little thought given to the entire faunal assemblage. For example, for years managers have attempted to provide large amounts of edge in the belief that edge would be beneficial to wildlife. More recently, however, managers have learned that large amounts of edge will not provide suitable habitat for some species; thus, this guideline alone will not meet all wildlife needs. Managers need guidelines that will ensure the retention of the entire faunal assemblage. The methods that Harris (1984) proposed to integrate principles of island biogeography into forest management should help solve some of these problems. Future research should build on this work to develop more precise methods and to extend guidelines to ecosystems other than forests.

Finally, we still have a need for continued work to evaluate the effects of fragmentation on wildlife in habitats other than forests. Gaps in our knowledge of such habitats are great, especially for amphibians, reptiles, and mammals in grasslands, deserts, shrublands, and woodlands. And, perhaps of greater concern is the utter lack of information about invertebrate communities. In most cases, we even lack information regarding local invertebrate faunal composition, so it is not surprising that we know almost nothing about the effects of habitat fragmentation on these taxa.

In conclusion, the understanding of habitat fragmentation has come a long way in the past 20 years or so, but in terms of management applications, we have just glimpsed the tip of the iceberg. Habitat fragmentation is and will continue to be an important concern to land managers, but it is hoped that better guidelines will soon be available to ameliorate its effects.

Acknowledgments

I thank H. D. Avant, D. H. Behrens, D. L. Connell, H. L. Greiman, J. H. Harn, G. E. Hartman, K. E. Mayer, and J. Verner for their helpful and insightful comments.

V

Linking Wildlife Models with Models of Vegetation Succession

Introduction:
Linking Wildlife Models with Models of Vegetation Succession

NORMAN L. CHRISTENSEN and LAWRENCE S. DAVIS

Preceding chapters have considered the nature and complexity of wildlife-habitat relationships and how these relationships can be used to develop predictive models. What has not been considered in great detail is the fact that habitats are not static. The areas of greatest habitat concern are often those that have been recently disturbed by man, as in the case of logging or road building, or by natural forces, as with fire. In the years following disturbance, ecosystems gradually succeed back toward their predisturbance status. The rate, mechanisms, and nature of such successional processes vary considerably among ecosystems and with the intensity and frequency of disturbance. It should be added that these are matters of considerable debate among plant ecologists (see, e.g., West et al. 1981).

Whatever their causes, this succession results in marked changes in habitat suitability. Habitat suitability for wildlife within a region or on a landscape is an integration of site-specific information coupled with an understanding of the role that patterning of habitats may play in determining suitability. Various management interventions will set back or deflect successional trajectories on particular sites. Thus, a habitat-suitability forecast for a managed landscape requires an understanding of habitat-vegetation relations, of vegetational change through time, and a clear picture of management objectives and their effects on the vegetation.

Several different models have been developed, using the digital computer, to predict successional change in various ecosystems. These models vary in several important features. The FORET–JABERWA-type models depict successional change in particular vegetation types. They are driven by biological properties of the component species, such as shade tolerance, growth rate, and dispersal. Horn (1975) used a Markov chain approach in which species replacement was based on species-by-species establishment probabilities to model forest change. Both of these approaches can be used to predict successional change on a particular site.

Management succession models (for example FORPLAN and DYNAST) are, for the most part, age-driven. That is, ecosystems are assumed to have particular properties that are predictable based on some measure of their age since disturbance. These properties have generally been derived empirically. Such models deal easily with large tracts of land. However, they cannot always provide details on variability from site to site. They are also regionally specific: empirical information (yield tables, compositional change, and so on) must be provided for each area in which the model is used. One of the great virtues of these particular regional models is that economic values can be integrated into them so that benefits of wildlife management can be weighed against their economic costs.

Assuming that a computer model (whatever its form) can provide a reasonable prediction of ecosystem change, the most challenging problem for the wildlife biologist is interpreting this change in terms of habitat suitability. Most successional models developed to date are timber-oriented. In some cases it may be a simple matter to turn information on the identities and size-class distributions of dominant trees into some measure of habitat suitability (as, for instance, when carrying capacity of a target species depends on tree mast). However, it is more usually the case that habitat suitability is more dependent on noneconomic aspects of the ecosystems that may not be easily predicted from stand-yield tables. These might include abundance of shrub and herb species, amount of dead wood and standing snags, vertical structure, and so on.

Three rather special challenges face us in effectively linking models of habitat suitability with models of vegetation succession. (1) Most habitat models previously discussed incorporate the notion that habitat suitability is something that varies in a gradient fashion over a landscape. However, most successional models are rather site-specific; that is, they do not incorporate the widely held notion that plant communities also vary continuously. This may severely affect the generality of many models. (2) As has been aptly demonstrated in preceding chapters, the mosaic nature or patterning of the landscape can, quite independently of average site conditions, influence habitat suitability. This is no less true for patterns of succession. We need to consider carefully how to incorporate such landscape patterns into models of both succession and habitat. (3) We need to develop models that recognize that the linkage between wildlife and succession is not a one-way street. The organisms we wish to understand cannot always be viewed as passive respondents to successional change. Such organisms may often slow, deflect, or halt succession.

We believe that the chapters that follow give us reason to be optimistic that effective succession-habitat models will be developed. Existing models already provide us with considerable power to predict consequences of management intervention on habitat in constantly changing landscapes.

NORMAN L. CHRISTENSEN: Department of Botany, Duke University, Durham, North Carolina 27706

LAWRENCE S. DAVIS: Department of Forestry and Resource Management, 145 Mulford Hall, University of California, Berkeley, California 94720

47

Predicting Canopy Cover and Shrub Cover with the Prognosis-COVER Model

MELINDA MOEUR

Abstract.—The combined Prognosis-COVER model predicts individual tree crown dimensions and shape, vertical and horizontal tree canopy closure, and the probability of occurrence, height, and cover of shrubs and other understory species in forest stands. The program may be used to produce a descriptive summary of a stand at the time of inventory or to project overstory and understory characteristics through time for natural and managed stands. Examples of simulations on 15 sites show how values obtained from the COVER program can be used to evaluate changes in wildlife habitat. Results point out the importance of considering unique site, vegetation, and treatment effects in making reliable shrub- and overstory-development predictions. Aggregate canopy-closure prediction is improved by a method that accounts for overlap of individual tree crown projections, assumed to be randomly distributed within the stand.

Forest managers in the Northern Rocky Mountains have extensively used the Prognosis Model for stand development (Stage 1973) to summarize current stand conditions and to predict future patterns of stand growth and the probable consequences of alternative management practices on stand development. Its primary application has been to evaluate silvicultural investments in growth and yield. The use of models such as Prognosis need not be restricted to timber-management applications. For example, forest managers must also consider how management may change the suitability of a stand for wildlife, the composition of the understory, or the sequence of succession. The COVER program extends Prognosis by modeling the structure of tree crowns and understory vegetation (Moeur 1985). COVER provides three types of information: (1) a description of the amount of cover and foliage in the tree canopy by height-class; (2) the height and cover of shrubs, forbs, and grasses in the understory; and (3) a summary of overstory and understory cover and biomass for the stand. Descriptions of stand structure produced by COVER are used to examine the probable effects of silvicultural treatments on forest-stand characteristics important to wildlife, such as thermal cover and hiding cover, browse production, and the interactions of shrubs and trees that determine vertical and horizontal stand structure.

Enhancements to Prognosis reflect a basic philosophy that alternative management prescriptions can be compared by examining simulated changes in the major vegetation components. Wildlife managers can use Prognosis-COVER to model the vegetation structure of single stands through time under different silvicultural prescriptions. The model can also estimate long-term, large-scale habitat changes for groups of stands arranged in time and space.

MELINDA MOEUR: USDA Forest Service, Intermountain Forest and Range Experiment Station, 1221 South Main Street, Moscow, Idaho 83843

How the Prognosis-COVER model works

Prognosis simulates stand development by modeling the growth of individual trees. Program versions containing tree-growth equations parameterized for different localities in the northwestern United States are accessible through the USDA Forest Service Computer Center in Fort Collins, Colorado, and elsewhere. Regional variants are presently available for portions of the Northern, Intermountain, and Pacific Northwest regions of the Forest Service. Prognosis incorporates submodels for individual tree diameter growth, height growth, mortality, and product volumes. Silvicultural activities that can be represented include thinning, site preparation, regeneration and establishment of seedlings, and pest management.

Each 10-year cycle in Prognosis starts with simulated thinning activities scheduled for that cycle. Next, diameter and height growth, change in crown ratio, and mortality rate are updated for each tree in the inventory, using prediction equations described by Wykoff et al. (1982). Then, COVER calculates crown dimensions for each tree and summarizes them for the stand. Finally, COVER computes shrub statistics, using site information and overstory density values.

Information needed to run Prognosis-COVER consists of a description of the inventory design used to measure the stand, a list of sampled trees for which species, diameter, and plot identification have been recorded, and site values for slope, aspect, elevation, habitat type, geographic location, time since stand disturbance, type of disturbance, and topographic position (Wykoff et al. 1982; Moeur 1985). Field measurements of shrub height and cover may be used to improve the understory predictions.

Overstory canopy models

The models constituting COVER are parameterized with data collected in the Inland Northwest and Northern Rocky Mountain forests. COVER can predict growth and crown development for 11 conifer species: western white pine

(*Pinus monticola*), western larch (*Larix occidentalis*), Douglas-fir (*Pseudotsuga menziesii*), grand fir (*Abies grandis*), western hemlock (*Tsuga heterophylla*), western redcedar (*Thuja plicata*), lodgepole pine (*Pinus contorta*), Engelmann spruce (*Picea engelmannii*), subalpine fir (*A. lasiocarpa*), ponderosa pine (*Pinus ponderosa*), and mountain hemlock (*Tsuga mertensiana*).

Tree crown width

Equations for crown width were derived from data on 370 trees on 14 sites in northern Idaho and western Montana. Sampled stand basal area ranged from 0.2 to 98 m^2/ha (1–426 ft^2/acre). Open-grown trees and trees that were obviously damaged or heavily defoliated were not sampled. Logarithmic regression equations in COVER predict individual tree crown width from species, dbh, height, crown length, and stand basal area (Moeur 1981).

Tree crown shape

Data for the crown-shape models came from 9800 trees on 12 sites in western Montana, the Blue Mountains of eastern Oregon, and the University of Idaho forest near Moscow, Idaho. Individual tree crown shape was predicted using a linear discriminant function (M. Moeur and E. O. Garton, pers. obs.). Each tree crown was classified into one of five shapes—circular, triangular, neiloid, parabolic, or elliptic—using species, height, dbh, crown length, crown radius, crown ratio, and trees/acre as discriminating variables.

In a projection, predicted tree crown width and shape are combined to display the structure of the canopy through time (Table 47.1). Crown cover values are also partitioned by 3-m (10-foot) height-classes within the tree canopy. Canopy closure is the percentage of ground area covered by the projections of individual crowns of trees whose total heights fall within a given height-class. Crown profile area is the area in ft^2/acre within height-classes occupied by crown profiles: this is the view one would have if the stand were flattened in a vertical plane. Also reported are conifer foliage biomass (Moeur 1981) and number of trees by height-class.

Shrub models

COVER predicts the probability of occurrence, height, and percent cover of understory vegetation in forest stands <40 years old. Thirty-one shrub, forb, grass, and fern species were selected for individual modeling because they provide an important source of browse or cover for wildlife or because they compete with establishing conifer seedlings. Equations were parameterized, using data from over 10,000 sample plots on 500 stands in Douglas-fir, grand fir, western redcedar, western hemlock, and subalpine fir forest habitat types. These data were from Idaho, northeastern Washington, northwestern Montana, and northwestern Wyoming.

Total shrub cover

Predictions for total cover, species cover, and species height for shrubs were taken from Laursen (1984). Laursen used a two-step method to predict total cover, first generating the probability of shrub cover through logistic regression on presence/absence data. Total shrub cover, conditional on the presence of shrubs computed in the first step, was predicted from logarithmic regression on the plots where cover was present. Variables in the model were slope, aspect, elevation, habitat type, overstory basal area, time since disturbance, type of disturbance, geographic location, and topographic position.

Species' occurrence

Probability of occurrence for 31 understory species and species groups was developed by S. Scharosch (pers. comm.). He used logistic regression to predict the proportion of 0.008-ha (1/300-acre) plots that contain the species from the site and stand variables listed above, including predicted total shrub cover.

Species' height

Laursen (1984) used logarithmic regression to model the average heights of shrub species on the plot, using the variables mentioned above, in addition to predicted total shrub cover.

Species' percent cover

Species' cover was predicted using either lognormal regression or logistic regression (Laursen 1984). Independent variables included predicted species' height and overtopping by taller species (the percent cover above current height), in addition to those used for total cover.

Inside the COVER program, total shrub cover on the site is predicted first. Then, probability of occurrence is predicted for individual species, using total shrub cover. Next, heights are predicted for these species, also using total shrub cover. The heights are arranged in order from tallest to shortest and, progressing down through the shrub list, individual species' cover is predicted, using predicted species' height. Finally, the cover values are summed and reported as total cover for the plot. A model run produces understory species' values (e.g., Table 47.2) and an understory and overstory summary (e.g., Table 47.3) displayed for each 10-year cycle. Although predictions are made for 31 understory species, only the nine species with greatest predicted cover, three each within low, medium, and tall height-classes, are displayed in Table 47.2. The remaining species within each height-class are combined into an "other" category. In Table 47.3, "successional stage" (Peterson 1982) is a classification computed from the vertical structure of both trees and shrubs in the predicted stand. It relates wildlife use to a particular vegetation life form.

Example simulations of managed stands

Many of the values produced by COVER can be related to wildlife. To illustrate understory and canopy development as predicted by Prognosis-COVER, simulations were made, using data from stands in Douglas-fir, grand fir, and subalpine fir habitat types from western Montana, Idaho, northeastern Oregon, and northeastern Washington. Within each stand, the species, dbh, total height, height of crown, crown

Table 47.1 Canopy-cover statistics for each 10-year projection cycle displayed by the Prognosis-COVER model.[a]

	Stand height-class (feet)																
Year	0.0–9.9	10.0–19.9	20.0–29.9	30.0–39.9	40.0–49.9	50.0–59.9	60.0–69.9	70.0–79.9	80.0–89.9	90.0–99.9	100.0–109.9	110.0–119.9	120.0–129.9	130.0–139.9	140.0–149.9	150.0+	Total
1981																	
Trees	68	142	107	67	47	45	32	21	13	17	17	1	0	4	1	1	582
% cover	2	9	12	9	6	6	4	3	3	10	8	1	0	5	2	2	82
Area	3,897	13,173	14,667	12,782	11,062	8,684	6,606	4,909	4,345	3,373	2,207	1,402	1,067	475	290	193	89,134
Biomass	286	975	1,111	1,072	1,070	991	879	851	889	689	399	178	110	50	30	15	9,597
1991																	
Trees	2	150	56	83	42	49	31	30	10	11	18	7	1	4	0	1	494
% cover	0	9	5	9	6	7	5	5	2	6	10	5	0	5	0	4	77
Area	2,309	12,544	13,088	13,352	11,638	9,017	6,859	4,764	3,945	3,461	2,664	1,390	999	436	148	33	86,648
Biomass	160	839	882	1,106	1,099	1,023	1,022	989	1,036	956	716	215	112	34	9	1	10,198
2001																	
Trees	0	59	115	37	60	45	11	39	24	8	12	14	2	3	1	1	432
% cover	0	3	11	4	8	8	1	7	6	5	7	8	1	4	1	4	79
Area	1,561	11,713	12,911	14,206	11,994	10,177	7,901	5,570	4,050	3,443	2,845	1,606	1,071	563	164	44	89,818
Biomass	108	811	943	1,182	1,142	1,141	1,088	990	995	902	727	258	122	42	10	2	10,461
2011																	
Trees	0	45	77	44	39	29	33	18	37	10	8	15	6	2	2	1	365
% cover	0	3	8	6	6	5	6	3	9	6	5	9	3	3	3	4	78
Area	774	8,864	12,494	13,538	12,552	11,111	9,014	6,970	4,594	3,525	2,967	2,022	1,189	698	196	62	90,571
Biomass	52	579	871	1,096	1,200	1,235	1,172	1,068	993	923	820	425	172	59	13	2	10,682
2011: Post-thin																	
Trees	0	2	4	2	2	1	2	1	2	1	0	1	0	0	0	0	18
% cover	0	0	0	0	0	0	0	0	0	0	0	0	0	0	0	0	4
Area	39	443	625	677	628	556	451	349	230	176	148	101	59	35	10	3	4,529
Biomass	3	29	44	55	60	62	59	53	50	46	41	21	9	3	1	0	534
2021																	
Trees	404	19	2	2	3	1	1	2	2	1	1	1	0	0	0	0	439
% cover	2	1	0	0	1	0	0	0	1	1	0	0	0	0	0	0	8
Area	1,305	735	1,059	1,076	966	878	754	578	404	242	184	137	93	53	22	16	8,501
Biomass	117	110	194	206	207	206	184	154	120	98	80	59	35	11	3	2	1,788
2031																	
Trees	471	64	0	3	3	1	2	2	2	2	1	1	0	0	0	0	553
% cover	6	3	0	0	1	0	1	1	0	1	1	1	0	0	0	0	15
Area	3,522	1,155	1,285	1,424	1,286	1,171	996	807	601	404	223	166	113	70	29	20	13,273
Biomass	335	144	243	276	276	272	243	214	172	129	88	74	40	19	4	2	2,529
2041																	
Trees	307	140	15	4	0	2	3	1	1	5	2	0	1	0	0	0	483
% cover	7	6	2	0	0	1	1	0	1	2	2	0	1	0	0	0	23
Area	6,408	2,196	1,764	1,893	1,965	1,817	1,456	1,297	1,014	597	325	190	105	62	27	22	21,137
Biomass	670	294	313	399	446	418	355	310	237	160	98	66	30	9	3	2	3,811
2051																	
Trees	80	261	51	14	4	0	5	0	1	3	4	1	1	0	0	0	426
% cover	3	14	6	2	1	0	2	0	0	1	3	1	1	0	0	0	33
Area	5,782	6,474	2,678	2,412	2,173	2,071	1,701	1,451	1,241	841	441	229	144	80	44	26	27,788
Biomass	782	780	403	393	375	377	329	295	247	176	116	71	49	14	6	3	4,417
2061																	
Trees	15	208	118	17	13	1	0	7	1	1	5	2	1	0	0	0	389
% cover	1	15	15	4	2	1	0	3	0	1	3	2	1	0	0	1	47
Area	5,038	10,783	6,429	3,278	2,788	2,358	2,203	1,663	1,333	1,076	648	273	187	94	56	31	38,239
Biomass	801	1,447	899	564	509	462	423	333	286	224	147	73	58	17	7	3	6,255
2071																	
Trees	0	117	103	67	44	10	0	3	4	1	3	5	0	1	0	0	359
% cover	0	10	19	12	10	2	0	1	2	0	3	3	0	1	0	1	64
Area	3,997	12,570	11,321	7,575	3,711	2,802	2,421	2,090	1,502	1,228	844	376	219	129	68	38	50,893
Biomass	884	2,286	1,658	1,212	579	468	426	368	292	243	170	86	65	33	8	3	8,782
2081																	
Trees	0	18	126	81	39	37	9	1	7	1	1	5	2	1	0	0	328
% cover	0	2	17	19	9	11	6	0	3	0	1	4	1	1	0	1	74
Area	1,236	10,020	15,288	10,737	7,709	3,889	2,597	2,226	1,663	1,214	982	539	219	162	75	48	58,603
Biomass	278	1,810	2,953	1,903	1,415	634	465	412	331	278	227	145	68	43	9	4	10,975

Note: Trees/acre, percent canopy closure, crown profile area (ft^2/acre), and foliage biomass (pounds/acre) predicted for site 2 (Table 47.4). In the simulation, the site was clearcut in the fourth projection cycle, when stand basal area exceeded 200 ft^2/acre.

[a] All Prognosis model output is generated in English units; metric equivalents are not supplied.

Table 47.2. Shrub-cover statistics displayed by the Prognosis-COVER model, predicting percent cover, height (feet), and probability of occurrence of individual shrub species for site 2[a]

Year	Low species[b]				Medium species[c]				Tall species[d]			
1981												
Species	BERB	SPBE	ARUV	OTHR	SYMP	ROSA	COMB	OTHR	HODI	AMAL	SALX	OTHR
% cover	3.2	2.9	2.3	0.1	8.2	1.1	0.8	0.5	6.5	0.8	0.0	0.1
Height	0.7	1.1	0.5		1.5	1.9	3.7		5.8	2.9	4.3	
Probability	29.8	29.5	8.5		33.1	11.4	5.0		32.9	7.0	0.3	
1991												
Species	SPBE	BERB	ARUV	OTHR	SYMP	ROSA	COMB	OTHR	HODI	AMAL	SALX	OTHR
% cover	2.7	2.5	2.4	0.1	7.4	1.0	0.6	0.4	5.2	0.7	0.1	0.1
Height	1.1	0.7	0.5		1.5	1.9	4.2		5.7	2.9	4.3	
Probability	26.7	23.2	8.5		30.2	10.3	3.5		29.7	5.9	0.4	
2001												
Species	SPBE	ARUV	BERB	OTHR	SYMP	ROSA	COMB	OTHR	HODI	AMAL	SALX	OTHR
% cover	2.5	2.4	2.1	0.1	6.9	0.9	0.5	0.3	4.1	0.6	0.1	0.0
Height	1.1	0.5	0.7		1.6	1.9	4.6		5.4	2.8	4.1	
Probability	24.8	8.4	19.1		28.3	9.7	2.7		27.6	5.1	0.4	
2011												
Species	ARUV	SPBE	BERB	OTHR	SYMP	ROSA	COMB	OTHR	HODI	AMAL	SALX	OTHR
% cover	2.4	2.3	1.8	0.1	6.4	0.8	0.4	0.3	3.2	0.5	0.1	0.0
Height	0.5	1.1	0.7		1.6	1.9	5.0		5.1	2.6	4.0	
Probability	8.2	23.3	16.3		26.9	9.3	2.2		26.0	4.6	0.5	
2011: Post-thin												
Species	ARUV	BERB	SPBE	OTHR	SYMP	COMB	FERN	OTHR	HODI	ACGL	AMAL	OTHR
% cover	16.3	6.9	5.1	0.0	24.6	14.9	13.9	13.6	12.4	4.5	0.4	0.6
Height	0.5	0.7	1.2		1.2	1.1	2.2		2.9	3.1	1.2	
Probability	99.9	63.3	72.9		95.9	99.7	66.4		97.6	97.0	7.6	
2021												
Species	ARUV	BERB	SPBE	OTHR	SYMP	COMB	ROSA	OTHR	HODI	AMAL	PRVI	OTHR
% cover	12.7	8.2	4.3	0.1	15.8	6.3	2.7	1.1	17.0	1.7	0.3	0.5
Height	0.5	0.7	1.2		1.5	2.4	1.9		4.9	1.8	10.2	
Probability	59.1	75.6	53.7		56.7	40.1	26.6		64.1	20.9	1.6	
2031												
Species	ARUV	BERB	SPBE	OTHR	SYMP	COMB	ROSA	OTHR	HODI	AMAL	PRVI	OTHR
% cover	10.5	8.1	4.2	0.1	14.4	4.9	1.9	0.9	16.5	1.8	0.3	0.4
Height	0.5	0.7	1.2		1.6	3.1	1.9		5.3	1.9	11.3	
Probability	47.3	74.5	51.5		51.8	30.1	19.3		58.9	21.0	1.2	
2041												
Species	ARUV	BERB	SPBE	OTHR	SYMP	COMB	ROSA	OTHR	HODI	AMAL	PRVI	OTHR
% cover	9.5	7.8	4.1	0.1	13.5	4.1	1.6	0.8	14.6	1.7	0.2	0.3
Height	0.5	0.7	1.2		1.7	3.7	1.9		5.3	1.9	11.7	
Probability	41.4	71.9	49.6		49.2	24.9	16.8		55.8	19.9	1.0	
2051												
Species	ARUV	BERB	SPBE	OTHR	SYMP	COMB	ROSA	OTHR	HODI	AMAL	SALX	OTHR
% cover	8.7	7.4	4.0	0.1	12.8	3.5	1.4	0.8	12.1	1.6	0.2	0.3
Height	0.5	0.7	1.2		1.7	4.1	1.9		5.1	1.8	4.4	
Probability	36.8	68.1	47.3		46.9	20.7	15.4		53.0	18.2	1.6	
2061												
Species	ARUV	BERB	SPBE	OTHR	SYMP	COMB	ROSA	OTHR	HODI	AMAL	SALX	OTHR
% cover	7.8	6.7	3.9	0.1	11.9	2.8	1.3	0.8	9.4	1.4	0.2	0.2
Height	0.5	0.7	1.2		1.7	4.6	1.9		4.9	1.8	4.7	
Probability	31.7	62.0	44.2		44.2	16.1	14.2		49.4	15.9	1.5	
2071												
Species	ARUV	BERB	SPBE	OTHR	SYMP	COMB	ROSA	OTHR	HODI	AMAL	SALX	OTHR
% cover	6.7	5.7	3.7	0.1	10.7	2.0	1.1	0.7	6.7	1.1	0.2	0.2
Height	0.5	0.7	1.1		1.7	5.0	1.9		4.6	1.8	5.0	
Probability	25.9	52.6	40.0		40.6	11.4	13.0		44.7	12.9	1.3	
2081												
Species	ARUV	BERB	SPBE	OTHR	SYMP	COMB	ROSA	OTHR	HODI	AMAL	SALX	OTHR
% cover	5.5	4.5	3.3	0.1	9.5	1.3	1.0	0.6	4.5	0.8	0.2	0.1
Height	0.5	0.7	1.1		1.7	5.4	1.9		4.2	1.7	5.2	
Probability	20.3	41.1	35.1		36.6	7.4	11.8		39.3	9.9	1.0	

[a] Prognosis model output is generated in English units; metric equivalents are not supplied.
[b] Low species (0–1.7 feet): ARUV = *Arctostaphylos uva-ursi*, BERB = *Berberis* spp., SPBE = *Spiraea betulifolia*, OTHR = other.
[c] Medium species (1.7–7 feet): ROSA = *Rosa* spp., SYMP = *Symphoricarpos* spp., FERN = ferns, COMB = other spp. combined.
[d] Tall species (7+ feet): ACGL = *Acer glabrum*, AMAL = *Amelanchier alnifolia*, HODI = *Holodiscus discolor*, PRVI = *Prunus virginiana*, SALX = *Salix* spp.

Table 47.3. Understory- and overstory-cover summary statistics for site 2[a]

	Understory				Overstory					
Year	Time since disturbance (years)	Average shrub height (feet)	Percent shrub cover	Successional stage code[b]	Stand age (years)	Average stand height (feet)	Percent canopy cover	Foliage biomass (pounds/acre)	Sum of stem diameters (feet)	Stems/acre
1981	20	2.3	27	7	20	102	82	9,597	238	582
1991	30	2.0	23	9	30	108	77	10,198	250	494
2001	40	2.0	20	8	40	111	79	10,461	251	432
2011	50	1.8	18	9	50	114	78	10,682	244	365
Thinned	1	0.8	113	2	50	52	4	534	12	18
2021	10	2.0	71	2	10	32	8	1,788	30	439
2031	20	2.2	64	2	20	40	15	2,529	50	553
2041	30	2.2	58	6	30	50	23	3,811	73	483
2051	40	2.3	53	6	40	57	33	4,417	102	426
2061	50	2.3	46	6	50	65	47	6,255	132	389
2071	60	2.3	39	8	60	72	64	8,782	164	359
2081	70	2.3	31	8	70	79	74	10,975	190	328

[a]Prognosis model output is generated in English units; metric equivalents are not supplied.
[b]Successional stage codes from Peterson (1982): 1 = recent disturbance; 2 = low shrub; 3 = medium shrub; 4 = tall shrub with no conifers; 5 = tall shrub with few conifers; 6 = tall shrub with mostly conifers; 7 = sapling timber; 8 = poletimber; 9 = mature timber; 10 = old-growth timber.

width, and crown profile form of all trees 2.54 cm (1 inch) dbh and larger were recorded on 25 plots of 0.04-ha (0.1-acre) each. Understory vegetation was measured on two 15-m (49.2-foot) transects within each plot by recording the species, length of intersection, and average height of each shrub intersecting the transect. Densiometer readings of overstory canopy closure were taken on five or six 0.1-m^2 (1.08-ft^2) microplots on each plot. From these observations, average observed shrub cover, shrub height, and percent canopy closure were calculated for each site (Garton and Langelier 1985).

Results and discussion

Total percent shrub cover, average shrub height, and species composition predicted at time 0 by Prognosis-COVER were compared to the values observed at the time of inventory. For the overstory, predicted and observed total percent canopy closure were compared. In addition to comparisons between observed values and predictions, qualitative comparisons of shrub cover, shrub height, and canopy closure were made on photographs taken from ground level at a number of places within each site.

Understory cover, height, and species composition

The difference in means between observed and predicted total shrub cover was 10%, for an overall underestimation of 33% (Table 47.4). Some of the variability in predicted shrub cover can be explained by considering individual site conditions. Even though each study site was treated as a single homogeneous unit in the projections, several of the sites were patchworks of treated and untreated areas, with a variety of shrub conditions. Sites 2, 3, 17, and 27 had combinations of dense, overstocked stands of trees interspersed with recent small patch-cuts or group-selection cuts. Average canopy conditions for each of these sites did not reflect the increased shrub response in openings, giving predicted shrub-cover values that were lower than observed. Sites 16 and 30 had areas of decadent, stagnating trees in which mortality created openings in the canopy. Values of 50–60 years for time since disturbance, an important predictor of shrub development, were misleading for these sites. In this case, stand age did not adequately represent the demise of the mature canopy. Understory removal on site 28 and high-grading on site 15 left practically no understory conifer component beneath the moderate-to-closed overstory of large trees. This condition produced greater shrub development than would be expected, given the canopy conditions.

Overprediction of shrub cover on sites 4 and 8 was due in part to heavy grazing pressures that reduced shrub cover below that expected on similar, undisturbed sites. Shrub cover was also overpredicted on site 12, where thin, rocky soils supported only grass. Visual comparisons of predicted shrub-cover values with stand photographs reinforced these conclusions.

Average species' height was overestimated 55% (the difference in means was 0.24 m [0.8 feet]). In general, the average rankings of predicted shrub heights from stand to stand corresponded well with average shrub heights observed on photographs of each stand. Height was underestimated or slightly overestimated on sites that had a well-developed tall-shrub component (1, 2, 3, and 26). Several stands had virtually no layer of tall shrubs (sites 4, 12, 15, 17, and 27–30), accounting for most of the overprediction of average height.

Predicted species composition closely matched observed on most sites. Eleven of the 15 sites were in the *Pseudotsuga menziesii/Symphoricarpos albus* habitat type. This tall-shrub-dominated ecosystem, described by Steele et al. (1983), is usually represented by *Symphoricarpos albus* (*SYAL*), *Spiraea betulifolia* (*SPBE*), *Amelanchier alnifolia* (*AMAL*), *Rosa gymnocarpa* (*ROGY*), *Berberis repens* (*BERE*), and *Prunus virginiana* (*PRVI*). An herbaceous

Table 47.4. Canopy-cover and understory-cover predictions made by the Prognosis-COVER model for 15 managed stands

Site	Time since disturbance (years)	Trees/acre	Basal area (ft²/acre)	Total percent shrub cover			Shrub height (feet)			Percent canopy closure[a]		
				Observed	Predicted	Bias	Observed	Predicted	Bias	Observed	Predicted	Bias
1	70	264	202	13	13	0	2.4	1.6	−0.8	51	61	+10
2	20	582	166	51	27	−24	1.9	2.3	+0.4	47	56	+ 9
3	15	479	139	55	20	−35	2.0	2.1	+0.1	56	52	− 4
4	30	278	111	0	15	+15	0.9	2.0	+1.1	33	47	+14
8	50	205	63	11	20	+ 9	2.2	2.1	−0.1	27	39	+12
11	50	281	85	9	12	+ 3	2.1	2.9	+0.8	42	43	+ 1
12	60	216	113	5	17	+12	0.7	1.5	+0.8	—	75	—
15	50	120	156	37	12	−25	1.1	1.5	+0.4	44	56	+12
16	60	515	133	29	16	−13	1.4	2.3	+0.9	75	64	−11
17	20	338	109	35	24	−11	0.9	2.7	+1.8	45	55	+10
26	10	189	16	43	43	0	2.3	2.0	−0.3	29	15	−14
27	6	169	112	37	20	−17	0.7	1.6	+0.9	33	44	+11
28	20	103	86	43	22	−21	0.7	2.3	+1.6	14	35	+21
29	6	124	40	39	27	−12	0.6	2.2	+1.6	18	24	+ 6
30	50	345	91	59	23	−36	0.7	2.8	+2.1	45	43	− 2
Mean				31	21	−10	1.4	2.2	+0.8	40	45	+ 5

[a] Adjusted for individual crown overlap (see text).

layer composed of *Calamagrostis rubescens* (*CARU*) or *Carex geyeri* (*CAGE*) often forms a layer beneath the shrubs. Site 17 was a good example of the correspondence of predicted and observed species. Species actually present were BERE, AMAL, SYAL, ROGY, and PRVI. Species predicted to occur with probability greater than 0.05 were SYAL (P = 0.46), CAGE (P = 0.34), SBPE (P = 0.31), AMAL (P = 0.18), ROGY (P = 0.10), and PRVI (P = 0.08).

Canopy closure

Canopy closure was overestimated by COVER by a factor of 1.5 on the example sites (Fig. 47.1). The model computed percent canopy closure by summing the cross-sectional areas of individual crowns, using their predicted crown widths. That is, unadjusted calculations did not account for overlap of crowns. The assumption that each tree crown was independent of all others became increasingly untenable as canopy density increased. A tentative method to correct for crown overlap assumed that if the ratio of crown cover of the first n trees was some amount called X (e.g., X = 0.4), the probability of overlap of the $n + 1$ tree was then equal to X (0.4). Thus, the quantity $[1 - X]$ (or 0.6) of the cover area of the $n + 1$ tree was added to the previous canopy cover (D. Satterlund, pers. comm.). This adjustment was made for each of the 15 example stands, computing an average tree crown area from the sum of individual crown areas in the stand and assuming that tree crowns were randomly distributed. Adjustment for crown overlap (Table 47.4, Fig. 47.1) produced a greatly improved correspondence between predicted and observed canopy closure ($\bar{x}$ overestimate = 13%).

Conclusions

The shrub and overstory models presented here have excellent potential for realistically representing existing vegetation conditions and for simulating how different treatment practices will change those conditions. Discrepancies between modeled values and observed values point out the need to consider each individual site as a product of unique influences—elevation, aspect, topography, soils, treatment history, and present species composition.

Most departures of understory predictions from observed values on the sample stands could be attributed to representing heterogeneous areas as a single stand (especially treated and untreated areas taken together), disturbance factors such as grazing, unexpected combinations of canopy and understory conditions (as in the case of severe understory

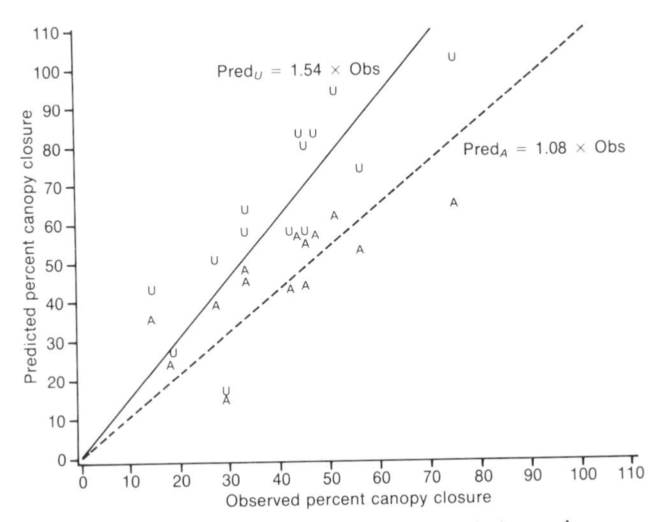

Figure 47.1. Relationship between predicted and observed canopy closure for the sites in Table 47.4. A simple regression of unadjusted, predicted canopy closure (U; solid line) on observed values, fit through the origin, produced the equation shown (root mean square error = 14.6, r^2 = 0.96). When predicted canopy closure was adjusted for individual crown overlap (A; dashed line), the relationship between predicted and observed was almost 1:1 (root mean square error = 11.9, r^2 = 0.95).

disturbance without removing the canopy), or using stand age as a predictor of shrub development in decadent stands. In addition, simply summing individual tree crown areas without considering the spatial distribution of trees on a site overestimated canopy density values. A method of adjusting canopy closure proposed here compensated for crown overlap for the example stands.

Planned improvements to COVER include making predictions on individual sample points within a stand, thus allowing a heterogeneous site to be represented in greater resolution. This will improve the prediction of shrub conditions, permit reporting of within-stand variance statistics, and provide a measure of the spatial distribution of overstory and understory cover.

Prognosis-COVER can provide information about the condition of the vegetation, including species composition, size, and distribution, status of the understory, and site productivity. The combination of shrub development and vertical and horizontal canopy development represents a framework with which changes that might affect wildlife habitat can be displayed in detail through time. The Prognosis-COVER program incorporates models that are specific to certain species and conditions prevalent in the Northern Rocky Mountains, but it is also a general system that can be calibrated to local conditions.

Acknowledgments

I am grateful to Edward O. Garton, Steve Laursen, and L. Jack Lyon for reviewing this manuscript and for making suggestions incorporated herein. Thanks also to Edward Garton for providing the stand-inventory data used to produce the sample simulations.

48

A Process for Integrating Wildlife Needs into Forest Management Planning

ROGER L. KIRKMAN, JODY A. EBERLY, WAYNE R. PORATH, and RUSSELL R. TITUS

Abstract.—A process for integrating information about wildlife habitat into forest land management is presented. Four steps are outlined that use state-of-the-art inventory information and analytical systems to evaluate present habitat conditions and compare management options. A multistand simulation model (DYNAST) was used to describe future forest structure, and species-capability models (PATREC) were used to estimate population responses of wildlife species. Economic factors were measured and displayed along with other resource parameters for consideration by administrators. The process assures better understanding of options and consequences and improves decision-making capability.

Comprehensive resource planning is becoming increasingly important as agencies face shrinking budgets, better informed and more diverse publics, new legislative mandates, and more complex agency policies. This trend has worked to the advantage of resource planners and managers by encouraging interagency development of integrated resource-planning systems.

Within the USDA Forest Service, current management policy was established by the National Forest Management Act of 1976 specifying that all resources shall be addressed in planning and management. Implementation of this policy requires that each Forest develop a management plan.

On the Mark Twain National Forest in Missouri, a plan was developed that identified and assigned hectares to one of several management prescriptions, each having a specified management emphasis (e.g., hardwood production). These prescriptions were then allocated to forest ranger districts and management areas. Management areas, units of 4000–8000 contiguous ha (10,000–20,000 acres) having similar geophysical and vegetative characteristics, were selected to meet management prescriptions. Standards and guidelines for land management were applied to these land units.

The purpose of this chapter is to describe the integration of one resource, terrestrial wildlife habitat, into the implementation of a multiresource plan. Although the example presented was applied to the Mark Twain National Forest in Missouri, the approach could have broad application elsewhere. The approach described here ensures that land capability for a broad spectrum of resources is considered in specifying the ultimate management emphases for a discrete unit of land. It deals with management areas of a size adequate to meet most resource needs, it can accommodate fragmented ownership patterns, and it is functional for data collection in the application of management practices.

Roger L. Kirkman: USDA Forest Service, Mark Twain National Forest, 401 Fairgrounds Road, Rolla, Missouri 65401

Jody A. Eberly: USDA Forest Service, Winona Ranger District, P.O. Box 182, Winona, Missouri 65588

Wayne R. Porath: Missouri Department of Conservation, P.O. Box 180, Jefferson City, Missouri 65102

Russell R. Titus: Missouri Department of Conservation, P.O. Box 509, Rolla, Missouri 65401

The process

Models and sources of information

Automated information systems and analytical tools defined below were used to integrate information about wildlife-habitat needs into land-management procedures on the forest.

DYNAST.—This is a systems-dynamic planning tool designed to aid in the evaluation of alternative silvicultural strategies and the quantification of interrelated forest benefits (Boyce 1980).

PATREC.—Pattern Recognition is a wildlife-habitat-evaluation system in the format of a descriptive word model that can be used to estimate present and to predict future population-density potentials (Williams et al. 1978).

MFWIS.—The Missouri Fish and Wildlife Information System is a data storage and retrieval system for 735 vertebrate species. Information includes classification, status, distribution, life history, food habits, and habitat requirements.

NHI.—The Natural Heritage Inventory is a data base detailing specific current locations of threatened, endangered, and rare animal and plant populations.

TMIS.—The Timber Management Information System is a data storage and retrieval system for information collected and maintained on timber stands. It is specifically designed to track accomplishments.

WMIS.—The Wildlife Management Information System is a data storage and retrieval system that coordinates with TMIS to provide descriptions of specific wildlife habitats.

Identification of present condition

This was the first of four steps in the process of integrating information about wildlife-habitat relationships into long-range plans for forest management. The three remaining steps described management options, evaluated those options, and developed a management program. In this first step, the present condition for all resources and programs had to be described and understood to establish a baseline from which future management options could be developed.

Habitat

Existing habitat was described from various inventories and information sources. Habitat data were organized by special habitats, target lands, and total area.

Special habitats.—Inventory data for riparian areas, caves, springs, glades, and fens were summarized from the WMIS, NHI, and Cave Inventory data bases. These areas were typically small but often provided habitat for endangered, threatened, or rare animals and plants. Actions needed for protection or management, or both, were identified and documented.

The status of all federal and state threatened, endangered, and rare species was determined. MFWIS provided information about species by counties. NHI contained information on specific locations of these species. Recovery plans, cave inventories, and other sources were used to assess actual and probable occurrence of these species.

Target ownership.—Target ownership included National Forest system lands subject to control and direction of the plan. Target-ownership lands ranged from 30% to 100% of all ownerships.

Vegetation data were available in TMIS and WMIS. For purposes of wildlife-habitat evaluation, these data were summarized into precise statements related directly to specific animal requirements (e.g., the percentage of target lands in oak-hickory and oak-pine forest types 50 years or older). These were used as variables in PATREC models.

All ownerships.—Selected information for all ownerships, public and private, was important for identifying total landscape patterns, particularly for species with large home ranges. The distribution and proportion of forest land to open and semiopen land, as well as the type (streams, lakes, etc.), size, and interspersion of surface water, were influenced by all ownerships within the boundaries of the management area. Forest fragmentation was quantified through the use of 10 descriptive patterns. Each pattern specified a range of proportions of open and semiopen land and visually displayed an interspersion pattern. The accompanying land uses were also determined. Data on water quality and availability were tabulated for rivers, streams, springs, lakes, and ponds.

Animals

Estimated population capability.—The capability to deal effectively with large numbers of species is lacking. The management indicator species (MIS) concept provided a realistic interim method to consider diverse life forms in land-management programs for a more manageable number of species, and 13 terrestrial vertebrate species were selected as MIS. In addition, all federal- and state-designated species and some unique life forms were considered individually. As examples for this study, we used the ruffed grouse (*Bonasa umbellus*), wild turkey (*Meleagris gallopavo*), pileated woodpecker (*Dryocopus pileatus*), and the white-tailed deer (*Odocoileus virginianus*). PATREC models were used to estimate population capability. Estimated populations were expressed as totals for the management area and in numbers/405 ha (1000 acres).

Description of management options

Each management area was assigned a specific land-management strategy (Management Prescription), directed by policies, standards, guidelines, and outputs in the higher-level plans (Forest Plan). These directional statements made it possible to describe a desired future condition for the management area. Because each management area was unique, its future condition had to be evaluated for reasonableness and attainability. The DYNAST system provided an excellent analytical tool to forecast future structure of the forest, and PATREC models could predict potential wildlife populations for selected species, given knowledge of future forest structure. Several rotation periods (1–200 years) are required to reach a desired future condition, so predictions were important only as a broad perspective.

Next, specific management programs for 10- or 20-year periods were formulated. Predictions of the effects of these short-term programs on long-term changes in vegetation were made using DYNAST. Several factors had to be considered in developing options for programs; for example: (1) the relationships between present and future conditions; (2) the opportunities to affect present conditions over the next 10–20 years; (3) cost-effectiveness and budget realities; and (4) markets for forest products, road systems in place, and public uses and interests.

Each option, or potential program, selected for analysis had to be described, including all affected resources and programs. TMIS and WMIS identified opportunities for treatments and conditions requiring protection. From PATREC models, insights into conditions of habitat parameters and their effects on populations of MIS were determined.

As an example, three options were described for the 3500-ha Carman Springs Management Area, located in Howell County, Missouri. The first option was characterized by complete implementation of standards and guides at the upper limit, with high investment. The second included a targeted emphasis of standards and guides, with medium investment. The third option reflected lower limits of standards and guides, with minimum investment.

Evaluation of options

Specifications were prepared for each option, outlining the proportion of each forest type to be harvested, applicable rotation ages, the amount of each forest type (conversion rates), the areas to be retained in old-growth, and the land base to be managed in early successional stages (open and semiopen habitats). These specifications were used with

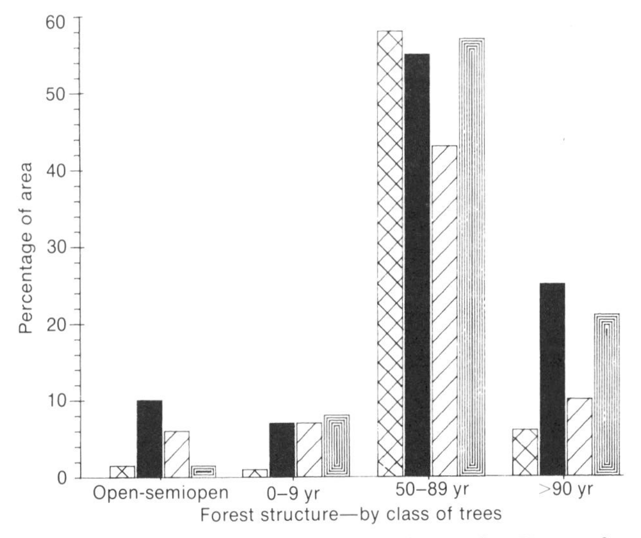

Figure 48.1. Forest structure by tree age-classes after 50 years for the present condition (cross-hatched), for option 1 (shaded), option 2 (diagonal lines), and option 3 (vertical lines).

DYNAST to simulate the effects of the three options on forest structure through time. Computer outputs were areas cut, volumes, forest land use as percent regeneration, poles, sawlogs, old-growth, and nonforested habitats. Differences between options at year 50 for several important forest-structure components were compared (Fig. 48.1).

Animal responses after 50 years of change in forest structure were measured by PATREC models. DYNAST outputs were used to estimate measures of habitat parameters required by PATREC models (Fig. 48.2). Changes in populations were used to estimate changes in harvest and other uses of wildlife. Estimates of recreation use were aggregated into visitor days (12-hour use periods) and assigned eco-

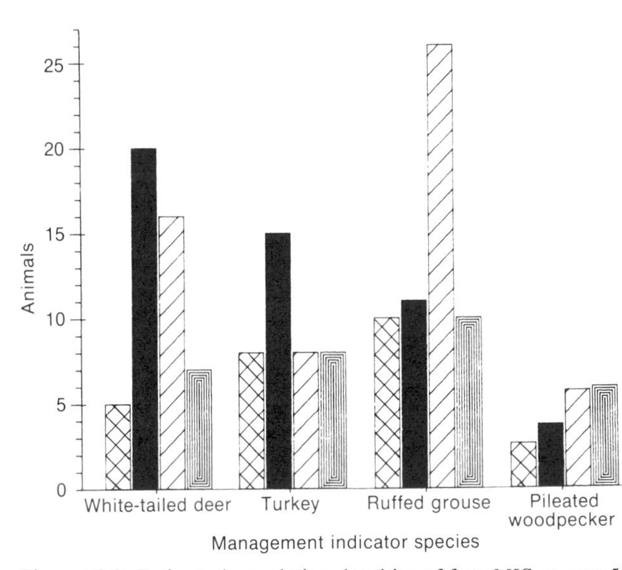

Figure 48.2. Estimated population densities of four MIS at year 50 for present conditions and options 1, 2, and 3 (histograms designated as in Fig. 48.1).

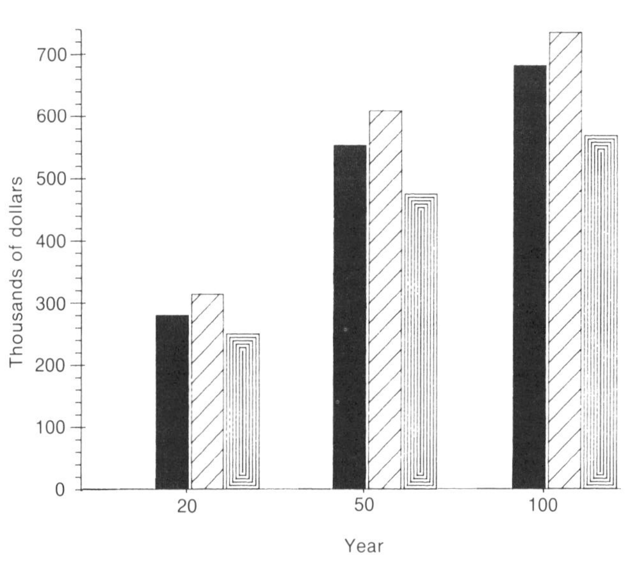

Figure 48.3. Net present worth at years 20, 50, and 100 for option 1 (shaded), option 2 (diagonal lines), and option 3 (vertical lines).

nomic values. Benefits (outputs) and costs for wildlife and timber, by time periods, were analyzed by the DYNAST economic program (Boyce 1982) to compare net present worth for three options at 20, 50, and 100 years (Fig. 48.3).

Development of a management program

Developing a program required preparing stand-selection criteria, selecting individual stands for treatment or protection, scheduling, budgeting, and field implementation.

Stand-selection criteria

Criteria for selecting individual stands were necessary because stand-treatment opportunities usually exceeded objectives or funds available to effect those treatments. Criteria were developed separately for old-growth, regeneration harvests, intermediate harvests, developing open and semiopen habitats, and water sources. Size, age-class, interspersion, location (e.g., proximity to water), and community type had to be considered for each set of selection criteria.

Selection criteria were developed independently for several resources, particularly wildlife habitat, timber management, and visual management. The final selection was made by a line officer, using recommendations from an interdisciplinary team.

Scheduling treatments and developing budgets

Selected stands were scheduled for treatment by fiscal year for the 10-year period. Information was available in TMIS and WMIS as the basis for constructing future budgets and was updated annually as schedules and budgets were adjusted.

Field implementation

It might appear that by the time of field implementation much of the need for resource professionals would have been met. Experience suggests that the opposite is true.

Field verification and on-the-ground adjustments are always necessary to meet stated objectives. The adjustments could best be made by professionals who understood and participated in the entire process. Improved information sources and better analytical tools do not mean that management is easier—they more often mean that managers are considering more factors in greater detail before making their decisions.

Discussion

Interdisciplinary land-management planning is in a fledgling stage and will continue to change and develop. Data bases and analytical tools will change as technology improves. New technology can easily be incorporated into the step-down process we have described. The framework of this process has potential for use in a variety of settings and is flexible enough to be adapted to varying operational goals, ownership patterns, available data, and analytical tools.

The system can be improved through refinements of input data. For example, the management indicator species we used may not have been the most appropriate and will probably change with time as information about species-habitat relationships improves, species-selection criteria are refined, and politics change. Modeling systems such as PATREC are, for the most part, first-generation tools. Research designed to refine parameters that drive these models will continue to be important. This research should include monitoring of both animals and habitats to improve predictive capabilities of such models. Also, the relationships between animal populations and the human dimensions of aesthetic and economic values need further definition.

In decision making, land managers in the past have been forced to use intuitive judgment rather than the evaluation of systematically organized data sets and processes. The process we have presented provides a systematic approach for evaluating resources, selecting options, and implementing a strategic plan. It will provide administrators with a better understanding of the range of outputs and economic values, and the consequences of their decisions. For these reasons, we believe better decisions will be made.

Acknowledgments

We appreciate the assistance provided by many employees of the Missouri Department of Conservation, the North Central Forest Experiment Station, and the Mark Twain National Forest in applying this process to the real world. We are also indebted to T. S. Baskett, L. E. Cambre, D. A. Hamilton, G. F. Houf, E. Morris, S. L. Sheriff, J. M. Sweeney, D. L. Urich, and O. Torgerson for reviewing the manuscript.

49

DYNAST: Simulating Wildlife Responses to Forest-Management Strategies

GARY L. BENSON and WILLIAM F. LAUDENSLAYER, JR.

Abstract.—A computer simulation approach (DYNAST) was used to evaluate effects of three timber-management alternatives on wildlife in a 2700-ha (6700-acre) study area located in the Sierra Nevada, California. Wildlife species selected to evaluate the effects of these alternatives were band-tailed pigeon (*Columba fasciata*), pileated woodpecker (*Dryocopus pileatus*), and mule deer (*Odocoileus hemionus*). Pileated woodpeckers responded most positively to the long-rotation alternative. Responses by band-tailed pigeons and mule deer indicated that either the mixed-rotation or optimal-timber alternative would be the most beneficial.

DYNAST was shown to be a useful tool for displaying the effects of various timber-management schemes. Program outputs were in a format that displayed trade-offs between timber and wildlife production. Because DYNAST simulations have not been tested for accuracy, the outputs should be used with caution. However, DYNAST is flexible, and forest dynamics can be simulated by this model.

The concept of multiple-use, "the balanced allocation of timber, water, range, recreation, wildlife, and other resources found on national forests" (Steen 1976:278), is the basis for integrating numerous resource concerns in National Forest management (Wengert et al. 1979). The DYNamically Analytic Silviculture Technique (DYNAST) (Boyce 1977, 1978; Biesterfeldt and Boyce 1978; Boyce and Cost 1978) was developed for comparing multiple-use resource productions under different management options. DYNAST simulates resource outputs associated with alternative land-management strategies (Boyce 1977). The model operates by combining resource models (e.g., wildlife, erosion, timber production) with models of natural succession for different land types (vegetation cover types, timber cover types, combinations of vegetation cover and slope types, and so on). Alternative timber prescriptions (rotation length for each land type, percent of each land type to be harvested, size of harvest units, and percent of land converted or reverted to another land type) are then developed and linked with DYNAST's successional model to simulate outputs on the areas to be manipulated. The DYNAST model operates through the use of the DYNAMO compiler, which is a FORTRAN source compiler or assembler language (Pugh 1980).

DYNAST outputs can be displayed in both tabular and graphic formats to show changes through time in the production of various benefits for each alternative. Outputs, especially the graphic ones, display trade-offs among the different alternatives. Decision makers then can view the DYNAST products for each alternative and select the option that achieves intended objectives for the land unit.

We selected DYNAST to display trade-offs among management alternatives developed specifically for the winter range of the Downieville (Sierra County) population of mule deer (*Odocoileus hemionus*) on the Tahoe National Forest. The most productive sites in this area are being converted to conifers by regeneration harvest methods that leave only remnant pockets of hardwoods for wildlife or other purposes.

Study area

The study area included 2700 ha (6700 acres) in eastern Yuba and western Sierra counties in the central Sierra Nevada of California. Elevations ranged from 600 to 1400 m (2000–4500 feet). The topography included steep canyons with mostly north- and south-facing slopes. The study area was dominated by mixed stands of conifers and hardwoods; however, small areas of pure hardwoods and conifers existed. Most of the stands were young, i.e., <80 years old. Principal trees were white fir (*Abies concolor*), Douglas-fir (*Pseudotsuga menziesii*), California black oak (*Quercus kelloggii*), and tanoak (*Lithocarpus densiflorus*) (Benson 1983). The study area included portions of the key winter range for the Downieville deer population.

Methods

Study-area characterization

Ten land types (combination of vegetation cover and slope types) and seven successional stages for each land type were defined for the study area, using information from the Tahoe National Forest land-management planning file and timber-type maps. Land types were (1) mixed-conifer on gentle

GARY L. BENSON: USDA Forest Service, Plumas National Forest, Oroville Ranger District, 875 Mitchell Avenue, Oroville, California 95965

WILLIAM F. LAUDENSLAYER, JR.: USDA Forest Service, Pacific Southwest Region, R-5, Tahoe National Forest, Highway 49 and Coyote Street, Nevada City, California 95959

slopes (<30%); (2) mixed-conifer on moderate slopes (30–50%); (3) mixed-conifer on steep slopes (>50%); (4) mixed-conifer–hardwood on gentle slopes; (5) mixed-conifer–hardwood on moderate slopes; (6) mixed-conifer–hardwood on steep slopes; (7) commercial hardwood on gentle or moderate slopes; (8) commercial hardwood on steep slopes; (9) noncommercial hardwood on gentle or moderate slopes; and (10) noncommercial hardwood on steep slopes. Habitat (successional) stages were seedling, sapling, small pole, medium pole, large pole, mature timber, and old-growth timber. Duration of successional stages varied by land type.

DYNAST simulation model

The DYNAST simulation model operates by moving the acreage of each successional stage within each forested land type through time. The simulation continuously moves all acreages through time, employing user-specified time intervals for successional stages (seven were established for each land type). Time intervals for successional stages vary, depending on the land type being simulated. When the acres of a land type reach the user-specified harvest age or completion of old-growth, they are "potentially" harvested or "die," and that number of acres is returned to the earliest successional stage. Meanwhile, all acres in the remaining younger successional stages are correspondingly "moved up" in age. In this way, the simulation model mimics natural vegetative succession.

The DYNAST simulation conducts potential harvesting in a study area which eventually brings the area to a steady-state or sustained-flow-of-timber condition. One of the ways it accomplishes this is to harvest the acreage of older timber at an accelerated rate, resulting in a peak of timber harvested before the sustained flow of timber is achieved.

DYNAST can operate using either English or metric units for the input and output values. English units were used for this study to facilitate prompt recognition and utilization of study results for resource agency personnel.

Management decisions that control the simulation sequence for DYNAST are rotation length for each land type, percentage of each land type to be harvested, and size range of harvest units. The size of harvest units stipulated for this simulation was expressed in terms of acres. Timber-stand-development information for the Tahoe Forest was used in DYNAST to simulate vegetation succession. Yield tables developed for the Tahoe National Forest Land Management Plan were used for mixed-conifer and mixed-conifer–hardwood land types and for maximum productivity or culmination of mean annual increment (with 10-year thinnings) of these timber types at 110 and 120 years, respectively.

Yield tables did not exist for hardwoods, so we developed them for tanoak and California black oak, using observations and measurements from regulated stands in the Challenge Experimental Forest, Challenge, California (P. McDonald, pers. comm.). Benson developed tables related to best growth, spacing, and hardwood thinning. The yield table for commercial hardwoods assumed that the hardwood land types were managed and the culmination of mean annual increment (with 10-year thinnings) occurred at 80 years of age.

Wildlife models

Initial development of wildlife models was accomplished by selecting those habitat variables that were (1) most critical to each wildlife species; and (2) could be linked to vegetation succession. Wildlife models to interface with DYNAST were constructed using information from the species-habitat matrices in Verner and Boss (1980). Coefficients inferred from the matrices rate the value of each successional stage for feeding, cover, and reproduction.

Models were developed for band-tailed pigeon (*Columba fasciata*), pileated woodpecker (*Dryocopus pileatus*), and mule deer. The values used indicated qualitative habitat suitability and ranged from 0 to 1. A value of 1 indicated that the simulated area had a combination of successional stages that provided the greatest habitat quality for that species; 0 represented the poorest combination of successional stages for that species. This qualitative range of values was used so that the effects on several wildlife species could be simultaneously displayed and compared on the same graph. Barrett and Salwasser (1982) described detailed approaches for building models to interface with DYNAST.

Alternatives

Three timber-management alternatives were developed to illustrate a range of management options for the study area. The number of acres in different successional stages within each land type varied considerably (e.g., from 0 to 1200 acres). All land types were assumed to be actively managed, and each land type was thinned every 10 years after the acreage for each successional stage reached or was grown up to 40 years of age. Brief descriptions of the timber alternatives are provided below (see Table 49.1 for more detailed information):

1. Long rotation. Potential harvest occurred within each land type whenever the acreage of the successional stages reached 200 years of age. All of the area within each land

Table 49.1. Timber-management alternatives for winter range of Downieville deer population

Alternative	Land type[a]	Rotation (years)	Opening size (acres)
Long rotation	C & HW/C	200	5 ± 1
	CHW	200	5 ± 1
	NHW	290[b]	1 ± 0.1
Mixed rotation (long/short)	C	120	25 ± 15
	HW/C	120/220[c]	25 ± 15
	CHW	200	15 ± 5
	NHW	290[b]	1 ± 0.1
Optimal-timber rotation	C	110	25 ± 15
	HW/C	120	25 ± 15
	CHW	80	10 ± 5
	NHW	290[b]	1 ± 0.1

[a] Land-type codes: C = conifers, HW/C = mixed hardwood and conifers, CHW = commercial hardwoods, NHW = noncommercial hardwoods.
[b] Stands were not managed at 290–300 years but were allowed to undergo natural regeneration.
[c] Year/year denotes that one-half of the land type was managed for each rotation.

type, except noncommercial hardwoods, was harvested when each successional stage reached 200 years of age.

2. Mixed rotation. Two potential harvest ages (120 and 200 years) were applied to the mixed-conifer–hardwood land types. One half of the acreage of each developing successional stage was harvested when it reached 120 years of age. The other half of that acreage was allowed to mature to 200 years of age before it was harvested. Half of the acreage in these land types reached the old-growth stage, and the other half was harvested near the optimal productivity level, i.e., near culmination of mean annual increment. Mixed-conifer land types were harvested when each successional stage reached 120 years of age, whereas commercial hardwood land types were harvested when the successional stages reached 200 years.
3. Optimal timber. Potential harvest of the land types occurred when each successional stage approached culmination of mean annual increment, which is as follows: mixed-conifer land types = 110 years; mixed-conifer–hardwood land types = 120 years; commercial hardwood land types = 80 years.

Each alternative made four additional assumptions: (1) reforestation was successful within a reasonable amount of time (<5 years); (2) control of competing growth (e.g., shrubs, forbs, and grasses) was largely ineffective; (3) removal of snags and dead material on the ground was minimal; and (4) no site conversions occurred.

Results

Annual timber-output levels for all alternatives were similar, approaching 14,000 m^3 (500,000 ft^3) for the study area 200 years after initial harvest. Although timber production was comparable among the three alternatives after 200 years, the time required to reach peak production and the total volume of timber produced through the simulation run varied considerably. Because much of the study area was dominated by young timber, little timber was produced early in the simulation; thus a lag period occurred before the timber harvest reached its maximum under each option. Maximal timber output appeared to depend on a combination of rotation age and number of land types harvested. Peak production was greater when more land types were managed and the rotation length closely approached the culmination of mean annual increment (point at which the rate of tree growth begins to decline) for each land type. The optimal-timber alternative produced the most timber of the three options because the timber-rotation lengths for mixed-conifer, mixed-conifer–hardwood, and commercial hardwood land types were at or near culmination of mean annual increment (see Table 49.1 and 49.2).

Wildlife responses to the three alternatives (Table 49.2) differed by the habitat quality, time necessary to reach greatest habitat quality, and long-term stability of the response curve (after curve reached the maximum habitat quality). Band-tailed pigeon habitat reached the highest point of habitat quality (i.e., 0.45–0.50) in the shortest time (i.e., approximately 60 years) under the mixed-rotation (Fig. 49.2) and optimal-timber (Fig. 49.1) alternatives. Both the mixed-

Table 49.2. Relative value of each alternative for each modeled resource

	Alternatives[a]		
Resource	Long rotation	Mixed rotation (long/short)	Optimal-timber rotation
Timber production	3	2	1
Band-tailed pigeon	3	1.5	1.5
Pileated woodpecker	1	2	3
Mule deer	3	1.5	1.5

Note: Table values based on our interpretation of the resource responses to each alternative when simulation reached steady-state conditions (approximately 200 years).

[a]Values represent the desirability of the alternative for the resource: 1 = high, 2 = moderate, 3 = low.

rotation (Fig. 49.2) and optimal-timber (Fig. 49.1) response curves also showed the greatest stability—that is, the highest point of habitat quality was reached in a relatively short time, and thereafter the curve remained relatively stable (habitat quality = 0.45).

Pileated woodpecker habitat reached the highest point of habitat quality (i.e., 1) under all three alternatives (Figs. 49.1, 49.2, and 49.3) in a short period of time, approximately 40 years. The long-rotation alternative (Fig. 49.3), however, provided the highest habitat quality (i.e., 1) from year 40 through the remainder of the rotation; hence, it was the most stable of the three alternatives. Both the mixed-rotation (Fig. 49.2) and optimal-timber (Fig. 49.1) alternatives showed a decline in habitat quality from age 40 to approximately 0.80 and 0.75, respectively, by year 200.

Mule deer habitat quality reached the highest point (0.75) under all three alternatives (Figs. 49.1, 49.2 and 49.3); however, the time required to attain that point varied among alternatives. The mixed-rotation (Fig. 49.2) and optimal-timber (Fig. 49.1) alternatives required a relatively short time to reach the highest point of habitat quality (50 and 60

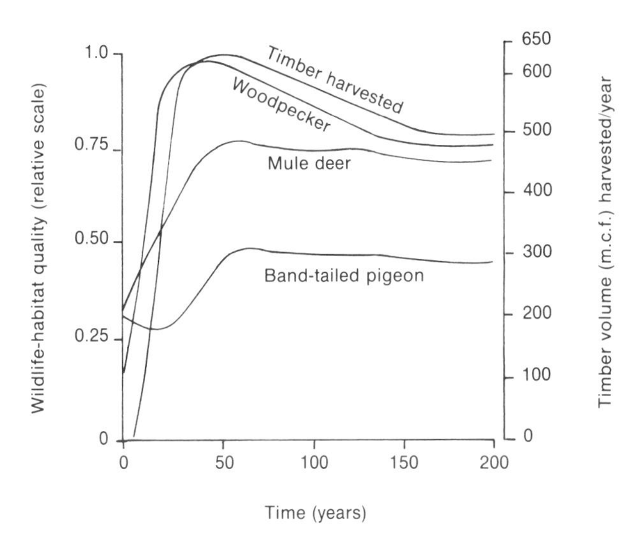

Figure 49.1. Wildlife-species and timber-harvest responses to the optimal-timber rotation DYNAST simulation. Timber-management information for this alternative provided in Table 49.1 (m.c.f. = thousands of cubic feet of harvested timber).

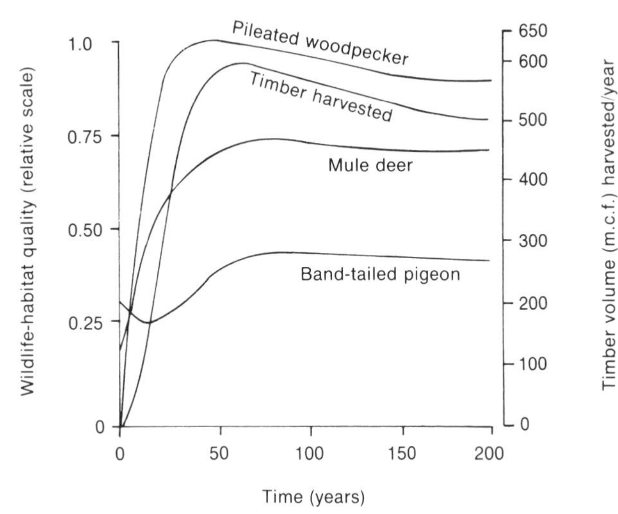

Figure 49.2. Wildlife-species and timber-harvest responses to the mixed (long/short) rotation DYNAST simulation. Timber-management information for this alternative provided in Table 49.1 (m.c.f. = thousands of cubic feet of harvested timber).

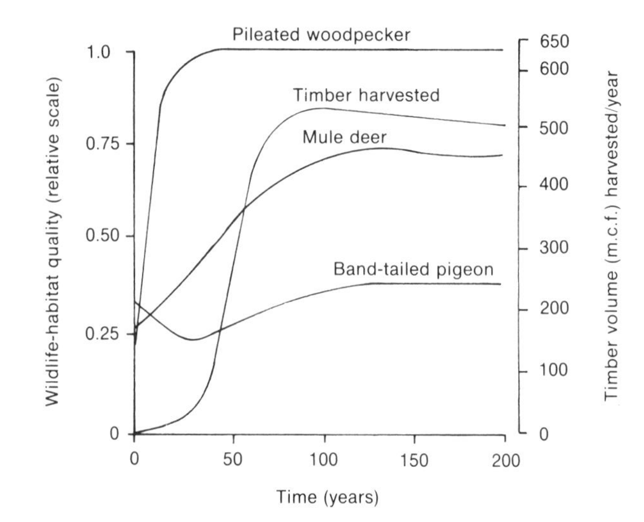

Figure 49.3. Wildlife-species and timber-harvest responses to the long rotation DYNAST simulation. Timber-management information for this alternative provided in Table 49.1 (m.c.f. = thousands of cubic feet of harvested timber).

years, respectively), whereas the long-rotation alternative (Fig. 49.3) showed a decline (to approximately 0.70) after attaining the highest point of habitat quality.

Discussion

Resource responses

The DYNAST simulation model displayed the potential effects of three different management approaches on a study area characterized by an uneven distribution of age-classes within the different land types. The model indicated that the time to reach peak timber production and total timber volume varied considerably according to the different alternatives. This variation is to be expected because the simulation model is rapidly harvesting the older successional stage timber at an accelerated rate to bring the entire study area to a steady-state or sustained-flow-of-timber condition.

The model harvests timber at all times whenever any amount of acreage reaches the specified harvest age. However, the study area was not initially in a steady-state or normal distribution of all age-classes. Consequently, the first 100–200 years of harvesting resulted in wide variations of timber production until steady-state was reached. If the user specifies a long rotation (e.g., 200 years before harvest), then it will take the model longer to reach steady-state than if a shorter rotation (e.g., 80 years before harvest) is used. Also if the user does not allow harvest to occur on certain land types, this will also affect the amount of timber that can be harvested.

Both the mixed-rotation and optimal-timber alternatives were very beneficial to band-tailed pigeons because of the resulting balance between the older mast-producing stands and younger stands supporting fruit- and berry-producing shrubs. Band-tailed pigeons feed extensively on mast during all seasons except summer (Grenfell et al. 1980). The mast is produced primarily by mature trees. In summer, forbs, fruits and berries, and grains become important diet items (Grenfell et al. 1980). Thus, the best habitat for band-tailed pigeons occurs where hardwood trees are harvested after mast production begins to decline and hardwoods are regenerated, producing younger tree stands supporting forbs and berry-producing shrubs.

Pileated woodpeckers found the most suitable habitat under the long-rotation alternative, in which trees and associated snags could grow to large sizes and provide the nesting sites that these birds need (McClelland 1979). Thus, under the long-rotation alternative, the habitat quickly reached maximum quality, which was retained throughout the 200-year simulation. Habitat quality declined as rotation lengths were shortened and large trees and snags were less readily available. Therefore, the optimal-timber alternative, which was based on relatively short rotation periods, provided the poorest-quality habitat for pileated woodpeckers.

Alternatives most beneficial to mule deer were the mixed-rotation and optimal-timber ones. Rotation age was also important to mule deer in relation to production of mast and to the amount of foraging habitat available. The shorter the rotation age, the greater the proportion of area suitable as foraging habitat (i.e., younger stands supporting shrubs, forbs, and grasses). Harvest-unit size also influenced deer habitat quality; very large and very small units lowered habitat values. Thus, those alternatives with relatively short rotation ages, yet retaining mast-producing trees, produced the best habitat for mule deer.

Value of DYNAST to decision makers

The greatest benefit of DYNAST to land and resource managers is its ability to portray management trade-offs in an understandable format, so that decision makers can select the alternative most closely meeting intended goals for the area. A second major benefit is the display of the cumulative effects of management decisions on resources. DYNAST can be altered to incorporate incremental, deleterious effects (e.g., soil loss and reduced productivity) or even thresholds for population viability. Simulations also will assist project-development teams by providing qualitative information on trade-offs for preparation of planning documents. Environmental documents that use graphics produced by DYNAST will enable the general public to visualize the consequences of actions.

Accuracy of DYNAST simulations

DYNAST outputs must be interpreted with caution. DYNAST was developed for use in the eastern United States and was later modified for western applications. The model uses information provided by the user concerning successional stages of vegetation. As a consequence, it can incorporate known successional information that is applicable to the specific area being modeled.

The DYNAST simulation is not a timber-stand model, because it does not operate on the basis of individual timber-stand development; instead it operates on the basis of land area occupied by user-specified vegetation cover (or land) types. The model also uses silvicultural controls, including rotation age, harvest-patch size, and percentage of land area converted or reverted from one cover (or land) type to another cover type. Wildlife and other resource models used within DYNAST may not relate directly to those variables, so resource responses may not be realistic. Finally, the model has not, to our knowledge, been tested for accuracy in the United States. Similarly, the wildlife models have not been verified. They are based on the literature and on professional judgment, but the accuracy of the proposed relationships has not been tested for many species.

Conclusion

Decisions to produce various timber and wildlife outputs are being made continually, and DYNAST simulations may provide decision makers with a better understanding of resource gains and losses. DYNAST provides managers of multiple-use forests with a tool to display quickly possible trade-offs under different timber-harvest options. Employment of this or other simulation procedures, in the absence of reasonable alternatives, has the potential to result in better informed and better documented decisions.

Acknowledgments

Harley Greiman, Dave Thomas, and Hal Salwasser initiated and supported the Downieville timber-wildlife-coordination project. Philip McDonald helped develop hardwood yield tables. Phil Aune, Dave Connell, Gary Hartman, Carol Malone, and Steve Underwood helped develop the data base. Stephen Boyce and Chuck Evans assisted with runs of the DYNAMO compiler to obtain DYNAST model outputs. Kathleen Franzreb, Harley Greiman, Gary Hartman, Ken Mayer, and Hal Salwasser provided critical and sensitive reviews of the manuscript. To all we express our sincere appreciation.

50
Refinement of DYNAST's Forest Structure Simulation

JAMES M. SWEENEY

Abstract.—DYNAST is a multiple-resource, systems-dynamic model that simulates normal forest succession from regeneration through old-growth. DYNAST includes an assumption that the inventory within a given age-class (or diameter-class) is equally distributed throughout that class. When the model is applied to a forest with a distinctly unbalanced distribution of age-classes (some age-classes absent), this assumption causes unrealistic movements of inventory through simulated time. As a result, the forest is prematurely driven to stability (balanced age-classes). This is important because all benefit algorithms for other forest resources are keyed to forest age-class structure. A change in the core model is presented that refines the simulation of forest structure, maintains model simplicity, and requires no additional user input.

In the face of increasing demands for all forest products, including timber, fiber, energy, wildlife, grazing, water, and recreation, managers of these resources can no longer afford the single-resource approach to management that was typical in the past. Management of any one forest resource must include consideration and adjustment for others. This managerial strategy was made binding in the 1970s by a series of laws aimed at resource integration: National Environmental Policy Act of 1969, Endangered Species Act of 1973, Renewable Natural Resources Planning Act of 1974, National Forest Management Act of 1976, Federal Land Policy and Management Act of 1976, and Fish and Wildlife Conservation Act of 1980 (Salwasser and Tappeiner 1981). As a result, we are now in an era of coordinated, intensive forest management in which interdisciplinary planning is an important element. In such circumstances decision makers need specific and accurate information on the relationships between differing land uses and probable consequences of alternative management prescriptions.

Over the past 5–10 years, numerous habitat evaluation and simulation models have been developed to help the resource manager simplify the maze of management alternatives. However, many of these models, originally designed as tools to aid in making decisions, have grown in complexity as they strive to quantify more accurately the ecological-sociological system they represent. As a result, some models have surpassed the point of practical day-to-day utility.

A useful model must strike a balance between complexity and simplicity to provide meaningful evaluations of alternative management strategies. Managing natural systems is not an exact science. Therefore, a model designed to assist forest managers with integrated management of alternative land uses need not be an exact replica of the forest ecosystem. Some concessions (assumptions) can and, indeed, should be made in going from a biological to a managerial system.

An example of a system that incorporates this philosophy is the dynamic simulation model DYNAST (Boyce 1980). This multiple-resource model simulates forest succession and provides a continuous inventory of all related forest benefits. The system is cybernetic in that it guides the modeled forest toward a steady state (constant annual output) through a series of feedback loops. DYNAST does not emphasize one benefit over another nor treat one benefit as arbitrarily constraining another. (For a more detailed discussion of DYNAST and its advantages and disadvantages, refer to Boyce [1977, 1978, 1980] and Sweeney and Bhullar [1983].)

JAMES M. SWEENEY: USDA Forest Service, North Central Forest Experiment Station, 1-26 Agriculture Building, University of Missouri, Columbia, Missouri 65211

The assumption

DYNAST develops and maintains a record of the simulated forest by keeping an inventory of hectares or acres by habitat type. It does not maintain individual stand records. A habitat type is either a diameter-class or an age-class within a given forest cover type, e.g., oak-hickory (*Quercus-Carya* spp.) poles or 50- to 90-year-old shortleaf pine (*Pinus echinata*). Inventory is therefore stored and moved by age- or diameter-class through the cells (habitat types) of this two-dimensional matrix of forest types. A first-order exponential delay is used to facilitate movement across these cells. The inventory within a given cell is assumed to be uniformly distributed across that cell.

The successional flow algorithm then consists of a simple movement of a proportional segment of the inventory from one cell to the next. For example, if age-class *C* was a 20-year age-class (year 20–39), each 1-year iteration of the model would move 1/20 of the inventory out of age-class *C* and into age-class *D*. Similarly, if age-class *D* was a 10-year age-class (year 40–49), the same 1-year iteration would move 1/10 of the initial inventory from age-class *D* to *E*.

The problem

In most cases, the assumptions of uniform distribution across an age-class and the proportional movement of inventory across age-classes provide a sufficiently accurate simulation of forest succession, while maintaining model simplicity (both in terms of input data needed and time and money required to run the model). However, when the initial forest inventory reflects a distinctly unbalanced distribution of age-classes (some cells with 0 ha), this assumption and its application cause two problems. For instance, if, in the previous example, age-class *C* had an initial inventory of 400 ha and age-class *D* was empty, the first 1-year iteration would move 20 ha from *C* into *D*. The 20 ha now in *D* is then assumed to be uniformly distributed across the 10-year span of *D*. As a result, the next 1-year iteration would move 2 ha out of *D* and into *E*. Thus, in only 2 years we have inventory moving into, across, and out of a 10-year age-class.

The second problem occurs when a lower age-class is empty and provides no input to the higher age-class for an extended number of years. For example, if age-class *B* remains empty for 20 years or more, age-class *C* should be empty by the end of 20 years (i.e., all inventory should have moved across and out of *C*). But this does not occur, because DYNAST moves the same set proportion (1/20 in this example) out each year. As a result, less and less is moved out, and *C* approaches zero inventory only after a long period of simulated time.

The ultimate effect of these two problems is the flattening of the age-class distribution curve. DYNAST prematurely drives the simulated forest to an early uniform distribution of inventory across age-classes (Fig. 50.1). This is a major concern because all benefit modules are driven by the stand-structure output of the core forest-succession model.

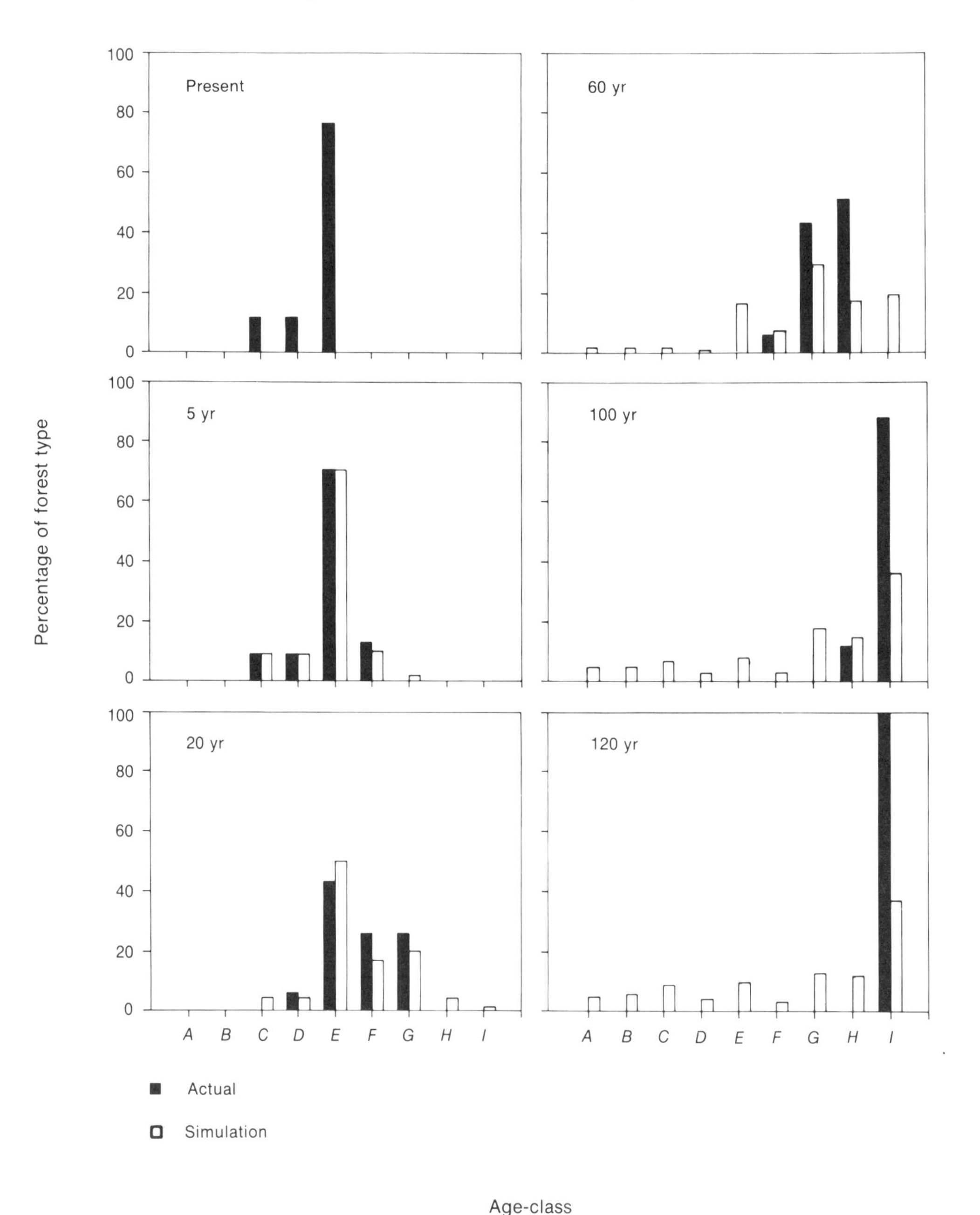

Figure 50.1. Stand structure as simulated by original succession algorithms in DYNAST–OB.

The solution

To be useful to resource managers, DYNAST must be able to simulate all possible conditions with the same level of reliability. One approach would be to maintain inventory records within DYNAST by stand. However, this would increase the model's complexity and associated computer time and costs. The challenge, then, is to improve the logic structure of forest-succession flow within DYNAST while maintaining model simplicity (time and cost per run) without requiring additional user input. The solution proposed meets these criteria.

A time-sensitive step function was inserted into the succession flow algorithms to account for the first problem. If at any time (either initially or later in the simulation) the inventory of an age-class goes to 0 (<0.1), a gate is closed and no outflow is allowed from that age-class. The gate remains closed until two conditions are met. First, there must be inflow from the age-class below. Second, a period of time must have elapsed, from the time inflow started, equal to the time span of the age-class. Thus, as long as inflow is 0, the gate remains closed. And, once inflow starts, outflow is allowed only after enough additional time has elapsed to allow the inventory to grow through the age-class. Then, as long as there is inventory in the age-class and inflow from the age-class below, outflow is determined by the orginal proportional method described above.

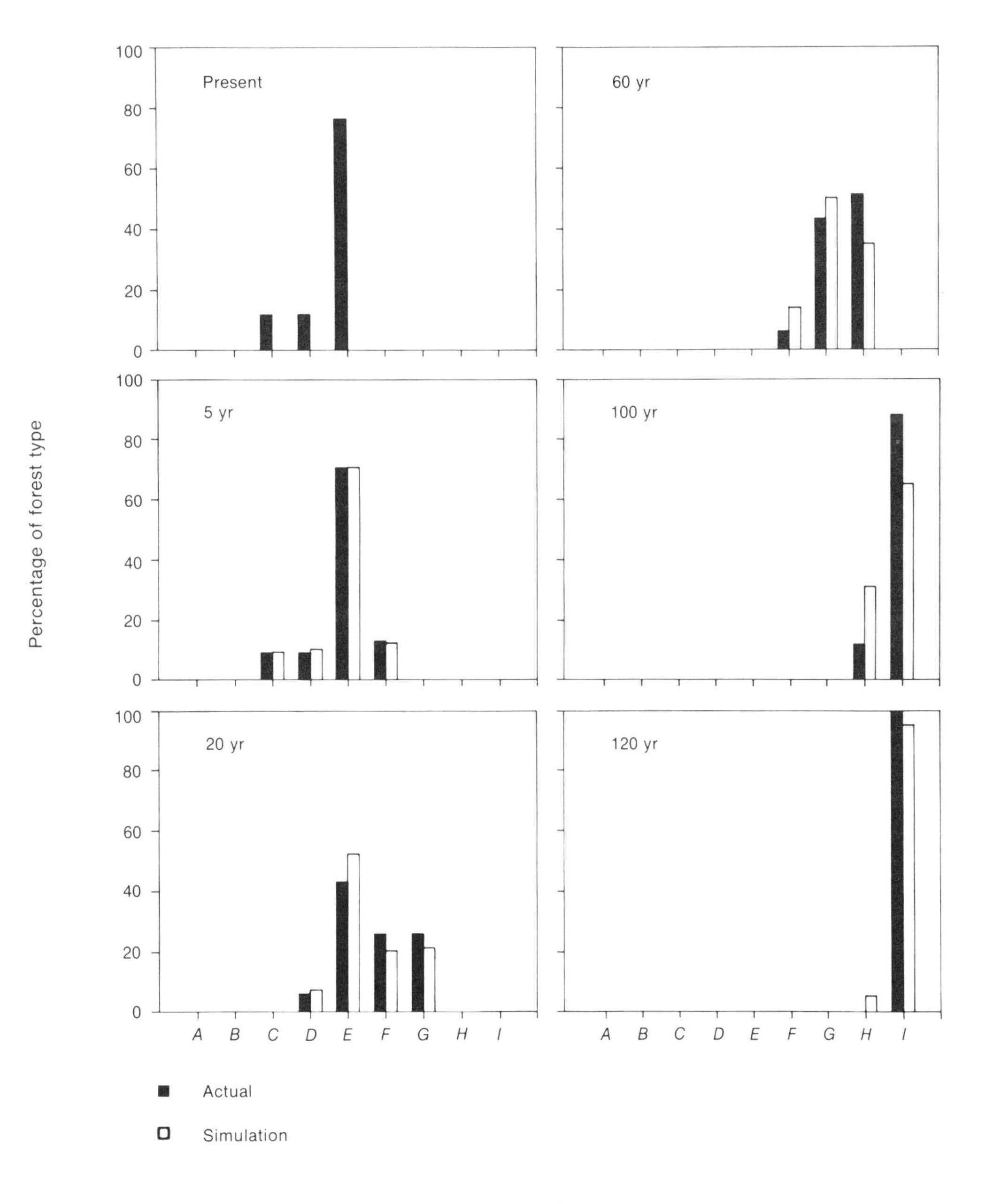

Figure 50.2. Stand structure as simulated by the revised succession algorithms presented in this chapter.

To meet the second problem, that of an age-class never going to 0, a counter was installed to modify the proportion removed each year. The counter remains at 0 as long as inflow occurs from the age-class below. When there is no inflow, the counter maintains a cumulative count of years without inflow. This count is used to increase the proportional outflow. For example, the normal outflow from a 20-year age-class is 1/20 of the inventory present. During the first year, no inflow occurs, the counter is incremented to 1, and the denominator is reduced by the count. Thus the outflow is increased to 1/19. Likewise, with no inflow for a second year the counter is incremented to 2, which results in an outflow of 1/18. As a result, if there is no inflow by the end of 20 years, all remaining inventory will have moved out, because the proportion will have increased to unity.

These two modifications, operating together, provide for a successful simulation of a forest, even when the initial age-class distribution is highly clumped (Fig. 50.2). The simulation is improved significantly, the cost per run is still below $5.00, and no additional user input is required. (A detailed listing of the revised coding and documentation is available upon request from the author.)

Conclusion

Besides the improvement in simulation of forest structure, perhaps the most important lesson to be derived from this exercise is that one should be aware of assumptions. Note that this is not a warning to "beware" of assumptions; these are a necessary part of wildlife-habitat models. However, along with the use of assumptions comes the responsibility on the part of the model designer to publish the underlying assumptions with the model. They must be clearly stated, and the implications of their use must be completely described. Likewise, the user of the habitat model has the responsibility for being fully aware of the assumptions operating within the model and of their influence under the specific conditions to which the model is being applied.

Acknowledgments

I thank Stephen Boyce, who assisted in the development stages of these refinements and reviewed the final manuscript. I also thank Steve Sheriff, Hal Salwasser, and Richard Holthausen for their critical reviews.

51

Linking Wildlife-Habitat Analysis to Forest Planning with ECOSYM

LAWRENCE S. DAVIS and LYNN I. DELAIN

Abstract.—The key to analytical linkup of mathematical wildlife-habitat models with forest planning, using systems such as FORPLAN or ECOSYM, is a suitable shared data base and an efficient computerized geographic-information system (GIS) to manipulate the data. To demonstrate how an integration of wildlife-habitat analysis into land-management planning can work, a mathematical habitat-suitability-index model for the spotted owl (*Strix occidentalis*) was applied to two spatial data bases. Although the model was implemented with no functional problems, many questions about model form and interpretation arose. A combination of geometric and arithmetic models may better evaluate home range. At issue is the literal interpretation of habitat index scores for suitability. A procedure for evaluating the potential impact of establishing habitat-management zones on other forest values such as timber is shown and provides the basis for trade-off analysis and identification of efficient management prescriptions.

Cutting trees and building roads to implement a forest plan are the dominant instrumental acts driving the structurally and spatially dynamic matrix of forest vegetation. This matrix is the basis of most wildlife habitat; thus, to influence current and future habitat, the best place to have input is in the forest-planning process.

Fortunately, the analytical sophistication of both forest planning and wildlife-habitat modeling has now progressed to the point where planners, wildlife biologists, timber managers, and other specialists can effectively work together—in fact, we seem poised to make some order-of-magnitude improvements. The key to the analytical linkup of mathematical wildlife-habitat models to forest planning, using systems such as FORPLAN, is a suitable shared data base and an efficient, computerized geographic-information system (GIS) to manipulate the data.

Our objectives in this chapter are threefold: (1) to outline the conceptual and procedural link of spatially explicit habitat analysis to the land strata used in planning; (2) to demonstrate this linkup, using ECOSYM data in a GIS called SPANMAP for two different data bases (the decision problem considered is identifying and selecting among candidate spotted owl [*Strix occidentalis*] habitat-management areas); and (3) to identify questions of interpretation when wildlife-habitat models are applied in a spatially explicit planning context.

ECOSYM data bases

ECOSYM (Henderson et al. 1978; Davis 1980) is an acronym for a concept and procedure for building a useful, spatially explicit land-management information system. The concept is based on a high-quality data base of objectively defined and mapped physical, vegetative, and human development attributes of the land resource. These basic attributes include a shared set of independent variables for hundreds of different models designed to predict everything from timber productivity to erosion rates and wildlife habitat. To achieve its maximum effect, the data base must be established in a GIS with the capability of complex mathematical manipulation, thereby allowing the models to produce spatially explicit dependent variables.

SPANMAP, an acronym for Spatial Analysis and Mapping, is a steadily evolving set of GIS software programs that originated at Harvard University. (SPANMAP is available on request from Davis.) It is a grid-based system designed for inexpensive and easy mathematical transformation of data, using equations, table models, or user-supplied FORTRAN programs. Tables and lineprinter maps are the current output forms, but pen plotters are an option. Input is terminal interactive, and modeling is done on a cell-by-cell basis. A recent linkup with the MAP software for grid data bases (Berry and Tomlin 1980) enables an efficient search over one data layer to calculate edge, diversity, clumping, and distance information.

Land stratification for forest planning

Forest planners first divide a forest into land strata, or units, and then make decisions about prescriptions under which each stratum will be managed over time. Prescriptions schedule the timing and kinds of management activity (e.g., fertilizing, timber harvesting, road building) but may also simply allocate land to natural reserve status. Most contemporary forest planning strives to optimally match land strata with prescriptions. Many possible prescriptions are defined for each land stratum, and the solution procedure simultaneously chooses a prescription for all land strata that best

LAWRENCE S. DAVIS and LYNN I. DELAIN: Department of Forestry and Resource Management, 145 Mulford Hall, University of California, Berkeley, California 94720

achieves the manager's goals. The USDA Forest Service does this with FORPLAN, a large-scale linear programming application.

Some land-stratification approaches do not identify the location of the land, and thus do not permit consideration or evaluation of the sites where activities take place on the ground. Because habitat for many wildlife species is strongly determined by the spatial arrangement of vegetation, water, topography, and roads, lack of information on location in land strata can make it difficult to formulate salient wildlife input to the planning process for these species. Four different approaches to land stratification—land types, land units, management units, and allocation and scheduling zones—are currently being used.

Land types

A land type includes all land in the planning area with the same user-defined subset of physical, biological, and human development attributes; for example, all ponderosa pine sawtimber above 1000 m (3280 feet) on south-facing slopes within 300 m (985 feet) of a road. When timber is a primary distinguishing attribute, we often use the term *stand type* for these strata. Land types respond in a predictable way to treatment because of their homogeneous character, so we can predict vegetation growth and yield with some confidence. Sample-plot data from forest surveys may tell us that 30,000 ha (74,100 acres) of this land type exist somewhere on the forest, but the plots do not tell where it is or how many parcels make up this land type. Planning decisions using these strata might say: Clearcut and regenerate 5500 ha (13,585 acres) in decade 2, 3000 ha (7410 acres) in decade 4, and so on. It is up to the managers in decades 2 and 4 to find and harvest the appropriate parcels (if they can). Because it is not known at the time of planning which specific areas will be cut in the future, wildlife, erosion, visual, and many other spatially dominated impacts cannot be accurately estimated and planning choices adjusted accordingly.

Land units

Land units (called "capability units" by the Forest Service) are contiguous parcels of the same land type that are larger than some specified minimum size. The forest is mapped, with the land-type classification used to identify land units. If the smallest area to be mapped is 5–10 ha (12.4–24.7 acres), many thousands of land units are found on a large forest. Because all units are of the same land type, they should respond in the same predictable way to treatment. But a large number of small land units can create a planning problem too great to solve. Additionally, the small-sized land units preclude good spatial analysis of large-area issues like preservation of raptors, maintenance of big game habitat, or control of stream sedimentation.

Management units

Management units are large contiguous parcels of land identified for administrative and planning purposes. Watersheds, ownership units, or administrative boundaries are used, and these units run from 1000 to 10,000 ha (2470–24,700 acres). Management units contain several land types and hundreds of individual land units; hence, they are heterogeneous. When treated as a single planning stratum with average attributes, management units do not predict or discriminate the response to selective treatments and management activities well. Alternatively, they are large enough so that meaningful, localized analysis of the effects of management decisions on wildlife habitat and other concerns can be carried out through the use of maps, photographs, and field inspection.

Allocation and scheduling zones

Allocation and scheduling zones (ALSCZ) are land strata that combine spatially defined management units, like watersheds, with homogeneous land types to achieve both the spatial-analysis capability of management units and the response predictions of land types. The distribution of area by land type within each management unit is determined. Prescriptions are written for both land types and management units in packages that are coordinated to achieve certain desired outcomes from the management unit. This allows the scheduling and location of ground activities for each package to be done on the input side of FORPLAN analysis to ensure that the prescription package can be implemented. Site visits, GIS, and models such as DYNAST can be used to prelocate or preevaluate the options for each management unit to ensure feasibility.

Integrating wildlife into forest planning: The spotted owl

Much of the integration of wildlife-habitat analysis into forest planning involves the Forest Service. FORPLAN I (Johnson et al. 1980) is the analytical model currently used to produce most forest plans; it normally stratifies land by land type and by a few selected special management units, e.g., big game winter range or wilderness-candidate areas. FORPLAN II (Johnson 1985) is just past the testing stage and has the ability to coordinate activities within and between ALSCZs.

Our examples assume that an ALSCZ approach to land stratification is used for analysis by FORPLAN II. We show how a GIS such as SPANMAP can be used to identify spotted owl habitat areas and prepare efficient prescriptions for spotted owl management.

Data bases

Two different ECOSYM-type spatial data bases—Blodgett Forest and Tanner Creek—are used for our examples and analysis. Vegetation, soils, topography, and other attributes are digitized and available for analysis by SPANMAP.

Blodgett Forest is a 1200-ha (3000-acre) mixed-conifer forest of the central Sierra Nevada, at an elevation of 1380 m (4500 feet). At least one pair of spotted owls occurs at Blodgett. High-resolution land-attribute data are digitized to grid cells of 0.25 ha (0.62 acre), and percent canopy cover of the tree vegetation is mapped at three canopy height levels.

Tanner Creek is a completely hypothetical data base with

physical and vegetative characteristics for this 10,000-ha (25,000-acre) watershed specified to represent those actually found on the Mount Hood National Forest in Oregon. This data base was constructed with 1-ha (2.47-acre) grid cells.

The Blodgett data base is used to investigate problems that arise when a typical mathematical habitat model is applied to spatially identify and evaluate habitat at a given point in time. The Tanner Creek data base demonstrates how habitat evaluations can integrate with timber-production goals to select an efficient habitat-management unit for an ALSCZ prescription package.

The spotted owl habitat model

A model of spotted owl habitat developed by Laymon et al. (1985) is used to illustrate application of habitat models. Their model takes the geometric mean of three variables: stand structure (number of canopy layers), percent canopy closure, and dbh of the larger stems. The functional form of the model is as follows:

$$HSI = (V_1 \times V_2 \times V_3)^{0.33}$$

where HSI = habitat-suitability index, which assigns an index value of 0–1.0 to each land unit of analysis; V_1 = tree diameter (diameters from 30 to 90 cm [12–36 inches] of the larger trees on a scale of 0–1.0); V_2 = percent canopy cover (from 20% to 80% are coded on a scale of 0–1); and V_3 = stand structure (codes 0–1.0 for single, double, or multiple canopy layers). The index value calculated for each geographic unit of analysis indicates its relative suitability. Laymon et al. suggest that the HSI for a large area can be calculated as the average of the individual stand- or unit-area HSIs, weighted by area. In addition to a suitable 60–120-ha (150–300-acre) roosting site, the literature suggests that an owl pair requires a blocky, contiguous home range of 400–3000 ha (1000–7500 acres).

To evaluate a planning area using a GIS, an operational interpretation of the habitat model is required. For our analysis, we made three normative assumptions: (1) a HSI of 1.0 was very good; (2) a HSI of 0 was nonhabitat; and (3) average HSI scores in a larger area were appropriate criteria for identifying habitat.

Given these assumptions, we somewhat arbitrarily defined minimum standards for a candidate habitat-management area: (1) at least 120 ha (295 acres) of contiguous core habitat with an average HSI > 0.5; and (2) a total home range of at least 2000 ha (4950 acres) with an average HSI > 0.3.

Example 1: Identifying spotted owl planning zones

Blodgett Forest's overstory vegetation data provided information to accurately quantify all three variables in the owl HSI equation for each 0.25-ha cell (Fig. 51.1). Four definite clusters of core owl habitat were identified, but considerable zero-value land (nonhabitat) occurred around the areas rated as suitable habitat. The observed spatial fragmentation of the habitat was a result of our using a small-cell cell data base and the geometric form of the model. If any factor in the equation had a zero value, the HSI was zero. And obviously the more factors used in the model, the greater the likelihood of a zero occurring. The geometric-mean model appeared to identify core habitat areas well, but the preponderance of very low or zero-value ratings seemed to us to understate the home range potential.

As an alternative, HSI was redefined as the arithmetic mean of the factors, with HSI = $(V_1 + V_2 + V_3)/3$ (Fig. 51.2). The core-area land units previously identified by high geometric scores were still intact, but the intervening areas were also rated as habitat by the model using arithmetic means, with scores from 0.2 to 0.4. We concluded that the geometric-mean HSI, with its critical factor properties, was good for identifying core roosting or breeding habitat, and the arithmetic-mean HSI was better for evaluating feeding and general home range territory. This implied that all factors must be present for core habitat but not for home range. In the Blodgett example, either model would have located the candidate management areas.

A cell near the center of each of four identifiable geometric HSI clusters was treated as the center of a superimposed, square habitat zone and used to define and numerically evaluate candidate management areas. On each of the four centers, evaluations were made for four habitat-zone sizes (50, 120, 240, and 400 ha), using SPANMAP to calculate five criteria: (1) average geometric HSI; (2) average arithmetic HSI; (3) percentage of the cells with nonzero scores; (4) percentage of the cells with HSI ≥ 0.4; and (5) the area of core habitat with HSI ≥ 0.5.

Results cautioned against using only geometric-mean models to evaluate home range. The average geometric HSIs for all zones were approximately three times smaller than the arithmetic HSI and had absolute values of approximately 0.1, which rated them as very poor habitat (Table 51.1). Both the geometric- and arithmetic-mean HSI values declined approximately 30–40% as the zone size was increased from 50 to 400 ha. This effect was logical when the zone was centered on the core habitat and low-valued or zero-valued cells outside the forest were included in the larger zones. The geometric scores showed much lower percentages of the area in suitable habitat and seemed to understate the size of the core area in zones 3 and 4 (Table 51.2).

By our standards, zone 2 was the best and zone 3 the poorest site (Table 51.3). If the habitat model had dominated the decision, zone 2 would have been selected; but zone 3 was where the owls roosted, and zone 2 was not used. Another owl pair had used zone 4 for foraging, although its nest site was just outside the forest. One important factor was missing from the analysis: the forest headquarters site was included in zone 2, and zones 3 and 4 had the least human activity. Moreover, zone 3 had a particularly desirable set of about a dozen very large roost trees.

All of Blodgett forest was smaller than our initial standard of a 2000-ha (5000-acre) minimum home range, and the core roosting areas were smaller than the 120 ha (300 acres) considered desirable. Because the owls were present and used Blodgett and surrounding areas, we must conclude that our initial operational criteria for implementing the model in a GIS were incorrect. We needed more guidance on accept-

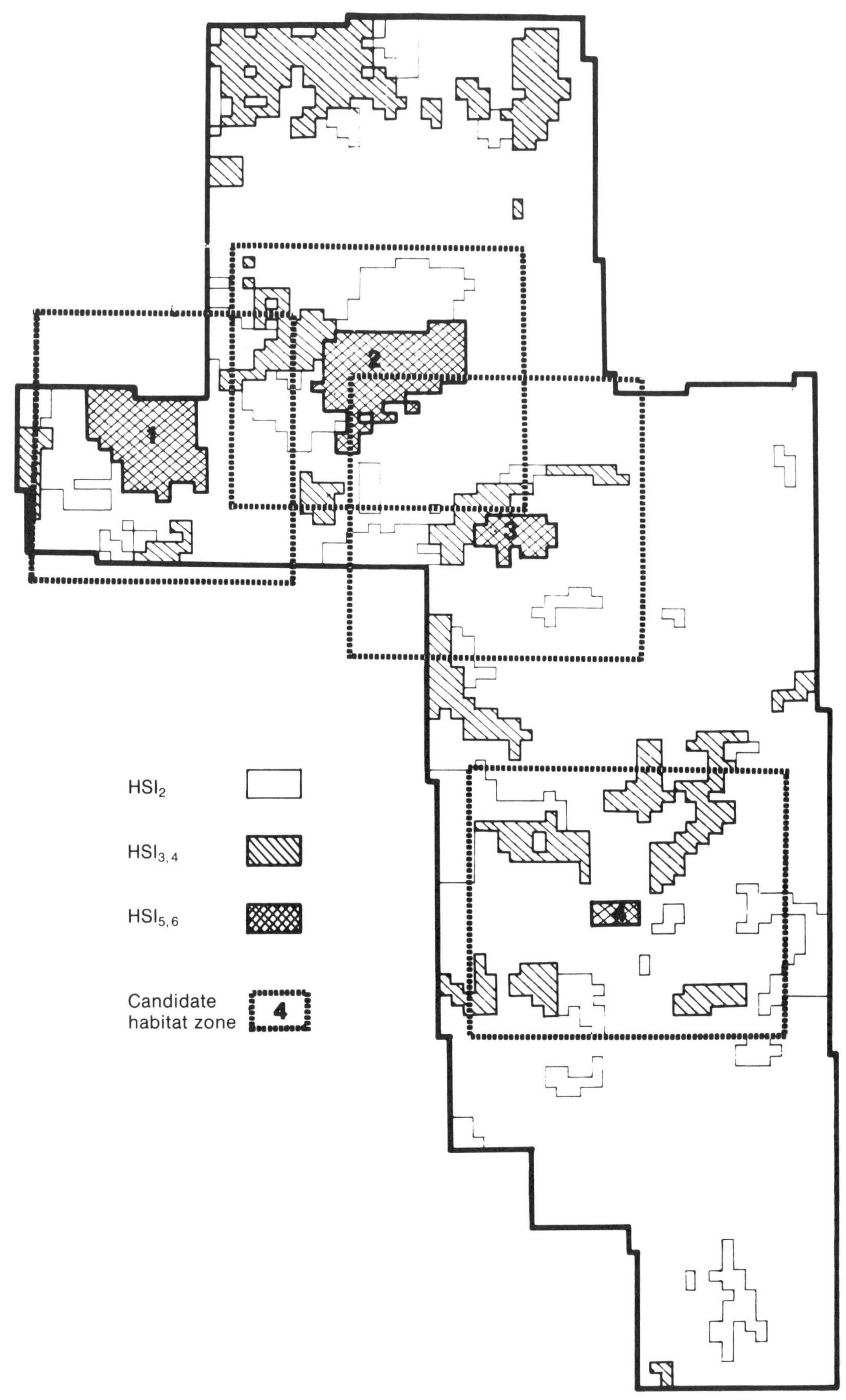

Figure 51.1. Spotted owl habitat index for Blodgett Forest, using the geometric-mean model.

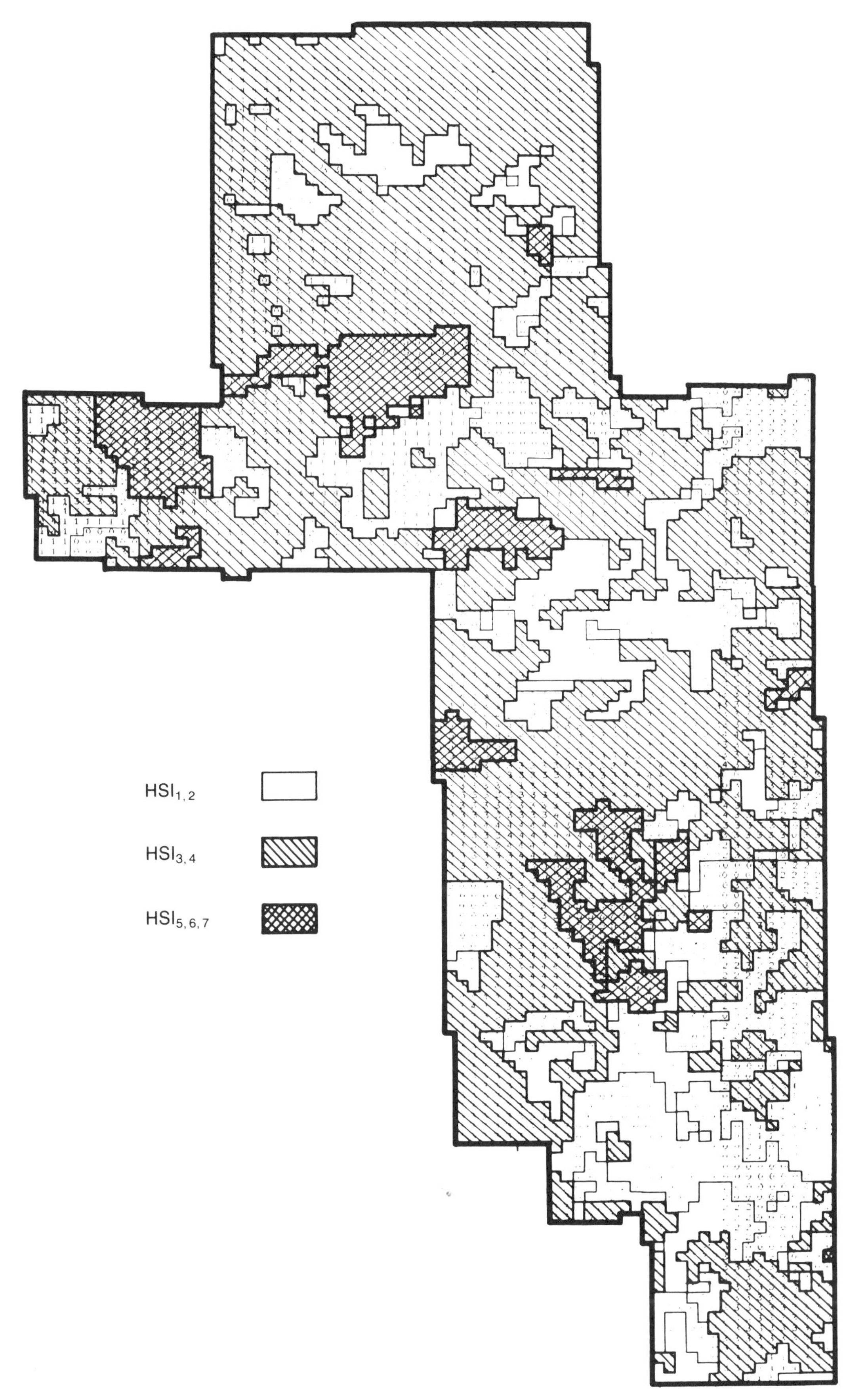

Figure 51.2. Spotted owl habitat index for Blodgett Forest, using the arithmetic-mean model.

Table 51.1. Mean value of owl habitat-suitability index for arithmetic (A) and geometric (G) HSI models applied to different sizes of four habitat-zone locations on Blodgett Forest

Area of habitat zone (ha)	Zone 1		Zone 2		Zone 3		Zone 4	
	G	A	G	A	G	A	G	A
50	0.191	0.337	0.305	0.424	0.123	0.305	0.093	0.324
120	0.154	0.288	0.169	0.322	0.072	0.231	0.072	0.315
240	0.129	0.231	0.117	0.304	0.076	0.233	0.067	0.284
400	0.113	0.217	0.096	0.298	0.077	0.231	0.061	0.256

able average habitat ratings and minimum core and home range sizes.

EXAMPLE 2: SELECTING A SPOTTED OWL MANAGEMENT AND PLANNING ZONE

Part of the assumed forestwide management direction for Tanner Creek included planning and managing for wildlife. This allocation and scheduling zone was selected as a potential site for managing spotted owl habitat. Spotted owls have occupied parts of the basin for some time, and at least one suitable 2000-ha (5000-acre) habitat area must be identified and preserved with appropriate prescription choices. Tanner Creek also had a timber-production objective, and the area selected for spotted owl habitat protection should have the minimum economic impact on the forest timber program. Four steps were followed, using SPANMAP to identify and select the zone with the least impact: (1) spotted owl habitat was evaluated, using the model of Laymon et al. on all land units in Tanner Creek; (2) candidate allocation zones that met minimum HSI and acreage requirements were identified; (3) these candidate allocation zones were evaluated for their relative contribution to timber production; and (4) a zone reserved for spotted owl habitat was selected that would minimize impact on the forestwide timber program.

Table 51.2. Comparison of four candidate management zones by three criteria based on geometric (G) and arithmetic (A) HSI models

Criterion	Zone size (ha)	Zone 1 G	Zone 1 A	Zone 2 G	Zone 2 A	Zone 3 G	Zone 3 A	Zone 4 G	Zone 4 A
1. Area (ha) in central core habitat with 0.5 ≤ HSI ≤ 1.0	All	18	19	23	24	6	12	2	19
2. Percentage of total zone area with 0 < HSI ≤ 1.0	50	42	88	70	99	33	92	28	86
	120	43	84	45	92	22	80	25	90
	240	30	68	33	93	21	18	24	87
	400	32	63	26	93	21	74	23	82
3. Percentage of total zone area in habitat with 0.4 ≤ HSI ≤ 1.0	50	35	43	44	47	21	36	10	48
	120	22	23	21	25	10	19	4	33
	240	16	16	13	22	10	17	3	27
	400	14	19	13	19	5	17	2	21

Table 51.3. Rank of candidate management zones by mean geometric (G) and arithmetic (A) HSI values for 120- and 400-ha zone sizes

Zone	Rank 120 ha G	Rank 120 ha A	Rank 400 ha G	Rank 400 ha A
1	2	3	1	4
2	1	1	2	1
3	3	4	4	3
4	4	2	3	2

Seventeen timber-defined analysis areas (A–Q) had been previously identified in the Tanner Creek basin (Fig. 51.3). We assumed that the geometric habitat model for spotted owls was valid and evaluated each analysis area for potential suitability as spotted owl habitat. Of the 17 analysis areas, only seven (K–Q) had mean HSI values > 0.3–suitable according to the operational habitat model. The seven analysis areas included 5980 ha (14,770 acres) of potential owl habitat with an average area HSI of 0.5.

Visual inspection of the suitable owl habitat suggested four candidate 2000-ha allocation zones (Fig. 51.4). The candidate allocation zones had approximately similar area HSI values, but they differed with respect to other resource values, such as timber yield. The zones overlapped, but this is acceptable for a mutually exclusive choice problem where only one zone is selected.

Tanner Creek apparently has a surplus of suitable spotted owl habitat. The candidate allocation zone with the least value in timber production could be identified in several

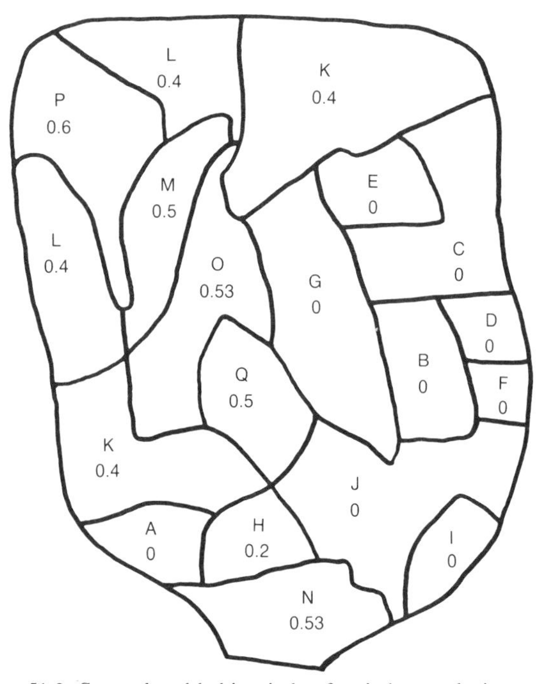

Figure 51.3. Spotted owl habitat index for timber-analysis areas on Tanner Creek, using the geometric-mean model.

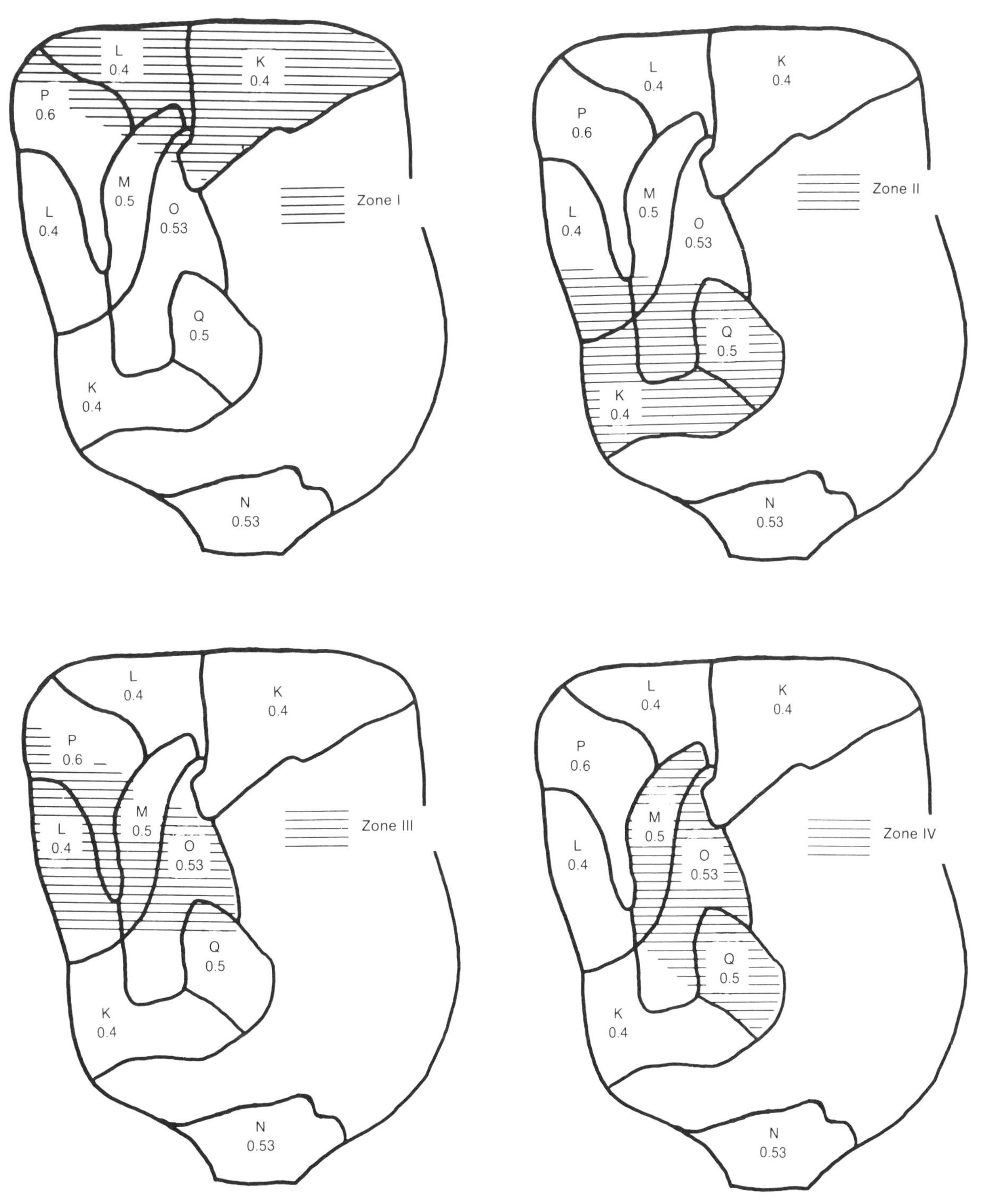

Figure 51.4. Four candidate spotted owl management zones in the Tanner Creek ALSCZ.

ways, ranging from present net-worth impact to impact on current harvest. Assuming that prescriptions for owl management would sharply curtail the volume of timber harvested in the management zone, we chose current timber inventory, distance to road, and logging cost as the criteria to illustrate the process. SPANMAP was used to calculate the distance to the nearest road for each cell in the data base. This variable, which related to road-building and skidding costs, was intersected with overstory vegetation to identify all cells with merchantable timber at least 60 years old and within 610 m (2000 feet) of a road (Fig. 51.5). These important lands had the greatest potential for supplying timber harvest in the near future.

All four zones contained accessible, merchantable timber.

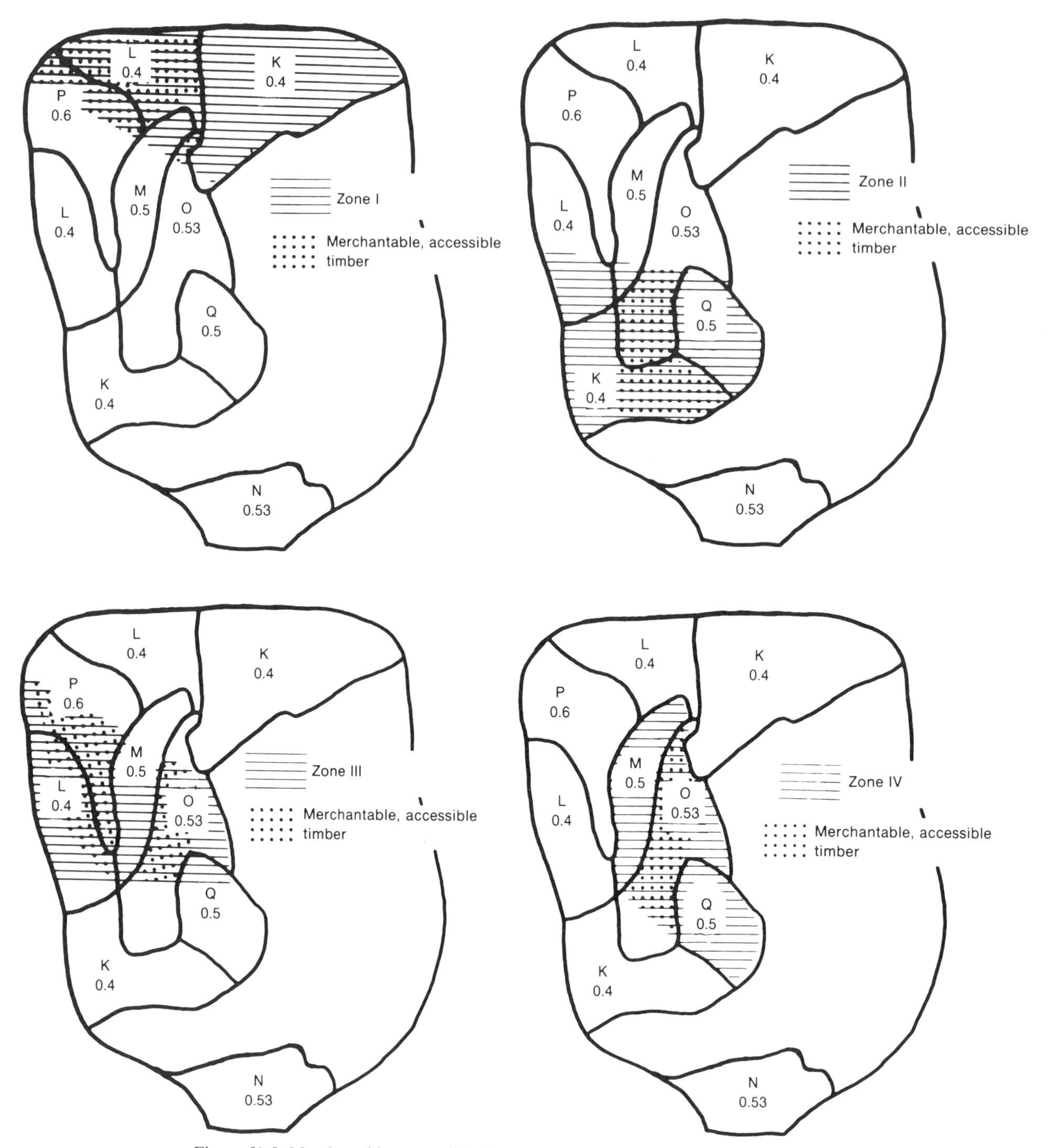

Figure 51.5. Merchantable, accessible timber on the candidate owl management zones.

Zones I and II contained 30–40% accessible, merchantable timber, and zone I had a large portion of lower HSI owl habitat. More than half of zone III was valuable timber land, but zone IV, composed largely of analysis areas M and Q, had only 20% of its area in valuable timber. This preliminary analysis suggested reserving either zone I or zone IV as owl habitat, based on minimum acreage of valuable timber. The choice between these two candidates was made by using the additional criterion of percent slope. Lands with slopes <40% are tractor-loggable and hence of greater value for economic timber production. The three-variable intersection of accessibility, merchantability, and percent slope (Fig. 51.6) indicated that zone I contained a much higher percentage of tractor land with slopes <40%. Zone IV was,

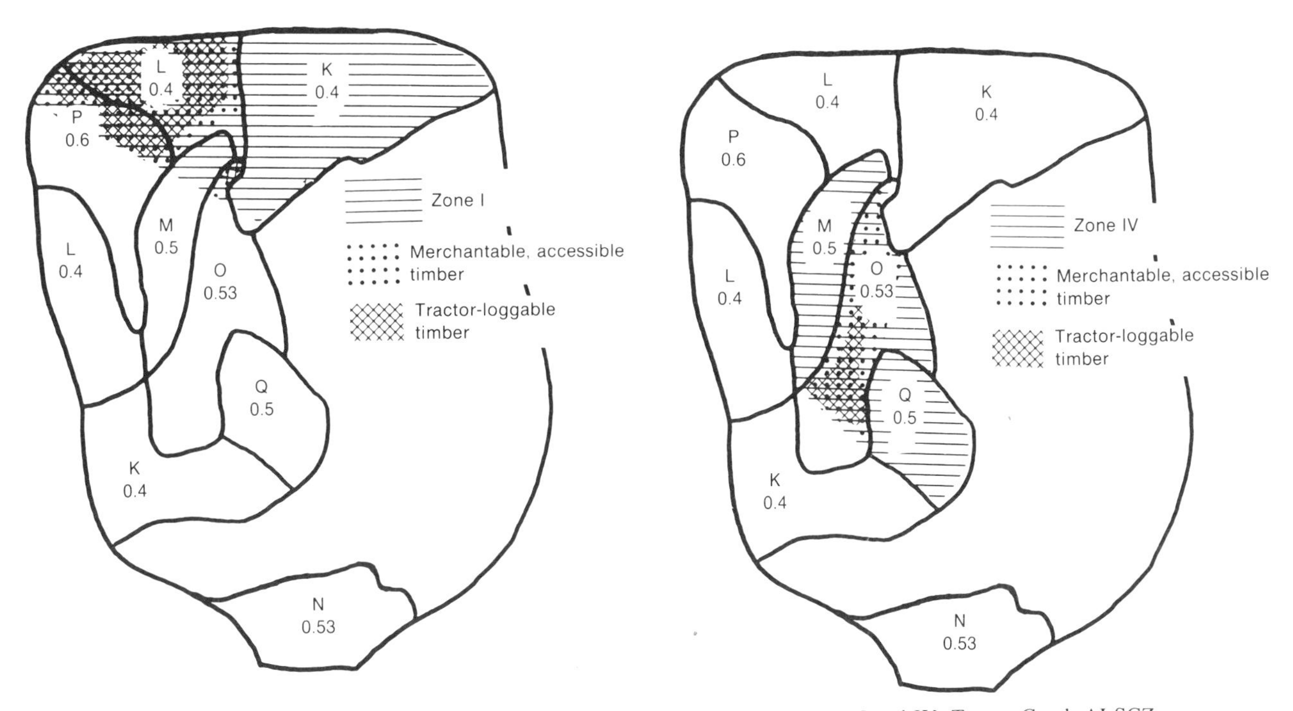

Figure 51.6. Tractor-loggable timberland in candidate owl management zones I and IV, Tanner Creek ALSCZ.

therefore, our best choice for an allocation zone reserved for spotted owl habitat.

The final step in linking the model to planning was to specify timber-management prescriptions that would maintain zone IV as suitable owl habitat. Constraints were added to the Tanner Creek ALSCZ owl prescription packages, requiring that area equivalent to habitat zone IV be managed under these prescriptions. If the forest plan selected Tanner Creek as an owl management area, then the maps and data from this analysis would be used to implement the prescription. Using an IBM 3081 mainframe computer, the analysis on the 15,000-cell Tanner Creek data base cost $4.25.

Discussion

The preceding examples show that if a good-quality spatial data base is available for an area, software like SPANMAP allows mathematical habitat and other models to be inexpensively and easily applied. Because they are inexpensive, many different model formulations and trial management areas can be examined before a recommendation is made. The front-end cost to accurately map, digitize, and maintain a spatial data base is, however, substantial. Such data bases need to support many kinds of forest analysis, from timber management to wildlife to recreation project planning, to lower the per-use cost to acceptable levels.

Our experience with the geometric-mean spotted owl model suggests that reliance on a single model may be unnecessarily limiting. Multiple models reflecting different needs and different periods in the species' life history permit more refined analysis. The combined results of both the arithmetic and geometric models provided better information than either one did separately. Furthermore, the spatial distribution of habitat conveyed additional information not captured by a simple numerical value of average HSI for a 400-ha (1000-acre) zone. For example, the actual location and size of core habitat may be considerably more important than the average HSI.

The spatial analysis presented in this chapter is static, in that we evaluated habitat at only one point in time. DYNAST and FORPLAN, in contrast, forecast the changing structure of vegetation and present a dynamic but spatially undefined estimate of habitat. The next step is to link a tree-growth model or some other succession model to the spatial data base and grow the vegetation in each cell. Alternative planned management activities, such as timber harvesting, can be mapped and digitized to present a spatial picture of future vegetation that can be evaluated for habitat. This would result in a spatially explicit and dynamic model.

These spatially explicit techniques of habitat analysis, when used to prepare the package prescriptions for ALSCZs, provide a potentially effective way to integrate wildlife on the input side of planning models like FORPLAN. Wildlife can then be considered and evaluated jointly with all other uses within the planning model, not as an after-the-fact reaction to the results of planning.

52

Use of Vegetation Projection Models for Management Problems

RICHARD S. HOLTHAUSEN

Abstract.—To be effective aids in reaching management decisions, wildlife-habitat-capability models must be linked to models that project changes in vegetation. This study compares two models that could be used for this purpose, the FORPLAN linear optimization model and the DYNAST simulation model. The models produced similar solutions to resource problems when objectives were well defined and the models were properly formulated. Each model has advantages for specific applications. Simulation models will generally be easier and less expensive to use than optimization models, and they are better suited to direct assessment of nonlinear wildlife functions. Linear optimization is required when the objective is to demonstrate that a resource solution produces optimum economic benefits.

Habitat-capability models are intended to assist with management decisions. Within the USDA Forest Service, those decisions concern habitats and how to manage them to create some set of desired conditions over time. To effectively support Forest Service decisions, habitat-capability models must be linked to models that predict changes in habitats. These models must operate at a geographic scale that allows meaningful analysis of wildlife habitat and is appropriate for the level of the decision being made.

Forest Service decisions are made at several geographic levels, and at each of these levels different elements of wildlife habitat are controlled. In this study, I have focused on the decisions and analysis that will be made at the level of a watershed. At this level, in units of roughly 400–8000 ha (1000–20,000 acres), our management actions must provide the amount and distribution of habitats to meet all seasonal requirements of wildlife species. The objective of management at this level is to implement direction from a National Forest plan. The controls available to the manager are the timing and location of management activities, and the decision to be made is how to best use those controls to achieve objectives.

Several models are available for analysis of vegetation dynamics at this scale: (1) the DYNAST model developed by Boyce (1977, 1978, 1980); and (2) the Direct-Entry version of FORPLAN (T. W. Stuart, pers. comm.). These two models present an interesting contrast. FORPLAN is a linear optimization model, representing the most recent evolution of models that were originally designed to solve the timber-harvest-scheduling and sustained-yield problem. FORPLAN finds a mathematically optimum solution to a problem defined in terms of desired resource outputs and constraints.

RICHARD S. HOLTHAUSEN: USDA Forest Service, Wildlife Fish and Ecology Unit, 3825 E. Mulberry Street, Fort Collins, Colorado 80524. *Present address:* Fish and Wildlife Staff, USDA Forest Service, Pacific Northwest Region, 319 SW Pine Street, Box 3623, Portland, Oregon 97208

DYNAST is a simulation model designed for analysis of the cumulative effects of multiple-resource management. Instead of seeking a mathematical optimum, it simply tracks the consequences of a proposed management action. Feedback loops are employed to help the user define management objectives and proposed actions.

DYNAST and other comparable simulation models are inherently simpler than the linear optimizations that are used for National Forest planning. Consequently, I expect that management biologists may use simulations for watershed-level analysis. Because simulation and optimization are different processes, it is important to look at the consequences of using simulation rather than optimization for this type of analysis.

To compare the two models, I did a case study of a watershed-level management problem. Three criteria were used for model evaluation: (1) potential for use in representing National Forest plan direction; (2) ease of use; and (3) the ability of the model to provide complete information on vegetation-wildlife interactions.

Study area

The case study was linked to the ongoing analysis of forest plan implementation on the Carman Springs area of the Mark Twain National Forest (R. L. Kirkman, pers. comm.). The study area encompasses 3500 ha (8640 acres) of forest cover types, including oak-hickory, oak-pine, pine, and bottomland hardwoods. Nonforest types constitute only about 2% of the area.

The forest vegetation is fairly well distributed across age-classes, but stands younger than 20 years or older than 90 years are scarce. Management objectives for the area include maintaining specified acreages of forest structural stages to provide sustained levels of habitat for indicator species. Habitat objectives used in model formulations were (1) to maintain 15% of the area as old-growth (defined as stands more than 90 years old); (2) on hardwood and hardwood-pine acres not managed for old-growth, to achieve balanced age-classes based on a 90-year rotation; (3) on pine acres not managed for old-growth, to achieve balanced age-classes based on a 60-year rotation; (4) to maintain 40% of the area as mast-producing stands (defined as hardwood stands greater than 50 years old); and (5) to convert 260 ha (640 acres) of forested area to permanent openings within 40 years.

Procedures

Habitat-capability models

Biologists from the Forest Service and the Missouri Department of Conservation identified 12 species as wildlife-management indicators for the study area, and they developed habitat-capability models for these species (J. M. Sweeney, pers. comm.). The habitat-capability models followed the Pattern Recognition (PATREC) form, using Bayesian probability functions to predict habitat capability under a given set of habitat conditions. I recognized that PATREC models could not be dynamically represented in FORPLAN because they reflect nonlinear relationships between habitat patterns and wildlife-habitat capability. Therefore, outputs in habitat capability for the indicators were not estimated. Instead, vegetation models were used to simulate the habitat components that are variables for the habitat-capability models. These habitat components are essentially identical to the structural stages used as habitat objectives, i.e., old-growth, hard mast, and openings.

Vegetation models

FORPLAN model

I used the Direct-Entry version of the FORPLAN model. It was better suited to this analysis than earlier versions of FORPLAN because of its enhanced abilities to represent outputs that are dependent on the age of vegetation (T. W. Stuart, pers. comm.). Within the FORPLAN model, the inventory for the Carman Springs area was classified by vegetation type, slope, and age. Ten-year age-classes were used for the classification. Outputs tracked in the model included cubic feet of wood fiber and the habitat components discussed earlier. The habitat components were modeled by using their relationship to vegetation age. For example, the old-growth output in the model was the proportion of forest area over 90 years old. Vegetation in the model was grown forward once in each decade by simply moving the area in each age-class forward to the next age-class. Management prescriptions used in the model were clearcuts, type conversions, and natural succession. Clearcuts could occur between the ages of 60 and 140 years for existing stands and regenerated pine stands, and between the ages of 90 and 140 years for regenerated hardwood and mixed conifer–hardwood stands. Forest vegetation that was not harvested was allowed to reach a maximum age of 180 years before it was assumed to regenerate naturally.

DYNAST model

The simulation model I used in this comparative analysis was a modified version of the DYNAST–OB model originally developed by Boyce (1980). J. M. Sweeney (pers. comm.) modified the model to make it specific to the Carman Springs area and to improve the vegetation-succession function. The first run reported here was taken directly from Sweeney's work, and a later analysis was made using a modification of the model that I developed.

The basic DYNAST structure consists of a simulator for forest growth and regeneration under a clearcutting regime. Feedback loops serve to bring the area analyzed into a steady state of balanced age-classes through forest management. The simulation is controlled by specifying rotation ages, time steps, opening sizes, and rates of conversion from one vegetation type to another (Boyce 1980).

Within DYNAST, vegetation was classified by using cover type, slope, and age. Nine structural stages were defined in the model for each cover type. The number of years represented varied from one age-class to another and ranged from 9 to 161 years. Vegetation was grown forward by taking a proportion of each age-class and moving it to the next age-class in each time step. This function produced some unrealistic growth projections in the original version of DYNAST but was improved by Sweeney's modifications. Wood fiber and habitat components were represented in DYNAST as described for FORPLAN. In addition, habitat-capability functions for the indicator species were included in the DYNAST formulation.

Formulation of the models for analysis

The comparison of the models focused on two strategies for achieving the habitat objectives listed above. The first strategy was to establish a strict even-flow of age-classes as rapidly as possible, and the second was to allow fluctuation of age-class proportions within limits set by the objectives. Different controls had to be used in the two models to represent each of these strategies. Controls used for the strict even-flow strategy in each model are described first.

In FORPLAN, the requirement for 15% old-growth was achieved by using a direct constraint on the old-growth output. However, this constraint could not be applied in the first two decades because the existing inventory did not contain 15% old-growth. The model formulation placed no constraint on old-growth in the first decade, a 12% minimum constraint in the second decade, and a 15% minimum constraint in all remaining decades.

The second important requirement was to achieve area regulation on a 90-year rotation in hardwood and hardwood-pine and on a 60-year rotation in pine. This objective could not be achieved in FORPLAN by simply assigning the appropriate rotation to each cover type. That strategy would cause each hectare to be harvested at the specified age, but the final age-class distribution would not look much different from the initial inventory. The solution to this problem was to constrain the area of each cover type in each age-class. For example, a balanced age-class distribution on a 90-year rotation would be achieved if each 10-year age-class contained $^{10}/_{90}$ or 11% of the total area. In practice, this constraint had to be applied only to the youngest and the oldest age-classes to achieve a balanced distribution. Some slack had to be allowed in the constraint to produce feasible solutions. This was achieved by allowing a 25% variation from desired area in the initial decades and a 10% variation in later decades. A final constraint was needed to produce 260 ha (640 acres) of type conversion within 40 years. This was used as a direct constraint on the type conversion output.

Besides constraints, FORPLAN requires an objective function to drive the solution of the matrix. The objective used most frequently is maximization of present net worth, but this could not be considered in my study because economic values were not included in the model. As a substitute, I used an objective of maximizing the total volume of timber output over the 200-year model run.

A different set of model specifications was needed to simulate the same set of resource objectives in DYNAST. The objectives for old-growth and for balanced age-classes based on specific rotations were simulated by using a mix of rotation ages for the cover types. In DYNAST, rotation age is used to determine both the appropriate age for harvesting and the area that should be harvested in each time step. Therefore, rotation can be used to achieve balanced age-classes and old-growth requirements in DYNAST. In this analysis, 60% of the hardwood and hardwood-conifer types were placed on a 90-year rotation, and 40% of those types were placed on a 140-year rotation. Specifications for the pine type placed 80% on a 60-year rotation and 20% on a 140-year rotation. Additional controls on the model were used to produce the desired type conversion of 260 ha (640 acres). No objective function is used in DYNAST because the model is driven by the specified rotation ages and conversion rates. Feedbacks drive the model toward a steady-state condition based on those specifications.

After concluding the initial runs, I reformulated the models to represent the strategy that allows fluctuation of age-class proportions within specified limits. In FORPLAN, I removed the constraint that set an upper limit on the area that could go into the youngest and oldest age-classes. This constraint had been intended to speed the achievement of the balanced age-class objective, but it was not needed to simply ensure that a minimum area would be provided in each age-class. The constraint on minimum area was adequate to achieve this objective.

Simulating the second strategy in DYNAST required the addition of some new simulation controls. The first of these allowed DYNAST to exceed the regulated rate of harvest during the initial decades. Harvest rate in DYNAST is normally determined by rotation age, but the added control allowed me to override this harvest rate. I also added controls that allowed me to specify the proportions of old-growth stands and mast-producing stands that would be maintained. These controls prevented the harvest of stands needed to meet those habitat requirements in each time step.

Results

Model results for the first strategy were closely matched (Fig. 52.1). Harvest in FORPLAN was about 60 ha (150 acres) greater than harvest in DYNAST in the first decade, but the DYNAST rate stayed at that level, while the FOR-

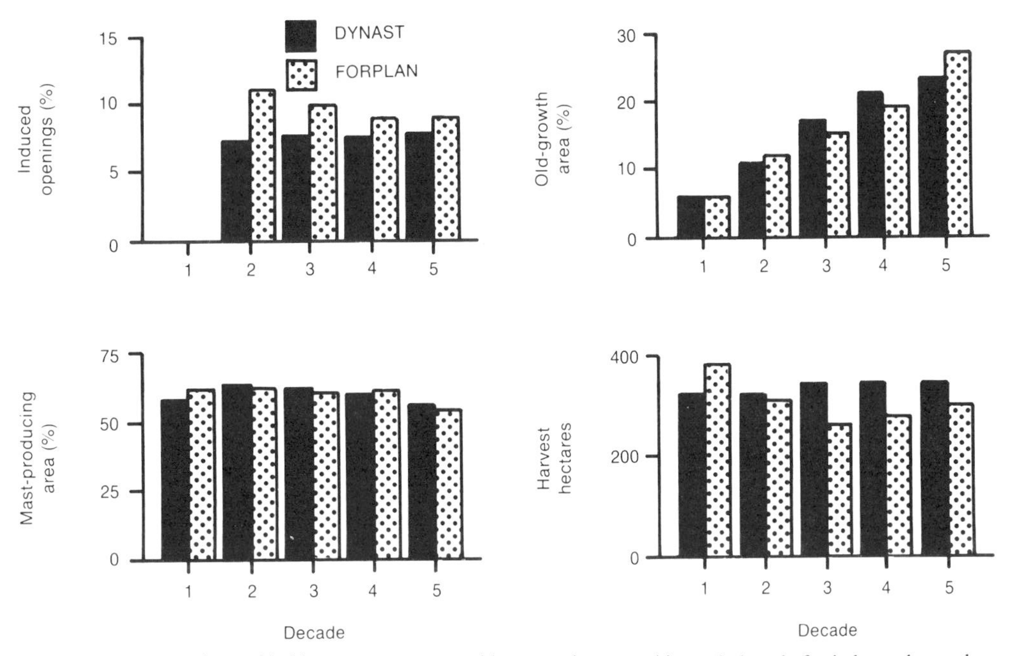

Figure 52.1. Proportions of habitat components and hectares harvested in each decade for balanced age-class strategy.

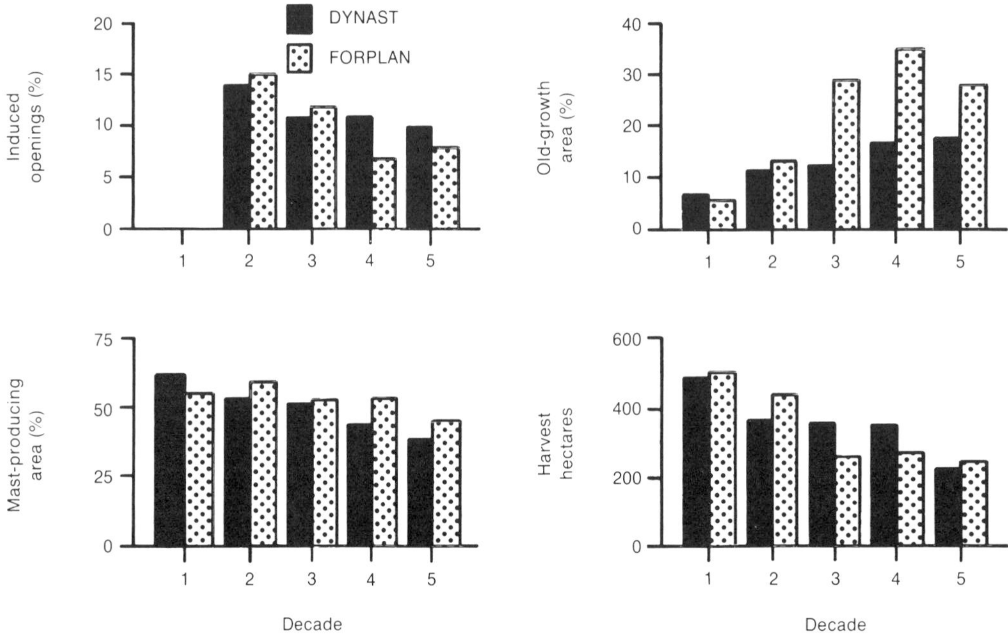

Figure 52.2. Proportions of habitat components and hectares harvested in each decade for strategy allowing fluctuations in age-classes.

PLAN rate declined slightly over the next 40 years. The different harvest rates produced some differences in induced openings, but there was generally good agreement on the percentages of old-growth and mast-producing areas.

There was also generally good correspondence in the results of the models for the second strategy (Fig. 52.2). Harvests in the first decade were nearly identical, and total harvests over five decades varied only by approximately 80 ha (200 acres). The only strong difference in model results occurred in the generation of old-growth. The old-growth maintained by FORPLAN greatly exceeded requirements in the third, fourth, and fifth decades, but DYNAST did not fully achieve the requirement until the fourth decade. The DYNAST model could probably be forced to achieve the requirement sooner by increasing the requirement for area in the 50- to 90-year age-classes. FORPLAN retained old-growth by harvesting almost exclusively from younger age-classes in the first four decades. This strategy may be related to the objective to maximize timber over the entire 200-year run. An objective for maximizing present net worth would encourage earlier harvest of these old-growth areas. Differences in assumptions about old-growth mortality also influenced the old-growth component of the model results, as discussed below.

Discussion

Model results and validity

FORPLAN and DYNAST produced similar solutions to the problem of representing National Forest plan direction on a single management area. The two models did not duplicate each other exactly, but either model could be used to generate a solution that would achieve the defined resource objectives for the area. When the models are heavily constrained by resource objectives, the solution space becomes very narrow and the distinction between optimization and simulation becomes less significant.

The strongest difference observed between the models was their calculation of old-growth area. Part of this difference can be traced to the assumptions used in the models for growth and mortality of old-growth. In DYNAST, the oldest timber stratum for hardwoods spanned 60 years and the oldest for pine spanned 40 years. Each year, a proportion (1/60 or 1/40) of this oldest stratum was assumed to die and revert to the youngest stratum. In FORPLAN, the initial inventory of old-growth was contained in the 90- to 99-year and 100- to 109-year age-classes, and none of these areas reverted to the youngest age-class until it had been harvested or grown forward for 80 years. Because information on mortality of older stands was lacking, it was not possible to determine which model formulation was more valid. However, with a few simple changes, both FORPLAN and DYNAST could be made to follow either set of assumptions.

Model controls

Controls available for the models play a very important role in determining model results. Work done in this study indicated that some controls are more useful than others. The control of harvest level in DYNAST, which bases harvest area on rotation age, is useful for achieving a balanced age-class distribution. In FORPLAN, the same objective required the use of constraints on area by age-class. This necessitated considerable pre-analysis, so it was more cumbersome to use.

The harvest-level and rotation-age control in DYNAST had limitations because it constrained initial harvest levels unnecessarily. This was improved by adding a control allowing the user to specify early-period harvests and then revert to the rotation-age control. This control was part of the final DYNAST formulation (Fig. 52.2) and proved very useful.

The other control needed to model habitat objectives was a control for area required in specific structural stages, e.g., old-growth. The direct controls available as constraints in FORPLAN were useful for modeling these objectives. A similar control was added to DYNAST, functioning through a feedback loop. This proved to be more convenient than simply using rotation to achieve the same objective.

Overall comparison of the models

Each model has advantages for specific applications. DYNAST would clearly be superior for rapid assessment of nonlinear habitat-capability functions, because those functions cannot be calculated with FORPLAN. DYNAST also has advantages in ease and economy of operation. The DYNAST runs reported here cost less than $5 each and were generally processed immediately. The FORPLAN runs cost $20–40 each and required 2–6 hours to process. The total cost of producing a FORPLAN solution was also increased by the number of infeasible runs that were made prior to generating an optimal solution. Each new constraint could potentially make the problem insolvable, so iterative runs were often necessary to generate a solution under a new set of conditions. DYNAST, by contrast, simulated the results of each new constraint, allowing the user to see the effect of the constraint and adjust it if necessary.

FORPLAN has distinct advantages when the objective is to demonstrate that a resource solution has optimum economic benefits. Economic optimization was not used in this study, but it is a very important concern in forest planning when timber harvests are being scheduled for entire national forests. It is a less important concern for the smaller-scale projects being considered here, because they will be driven by resource objectives identified in the forest-planning process.

Recommendations

The following specific recommendations are offered in regard to the use of vegetation models to support habitat-capability models in decision making.

1. Use the simplest tool that does the job. At the level of individual management areas, simulations will generally be adequate to determine management actions that achieve defined objectives.
2. Provide model controls that emulate management controls. Using controls that are similar to actual management controls makes models more accessible to managers and helps ensure that the model solution can be understood and implemented. Examples of such controls are direct control over area to be harvested and direct control over area required to be in the old-growth and mast-producing condition.

Acknowledgments

I thank Brad Gilbert and Tom Stuart for assistance with FORPLAN modeling. Gilbert also reviewed and helped clarify the manuscript. James Sweeney produced the initial DYNAST results reported here and provided a helpful review of the manuscript. Finally, Hal Salwasser helped to define the study and assisted in a thorough review of the manuscript.

53

A Simulation Procedure for Modeling the Relationships Between Wildlife and Forest Management

KENNETH J. RAEDEKE and JOHN F. LEHMKUHL

Abstract.—A habitat-simulation procedure (HABSIM) was developed to model the response of wildlife populations to changes in their habitat resulting from forest-management activities. HABSIM combines a wildlife-habitat-relationships model with a model of forest succession to calculate potential wildlife carrying capacity. HABSIM reconstructs changes in the carrying capacity of the habitat from former times to the present, calculates future carrying capacity, and simulates effects of changes in forest management, such as thinning, cutting rates, and so on. The HABSIM model has been employed to model the responses of black-tailed deer (*Odocoileus hemionus columbianus*) and elk (*Cervus elaphus*) to forest management on USDA Forest Service lands, National Park Service lands, and also private lands. The procedure was found to be comparable to other simulation techniques and to agree with available historical data.

Resource managers are often required to predict the effects of land-management practices on wildlife habitat and wildlife species. We developed a simulation procedure, HABSIM, that links a model of wildlife-habitat carrying capacity with a model of habitat succession. HABSIM gives the land manager a simulation model that is conceptually tractable, uses data that are generally available, is cost-effective, and can simulate the results of alternative habitat-management practices. HABSIM also allows reconstruction of the status of wildlife habitat during historical time and contrasts future changes to either current or original conditions.

The underlying assumptions of the model are that animal population status is correlated with changes in carrying capacity and that carrying capacity can be estimated for different forest successional stages. A further assumption is that a predictable pattern of forest succession exists with only minor deviations. Total potential carrying capacity is then determined by the proportions of each stage existing in the area at a given time interval.

This is an arguably simple view of carrying capacity, but elk (*Cervus elaphus*) habitat relationships for western Washington are known in only a fragmentary manner. A model to examine past and projected changes in carrying capacity that incorporated the probable effects of road density, patch size, within-habitat variability, and weather certainly would be neither simple nor particularly tractable. These variables are either stochastic or are decision variables that may be predictable but are not easily modeled.

HABSIM was written in FORTRAN-77 for use on the University of Washington CDC CYBER 170–750 computer. We used the program successfully to model responses of black-tailed deer (*Odocoileus hemionus columbianus*) and elk populations to forest management at the ranger district level in the Pacific Northwest for the USDA Forest Service and for the National Park Service.

KENNETH J. RAEDEKE and JOHN F. LEHMKUHL: Wildlife Science Group, College of Forest Resources, AR-10, University of Washington, Seattle, Washington 98195

Description of the model

The simulation model HABSIM combines a model of habitat succession, in this case a forest-succession model, with a model of wildlife response to habitat succession. HABSIM is constructed to allow the user to vary parameters interactively in either the succession or wildlife-response model to simulate different management situations.

Succession simulation

The core of the model is a simple algorithm that creates an $n \times n$ time-interval matrix of forest age-class rows and year-interval columns. The time-interval length represented by each column or row is a matter of convenience and of the length of the simulation period. For example, 5 years would be a useful time interval represented by each matrix cell if the simulation period was the 100 years, from, say, 1931 to 2030. The matrix would be 20×20, the rows being stand age (0–5, 6–10, and so on) and the columns being time (1931–1935, 1936–1940, and so on). The program can use only 40 intervals.

The first row of the matrix consists of (1) the areas of forest harvested during each time interval from the start of the simulation period to the present, or base year; (2) the area to be harvested during each interval from the base year to the end of the simulation period; and (3) the total area of commercial old-growth forest. These data are input from a data file. The core algorithm constructs the successional matrix by successively transferring the area of forest input as the first cell diagonally to the adjacent lower-right cell in the next age and time cell until the triangular matrix is filled. Successional stage areas are calculated as the sum of cell areas within predetermined age limits.

Noncommercial forest habitat types were not included in the model, because in the Pacific Northwest, noncommercial forest lands are a minor part of all forest lands. For example, in the Soleduck and Quinault ranger districts of the Olympic National Forest, only 1.9% of the landscape is classified as noncommercial forest lands. This includes lands in administrative uses, meadows, nonforested lands, subalpine, and water.

Projected timber harvest can be estimated by planned harvest of old-growth or virgin timber, or by total or partial harvest of second-growth stands in rotation. Rotation age and proportion of area to be harvested at rotation can be varied with each run on a particular data set. A constant proportion of timber in each interval past rotation is harvested, with partial harvest at rotation.

Wildlife response

HABSIM requires the user to provide a model of wildlife response to habitat succession. The model must specify animal density, numbers, or relative carrying capacity for each successional stage identified in the successional model. Brown's (1961) model of the response of deer populations to coniferous forest succession after timber harvests is the type of model required by HABSIM. Coefficients of habitat suitability derived from regression analysis could also be used as the wildlife-response model values in the place of density values, as in Brown's (1961) model.

Program execution

Input data and parameters are read from a data file and entered interactively from the terminal. The data file contains a title, starting year, ending year, and base year, the total area of commercial forest, the past area harvested during each time interval, and the area of planned harvest during each interval. Interactive input is (1) name of run with the data set; (2) potential density of animals for each successional stage; (3) the rotation length; and (4) proportion of rotation to be harvested.

After filling the succession matrix, the program calculates, for each time interval, the area in predetermined successional age-classes, multiplies by the corresponding wildlife-density value, and calculates the total potential carrying capacity (PCC) during the time interval. The relative change in PCC from the base year is also calculated. Successional age-class boundaries are written into the program but can be changed easily. The wildlife-density parameters are entered at the beginning of each run, which allows for a sensitivity analysis of changes in PCC with changes in the value of successional stages.

Output from HABSIM displays the habitat-succession matrix for the simulation period, the amount of virgin or unaltered habitat remaining at the end of each time interval, a summary of the PCC of the habitat for each time interval, and the relative change in the PCC of the habitat by time interval over base-year conditions (= 1980).

Assumptions

The simulation of wildlife PCC as a function of successional change as described above requires a number of assumptions: (1) the spatial distribution of required forage and cover resources will not change over the simulation period (i.e., current timber-harvesting practices that determine patch size and interspersion will be continued); (2) the effects of road density and other factors that might reduce the availability of resources will remain constant; (3) habitat within each successional stage in the area under consideration is uniform in its PCC and changes in a consistent successional sequence; and (4) the effects of weather are constant over the simulation period.

Any attempt to model future habitat conditions requires assumptions such as the first two listed above. We have used the most conservative assumptions, i.e., those permitting a continuation of current management. These assumptions are most likely to be violated if the area under consideration is not being managed under sustained-yield management. The third assumption is common to deterministic models. This assumption simply means that sampling units are homogeneous.

Application of HABSIM: An example

We used HABSIM to simulate the response of elk populations to past and future forest management on forested lands surrounding Mount Rainier National Park and in the Olympic National Forest. The forest-succession model (Table 53.1) was supplied by the Olympic National Forest staff, and the elk-response model (Table 53.1) was based on Taber and Raedeke (1980) and Hett et al. (1978). Separate simulations were done for summer and winter ranges, as defined by agency biologists. Winter and summer ranges were defined by using snow-accumulation patterns.

Data on area harvested by year, with even-age silvicultural objectives from 1900 to 1980, and on actual area of commercial forest lands were provided by the Forest Service and other landowners. These were tallied into 5-year-interval classes for input into the computer model. Rotation length and projected timber-harvest rates were also provided by landowners. HABSIM was used to simulate the PCC of the various Forest Service ranger districts, starting from the time of initial timber harvesting to the current year, and then for a 50-year planning period. Results of proposed forest-management practices were contrasted with various management alternatives.

Table 53.1. Estimated potential carrying capacities (PCC), given in elk/km^2, for different forest successional stages on lands surrounding Mount Rainier National Park, Washington

Successional stage	Dominant vegetation	Stand age (years)	Elk carrying capacities	
			Summer	Winter
1	Grasses-forbs	0–5	3.2	0.8
2	Forbs-shrubs	6–10	6.0	2.0
3	Shrubs–low trees	11–30	4.8	2.4
4	Tall trees	31–60	0.8	0.8
5	Mature trees	60–150	3.2	3.2
6	Virgin timber	150+	3.2	3.2

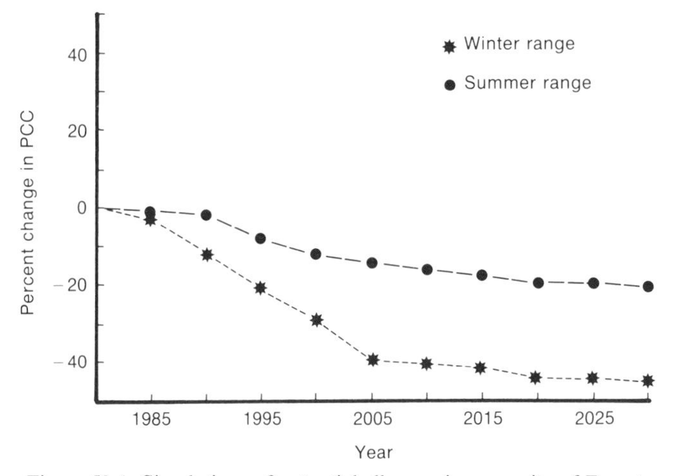

Figure 53.1. Simulations of potential elk carrying capacity of Forest Service lands north of Mount Rainer National Park, using HABSIM and assuming current forest-management practices. Summer and winter ranges were treated separately.

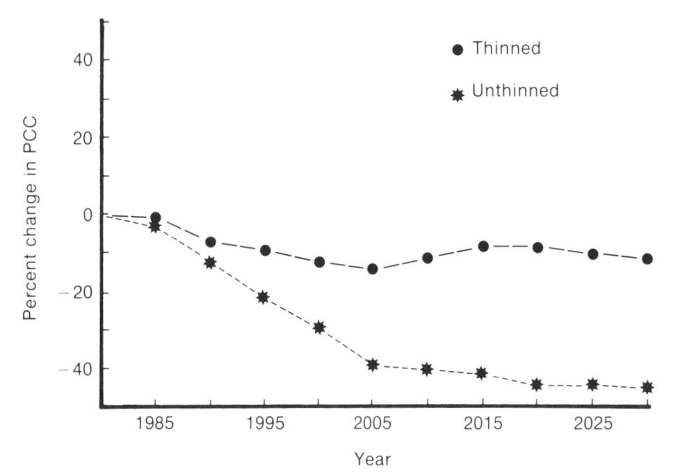

Figure 53.2. HABSIM simulations of potential elk carrying capacity on Forest Service lands north of Mount Rainier National Park with and without thinning of pole-stage stands. Only the winter range was modeled in this example.

The HABSIM simulations of PCC for elk habitat on Forest Service lands north of Mount Rainier National Park indicated that elk PCC would decline on both the summer and winter ranges over the simulation period, with proposed management practices. The projected decline was 45% on the winter range and 18% on the summer range (Fig. 53.1). The decline on the winter range was due to the loss of old-growth forest, and on the summer range it was due to conversion of the commercial old-growth forest to second-growth forest (Table 53.2), which is less productive elk habitat in the Pacific Northwest. These reductions in summer- and winter-range PCC were not additive because the elk used these areas at separate times of the year. The actual magnitude of the impact on the elk population would depend on the relative amounts of summer- and winter-range types currently available, and on changes in the habitats with forest management. In the example, the reduction of the quality of the winter range was probably more important than impacts on the summer range.

Various levels of thinning, as well as other management practices, can be simulated by expressing the effects of thinning in terms of animal density for the successional stage in which thinning occurs. We assumed that thinning would

Table 53.2. The habitat matrix produced by HABSIM for Forest Service lands north of Mount Rainier National Park, Washington, given in hectares by successional stage. (The simulation assumes current timber-harvest rates, and no thinning.)

Successional stage	Year 1940	1960	1980	2000	2020
1	0	59	112	100	0
2	0	22	137	100	0
3	0	22	304	449	352
4	0	0	22	326	671
5	0	0	0	0	104
6	1127	1022	575	152	0

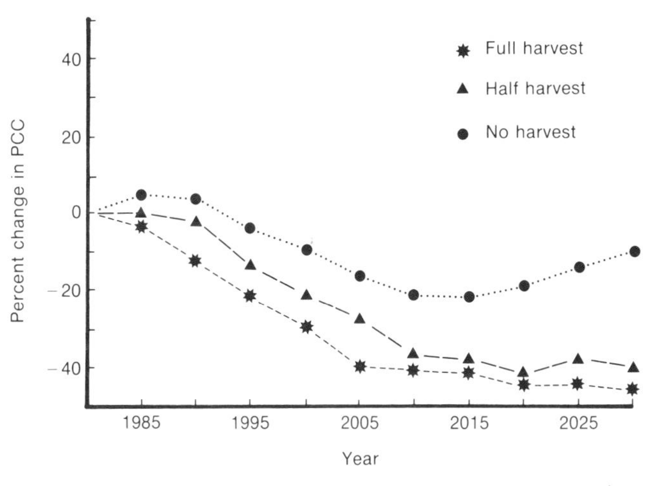

Figure 53.3. HABSIM simulation of potential elk carrying capacity on Forest Service lands north of Mount Rainier National Park with three different timber-harvest rates. Only the winter range was modeled in this example.

maintain carrying-capacity values of the previous successional stage (Taber and Raedeke 1980; Hungerford 1969). The increase in elk PCC from 1980 to 2030 as a result of thinning stands of pole-sized trees on winter range was substantial relative to the unthinned alternative (Fig. 53.2). However, even with the thinned alternative, PCC for elk declined over the 50-year simulation period. Changes in projected timber-harvest levels caused variable results (Fig. 53.3).

Discussion

Validation of the results of the HABSIM simulations is possible through validation of the habitat-succession model and the wildlife-response model or through analysis of the reconstruction of changes in PCC to present time. The results of HABSIM should be valid if both the habitat-succession model and the wildlife-response model are valid. Because both models used in the present example were based on empirical data, the HABSIM simulations should reflect our current understanding of elk-habitat relationships within the constraints of the assumptions listed.

The close correspondence between the results of the HABSIM simulations and reported elk population trends over historical time provides circumstantial evidence that the HABSIM model is valid. The simulations of elk PCC for the Olympic National Forest over the period from initiation of timber harvests to 1980 (based on the HABSIM model) reflected the reported trends in elk population numbers in the area studied (Taber and Raedeke 1980). Elk population numbers of the Olympic Peninsula were reported to be gradually increasing, from 7000 in the 1910s to a peak of unknown magnitude in the mid-1950s. Elk numbers had declined to 22,550 by 1968 and continued to decline to 15,000 by 1978. The simulations of elk PCC by HABSIM showed an increase in elk PCC from the mid-1930s to a peak in the mid-1960s, approximately 95% above the level of the 1930s. This was followed by a decline in elk PCC through 1980 to a level 35% below that of 1930. Thus, in a general way, the results of the HABSIM simulations of PCC are consistent with the present data on elk numbers. However, a precise comparison of the two data sets is not possible, because accurate estimates of elk numbers are not available.

A one-to-one correspondence between simulations and actual population numbers should not be expected, because actual population numbers are dependent on a variety of additional factors not included in HABSIM, such as hunting and periodic severe winter snow conditions. Furthermore, HABSIM models the potential habitat carrying capacity, not population numbers.

HABSIM produces results similar to a linear compartment model, such as the one developed by Hett et al. (1978) for elk in western Washington. Both types of models simulate successional change and estimate potential carrying capacity by using estimates of successional-stage carrying capacity. Compartment models, however, use a system of linear differential equations to simulate regional succession (Bledsoe and Van Dyne 1971; Shugart et al. 1973). Another basic difference is the form of input data. HABSIM uses the past or projected area harvested during each time interval for a finite time period and the total harvestable area to determine the areas of successional stages. Thus, the simulation period is always constant and is only as long as the period from which harvest data were recorded or projected. The compartment model requires an initial area for each stage (compartment) and transfer rates (including harvest rates) between stages. The length of the simulation period can vary. A key difference is the requirement of a constant harvest rate in the compartment model and a variable estimate of actual harvest for HABSIM. Some variability in harvest rate could be introduced in a compartment model if a suitable distribution could be found to represent such a decision variable.

The continuous nature of transfer, or successional change, in the compartment model introduces a slight bias, because an area entering a new stage, or age-class, immediately has some portion of its area transferred to the next stage. For example, a 30-year-old 1500-ha (3700-acre) stand entering a pole size-class, defined as 30–60 years old, has some portion of its area transferred to the next size- or age-class, defined as 60–150 years old, during the next time step, despite the fact that the stand may then be only 31 years old. This could be a reasonably accurate process if harvest, or the initiation of the successional process, was relatively constant and extensive. In reality, however, harvest varies unpredictably. The ability of HABSIM to incorporate variable harvest is a strong asset in providing accurate simulation of changing stage proportions. Projected harvest data with which we have worked have been in the form of area/year as opposed to a proportionality coefficient.

A direct comparison of our model with the similar compartment model of Hett et al. (1978) was not possible because Hett broke the successional-model sequence down into more stages than were used in our model. Therefore, we constructed a compartment model, using the method described by Shugart et al. (1973), that incorporated the state variables and transition rates used in HABSIM.

A comparison of PCC estimates from 1980 to 2030 for HABSIM and for this compartment model showed little difference, despite differences in actual area proportions (Table 53.3). The compartment model initially estimated lower PCC values than HABSIM but predicted higher values at the end of the simulation. This was the result of using an average harvest rate, rather than actual estimates, that transferred relatively more high-quality old-growth to clearcut stage at the beginning of the simulation period than at the end.

DYNAST–MB (Boyce 1977; Barrett and Salwasser 1982) is similar to HABSIM in that it links proportions of successional stages to algorithms that calculate wildlife-habitat-suitability indices. DYNAST, in contrast to HABSIM, incorporates stand-size information in calculating the habitat-

Table 53.3. Potential elk carrying capacity values (PCC) for Forest Service summer-range lands north of Mount Rainier National Park, Washington, as calculated by HABSIM and a linear differential equation compartment model. (Full projected harvest and no thinning were assumed.)

	HABSIM		Compartment Model	
Year	PCC	% change[a]	PCC	% change[a]
1980	596	0	596	0
1985	590	−1	582	−2
1990	582	−2	570	−4
1995	547	−8	558	−6
2000	526	−12	546	−8
2005	512	−14	534	−10
2010	502	−16	523	−12
2015	492	−17	513	−14
2020	484	−19	503	−16
2025	483	−19	494	−17
2030	478	−20	486	−18

[a] 1980 used as base year.

suitability index. Management activities, such as thinning, have not been used in DYNAST (Boyce 1977), but they can be readily incorporated into the HABSIM model.

HABSIM is a simulation procedure, not directly comparable to FORPLAN or other optimization models with constraints that define a feasible domain of solutions. Simulation models such as HABSIM could best be used to develop constraints for optimization techniques, or to analyze the effects of management activities in habitats with a relatively homogeneous successional pathway and animal response.

HABSIM is most useful for providing a quick estimate of the long-term implications of management actions. The conceptual basis, procedure, data requirements, and implementation of HABSIM are simple and easily understood. The technique works best in fairly homogeneous plant communities, in terms of value for wildlife, as found in western Washington. Complex system topology (Shugart et al. 1973) is best handled by compartment models (Bledsoe and Van Dyne 1971; Shugart et al. 1973). HABSIM effectively incorporates pulses of harvest input into the successional system by using actual harvest data, rather than constant rates. A single HABSIM run on the University of Washington CDC CYBER 170–750 computer cost approximately $0.50, compared to approximately $0.80 for a run of a similar compartment model. A compiled version of HABSIM could easily be run on a microcomputer without modification or additional software.

Acknowledgments

The HABSIM model was developed as part of a research project funded by the National Park Service, and the procedure was developed with funding from the USDA Forest Service. The first draft of this manuscript was reviewed by Joan Hett, Thomas Hanley, David Patton, Reuben Weisz, and Rick Wadleigh. We are grateful for the assistance provided by these agencies and individuals.

54

Linking Wildlife and Vegetation Models to Forecast the Effects of Management

GARY J. BRAND, STEPHEN R. SHIFLEY, and LEWIS F. OHMANN

Abstract.—Assessment of forest vegetation is essential for evaluating wildlife habitat. TWIGS, a forest-growth projection model based on individual trees, is a useful tool for assessing the effect of management on the forest overstory. Combining models capable of forecasting overstory-vegetation dynamics with models of wildlife-habitat suitability enables managers to quantify and evaluate the probable consequences of management alternatives. We developed a linkage between TWIGS and a gray squirrel (*Sciurus carolinensis*) habitat-suitability model to illustrate the value of such combinations, using three management options for a stand in southern Indiana. Do-nothing management provided the best gray squirrel habitat but the worst economic return; the diameter-limit alternative produced poorer squirrel habitat but a better financial return; and intensive management provided the highest economic return but produced the poorest squirrel habitat.

Forecasting forest overstory change is essential to forecasting habitat suitability for many forest wildlife species. Although forest overstory is only one of several important pieces of information needed to evaluate wildlife habitat, its dynamics have been modeled extensively and can often be linked to other changes in the forest community that affect habitat. Combining models that forecast vegetation changes with models that convert those changes to evaluations of wildlife habitat is an essential step toward integrated forest management. Linking these models allows the trade-offs implicit in management alternatives to be explored.

In this chapter we demonstrate how a widely used tree-growth projection model can be coupled with a habitat-suitability model for gray squirrels (*Sciurus carolinensis*). The combined model can run on a microcomputer and uses common forest-inventory information. In addition, an integrated economic-analysis routine allows the forest manager to determine the probable consequences of each management strategy in terms of wood production, habitat production, and economic return. Here we describe the composite model, illustrate its application, and discuss important aspects of evaluating linked vegetation and wildlife models.

The linked model

The forest-overstory model drives the linked model by projecting the state of the forest overstory to any future time. The projected condition of the forest overstory is converted into variables required by the habitat model, and the corresponding habitat-suitability index is computed.

The overstory model

Overstory dynamics are projected using the TWIGS simulator (Belcher 1982). TWIGS is the microcomputer version of STEMS (the Stand and Tree Evaluation and Modeling System) (Belcher et al. 1982). Both models are widely used by the USDA Forest Service, industry, and consultants for inventory projection, forest planning, and forest management. TWIGS uses individual-tree-based, distance-independent projection models (Munro 1974). Growth and mortality for each tree are estimated using generalized model forms with species-specific regression coefficients. Equations have been calibrated and independently tested (Holdaway and Brand 1983; Forest Service 1983, unpubl. report) for important tree species indigenous to the Lake States and the Central States.

Information required to apply the overstory model is typically collected during a timber inventory. Each inventory plot is represented by a list of sampled trees, including species, diameter, and crown ratio. Using this tree list and the estimated site index, TWIGS applies the appropriate growth and mortality estimates for a given number of years. Estimates of growth and mortality depend on tree size, crown ratio, size relative to competitors, and stand density. The projected tree list is identical in form to one that could be obtained by a future inventory of the plot. At any time during the course of the overstory projection, the user may opt to simulate tree cutting by implementing any desired cutting prescription. Product values for cut trees are automatically recorded in a computerized ledger as trees are removed. When cutting is complete, the value of harvested products, along with annual and one-time costs or revenues, is used to automatically compute an internal rate of return.

The gray squirrel habitat-suitability model

Allen's (1982b) model estimates a year-round habitat-suitability index (HSI) for gray squirrels, in terms of winter food availability and needs for cover and reproduction, in

Gary J. Brand, Stephen R. Shifley, and Lewis F. Ohmann: USDA Forest Service, North Central Forest Experiment Station, 1992 Folwell Avenue, St. Paul, Minnesota 55108

deciduous forests larger than 0.4 ha (1 acre). Habitat suitability is expressed as the more limiting of these two factors. Five variables are used to compute gray squirrel HSI: (1) percent canopy closure of hard-mast-producing trees 25.4 cm (10 inches) dbh or larger; (2) number of species producing hard mast; (3) percent canopy closure of trees taller than 5 m (16.5 feet); (4) mean dbh of trees that are at least 80% as tall as the tallest tree in the stand; and (5) percent crown cover of woody vegetation shorter than 5m (16.5 feet) tall. The winter food index ranges from 0 (unsuitable) to 1 (optimum) and is a function of the first two variables. The cover-reproduction index also ranges from 0 to 1 and is a function of the remaining three variables.

Allen (1982b) suggested ways to sample forests directly for these variables, but it would be too costly and tedious to do this solely to evaluate present squirrel habitat. Direct sampling also does not provide values for estimating future habitat suitability. The vegetation model, therefore, must be able to forecast the variables required by the gray squirrel model.

THE LINKAGE

Morrison (1983) stated that even though both wildlife-habitat and timber surveys measure vegetation, it is only by chance that they measure the same variables. The incompatibility in survey methods is usually carried over into wildlife and vegetation models. Rarely are the inputs and outputs of vegetation models fully compatible with the inputs and outputs of wildlife models. Yet, compatibility is crucial if models are to be linked. Smith et al. (1981b) listed two requirements for linking vegetation models and wildlife-habitat models: (1) relate vegetation variables to wildlife habitat; and (2) generate these vegetation variables with the vegetation model. Careful planning early in model development can greatly simplify future efforts to link wildlife and vegetation models.

The five variables in the squirrel model provide the linkage between the vegetation and wildlife models and exemplify direct, indirect, and subjective model links. The number of mast species is a direct link. TWIGS records individual trees, so it is easy to count the number of hard-mast-producing species in the sample. Canopy closure of mast-producing trees, canopy closure of all trees, and mean dbh of overstory trees are indirect links. The crown area and height of each tree are required to determine canopy closure for overstory trees; neither is available in TWIGS. However, it is possible to obtain crown diameter (and therefore crown area) and height indirectly from other variables in TWIGS. In this case, we used species-specific regression equations (Minckler and Gingrich 1970; Ek 1974) to predict crown diameter and height from dbh. The final variable, shrub crown cover, is a subjective link. Although the condition of the forest overstory strongly influences the amount of shrub crown cover, we cannot estimate shrub cover from variables in TWIGS. Therefore, we can only estimate squirrel habitat suitability for specified levels of shrub crown cover and make an educated guess about how shrub crown cover will change as the overstory changes.

Results of the linked model: An example

The following example presents the results of a 35-year simulation in an oak-hickory stand. It shows how forest managers can use the TWIGS-squirrel model to study and compare future timber yields, economic returns, and gray squirrel habitat suitability expected under a variety of management alternatives.

Our sample stand came from a 1977 inventory of a 45-year-old oak-hickory forest located in the Hoosier National Forest in southern Indiana. The site index for white oak was 21.3 m (70 feet) at age 50. Although a few large white oaks (*Quercus alba*), black oaks (*Q. velutina*), and pignut hickories (*Carya glabra*) were in the overstory, the stand consisted predominantly of 16–30 cm (6.3–11.8 inches) dbh white oaks, black oaks, scarlet oaks (*Q. coccinea*), and northern red oaks (*Q. rubra*) (Table 54.1). The understory contained white oaks and pignut hickories. According to an upland oak stocking guide (Gingrich 1967), the initial inventory of 1700 trees/ha (688 trees/acre) with a basal area of 24.8 m^2/ha (108 ft^2/acre) indicates that the stand was overstocked.

We considered three management alternatives: (1) do nothing; (2) remove oak sawtimber in periodic diameter-limit cuts; and (3) intensively manage for high-quality sawtimber at rotation. With the do-nothing management option, the stand was clearcut in the year 2012 at age 80. In the diame-

Table 54.1. Inventory of all trees 2.5 cm (1 inch) dbh and larger in a 45-year-old southern Indiana oak-hickory stand in 1977

	Live trees		Basal area		Average dbh		Sawtimber[a]	
Species	no./ha	no./acre	m^2/ha	ft^2/acre	cm	inches	m^3/ha	(mbf/acre)
Pignut hickory	504	204	1.5	6.5	4.0	1.6	3.0	0.51
White oak	894	362	13.2	57.6	12.4	5.0	5.9	1.01
Northern red oak	22	9	1.1	4.8	25.2	10.2	0	0
Black oak	124	50	4.9	21.5	20.7	8.4	6.2	1.06
Scarlet oak	163	66	4.1	18.0	17.5	7.1	0	0
Total	1707	691	24.8	108.4			15.1	2.58
Combined mean					11.2	4.5		

[a]Sawtimber (Doyle log rule) minimum dbh is 28 cm (11 inches); mbf = thousands of board feet.

Table 54.2. Simulated timber yields 1977–2012 for three management options

	Do nothing				Diameter limit				Intensive			
	Sawtimber[a]		Firewood[b]		Sawtimber[a]		Firewood[b]		Sawtimber[a]		Firewood[b]	
Year	m^3/ha	mbf/acre	m^3/ha	ft^3/acre	m^3/ha	mbf/acre	m^3/ha	ft^3/acre	m^3/ha	mbf/acre	m^3/ha	ft^3/acre
1977	...	...	...	...	...	...	...	...	6	1.0	20	290
1987	...	...	...	...	...	...	...	...	0	0	31	440
1997	...	...	...	...	17	2.8	20	278	...	...	...	...
2012 (cut)	29	4.9	107	1830	...	...	...	...	29	4.9	49	700
2012 (residual)	...	...	...	...	20	3.4	94	1344	...	...	...	...
Total	29	4.9	107	1830	37	6.2	114	1622	35	5.9	100	1430

[a]Minimum sawtimber (Doyle log rule) dbh is 28 cm (11 inches); mbf = thousands of board feet.
[b]Firewood is the volume of wood in the portion of the stem above the sawlog and the volume of wood in trees between 13 cm (5 inches) and 28 cm (11 inches) dbh.

ter-limit cut (similar to high-grading), all merchantable trees at least 36 cm (14 inches) in diameter were removed when the combined harvest volume exceeded 15 m^3/ha (2.5 mbf [thousands of board feet]/acre). Simulated harvests were made without regard for the stocking or species composition of the residual stand; the only harvest was in 1997. Because of the many small white oak and hickory trees in this stand, the stocking remained above 100%, even after the cut in 1997. We followed Sander's (1977) guidelines to simulate intensive oak management. In 1977, we simulated the removal of large white and black oaks and hickories, as well as white oaks with dbh from 12.5 to 16.2 cm (4.9–6.4 inches). The residual stocking was 73%. Simulated cutting of white, black, and scarlet oaks 19.0–22.4 cm (7.5–8.8 inches) dbh in 1987 thinned the stand to 63% stocking, with no further cutting until the clearcut in 2012.

Timber yields

Diameter-limit cutting produced the largest total volume of sawtimber and firewood (including the standing crop in 2012) (Table 54.2). By periodically removing trees and increasing growing space for the residual stand, diameter-limit cutting and intensive management produced 28% and 21% more sawtimber volume, respectively, than the do-nothing option.

Financial yields from timber

Another way to evaluate these options is to compare their financial yields. For each species in this stand we assigned stumpage values based on average values reported for Indiana in the third quarter of 1983 (Hoover and Park 1983). We used the average stumpage for number 1 and number 2 sawlogs: hickory \$14/m^3 (\$33/mbf), white oak \$46/m^3 (\$109/mbf), black and scarlet oak \$48/m^3 (\$114/mbf), and red oak \$56/m^3 (\$133/mbf). We estimated stumpage for firewood to be \$5/m^3 (\$0.15/ft^3). We also assumed an annual cost of \$4.94/ha (\$2/acre) for taxes and administration from 1932 to 2012. Given these assumptions and the yields from Table 54.2, do-nothing management yielded an internal rate of return (IRR) of 2.3%; diameter-limit gave 3.4%; and intensive management gave 3.9%. However, this analysis did not include the effect of management on tree quality and value. The diameter-limit option should produce low-quality sawlogs, and intensive management should produce high-quality sawlogs at rotation (Sander 1977). Therefore, we added a quality bonus of \$42/m^3 (\$100/mbf) to trees harvested in 2012 under intensive management. This corresponded to an average increase of about one quality grade per log. With this adjustment, the IRR for intensive management was 5.6%.

Squirrel-habitat yields

The intensive-management option consistently produced the poorest squirrel habitat (Table 54.3). All three options produced high winter food indices, and only in a few cases was this the limiting resource (recall that the habitat-suitability index is the lesser of the winter food index and the cover-reproduction index).

To determine the cover-reproduction index, we needed to estimate the shrub crown cover. Shrub crown cover should be low for the do-nothing and diameter-limit options, because both produced overstocked stands throughout the period (R. Burt and J. R. Probst, pers. comm.). For these two

Table 54.3. Projected squirrel habitat-suitability index (1 = optimum, 0 = unsuitable) for three management options, 1977–2012. (The HSI value used for modeling was the lower of the two values—food or cover—shown for each of the three options.)

	Do nothing		Diameter limit		Intensive	
Year	Food	Cover[a]	Food	Cover[a]	Food	Cover[b]
1977	0.97	0.85	0.97	0.85	0.97	0.85
1977 postcut	No cut		No cut		0.70	0.60
1987	1.00	0.85	1.00	0.85	0.95	0.43
1987 postcut	No cut		No cut		0.95	0.35
1997	0.92	0.85	0.92	0.85	0.86	0.49
1997 postcut	No cut		0.81	0.59	No cut	
2007	0.77	0.85	0.86	0.67	0.78	0.53
2012	0.77	0.85	0.77	0.71	0.77	0.55
2012 postcut	0	0	0.77	0.71	0	0

[a]Assumes 15% shrub crown cover; stand is overstocked throughout the period.
[b]Assumes 60% shrub crown cover; stocking is 60–90%.

options, we assumed that shrub crown cover was 15%. Stocking of the intensively managed stand ranged from 60% to 90% because of the regular, heavy thinnings. Shrub crown cover should be highest with this option; we used a figure of 60% (R. Burt and J. R. Probst, pers. comm.).

Under intensive management, high shrub crown cover was the limiting factor in the gray squirrel HSI. We believe that the 60% shrub crown cover used was characteristic of stands managed in this way. However, when the limiting effect of shrub crown cover was removed from the cover-reproduction index, the HSI for intensive management after 1997 was as high as that for the other management options. The importance of the shrub crown cover estimate in the HSI model highlights the need for additional research to quantify the relations between shrub cover, tree overstory density, and squirrel habitat.

Selecting among the alternatives

In this example, the best option was not obvious. Do-nothing management favored squirrel habitat at the expense of future revenue. Diameter-limit cutting generated more revenue from timber but poorer squirrel habitat. However, because an overstory still remained in 2012, squirrel habitat would be better than under the other options for some years after 2012. If log grade could be improved through management, intensive management would give by far the best financial return from timber, with either a slight or a dramatic decrease in squirrel habitat, depending on the true level and effect of shrub crown cover.

As is usual in resource-management decisions, there are no simple answers, only trade-offs. We think the TWIGS-squirrel model emphasizes the relevant trade-offs effectively and provides managers with basic information necessary to guide decision making.

Evaluating the linked model

We have demonstrated that a forest-overstory model and a habitat model can be linked to help land managers forecast the impact of alternative actions. Ultimately it is the user who must evaluate any model in light of the particular application. A model may perform well for one purpose but poorly for another.

Buchman and Shifley (1983) presented three main criteria for model evaluation: (1) application environment; (2) model performance; and (3) model design. Application environment deals with the ease of operating the model. Model performance focuses on the accuracy and precision of the model. Model design assesses the biological realism incorporated into the model and the model flexibility. In this section, we use these criteria to help evaluate the utility of the TWIGS-squirrel model for comparing management alternatives in oak-hickory stands.

Application environment

TWIGS is easy to use and can be run on microcomputers. With the exception of percent shrub crown cover, the linked model can be implemented with data collected in a traditional inventory that records individual tree data.

Model performance

The growth and mortality components of TWIGS have been thoroughly tested with independent data (Holdaway and Brand 1983; Forest Service 1983, unpubl. report). For a sample of 13 oak-hickory stands on the Hoosier National Forest, the mean basal area error was -1.3 m^2/ha (-5.8 ft^2/acre) after 10 years, and the mean error in predicting number of trees was -57 trees/ha (-23 trees/acre). The mean errors in dbh prediction after 10 years for each species occurring in this sample ranged from -0.20 cm (-0.08 inches) for red oak to 0.13 cm (0.05 inches) for scarlet oak. The other important vegetation variables (volume, height, and crown diameter) are functions of dbh and tree density and will closely follow the above errors.

Evaluating wildlife-model performance is more difficult because data to develop the model are usually scarce (Euler 1975). The need to evaluate performance with an independent data set enlarges the problem. We did not have independent estimates of squirrel HSI. We can only say that the squirrel model is consistent with our expectations; we have no quantitative evaluations of this component.

We compared the projected 1967–1977 changes in our sample stand with the observed changes for the same period. Projected basal area was 0.7 m^2/ha (3 ft^2/acre) lower, and the projected tree density was 62 trees/ha (25 trees/acre) higher, than actually observed in 1977. The similarity between observed and simulated 10-year changes supports our confidence in the ability of the forest component of the TWIGS-squirrel model to simulate this stand. Observed and predicted HSI values were identical for the 1967–1977 interval.

Model design

TWIGS components are designed to interact in a manner that is biologically realistic and also logical. For example, large trees compete more effectively than small trees, and removing trees in a stand allows the remaining trees to grow faster and live longer. Both the TWIGS and the squirrel models are designed to produce reasonable results. Stand basal area and tree diameter cannot increase indefinitely, but are constrained by the equations. The squirrel model constrains habitat suitability to values between 0 and 1.

The logic of the TWIGS-squirrel model has two major weaknesses. The first is the lack of a small-plant model. TWIGS allows the user to add new trees, but it is up to the user to know the characteristics (species, dbh, and number) of the new trees. Short vegetation (e.g., tree seedlings, shrubs, and herbs) is not modeled. This vegetative layer strongly influences the estimate of HSI. It also is the layer most likely to be affected by vertebrate herbivores. The second weakness is HSI itself. HSI is intangible; it has not been directly tied to squirrel population size, recreation value, or to any other forest output that can be evaluated from an economic perspective. HSI is a dimensionless scale measuring relative improvement in habitat quality. We use it for lack of a better alternative.

Model flexibility is another important aspect of model design. TWIGS is programmed in modules, so adding or changing parts is easy. Habitat models for other species (or species' occurrence) or wildlife-population models could be added as long as the important vegetation variables are generated by TWIGS. The management module in TWIGS is also flexible. Any tree-cutting action based on species, tree quality, or dbh can be implemented. Tree cavities, important for gray squirrels, are more likely to occur as the average diameter of overstory trees increases (Allen 1982b). Although we lacked information about which trees in this stand, if any, were cull or had cavities, TWIGS offers the opportunity to record cull trees for a more detailed evaluation of den-tree availability.

Conclusions

Considerations of model evaluation can be summarized in the following two questions: (1) Is this model better than the alternative models available? (An alternative model can always be found, even if only as an educated guess.) (2) Are the estimates given by this model close enough for the anticipated application? Judged in this context, we think that the combined model, despite its limitations, has forest-management applications. It provides a convenient method to assimilate information about the impact of tree growth and management on gray squirrel habitat suitability. If the user desires, the economic implications of alternative actions can also be explored. This methodology can be extended to other wildlife models for which appropriate linkages can be developed. Although the TWIGS and gray squirrel models have shortcomings, improvements in each will come in the future. Meanwhile, the models provide an easy-to-use tool enabling forest managers to study the potential impact of their management alternatives.

Acknowledgments

We thank the Forest Inventory and Analysis work unit at the North Central Forest Experiment Station for providing inventory data for our sample stand. The expert opinions of Richard Burt of the Hoosier National Forest and John Probst of the North Central Forest Experiment Station helped us estimate present shrub crown cover. We also appreciate the helpful comments of Richard Buech, Richard Burt, William McComb, Ivan Sander, and James Sweeney, who reviewed the manuscript.

55

Habitat-Simulation Models: Integrating Habitat-Classification and Forest-Simulation Models

THOMAS M. SMITH

Abstract.—Habitat-simulation models integrate habitat-classification and forest-simulation models to predict the effects of forest-management practices on the availability of habitat for wildlife species. The classification of habitat must consider both the level of habitat resolution and the spatial scale at which habitat patches are considered to be homogeneous units. The forest-simulation model must be able to predict changes in the variables on which the classification is based at an appropriate spatial scale. FORHAB (a deciduous-forest habitat-simulation model) is presented as an example of this methodology. The model is used to predict the effects of timber harvest on habitat availability for two nongame bird species on the Walker Branch Watershed in eastern Tennessee, the red-eyed vireo (*Vireo olivaceus*) and the downy woodpecker (*Picoides pubescens*).

With ever-increasing demands being placed on our natural resources and the continuing call for multiple-land-use management, wildlife biologists are having to reevaluate their general approach to wildlife management. In the past, emphasis was placed on estimating population numbers and actively regulating those numbers through stocking and hunting programs. The general approach to habitat management was an attempt to maintain the status quo.

With the need for multiple-land-use programs in our natural areas and the growing interest in nongame species, a shift has taken place. Wildlife biologists are placing more emphasis on habitat management and on integrating wildlife management with forest management and timber harvest. This shift has required an increased information base relating to species' habitat requirements and the effects of various management practices on the availability of that habitat. At first, the habitat requirements of species were based on qualitative descriptions relating the presence or absence of species to the general forest type or structure of the vegetation. This information, coupled with a knowledge of the effects of various silvicultural practices on the structure of forests, was used to predict the effects of forest-management practices on wildlife habitat. In recent years, however, there has been a growing interest in the use of more quantitative techniques to describe the habitat-selection patterns of animals (Capen 1981). An increase in the development and use of forest-simulation models (Shugart and West 1980) by both plant ecologists and foresters, for the purpose of predicting the temporal dynamics of forested systems, has accompanied the development of statistical approaches to habitat classification. The development of these two techniques, habitat-classification and vegetation-simulation models, has provided wildlife biologists and managers with the new potential to move habitat management into a dynamic framework. The purpose of this chapter is to discuss a method for predicting the dynamics of habitat availability under various management practices on specific sites through the use of habitat-simulation models.

THOMAS M. SMITH: Environmental Sciences Division, Building 1505, Oak Ridge National Laboratory, P.O. Box X, Oak Ridge, Tennessee 37831

Dynamic habitat models

Habitat-simulation models integrate habitat classification with the ability of vegetation-simulation models to project changes in vegetation through time. This process involves (1) the classification of vegetation in terms of suitability to provide habitat for a given animal species; and (2) a vegetation simulator with the ability to generate specific structural or compositional variables on which the classification is based. By introducing vegetation-management scenarios into the model, we can predict the long-term effects of various alternatives on the availability of habitat for a particular species or for community characteristics, such as species diversity. The structural classification of habitat must consider both the level of habitat resolution and the spatial scale at which habitat patches are considered to be homogeneous units. The level of habitat resolution is the scale at which the vegetation structure must be defined to differentiate between suitable and unsuitable (unoccupied) areas. For example, if all pine forests were suitable habitat for a given animal species, then it would be sufficient to consider only taxonomic composition of the stand to predict the potential of a site to provide habitat. However, if the species occupies only old-age pine stands, it would be necessary to consider not only the species composition, but also some measure of stand structure and/or age, such as basal area or average tree size.

The second important feature of the habitat classification is the spatial scale at which the habitat is defined. This will be related to the size of the territory or home range of the species, to the degree to which the area can be partitioned into homogeneous units, and to the importance of the spatial arrangement of these homogeneous units. For example, the territory of a species of warbler may correspond to an area <1 ha and be considered relatively homogeneous with respect to the variables defining habitat suitability. In contrast, certain large mammals may require relatively large home ranges with various habitat types interspersed over the landscape at a spatial scale of hundreds of hectares. Therefore, to determine the form of the vegetation model to be used to simulate changes in habitat availability, one must consider both the variables required to differentiate between suitable and unsuitable vegetation structures and the scale at which these structures must appear as homogeneous units on the landscape. When a variety of defined structural units are required, it will also be necessary to define any restrictions on their spatial pattern on the landscape.

A case study

Much of the research in the development of statistical methods for identifying patterns of habitat selection has focused on determining the structural characteristics of vegetation associated with the breeding territories of bird species. The homogeneous sampling unit used to describe microhabitat-selection patterns (James and Shugart 1970; James 1971) corresponds to the spatial unit generally used as a statistical sample for the description of forest stands (Cottam and Curtis 1956). This same spatial unit (0.04–0.10 ha) has been described as the forest "gap," a functional unit of forest dynamics (Watt 1925, 1947; Shugart 1984). As a result, numerous forest-simulation models use the forest gap as their functional spatial unit (Shugart and West 1981). Therefore, territorial songbirds provide an excellent example for exploring habitat simulation, because the spatial scale needed to define the required level of habitat resolution corresponds to a meaningful homogeneous unit in terms of forest dynamics. When vegetation structure is classified in terms of providing suitable territory sites at the spatial scale of the forest stand and that classification is merged with forest-stand-simulation models, we can project changes in habitat availability as a function of vegetation dynamics. The present example is a forest-habitat-simulation model, FORHAB (Smith et al. 1981a, 1981b), which was used to predict the impact of certain forest-management decisions on the availability of breeding habitat for the bird community inhabiting the Walker Branch Watershed in eastern Tennessee. Results for two species, the red-eyed vireo (*Vireo olivaceus*) and the downy woodpecker (*Picoides pubescens*), are illustrated in this example.

Habitat classification used the statistical procedure of two-group discriminant function analysis (Morrison 1967). Classification criteria were constructed, using the vegetation data collected on 298 permanent census plots 0.08 ha in size located on the Walker Branch Watershed (Curlin and Nelson 1968; Anderson and Shugart 1974). Breeding territories of the various bird species that either contained or overlapped any of the 298 plots were located and mapped. If a plot was located within the territory of a breeding pair, the plot was considered as potential habitat for the species. Conversely, if a given plot was not within the boundary of a territory of that species, the plot was classified as not providing habitat for that species. Thus, data were obtained both on areas of suitable habitat and on areas that were not used by the various bird species. Data collected from the census plots (species and diameter of each individual tree) were then used to generate biomass variables (Smith et al. 1981a), using allometric regression equations specific to the Walker Branch Watershed (Harris et al. 1973; see description of subroutine HABIT). This data set (biomass variables) was used to construct linear decision scales, a classification scheme using the Bayes, or minimum-loss classification, rule based on two-group discriminant function analysis (Smith et al. 1981a).

The forest-habitat-simulation model, FORHAB, is a modified version of FORET (Shugart and West 1977), an Appalachian deciduous-forest-stand simulator. (A detailed description of the FORET model can be found in Shugart and West [1977].) FORHAB simulates the annual change of a forest stand (0.08 ha) by calculating the growth increment of each tree on the stand (subroutine GROW), by tabulating the addition of new saplings to the stand (subroutines BIRTH and SPROUT), and by tabulating the death of trees present on the stand (subroutine KILL). These processes are simulated on the basis of general silvicultural information. The maximum growth of each tree on a plot is computed from the inherent radial- and height-increment potentials and longevity of the tree species. The maximum growth is modified by the response of the tree to such environmental conditions as light availability, annual temperature, growing degree-days, and spatial crowding. These environmental factors are considered to be homogeneously distributed over the plot. Seed and sprout establishment are based on the availability of light, proper temperatures, substrate requirements for germination, and sprouting capabilities of dying trees. Mortality is a function of the growth and expected maximum age of the tree. Each growth response to environmental conditions is described as a probability function. Therefore, the model is stochastic, allowing for a wide variation in the response of any individual tree. Provided that the mean and variance of the biological response of an individual tree to the management technique is known, the model can be used to simulate the subsequent effects of management practices on forest structure. Those subroutines involved in the management and classification of simulated stands will be discussed separately.

Subroutines DISCRIM and HABIT

The process of classifying simulated stands according to their potential to provide habitat for a given species is carried out in subroutine DISCRIM. The classification uses the linear decision scales described above. Therefore, model output, in the form of species and tree diameters for each simulated plot, must be converted to the biomass variables on which the classification is based. Subroutine HABIT divides all trees on the simulated plot into two groups, conifer and deciduous, and then into three size-classes within each of these two groups (Smith et al. 1981a). The foliage, branch, and bole biomass for each tree are then calculated using the site-specific biomass regression equations (Harris et al. 1973). Values are then summed for all trees on the plot for each class to provide the above-mentioned variables. Data are then input into subroutine DISCRIM, where a decision is made as to whether the plot provides potential breeding habitat for each of the species.

Subroutine CUT

The CUT subroutine simulates various forest-management practices that are applicable to the southeastern deciduous-forest type. The version of this subroutine used for the following analysis was a diameter-limit cut. In this subroutine, all commercially valuable species for sawtimber > 28 cm dbh were removed from the plot on a 60-year rotation. The rotation period of 60 years was determined by analysis of stem and basal area curves generated by FORHAB after initial simulations of logging on the watershed.

Methods

Two 120-year model runs were made, one each to represent undisturbed and managed conditions. Each model run consisted of 298 simulated plots. Model runs were initiated using data from the 298 census plots on which the habitat classification was based. Therefore, year 0 of the simulations represents the present structure and composition of the watershed. For the simulation of managed plots, sawtimber cuts were made at years 1 and 60. For each model run, results from the 298 plots for each year were averaged to give a composite picture of stand structure (Fig. 55.1). For each year of the simulations, the 298 plots were individually classified as to their potential to provide suitable habitat. Habitat availability through time was then expressed as the percentage of the 298 plots classified as potential breeding habitat for each of the species (Fig. 55.2).

Results and discussion

Timber harvest resulted in a general increase in available habitat for both bird species (Fig. 55.2). This increase contrasted with the decline in habitat availability for the undisturbed simulations. Both the red-eyed vireo (Williamson 1971) and downy woodpecker (Conner and Adkisson 1977) are associated with a well-developed understory. The red-eyed vireo characteristically feeds by gleaning insects from the foliage where the canopy is open and the understory is dense. The downy woodpecker forages by probing the bark, focusing its activity on small trees in the understory and peripheral branches of large trees. The general decline in habitat was a function of canopy closure and the lack of a well-developed understory as the forest increased in age over the simulation. When the canopy was opened after thinning, suitable habitat increased with the development of the understory. This can be seen in the increase of stems, foliage, and branch biomass following the thinnings at years 1 and 60 (Fig. 55.1).

Considerations and conclusions

It should be noted that the model simulated the availability of potential habitat expressed as a percentage of the total land area under consideration. The model did not simulate the population dynamics of the bird species per se. The ability of the populations to track changes in habitat availability in time (Seagle, Chapter 40) or in space (Urban and Shugart, Chapter 39), or to reinvade after elimination of potential breeding habitat, was not considered in the model. Likewise, the model did not consider the quality of habitat provided by a given stand. Some marginal areas may be used, depending on the size of the population (Fretwell and Lucas 1969). Further work is needed in the area of linking animal-population models to habitat-simulation models. This will be especially important for animal species that have a significant influence on vegetation dynamics, therefore providing a feedback between animal and plant populations.

Although the model examined microhabitat requirements at the level of the forest stand or gap, there are various possible scales at which habitat can be defined for any given species (see Shugart and Urban, Chapter 60). However, as one decreases the level of resolution at which the habitat is defined, the probability of error associated with predicting the potential suitability or occupancy of a given area increases. This increasing probability of misclassification is a function of aggregating vegetation attributes to a level of resolution at which the categories or variables defining the state of the vegetation no longer correspond to the structural characteristics that define habitat preference. Therefore, the investigator faces the dilemma of choosing between modeling vegetation dynamics at a spatial scale appropriate for distinguishing those structural characteristics that define habitat suitability, or modeling vegetation dynamics at a level of structural resolution that is most appropriate (in terms of time and money) for examining vegetation dynamics over large tracts of land. However, regardless of which approach is used, the model must consider vegetation structure at a level of resolution that is able to incorporate changes in vegetation structure and dynamics resulting from management practices, because those practices may influence habitat suitability.

The habitat models presented in this volume represent these two different approaches to simulating habitat dynamics in managed forests. The DYNAST model (Holthausen, Chapter 52; Benson and Laudenslayer, Chapter 49) simulates vegetation dynamics using a large-scale, coarse-resolution, deterministic approach. This approach is best adapted for use in simulating vegetation dynamics for large, even-aged forests, where timber harvest is done with clear-cut techniques. In contrast, both TWIGS (Brand et al., Chapter 54) and FORHAB use small-scale, fine-resolution approaches to simulating vegetation dynamics. As a result, both models are able to incorporate changes in structure and dynamics at the spatial scale of the forest stand, resulting from selective cutting practices. Therefore, these modeling approaches are better suited to many of the management practices used in eastern deciduous forests. However, TWIGS and FORHAB differ in their basic structure. TWIGS

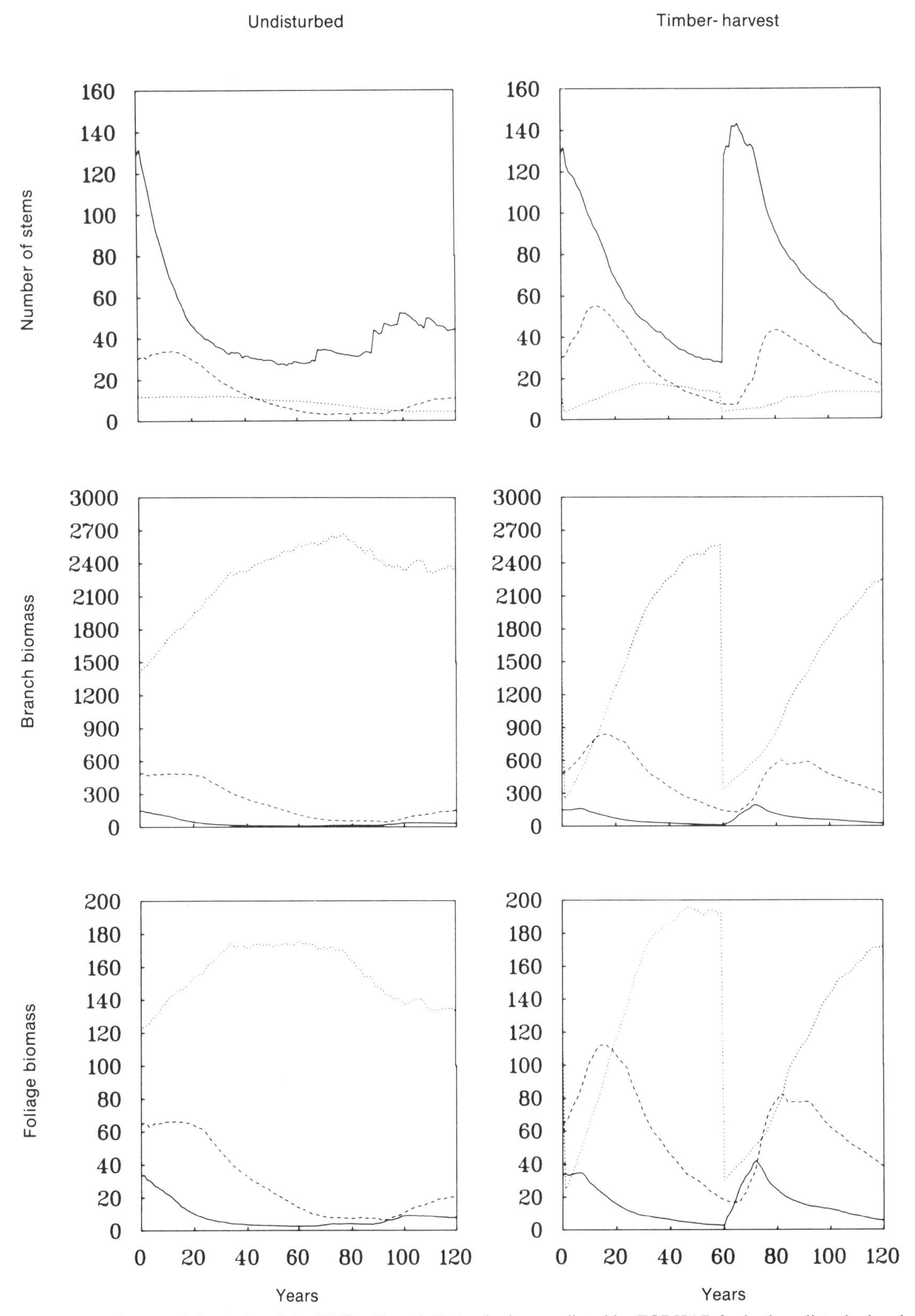

Figure 55.1. Structural dynamics of the Walker Branch Watershed as predicted by FORHAB for both undisturbed and timber-harvest conditions. Biomass measurements are expressed as kilograms dry weight/0.08 ha. Results are grouped into three size-classes: solid line = < 8 cm dbh; dashed line = 8–23 cm dbh; dotted line = > 23 cm dbh (from Smith et al. 1981b).

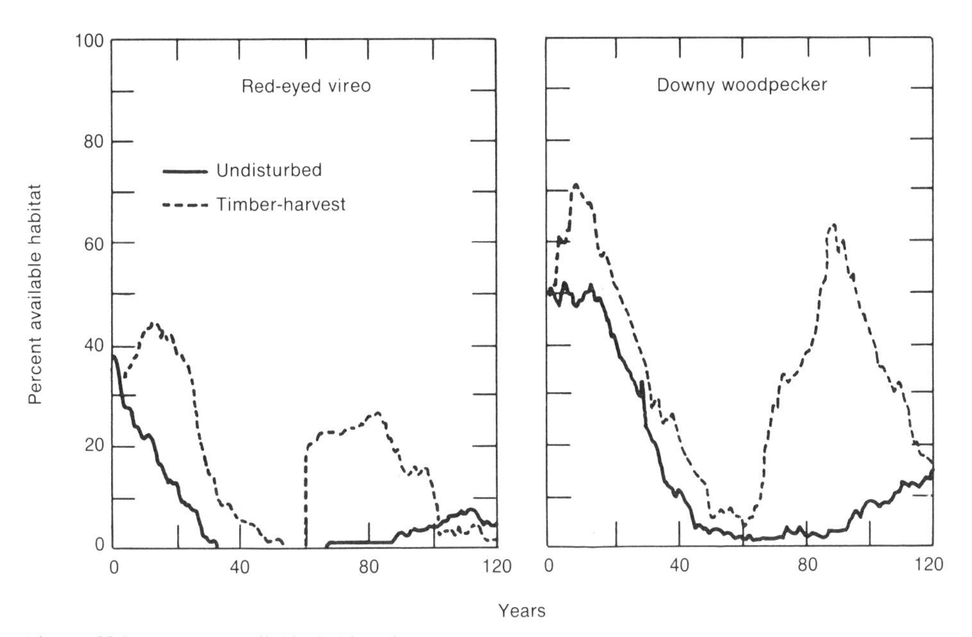

Figure 55.2. Percent available habitat for red-eyed vireo and downy woodpecker as predicted by FORHAB for both undisturbed and timber-harvest conditions. Percent available habitat is expressed as percentage of the total land area of the watershed.

is a data-rich, deterministic modeling approach based on species-specific regressions. As such, it is difficult to adapt the model to different forested areas, because of the data requirements needed for parameterization. The parameters of FORET, the vegetation simulator for the FORHAB model, are based on easily obtained, species-specific silvicultural data (see Shugart and West 1977). As a result, the FORET modeling approach has been widely adapted and is used in a variety of forested systems, both in the United States and around the world (Shugart and West 1980), making it a potentially important tool for future applications integrating timber and wildlife management.

Acknowledgments

The manuscript was reviewed by Jared Verner and Larry Davis. Dean Urban and Hank Shugart provided helpful comments and discussion on the development of habitat-simulation models. This research was supported by the National Science Foundation's Ecosystem Studies Program under Interagency Agreement No. BSR–8315185 with the U.S. Department of Energy, under contract No. DE–AC05–84OR21400 with Martin Marietta Energy Systems, Inc.

56

Dealing with Wicked Problems: A Case Study of Models and Resource Management in Southeast Alaska

PETER J. McNAMEE, PILLE BUNNELL, and NICHOLAS C. SONNTAG

Abstract.—Resource and environmental issues are difficult to resolve because the issues cross a multitude of scientific disciplines and institutional concerns. This chapter describes an ongoing simulation-modeling project within the USDA Forest Service for research and management of the timber, wildlife, and fisheries resources of the Tongass National Forest in Southeast Alaska. The procedures for model development used are different from those in most traditional modeling efforts aimed at resource-management issues and are specifically designed to overcome the difficulties mentioned above. The emphasis has been on achieving interagency consensus for a model as a practical tool. Key features of model development are explained, and the results of the modeling project are presented.

Resource management is difficult because of the inherent complexity of biophysical, socioeconomic, and management systems, and because of our incomplete understanding of these processes. The multitude of disciplinary and institutional concerns, combined with ecosystem interactions, make resource managment a complex or "wicked" problem, as defined by Rittel (1972). Mason and Mitroff (1981) presented five basic characteristics of wicked problems: (1) interconnectedness and complexity: wicked problems have numerous important elements with relationships among them, including feedback loops through which management actions can have unexpected impacts (Walters 1977); (2) uncertainty: our understanding of biophysical systems will always be incomplete— management decisions, however, are always made in the face of uncertain future priorities and demands; (3) ambiguity: no single correct view of a wicked problem exists—the problem can be seen in many different ways depending on the viewer's personal characteristics, loyalties, past experience, and circumstances of involvement; (4) controversy: conflicts among interpretation, jurisdiction, and values are the norm with wicked problems; and (5) societal constraints: organizational, political, and cultural constraints, together with technical capabilities, limit the feasibility of solutions to wicked problems.

Given the wicked nature of resource-management problems, it is not surprising that managers have found models to be of limited utility in providing solutions (Lee 1973; Watt 1977; Hilborn 1979). Statistically sound models based on carefully collected and accurate data often contribute little to the resolution of environmental problems.

This lack of utility is not inherent in the models per se. Many model attributes, such as their ability to unambiguously represent complex interactions among system components and to form predictions that allow comparison of management and policy alternatives, make them potentially powerful management tools (Walters 1974; Gilbert et al. 1976). Furthermore, if these special attributes of models are embedded in a process capable of dealing with some of the concerns mentioned above, then new opportunities occur. The major objective of this chapter is to present a case study of the Southeast Alaska Multiresource Model (SAMM), demonstrating a process by which modeling and models can aid in difficult resource-management problems.

PETER J. MCNAMEE: ESSA Environmental and Social Systems Analysts Ltd., 102-66 Isabella Street, Toronto, Ontario, Canada M4Y 1N3

PILLE BUNNELL and NICHOLAS C. SONNTAG: ESSA Environmental and Social Systems Analysts Ltd., Box 12155, Nelson Square, #705, 808 Nelson Street, Vancouver, B.C., Canada V6Z 2H2

Case-study description

The management of the natural resources within the Tongass National Forest in Southeast Alaska is a wicked problem. State agencies manage anadromous fish and wildlife populations, and the USDA Forest Service manages fish and wildlife habitat. Historically, opinion has been divided on how timber harvest affects the land base. There have been differences in perspective and even bitter conflicts regarding management of the land base for the various resources. Most Forest Service policy documents for the Tongass National Forest, including the 1977 Southeast Alaska Area Guide, the Tongass Land Management Plan, and the Southeast Alaska Regional Guide (Forest Service Region 10, Juneau, Alaska) state that the disputed effect of timber-harvest activities on fish and wildlife is a major issue and, in fact, an impediment to the satisfactory resolution of land allocation and management in Southeast Alaska.

The management situation in Southeast Alaska is made difficult by the lack of a biophysical understanding of the natural resources. The area has only recently come under intensive management and, as a result, research data to support management decisions are meager (see McKnight 1979). This lack of information has aggravated the difficulties of reconciling different constituencies.

The Forest Service took the initiative toward developing a unified management perspective by sponsoring the development of a simulation model. Clearly, the primary purpose of the model was not to provide accurate predictions of a poorly understood ecological system. Rather, the modeling process was needed to promote interagency cooperation, to provide a better understanding of important issues, and to aid in the resolution of these issues. A simulation model firmly grounded in existing biophysical understanding provided the unifying force.

Methods

The Forest Service used Adaptive Environmental Assessment and Management (AEAM) for model development. AEAM (Holling 1978; Anonymous 1982) is a collection of concepts, techniques, and procedures intended to aid in the design of creative resource-management and policy alternatives. Through its range of applications, there have evolved a number of distinctive features that take into account many of the interdisciplinary and institutional problems often found when management policies are being developed. Its most important feature is that it uses the development of a computer simulation model of the resource system as a mechanism for resolving many of the difficulties associated with wicked problems.

A computer simulation model, once constructed, can be used as a kind of laboratory world in which alternative management strategies can be examined. What distinguishes AEAM models from other models is the forum in which the models are designed and used. The simulation model is not constructed by modelers in isolation. Short, intensive workshops, conducted by a workshop staff trained in simulation modeling and group dynamics, are used to design the simulation model. These workshops are attended by key scientists, managers, and decision makers chosen from all disciplines and agencies working on the problem. The modeling workshops use a structured set of exercises conducted by workshop staff to guide the diverse set of participants through the process of model building.

Model development began in March 1982 and was completed in September 1984. Thirty-two participants (with direct involvement in model construction) and 34 collaborators (providing direction, advice, and agency commitment for model development) from seven agencies or constituencies assisted in model development. A three-person core team, consisting of one member of the modeling staff plus one representative each from the Forest Service Regional Office and the Research Experiment Station, took major responsibility for overseeing model development. This was a cyclical process of model construction, refinement and analysis, and implementation (Fig. 56.1). Three major modeling workshops were conducted; shorter technical meetings in the intervals between them were used by participants to refine the model. The understanding of model capabilities and limitations was improved through model simplification and sensitivity analyses. Flexible audiovisual programs and demonstration workshops were used to involve new participants, particularly managers, in model implementation and refinement.

The major objective of the core team throughout the project was to use the model to build consensus among the constituencies. Because there was disagreement among various agencies on appropriate managment direction for the Tongass National Forest, the core team did not attempt to start with management policy. Rather, it sought to develop a noncontroversial, interdisciplinary biophysical representation of timber, water, anadromous fisheries, and Sitka black-tailed deer (*Odocoileus hemionus sitkensis*) resources on a watershed over two forest rotations (Fig. 56.2). The core team ensured that the model included management actions available to all agencies. Therefore, each agency could use the model, which it had agreed on as a credible representation of

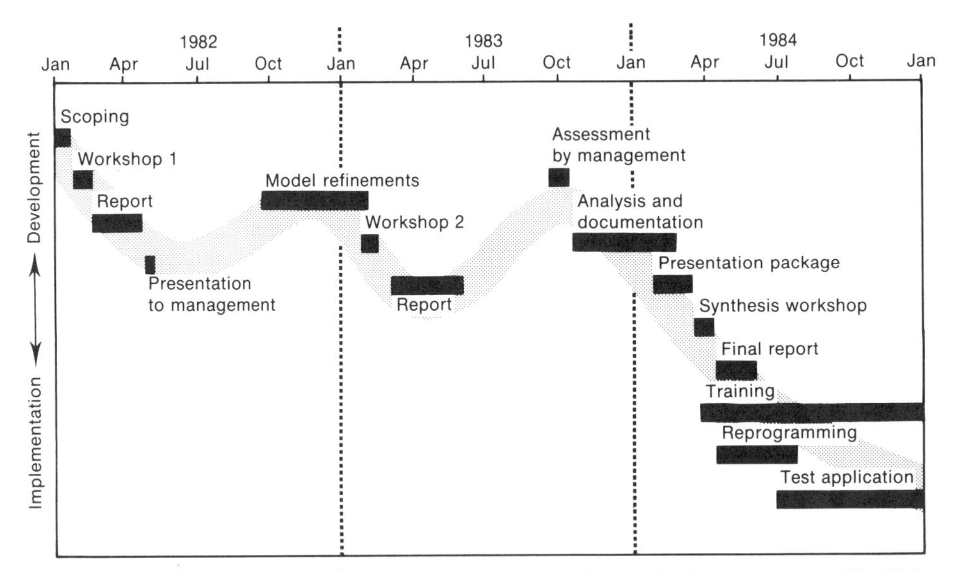

Figure 56.1. History of the development of the Southeast Alaska Multiresource Model (SAMM).

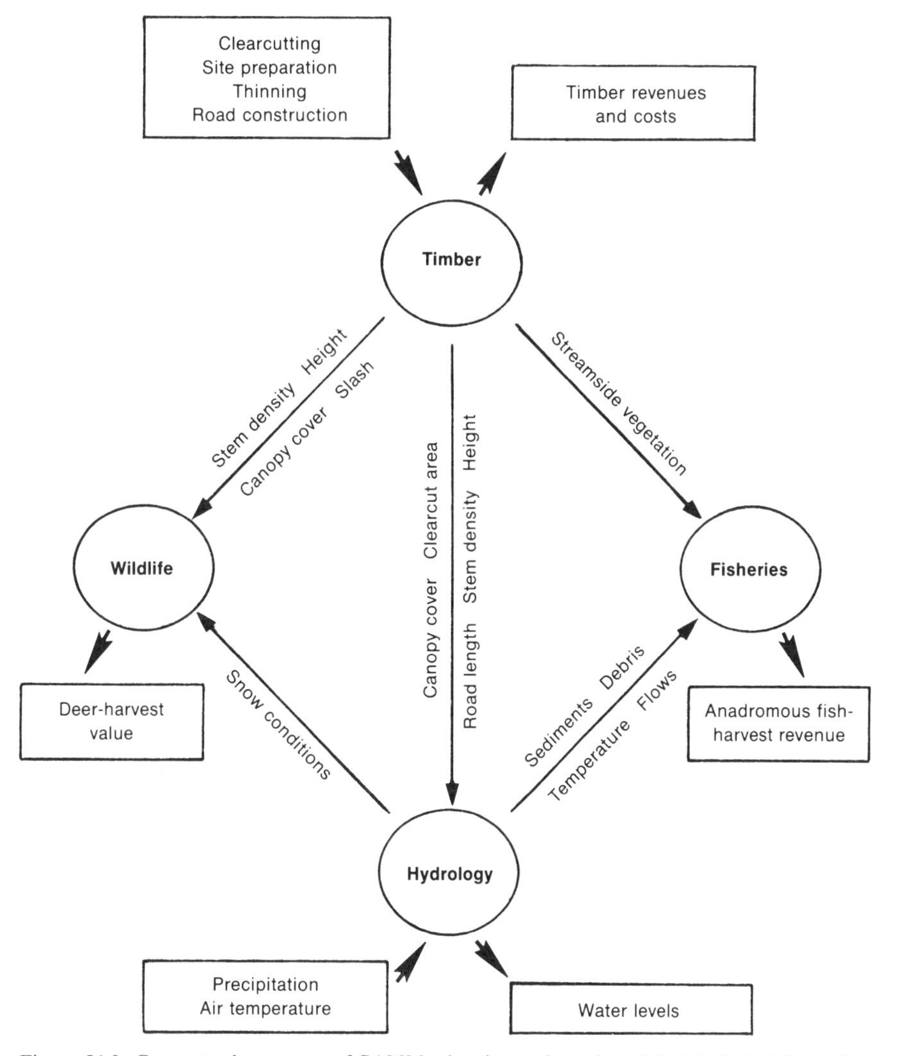

Figure 56.2. Conceptual structure of SAMM, showing major submodels (circled), information transfers between submodels (arrows), and management actions and indicators (boxes).

the resource system, to examine consequences of its preferred management strategy.

This approach meant that understanding and support were built from the bottom up. The core team worked with the technical experts and research scientists to achieve a common understanding of the resource system. The entire modeling group was then able to go to senior management groups and present the results as an interagency, interdisciplinary consensus. Such a consensus, so rarely achieved in Southeast Alaska, was met with enthusiasm and commitment.

There was an explicit effort to avoid assembling a highly trained group of modelers and programmers in isolation from the technical experts and resource managers. The core team kept the model open to scrutiny throughout its development. Approximately 90% of model development took place in a workshop forum, and all participants were made aware of any model changes, even in parts of the model that were not in their areas of specialty. This lack of compartmentalization promoted continuing interdisciplinary communication and interagency cooperation.

Results

Demonstration scenarios

We present five demonstration scenarios intended to illustrate how SAMM can be used for managing the natural resources of Southeast Alaska. Three different combinations of logging activities are compared to a base scenario of no action, and deer predation effects are examined for one logging option (Fig. 56.3). We stress that these scenarios were chosen to illustrate the temporal consequences of alternate management strategies rather than to produce quantitative predictions of natural resource levels.

Salmonids respond to logging activities in two distinct phases. The first is a decrease in juvenile production im-

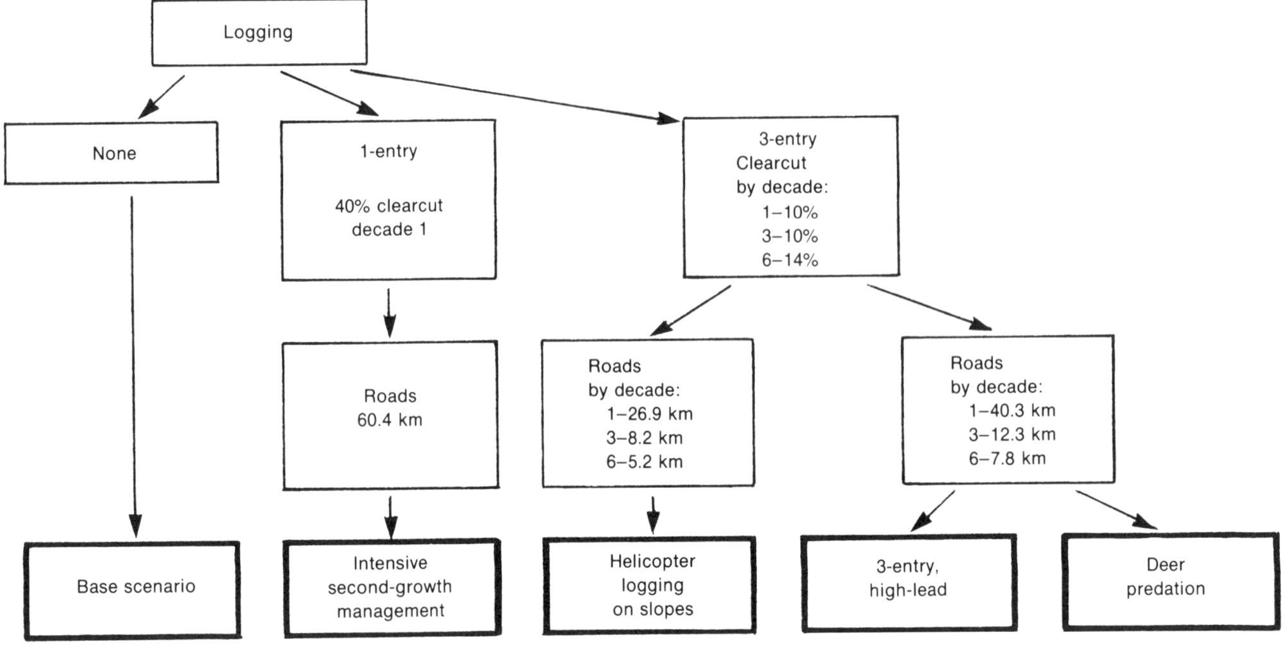

Figure 56.3. Scenario descriptions.

mediately after an entry; the second is an increase in production approximately 10 years after the entry (Fig. 56.4*a*). The initial declines are caused by the heavy siltation of streams from roads built into the watershed, whereas the ensuing rise in juvenile production occurs because of increased streamflows from higher runoff in clearcuts. Increased streamflow improves spawning habitat and winter egg survival. The model predicted no substantial difference in salmonid response for the helicopter logging option. Accordingly, the increased costs of helicopter logging (Fig. 56.4*d*) cannot be justified by improved fisheries production.

Deer populations in the watershed undergo a long-term decline in all the logging scenarios (Fig. 56.4*b*). Intensive second-growth management does not significantly improve deer population levels, though this option does provide for higher revenues in the first half-century due to the higher proportion of clearcut in the single entry (Fig. 56.4*c*). After this time, the increased costs of second-growth management (Fig. 56.4*d*) are greater than the revenues gained. Predation dramatically reduces deer populations (Fig. 56.4*b*), especially when the adverse effects of a reduced habitat, predation, and heavy snowfall occur concurrently in about year 60. Once the decline has occurred, predation prevents the deer populations from recovering.

Benefits

Interagency coordination

The most significant benefit of the model-development process has been the quantum leap in interagency communication and cooperation in Southeast Alaska. The primary evidence of this is the consensus on a biophysical representation of the resource system in the form of a simulation model (Fight et al., in press). This biophysical representation integrates four resources and disciplines—timber, hydrology, fisheries, and wildlife. The agencies involved in model development are committed to continued model use. The model will be used to identify research required, to evaluate existing data, and to test model utility in an actual management context, as discussed below.

Management application

The first model test in a real management situation has begun. The Forest Service initiated a management-area analysis on the Cleveland Peninsula near Ketchikan in early 1984. This analysis involves 16 watersheds over approximately 50,000 ha (120,000 acres) to determine appropriate allocation of the land base to timber harvest, fisheries, and wildlife use.

SAMM will be used in parallel with a traditional management-area analysis to evaluate its usefulness. The interdisciplinary team and its consultants will be able to specify different combinations of management actions and to view the simulated consequences of these scenarios on the timber, hydrology, fisheries, and wildlife resources over two forest rotations.

Research direction

SAMM has generated clear priorities for natural resource research in Southeast Alaska. The model-building process identified deficiencies in the collective understanding of the biophysical system, thereby indicating which data should be collected, and suggested novel ways of analyzing existing data. The process of model improvement and research should be recursive. Model construction and gaming identify the research needed to improve understanding, and research results improve the capabilities of the model.

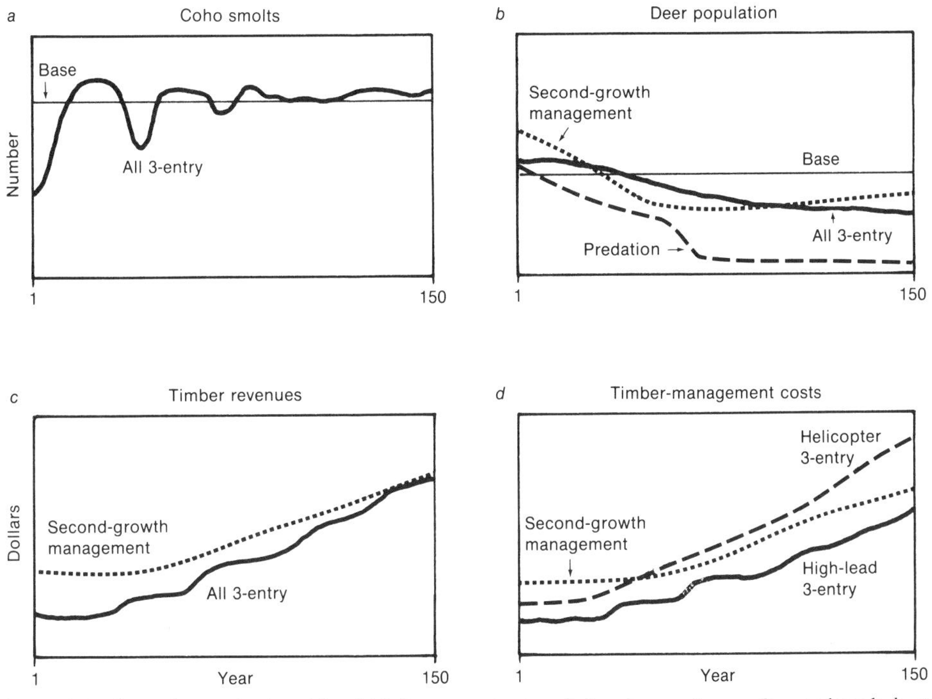

Figure 56.4. Scenario results (see Fig. 56.3 for scenario descriptions): *a*. coho smolts produced; *b*. estimated deer population; *c*. undiscounted, cumulative timber revenues; *d*. undiscounted, cumulative timber-management costs.

Several broad classes of research needs have been identified and now form the basis of agency research programs: (1) snow interception processes, including effects of stand and site characteristics on the levels of snow reaching the ground; (2) factors and processes contributing to the growth and survival of anadromous fish species; (3) growth responses of forests to thinning; (4) sediment loading and transport processes; (5) effects of predation and hunting on deer population dynamics; and (6) influence of lichen as a food refuge for deer in the winter.

Discussion and conclusions

Management of timber, fisheries, and wildlife in Southeast Alaska is a classic example of a wicked problem. It has all the characteristics of such problems: a complicated and interconnected system, with inherent uncertainties and ambiguities, set in an environment exhibiting social constraints and institutional conflicts. Attributes of both the simulation model and the AEAM process have helped to address these concerns.

The development of SAMM has not been without difficulties. Indeed, uncertainty and skepticism still exist concerning the potential use of SAMM in a management context. Also, the transfer of the necessary skills for using SAMM within the resource agencies has been slow. However, interagency and interdisciplinary conflicts have been reduced, research priorities have been clearly defined, and a commitment has been made to management evaluation. Embedding resource simulation model development within a proper development process does work toward resolving wicked problems.

The necessary attributes of an effective resource model development process are (1) research scientists, managers, and decision makers from all affected agencies and constituencies should be involved in model development; (2) the biophysical linkages between the natural resource components should be defined early in the process to develop a common interdisciplinary and interagency understanding about system structure and function; (3) model development should be used to identify critical uncertainties in the understanding of system dynamics; (4) the model-development process should be iterative so that new research results are incorporated to maintain it as an up-to-date management tool; and (5) the model-building process and the model should be kept open to scrutiny from all affected agencies and constituencies to maintain consensus.

Acknowledgments

It is a pleasure to thank the following persons for their perseverance and patience in the development of SAMM: L. Bartos, R. Benda, C. Bey, S. Boyle, R. Clark, R. Dewey, J. Edgington, R. Everitt, W. Farr, R. Fight, D. Flora, R. Flynn, L. Garrett, R. Gerdes, D. Gibbons, T. Hanley, R. Hilborn, P. Huberth, M. Jones, M. Kirchoff, K. Koski, D. Larson, J. Matthews, W. Meehan, J. Mehrkens, E. Merrell, M. Murphy, M. Prather, A. Puffer, J. Schoen, T. Sheehy, D. Swanston, M. Thomas, T. Webb, W. Wilson, and K. Wright. We thank R. Ellis, T. Hall, and D. Tait for their constructive reviews of the manuscript.

57

Wildlife-Habitat Planning Demonstration: Sierra National Forest

MICHAEL T. CHAPEL

Abstract.—A wildlife-habitat planning demonstration was developed for a forest-management unit in the central Sierra Nevada. A primary objective of the demonstration was to improve methods for using timber management as a tool for managing wildlife habitat. Timber-sale planning was used to demonstrate coordination between timber and wildlife habitat. A management prescription was prepared to detail general and specific direction for managing timber, visual resources, and wildlife resources in the project area. A timber-harvest plan was developed for six decades, and resource objectives were found to be compatible with careful planning. The planning demonstration establishes a method for coordinating timber, viewshed, and wildlife-habitat management within the framework of the current Forest Service timber-sale planning process. The time needed to develop succession and species-habitat models was excessive, suggesting that access to standardized models may be a key to widespread use of this approach by the USDA Forest Service.

The high costs of managing programs of direct wildlife-habitat improvement on national forest lands has stimulated several recent efforts to use timber harvest as a tool for habitat management. Holbrook (1974) related forest succession to various wildlife species and created a logging system that manipulated rotation ages of even-aged stands to manage desirable habitat. Roach (1974) recognized the relationship between deer habitat quality and forest succession. He described a general logging plan to eliminate the "feast or famine" in deer habitat for an eastern hardwood forest. Recently, Salwasser and Tappeiner (1981) produced a system for integrating timber and wildlife-habitat management for a land unit on the Tahoe National Forest. Their system produced sustained and acceptable yields of timber and habitat for five management indicator species through a five-decade planning period.

The Sierra National Forest began a demonstration project for the Rancheria Planning Unit (Kings River Ranger District) in 1981 to improve USDA Forest Service procedures for coordinating objectives for timber and wildlife resources. Specific project objectives were (1) to develop site-specific methods for using timber harvest and regeneration as long-term tools for habitat management; and (2) to test the on-the-ground application of the Forest Service's Land and Resource Management Plan. Forest Service timber-sale planning was used to address these project objectives.

In this chapter, I summarize the wildlife-habitat planning demonstration and focus on the process developed for integrating wildlife-habitat objectives into the timber-sale planning program. Specific details regarding habitat analyses and management recommendations can be found in Chapel et al. (1983).

MICHAEL T. CHAPEL: USDA Forest Service, Sierra National Forest, 1130 O Street, Fresno, California 93721

Project area

The 2600-ha (6400-acre) Rancheria Planning Unit was selected for the planning demonstration because it was an unlogged area under consideration for future timber harvest. The planning unit is bordered by the John Muir Wilderness on the east, commercial forest lands on the north and west, and dense chaparral to the south. Project-area slopes are gentle to moderate, and elevations range from 2100 to 2800 m (6900–9200 feet).

California red fir (*Abies magnifica*) dominates the timber stands in the Rancheria Planning Unit, but small tracts of mixed-conifer and Jeffrey pine (*Pinus jeffreyi*) occur at lower elevations. A jeep trail forms the southern boundary and provides the only vehicle access into the planning area. Two important deer fawning areas are located in the project area. Numerous meadows, riparian areas, and large tracts of old-growth forest provide excellent habitat for many other wildlife species.

Project development

THE MANAGEMENT PRESCRIPTION

The management prescription lists general and specific direction for managing resources in the project area. The prescription was developed by researching management direction for the project area, determining the principal resources there, and coordinating resource needs through an interdisciplinary planning approach. Integration of wildlife-habitat objectives into the management prescription was accomplished through the five tasks listed in sequence below.

Task 1: Assessment of management direction

An assessment of management direction consisted of a review of prevailing forestwide and regionwide documents that applied to the project area. The assessment of management direction produced an understanding of current policy and standards for the project area. Although detailed direction for managing timber and visual resources was found during the review, specific direction for project-area wildlife was lacking.

Task 2: Project-area inventory

An inventory of project-area resources was completed concurrently with the assessment of management direction. A base map and a series of overlays were prepared to display the characteristics of various resources in the project area. Included on overlays were timber-stand information (e.g., age, type, and size), watercourses, wildlife habitats, key areas for mule deer (*Odocoileus hemionus*), recreation trails, the off-road vehicle routes, and viewsheds from important trails and roads. Wildlife-habitat overlays were made with the labels of Verner and Boss (1980), as interpreted from timber-type labels. This effort produced an understanding of the quantities, qualities, and distribution of various resources in the project area.

Task 3: Determination of targets for resource management

Because prevailing planning direction established timber and visual resource outputs to the project area, both resources were included as management targets for the demonstration project. This task was necessary because no direction for selecting wildlife-management targets was presented in the forest plan or elsewhere.

The planning group used results from the inventory of wildlife resources to divide the project area into sections with two different wildlife targets. The mule deer was selected as a management target in the two deer fawning areas because the fawning areas were regarded as significantly important habitat for a subunit of the North Kings deer herd. Local wildlife biologists also considered mule deer to be good indicators of habitat quality for many wildlife species associated with early and middle stages of forest succession.

In the remaining area, a guild of birds associated with mature forests was chosen as the management target. Mature forest species were selected to balance wildlife targets between animals associated with early and mature stages of forest succession. A management guild was chosen in favor of a single management indicator species because of concern over the ability to choose an appropriate indicator species and the comparable costs of monitoring single vs multiple species of birds (Verner 1984).

The guild for this project was composed of the red-breasted sapsucker (*Sphyrapicus ruber*), hairy woodpecker (*Picoides villosus*), white-headed woodpecker (*P. albolarvatus*), black-backed woodpecker (*P. arcticus*), northern flicker (*Colaptes auratus*), pileated woodpecker (*Dryocopus pileatus*), mountain chickadee (*Parus gambeli*), red-breasted nuthatch (*Sitta canadensis*), brown creeper (*Certhia americana*), golden-crowned kinglet (*Regulus satrapa*), and western tanager (*Piranga ludoviciana*). The species chosen for the guild are those that find optimum breeding conditions in mature mixed-conifer forests, red fir forests, or both (Verner and Boss 1980).

Task 4: Development of habitat objectives for target species

Since forest planning direction for the Sierra National Forest adequately detailed management objectives for timber and visual resources, this task was limited to developing habitat objectives for wildlife-management targets. Literature regarding species-habitat relationships was combined with expert knowledge about target wildlife by the planning-team biologists. An extensive literature search was not done because of time and manpower constraints. Some of the information was used to make direct management recommendations about song posts, snag characteristics, and other habitat elements. Other information was organized into graphic displays showing reported or expected relationships between habitat parameters and their effect on habitat value for target species (e.g., Fig. 57.1). Expansion of these data to population responses was not attempted.

The graphic displays for deer and the bird guild were summarized into a one-page display of relationships between important habitat parameters and habitat values for deer (Table 57.1) and the guild of birds. Similar habitat summaries have been developed and termed "habitat-capability models" by Hurley et al. (1982). As an example of this process, the data for herbaceous, shrub, and tree habitat components in Figure 57.1 were summarized as the last item in Table 57.1. Habitat values ranging from 0.8 to 1.0 on the graphic displays were considered optimum quality and served as habitat objectives (Table 57.1). Values of 0–0.3 and 0.4–0.7

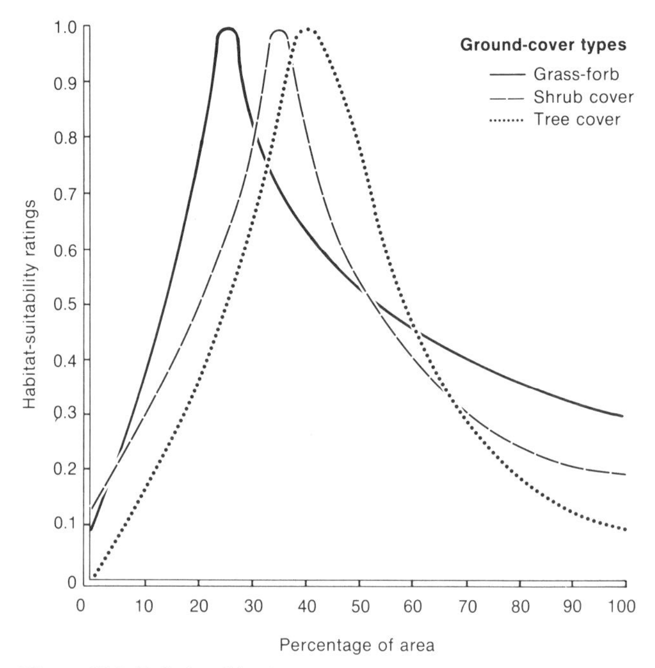

Figure 57.1. Relationships between ground-cover types and habitat value for mule deer.

Table 57.1. Summer-range habitat-capability model for mule deer on the Sierra National Forest

Habitat parameter	Optimum	Moderate	Low
Aspect (degrees)	0–60; 300–360	61–120; 240–299	121–239
Slope (%)	0–15	16–60	61–100
Distance from forage area to escape cover (m)	0–50	51–100	> 100
Distance to water (km)	0–0.8	0.9–1.6	> 1.6
Patch size for escape cover patch (ha)	8–12	4–7; 13–17	0–3; > 17
Patch size for forage areas (ha)	0.4–1.2	1.5–4	> 4
Percent forage and cover areas			
Herbaceous forage (% area)	20–30	10–19; 31–50	0–9; > 50
Shrub forage/cover (% area)	30–40	15–29; 41–60	0–14; > 60
Tree cover (% area)	30–50	15–29; 51–60	0–14; > 60

were considered low and moderate quality, respectively, and were not used as habitat objectives. Therefore, the habitat objectives in deer fawning areas were roughly 25% herbaceous, 35% shrub, and 40% tree cover.

Task 5: Interfacing resource needs

The principal tools for developing site-specific wildlife-habitat recommendations for the management prescription were the habitat objectives from task 4. The objectives were discussed by an interdisciplinary team, and management prescriptions were negotiated to balance habitat objectives with management objectives for other resources. For example, the data for escape cover patch size in Table 57.1 were used to prescribe management that provides 8–12-ha (20-30-acre) cover blocks near the end of the rotation. To develop management direction that yielded long-term balance of herbaceous, shrub, and tree components, forest-succession models were also developed and evaluated by the planning team. Direction was developed to address all habitat objectives for wildlife-management targets. The management direction for all resources was then combined to form the management prescription.

THE HARVEST PLAN

The harvest plan was developed in two phases intended to simulate application of the management prescription to an on-the-ground timber-sale planning project.

Phase 1: The unit layout

Using direction in the management-prescription and resource-inventory data, transportation engineers planned a road system and logging engineers prepared a layout of all potential harvest units for the project area (Fig. 57.2). The resource-inventory data yielded information on quantities and distribution of target resources. The management prescription provided direction for planning roads, harvest-unit sizes, and harvest-unit shapes in a manner consistent with the combined objectives for target resources. Two of the most important wildlife objectives considered during this work were (1) to emphasize large (8–12-ha) cut blocks, where the guild of birds was the management target; and (2) to provide small (1–4-ha) cut blocks that can later aggregate to large (8–12-ha) cover blocks for deer. The unit layout provided the basis for development of the harvest schedule.

Phase 2: The harvest schedule

To develop a harvest schedule, the planning group first determined the size, timber type, and wildlife-management target for all potential harvest units in the planning area. Selection of harvest units for the first logging entry was then made by an interdisciplinary team, using the project management prescription and timber-output targets as planning direction. Timber outputs were expressed as priority areas for harvest and total acreage targets. The management prescription provided direction for dispersing harvest areas and choosing treatments that were consistent with visual and wildlife-resource objectives.

The interdisciplinary team planned the first harvest entry by selecting an array of treatment areas that met the direction for timber outputs and habitat objectives in the management prescription. Examples of wildlife-habitat objectives that were used for harvest-unit selection are (1) to spread treatment units rather evenly across the planning area where the guild of birds was the management target; and (2) to harvest small cut blocks in groups that mature into aggregates of large cover blocks for deer.

To evaluate the application of the management prescription over a longer planning horizon, the planning group also developed a harvest plan for six decades (Fig. 57.3). During this effort, the management prescription and anticipated timber-output targets were used to develop consecutive timber sales for the planning area. In each of the six decades, the timber output targets of 22–36 mmbf (million board feet) timber and 162–263 regenerated hectares (400–650 acres) were met. The wildlife-habitat objectives described in the prescription were also met during each decade. The planning demonstration was not extended beyond six decades because the planning team agreed that the pattern observed could continue indefinitely.

Project evaluation

The planning demonstration suggests that coordinated timber management can produce desirable habitat quality for target wildlife because prescribed objectives were maintained for both resources. Further, coordination can occur without changing the Forest Service timber-sale planning process. Changing the role of the wildlife manager on the

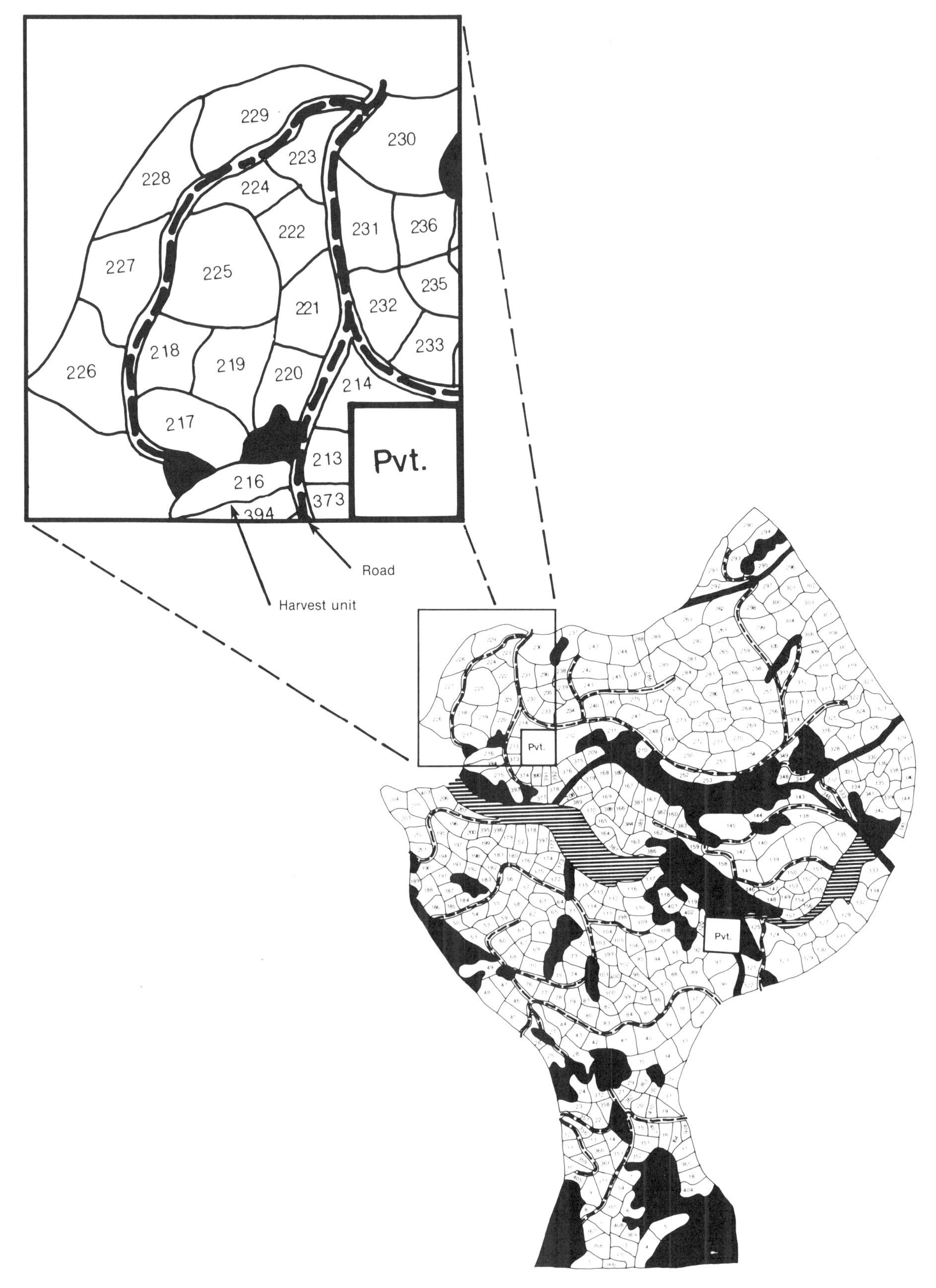

Figure 57.2. The unit layout for the Rancheria Planning area.

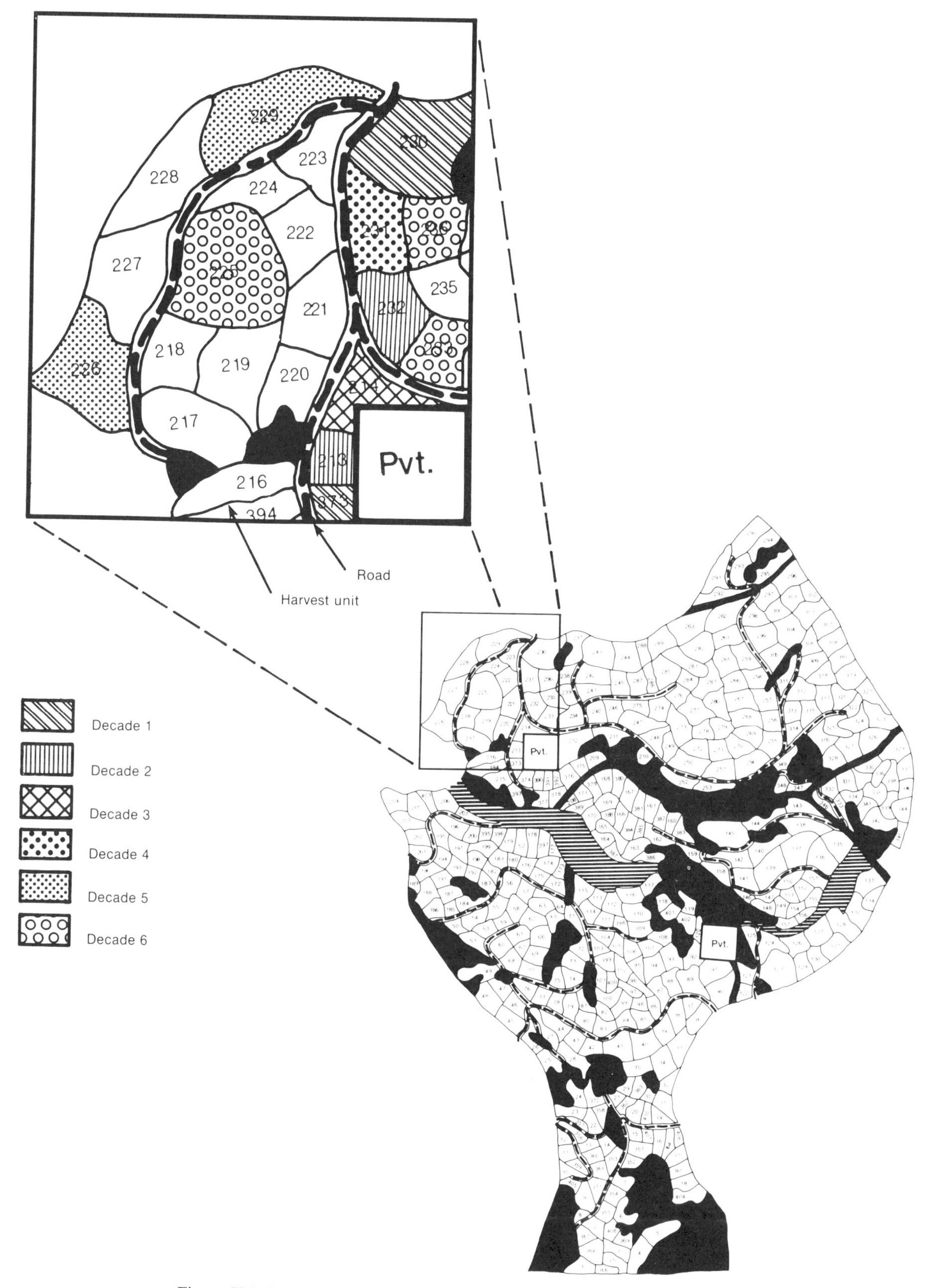

Figure 57.3. The harvest plan for six decades in the Rancheria Planning Area.

planning team was critical for integrating wildlife-habitat needs with timber-sale planning.

On typical Forest Service planning teams, the wildlife biologist's input is limited to discussions of impacts on associated species. For this planning demonstration, however, wildlife biologists assumed a new role by (1) systematically selecting wildlife-management targets for the project area; (2) determining habitat needs for target wildlife; and (3) finding methods for achieving habitat needs through coordinated timber-management activity.

These changes appeared to be essential for proper resource coordination, and biologists on the Sierra National Forest have adopted this new approach to timber-sale planning. Unfortunately, Forest Service biologists must assume a substantial workload to adopt this new planning approach. For example, wildlife biologists typically find that regional or forest direction for selecting specific wildlife targets for most areas is lacking. For the Rancheria Planning Unit, the results of local research and the habitat matrices in Verner and Boss (1980) provided the necessary species-habitat-relationships information for selecting appropriate management targets. The habitat matrices are applicable for most areas on the west slope of the Sierra and elsewhere, and their use is recommended if adequate local study data are lacking.

Another essential element for changing the role of wildlife planners is the development of habitat objectives for target species. Direction for making such assessments is also usually lacking. For this project, a cursory literature review and various personal communications yielded information regarding wildlife targets and habitat requirements. Organization of the information into simple graphic displays and habitat-capability models provided yardsticks from which wildlife-habitat objectives were developed. The development of habitat-capability models is highly recommended for establishing wildlife-habitat management objectives elsewhere.

The final essential step for changing the role of wildlife biologists on timber-sale planning teams is the integration of habitat needs with timber management. For this project, methods for managing desirable wildlife habitat were developed by presenting and discussing organized habitat objectives with the interdisciplinary planning team. This approach can easily be incorporated into other Forest Service timber-sale planning efforts because the interdisciplinary planning approach has been well established. The interdisciplinary team's success at coordinating timber and wildlife-habitat management will largely depend on the quality of habitat objectives provided by the wildlife biologist.

Although the project demonstrated how habitat objectives can be developed and linked with timber management, our process requires at least a month of intensive work by a wildlife biologist. Since few biologists can afford this time commitment for individual planning projects, further modification of the process is needed.

Two principal modifications would significantly improve the planning process. Development of standard species-habitat-relationships models for indicator species and management guilds would greatly enhance planning work. The use of standard models would eliminate time-consuming model-development work for individual projects and would provide the baseline information for developing habitat objectives. Development of habitat-succession models for the entire Sierra National Forest would also improve coordinated resource-management efforts. Standard succession models for each major habitat type would eliminate the need to develop models for individual projects. Succession models could also be integrated with species-habitat-relationships models to estimate population responses to proposed projects. The development of standard succession and species-habitat-relationships models may be the key to widespread habitat coordination in the Forest Service.

Acknowledgments

Hal Salwasser provided vital assistance in organizing and supporting this project. Mark Smith, Ken Sonksen, Grace Terrazas, and John Lorenzana were the other principal members of the planning team. Jerry Verner helped select target species for wildlife management; Rod Kindlund prepared the graphics; and Bill Laudenslayer, Dan Airola, and Clint McCarthy, reviewed the manuscript. My sincere thanks to all.

58

Habitat Evaluation Procedures as a Method for Assessing Timber-Sale Impacts

JOHN P. DOERING III and MARIANN B. ARMIJO

Abstract.—The Habitat Evaluation Procedures (HEP) were evaluated for their suitability to assess timber-sale impacts on wildlife habitat. Habitat Suitability Index (HSI) values used in HEP were determined at three intensities of inventory: aerial photo interpretation, field estimation, and field measurement of habitat variables. Aerial photo interpretation showed less deviation (5%) from field measurement than did field estimation (12%). Developing HSI models and performing calculations consumed 56% of the time required to implement HEP. New computer software can reduce the analysis phase to <2% of the time required by this study. Models developed for one study can be used on subsequent studies in the same area. HEP performed satisfactorily as an assessment methodology.

The need for rapid and accurate assessment of timber-sale impacts on wildlife habitat has grown over the years as demand for timber products has increased. A standardized, quantitative, and comprehensive assessment methodology is mandated by the National Environmental Policy and National Forest Management Acts. The methodology should display effects of each timber-harvest alternative through time in a manner easily understood by resource managers. Biologists currently use a variety of systems that may be effective but are not standardized. A standardized methodology would provide a common understanding of recommendations and would improve communication among biologists and decision makers.

Habitat Evaluation Procedures (HEP) (Fish and Wildlife Service 1980b), constitute an impact-assessment methodology for evaluating and predicting effects of resource projects on wildlife habitat. The HEP methodology uses species-habitat models to generate a Habitat Suitability Index (HSI), which ranges from 0 for valueless habitat to 1 for optimal habitat. Habitat Suitability Index values for each species of concern are multiplied by the area of each habitat inventoried to yield habitat units (HUs), a measure of habitat quality and quantity. Habitat units are assumed to be related directly to the carrying capacity for each species. Annualization of HUs allows for prediction of average annual changes in habitat values over the project life.

HEP has been used for assessing water-resource projects (Schamberger and Farmer 1978), surface mining (Rhodes et al. 1983), and project planning on wildlife areas (Urich and Graham 1983). This study evaluated the application of HEP for a USDA Forest Service timber sale.

JOHN P. DOERING III: USDA Forest Service, Eldorado National Forest, 100 Forni Road, Placerville, California 95667. *Present address:* Placerville Ranger District, 3491 Carson Court, Placerville, California 95667.

MARIANN B. ARMIJO: USDA Forest Service, Eldorado National Forest, 100 Forni Road, Placerville, California 95667.

Study area and methods

The Lincoln Timber Sale on the Eldorado National Forest was chosen to evaluate HEP. The sale was situated on the west slope of the Sierra Nevada approximately 27 km (17 miles) southeast of Placerville, California, at elevations from 1260 m to 1530 m (4200–5100 feet). Field data were collected during spring of 1984. The 846-ha (2115-acre) sale area was classified into five habitat types by aerial photo interpretation (see Verner and Boss [1980] for habitat descriptions and Hurley and Asrow [1980] for habitat codes): (1) mixed-conifer stands >15 m (50 feet) in height with 40–100% canopy closure (MC 4B & 4C); (2) mixed-conifer stands 6–15 m (20–50 feet) in height with 40–100% canopy closure (MC 3B & 3C); (3) mixed-conifer stands >15 m (50 feet) in height with 0–39% canopy closure (MC 4A); (4) plantations <6 m (20 feet) in height (MC 2); and (5) riparian deciduous stands at low elevation (RIPR-L).

Lists of candidate evaluation species were derived from Verner and Boss (1980), and a guilding process (Short and Burnham 1982) was used to select 13 evaluation species (Table 58.1). Species were selected for each habitat type according to the following criteria: (1) numbers of feeding and reproductive guilds represented; (2) limitation of geographic range; (3) potential sensitivity to project impacts; (4) availability of habitat data; and (5) emphasis status, e.g., whether endangered or threatened (Verner and Boss 1980). Of the 13 HSI models used, seven new models were developed, four existing models were modified to fit local conditions, and two existing models (Laymon et al. 1985; Schroeder 1983) were used without modification. Model development followed standards established by the Fish and Wildlife Service (1981a). None of the models was tested against field data.

Mean HSI values for each evaluation species were determined using three intensities of inventory: (1) estimation by interpretation of 1:24,000-scale color aerial photographs dated 1980; (2) visual estimation while traversing the study area; and (3) field measurement of specific habitat parameters. Sample plots were selected for the third level of inven-

Table 58.1. Evaluation species for the Lincoln Timber Sale study area and the habitats (indicated by X) in which they were evaluated, showing height and canopy cover, with the evaluation codes of Hurley and Asrow (1980) given in parentheses

	Mixed conifer				
Species	<6 m (MC 2)	6–15 m 40–100% cover (MC 3B & 3C)	>15 m 0–39% cover (MC 4A)	>15 m 40–100% cover (MC 4B & 4C)	Riparian
Long-toed salamander (*Ambystoma macrodactylum*)					X
Pacific tree frog (*Hyla regilla*)	X	X	X	X	X
Mountain quail (*Oreortyx pictus*)	X		X		
Spotted owl (*Strix occidentalis*)		X		X	
Calliope hummingbird (*Stellula calliope*)			X		
Northern flicker (*Colaptes auratus*)	X		X		
Pileated woodpecker (*Dryocopus pileatus*)		X		X	
Trowbridge's shrew (*Sorex trowbridgii*)		X		X	
Douglas' squirrel (*Tamiasciurus douglasii*)		X		X	
Mountain pocket gopher (*Thomomys monticola*)	X		X		
Long-tailed weasel (*Mustela frenata*)	X				X
Striped skunk (*Mephitis mephitis*)	X	X	X	X	X
Mule deer (*Odocoileus hemionus*)	X	X	X	X	X

tory, using stratified random sampling without replacement. Plot size was negatively correlated with density of understory vegetation in each habitat. Plots were 0.12 ha (0.3 acre) in more open MC 4B & 4C, MC 3B & 3C, and MC 4A habitats; 0.04 ha (0.1 acre) in MC 2 habitat; and 0.004 ha (0.01 acre) in RIPR–L habitat. Habitat parameters were measured as recommended in the HSI models.

Field data were used to calculate HSI values manually for each habitat to provide the assessment for target year 0. Changes in habitat variables resulting from harvest treatments were predicted for target years 1, 5, 10, 20, and 40. Habitat Suitability Index values and HUs were computed from predictions based on examples of actual stands reflecting each target year. Net change in habitat suitability was determined for each alternative treatment by comparing baseline habitat conditions with predicted conditions.

Of the seven harvest alternatives evaluated, the first alternative, "no action," was the baseline assessment. Although no harvest activity occurs with this alternative, predictions were made for successional changes in habitat parameters for each target year. The six remaining alternatives were combinations of three treatments: (1) clearcutting with herbicide use to release seedlings from vegetative competition; (2) clearcutting without herbicide use; and (3) overstory removal without herbicide use. Several assumptions were made to predict changes in habitat parameters associated with these treatments: (1) target year 1 commenced with road construction; (2) logging, site preparation, and planting would occur during target years 1–4; (3) herbicide use would start in target year 5; (4) use of herbicides would yield an 80% kill of competing vegetation; (5) soil duff layer would be removed during site preparation; and (6) riparian habitat would not be altered by any treatment.

Results

Field measurement of habitat parameters was assumed to reflect habitat conditions accurately and was used as the standard of comparison for aerial photo interpretation and field estimation (Table 58.2). Deviation in HUs from field measurement values was computed for each evaluation species. Average species deviation was 37% for field estimation and 16% for aerial photo interpretation. Deviation dropped to 12% for field estimation and 5% for aerial photo interpretation when HUs for all species were totaled and compared with the field measurement value.

Time required to assess the Lincoln Timber Sale was divided into three activities: pre-field, field, and analysis. Pre-field activities consumed 64% of the time required to implement HEP, compared with 14% for field activities and 22% for analysis activities. Model development constituted 58% of the pre-field activity, compared with 20% for study design, 19% for selection of evaluation species, and 3% for defining the study area. Field measurement of habitat variables constituted 72% of the field activities, compared with 23% for field estimation and 5% for photo interpretation. Computing weighted HSI values constituted 53% of the analysis activity, compared with 28% for computing HSI

Table 58.2. Display of habitat units for evaluation species of the Lincoln Timber Sale for three intensities of analysis in target year 0

Evaluation species	Photo interpretation	Field estimation	Field measurement
Long-toed salamander	11	35	43
Pacific tree frog	444	783	783
Mountain quail	524	378	514
Spotted owl	870	903	958
Calliope hummingbird	332	354	547
Northern flicker	825	572	815
Pileated woodpecker	870	892	1046
Trowbridge's shrew	760	782	694
Douglas' squirrel	980	859	793
Mountain pocket gopher	611	388	660
Long-tailed weasel	29	107	27
Striped skunk	1354	1291	1375
Mule deer	1481	1100	1312
Total	9091	8444	9567

Table 58.3. Predicted changes in habitat units resulting from alternative 1 (clearcut with herbicide treatment) for evaluation species on the Lincoln Timber Sale

	Target year					
	0	1	5	10	20	40
Long-toed salamander	43	1	1	1	1	1
Pacific tree frog	783	741	741	762	790	783
Mountain quail	514	520	520	712	797	797
Spotted owl	958	870	870	870	969	968
Calliope hummingbird	547	531	572	555	459	443
Northern flicker	815	712	531	393	775	733
Pileated woodpecker	1046	958	958	958	958	947
Trowbridge's shrew	694	639	639	639	804	848
Douglas' squirrel	793	749	749	760	771	782
Mountain pocket gopher	660	701	733	797	775	701
Long-tailed weasel	27	27	28	28	28	28
Striped skunk	1375	1566	1566	1566	1566	1587
Mule deer	1312	1270	1289	1291	1312	1375

baseline values, 13% for predicting changes in habitat parameters, and 6% for computing average annual HUs.

The application of HEP to the Lincoln Timber Sale required 78 working days. Development of HSI models and calculation of HSI values, HUs, and average annual HUs accounted for 56% of the total time required to implement HEP. Developing HSI models was the single most time-consuming task, requiring 28.6 working days to develop or modify 11 models. Calculations involved in the analysis of field data required 14.8 days for completion.

Changes in HUs over the project life were determined for all alternatives. An example of results is shown for one alternative, clearcutting with herbicide use (Table 58.3). Most harvest activity (61%) would occur in mature, mixed-conifer, sparsely stocked (MC 4A) habitat. Species using MC 4A habitat, such as the calliope hummingbird, common flicker, and mountain quail showed a greater change in HUs than species using other habitat types, e.g., mule deer and spotted owl. The long-term effects of the alternatives on mule deer and spotted owl are shown by calculating average annual HUs (Table 58.4) The spotted owl showed a more variable response that did mule deer.

Forest Service personnel, including the interdisciplinary team evaluating the timber sale, were given a training session to provide an overview of HEP and its application. The results of the HEP analysis were offered for inclusion in the environmental assessment of the timber sale. Habitat units were described as representing both habitat quality and quantity and were assumed to be related directly to habitat carrying capacity. Resource managers felt that relating HUs to potential animal abundance was an important aspect of HEP. Consensus was that the analysis was understandable and that the comparison of HUs would provide a clear choice for the decision maker.

Discussion

The Fish and Wildlife Service (1980b) estimates that aerial photo interpretation would require 0.5 working day/800 ha. Our result of 0.5 day for 846 ha supports this estimate. Results obtained through aerial photo interpretation were close to those obtained by measuring habitat variables. This level of inventory appears adequate for preliminary assessments when time is limited. Caution is required, however, when using aerial photographs to assess wildlife habitats. Some habitat variables, such as litter depth or downed-log density, cannot be estimated from aerial photographs and estimates may be inaccurate. Accuracy of aerial photo interpretation can be increased by using recent, large-scale photographs and, in some instances, color infrared photographs. Estimating habitat values as a function of structure may be more accurate than estimating habitat parameters and may improve the utility of aerial photo interpretation.

The Fish and Wildlife Service (1980b) suggests a minimum of 2.0 working days to develop a species model. Model development averaged 2.6 days/model in our study. Although model development took a large part of the time required to implement HEP, this is a minimal amount of time to develop an adequate or reliable HSI model. Managers often must develop models by literature review, and model verification is limited because of time constraints. For HEP to be a useful tool, HSI models need to be available and ready to use.

Urich and Graham (1983) reported that calculations for their application of HEP required 28 working days and suggested that the process could be streamlined. HEP is an accounting system for HSI values, and a computer program exists to perform some of the required calculations. However, existing software requires manual calculation of all values except average annual HUs. Software capable of reducing analysis to 1 day of data entry is being developed by

Table 58.4. Comparison of average annual habitat units for mule deer and spotted owl on the Lincoln Timber Sale: a no-action alternative (without harvest) compared with each of six harvest alternatives. (Net changes in habitat units are given in parentheses.)

	Mule deer	Spotted owl
Without harvest	1333	1020
With harvest		
Alternative 1: clearcut with herbicides (94 ha)	1319 (−14)	934 (−86)
Alternative 2: partial cut without herbicides (203 ha)	1338 (+5)	809 (−211)
Alternative 3: clearcut with herbicides (129 ha)	1302 (−31)	841 (−179)
Alternative 4: clearcut without herbicides (101 ha); partial cut without herbicides (559 ha)	1340 (+7)	706 (−314)
Alternative 5: clearcut with herbicides (203 ha)	1324 (−9)	666 (−354)
Alternative 6: clearcut without herbicides (203 ha)	1353 (+20)	637 (−383)

the Fish and Wildlife Service (R. Easterbrooks, pers. comm.) and independently by Pacific Gas and Electric Co. (M. Fry, pers. comm.). The development of improved software will greatly increase the utility of HEP.

We found HEP compatible with the Forest Land Management Plan because existing timber and wildlife data bases can be used with little or no modification. Moreover, the results of HEP analyses can be expressed in terms of carrying capacity and economics, which are important when considering the long-term consequences of a particular plan alternative. HEP has the potential to enhance future Land Management Plan analyses by providing the framework for monitoring habitat structure and diversity, an ability that is not currently available.

Our analysis provided reasonable results. For example, it indicated that mule deer are not as sensitive to the harvest alternatives as spotted owls. Mule deer are habitat generalists, and alteration of one habitat type can be compensated for by their ability to use other habitats. Spotted owls specialize in using older, mature stands and are less adaptable to the harvest alternatives. These findings are compatible with current knowledge of these species and are an indication that HEP can accurately display animal response as predicted by HSI models (Laymon et al. 1985; H. Salwasser, pers. comm.). One advantage of using HEP is the ability to display impacts on one species or group of species through time. The predictive power of HEP can be improved by integrating vegetation-succession models, such as Prognosis (Wykoff et al. 1982) or DYNAST (Boyce 1978).

We found HEP to be a useful application of HSI models. Some caution in the use of HEP is warranted, however. Species HSI models are the basis of HEP, and they predict habitat use. Reliability of results in an assessment is related directly to the validity of the HSI models. In addition to model validity, the outcome of an evaluation is greatly influenced by characteristics of the species selected.

Development of habitat criteria, stand-management prescriptions, and monitoring are three principles needed for integrating wildlife management into forest management (Salwasser and Tappeiner 1981). HEP offers a framework for accomplishing all three. Species HSI models are used in HEP to document and quantify habitat criteria and are the basis for generating stand prescriptions to meet wildlife needs. Monitoring wildlife habitats can be accomplished by using HEP accounting software. The optimum decision in resource management is one that maintains the greatest range of future options within the constraints of current needs. The current needs of wildlife can be quantified with HEP to ensure that meaningful recommendations are offered to managers.

Acknowledgments

We wish to thank J. Tartaglia for her support and encouragement throughout this project. Special thanks to B. Tannehil, P. Minton, M. Halterman, and D. Fortna for aiding in model development and performing calculations and to C. Bonser for editorial help. The authors also acknowledge the following people for reviewing the manuscript: W. Laudenslayer, M. Fry, R. Risser, B. Mapes, and R. Morat.

Summary: Linking Wildlife Models with Models of Vegetation Succession—The Manager's Viewpoint

KENNETH E. MAYER

Modeling wildlife responses to changes in habitats involves a marriage between science and management. Pressures on wildland resources resulting from human population expansion into rural areas and increased economic exploitation have created serious problems for today's wildlife managers. Therefore, predictive tools that can provide a realistic view of vegetation change through time and space, and any associated effects on wildlife, are keys to successful resource-management planning.

The chapters in this part suggest that systems developed to make wildlife predictions should consist of five major components: (1) habitat-vegetation data (inventory information); (2) vegetation-succession models; (3) an information base upon which to develop wildlife models; (4) models that relate wildlife to vegetation and seral stages; and (5) integrated computer systems for data manipulation and predictions using models. It is not enough, however, to simply adopt a system and apply it to management questions without first evaluating the assumptions upon which the various components are based. Furthermore, one should not assume that each component can be easily or reliably linked. Linkage problems can be the overriding factor in determining the reliability of such a system.

So often managers, in their haste to adopt a tool to solve a pressing problem, forget (or are blind to) underlying assumptions or problems involved with linking the system, or both. Furthermore, when problems of component integration are identified, such as the incompatibility between wildlife models and inventory data, steps are seldom taken to collect data according to the proper assumptions and definitions useful to all disciplines (e.g., forestry, range, wildlife, and so on). Often, interpretations of data are made in spite of problems with component incompatibility; indeed, examples may be found in this part of the present volume. Thus, little has been offered in the chapters in Part V to solve "linkage" problems or to improve approaches for collecting and analyzing compatible inventory data. Instead, the various chapters generally addressed how existing tools might be used, without taking into consideration the associated problems.

Some discussion is needed on each component of the process in order for this part to be complete. In this summary, I address other management tools available for developing an integrated system and the major problems and cautions we managers must recognize. Specific topics include ideas and cautions for collecting, analyzing, and linking data and wildlife-modeling efforts, and a discussion of the relative strengths and weaknesses of the chapters presented in Part V.

KENNETH E. MAYER: California Department of Forestry, 1416 Ninth Street, Sacramento, California 95814

System description

Inventory data

Habitat-vegetation data or land-type-inventory data must be similar in definition and collected with standard techniques. Single-purpose vegetation data (forest-inventory data extrapolated for use in wildlife assessments) are often inappropriate inputs for predictive modeling for management purposes. One cannot expect succession or Habitat Suitability Index (HSI) models to perform well when the vegetation variables are not appropriate for the intended model. Furthermore, inventory information needs to be integrated for a number of resources in order that data can be more efficiently gathered. Therefore, it is imperative that thought be given to the development of a list of standard definitions useful for the forester and the wildlife biologist. Finally, data need to be collected and analyzed, using standard techniques to ensure compatibility. For example, foresters may use quadratic-mean equations to calculate mean tree sizes, and wildlife biologists may use arithmetic-mean equations to calculate the same structural classes—an obvious discrepancy that may create linkage problems.

Vegetation models

Models of vegetation succession have been available in the literature for many years (Hawkes et al. 1983). These vary in structure and purpose. Each is unique and must be evaluated (i.e., in terms of assumptions, production functions, report writing, and so on) before being used by the manager. Furthermore, the scale of the management problem will also dictate which successional model will be most useful. As an example, Benson and Laudenslayer (Chapter 49) discuss the DYNAST model and its use as input to forest planning in California. DYNAST, an extensive model as described by the authors, provides predictions of future forest conditions. The model, as adapted, may provide reasonable predictions and may be adequate for their scale of planning, but problems may exist because of the underlying assumptions and the quality of data used as input for the model. Problems may include the lack of good growth and yield equations for various vegetation types, as well as the fact that the DYNAST model was developed for an even-aged management system in eastern forests. Sweeney (Chapter 50) aptly points out that with an unbalanced distribution of tree age-classes (which is common in California forests) the model unrealistically moves a timber inventory through simulated time. These problems are not insignificant when the results of the simulation run are to be applied or to be used as guidance for management actions on the ground, particularly for small-scale resource decisions. Fortunately, none of these authors has used the model for this purpose or at this scale.

As a further example, the study by Brand et al. (Chapter 54) illustrates the problem of using a forest overstory model to predict wildlife species' occurrence. Brand et al. point out that three types of model links can be used to combine succession and HSI models: (1) direct; (2) indirect; and (3) subjective. The forest-overstory model they used (TWIGS) had only one directly linked variable (number of mast-producing trees) for running the gray squirrel (*Sciurus carolinensis*) HSI model. They acknowledge that this factor could cause serious problems if results are important as accurate predictions of the presence of gray squirrels. Special caution is particularly warranted when interpretations are made from indirect and subjective model links. Model users should always describe the reliability of these model linkages, with a description of their limitations, as part of the published results.

Species-habitat-relationships models

Species-habitat-relationships models vary in complexity and appear to produce results of questionable reliability, perhaps primarily as a result of a lack of field validation. This does not mean, however, that managers should avoid using these models in their management efforts. Rather, they should be used as estimates (objective guidance) to strengthen subjective decisions (Barrett and Salwasser 1982).

Most of the studies in Part V are based upon the use of such a model, and it is important that the differences between these models be identified and that the models be used in the appropriate management situation. As a guide for managers, I suggest that we recognize three model levels, as follows:

Level 1. Relative habitat value of a habitat patch, based on species' life requisites, as met by stand conditions. The primary use is in estimates of species richness. The California Wildlife-Habitat Relationships Models described by Chapel (Chapter 57) are good examples.

Level 2. Relative capability of a mix of habitat patches and associated specific habitat elements to support a species or species group, based on all life requisites and mean home range sizes of the species. The primary use would be in featured species–habitat evaluations. HSI models, such as that for the gray squirrel used by Brand et al. (Chapter 54), represent this level.

Level 3. Habitat capability of the aggregation and distribution of capable habitats for all population subgroups within an area. Primary use would be for determining wildlife-production capability over the range from minimum viable populations to k (carrying capacity) for featured species. The PATREC model described by Kirkman et al. (Chapter 48) is the first generation of such a level 3 model.

Adequate predictions for an entire forest may be made by using level 1 models (Mayer 1984). The accuracy of these models can be high enough to provide insight into the effects on wildlife due to human-caused changes in the forest environment. Models such as these, however, are inappropriate for project-level predictions of wildlife use on a multiple-stand basis.

The level 2 models have been developed to predict more accurately the impacts of management decisions on wildlife. As many chapters in this volume show, few of these models perform as well as one would like. This should not be too disturbing to either the researcher or the manager, however, as modeling efforts are just beginning. The accuracy of many management decisions is open to question as well, and any assistance that strengthens the decision-making process is useful. It is important that the proper model (level 1 or level 2) be linked to the appropriate vegetation model and that the results be used as indications, not absolute predictions.

Finally, level 3 models, such as PATREC, which make population estimates, are precisely what the manager needs for management in the 1980s. Kirkman et al. (Chapter 48) describe how future population-density potentials can be estimated and used within a management framework. These models, however, still need much refinement, and we must have better population data to use them accurately.

Integrated computer systems

Integrated computer systems for data manipulation have been described briefly in Part V. A Geobased Information System (GIS) is a commonly recognized system with a computerized resource data base that is derived from and can be related back to a ground base (Mead et al. 1981). A GIS includes a set of procedures for assessing and modeling data to describe and predict resource dynamics. The power of a GIS is its ability to spatially track vegetation through time and apply various models to spatial data. Furthermore, cumulative effects of management are more easily evaluated with a GIS. As managers, we must move toward this technology as an answer to many of our "real time" resource problems. For example, the chapter by Davis and DeLain (Chapter 51) discusses an approach to a GIS and describes the possible benefits to wildlife management. The ability to ask the "what if" questions, with answers in the form of maps and statistics, gives credibility to management decisions. In addition, benefits such as computing speed, repeatability, and ease of updating provide additional power.

Shortcomings of Part V

Part V and this volume as a whole suffer from a lack of information on remote sensing and its applicability to wildlife-vegetation modeling. Remote sensing technology can be an integral part of either developing, using, or testing wildlife models. Remote sensing has been used extensively in many resource fields for many years as a tool to provide invaluable inventory data (e.g., Heller et al. 1964; Aldrich and Heller 1969; Walsh 1980). This tool, however, has not been implemented to any great degree by wildlife managers, especially in the development or use of models. Technologies such a Landsat multispectral scanner (MSS) or thematic mapper (TM) data may provide clues to how wildlife perceive habitats (Mayer 1984). The objective nature of MSS and TM data may allow us to go beyond the human limitation of biased interpretations. For example, using home range data as a guide for clustering multispectral data may allow the biologist to classify habitat differently, and perhaps more realistically, than traditional approaches permit (Trivedi et al. 1982). The California Department of Fish and Game, through its big game program, has had reasonable success in identifying fawn habitat from radio-telemetry home range data in conjunction with preprocessed Landsat data (Benson 1984). Interestingly, the Landsat data had been previously classified solely for forest-inventory purposes. Finally, either aerial or satellite-derived remote sensing data can provide vegetation information useful as input to wildlife-habitat models. Therefore, model development should seek to identify vegetation variables that are easily assessed or inventoried with remote sensing tools.

Conclusion

Each chapter presented in Part V gives bits and pieces that collectively describe a complete system. Ideas have been presented that provide insight and possible tools for making better informed resource-management decisions. Unfortunately, but not unexpectedly, none of the chapters has offered a cohesive answer or integrated approach to wildlife assessments. Many models for wildlife assessments have been, and continue to be, developed independently. Most of these models have used similar approaches and suffer from the same deficiencies. I believe it is time to step back and evaluate these various models. A pooling of unique ideas into a system that functions properly should be our next step.

Summary: Linking Wildlife Models with Models of Vegetation Succession— The Researcher's Viewpoint

JAMES M. SWEENEY

In a very real sense, Part V is the capstone to the theme of this volume. Previous sections addressed the various parts of an integrated modeling system. Chapters in this section address the challenge of combining those parts into an operative whole. The goal for this operative whole is to permit long-term forecasting of wildlife populations by using inferences from our knowledge of species' habitat requirements. This assumes that in our simulations of plant communities, the wildlife-habitat relationships embodied in the system are accurate enough for managers to reach reasonably accurate conclusions about the viability of future populations. This goal is obviously an ambitious one, and reliance on models to achieve it represents a bold step. Given the nascent state of this effort, perhaps the most we can expect from chapters in this part is (1) an indication that linking wildlife models with models of vegetation succession holds some promise for achieving the goal; and (2) some reasonable directions for future work in this important arena.

Chapters in this section amply demonstrate ways to link existing forest-succession models with wildlife-capability models. Brand et al. (Chapter 54) show how a widely used tree-growth projection model (TWIGS) could be coupled with an HSI model for gray squirrels (*Sciurus carolinensis*) to predict trends in squirrel populations well into the future. Raedeke and Lehmkuhl (Chapter 53) describe a model (HABSIM) that combines wildlife-habitat relationships with forest succession to produce outputs in general agreement with the results of independent studies. In addition, chapters by Benson and Laudenslayer (Chapter 49) and by Kirkman et al. (Chapter 48) report actual case histories in which DYNAST (an integrated resource-simulation model) was used to predict future values of timber and wildlife to help forest managers make better decisions. The results from these models were apparently consistent with biological judgment, as line officers accepted the procedures as a way to improve long-term planning and to give wildlife resources more attention in that planning.

Although these chapters show that physical linkage of different models is possible, and indeed valuable to managers and researchers alike, they also clearly show that information has often been lacking for the optimum linkage of forest-succession simulators to wildlife-habitat-capability models.

JAMES M. SWEENEY: USDA Forest Service, North Central Forest Experiment Station, 1-26 Agriculture Building, University of Missouri, Columbia, Missouri 65211

This is especially clear in the chapter by Brand et al. (Chapter 54). Because shrub cover is a critically important habitat variable in the squirrel HSI model, and the TWIGS model does not simulate shrub cover, Brand et al. inserted an assumed shrub cover value, which was static through time. Consequently, although they show the value of the linkage, they are uncertain about the accuracy of their results. The direction taken by Moeur (Chapter 47) to model shrub cover may provide an answer to this shortcoming of current succession models.

Although general habitat requirements and population characteristics are well documented, few studies provide cause-and-effect data to permit predictions of population numbers from forest-vegetation variables. Most models simulate the availability of potential habitat, not the population dynamics of the wildlife species per se. Such factors as the ability of wildlife populations to track changes in habitat availability in time and space, or their ability to reinvade potential habitat, are often not included (as Smith suggests in Chapter 55). The wildlife-research community must not allow the tool (modeling) to become an end in itself, but instead must continue to explore and test the underlying biological principles of wildlife-habitat relationships.

Each model represents a compilation of knowledge of the relationship modeled and as such represents a hypothesis suitable for testing. Use of the models by managers identifies those variables that most strongly influence model outcomes and that need immediate detailed research attention. Model use also defines the range of values over which the variable needs to be addressed. Testing and refinement will be complex and costly, and will probably be accomplished only through a cooperative effort between land managers and wildlife researchers. Brand et al. (Chapter 54), Moeur (Chapter 47), Raedeke and Lehmkuhl (Chapter 53), and many authors in previous parts have pointed out the difficulties associated with testing habitat models. These difficulties range from data sets that are small and/or highly variable, to differential responses to location, to the presence of independent variables (such as disease and poaching) that directly influence populations. The problems inherent in model linkage discussed above are equally important in model testing.

Not all environmental features useful for describing wildlife habitat lend themselves to mathematical modeling. Raedeke and Lehmkuhl (Chapter 53) point out that external decisions or stochastic variables, such as road densities and weather, are not particularly tractable. The habitat indices generated are themselves intangible and, in most cases, not tied to population size (Brand et al., Chapter 54). Model form and structure also pose problems. For example, Davis and DeLain (Chapter 51) show the need to consider carefully the relative merits of using geometric means or arithmetic means. Raedeke and Lehmkuhl and also Sweeney (Chapter 50) point out the bias in forest-succession simulators based on the continuous simulation approach.

An important point surfaces repeatedly in discussions about the application of these models. Model developers, usually researchers, are responsible for clearly stating all assumptions and limitations affecting each model. The chapters by Benson and Laudenslayer (Chapter 49), by Raedeke and Lehmkuhl (Chapter 53), and by Smith (Chapter 55) provide good examples of concise statements of the assumptions and limitations to consider with their models. At the same time, users are responsible for reading the assumptions, and for understanding fully the impacts and limitations on the particular use intended for any given model. Holthausen (Chapter 52), in comparing FORPLAN and DYNAST, has given an excellent example of the importance of understanding the workings within a model (assumptions and limitations) and the proper use (control) of the model.

Along with model refinement should also come continued development. Models of forest succession give the time dimension needed for our wildlife-habitat-capability models. But the important spatial aspects—patch size and juxtaposition—have been generally ignored by modeling efforts. It is commonly accepted that the juxtaposition of vegetation with other land features, such as water and topography, is an important attribute. The analytical linkup of a geographic information system will provide the spatial evaluation. Davis and DeLain (Chapter 51) argue strongly for the addition of spatially explicit variables, and in Part IV Urban and Shugart (Chapter 39) and Seagle (Chapter 40) have shown ways by which this may be possible. In the end, however, the need for spatially defined model outputs must be weighed against the significant cost of establishing and maintaing a complete, digitized, geographic-information data base.

Another underlying concern relates to this part: Model developers should produce programs that are interactive, user-friendly, and compatible with the more common microcomputers. Although researchers must elucidate the true biological, wildlife-to-habitat relationships, they should also produce a usable research product. As our data base increases, so often does the complexity of our models. Researchers must strive to maintain model simplicity and, where possible, to use the more easily obtainable habitat variables. Where complexity becomes necessary, researchers must provide interactive menus and detailed documentation to help users correctly apply models and correctly interpret their results. Successful model development, therefore, will require continued communication and cooperation between researchers and managers.

VI *Synopsis*

59

Modeling Habitat Relationships of Terrestrial Vertebrates—The Manager's Viewpoint

HAL SALWASSER

The goal of the Wildlife 2000 Symposium was to present an up-to-date synthesis of models that predict the responses of wildlife to habitat change. Specific objectives were to identify the types of models in use, explore ways to test them, review the accuracy of existing models, and examine the application of models in conservation decisions. My goals in this chapter are to provide a manager's perspective on the degree to which the studies in this volume accomplish the stated objectives and what they show about the state of the art. I will do this in three parts: (1) roles for models in the conservation decision-making process; (2) a management view of the utility of models; and (3) an assessment of the status of wildlife-habitat-relationships models, as reflected by this volume.

Roles for models in making conservation decisions

Making conservation decisions is a bit like traversing a maze. It is complex, uncertain, and full of blind alleys and intricate passageways. Players need good rules for making turns in order to solve the maze efficiently. One role of simple wildlife-habitat-relationships models is to serve as biological rules that help in making prudent conservation decisions. Models serve to improve both communication and the body of knowledge when they are expressed as explicit hypotheses rather than implicit assumptions. Another role of models is to aid in the synthesis of many parts into a whole. At any specific point in time, our activities in conservation might appear to be frantic expenditures of energy on lots of little things—like running around in the mazes: probing, testing, changing directions, without knowing, until the end, if the path we are on is the solution to the maze. Periodically someone steps back (or above the maze) and makes a logical interpretation of the new facts learned from the little actions. Such a synthesis provides a better set of rules to help in our probings. For a while our actions are better focused and more efficient. Leopold's (1933) text on game management, Thomas's (1979) book on wildlife habitats in managed forests, Soulé and Wilcox's (1980) treatment of principles of conservation biology, and Harris's (1984) application of biogeographic theory to fragmented forests are good examples of useful syntheses in our business. In this role, models are a synthesis of knowledge applied to specific kinds of issues and decisions.

HAL SALWASSER: USDA Forest Service, P.O. Box 2417, Washington, D.C. 20013

Models are useful, but they must not be viewed as permanent expressions of truth. Occasionally someone steps even further back from the conservation mazes and, looking more closely at our knowledge in relation to biological and social changes, tells us the point of exit—the goal—for the maze is changing. Of course, that means that the solution and some of the rules for reaching the goal are probably also changing. Aldo Leopold (1949) did that with *A Sand County Almanac*. Society, through Congress, validated his perception with the environmental legislation of the 1960s and 1970s. Resource conservation goals were enlarged from a concern with game and commodities to include diversity and nonconsumptive uses. The world is dynamic, and our models of how it works must be responsive to many kinds of change.

Our perception of the prevailing rules changes in proportion to the state of our knowledge. Consider two familiar reversals—paradigm shifts—in what we once thought were valid conservation rules: the wisdom of exterminating predators and the suppression of fire for ecosystem health. Although we strive to base our rules on empirical evidence and logical concepts, we occasionally make incorrect interpretations because the evidence is incomplete or our logic is faulty. Thus, the actual utility of a model may become clear only when tested over a sufficiently long period of time (Rice et al., Chapter 13), or on a large enough scale (Gaud et al., Chapter 32), or in the context of a specific management application (Kirkman et al., Chapter 48).

Extending the maze metaphor further, envision a hierarchy of mazes, with lower-level mazes being more detailed subsets of higher-level ones. Wildlife management is a lower-level maze nested within a resource conservation maze, that is nested within a maze of social and economic systems, that are ultimately nested within the biosphere maze. A single turn in a higher maze is the "effect" that results from a full set of turns or "causes" in a lower-level maze. Models depict relationships, that is, how to link things like causes and effects or inputs and outputs. They help us describe how we think the biological world functions. To be useful, the models must, in some way, conform to the rules of the biosphere and how man fits into it; that is, the search for new knowledge (science) guides the development of models, while a land ethic and politics guide their application.

I have noted three roles of models in conservation: (1) as explicit rules for making decisions; (2) as applied syntheses of the state of knowledge; and (3) as expressions of the linkages between causes and effects. How, then, should we judge the utility of a model for these roles? Is accuracy the most important rule for determining utility? Does a particular model represent the full solution of a maze or merely a decision rule for making turns? Is a particular model, in the context of possible applications, likely to lead to the solution or to a blind alley?

The utility of models

Accuracy and practicality

One of the struggles of an emerging discipline is sorting out the accomplishments—distinguishing real progress from the cloud of dust. Many chapters in this volume focused on the validity (or, better, veracity) of a particular model, or how validity can be determined. The criteria were predominantly biological or statistical. From a manager's perspective, utility is really the key issue. And utility for management must reflect both biological accuracy and immediate usefulness. Models based on obscure variables for which inventories are not available, or which derive results unrelated to the needs of a decision, are not practical.

Consider this simple example. Someone builds a logical model designed to portray the relationships of a small bird to its habitats at a watershed scale. A scientist finds that the model explains only 50% of the actual variation in occurrence of the bird. It could have been 15% or 75%. So far, so good; but do not rule out the possibility that the empirical evidence may be incomplete or at the wrong scale. The scientist then concludes that the model is not a good rule for determining the course of habitat management. He or she may even go so far as to say it is dangerous to use the model. But this might be going *too* far. The scientist used only one criterion—a good one for determining accuracy, but not for inferring utility. A valid, logically correct model with the ability to explain 15%, 50%, or 75% of the variation in occurrence or abundance of the bird on the basis of habitat conditions may be useful for convincing people to provide habitats for the bird. So a manager uses the model. Now, if the manager assumes that the model is the whole truth about how nature works because it works to effect a favorable decision, this would again be going too far. That kind of conclusion requires scientific rather than utilitarian criteria.

The point is that the utility of a model depends on many factors, including, but not limited to, biological accuracy. One major step in settling the dust and getting us on the right path is to specifically define the utility of a model, not imply that judgments about accuracy are all there is to it. Determining accuracy is the purview of the scientist; practicality, that of the manager.

If application to conservation decisions is the intent of a model, utility is the ultimate concern. We all have a stake in and a responsibility for the determination and improvement of model utility. Both scientist and manager are capable of logical thought, so building useful models should be a joint responsibility. What the models are used for also has a bearing on utility.

Use of models in management decision and control

Wildlife-management decisions aim at two kinds of broad goals: to maintain full biotic diversity or to produce specific resources for people to use and enjoy, or both. These goals must be met largely by managing wildland ecosystems, most of which have additional, nonwildlife purposes as well. Yet we can control relatively few things to reach those goals. We can exert an influence on animal populations through their vegetative habitat resources and, in some cases, through direct population management. Climate and geology are largely beyond our control, as are many of the natural competitors and predators of the resources of concern.

Management decisions have long been made under these circumstances without the assistance of quantitative wildlife-habitat models. Such models are *needed* by managers only if their efficiency in identifying opportunities or explaining the consequences of decisions offsets the cost of building and using them. All other uses or roles of models in the management arena are luxuries, not necessities.

The most important assumption, often an implied one, when modeling wildlife-habitat relationships is that some aspect of populations can be predicted from some aspect of habitats. We have known for a long time that animal populations are affected by many things other than their habitats. As chapters in this volume repeatedly document, one can rarely, if ever, be 100% accurate in predicting population levels or reproductive success of a species by using habitat variables alone. It is also axiomatic that without suitable habitats populations cease to exist. So we should always be able to explain something about populations from knowledge of habitat variables. The key question in modeling would seem to be, given various kinds of information about habitats, how much can we explain about associated wildlife populations? I recognize four criteria for the utility of models employed to aid decisions: (1) they must relate species to habitats at a geographic scale appropriate to the species' home range; (2) they must encompass habitat variables that are related to population attributes; (3) the habitat variables must include those that the management decision will affect; and (4) the degree to which population changes can be inferred from changes in the habitat variables must be specified. Obviously the higher the predictability the better, but no one expects it to be 100%, nor will a certain "validity threshold" be most efficient to all applications.

Species' occurrence models

The simplest habitat-relationships model should explain the likelihood of a species' being present in different kinds of habitats. The habitats must be classified in ways that make inventories feasible. This model would have to account for the biology of the species, especially its life needs, home range area, and seasonal movements. Many of these types of models have been developed and put into use since the mid-1970s (e.g., Patton 1978; Thomas 1979; Verner and Boss 1980; Sheppard et al. 1982). Hamel et al. (Chapter 20) and Raphael and Marcot (Chapter 21) indicate that models can be quite reliable when used to predict species' occurrences at sufficiently large geographic and temporal scales. They do not work well to predict species' occurrence in a specific site at a specific time. That is not a significant drawback for management uses of occurrence models.

Habitat-capability models

Beyond the simple predictions provided by a basic occurrence model, managers will want to know the extent to which additional population attributes can be explained by additional information on habitats and the changes that man and nature will make in them. Although managers need not, and should not, try to predict populations precisely from habitat variables alone, they do need to be able to estimate the probable changes in populations that their actions on habitats might cause. This requires a more complicated model than is needed to predict occurrence. Examples of such models, under the names Habitat Suitability Index and Habitat-Capability Model, are presented in many of the chapters in this volume.

THE NEED FOR CONSISTENCY IN DEFINING HABITATS

It is unsettling that 50 years into the business of wildlife management people still argue about what habitat is—especially because employing models to predict the effect of habitat changes on wildlife requires that people first classify habitat. Drawing from the chapters in this volume, it appears that habitat variables in three categories could be considered as a basic set for building models. First are kinds of HABITAT ELEMENTS (Short and Williamson [Chapter 16] call them substrates) that could occur on sites, e.g., snags, talus, hardwood mast, litter, large tree diameters, and shrub layer. Species could be related to each habitat element in proportion to its value in meeting life needs. The main uses of these models would be in developing detailed prescriptions for managing sites for particular kinds of species (alpha diversity), determining groups of species with similar habitat affinities (Short and Williamson's guilds, Thomas's [1979] life forms), and building more detailed species-habitat-suitability/capability models useful in assessing the potential productivity of a site.

Second are HABITAT CLASSES that represent major differences in floristic and structural characteristics of the habitat elements on sites, e.g., successional stages or habitat classes as described in Thomas (1979) and Verner and Boss (1980). Species would be related to these classes qualitatively or quantitatively as appropriate for rapid assessment of potential species richness (beta diversity) across a landscape.

Third are LOCATIONAL VARIABLES that describe area, shape, and pattern of different habitat sites. Models developed from these variables would be useful for considering the effects of biogeography on species diversity and productivity at the watershed or larger scale. If time dynamics is an issue in a decision, then successional pathways and rates, and mechanisms to deal with changes in locational variables, would need to be included in models. For certain species, it will also be necessary to know about the degree to which the presence of human activities (see, e.g., Lyon 1983) or of other animals (see, e.g., Rothstein 1982) results in interference with access to habitats.

The basic habitat variables, however they are eventually defined, will not always be equally important. Some will be irrelevant in making certain decisions or for certain species. The biology, life history, and population status of a species will determine which habitat variables are most important to that species in a certain place at a certain time. And the kinds of decisions and their geographic scope will determine which habitat variables to include in a model as the elements of management control. It should not surprise anyone that a model built without a specific use in mind may not work well for any use. Nor should it surprise anyone that a model built for use at one geographic or decision level, say, an entire watershed, may not be of much use at another, say, a single-vegetation stand (Laymon and Reid, Chapter 15).

AN ADAPTIVE MANAGEMENT FRAMEWORK FOR USE OF MODELS

So much for wildlife-habitat-relationships models per se. The context in which they will be applied is also an important determinant of utility. McNamee et al. (Chapter 56) characterize resource-management issues as wicked problems. In addition to their description of wicked problems, in the typical resource conservation maze: (1) demand for knowledge always exceeds supply; (2) future uncertainty is the rule; (3) no one factor uniquely controls the goals; (4) some cause- and-effect relationships are beyond human control; (5) nonwildlife goals may at times or places override wildlife goals; (6) limited dollars and technical resources are available to affect controls; (7) the most important effects on diversity and even single-species production often occur at landscape or larger scales; and (8) past events limit short-term opportunities to meet goals.

It would be folly to try to reduce such resource-management problems to a single model. It may be possible, but no one could comprehend it. Nor is it realistic to assume that any or all aspects of models designed to serve such problems will be linear and deterministic. The result is that models will generally portray only parts of complicated decisions. They are means to an end, but the decisions themselves are also only means. A decision is only one step in a dynamic management process—traversing the resource conservation maze.

The concept of adaptive management (Holling 1978) describes a systematic approach to these wicked problems. I suspect that it is the most rational approach. Adaptive management starts with defining specific issues and opportunities. This provides a goal-driven focus to both assessments and management. Inventories and state-of-the-art models are then used to explore opportunities and the consequences of possible decisions. The management actions themselves are considered to be management experiments. Monitoring is used to "read" the experiment, while research focuses on assumptions and relationships that require more rigorous experimental design and sampling. Finally, to complete the cycle, the course of management adapts to new knowledge gained from experience, monitoring, or research, and to changing goals.

Adaptive management does not assume that the models used in planning analyses provide definitive statements of truth. Instead, models are used to help synthesize knowledge and apply it systematically to decisions that guide management toward goals. The decisions are not considered inviolate for all times, nor are the models that are used to help make those decisions.

At least five questions must be considered when building or selecting a model for use in adaptive management. How much uncertainty can the managers accept? Which of the factors that control the goals are most significant in a specific time or place? What resource and economic trade-offs are associated with these factors? What are the technical and financial capabilities of the people and agencies making and implementing the decisions? And what proportion of the goal can be controlled by human actions? This last question is the converse of, How much of the goal is beyond our control?

The resulting models should (1) be easy to understand and communicate with (i.e., user friendly); (2) represent the major elements of control; (3) run on commonly collected inventory data; (4) provide information in terms of the outputs of concern; (5) be reliable in proportion to risks and values involved in the decisions; and (6) allow the estimation of incremental movement toward goals that is likely to result from a specific set of actions on a specific piece of land.

A view on the state of the art

Are we there? Not yet, but the chapters in this volume show that we are definitely on the way. The mere existence of the volume makes trivial the conclusion that modeling of wildlife-habitat relationships has come a long way. But we obviously still have a long way to go. In this section, I examine our progress in developing accurate models that meet a manager's need for practical tools at a scale commensurate with typical decisions.

This assessment has a definite management-application bias. I consider examples of how models were used to shape a specific conservation decision as the endpoint of a spectrum that starts with model building and proceeds through verification testing and use in environmental assessments. I have somewhat arbitrarily categorized the papers into a $3 \times 3 \times 3 + 1$ matrix. The first axis is building-testing models, use in environmental analyses, and use in shaping conservation decisions. The second axis is time- and space-static models, time- or space-dynamic models, and time- and space-dynamic models. The third axis is community models in which no species detail is present, single-species or indicator-species models that may deal with one or more species, and comprehensive community models that include detail on all species possibly present. The "1" is for papers that are purely conceptual or technical without reference to particular species or kinds of habitats.

I view the building of models of community richness, static in time and space, as the simplest task, and of models of comprehensive communities, dynamic in time and space, to shape land-use decisions as the most complex of tasks in this matrix. I do not view any particular block in the matrix as less important than any other. The number of chapters falling in each block, and their content, tell us much about the state of this modeling business. Let me start with what I believe managers would consider the ultimate goal of modeling wildlife-habitat relationships.

Uses of models to shape decisions

I categorized only four chapters as examples of using models to shape decisions. Each is a significant example of where habitat-relationships modeling must be going. Both Kirkman et al. (Chapter 48) and Toth et al. (Chapter 22) used models to project future conditions for wildlife, and then used these estimations to shape the course of intended management. The former chapter does this using indicator species and a multistand habitat-simulation model. The latter uses a comprehensive community-richness approach to specify the ultimate pattern of habitats that enhances full diversity. Chapel (Chapter 57) shows the use of a model to help shape both temporal and spatial habitat goals. Davis and DeLain (Chapter 51) have taken this a step further by discussing the use of geographic-information systems (GIS) and habitat-relationships models to determine an optimum course of management actions. Each of these chapters yields a glimpse of what are, at present, some of the most challenging and potentially rewarding developments in the modeling of wildlife-habitat relationships.

Uses of models in environmental analyses

Only a notch back of the four chapters just discussed are seven chapters that illustrate the use of models in environmental analyses. It is an easy step from these chapters to using their models or techniques to shape decisions. McLellan et al. (Chapter 44), working from a sound empirical basis and blessed with excellent analytical skills, have shown how temporal and spatial habitat dynamics can affect wildlife communities, and have derived rule-of-thumb principles for regional land-use decisions. Lancia et al. (Chapter 11) appear headed in the same direction as Davis and DeLain (Chapter 51); moreover, they offer useful tips on ways to verify models. As an aside, R. H. Barrett (pers. comm.) at the University of California, Berkeley, and D. Winn (pers. comm.) of the USDA Forest Service in Ogden, Utah, are also actively engaged in developing the linkage of wildlife with temporal and spatial dynamics models of habitats. Both of these efforts were in their infancy when the call for papers for Wildlife 2000 went out and were operational by the time the symposium was held—a good indication of the lag between actual state of the art and what appears in volumes such as this.

Two chapters in this group of seven are especially noteworthy from the manager's perspective. The aforementioned temporal and spatial dynamics models can require expensive computer hardware and software that may not be available to all potential users. The chapters by Benson and

Laudenslayer (Chapter 49) and Raedeke and Lehmkuhl (Chapter 53) illustrate time-dynamic models that can be run on personal or microcomputers. Both show the linkage of wildlife-habitat-capability/suitability models with habitat simulators. This technology is growing rapidly and is now used routinely by the Forest Service in cumulative effects analyses (R. S. Holthausen, pers. comm.). Doering and Armijo (Chapter 58) show how Habitat Evaluation Procedures of the U.S. Fish and Wildlife Service can be used when there is no access to computerized simulation programs.

The final two chapters in the group covering employment of models in environmental analyses illustrate uses of models in selecting indicator species (Fry et al., Chapter 17) and management guilds (Short and Williamson, Chapter 16) for habitat evaluations. They are both useful techniques for management biologists.

Before discussing the set of chapters dealing with building or testing models, we should think about what this number of contributions (11 out of 60) says about the symposium's objective of examining ways to apply models to conservation decisions. The chapters are enlightening but make up a rather small proportion of the volume. One might conclude that (1) managers are not spending much energy on developing ways to apply models; or (2) they are not documenting their efforts; or (3) they do not perceive much value in sharing their experiences with other practitioners through media such as this volume. I find any one of these three possibilities, and certainly any combination of them, a distressing sign of failure on the part of managers to fulfill their part of the bargain in advancing the state of the art.

Building and testing models

The bulk of this volume—44 chapters—addresses building or testing different kinds of models. This is largely a research endeavor, so I highlight here only what I believe to be significant advances from a management perspective. (Scientific importance is not implied by this assessment.) Hamel et al. (Chapter 20) show that a habitat inventory at the habitat-class level can be used over a broad geographic area, though within limits, to infer bird species richness. This is a most useful application for habitat planners. Marcot's (Chapter 23) introduction to expert systems provides a peek into the near future on user-friendly models. Raphael and Marcot (Chapter 21) provide precisely the kind of information on model veracity that a manager wants: how well the model works and where it is likely to be weak. Their empirical data base is, I suspect, the minimum in terms of time, space, and size to allow such inferences, though many chapters in this group of 44 have attempted to assess accuracy with fewer data. I would tend to discount the conclusions from impauperate data bases, regardless of whether they confirm preconceptions.

Haila (Chapter 45) gives sobering evidence of the need to consider scale in understanding the effects of habitat fragmentation. Rotenberry's study (Chapter 31) provides a fine and lucid illustration of why managers need to consider more than habitat in predicting future populations. Gaud et al. (Chapter 32) show the importance of both time and scale in verification studies, an illustration possible only with the kind of data sets available to precious few authors in this volume.

Because of my concern for the need to portray both the time and space effects of habitat change on wildlife, I am particularly impressed by the contributions of Seagle (Chapter 40), Urban and Shugart (Chapter 39), O'Connor (Chapter 34), and Holthausen (Chapter 52). These chapters clearly indicate that modeling technology and, in O'Connor's case, empirical evidence will soon allow routine analysis of time and space effects. When species-habitat-relationships models are linked to these habitat-dynamics simulators, models are downsized to interactive micro- and personal computer programs, and nonhabitat factors are considered, we will have available practical tools to analyze cumulative effects. I have no doubt that someone, somewhere, is currently developing or using such models.

Conceptual aspects

I considered four chapters to be primarily conceptual; of these, two are noteworthy for managers. Schamberger and O'Niel (Chapter 1) make excellent points about differentiating the scientific and planning uses of models and recognizing that planning utility will probably be shown only through planning efficiency and by monitoring the attainments of plan implementation. Management uses of models may well be too complicated to simply test the hypotheses to elucidate veracity. McNamee et al. (Chapter 56) describe just what the model-application environment is like; certainly it is not controlled by scientific veracity, though one would like to have the most accurate models affordable.

Conclusions

Success of the symposium

Finally, a few words on the success of the symposium and on the future. I am a biased observer; I wanted the symposium to succeed, and I am excited by the advances shown. I believe this volume documents well the diversity of models in use and under development. It does not, by any means, show a consensus or consistency in testing models or describing their accuracy. Much work, both conceptual and empirical, remains. Even more work remains on the fourth objective: application of models in conservation decisions. This is a critical need, because the applications will eventually clarify what the utility criteria should be, and thereby resolve issues about accuracy and practicality.

A peek at the future

I think we are at the base of a new plateau in wildlife-habitat-relationships modeling: the routine use of appropriate-scale interactive models in a goal-driven, adaptive management process. At the current rate of real progress, I believe we can reach this plateau in 5–10 years. The next generation of models will portray the relationships of species to habitats and human uses at the same scale as the kinds of decisions being made and with reliability commensurate with the "stakes" in the decision. This means that managers will need a toolbox full of models of wildlife-habitat relationships that (1) span the geographic scale from sites, to landscapes, to whole species' ranges; and (2) span the population scale from individual life needs to regional population viability or productivity.

Expert systems and geographic information systems will be among the major new tools in the box. Considerations of populations across landscapes will augment our current preoccupation with the value of individual sites for a species. Valid and practical models are a must, but empirical verification will be needed only in proportion to the importance of a model in management decisions, that is, its role in assessing the risks, trade-offs, and benefits of investments. Managers and scientists will work together to integrate model building, application, monitoring, and hypothesis-testing research into adaptive management.

The next plateau is not like our present situation, and we may not get there if we rely upon the largely uncoordinated, piecemeal efforts that got us here. The development and adaptive use of reliable, appropriate-scale models to solve wicked problems will require no small amount of organization and coordination. It necessitates a teamwork effort—shared responsibility among individuals from all points along the spectrum of scientists, developers, extension agents, educators, and managers. To have individuals from each group is of critical importance because collectively they provide the full array of knowledge and technical skills needed to successfully negotiate increasingly complex resource conservation mazes.

The teamwork must be encouraged to be multiagency and multifunctional (see the discussion of "skunkworks" in Peters and Waterman [1982]) and focused on major, long-term management issues. Only the ongoing tasks of resource management are likely to provide the range of opportunity and the financial resources needed to dig beneath superficial relationships (Holling 1978).

Most important, we need to implement a formal role for technology developers and extension agents to serve as links between science and its application (Callaham 1984). The historical time-lag between new knowledge and its application is no longer acceptable. Models can become the common language to effect that linkage and reduce the wait.

No focus for wildlife research and management is more legitimate than maintaining full faunal diversity in the face of large-scale resource development. And no set of hypotheses is more legitimate than the models and management plans of state and federal conservation agencies. Finally, no arena is more legitimate for testing the hypotheses than the lands and waters of this nation's remaining wildland ecosystems. But make no mistake: we are dealing with wicked problems that are beyond the scope of traditional science or management to tackle alone. If we do not come together and function as a team, decision makers will get through the maze with bad rules for wildlife conservation or, even worse, without any wildlife rules at all.

Acknowledgments

I want to thank R. Ellis, J. Harn, R. Holthausen, J. Thomas, W. Laudenslayer, F. Samson, M. Schamberger, and J. Verner for their assistance with the manuscript.

60

Modeling Habitat Relationships of Terrestrial Vertebrates—The Researcher's Viewpoint

HERMAN H. SHUGART, JR., and DEAN L. URBAN

It is becoming increasingly obvious that the development of computer models of the dynamics of wildlife populations in time and space, and in response to vegetative change, is a necessary next step in the development of wildlife management. This volume documents some of the optimistic starts in the direction of developing and sharpening these management tools. As is often the case in any initial entry into a new area of research, there are differences in philosophies, approaches, and methods, and that is true of the studies presented in this volume. Yet, in this diversity is the eventual hope of finding satisfactory methods for the difficult problem of managing dynamic populations in an interactive and changing environment.

Rather than discuss the details of all the chapters that have been presented, we will discuss in a general sense developments in modeling the dynamics of habitats. The most basic model of habitat dynamics consists of two parts. First, a model must include a functional relation between the nature of the environment and the animal population (see Schamberger and O'Neil, Chapter 1; Hamel et al., Chapter 20; Capen et al., Chapter 26). Second, it must simulate the dynamics of the environment (see Sweeney, Chapter 50; Brand et al., Chapter 54). This sort of model is the most elementary dynamic habitat model, because the effects of environmental dynamics are manifested on wildlife without feedback from the animals to the environment, nor are there other complexities (such as spatial effects) on the internal dynamics of the animal populations. Recognizing that this case is a simple one, given the potential range of complexity in the formation of habitat models, it is appropriate to discuss the details of developing such simple models as a preamble to considering more complex ones.

Associating animals with habitats

As is the case with computer models of habitat dynamics, models relating animals with habitats form a diverse array of methodological approaches and a correspondingly large literature. The extant cases are dominated by nondynamic, statistical approaches. They differ most clearly with respect to the scale in space to which the methods can be expected to apply. For purposes of discussion, the dichotomy of large- and small-scale methods will be used to frame a general discussion.

Large-scale approaches

Most population-survey work on nongame birds has been at spatial scales of the order of 40 ha (100 acres) in areas considered to represent a single plant community. These surveys are typically directed toward breeding birds (usually sampled as territorial males). The association of bird species to habitats is in terms of the number of territorial males found on an area of a given size with a given vegetative character (see Rotenberry, Chapter 31).

Survey work for game animals is most intense in seasons in which the animals are most easily censused. As is the case for nongame birds, this may be the breeding season, when animals are in territorial display. The populations may be estimated at the time of harvest or at other times when large volumes of data can be obtained with relative ease. Game biologists are more aware of the use of multiple habitats by animals, and the habitat is often assessed with respect to food, resting places, water availability, or other animal requirements. Game surveys are often done at a spatial scale larger than 100 ha.

For nongame animals other than birds, the spatial scale of the survey is often determined by sampling considerations and logistics. For example, there is a limit to how many livetraps one can haul to the field and run on a daily basis. Such logistic considerations set a limit on the spatial extent of many surveys of small mammals and other nongame animals. With such surveys, the tendency is to associate a density with a plant community as a basis for quantifying habitat affinities.

Thus, for both game and nongame animals, a large repository of data and insights involves a relatively large spatial scale and is either associated with a plant-community concept or an even more general concept of habitat types. These sorts of information are also encoded into the habitat descriptions in field guides. When such information is used to perform habitat evaluations for actively managed land, the nature of the relationships that are being used is inherently correlative. The fact that an animal occurs in a habitat called "hardwood forest" provides a reasonable expectation that it might inhabit a managed hardwood forest, but this expectation is based on correlations between apparent animal abundance and the gross nature of the habitat. (In this volume see Laymon and Barrett [Chapter 14], Stauffer and Best

HERMAN H. SHUGART, JR.: Department of Environmental Sciences, University of Virginia, Charlottesville, Virginia 22903

DEAN L. URBAN: Graduate Program in Ecology, University of Tennessee, Knoxville, Tennessee 37916

[Chapter 12], Raphael and Marcot [Chapter 21], Dedon et al. [Chapter 19], and Doering and Armijo [Chapter 58] for a range of examples of large-scale approaches to the association of animal density with habitat features.)

Fine-scale approaches

An alternate method of associating animals with habitat features relates the abundance (or presence) of the animal to habitat features measured at a very fine scale (0.1 ha or less) and explores the correlations among the animal's abundance and the quantitative measures of the habitat (Larson and Bock, Chapter 7). Such statistical methods are often based on the use of a multivariate statistical model that takes into account the intercorrelations that are a prominent feature of habitat-variable data sets, but they also produce results in a form that is inscrutable to land managers. Fine-scale approaches to habitat evaluation have been the topic of a recent symposium (Capen 1981), whose proceedings provide a sample of the diversity of approaches in this area. Several of the chapters in the present volume also give a sample of fine-scale methodologies (e.g., Mosher et al., Chapter 6; Brennan et al., Chapter 27; Capen et al., Chapter 26; Smith and Conners, Chapter 8). Here it is appropriate to take a broader view of these methods as they contrast with the more classical methods categorized as large-scale in the previous section. The essential observation behind fine-scale approaches is that animals found in, say, oak-hickory forest are not necessarily found in all parts of such a forest. The habitat of the animal is characterized by the unique features of the subset of the area where the animal is most frequently found. An element of capitalizing on the heterogeneity of patterns exists, as if to analyze a flawed natural experiment. The hypothesis in this case might be something to the effect: "Given that species *A* occurs in this area, that it could (in theory) be found anywhere in the area, and that it (in fact) is only found in part of this area, the common features of the places where species *A* is found are important to the species." Even if such analyses resemble experiments in some senses, they are, nonetheless, correlational in nature. The level of detail in the analysis does not substitute for experimentation as a means to indicate the actual importance of the habitat features to the animal. In the use of these methods in management, such experiments may actually be done.

Computer models of vegetation dynamics

Simulation of the dynamics of habitats in response to management, disturbance, or succession is a complex modeling problem. Wildlife scientists can use models of vegetation dynamics as a starting point for the development of habitat-dynamics models. Because we are more familiar with models of forest systems, we will focus this discussion on those models. As early as the mid-1960s, foresters and ecologists independently began to implement digital computer models of forest growth and development. These were fairly detailed models that attempted to incorporate a wide variety of natural processes. At the time that they were developed, they were considered to be very complex. However, because of the rapid rate at which computer technology has developed, many of these models would easily fit on today's personal computer.

In forestry, model development was motivated by a concern that changes in forestry practice (e.g., change in trees due to genetic improvement, the use of forest fertilization) would render the stand-yield tables that had been so laboriously developed over prior decades less useful. Foresters began to develop models of forest growth and yield models that could be calibrated from the extant, stand-table data sets and, at the same time, could be used to incorporate some of the changes in forestry practice.

At about this same time, ecologists had become dissatisfied with the static notions of forest typology and had embarked on detailed studies of the internal dynamics of natural ecosystems. A most conspicuous effort of this sort was the International Biological Program. The interest in ecosystem dynamics led naturally to the development of computer models of vegetation dynamics.

By the mid-1970s, a general convergence occurred in the approaches used to simulate the dynamics of forests. Efforts could generally be categorized (Munro 1974; Shugart 1984) as:

1. Tree models. These take the individual tree as the basic unit of the simulation model. The degree of complexity of these models ranges from simple tabulations of the probabilities of an individual species replacement to extremely detailed models that include the three-dimensional geometry of the crown overlap of different species of trees of different sizes at different locations.
2. Forest models. These consider the forest as the focal point of the simulation. Forest yield tables are one subset of these models, as are several ecological models of element-cycling and ecosystem energetics.

These two sorts of models will be discussed in some detail to provide a background for understanding how such formulations can be used in wildlife-habitat simulation.

Tree models

There are several hundred digital computer models that project changes in forest stands by simulating the growth (and sometimes the birth and death) of individual trees on small areas. These models have recently been reviewed from both a forestry (Munro 1974) and an ecological (Shugart 1984) perspective and are often referred to as "individual tree-based models" or, simply as "tree models." A fair degree of convergence occurs in the approaches of these models, but the details of the exact functional forms used vary considerably. The models use the size of the individual trees as the set of state variables of interest. Usually, these sizes are expressed as the diameters of the trees, but some of the models also use an expanded set of state variables that may include each individual tree's height, the diameters of sections of trees at different heights, or the sizes (volume or diameter) of the crowns of the trees.

The equations that increment these state variables to simulate tree growth are almost always at annual or greater than annual time-scales. Simulation of the annual diameter increment of each tree is by far the most common case. Tree models vary with respect to several fundamental assumptions used to develop the model equations, and some of these assumptions are directly involved in the formulation of the tree-growth equations. The fundamental structural assumptions of the models revolve around the following considerations:

1. Stand diversity. Models can simulate either mixed- or single-species stands. Whether or not more than one species is simulated in the model influences the degree of detail that is used in the growth equations, as well as the functional complexity of the formulation. The models of single-species stands are generally more complex and more likely to grow the trees by incrementing height, crown size, and diameter than the models of mixed-species stands, which typically grow trees by increasing tree diameters. Model developers have a predilection for using single-species approaches in plantations and for commercially important species. The considerable amount of growth-rate data for such species has resulted in a tendency to incorporate rather elaborate statistical analyses into the models to capitalize on these large data sets.
2. Stand age structure. Many models simulate even-aged stands and are used for the commercially important species that are found in plantations. In addition, several species naturally regenerate in cohorts (e.g., Douglas-fir [*Pseudotsuga menziesii*]), and these species are frequently simulated with models that assume an even age structure. The models of uneven age structure are normally used for natural forests that typically have less commercial importance. The principal effect of using a model applicable to stands with an uneven age structure is that the simulation of the births and deaths of trees is important to raise the model to a higher level of detail. The growth equations in uneven-aged models generally incorporate tree interaction equations that are responsive to the range of sizes of trees that may occur in simulated stands of this type.

Forest models

Forest models are necessarily concerned with scales that are larger than those of tree models. In the latter case, the dichotomies used to categorize the models were formed with respect to the nature of the tree populations constituting the forests. It is easy to categorize forest models with respect to the form of the mathematical equations used in the model formulation.

In the first such category, several forest models take a holistic view of the response of a forest and are not concerned with the spatial scale of the forest. Such models can be formed by statistical regression (or related techniques) that might relate the yield of timber over time to site conditions. These approaches can be extremely sophisticated in their development, or they can consist of computer-efficient tabulations of data sets. For example, Benson and Laudenslayer (Chapter 49) used the DYNAST Model (Boyce 1978) to project the change expected for a forested landscape following three different management alternatives. By associating several wildlife species with forest-level variables (e.g., density of openings, proportions of forest in various successional states), the authors were able to make large-scale predictions about the change of habitat. Such an approach has the advantage of relating the large-scale changes in a forest landscape with equivalently large-scale changes in wildlife habitat.

An alternate approach is to view the change of the forest as the summation of the changes in its small pieces. Such approaches categorize the habitat on a grid cell, a landscape mosaic element, or a sample plot. The models then tabulate the fates of hundreds of such pieces of the forest to determine the successional change of the entire system. Examples of such models include studies by Kessell (1979), Kessell et al. (1982), Hett et al. (1978), Raedeke and Lehmkuhl (Chapter 53), Rice et al. (Chapter 13), and Smith (Chapter 55). Seagle (Chapter 40) has also explored the theoretical consequences of this sort of model to find that the species-area curves used in biogeographic theory appear to be derivable from formulations of this kind.

Synthesis

From the previous discussion, it appears that the two scales of modeling forest dynamics and the two scales at which vertebrates are typically surveyed and associated with vegetation are roughly compatible. In a management context, understanding the temporal dynamics, and hence the long-term consequences of the larger landscape-scale phenomena, may be the ultimate goal of scientists developing research in support of management objectives. One way to frame the problems and the accomplishments in this direction is to consider the progress that has been made to date in the simulation of landscapes by using computer models like the ones described earlier for forests. Weinstein and Shugart (1983) provided an overview of some of the approaches to modeling landscape dynamics using succession models. A modified version of their categorization is shown in Figure 60.1. In this figure, four modeling approaches are illustrated:

1. Area-flow models (a category melding the ordinary differential and difference equation categories of Weinstein and Shugart [1983]). These models are of the forest model type. Examples of their use in simulating change in large areas of forests may be found in Shugart et al. (1973) and Johnson and Sharpe (1976).
2. Gap models. These models are of the tree model type. The basic equations used to compute the competition between each of the individual trees in the simulated stand are of a mathematical form that is appropriate to simulating the population of trees on a sample plot. The first such model was developed by Botkin et al. (1972), and a review of the models, including examples of applications as habitat simulators, is found in Shugart (1984).
3. Spatially explicit, individual-tree models. These models simulate the size, location, and growth over time of each tree in an area. The output of such models is in a form that

Attribute \ Model:	Gap models		Compartment	
	Nonspatial	Spatial	Markov	Flow
Gradients	↑	↑	?	?
Diversity	↑	↑ $	? $	? $
Contagion	? $	↑ $	? $	↓
Largeness	$	$	↑	↑
Contrast	↓	↓	↑	↑

↑ Can be or has been done
↓ Cannot be done
? Difficult to assess; uncertain parameters
$ Considerable time, effort, or computer cost

Figure 60.1. Current status of four different approaches to modeling habitat dynamics with respect to five classes of modeling problems associated with landscapes. Downward arrows ("cannot be done") signify problems that are precluded from solution by the fundamental assumptions underlying the particular model.

resembles a map of each tree's size and location that is updated each simulated year. Most such models are of plantations or of even-aged forests (e.g., Mitchell 1975), but Ek and Monserud (1974) have a very detailed model of uneven-aged, mixed-species forests for Wisconsin.

4. Markov models. These models simulate change in habitat by computing the likelihood that a small area of a given vegetation character will either remain the same or be transformed to some new vegetation type over a given time step. Such models are fairly simple to implement on a computer, and they have been used to simulate habitat dynamics of landscapes. Kessell's (1979) study is a particularly good example of such an application.

All of these modeling approaches have been used in applications involving the simulation of landscape dynamics, and several have been tried in applications that involve simulation of animal habitat. The first model category would be most likely to be implemented by using large-scale habitat data. Examples in this book are Benson and Laudenslayer (Chapter 49), Holthausen (Chapter 52), and Kirkman et al. (Chapter 48). The last three categories would use the detailed information provided by the models in conjunction with the detailed analysis used in fine-scale habitat work. (See work in this volume by Brand et al., [Chapter 54] and Moeur [Chapter 47] for examples.) The finer-scale, more detailed approaches would represent the change of a landscape as the summation of the changes of small pieces of the landscape (Seagle, Chapter 40; Urban and Shugart, Chapter 39). This dynamic-mosaic concept of landscape has recently been discussed by Bormann and Likens (1979) and has been an important concept in ecology since the classic papers of Watt (1925, 1947).

Area-flow models also are based on the assumption that the landscape can be thought of as a great number of small patches with the sum of their areal extent being equivalent to the area of the landscape. Landscape change is the change of these patches from one category (such as land-use type or some community classification) to others. If the number of patches is large, this change can be thought of as being continuous, and the dynamics of the landscape can be abstracted as a flow of area from one landscape category to another over time. Usually, differential equations are used if this change is continuous in time, and difference equations are used when the change over time is at intervals (such as the change in a landscape seen as the difference between sample intervals).

Each of these modeling approaches has strengths and weaknesses in terms of its actual application to the simulation of real landscapes. To illustrate this point, the following potential problems in simulating the dynamics of landscapes are considered relative to the four modeling approaches discussed above:

1. Gradients. Certain landscapes exhibit a pronounced heterogeneity of substrates, slope, aspect, micrometeorology, and so on that must be taken into account to understand the landscape. This feature of landscapes can create parameter-estimation problems in some model formulations.
2. Diversity. When considering the pattern of landscapes, it is frequently necessary to treat a large number of species (or other functional entities). In some formulations (e.g., Markov models), the number of states increases the number of parameters as a square. In most formulations, the inclusion of more entities increases the effort in parameter estimation, and hence the cost.
3. Contagion. The proximity of entities on the landscape may affect their behavior. For example, successful tree regeneration may require a seed tree within a given area. If the contagion effects are dynamic, the complexity of the modeling task may be greatly increased.
4. Largeness. The actual size of landscapes may proscribe certain modeling approaches and may make others prohibitively expensive.
5. Patch contrast. Landscapes may be composed of patches dominated by one life form in some cases and other life forms in others. Some natural landscapes, such as savannas, are locally dominated by trees, shrubs, or grasses, and such mixtures are frequent on human-altered landscapes. In some human-controlled landscapes, the basic rules that determine change may vary. For example, the woodlots in an agricultural matrix may behave according to ecological rules, but the agricultural land may change in response to the price of commodities. The higher the contrast among patches, the greater the difficulty in constructing submodels of equal realism to simulate the range of behaviors.

When one forms a matrix of modeling approaches vs potential modeling problems (Fig. 60.1), it is clear that different methodologies have different limitations and strengths, depending upon the context of the problem. This may be a discouraging situation for those hoping for a single "best" modeling approach to simulating the dynamics of animal habitat over landscape scales. Nevertheless, it seems to be a fair assessment of the situation today. The use of models as tools to assess the response of wildlife to habitat alteration by humans and by natural processes over time will profit from looking for the robust results that are produced by a range of models with different underlying assumptions and with differing strengths and weaknesses. In this sense, the diversity of approaches presented in this book is an indication of health in wildlife science. The professional challenge will be to select wisely from this diverse menu of methods to suit the particular problem at hand.

Acknowledgments

Research was supported by the National Science Foundation's Ecosystem Studies Program under Interagency Agreement No. BSR 83–15185, under contract W–7405–eng–26 with the Martin Marietta Corporation.

References
Index

References

Compiled by JARED VERNER and CLARA L. SLATTERY

Abele, L. G., and E. F. Connor. 1979. Application of island biogeography theory to refuge design: Making the right decision for the wrong reasons. Pp. 89–94 *in* R. M. Linn (ed.), Proceedings of the First Conference on Scientific Research in National Parks. U.S. Department of Interior, National Park Service, Washington, D.C.

Adams, D. A. 1980. Wildlife habitat models as aids to impact evaluation. The Environmental Professional 2:253–262.

Adams, D. A., M. F. Overton, T. C. Gopalkrishnan, J. S. Wei, and G. M. Cressman. 1983. Solving species-habitat functions with georeferenced environmental impact assessment. Pp. 399–408 *in* W. K. Laurenroth, G. V. Skogerboe, and M. Flug (eds.), Analysis of Ecological Systems: State-of-the-Art in Ecological Modelling. Elsevier, New York.

Adler, P. E., and D. L. Pearson. 1982. Why do male butterflies visit mud puddles? Canadian Journal of Zoology 60:322–325.

Ahlén, I., and S. G. Nilsson. 1982. Species richness and area requirements of forest bird species on islands with natural forests in Lake Malaren and Hajalmaren. Vår Fågelvärld 41:161–184.

Airola, D. A. (ed.). 1980. California Wildlife/Habitat Relationships Program: Northeast Interior Zone. Volume 3. Birds. U.S. Department of Agriculture, Forest Service, Lassen National Forest, Susanville, CA. 590 pp.

Alatalo, R. V. 1981. Habitat selection of forest birds in the seasonal environment of Finland. Annales Zoologici Fennici 18:103–114.

Aldous, S. E. 1938. Beaver food utilization studies. Journal of Wildlife Management 2:215–222.

Aldrich, R. C., and R. C. Heller. 1969. Large-scale color photography reflects changes in a forest community during a spruce budworm epidemic. Pp. 30–45 *in* P. L. Johnson (ed.), Remote Sensing in Ecology: Proceedings of the Symposium. University of Georgia Press, Athens, GA.

Aleksiuk, M. 1970. The seasonal food regime of arctic beavers. Ecology 51:264–270.

Allen, A. W. 1982a. Habitat Suitability Index Models: Marten. FWS/OBS–82/10.11. U.S. Department of Interior, Fish and Wildlife Service, Biological Services Program, Washington, D.C. 9 pp.

Allen, A. W. 1982b. Habitat Suitability Index Models: Gray Squirrel. FWS/OBS–82/10.19. U.S. Department of Interior, Fish and Wildlife Service, Biological Services Program, Washington, D.C. 11 pp.

Allen, T. F. H., and T. B. Starr. 1982. Hierarchy: Perspectives for Ecological Complexity. University of Chicago Press, Chicago. 310 pp.

Ambuel, B., and S. A. Temple. 1982. Songbird populations in southern Wisconsin forests: 1954 and 1979. Journal of Field Ornithology 53:149–158.

Ambuel, B., and S. A. Temple. 1983. Area-dependent changes in the bird communities and vegetation of southern Wisconsin forests. Ecology 64:1057–1068.

American Geological Institute. 1976. Dictionary of Geological Terms. Revised edition. Anchor, New York. 472 pp.

Anderson, B. W., R. W. Engel-Wilson, D. Wells, and R. D. Ohmart. 1977. Ecological study of southwestern riparian habitats: Techniques and data applicability. Pp. 146–155 *in* R. R. Johnson and D. A. Jones (tech. coords.), Importance, Preservation and Management of Riparian Habitat: A Symposium. U.S. Department of Agriculture, Forest Service, General Technical Report RM–43. Rocky Mountain Forest and Range Experiment Station, Fort Collins, CO.

Anderson, B. W., R. D. Ohmart, and J. Rice. 1983. Avian and vegetation community structure and their seasonal relationships in the lower Colorado River Valley. Condor 85:392–415.

Anderson, D. R. 1975a. Optimal exploitation strategies for an animal population in a Markovian environment: A theory and an example. Ecology 56:1281–1297.

Anderson, D. R. 1975b. Population Ecology of the Mallard: Part V. Temporal and Geographic Estimates of Survival, Recovery, and Harvest Rates. Resource Publication 125. U.S. Department of Interior, Fish and Wildlife Service, Washington, D.C. 110 pp.

Anderson, S. H. 1981. Correlating habitat variables and birds. Pp. 538–542 *in* C. J. Ralph and J. M. Scott (eds.), Estimating Numbers of Terrestrial Birds. Studies in Avian Biology 6. Cooper Ornithological Society.

Anderson, S. H., and H. H. Shugart, Jr. 1974. Habitat selection of breeding birds in an east Tennessee deciduous forest. Ecology 55:828–837.

Andrews, T. L., R. H. Harms, and H. R. Wilson. 1973. Protein requirement of the bobwhite chick. Poultry Science 52:2199–2201.

Anonymous. 1982. Review and Evaluation of Adaptive Environmental Assessment and Management. Environment Canada, Vancouver, British Columbia. 116 pp.

Anonymous. 1983. International Mathematical and Statistical Libraries [computer software]. International Mathematical and Statistical Libraries, Incorporated, Houston, TX. Four volumes, unpaged.

Arms, K., P. Feeney, and R. Lederhouse. 1974. Sodium: Stimulus for puddling behavior by tiger swallowtail butterflies (*Papilio glaucus*). Science 185:373–374.

Army Corps of Engineers. 1980. A Habitat Evaluation System for Water Resources Planning. U.S. Army Corps of Engineers, Lower Mississippi Division, Planning Division, Environmental Analysis Branch, Vicksburg, MS. 89 pp. + appendices.

Austin, M. P. 1977. Use of ordination and other multivariate descriptive methods to study succession. Vegetatio 35:165–175.

Bailey, R. S. 1967. An index of bird population changes on farmland. Bird Study 14:195–209.

Balda, R. P. 1975a. The Relationship of Secondary Cavity Nesters to Snag Densities in Western Coniferous Forests. U.S. Department of Agriculture, Forest Service, Wildlife Habitat Technical Bulletin 1. Southwestern Region, Albuquerque, NM. 27 pp.

Balda, R. P. 1975b. Vegetation structure and breeding bird diversity. Pp. 59–80 *in* D. R. Smith (tech. coord.), Management of Forest and Range Habitats for Nongame Birds: Proceedings of the Symposium. U.S. Department of Agriculture, Forest Service, General Technical Report WO-1. Washington, D.C.

Balda, R. P., W. S. Gaud, and J. D. Brawn. 1983. Predictive models for snag nesting birds. Pp. 216–222 *in* J. W. Davis, G. A. Goodwin, and R. A. Ockenfels (tech. coords.), Snag Habitat Management: Proceedings of the Symposium. U.S. Department of Agriculture, Forest Service, General Technical Report RM–99. Rocky Mountain Forest and Range Experiment Station, Fort Collins, CO.

Balser, D., A. Bielak, G. DeBoer, T. Tobias, G. Adindu, and R. S. Corney. 1981. Nature reserve designation in a cultural landscape, incorporating island biogeography theory. Landscape Planning 8:329–347.

Barbour, M. G., and J. Major. 1977. Terrestrial Vegetation of California. John Wiley and Sons, New York. 1002 pp.

Barr, A. J., J. H. Goodnight, J. P. Sall, and J. T. Helwig. 1976. A User's Guide to SAS. SAS Institute, Raleigh, NC. 329 pp.

Barrett, R. H. 1983. Smoked aluminum track plots for determining furbearer distribution and abundance. California Fish and Game 69:188–190.

Barrett, R. H., and H. Salwasser. 1982. Adaptive management of timber and wildlife habitat using DYNAST and wildlife habitat relationships models. Proceedings of the Annual Conference of the Western Association of Fish and Wildlife Agencies 62:350–365.

Barstow, D. R., N. Aiello, R. O. Duda, L. D. Erman, C. L. Forgy, D. Gorlin, R. D. Greiner, D. B. Lenat, R. E. London, J. McDermott, H. P. Nii, P. Politakis, R. Reboh, S. Rosenschein, A. C. Scott, W. Van Melle, and S. M. Weiss. 1983. Expert system tools. Pp. 283–348 *in* F. Hayes-Roth, D. A. Waterman, and D. B. Lenat (eds.), Building Expert Systems. Addison-Wesley, Reading, MA.

Bart, J., and J. D. Schoultz. 1984. Reliability of singing bird surveys: Changes in observer efficiency with avian density. Auk 101:307–318.

Baskett, T. S., D. A. Darrow, D. L. Hallett, M. J. Armbruster, J. A. Ellis, B. F. Sparrowe, and P. A. Korte. 1980. A Handbook for Terrestrial Habitat Evaluation in Central Missouri. Resource Publication 133. U.S. Department of Interior, Fish and Wildlife Service, Washington, D.C. 155 pp.

Beasom, S. L., and R. A. Moore. 1977. Bobcat food habit response to a change in prey abundance. Southwestern Naturalist 21:451–457.

Beck, J. R., and D. O. Beck. 1955. A method for nutritional evaluation of wildlife foods. Journal of Wildlife Management 19:198–205.

Beedy, E. C. 1982. Bird Community Structure in Coniferous Forests of Yosemite National Park, California. Ph.D. dissertation. University of California, Davis, CA. 167 pp.

Beissinger, S. R., and D. R. Osborne. 1982. Effects of urbanization on avian community organization. Condor 84:75–83.

Belcher, D. M. 1982. TWIGS: The Woodsman's Ideal Growth Projection System. Pp. 70–95 *in* J. W. Moser, Jr. (ed.), Microcomputers: A New Tool for Foresters. Purdue University Press, West Lafayette, IN.

Belcher, D. M., M. R. Holdaway, and G. J. Brand. 1982. A Description of STEMS—the Stand and Tree Evaluation and Modeling System. U.S. Department of Agriculture, Forest Service, General Technical Report NC–79. North Central Forest Experiment Station, St. Paul, MN. 18 pp.

Bella, D. A. 1970. Simulating the effect of sinking and vertical mixing on algal population dynamics. Journal of the Water Pollution and Control Federation 42 (Number 5, Part 2):R140–R152.

Bellrose, F. C. 1976. Ducks, Geese and Swans of North America. Stackpole, Harrisburg, PA. 543 pp.

Bengtson, S. -A., and D. Bloch. 1983. Island land bird population densities in relation to island size and habitat quality on the Faroe Islands. Oikos 41:507–522.

Benson, A. S. 1984. Characterization of Spotted Fawn Habitat for the Yolla Bolly Deer Herd in Northern California. Final Report IA 83/84–C–718, Space Sciences Laboratory, University of California, Berkeley, CA. 20 pp.

Benson, G. L. 1983. Modeling forest resource responses to various timber management strategies. Pp. 305–309 *in* Proceedings of the Annual Convention of the Society of American Foresters, 16–20 October 1983, Portland, Oregon. Publication 84–03, Society of American Foresters, Bethesda, MD.

Bergerud, A. T., W. Wyett, and B. Snider. 1983. The role of wolf predation in limiting a moose population. Journal of Wildlife Management 47:977–988.

Bergman, G. 1939. Untersuchungen über die Nistvogelfauna in einem Schärengebiet westlich von Helsingfors. Acta Zoologica Fennica 23:1–134.

Berry, J. K., and C. D. Tomlin. 1980. Geographic Information Analysis Workbook. Papers in Spatial Information Systems. Yale School of Forestry and Environmental Studies, New Haven, CT. 140 pp.

Best, L. B. 1981. Seasonal changes in detection of individual bird species. Pp. 252–261 *in* C. J. Ralph and J. M. Scott (eds.), Estimating Numbers of Terrestrial Birds. Studies in Avian Biology 6. Cooper Ornithological Society.

Best, L. B. 1983. Bird use of fencerows: Implications of contemporary fencerow management practices. Wildlife Society Bulletin 11:343–347.

Biesterfeldt, R. C., and S. G. Boyce. 1978. Systematic approach to multiple-use management. Journal of Forestry 76:342–345.

Billingsley, B. B., Jr., and D. H. Arner. 1970. The nutritional value and digestibility of some winter foods of the eastern wild turkey. Journal of Wildlife Management 34:176–182.

Bledsoe, L. J., and S. M. Van Dyne. 1971. A compartment model of secondary succession. Pp. 480–513 *in* B. C. Patten (ed.), Systems Analysis and Simulation in Ecology. Academic, New York.

Blenden, M. D. 1982. Relation of Arthropod Abundance to Avian Habitat in Central Missouri. M.S. thesis. University of Missouri, Columbia, MO. 161 pp.

Blondel, J. 1979. Biogeographie et Ecologie. Collection d'Ecologie 15. Masson, Paris, New York, Barcelon, and Milan. 173 pp.

Blonski, K. S., and J. L. Schramel. 1981. Photo Series for Quantifying Natural Forest Residues: South Cascades, North Sierra Nevada. U.S. Department of Agriculture, Forest Service, General Technical Report PSW–56. Pacific Southwest Forest and Range Experiment Station, Berkeley, CA. 145 pp.

Bock, C. E., and J. F. Lynch. 1970. Breeding bird populations of burned and unburned conifer forests in the Sierra Nevada. Condor 72:182–189.

Bock, C. E., and B. Webb. 1984. Birds as grazing indicator species in southeastern Arizona. Journal of Wildlife Management 48:1045–1049.

Bormann, F. H., and G. E. Likens. 1979. Pattern and Process in a Forested Ecosystem. Springer-Verlag, New York. 253 pp.

Bostrom, U., and S. G. Nilsson. 1983. Latitudinal gradients and local variations in species richness and structure of bird communities of raised peat-bogs in Sweden. Ornis Scandinavica 14:213–226.

Botkin, D. B., J. F. Janak, and J. R. Wallis. 1972. Some ecological consequences of a computer model of forest growth. Journal of Ecology 60:849–873.

Bovee, K. D. 1982. A Guide to Stream Habitat Analysis Using the Instream Flow Incremental Methodology. Instream Flow Information Paper 12. FWS/OBS–82/26. U.S. Department of Interior, Fish and Wildlife Service, Western Energy and Land Use Team, Fort Collins, CO. 268 pp.

Bovee, K. D., and R. T. Milhouse. 1978. Hydraulic Simulation in Instream Flow Studies: Theory and Techniques. Instream Flow Information Paper 5. FWS/OBS–78/83. U.S. Department of Interior, Fish and Wildlife Service, Western Energy and Land Use Team, Fort Collins, CO. 141 pp.

Box, G. E. P., W. G. Hunter, and J. S. Hunter. 1978. Statistics for Experimenters: An Introduction to Design, Data Analysis, and Model Building. John Wiley and Sons, New York. 653 pp.

Boyce, S. G. 1977. Management of Eastern Hardwood Forests for Multiple Benefits (DYNAST–MB). U.S. Department of Agriculture, Forest Service, Research Paper SE–168. Southeastern Forest Experiment Station, Asheville, NC. 116 pp.

Boyce, S. G. 1978. Management of Forests for Timber and Related

Benefits (DYNAST–TM). U.S. Department of Agriculture, Forest Service, Research Paper SE–184. Southeastern Forest Experiment Station, Asheville, NC. 140 pp.

BOYCE, S. G. 1980. Management of Forests for Optimal Benefits (DYNAST–OB). U.S. Department of Agriculture, Forest Service, Research Paper SE–204. Southeastern Forest Experiment Station, Asheville, NC. 92 pp.

BOYCE, S. G. 1982. Increasing the joint production of wildlife and timber. Pp. 204–208 *in* Proceedings of the 1981 Convention of the Society of American Foresters, Orlando, Florida. Society of American Foresters, Bethesda, MD.

BOYCE, S. G., and N. D. COST. 1978. Forest Diversity, New Concepts and Applications. U.S. Department of Agriculture, Forest Service, Research Paper SE–194. Southeastern Forest Experiment Station, Asheville, NC. 36 pp.

BOYD, R. L., and C. L. CINK. 1980. Breeding bird populations of selected oak-hickory forests in northeastern Kansas. American Birds 34:104–106.

BOYD, R. L., C. L. CINK, D. BRYAN, and M. JOYCE. 1982. Breeding bird populations of selected oak-hickory forests in northeastern Kansas. American Birds 36:85–86.

BOYNTON, A. C. 1979. A Multivariate Analysis of Vegetative Structure of Songbird Territories in a Northern Hardwoods Forest. M.S. thesis. University of Vermont, Burlington, VT. 66 pp.

BRACHMAN, R. J., S. AMAREL, C. ENGELMAN, R. S. ENGELMORE, E. A. FEIGENBAUM, and D. E. WILKINS. 1983. What are expert systems? Pp. 31–58 *in* F. Hayes-Roth, D. A. Waterman, and D. B. Lenat (eds.), Building Expert Systems. Addison-Wesley, Reading, MA. 444 pp.

BRAUN, C. E., M. F. BAKER, R. L. ENG, J. S. GASHWILLER, and M. H. SCHROEDER. 1976. Conservation committee report on effects of alteration of sagebrush communities on the associated fauna. Wilson Bulletin 88:165–171.

BRAUN, E. L. 1950. Deciduous Forests of Eastern North America. Hafner, New York, 596 pp.

BRAY, J. R., and J. T. CURTIS. 1957. An ordination of the upland forest communities of southern Wisconsin. Ecological Monographs 27:325–349.

BRENNAN, L. A. 1984. Summer Habitat Ecology of Mountain Quail in Northern California. M.S. thesis. Humboldt State University, Arcata, CA. 71 pp.

BRITTINGHAM, M. C., and S. A. TEMPLE. 1983. Have cowbirds caused forest songbirds to decline? BioScience 33:31–35.

BROWN, E. R. 1961. The Black-tailed Deer of Western Washington. Biological Bulletin Number 13. Washington State Game Department, Olympia, WA. 124 pp.

BROWN, J. H. 1971. Mammals on mountaintops: Nonequilibrium insular biogeography. American Naturalist 105:467–478.

BROWN, J. H. 1978. The theory of insular biogeography and the distribution of boreal birds and mammals. Great Basin Naturalist Memoirs 2:209–227.

BROWN, J. H., and A. KODRIC-BROWN. 1977. Turnover rates in insular biogeography: Effect of immigration on extinction. Ecology 58:445–449.

BROWN, J. L. 1969. Territorial behavior and population regulation in birds. Wilson Bulletin 81:293–329.

BRUNSWIG, N. L., and A. S. JOHNSON. 1972. Bobwhite quail foods and quail populations on pine plantations in the Georgia Piedmont during the first seven years following site preparation. Proceedings of the Annual Conference of the Southeastern Association of Game and Fish Commissioners 26:96–107.

BUCHANAN, B. G., and E. H. SHORTLIFFE. 1983. Rule-based Expert Systems: The MYCIN Experiments of the Heuristic Programming Project. Addison-Wesley, Reading, MA. 748 pp.

BUCHANAN, B. G., D. BARSTOW, R. BECHTEL, J. BENNETT, W. CLANCEY, C. KULIKOWSKI, T. MITCHELL, and D. A. WATERMAN. 1983. Constructing an expert system. Pp. 127–168 *in* F. Hayes-Roth, D. A. Waterman, and D. B. Lenat (eds.), Building Expert Systems. Addison-Wesley, Reading, MA.

BUCHMAN, R. G., and S. R. SHIFLEY. 1983. Guide to evaluating forest growth projection systems. Journal of Forestry 81:232–234, 254.

BUREAU OF LAND MANAGEMENT. 1982. Integrated Habitat Inventory and Classification System. Manual Section 6602. U.S. Department of Interior, Bureau of Land Management, Washington, D.C. Unpaged.

BURGESS, R. L., and D. M. SHARPE (eds.). 1981. Forest Island Dynamics in Man-Dominated Landscapes. Springer-Verlag, New York, Heidelberg, and Berlin. 310 pp.

BURNHAM, K. P., D. R. ANDERSON, and J. L. LAAKE. 1980. Estimation of density from line-transect sampling of biological communities. Wildlife Monographs 72:1–202.

BURY, R. B., and M. G. RAPHAEL. 1983. Inventory methods for amphibians and reptiles. Pp. 416–419 *in* J. F. Bell and T. Atterbury (eds.), Renewable Resource Inventories for Monitoring Changes and Trends. College of Forestry, Oregon State University, Corvallis, OR.

BUTCHER, G. S., W. A. NIERING, W. J. BARRY, and R. H. GOODWIN. 1981. Equilibrium biogeography and the size of nature preserves: An avian case study. Oecologia 49:29–37.

CADZOW, J. A. 1973. Discrete-Time Systems. Prentice-Hall, Englewood Cliffs, NJ. 440 pp.

CALLAHAM, R. Z. 1984. Managing for applications, not just for research and development. Journal of Forestry 82:224–227.

CAMPBELL, H. W., and S. P. CHRISTMAN. 1982. Field techniques for herpetofaunal community analysis. Pp. 193–200 *in* N. J. Scott (ed.), Herpetological Communities. Research Paper 13. U.S. Department of Interior, Fish and Wildlife Service.

CANNINGS, R. J., and W. THRELFALL. 1981. Horned lark breeding biology at Cape St. Mary's, Newfoundland. Wilson Bulletin 93:519–530.

CAPEN, D. E. (ed.). 1981. The Use of Multivariate Statistics in Studies of Wildlife Habitat. U.S. Department of Agriculture, Forest Service, General Technical Report RM–87. Rocky Mountain Forest and Range Experiment Station, Fort Collins, CO. 249 pp.

CAREY, A. B. 1984. A critical look at the issue of species-habitat dependency. Pp. 346–351 *in* Proceedings of the 1983 Convention of the Society of American Foresters, Portland, Oregon. Society of American Foresters, Bethesda, MD.

CARNES, B. A., and N. A. SLADE. 1982. Some comments on niche analysis in canonical space. Ecology 63:888–893.

CARRIER, W. D., D. M. SOLIS, and E. F. TOTH. 1985. Providing for spotted owls through a managed forest ecosystem in California's national forests. Pp. 283–291 *in* W. McComb (ed.), Management of Nongame Species and Ecological Communities: Proceedings of the Workshop. College of Agriculture, University of Kentucky, Lexington, KY.

CASE, R. M. 1972. Energetic requirements for egg-laying bobwhites. Proceedings of National Bobwhite Quail Symposium 1:205–212.

CASE, R. M., and R. J. ROBEL. 1974. Bioenergetics of bobwhites. Journal of Wildlife Management 38:638–652.

CASWELL, H. 1976a. Community structure: A neutral model analysis. Ecological Monographs 46:326–354.

CASWELL, H. 1976b. The validation problem. Pp. 313–325 *in* B. C. Patten (ed.), Systems Analysis and Simulation in Ecology. Volume 4. Academic, New York.

CATZEFLIS, F. 1978. Sur la biologie du reproduction du pipit spioncelle alpin. Nos Oiseaux 34:287–302.

Caughley, G. 1977. Analysis of Vertebrate Populations. John Wiley and Sons, New York. 234 pp.

Cawthorne, R. A., and J. H. Marchant. 1980. The effects of the 1978/79 winter on British bird populations. Bird Study 27:163–172.

Chapel, M. T., M. Smith, K. Sonksen, G. Terrazas, J. Lorenzana, and R. Kindlund. 1983. Wildlife Habitat Planning Demonstration—Rancheria Planning Unit. U.S. Department of Agriculture, Forest Service, Sierra National Forest, Fresno, CA. 50 pp.

Chatterjee, S., and B. Price. 1977. Regression Analysis by Example. John Wiley and Sons, New York. 228 pp.

Cink, C. L., and R. L. Boyd. 1979. Breeding bird populations of selected oak-hickory forests in northeastern Kansas. American Birds 33:66–67.

Cink, C. L., and R. L. Boyd. 1981. Breeding bird populations of selected oak-hickory forests in northeastern Kansas. American Birds 35:61–62, 112.

Clancy, W. J. 1983. The epistomology of a rule-based expert system—a framework for explanation. Artificial Intelligence 20:205–251.

Clark, J. D., and J. C. Lewis. 1983. A validity test of a Habitat Suitability Index model for clapper rail. Proceedings of the Annual Conference of the Southeastern Association of Fish and Wildlife Agencies 37:95–102.

Clawson, M. E., T. S. Baskett, and M. J. Armbruster. 1984. An approach to habitat modeling for herpetofauna. Wildlife Society Bulletin 12:61–69.

Cochran, W. G. 1963. Sampling Techniques. Second edition. John Wiley and Sons, New York. 413 pp.

Cole, B. J. 1981. Colonizing abilities, island size, and the number of species on archipelagoes. American Naturalist 117:629–638.

Cole, C. A., and R. L. Smith. 1983. Habitat suitability indices for monitoring wildlife populations—an evaluation. Transactions of the North American Wildlife and Natural Resources Conference 48:367–375.

Coleman, B. D. 1981. On random placement and species-area relations. Mathematical Biosciences 54:191–215.

Collins, S. L. 1981. Habitat Relationships and Habitat Variability of the Wood Warblers. Ph.D. dissertation. University of Oklahoma, Norman, OK. 100 pp.

Collins, S. L. 1983a. Geographic variation in habitat structure for the wood warblers in Maine and Minnesota. Oecologia 59:246–252.

Collins, S. L. 1983b. Geographic variation in habitat structure of the black-throated green warbler (*Dendroica virens*). Auk 100:382–389.

Commoner, B. 1971. The Closing Circle: Nature, Man, and Technology. Alfred A. Knopf, New York. 326 pp.

Connell, J. H. 1983. On the prevalence and relative importance of interspecific competition: Evidence from field experiments. American Naturalist 122:661–696.

Conner, R. N., and C. S. Adkisson. 1976. Discriminant function analysis: A possible aid in determining the impact of forest management on woodpecker nesting habitat. Forest Science 22:122–127.

Conner, R. N., and C. S. Adkisson. 1977. Principal component analysis of woodpecker nesting habitat. Wilson Bulletin 89:122–129.

Conner, R. N., and J. G. Dickson. 1980. Strip transect sampling and analysis for avian habitat studies. Wildlife Society Bulletin 8:4–10.

Conner, R. N., J. G. Dickson, B. A. Locke, and C. A. Segelquist. 1983. Vegetation characteristics important to common songbirds in east Texas. Wilson Bulletin 95:349–361.

Connor, E. F., and E. D. McCoy. 1979. The statistics and biology of the species-area relationship. American Naturalist 113:791–833.

Connor, E. F., and D. Simberloff. 1979. The assembly of species communities: Chance or competition? Ecology 60:1132–1140.

Connors, P. G., C. S. Connors, and K. G. Smith. 1984. Shorebird littoral zone ecology of the Alaskan Beaufort coast. U.S. Department of Commerce, National Oceanic and Atmospheric Administration, Boulder, CO. Outer Continental Shelf Environmental Assessment Program. Final Report 23:295–396.

Conover, W. J. 1971. Practical Nonparametric Statistics. John Wiley and Sons, New York. 462 pp.

Cook, J. G., L. L. Irwin, A. W. Allen, and M. J. Armbruster. 1984. Field test of a winter pronghorn habitat suitability index model. Pp. 207–224 *in* C. K. Winkler (chairman), Proceedings of the Eleventh Biennial Pronghorn Antelope Workshop. Texas Parks and Wildlife Department, Austin, TX.

Cooley, W. W., and P. R. Lohnes. 1971. Multivariate Data Analysis. John Wiley and Sons, New York. 364 pp.

Cottam, G., and J. T. Curtis. 1956. The use of distance measures in phytosociological sampling. Ecology 37:451–460.

Council on Environmental Quality. 1980. Environmental Quality—1980: Eleventh Annual Report of the Council on Environmental Quality. U.S. Government Printing Office, Washington, D.C. 497 pp.

Cowardin, L. M., and D. H. Johnson. 1979. Mathematics and mallard management. Journal of Wildlife Management 43:18–35.

Cowardin, L. M., D. H. Johnson, A. M. Frank, and A. T. Klett. 1983. Simulating results of management actions on mallard production. Transactions of the North American Wildlife and Natural Resources Conference 48:257–272.

Cowardin, L. M., D. S. Gilmer, and C. W. Shaiffer. 1985. Mallard recruitment in the agricultural environment of North Dakota. Wildlife Monographs 92:1–37.

Cox, D. R. 1970. The Analysis of Binary Data. Methuen, London. 142 pp.

Crissey, W. F. 1969. Prairie potholes from a continental viewpoint. Pp. 161–191 *in* Saskatoon Wetlands Seminar. Report Series 6. Canadian Wildlife Service, Ottawa, Ontario.

Crocker, D. C. 1972. Some interpretations of the multiple correlation coefficient. The American Statistician 26:31–32.

Cronquist, A., A. H. Holmgren, N. H. Holmgren, and J. L. Reveal. 1972. Intermountain Flora. Volume 1. Hafner, New York. 270 pp.

Csaki, F. 1977. Space-State Methods for Control Systems. Academia Kiaedo, Budapest, Hungary. 672 pp.

Cunningham, J. B., R. P. Balda, and W. S. Gaud. 1980. Selection and Use of Snags by Secondary Cavity Nesting Birds of the Ponderosa Pine Forest. U.S. Department of Agriculture, Forest Service, Research Paper RM–222. Rocky Mountain Forest and Range Experiment Station, Fort Collins, CO. 15 pp.

Curlin, J. W., and D. J. Nelson. 1968. Walker Branch Watershed Project: Objectives, Facilities, and Ecological Characteristics. ORNL TM 2271. Oak Ridge National Laboratory, Oak Ridge, TN. 188 pp.

Currier, P. J. 1984. Woody vegetation clearing on the Platte River: Restoration of sandhill crane roosting habitat (Nebraska). Restoration and Management Notes 2 (1):38.

Danilov, N. N. 1980. Formirovanie prostranstvennoj struktury naseleniya ptits. Pp. 113–120 *in* I. A. Nejfel'dt (ed.), Ekologiya, geografiya i ohrana ptits. Zoologicheskij Institut, Akademij Nauk, SSSR, Leningrad.

Daruna, J. H., and R. Karrer. 1981. On the validation of discriminant functions: An empirical analysis using event related potentials. Psychophysiology 18:82–87.

DAUBENMIRE, R. 1959. A canopy-coverage method of vegetational analysis. Northwest Science 33:43–64.

DAVIS, A. M., and T. F. GLICK. 1978. Urban ecosystems and island biogeography. Environmental Conservation 5:299–304.

DAVIS, J. 1957. Comparative foraging behavior of the spotted and brown towhees. Auk 74:129–166.

DAVIS, L. S. 1980. Strategy for building a location-specific multipurpose information system for wildland management. Journal of Forestry 78:402–408.

DAWSON, D. G. 1981. Counting birds for a relative measure (index) of density. Pp. 12–16 *in* C. J. Ralph and J. M. Scott (eds.), Estimating Numbers of Terrestrial Birds. Studies in Avian Biology 6. Cooper Ornithological Society.

DEDON, M. F. 1982. A Test of Wildlife Habitat Relationship Models for Black Oak and Mixed Conifer Habitats in California. M.S. thesis. University of California, Berkeley, CA. 175 pp.

DEDON, M. F., and R. H. BARRETT. 1982. An inventory system for assessing wildlife habitat relationships in forests. Cal-Neva Wildlife Transactions 1982:55–60.

DEGRAAF, R. M., G. M. WITMAN, J. W. LANIER, B. J. HILL, and J. M. KENISTON. 1980. Forest Habitat for Birds of the Northeast. U.S. Department of Agriculture, Forest Service, Northeastern Forest Experiment Station and Eastern Region, Amherst, MA. 598 pp.

DELONG, D. M. 1932. Some problems encountered in the estimation of insect populations by the sweeping method. Annals of the Entomological Society of America 25:13–17.

DEVOS, A., and H. S. MOSBY. 1969. Habitat analysis and evaluation. Pp. 135–172 *in* R. H. Giles (ed.), Wildlife Management Techniques. Third edition. The Wildlife Society, Washington, D.C.

DEWITT, J. B., R. B. NESTLER, and J. V. DERBY, JR. 1949. Calcium and phosphorus requirements of breeding bobwhite quail. Journal of Nutrition 39:567–577.

DIAMOND, J. M. 1972. Biogeographic kinetics: Estimation of relaxation times for avifaunas of Southwest Pacific islands. Proceedings of the National Academy of Sciences (USA) 69:3199–3203.

DIAMOND, J. M. 1975a. The island dilemma: Lessons of modern biogeographic studies for the design of natural preserves. Biological Conservation 7:129–146.

DIAMOND, J. M. 1975b. Assembly of species communities. Pp. 342–444 *in* M. L. Cody and J. M. Diamond (eds.), Ecology and Evolution of Communities. Harvard University Press, Cambridge, MA.

DIAMOND, J. M. 1976. Island biogeography and conservation: Strategy and limitations. Science 193:1027–1029.

DIAMOND, J. M. 1984a. "Normal" extinctions of isolated populations. Pp. 191–246 *in* M. H. Nitecki (ed.), Extinctions. University of Chicago Press, Chicago.

DIAMOND, J. M. 1984b. Historic extinctions: A Rosetta Stone for understanding prehistoric extinctions. Pp. 824–862 *in* P. S. Martin and R. G. Klein (eds.), Quaternary Extinctions: A Prehistoric Revolution. University of Arizona Press, Tucson, AZ.

DIAMOND, J. M., and M. E. GILPIN. 1982. Examination of the "null" model of Connor and Simberloff for species co-occurrence on islands. Oecologia 52:64–74.

DIAMOND, J. M., and R. M. MAY. 1982. Island biogeography and the design of natural reserves. Pp. 228–252 *in* R. M. May (ed.), Theoretical Ecology. Second edition. Sinauer Associates, Sunderland, MA.

DIAMOND, J. M., and E. MAYR. 1976. Species-area relation for birds of the Solomon Archipelago. Proceedings of the National Academy of Sciences (USA) 73:262–266.

DICKSON, J., R. N. CONNER, and J. H. WILLIAMSON. 1980. Relative abundance of breeding birds in forest stands in the Southeast. Southern Journal of Applied Forestry 4:175–179.

DIEHL, B. 1985. A 20-year study of bird communities in a heterogeneous and changing habitat. Pp. 213–230 *in* G. K. Taylor, R. J. Fuller, and P. C. Lack (eds.), Bird Census and Atlas Studies: Proceedings of the VIII International Conference on Bird Census Work. British Trust for Ornithology, Tring, England.

DIXON, W. J. (ed.). 1983. BMDP Statistical Software. University of California Press, Los Angeles. 735 pp.

DIXON, W. J., and M. B. BROWN (eds.). 1979. Biomedical Computer Programs, P-Series. University of California Press, Berkeley, CA. 880 pp.

DOLNIK, V. R. 1982. Populyatsionnaya ekologiya zyablika. Nauka, Leningradskoe Otd., Leningrad, USSR. 354 pp.

DOTY, H. A., and F. B. LEE. 1974. Homing to nest baskets by wild female mallards. Journal of Wildlife Management 38:714–719.

DOUGENIK, J. A., and D. E. SHEEHAN. 1979. SYMAP User's Reference Manual. Laboratory for Computer Graphics and Spatial Analysis, Graduate School of Design, Harvard University, Cambridge, MA. 189 pp.

DOW, D. D. 1969. Home range and habitat of the cardinal in peripheral and central populations. Canadian Journal of Zoology 47:103–114.

DRAPER, N. R., and H. SMITH. 1981. Applied Regression Analysis. Second edition. John Wiley and Sons, New York. 709 pp.

DUDA, R. O., and J. G. GASCHNIG. 1981. Knowledge-based expert systems come of age. Byte 6:238–248, 281.

DUDA, R. O., and E. H. SHORTLIFFE. 1983. Expert systems research. Science 220:261–268.

DUDA, R. O., J. G. GASCHNIG, and P. E. HART. 1981. Model design in the Prospector consultant system for mineral exploration. Pp. 153–167 *in* D. Michie (ed.), Expert Systems in the Microelectronic Age. Edinburgh University Press, Edinburgh, Scotland.

DUESER, R. D., and W. C. BROWN. 1980. Ecological correlates of insular rodent diversity. Ecology 61:50–56.

DUESER, R. D., and H. H. SHUGART, JR. 1979. Niche pattern in a forest floor small mammal fauna. Ecology 60:108–118.

DUESER, R. D., and H. H. SHUGART, JR. 1982. Reply to comments by Van Horne and Ford and by Carnes and Slade. Ecology 63:1174–1175.

DUESER, R. D., H. H. SHUGART, JR., and J. C. RANDOLPH. 1976. Structural Niches in a Forest Floor Small Mammal Community. Environmental Sciences Division Publication Number 752. Oak Ridge National Laboratory, Oak Ridge, TN. 118 pp.

DUNKELGOD, K. E. 1961. Practical Nutrient Intake Standards and Feed Formulas for Growing Market Turkeys. Oklahoma Agricultural Experiment Station, Stillwater, OK. 39 pp.

DWERNYCHUK, L. W., and D. A. BOAG. 1973. Effect of herbicide-induced changes in vegetation on nesting ducks. Canadian Field-Naturalist 87:155–165.

ECKHARDT, R. C. 1979. The adaptive syndromes of two guilds of insectivorous birds in the Colorado Rocky Mountains. Ecological Monographs 49:129–149.

EDWARDS, R. Y., and C. D. FOWLE. 1955. The concept of carrying capacity. Transactions of the North American Wildlife and Natural Resources Conference 20:589–602.

EFRON, B. 1975. The efficiency of logistic regression compared to normal discriminant analysis. Journal of the American Statistical Association 70:892–898.

EHRLICH, P. R., and H. A. MOONEY. 1983. Extinction, substitution, and ecosystem services. BioScience 33:248–254.

EK, A. R. 1974. Dimensional Relationships of Forest and Open Grown Trees in Wisconsin. Forestry Research Note 181. School of Natural Resources, Department of Forestry, University of Wisconsin, Madison, WI. 7 pp.

EK, A. R., and R. A. MONSERUD. 1974. FOREST: A Computer

Model for the Growth and Reproduction of Mixed Species Forest Stands. Research Report A2635. College of Agricultural and Life Sciences, University of Wisconsin, Madison, WI. 90 pp.

Elton, C. S., and R. S. Miller. 1954. The ecological survey of animal communities: With a practical system of classifying habitats by structural characters. Journal of Ecology 42:460–496.

Emlen, J. T. 1971. Population densities of birds derived from transect counts. Auk 88:323–342.

Emlen, J. T. 1977. Estimating breeding season bird densities from transect counts. Auk 94:455–468.

Emlen, J. T., and M. J. DeJong. 1981. The application of song detection threshold distance to census operations. Pp. 346–352 *in* C. J. Ralph and J. M. Scott (eds.), Estimating Numbers of Terrestrial Birds. Studies in Avian Biology 6. Cooper Ornithological Society.

Enemar, A. 1959. On the determination of the size and composition of a passerine bird population during the breeding season. Vår Fågelvärld, Supplement 2:1–114.

Engelman, L. 1981. Stepwise logistic regression. Pp. 330–344 *in* W. J. Dixon, L. Engelman, J. W. Frane, M. A. Hill, R. I. Jennrich, and J. D. Toporek (eds.), Biomedical Computer Programs, P-Series. University of California Press, Los Angeles.

Engelman, L. 1983. Stepwise logistic regression. Pp. 330–344 *in* W. J. Dixon (ed.), BMDP Statistical Software. University of California Press, Berkeley, CA.

Engen, S. 1977. Comments on two different approaches to the analysis of species frequency data. Biometrics 33:205–213.

Engstrom, R. T., and F. C. James. 1981. Plot size as a factor in winter bird population studies. Condor 83:34–41.

Euler, O. L. 1975. Inventory of wildlife habitat. Pp. 103–108 *in* V. G. Smith and P. L. Aird (eds.), Canadian Forest Inventory Methods. University of Toronto Press, Toronto, Ontario.

Evans, K. E., and R. N. Conner. 1979. Snag management. Pp. 214–255 *in* R. M. DeGraaf (tech. coord.), Management of Northcentral and Northeastern Forests for Nongame Birds: Proceedings of the Workshop. U.S. Department of Agriculture, Forest Service, General Technical Report NC–51. North Central Forest Experiment Station, St. Paul, MN.

Evans, K. E., and R. A. Kirkman. 1981. Guide to Bird Habitats of the Ozark Plateau. U.S. Department of Agriculture, Forest Service, General Technical Report NC–68. North Central Forest Experiment Station, St. Paul, MN. 79 pp.

Farmer, A. H., M. J. Armbruster, J. W. Terrell, and R. L. Schroeder. 1982. Habitat models for land-use planning: Assumptions and strategies for development. Transactions of the North American Wildlife and Natural Resources Conference 47:47–56.

Farr, W. A. 1977. Interim Thinning Guides for Even-aged Stands of Western Hemlock-Spruce in Southeast Alaska. Unpublished typescript. U.S. Department of Agriculture, Forest Service, Region 10, Juneau, AK. 7 pp.

Fasham, M. J. R. 1977. A comparison of nonmetric multidimensional scaling, principal components and reciprocal averaging for the ordination of simulated coenoclines and coenoplanes. Ecology 58:551–561.

Feeny, P. 1970. Seasonal changes in oak leaf tannins and nutrients as a cause of spring feeding by winter moth caterpillars. Ecology 51:565–581.

Fenwick, J. W. 1983. Quantifying Songbird Habitat: A Comparison of Discriminant Analysis and Logistic Regression. M.S. paper. Biostatistics Program, University of Vermont, Burlington, VT. 47 pp.

Ferns, P. N. 1980. Energy flow through small mammal populations. Mammal Review 10:165–188.

Fienberg, S. E. 1980. The Analysis of Cross-classified Categorical Data. Massachusetts Institute of Technology Press, Cambridge, MA. 198 pp.

Fight, R. D., L. D. Garrett, P. J. McNamee, and N. C. Sonntag. In press. SAMM: A Conceptual Model for Projecting Resource Impacts of Management Actions in Southeast Alaska. U.S. Department of Agriculture, Forest Service, General Technical Report series PNW. Pacific Northwest Forest and Range Experiment Station, Portland, OR.

Fish and Wildlife Service. 1980a. Habitat as a Basis for Environmental Assessment. Ecological Services Manual 101. U.S. Department of Interior, Fish and Wildlife Service, Division of Ecological Services. Government Printing Office, Washington, D.C. 28 pp.

Fish and Wildlife Service. 1980b. Habitat Evaluation Procedures (HEP). Ecological Services Manual 102. U.S. Department of Interior, Fish and Wildlife Service, Division of Ecological Services. Government Printing Office, Washington, D.C. 84 pp. + appendices.

Fish and Wildlife Service. 1980c. Human Use and Economic Evaluation (HUEE). Ecological Services Manual 104. U.S. Department of Interior, Fish and Wildlife Service, Division of Ecological Services. Government Printing Office, Washington, D.C. 26 pp. + appendices.

Fish and Wildlife Service. 1981a. Standards for the Development of Suitability Index Models. Ecological Services Manual 103. U.S. Department of Interior, Fish and Wildlife Service, Division of Ecological Services. Government Printing Office, Washington, D.C. 68 pp. + appendices.

Fish and Wildlife Service. 1981b. The Platte River Ecology Study: Special Research Report. U.S. Department of Interior, Fish and Wildlife Service, Northern Prairie Wildlife Research Center, Jamestown, ND. 203 pp.

Fisher, R. A., A. S. Corbet, and C. B. Williams. 1943. The relation between the number of species and the number of individuals in a random sample from an animal population. Journal of Animal Ecology 12:42–58.

Flather, C. H., and T. W. Hoekstra. 1985. Evaluating population-habitat models using ecological theory. Wildlife Society Bulletin 13:121–130.

Folse, L. J., Jr. 1982. An analysis of avifauna-resource relationships on the Serengeti Plains. Ecological Monographs 52:111–127.

Forest Service. 1977a. The Nation's Renewable Resources: An Assessment, 1975. Research Report 21. U.S. Department of Agriculture, Forest Service, Washington, D.C. 243 pp.

Forest Service. 1977b. Field Instructions for South Carolina, 1977. U.S. Department of Agriculture, Forest Service, Southeastern Forest Experiment Station, Asheville, NC. 36 pp.

Forgy, C. L. 1981. The OPS5 User's Manual. Technology Report CMU–CS–81–135. Computer Science Department, Carnegie-Mellon University, Pittsburgh, PA. Unpaged.

Forman, R. T. T. (ed.). 1979. Pine Barrens: Ecosystem and Landscape. Academic, New York. 601 pp.

Forman, R. T. T. 1983. Corridors in a landscape: Their ecological structure and function. Ekologia 2:375–387.

Forman, R. T. T., and R. E. Boerner. 1981. Fire frequency and the Pine Barrens of New Jersey. Bulletin of the Torrey Botanical Club 108:34–50.

Forman, R. T. T., and M. Godron. 1981. Patches and structural components for a landscape ecology. BioScience 31:733–740.

Forman, R. T. T., A. E. Galli, and C. F. Leck. 1976. Forest size and avian diversity in New Jersey woodlots with some land-use implications. Oecologia 26:1–8.

Forsman, E. D., E. C. Meslow, and H. M. Wight. 1984. Distribution and biology of the spotted owl in Oregon. Wildlife Monographs 87:1–64.

Fox, J. 1984. Linear Statistical Models and Related Methods: With Applications to Social Research. John Wiley and Sons, New York. 449 pp.

Frane, J. W. 1981. All possible subsets regression. Pp. 264–277 *in* W. J. Dixon, L. Engelman, J. W. Frane, M. A. Hill, R. I. Jennrich, and J. D. Toporek (eds.), Biomedical Computer Programs, P-Series. University of California Press, Los Angeles, CA.

Frankel, O. H., and M. E. Soulé. 1981. Conservation and Evolution. Cambridge University Press, Cambridge, England, 327 pp.

Franklin, J. F., and C. T. Dyrness. 1973. Natural Vegetation of Oregon and Washington. U.S. Department of Agriculture, Forest Service, General Technical Report PNW-8. Pacific Northwest Forest and Range Experiment Station, Portland, OR. 417 pp.

Fretwell, S. D. 1968. Habitat distribution and survival in the field sparrow (*Spizella pusilla*). Bird-Banding 34:293–306.

Fretwell, S. D. 1970. On territorial behavior and other factors influencing habitat distribution in birds. Part III. Breeding success in a local population of field sparrows (*Spizella pusilla* Wils.). Acta Biotheoretica 19:210–221.

Fretwell, S. D. 1972. Populations in a Seasonal Environment. Princeton University Press, Princeton, NJ. 217 pp.

Fretwell, S. D. 1980. Evolution of migration in relation to factors regulating bird numbers. Pp. 517–527 *in* A. Keast and E. S. Morton (eds.), Migrant Birds in the Neotropics. Smithsonian Institution, Washington, D.C.

Fretwell, S. D., and H. L. Lucas, Jr. 1970. On territorial behavior and other factors influencing habitat distribution in birds. Part I. Theoretical development. Acta Biotheoretica 19:16–36.

Freund, R. J., and R. C. Littell. 1981. SAS for Linear Models: A Guide to the ANOVA and GLM Procedures. SAS Institute, Cary, NC. 231 pp.

Freund, R. J., and P. D. Minton. 1979. Regression Methods. Marcel Dekker, New York, 261 pp.

Frith, C. R. 1974. The Ecology of the Platte River as Related to Sandhill Cranes and Other Waterfowl in South Central Nebraska. M.S. thesis. Kearney State College, Kearney, NB. 115 pp.

Fuller, M. R., and J. A. Mosher. 1981. Methods of detecting and counting raptors: A review. Pp. 235–246 *in* C. J. Ralph and J. M. Scott (eds.), Estimating Numbers of Terrestrial Birds. Studies in Avian Biology 6. Cooper Ornithological Society.

Fuller, R. J. 1982. Bird Habitats in Britain. T. and A. D. Poyser, Calton, England. 320 pp.

Galli, A. E., C. F. Leck, and R. T. T. Forman. 1976. Avian distribution patterns in forest islands of different sizes in central New Jersey. Auk 93:356–364.

Garton, E. O., and L. A. Langelier. 1985. Effects of stand characteristics on avian predators of western spruce budworm. Pp. 65–72 *in* L. Safranyik (ed.), Proceedings of the International Union of Forestry Research Organizations. Canadian Forestry Service and USDA Forest Service, Vancouver, B.C., Canada.

Gaschnig, J., P. Klahr, H. Pople, E. Shortliffe, and A. Terry. 1983. Evaluation of expert systems: Issues and case studies. Pp. 241–280 *in* F. Hayes-Roth, D. A. Waterman, and D. B. Lenat (eds.), Building Expert Systems. Addison-Wesley, Reading, MA.

Gates, C. E. 1981. Optimizing sampling frequency and numbers of transects and stations. Pp. 399–404 *in* C. J. Ralph and J. M. Scott (eds.), Estimating Numbers of Terrestrial Birds. Studies in Avian Biology 6. Cooper Ornithological Society.

Gates, J. E., and L. W. Gysel. 1978. Avian nest dispersion and fledging success in field-forest ecotones. Ecology 59:871–883.

Gauch, H. G. 1982a. Multivariate Analysis in Community Ecology. Cambridge University Press, Cambridge, England. 298 pp.

Gauch, H. G. 1982b. Noise reduction by eigenvalue ordinations. Ecology 63:1643–1649.

Geibert, E. H. 1979. Songbird diversity along a powerline right-of-way in an urbanizing Rhode Island environment. Transactions of the Northeastern Section of The Wildlife Society 36:32–44.

Geis, A. D., R. K. Martinson, and D. R. Anderson. 1969. Establishing hunting regulations and allowable harvest of mallards in the United States. Journal of Wildlife Management 33:848–859.

Gentry, J. B., E. P. Odum, M. Mason, V. Nabholz, S. Marshall, and J. T. McGinnis. 1968. Effects of altitude and forest manipulation on relative abundance of small mammals. Journal of Mammalogy 49:539–541.

Gibb, J. A. 1950. The breeding biology of the great and blue titmice. Ibis 92:507–539.

Gibb, J. A., and M. M. Betts. 1963. Food and food supply of nestling tits (Paridae) in Breckland Pine. Journal of Animal Ecology 32:489–533.

Gilbert, N. A., P. Gutierrez, B. D. Frazer, and R. E. Jones. 1976. Ecological Relationships. W. H. Freeman, San Francisco. 157 pp.

Giles, R. H., Jr. 1969. Wildlife Management Techniques. Third edition. The Wildlife Society, Washington, D.C. 623 pp.

Giles, R. H., Jr. 1978. Wildlife Management. W. H. Freeman, San Francisco. 416 pp.

Gillard, P. 1976. Classification of sequential data: A three-dimensional approach. Pp. 259–266 *in* W. T. Williams (ed.), Pattern Analysis in Agricultural Science. Elsevier, New York.

Gilpin, M. E., and R. A. Armstrong. 1981. On the concavity of island biogeographic rate functions. Journal of Theoretical Biology 20:209–217.

Gilpin, M. E., and J. M. Diamond. 1976. Calculation of immigration and extinction curves from the species-area-distance relation. Proceedings of the National Academy of Sciences (USA) 73:4130–4134.

Gilpin, M. E., and J. M. Diamond. 1981. Immigration and extinction probabilities for individual species: Relation to incidence functions and species colonization curves. Proceedings of the National Academy of Sciences (USA) 78:392–396.

Gingrich, S. F. 1967. Measuring and evaluating stocking and stand density in upland hardwood forests in the central states. Forest Science 13:38–53.

Gleason, H. A. 1926. The individualistic concept of the plant association. Bulletin of the Torrey Botanical Club 53:7–26.

Gleason, H. A. 1939. The individualistic concept of plant association. American Midland Naturalist 21:92–110.

Gluesing, E. A., and D. M. Field. 1982. Forest-Wildlife Relationships: An Assessment of the Biological State-of-the-Art. Final report, Project 0608, Mississippi Agricultural and Forestry Experiment Station. Department of Wildlife and Fisheries, Mississippi State University, Mississippi State, MS. 135 pp.

Gnanadesikan, R. 1977. Methods for Statistical Data Analysis of Multivariate Observations. John Wiley and Sons, New York. 311 pp.

Goeden, G. B. 1979. Biogeographic theory as a management tool. Environmental Conservation 6:27–32.

Goh, B. S. 1980. Management and Analysis of Biological Populations. Elsevier, New York, 288 pp.

Goldstein, E. L., M. Gross, and R. M. DeGraaf. 1981. Explorations in bird-land geometry. Urban Ecology 5:113–124.

Goldstein, E. L., M. Gross, and R. M. DeGraaf. 1983. Wildlife and greenspace planning in medium scale residential developments. Urban Ecology 7:201–214.

Goldstein, R. A. 1977. Reality and models: Difficulties associated with applying general ecological models to specific situations. Pp. 207–215 *in* D. L. Solomon and C. F. Walter (eds.), Mathematical Models in Biological Discovery. Springer-Verlag, New York.

Golley, F. B., K. Petrusewiez, and L. Ryszkowski (eds.). 1975.

Small Mammals: Their Productivity and Population Dynamics. Cambridge University Press, Cambridge, England, 451 pp.

Gotfryd, A. 1984. Urban woodlots in Ontario. American Birds 38:78–84.

Gotfryd, A., and P. Smith (eds.). 1980. Cedarvale Ravine: An Ecological and Human Use Study. Ontario Ministry of Environment, Toronto, Ontario. 211 pp.

Gould, S. J. 1979. An allometric interpretation of species-area curves: The meaning of the coefficient. American Naturalist 114:335–343.

Graber, J. W., and R. R. Graber. 1979. Severe winter weather and bird populations in southern Illinois. Wilson Bulletin 91:88–103.

Graber, J. W., and R. R. Graber. 1983a. Expectable decline of forest bird populations in severe and mild winters. Wilson Bulletin 95:682–689.

Graber, J. W., and R. R. Graber. 1983b. Feeding rates of warblers in spring. Condor 85:139–150.

Green, B. F. 1977. Parameter sensitivity in multivariate methods. Journal of Multivariate Behavioral Research 14:163–187.

Green, P. E. 1978. Analyzing Multivariate Data. Dryden, Hinsdale, IL. 519 pp.

Green, R. H. 1971. A multivariate statistical approach to the Huchinsonian niche: Bivalve molluscs in central Canada. Ecology 52:543–556.

Green, R. H. 1974. Multivariate niche analysis with temporally varying environmental factors. Ecology 55:73–83.

Green, R. H. 1979. Sampling Design and Statistical Methods for Environmental Biologists. John Wiley and Sons, New York. 257 pp.

Grenfell, W. E., B. M. Browning, and W. E. Steinecker. 1980. Food Habits of California Upland Game Birds. California Department of Fish and Game, Wildlife Management Branch Administrative Report No. 80–1, Sacramento, CA. 130 pp.

Grinnell, J., and A. H. Miller. 1944. The Distribution of the Birds of California. Pacific Coast Avifauna, Number 27. Cooper Ornithological Society. 608 pp.

Grue, C. E., R. R. Reid, and J. J. Silvy. 1983. Correlation of habitat variables with mourning dove call counts in Texas. Journal of Wildlife Management 47:186–195.

Grzybowski, J. A. 1983. Patterns of space use in grassland bird communities in winter. Wilson Bulletin 95:591–602.

Gula, T. 1977. Foraging Patterns of the Carolina Chickadee (*Parus carolinensis*) in Oak and Pine Forests of the New Jersey Pine Barrens. M.S. thesis. Rutgers University, New Brunswick, NJ. 50 pp.

Gullion, G. W. 1977. Forest manipulation for ruffed grouse. Transactions of the North American Wildlife and Natural Resources Conference 42:449–458.

Gullion, G. W., and F. J. Svoboda. 1972. Aspen—The basic habitat resource for ruffed grouse. Pp. 113–119 *in* Aspen Symposium Proceedings. U.S. Department of Agriculture, Forest Service, General Technical Report NC–1. North Central Forest Experiment Station, St. Paul, MN.

Gutiérrez, R. J. 1975. Literature Review and Bibliography of the Mountain Quail (*Oreortyx pictus*). U.S. Department of Agriculture, Forest Service, California Region, San Francisco. 33 pp.

Gutiérrez, R. J. 1980. Comparative ecology of the mountain and California quail in the Carmel Valley, California. Living Bird 18:71–93.

Haefner, J. W. 1981. Avian community assembly rules: The foliage-gleaning guild. Oecologia 50:131–142.

Haila, Y. 1983a. Land birds on northern islands: A sampling metaphor for insular colonization. Oikos 41:334–351.

Haila, Y. 1983b. Colonization of islands in a north-boreal Finnish lake by land birds. Annales Zoologici Fennici 20:179–197.

Haila, Y., and I. K. Hanski. 1984. Methodology for studying the effect of habitat fragmentation on land birds. Annales Zoologici Fennici 21:393–397.

Haila, Y., and O. Järvinen. 1981. The under-exploited potential of bird censuses in insular ecology. Pp. 559–565 *in* C. J. Ralph and J. M. Scott (eds.), Estimating Numbers of Terrestrial Birds. Studies in Avian Biology 6. Cooper Ornithological Society.

Haila, Y., and O. Järvinen. 1983. Land bird communities on a Finnish island: Species impoverishment and abundance patterns. Oikos 41:255–273.

Haila, Y., O. Järvinen, and R. A. Väisänen. 1980. Effects of changing forest structure on long-term trends in bird populations in SW Finland. Ornis Scandinavica 11:12–22.

Haila, Y., O. Järvinen, and S. Kuusela. 1983. Colonization of islands by land birds: Prevalence functions in a Finnish archipelago. Journal of Biogeography 10:499–531.

Hairston, N. G. 1980. The experimental test of an analysis of field distributions: Competition in terrestrial salamanders. Ecology 61:817–826.

Hale, B. E., A. S. Johnson, and J. L. Landers. 1982. Characteristics of ruffed grouse drumming sites in Georgia. Journal of Wildlife Management 46:115–123.

Hall, C. A. S., and J. W. Day. 1977. Systems and models: Terms and basic principles. Pp. 5–36 *in* C. A. S. Hall and J. W. Day (eds.), Ecosystem Modeling in Theory and Practice: An Introduction with Case Histories. John Wiley and Sons, New York.

Hall, G. A. 1984. A long-term bird population study in an Appalachian spruce forest. Wilson Bulletin 96:228–240.

Hall, J. G. 1960. Willow and aspen in the ecology of beaver on Sagehen Creek, California. Ecology 41:484–494.

Halperin, M., W. C. Blanckwelder, and J. I. Verter. 1971. Estimation of the multivariate logistic risk function: A comparison of the discriminant function and maximum likelihood approaches. Journal of Chronic Diseases 24:125–158.

Hamel, P. B. 1983. Breeding bird censuses 28–37. American Birds 37:59–65.

Hamel, P. B., and M. O. Efird. 1985. Wildlife and fish habitat relationships data base for the Forest Service southern region. Pp. 44–52 *in* W. C. McComb (ed.), Management of Nongame Species and Ecological Communities: Proceedings of the Workshop. College of Agriculture, University of Kentucky, Lexington, KY.

Hamel, P. B., H. E. LeGrand, Jr., M. R. Lennartz, and S. A. Gauthreaux, Jr. 1982. Bird Habitat Relationships on Southeastern Forest Lands. U.S. Department of Agriculture, Forest Service, General Technical Report SE–22. Southeastern Forest Experiment Station, Asheville, NC. 417 pp.

Hamilton, D. A. 1974. Event Probabilities Estimated by Regression. U.S. Department of Agriculture, Forest Service, Research Paper INT–152. Intermountain Forest and Range Experiment Station, Ogden, UT. 18 pp.

Hamilton, W. J., and K. E. F. Watt. 1970. Refuging. Annual Review of Ecology and Systematics 1:263–286.

Hammack, J., and G. M. Brown, Jr. 1974. Waterfowl and Wetlands: Toward Bioeconomic Analysis. Resources for the Future, Washington, D.C. 95 pp.

Hammill, J. H., and L. G. Visser. 1984. Status of Aspen in Northern Michigan as Ruffed Grouse Habitat. Michigan Department of Natural Resources, Wildlife Division Report 2976. Lansing, MI. 14 pp.

Hansson, L. 1983. Bird numbers across edges between mature conifer forest and clearcuts in central Sweden. Ornis Scandinavica 14:97–103.

Hardy, T. B., C. G. Prewitt, and K. A. Voos. 1982. Application of a physical habitat usability model to the fish community in a spring-fed desert stream. Pp. 391–397 *in* W. K. Lavenroth, G. V. Skogerboe, and M. Fug (eds.), Analysis of Ecological Systems: State-of-the-Art in Ecological Modelling. Elsevier, New York.

Harris, L. D. 1984. The Fragmented Forest: Island Biogeography Theory and the Preservation of Biotic Diversity. University of Chicago Press, Chicago. 211 pp.

Harris, W. F., R. A. Goldstein, and D. Sollins. 1973. Net aboveground production and estimates of standing biomass on Walker Branch Watershed. Pp. 71–80 *in* H. E. Young (ed.), Proceedings of the IUFRO Conference on Forest Biomass. University of Maine Press, Orono, ME.

Hatt, R. T. 1929. The red squirrel: Its life history and habits with special reference to the Adirondacks of New York and the Harvard Forest. Roosevelt Wildlife Annals 2:10–46.

Havens, A. V. 1979. Climate and microclimate of the New Jersey Pine Barrens. Pp. 113–131 *in* R. T. T. Forman (ed.), Pine Barrens: Ecosystem and Landscape. Academic, New York.

Hawkes, C. L., D. E. Chalk, T. W. Hoekstra, and C. H. Flather. 1983. Prediction of Wildlife and Fish Resources for National Assessments and Appraisals. U.S. Department of Agriculture, Forest Service, General Technical Report RM–100. Rocky Mountain Forest and Range Experiment Station, Fort Collins, CO. 21 pp.

Hayes-Roth, F., D. A. Waterman, and D. B. Lenat (eds.). 1983. Building Expert Systems. Addison-Wesley, Reading, MA. 444 pp.

Hayward, C. L., C. Cottam, A. W. Woodbury and H. H. Frost. 1976. Birds of Utah. Great Basin Naturalist Memoirs, Number 1. 229 pp.

Heatwole, H. 1982. A review of structuring in herpetofaunal assemblages. Pp. 1–19 *in* N. J. Scott, Jr. (ed.), Herpetological Communities. Wildlife Research Report 13. U.S. Department of Interior, Fish and Wildlife Service. Washington, D.C.

Helle, E., and P. Helle. 1979. Changes in land bird populations on the Krunnit Islands in the Bothnian Bay, 1939–77. Ornis Fennica 56:137–147.

Helle, E., and P. Helle. 1982. Edge effect on forest bird densities on off-shore islands in the northern Gulf of Bothnia. Annales Zoologici Fennici 19:165–169.

Helle, P. 1984. Effects of habitat area on breeding bird communities: A study in northeastern Finland. Annales Zoologici Fennici 21:421–425.

Heller, R. C., G. E. Doverspike, and R. C. Aldrich. 1964. Identification of Tree Species on Large Scale Panchromatic and Color Aerial Photography. U.S. Department of Agriculture, Forest Service, Agricultural Handbook Number 261. U.S. Government Printing Office, Washington, D.C. 17 pp.

Helliwell, D. R. 1976. The effects of size and isolation on the conservation value of wooded sites in Britain. Journal of Biogeography 3:407–416.

Helwig, J. T., and K. A. Council (eds.). 1979. SAS User's Guide. 1979 edition. SAS Institute, Cary, NC. 494 pp.

Henderson, J. A., L. S. Davis, and E. M. Ryberg. 1978. ECOSYM, A Classification and Information System for Wildland Management. Department of Forest Resources, Utah State University, Logan, UT. 30 pp. + appendices.

Henny, C. J., W. S. Overton, and H. M. Wight. 1970. Determining parameters for populations by using structural models. Journal of Wildlife Management 34:690–703.

Hett, J., R. D. Taber, J. Long, and J. W. Schoen. 1978. Forest management policies and elk summer range carrying capacity in the *Abies amabalis* forest, western Washington. Environmental Management 2:56–66.

Higgs, A. J. 1981. Island biogeography theory and nature reserve design. Journal of Biogeography 8:117–124.

Higgs, A. J., and M. B. Usher. 1980. Should nature reserves be large or small? Nature 285:568–569.

Hilborn, R. 1979. Some failures and successes in applying systems analysis to ecological systems. Journal of Applied Systems Analysis 6:25–31.

Hildén, O. 1965. Habitat selection in birds: A review. Annales Zoologici Fennici 2:53–75.

Hill, M. O. 1973. Reciprocal averaging: An eigenvector method of ordination. Journal of Ecology 61:237–241.

Hill, M. O. 1974. Correspondence analysis: A neglected multivariate method. Journal of the Royal Statistical Society, Series C, 23:340–354.

Hill, M. O. 1979a. DECORANA—A FORTRAN Program for Detrended Correspondence Analysis and Reciprocal Averaging. Cornell University Press, Ithaca, NY. 52 pp.

Hill, M. O. 1979b. TWINSPAN—A FORTRAN Program for Arranging Multivariate Data in an Ordered Two-way Table by Classification of the Individuals and Attributes. Cornell University Press, Ithaca, NY. 47 pp.

Hinchen, J. D. 1970. Multiple regression with unbalanced data. Journal of Quality Technology 2:22–29.

Hiner, L. E. 1938. Observations on the foraging habits of beavers. Journal of Mammalogy 19:317–319.

Hochbaum, G. S., and F. D. Caswell. 1978. A Forecast of Long-term Trends in Breeding Mallard Populations on the Canadian Prairies. Canadian Wildlife Service Progress Notes 90. Canadian Wildlife Service, Ottawa, Ontario. 8 pp.

Hoekstra, T. W., and C. T. Cushwa. 1979. Compilation of avian information in computerized data storage and retrieval systems of the north central and northeastern United States. Pp. 238–244 *in* R. M. DeGraaf (tech. coord.), Management of Northcentral and Northeastern Forests for Nongame Birds: Proceedings of the Workshop. U.S. Department of Agriculture, Forest Service, General Technical Report NC–51. North Central Forest Experiment Station, St. Paul, MN.

Hogstad, O. 1967. The edge effect on species and population density of some passerine birds. Nytt Magasin for Zoologi 15:40–43.

Hohlova, T. Yu. 1977. Ekologo-faunisticheskaya charakteristika gnesdovoi ornitofaunij Zaonezhya. Vestnik LGU, Seriya Biologii 15:22–30. Leningrad, USSR.

Holbrook, H. L. 1974. A system of wildlife habitat management on southern national forests. Wildlife Society Bulletin 2:119–123.

Holdaway, M. R., and G. J. Brand. 1983. An Evaluation of the STEMS Tree Growth Projection System. U.S. Department of Agriculture, Forest Service, Research Paper NC–234. North Central Forest Experiment Station, St. Paul, MN. 20 pp.

Holling, C. S. (ed.). 1978. Adaptive Environmental Assessment and Management. John Wiley and Sons, New York. 377 pp.

Holmes, R. T. 1981. Theoretical aspects of habitat use by birds. Pp. 33–37 *in* D. E. Capen (ed.), The Use of Multivariate Statistics in Studies of Wildlife Habitat. U.S. Department of Agriculture, Forest Service, General Technical Report RM–87. Rocky Mountain Forest and Range Experiment Station, Fort Collins, CO.

Holmes, R. T., R. E. Bonney, Jr., and S. W. Pacala. 1979. Guild structure of the Hubbard Brook bird community: A multivariate approach. Ecology 60:512–520.

Hooper, M.D. 1971. The size and surroundings of nature reserves. Pp. 555–561 *in* E. Duffey and A. S. Watt (eds.), The Scientific Management of Animal and Plant Communities for Conservation. Blackwell, Oxford, England.

Hoover, W. H., and E. L. Park. 1983. 1983 Price report—third quarter. Pp. 16–18 *in* B. C. Fischer (ed.), Indiana Forest Products Marketing and Wood Utilization Report. Purdue University Press, West Lafayette, IN.

HORN, H. S. 1975. Markovian properties of forest succession. Pp. 196–211 *in* M. L. Cody and J. M. Diamond (eds.), Ecology and Evolution of Communities. Harvard University Press, Cambridge, MA.

HOWE, R. W., and G. JONES. 1977. Avian utilization of small woodlots in Dane County, Wisconsin. Passenger Pigeon 39:313–319.

HUBBARD, J. P. 1970. Checklist of the Birds of New Mexico. New Mexico Ornithological Society Publication Number 3. New Mexico Ornithological Society, Albuquerque, NM. 108 pp.

HUEY, W. S. 1956. New Mexico Beaver Management. New Mexico Department of Game and Fish Bulletin Number 4. New Mexico Department of Game and Fish, Sante Fe, NM. 49 pp.

HULL, C. H., and N. H. NIE. 1981. SPSS Update 7–9. McGraw-Hill, New York. 402 pp.

HUNGERFORD, K. E. 1969. Influence of forest management on wildlife. Pp. 39–41 *in* H. C. Black (ed.), Wildlife and Reforestation in the Pacific Northwest. School of Forestry, Oregon State University, Corvallis, OR.

HURLBERT, S. H. 1984. Pseudoreplication and the design of ecological field experiments. Ecological Monographs 54:187–211.

HURLEY, J. F., and E. S. ASROW. 1980. Western Sierra Nevada Wildlife/Habitat Relationships Computer Use Manual. Wildlife Habitat Relationships Applications Note 80–1. U.S. Department of Agriculture, Forest Service, Pacific Southwest Region, San Francisco. 155 pp.

HURLEY, J. F., H. SALWASSER, and K. SHIMAMOTO. 1982. Fish and wildlife habitat capability models and special habitat criteria. Cal-Neva Wildlife Transactions 1982:40–48.

HURST, G. A. 1972. Insects and bobwhite quail brood habitat management. Proceedings of the National Bobwhite Quail Symposium 1:65–82.

HUTCHINSON, G. E. 1957. Concluding remarks. Cold Spring Harbor Symposium on Quantitative Biology 22:415–427.

HUTCHINSON, G. E. 1978. An Introduction to Population Ecology. Yale University Press, New Haven, CT. 260 pp.

INKLEY, D. B. 1980. A Quantitative Analysis of Forest Habitat Selected by Breeding Songbirds in Grafton, Vermont. M.S. thesis. University of Vermont, Burlington, VT. 68 pp.

JACCARD. P. 1912. The distribution of flora in the alpine zone. New Phytologist 11:37–50.

JAKSIĆ, F. M. 1981. Abuse and misuse of the term "guild" in ecological studies. Oikos 37:397–400.

JAMES, F. C. 1971. Ordinations of habitat relationships among breeding birds. Wilson Bulletin 83:215–236.

JAMES, F. C., and C. E. MCCULLOCH. 1985. Data analysis and the design of experiments in ornithology. Pp. 1–102 *in* R. F. Johnston (ed.), Current Ornithology. Volume 2. Plenum, New York and London.

JAMES, F. C., and H. H. SHUGART, JR. 1970. A quantitative method of habitat description. Audubon Field Notes 24:727–736.

JAMES, F. C., R. F. JOHNSTON, N. O. WAMER, G. J. NIEMI, and W. J. BOECKLEN. 1984. The Grinnellian niche of the wood thrush. American Naturalist 124:17–47.

JÄRVINEN, A. 1983. Breeding strategies of hole-nesting passerines in northern Lapland. Annales Zoologici Fennici 20:129–149.

JÄRVINEN, O. 1980. Dynamics of North-European bird communities. Pp. 770–776 *in* R. Nöhring (ed.), Acta XIII Congressus Internationalis Ornithologici. Verlag der Deutschen Ornithologen-Gesellschaft, Zoologischer Garten, Berlin.

JÄRVINEN, O., and Y. HAILA. 1984. Assembly of land bird communities on northern islands: A quantitative analysis of insular impoverishment. Pp. 138–147 *in* D. R. Strong, S. Simberloff, L. G. Abele, and A. B. Thistle (eds.), Ecological Communities: Conceptual Issues and the Evidence. Princeton University Press, Princeton, NJ.

JÄRVINEN, O., and R. A. VÄISÄNEN. 1979. Changes in bird populations as criteria of environmental changes. Holarctic Ecology 2:75–80.

JÄRVINEN, O., and R. A. VÄISÄNEN. 1981. Methodology for censusing land bird faunas in large regions. Pp. 146–151 *in* C. J. Ralph and J. M. Scott (eds.), Estimating Numbers of Terrestrial Birds. Studies in Avian Biology 6. Cooper Ornithological Society.

JENKINS, R. E. 1977. Classification and inventory for the perpetuation of ecological diversity. Pp. 41–51 *in* A. Marmelstein (ed.), Classification, Inventory, and Analysis of Fish and Wildlife Habitat. FWS/OBS–78/76. U.S. Department of Interior, Fish and Wildlife Service, Office of Biological Services, Washington, D.C.

JENKINS, S. H. 1975. Food selection by beavers: A multidimensional contingency table analysis. Oecologia 21:157–173.

JENKINS, S. H. 1979. Seasonal and year-to-year differences in food selection by beavers. Oecologia 44:112–116.

JENKINS, S. H. 1980. A size-distance relation in food selection by beavers. Ecology 61:740–746.

JENNRICH, R. I., and P. SAMPSON. 1981. Stepwise discriminant analysis. Pp. 519–537 *in* W. J. Dixon, L. Engelman, J. W. Frane, M. A. Hill, R. I. Hennrich, and J. D. Toporek (eds.), Biomedical Computer Programs, P-Series. University of California Press, Los Angeles.

JOHNSGARD, J. S. 1963. Temperature and Water Balance for Oregon Weather Stations. Special Report 150. Agricultural Experiment Station, Oregon State University, Corvallis, OR. 127 pp.

JOHNSON, D. H. 1979. Estimating nest success: The Mayfield method and an alternative. Auk 96:651–661.

JOHNSON, D. H. 1981a. The use and misuse of statistics in wildlife habitat studies. Pp. 11–19 *in* D. E. Capen (ed.), The Use of Multivariate Statistics in Studies of Wildlife Habitat. U.S. Department of Agriculture, Forest Service, General Technical Report RM–87. Rocky Mountain Forest and Range Experiment Station, Fort Collins, CO.

JOHNSON, D. H. 1981b. How to measure habitat—a statistical perspective. Pp. 53–57 *in* D. E. Capen (ed.), The Use of Multivariate Statistics in Studies of Wildlife Habitat. U.S. Department of Agriculture, Forest Service, General Technical Report RM–87. Rocky Mountain Forest and Range Experiment Station, Fort Collins, CO.

JOHNSON, K. N. 1985. FORPLAN (Version II): A Mathematical Programmer's Guide. U.S. Department of Agriculture, Forest Service, Systems Applications Unit for Land Management Planning, Fort Collins, CO. Unpublished review draft.

JOHNSON, K. N., D. P. JONES, and B. KENT. 1980. Forest Planning Model (FORPLAN): User's Guide and Operations Manual. U.S. Department of Agriculture, Forest Service, Systems Applications Unit for Land Management Planning, Fort Collins, CO. 258 pp.

JOHNSON, M. P., and D. S. SIMBERLOFF. 1974. Environmental determinants of island species number in the British Isles. Journal of Biogeography 1:149–154.

JOHNSON, N. K. 1975. Controls of number of bird species on montane islands in the Great Basin. Evolution 29:545–567.

JOHNSON, W. C., and D. M. SHARPE. 1976. Forest dynamics in the northern Georgia Piedmont. Forest Science 22:307–322.

JOHNSTON, D. W., and E. P. ODUM. 1956. Breeding bird populations in relation to plant succession on the Piedmont of Georgia. Ecology 37:50–62.

JONES, C. A. 1973. The Conservation of Chalk Downland in Dorset. Report. Dorset County Council, Dorchester, Dorset, England. 48 pp.

JONES, J. H., and N. S. SMITH. 1979. Bobcat density and prey selection in central Arizona. Journal of Wildlife Management 43:666–672.

Kaminski, R. M., and H. H. Prince. 1981. Dabbling duck and aquatic macroinvertebrate responses to manipulated wetland habitat. Journal of Wildlife Management 45:1–15.

Karr, J. R. 1971. Structure of avian communities in selected Panama and Illinois habitats. Ecological Monographs 41:207–233.

Karr, J. R. 1980. Geographical variation in the avifaunas of tropical forest undergrowth. Auk 97:283–298.

Karr, J. R. 1982. Avian extinction on Barro Colorado Island, Panama: A reassessment. American Naturalist 119:220–239.

Karr, J. R., and K. E. Freemark. 1983. Habitat selection and environmental gradients: Dynamics in the "stable" tropics. Ecology 64:1481–1494.

Keast, A., and E. S. Morton (eds.). 1980. Migrant Birds in the Neotropics: Ecology, Behavior, Distribution, and Conservation. Smithsonian Institution, Washington, D.C. 576 pp.

Kelker, G. H. 1964. Appraisal of ideas advanced by Aldo Leopold thirty years ago. Journal of Wildlife Management 28:180–185.

Kendall, M., and A. Stuart. 1979. The Advanced Theory of Statistics. Volume 2. Inference and Relationship. Fourth edition. Macmillan, New York. 748 pp.

Kendeigh, S. C. 1944. Measurement of bird populations. Ecological Monographs 14:67–106.

Kendeigh, S. C. 1946. Breeding birds of the beech-maple-hemlock community. Ecology 27:226–245.

Kessell, S. R. 1979. Gradient Modelling. Resource and Fire Management. Springer-Verlag, New York. 432 pp.

Kessell, S. R., R. B. Good, and M. W. Potter. 1982. Computer Modeling in Natural Area Management. Special Publication Number 9. Australian National Parks and Wildlife Service, Canberra, Australia. 45 pp.

Kie, J. G., M. White, and D. L. Drawe. 1983. Condition parameters of white-tailed deer in Texas. Journal of Wildlife Management 47:583–594.

Kincaid, W. B., and E. H. Bryant. 1983. A geometric method for evaluating the null hypothesis of random habitat utilization. Ecology 64:1463–1470.

Kincaid, W. B., G. N. Cameron, and B. A. Carnes. 1983. Patterns of habitat utilization in sympatric rodents on the Texas coastal prairie. Ecology 64:1471–1480.

Kirby, P. C. 1984. Natural diversity requirements in environmental legislation affecting National Forest planning (except the National Forest Management Act). Pp. 11–20 *in* J. L. Cooley and J. H. Cooley (eds.), Natural Diversity in Forest Ecosystems: Proceedings of the Workshop. Institute of Ecology, Athens, GA.

Kirsch, L. M., H. F. Duebbert, and A. D. Kruse. 1978. Grazing and haying effects on habitats of upland nesting birds. Transactions of the North American Wildlife and Natural Resources Conference 43:486–497.

Kleinbaum, D. G., L. L. Kupper, and H. Morgenstern. 1982. Epidemiological Research: Principals and Quantitative Methods. Lifetime Learning, Belmont, CA. 529 pp.

Klopfer, P. 1963. Behavioral aspects of habitat selection: The role of early experience. Wilson Bulletin 75:15–22.

Kluyver, H. N., and L. Tinbergen. 1953. Territory and the regulation of density in titmice. Archives Neerlandaises de Zoologie 10:265–289.

Koford, R. R. 1979. Behavior and Ecology of a Population of *Tamiasciurus douglasii*. Ph.D. dissertation. University of California, Berkeley, CA. 129 pp.

Konopasek, M., and S. Jayaraman. 1984. Expert systems for personal computers. Byte 9:137–156.

Kotler, B. P. 1984. Risk of predation and the structure of desert rodent communities. Ecology 65:689–701.

Kozlowski, T. T., and J. J. Clausen. 1966. Shoot growth characteristics of heterophyllous woody plants. Canadian Journal of Botany 44:826–843.

Krapu, G. L., A. T. Klett, and D. G. Jorde. 1983. The effect of variable spring water conditions on mallard reproduction. Auk 100:689–698.

Krapu, G. L., D. E. Facey, E. K. Fritzell, and D. H. Johnson. 1984. Habitat use by migrant sandhill cranes in Nebraska. Journal of Wildlife Management 48:407–417.

Krebs, J. R. 1971. Territory and breeding density in the great tit, *Parus major* L. Ecology 52:1–22.

Krebs, J. R. 1980. Ornithologists as unconscious theorists. Auk 97:411.

Kroodsma, R. L. 1982. Bird community ecology on power-line corridors in eastern Tennessee. Biological Conservation 23:79–84.

Küchler, A. W. 1977. Potential Natural Vegetation of California. (Map) *in* M. G. Barbour and J. Major (eds.), Terrestrial Vegetation of California. John Wiley and Sons, New York. 1002 pp.

Laake, J. L., K. P. Burnham, and D. R. Anderson. 1979. User's Manual for the Program TRANSECT. Utah State University Press, Logan, UT. 26 pp.

Lachenbruch, P. A. 1975. Discriminant Analysis. Hafner, New York. 128 pp.

Lachenbruch, P. A., and M. R. Mickey. 1968. Estimation of error rates in discriminant analysis. Technometrics 10:1–11.

Lack, D. 1937. The psychological factor in bird distribution. British Birds 31:130–136.

Lack, D. 1942. Ecological features of bird faunas of small British islands. Journal of Animal Ecology 11:9–36.

Lack, D. 1966. Population Studies of Birds. Oxford University Press, Oxford, England. 341 pp.

Lack, D. 1976. Island Biology: Illustrated by the Land Birds of Jamaica. Blackwell, Oxford, England. 445 pp.

Lance, G. N., and W. T. Williams. 1967. A general theory of classificatory sorting strategies. Part I. Hierarchical systems. Computer Journal 9:373–380.

Lancia, R. A., D. W. Hazel, S. D. Miller, and J. D. Hair. 1981. Computer mapping of potential habitat quality for bobcat based on digital LANDSAT imagery. Pp. 766–770 *in* T. B. Brann, L. O. House, and H. G. Lund (eds.), In-place Resources Inventories: Principles and Practices. Society of American Foresters. Bethesda, MD.

Lancia, R. A., S. D. Miller, D. A. Adams, and D. W. Hazel. 1982. Validating habitat quality assessment: An example. Transactions of the North American Wildlife and Natural Resources Conference 47:96–110.

Landers, J. L., and A. S. Johnson. 1976. Bobwhite Quail Food Habits in the Southeastern United States with a Seed Key to Important Foods. Miscellaneous Publication Number 4. Tall Timbers Research Station, Tallahassee, FL. 90 pp.

Landwehr, J. M., D. Pregibon, and A. C. Shoemaker. 1984. Graphical methods for assessing logistic regression models. Journal of the American Statistical Association 79:61–71.

Larson, D. L. B. 1984. Bird-centered Habitat Analysis of Shrubsteppe Birds in Northwestern New Mexico. M.A. thesis. University of Colorado, Boulder, CO. 72 pp.

Larsson, S., and O. Tenow. 1979. Utilization of dry matter and bioelements in larvae of *Neodiprion sertifer* Geoffr. (Hymenoptera: Diprionidae) feeding on Scotch pine (*Pinus sylvestris* L.). Oecologia 43:157–172.

Laudenslayer, W. F., Jr., and W. E. Grenfell, Jr. 1983. A list of amphibians, reptiles, birds and mammals of California. Outdoor California 44(1):5–14.

LAURSEN, S. 1984. Predicting Shrub Community Composition and Structure Following Management Disturbance in Forest Ecosystems of the Intermountain West. Ph.D. dissertation. University of Idaho, Moscow, ID. 261 pp.

LAYMON, S. A., H. SALWASSER and R. H. BARRETT. 1985. Habitat Suitability Index Models: Spotted Owl. Biological Report 82(10.113). U.S. Department of Interior, Fish and Wildlife Service, Biological Services Program, Washington, D.C. 14 pp.

LEE, D. B., JR. 1973. Requiem for large scale models. Journal of the American Institute of Planners 34:163–177.

LEE, J. L., C. N. REED, R. L. FROSH, L. C. SULLIVAN, and F. D'ERCHIA. 1985. MOSS Users Manual (Version 85.01) WELUT-85/W04. Western Energy and Land Use Team. U.S. Fish and Wildlife Service. Fort Collins, CO. 70 pp. + tables.

LEGENDRE, L., and P. LEGENDRE. 1983. Numerical Ecology. Elsevier, New York. 419 pp.

LEHMANN, V. W. 1953. Bobwhite population fluctuations and vitamin A. Transactions of the North American Wildlife and Natural Resources Conference 18:199–245.

LEHMKUHL, J. F., and D. R. PATTON. 1982. User's Manual for the RUN WILD III Data Storage and Retrieval System. Unpublished typescript. U.S. Department of Agriculture, Forest Service, Southwestern Region, Wildlife Unit, Albuquerque, NM. 68 pp.

LEOPOLD, A. 1933. Game Management. Charles Scribner's Sons, New York. 481 pp.

LEOPOLD, A. 1949. A Sand County Almanac. Oxford University Press, New York. 226 pp.

LEOPOLD, A. S., R. J. GUTIÉRREZ, and M. T. BRONSON. 1981. North American Gamebirds and Mammals. Charles Scribner's Sons, New York. 198 pp.

LEVINS, R., and D. CULVER. 1971. Regional coexistence of species and competition between rare species. Proceedings of the National Academy of Sciences (USA) 68:1246–1248.

LEWIS, J. C. 1974. Ecology of the Sandhill Crane in the Southeastern Central Flyway. Ph.D. dissertation. Oklahoma State University, Stillwater, OK. 213 pp.

LINDSAY, R. K., B. G. BUCHANAN, E. A. FEIGENBAUM, and J. LEDERBERG. 1980. Applications of Artificial Intelligence for Organic Chemistry: The DENDRAL Project. McGraw-Hill, New York. 194 pp.

LINES, I. L., JR., and C. J. PERRY. 1978. A numerical wildlife habitat evaluation procedure. Transactions of the North American Wildlife and Natural Resources Conference 43:284–301.

LINSDALE, J. M. 1936. The Birds of Nevada. Pacific Coast Avifauna, Number 23. Cooper Ornithological Society. 145 pp.

LINSDALE, J. M. 1938. Environmental responses of vertebrates in the Great Basin. American Midland Naturalist 19:1–206.

LOCKERD, M. J. 1980. Small Mammal Niche Patterns in Old Field Communities. M. S. thesis. University of Arkansas, Fayetteville, AR. 33 pp.

LOCKERD, M. J., D. JAMES, and J. E. DUNN. 1979. Estimating small mammal populations at the Buffalo National River in Arkansas. Pp. 243–257 *in* Proceedings of the Second Conference on Scientific Research in National Parks, Volume 12. National Park Service, Washington, D.C.

LOVEJOY, T. E. 1975. Bird diversity and abundance in Amazonian forest communities. Living Bird 13:127–191.

LOVEJOY, T. E., J. M. RANKIN, R. O. BIERREGAARD, JR., K. S. BROWN, JR., L. H. EMMONS, and M. E. VAN DER VOORT. 1984. Ecosystem decay of Amazon forest fragments. Pp. 296–325 *in* M. H. Nitecki (ed.), Extinction. University of Chicago Press, Chicago.

LOVVORN, J. R., and C. M. KIRKPATRICK. 1981. Roosting behavior and habitat of migrant greater sandhill cranes. Journal of Wildlife Management 45:842–857.

LYNCH, J. F., and N. K. JOHNSON. 1974. Turnover and equilibria in insular avifaunas, with special reference to the California Channel Islands. Condor 76:370–381.

LYNCH, J. F., and D. F. WHIGHAM. 1984. Effects of forest fragmentation on breeding bird communities in Maryland, U.S.A. Biological Conservation 28:287–324.

LYNCH, J. F., and R. F. WHITCOMB. 1978. Effects of the insularization of the eastern deciduous forest on avifaunal diversity and turnover. Pp. 461–489 *in* A. Marmelstein (ed.), Classification, Inventory, and Analysis of Fish and Wildlife Habitat. FWS/OBS–78/76. U.S. Department of Interior, Fish and Wildlife Service, Office of Biological Services, Washington, D.C.

LYON, L. J. 1983. Road density models describing habitat effectiveness for elk. Journal of Forestry 81:592–595.

LYON, L. J., and J. M. MARZLUFF. 1985. Fire effects on a small bird population. Pp. 16–22 *in* J. E. Lotan and J. K. Brown (tech. compilers), Fire Effects on Wildlife Habitat: Symposium Proceedings. U.S. Department of Agriculture, Forest Service, General Technical Report INT-186. Intermountain Research Station, Ogden, UT.

MACARTHUR, R. H. 1972. Geographical Ecology. Harper and Row, New York. 269 pp.

MACARTHUR R. H., and J. W. MACARTHUR. 1961. On bird species diversity. Ecology 42:594–598.

MACARTHUR, R. H., and E. O. WILSON. 1963. An equilibrium theory of insular biogeography. Evolution 17:373–387.

MACARTHUR, R. H., and E. O. WILSON. 1967. The Theory of Island Biogeography. Princeton University Press, Princeton, NJ. 203 pp.

MACARTHUR, R. H., H. RECHER, and M. CODY. 1966. On the relation between habitat selection and species diversity. American Naturalist 100:310–332.

MCCLELLAND, B. R. 1979. The pileated woodpecker in forests of the northern Rocky Mountains. Pp. 283–299 *in* J. G. Dickson, R. N. Conner, R. R. Fleet, J. C. Kroll, and J. A. Jackson (eds.), The Role of Insectivorous Birds in Forest Ecosystems. Academic, New York.

MACCLINTOCK, L., R. F. WHITCOMB, and B. L. WHITCOMB. 1977. Island biogeography and "habitat islands" of eastern forest. Part II. Evidence for the value of corridors and minimization of isolation in preservation of biotic diversity. American Birds 31:6–16.

MCCOY, E. D. 1982. The application of island-biogeographic theory to forest tracts: Problems in the determination of turnover rates. Biological Conservation 22:217–227.

MCCULLAGH, P., and J. A. NELDER. 1983. Generalized Linear Models. Chapman and Hall, New York. 261 pp.

MCDERMOTT, J. 1982. RI: A rule-based configurer of computer systems. Artificial Intelligence 19:39–88.

MCDIARMID, R. W., R. E. RICKLEFS, and M. S. FOSTER. 1977. Dispersal of *Stemmadenia donnel-smithii* (Apocynaceae) by birds. Biotropica 9:9–25.

MCGUINNESS, K. A. 1984. Equations and explanations in the study of species-area curves. Biological Review 59:423–440.

MCKNIGHT, D. E. 1979. The history of deer research in Alaska. Pp. 2–10 *in* O. C. Wallmo and J. W. Schoen (eds.), Sitka Black-tailed Deer: Proceedings of a Conference in Juneau, Alaska. Series Number R10–48. U.S. Department of Agriculture, Forest Service, Region 10, in cooperation with Alaska Department of Fish and Game. Juneau, AK.

MAGNUSSON, W. E. 1983. Use of discriminant function to characterize ruffed grouse drumming sites in Georgia: A critique. Journal of Wildlife Management 47:1151–1152.

MAGUIRE, B. 1973. Niche response structure and the analytical potentials of its relationship to the habitat. American Naturalist 107:213–246.

MANNAN, R. W., M. L. MORRISON, and E. C. MESLOW. 1984. Comment: The use of guilds in forest bird management. Wildlife Society Bulletin 12:426–430.

MARASCUILO, L. A., and J. R. LEVIN. 1983. Multivariate Statistics in the Social Sciences: A Researcher's Guide. Brooks-Cole, Monterey, CA. 580 pp.

MARASCUILO, L. A., and M. MCSWEENEY. 1977. Non-parametric and Distribution-free Methods for the Social Sciences. Brooks-Cole, Monterey, CA. 556 pp.

MARCHANT, J. H. 1981. Residual edge effects with the mapping bird census method. Pp. 488–491 *in* C. J. Ralph and J. M. Scott (eds.), Estimating Numbers of Terrestrial Birds. Studies in Avian Biology 6. Cooper Ornithological Society.

MARCOT, B. G. (ed.). 1979. California Wildlife/Habitat Relationships Program: North Coast/Cascades Zone. Volumes 1–4. U.S. Department of Agriculture, Forest Service, Six Rivers National Forest, Eureka, CA. 899 pp.

MARCOT, B. G. 1980. Use of a habitat-niche model for old growth management: A preliminary discussion. Pp. 390–402 *in* R. M. DeGraaf (ed.), Management of Western Forests and Grasslands for Nongame Birds. U.S. Department of Agriculture, Forest Service, General Technical Report INT–86. Intermountain Forest and Range Experiment Station, Ogden, UT.

MARCOT, B. G., M. G. RAPHAEL, and K. H. BERRY. 1983. Monitoring wildlife habitat and validation of wildlife-habitat relationships models. Transactions of the North American Wildlife and Natural Resources Conference 48:315–329.

MARKS, J. S., J. H. DOREMUS, and A. R. BAMMAN. 1980. Black-throated sparrows breeding in Idaho. The Murrelet 61:112–113.

MARRIOT, F. H. C. 1974. The Interpretation of Multiple Observations. Academic, New York and London. 117 pp.

MARSDEN, S. J., and J. H. MARTIN. 1949. Turkey Management. Fifth edition. The Interstate, Danville, IL. 774 pp.

MARSTON, N. 1965. Recent modifications in the design of Malaise insect traps with a summary of the insects represented in collections. Journal of the Kansas Entomological Society 38:154–162.

MARTIN, F. W., R. S. POSPAHALA, and J. D. NICHOLS. 1979. Assessment and population management of North American migratory game birds. Pp. 187–239 *in* J. Cairns, G. P. Patil, and W. E. Waters (eds.), Environmental Biomonitoring, Assessment, Prediction and Management—Certain Case Studies and Related Quantitative Issues. International Co-operative Publishing House, Fairland, MD.

MARTIN, N.D. 1960. An analysis of bird populations in relation to forest succession in Algonquin Provincial Park, Ontario. Ecology 41:126–140.

MARTIN, T. E. 1981a. Species-area slopes and coefficients: A caution on their interpretation. American Naturalist 118:823–837.

MARTIN, T. E. 1981b. Limitation in small habitat islands: Chance or competition? Auk 98:715–734.

MARTINKA, R. R. 1972. Structural characteristics of blue grouse territories in southwestern Montana. Journal of Wildlife Management 36:498–510.

MARZLUFF, J. M., and L. J. LYON. 1983. Snags as indicators of habitat suitability for open-nesting birds. Pp. 140–146 *in* J. W. Davis, G. A. Goodwin, and R. A. Ockenfels (tech. coords.), Snag Habitat Management: Proceedings of the Symposium. U.S. Department of Agriculture, Forest Service, General Technical Report RM–99. Rocky Mountain Forest and Range Experiment Station, Fort Collins, CO.

MASER, C., B. R. MATE, J. F. FRANKLIN, and C. T. DYRNESS. 1981. Natural History of Oregon Coast Mammals. U.S. Department of Agriculture, Forest Service, General Technical Report PNW–133. Pacific Northwest Forest and Range Experiment Station, Portland, OR. 496 pp.

MASON, R. O., and I. I. MITROFF. 1981. Challenging Strategic Planning Assumptions: Theory, Cases, and Techniques. John Wiley and Sons, New York. 324 pp.

MATTHEWS, R. W., and J. R. MATTHEWS. 1970. Malaise trap studies of flying insects in a New York mesic forest. Part I. Ordinal composition and seasonal abundance. Journal of the New York Entomological Society 78:52–59.

MAY, R. M. 1975. Patterns of species abundance and diversity. Pp. 81–120 *in* M. L. Cody and J. M. Diamond (eds.), Ecology and Evolution of Communities. Belknap (Harvard University), Cambridge, MA.

MAY, R. M. 1976a. Patterns in multi-species communities. Pp. 142–162 *in* R. M. May (ed.), Theoretical Ecology: Principles and Applications. Saunders, Philadelphia, PA.

MAY, R. M. (ed.). 1976b. Theoretical Ecology: Principles and Applications. Saunders, Philadelphia, PA. 489 pp.

MAY, R. M. 1981. Modeling recolonization by neotropical migrants in habitats with changing patch structure, with notes on the age structure of populations. Pp. 207–213 *in* R. L. Burgess and D. L. Sharpe (eds.), Forest Island Dynamics in Man-Dominated Landscapes. Springer-Verlag, New York, Heidelberg, and Berlin.

MAYER, K. E. 1984. A review of selected remote sensing and computer technologies applied to wildlife habitat inventories. California Fish and Game 70:101–112.

MAYER-GROSS, H. 1970. The Nest Record Scheme. Field Guide Number 12. British Trust for Ornithology, Tring, England. 36 pp.

MAYFIELD, H. 1961. Nesting success calculated from exposure. Wilson Bulletin 73:255–261.

MAYFIELD, H. 1975. Suggestions for calculating nest success. Wilson Bulletin 87:456–466.

MAYNARD SMITH, J. 1974. Models in Ecology. Cambridge University Press, Cambridge, England. 146 pp.

MEAD, R. A., T. L. SHARIK, S. P. PRESLEY, and J. T. HEINEN. 1981. A computerized spatial analysis system for assessing wildlife habitat from vegetation maps. Canadian Journal of Remote Sensing 7:34–40.

MEALEY, S. P., J. F. LIPSCOMB, and K. N. JOHNSON. 1982. Solving the habitat dispersion problem in forest planning. Transactions of the North American Wildlife and Natural Resources Conference 47:142–153.

MEENTS, J. K., J. RICE, B. W. ANDERSON, and R. D. OHMART. 1983. Nonlinear relationships between birds and vegetation. Ecology 64:1022–1027.

MIDDLETON, J., and G. MERRIAM. 1981. Woodland mice in a farmland mosaic. Journal of Applied Ecology 18:703–710.

MIDDLETON, J., and G. MERRIAM. 1983. Distribution of woodland species in farmland woods. Journal of Applied Ecology 20:625–644.

MILHOUSE, R. T., D. L. WEGNER, and T. WADDLE. 1984. User's Guide to the Physical Habitat Simulation System (PHABSIM). Instream Flow Information Paper 11. FWS/OBS–81/43 (revised). U.S. Department of Interior, Fish and Wildlife Service, Western Energy and Land Use Team, Fort Collins, CO. 320 pp.

MILLER, S. 1984. Estimation of Animal Production Numbers for National Assessments and Appraisals. U.S. Department of Agriculture, Forest Service, General Technical Report RM–105. Rocky Mountain Forest and Range Experiment Station, Fort Collins, CO. 23 pp.

MINKLER, L. S., and S. F. GINGRICH. 1970. Relation of Crown Width to Tree Diameter in Some Upland Hardwood Stands of Southern

Illinois. U.S. Department of Agriculture, Forest Service, Research Note NC–99. North Central Forest Experiment Station, St. Paul, MN. 4 pp.

Mitchell, K. J. 1975. Dynamics and Simulated Yield of Douglas-fir. Forest Science Monographs 17. Society of American Foresters, Bethesda, MD. 39 pp.

Moen, A. N. 1973. Wildlife Ecology. W. H. Freeman, San Francisco. 458 pp.

Moeur, M. 1981. Crown Width and Foliage Weight of Northern Rocky Mountain Conifers. U.S. Department of Agriculture, Forest Service, Research Paper INT–283. Intermountain Forest and Range Experiment Station, Ogden, UT. 14 pp.

Moeur, M. 1985. COVER: A User's Guide to the CANOPY and SHRUBS Extension of the Stand Prognosis Model. U.S. Department of Agriculture, Forest Service, General Technical Report INT–190. Intermountain Forest and Range Experiment Station, Ogden, UT. 49 pp.

Monson, G., and A. R. Phillips. 1981. Annotated Checklist of the Birds of Arizona. University of Arizona Press, Tucson, AZ. 240 pp.

Montgomery, D. C., and E. A. Peck. 1982. Introduction to Linear Regression Analysis. John Wiley and Sons, New York. 504 pp.

Moore, N. W. 1962. The heaths of Dorset and their conservation. Journal of Ecology 50:369–391.

Moore, N. W., and M. D. Hooper. 1975. On the number of bird species in British woods. Biological Conservation 8:239–250.

Morgan, R. A., and R. J. O'Connor. 1980. Farmland habitat and yellowhammer distribution in Britain. Bird Study 27:155–162.

Morin, P. J. 1981. Predatory salamanders reverse the outcome of competition among three species of anuran tadpoles. Science 212:1284–1286.

Morrison, D. F. 1967. Multivariate Statistical Methods. McGraw-Hill, New York. 338 pp.

Morrison, D. G. 1969. On the interpretation of discriminant analysis. Journal of Marketing Research 6:156–163.

Morrison, M. L. 1983. Assessing changes and trends in wildlife habitat in a forest management context. Pp. 101–103 *in* J. F. Bell and T. Atterbury (eds.), Renewable Resource Inventories for Monitoring Changes and Trends. College of Forestry, Oregon State University, Corvallis, OR.

Morton, E. S. 1978. Reintroducing recently extirpated birds into a tropical forest preserve. Pp. 379–384 *in* S. A. Temple (ed.), Endangered Birds: Management Techniques for Preserving Threatened Species. University of Wisconsin Press, Madison, WI.

Mueller-Dombois, D., and H. Ellenberg. 1974. Aims and Methods of Vegetation Ecology. John Wiley and Sons, New York. 547 pp.

Munro, D. D. 1974. Forest growth models—a prognosis. Pp. 7–21 *in* J. Fries (ed.), Growth Models for Tree and Stand Simulation. Research Note 30. Skogshoskolan, Royal College of Forestry, Stockholm, Sweden.

Murphy, D. D., A. E. Launer, and P. R. Ehrlich. 1984. The role of adult feeding in egg production and population dynamics of the checkerspot butterfly, *Euphydryas editha*. Oecologia 56:257–263.

Murphy, D. D., M. S. Menninger, and P. R. Ehrlich. 1984. Nectar source distribution as a determinant of oviposition host species in *Euphydryas chalcedona*. Oecologia 62:269–271.

Myers, N. 1983. A Wealth of Wild Species: Storehouse for Human Welfare. Westview, Boulder, CO. 274 pp.

Myers, R. H. 1971. Response Surface Methodology. Allyn and Bacon, Boston, MA. 246 pp.

National Research Council. 1971. Nutrient Requirements of Domestic Animals: Number 1, Nutrient Requirements of Poultry. Sixth revised edition. National Academy of Sciences, Washington, D.C. 54 pp.

Nau, D. S. 1983. Expert computer systems. Computer (February):63–85.

Nelson, R. D., and H. Salwasser. 1982. The Forest Service wildlife and fish habitat relationship program. Transactions of the North American Wildlife and Natural Resources Conference 47:174–183.

Nestler, R. B. 1940. Feeding requirements of gallinaceous upland game birds. Pp. 893–924 *in* Food and Life. Yearbook of Agriculture, 1939. U.S. Department of Agriculture, Government Printing Office, Washington, D.C.

Nestler, R. B. 1946. Vitamin A, vital factor in the survival of bobwhite. Transactions of the North American Wildlife Conference 11:176–192.

Nestler, R. B. 1949. Nutrition of bobwhite quail. Journal of Wildlife Management 13:342–358.

Nestler, R. B., and W. W. Bailey. 1943. Vitamin A deficiency in bobwhite quail. Journal of Wildlife Management 7:170–173.

Nestler, R. B., W. W. Bailey, L. M. Llewellyn, and M. J. Rensberger. 1944. Winter protein requirements of bobwhite quail. Journal of Wildlife Management 8:218–222.

Nestler, R. B., W. W. Bailey, M. J. Rensberger, and M. Benner. 1944. Protein requirements of breeding bobwhite quail. Journal of Wildlife Management 8:284–289.

Nestler, R. B., W. W. Bailey, and H. E. McClure. 1942. Protein requirements of bobwhite quail chicks for survival, growth, and efficiency of feed utilization. Journal of Wildlife Management 6:185–193.

Nestler, R. B., J. V. Derby, and J. B. DeWitt. 1948. Storage by bobwhite quail of vitamin A fed in various forms. Journal of Nutrition 36:323–329.

Nestler, R. B., J. V. Derby, and J. B. DeWitt. 1949a. Vitamin A and carotene content of some wildlife foods. Journal of Wildlife Management 13:271–274.

Nestler, R. B., J. V. Derby, and J. B. DeWitt. 1949b. Vitamin A storage in wild quail and its possible significance. Journal of Wildlife Management 13:265–271.

Neter, J., and W. Wasserman. 1974. Applied Linear Statistical Models. Richard D. Irwin, Homewood, IL. 842 pp.

Nice, M. M. 1937. Studies in the Life History of the Song Sparrow. 1964 reprint. Dover, New York. 328 pp.

Nicholls, C. F. 1960. A portable mechanical insect trap. The Canadian Entomologist 92:48–51.

Nickell, W. P. 1965. Habitats, territory and nesting of the catbird. American Midland Naturalist 73:433–478.

Nie, N. H., C. H. Hull, J. G. Jenkins, K. Steinbrenner, and D. H. Bent. 1975. Statistical Package for the Social Sciences. Second edition. McGraw-Hill, New York. 675 pp.

Niemela, P., and E. Haukioja. 1982. Seasonal patterns in species richness of herbivorous macrolepidopteran larvae on Finnish deciduous trees. Ecological Entomology 7:169–175.

Niemi, G. J. 1982. Determining priorities in nongame management. Loon 54:28–36.

Nilsson, S. G. 1977. Density compensation and competition among birds breeding on small islands in a south Swedish lake. Oikos 28:170–176.

Nilsson, S. G. 1978. Fragmented habitats, species richness and conservation practice. Ambio 7:26–27.

Nilsson, S. G., and B. Ebenman. 1981. Density changes and niche differences in island and mainland willow warblers, *Phylloscopus trochilus*, at a lake in southern Sweden. Ornis Scandinavica 12:62–67.

Nisbet, R. M., and W. S. C. Gurney. 1982. Modelling Fluctuating Populations. John Wiley and Sons, New York. 379 pp.

Nixon, C. M., and J. Ely. 1969. Foods eaten by a beaver colony in southeast Ohio. Ohio Journal of Science 69:313–319.

Nixon, C. M., S. P. Havera, and L. P. Hansen. 1980. Initial response of squirrels to forest changes associated with selection cutting. Wildlife Society Bulletin 8:298–306.

Noble, I. R., and R. O. Slatyer. 1980. The use of vital attributes to predict successional changes in plant communities subject to recurrent disturbances. Vegetatio 43:5–21.

Noon, B. R. 1981a. Techniques for sampling avian habitats. Pp. 42–52 *in* D. E. Capen (ed.), The Use of Multivariate Statistics in Studies of Wildlife Habitat. U.S. Department of Agriculture, Forest Service, General Technical Report RM–87. Rocky Mountain Forest and Range Experiment Station, Fort Collins, CO.

Noon, B. R. 1981b. The distribution of an avian guild along a temperate elevational gradient: The importance and expression of competition. Ecological Monographs 51:105–124.

Noon, B. R., D. K. Dawson, D. B. Inkley, C. S. Robbins, and S. H. Anderson. 1980. Consistency in habitat preference of forest bird species. Transactions of the North American Wildlife and Natural Resources Conference 45:226–244.

Norman, G. W., and R. L. Kirkpatrick. 1981. Estimating carcass fat levels in ruffed grouse from wing fat. Proceedings of the Annual Conference of the Southeastern Association of Fish and Wildlife Agencies 35:211–215.

Norris, R. A. 1968. Green-tailed towhee. Pp. 547–562 *in* A. C. Bent (ed.), Life Histories of North American Cardinals, Grosbeaks, Towhees, Finches, Sparrows and Their Allies. U.S. National Museum Bulletin 237. Government Printing Office, Washington, D.C.

Noss, R. 1983. A regional landscape approach to maintain diversity. BioScience 33:700–706.

Nuorteva, P. 1971. The synanthropy of birds as an expression of the ecological cycle disorder caused by urbanization. Annales Zoologici Fennici 8:547–553.

O'Connor, R. J. 1980. Population regulation in the yellowhammer, *Emberiza citrinella*. Pp. 190–200 *in* H. Oelke (ed.), Bird Census Work and Nature Conservation. Proceedings of the VI International Conference on Bird Census Work. University of Göttingen, West Germany.

O'Connor, R. J. 1981. Comparisons between migrant and non-migrant birds in Britain. Pp. 167–195 *in* D. J. Aidley (ed.), Animal Migration. Cambridge University Press, Cambridge, England.

O'Connor, R. J. 1982. Habitat occupancy and regulation in clutch size in the European kestrel, *Falco tinnunculus*. Bird Study 29:17–26.

O'Connor, R. J. 1984. The importance of hedges to songbirds. Pp. 117–123 *in* D. Jenkins (ed.), Agriculture and the Environment. ITE Symposium Number 13. Institute of Terrestrial Ecology, Cambridge, England.

O'Connor, R. J. 1985. Behavioural regulation of bird populations: A review of habitat use in relation to migration and residency. Pp. 105–142 *in* R. M. Sibley and R. H. Smith (eds.), Behavioural Ecology: Ecological Consequences of Adaptive Behaviour. British Ecological Symposium Number 25. Blackwell Scientific Publications, Oxford, England.

O'Connor, R. J., and R. J. Fuller. 1985. Bird population responses to habitat. Pp. 197–211 *in* K. Taylor, R. J. Fuller, and P. C. Lack (eds.), Bird Census and Atlas Studies: Proceedings of the VIII International Conference on Bird Census Work. British Trust for Ornithology, Tring, England.

Oelke, N. 1966. Thirty-five years of breeding bird census work in Europe. Audubon Field Notes 20:635–642.

Ohmann, J. L. 1983. Evaluating wildlife habitat as part of continuing, extensive forest inventory. Pp. 623–627 *in* J. F. Bell and T. Atterbury (eds.), Renewable Resource Inventories for Monitoring Changes and Trends. College of Forestry, Oregon State University, Corvallis, OR.

Osborne, P. E. 1982. The Effects of Dutch Elm Disease on Farmland Bird Populations. Ph.D. dissertation. Oxford University, Oxford, England. 358 pp.

Otis, D. L., K. P. Burnham, G. C. White, and D. R. Anderson. 1978. Statistical inference from capture data on closed animal populations. Wildlife Monographs 62:1–135.

Overton, W. S. 1965. A modification of the Schnabel estimator to account for removal of animals from the population. Journal of Wildlife Management 29:392–395.

Overton, W. S. 1977. A strategy of model construction. Pp. 49–74 *in* C. A. S. Hall and J. W. Day (eds.), Ecosystem Modeling in Theory and Practice. An Introduction with Case Histories. John Wiley and Sons, New York.

Palmgren, P. 1935. Über die Vogelfauna des Kulturgeländes auf Åland. Ornis Fennica 12:4–22.

Parker, I. E., and W. J. Matyas. 1978. CALVEG—Vegetation Classification and Mapping of California. U.S. Department of Agriculture, Forest Service, Region 5, San Francisco. 158 pp.

Patterson, J. H. 1976. The role of environmental heterogeneity in the regulation of duck populations. Journal of Wildlife Management 40:22–32.

Patton, D. R. 1978. RUN WILD: A Storage and Retrieval System for Wildlife Habitat Information. U.S. Department of Agriculture, Forest Service, General Technical Report RM–51. Rocky Mountain Forest and Range Experiment Station, Fort Collins, CO. 8

Pearson, I. 1980. The Prediction of Bird Species Diversity and Territory Density from Foliage Profile Measurements in Southern English Woodlands. B. S. thesis. University of East Anglia, Norwich, England. 80 pp.

Peters, T. J., and R. H. Waterman. 1982. In Search of Excellence: Lessons from America's Best Run Companies. Harper and Row, New York. 360 pp.

Peterson, S. R. 1982. A preliminary survey of forest bird communities in northern Idaho. Northwest Science 56:287–298.

Pianka, E. R. 1974. Evolutionary Ecology. Harper and Row, New York. 356 pp.

Pickett, S. T. A., and J. N. Thompson. 1978. Patch dynamics and the design of nature reserves. Biological Conservation 13:27–37.

Picton, H. D. 1979. The application of insular biogeographic theory to the conservation of large mammals in the northern Rocky Mountains. Biological Conservation 15:73–79.

Pielou, E. C. 1969. An Introduction to Mathematical Ecology. John Wiley and Sons, New York. 286 pp.

Pielou, E. C. 1975. Ecological Diversity. John Wiley and Sons, New York. 165 pp.

Pielou, E. C. 1984. The Interpretation of Ecological Data. John Wiley and Sons, New York. 263 pp.

Pienkowski, M., and P. R Evans. 1982. Breeding behaviour, productivity and survival of colonial and non-colonial shelducks *Tadorna tadorna*. Ornis Scandinavica 13:101–116.

Pimentel, R. A. 1979. Morphometrics, the Multivariate Analysis of Biological Data. Kendall/Hunt, Dubuque, IA. 276 pp.

Pitelka, F. A. 1951. Speciation and distribution of American jays of the genus *Aphelocoma*. University of California Publications in Zoology 50:194–464.

Press, S. J., and S. Wilson. 1978. Choosing between logistic regression and discriminant analysis. Journal of the American Statistical Association 73:699–705.

Preston, F. W. 1948. The commonness and rarity of species. Ecology 29:254–283.

Preston, F. W. 1960. Time and space and the variation of species. Ecology 41:611–627.

Preston, F. W. 1962. The canonical distribution of commonness and rarity. Part I. Ecology 43:185–215.

PRESTON, F. W., and R. T. NORRIS. 1947. Nesting heights of breeding birds. Ecology 28:241–273.

PUGH, A., III. 1980. DYNAMO User's Manual. Fifth edition. Massachusetts Institute of Technology Press, Cambridge, MA. 131 pp.

PYLE, R. M., M. BENTZIEN, and P. OPLER. 1981. Insect conservation. Annual Reviews of Entomology 26:233–258.

RABENOLD, K. N. 1978. Foraging strategies, diversity, and seasonality in bird communities of Appalachian spruce-fir forests. Ecological Monographs 48:396–424.

RACKHAM, O. 1976. Trees and Woodland in the British Landscape. Dent, London. 204 pp.

RAILE, G. K., and W. B. SMITH. 1983. Michigan Forest Statistics, 1980. U.S. Department of Agriculture, Forest Service, Resource Bulletin NC–67. North Central Forest Experiment Station, St. Paul, MN. 101 pp.

RALPH, C. J., and J. M. SCOTT (eds.). 1981. Estimating Numbers of Terrestrial Birds. Studies in Avian Biology 6. Cooper Ornithological Society. 630 pp.

RANDS, M. R. W. 1982. The importance of nesting cover quality to partridge. Annual Review of the Game Conservancy 13:58–64.

RANNEY, J. W., M. C. BRUNER, and J. B. LEVENSON. 1981. The importance of edge in the structure and dynamics of forest islands. Pp. 67–95 *in* R. L. Burgess and D. M. Sharpe (eds.), Forest Island Dynamics in Man-Dominated Landscapes. Springer-Verlag, New York.

RAPHAEL, M. G. 1984. Wildlife populations in relation to stand age and area in Douglas-fir forests of northwestern California. Pp. 259–274 *in* W. R. Meehan, T. R. Merrell, Jr., and T. A. Hanley (eds.), Fish and Wildlife Relationships in Old-Growth Forests: Proceedings of a Symposium. American Institute of Fishery Research Biologists.

RAPHAEL, M. G., and R. H. BARRETT. 1981. Methodologies for a comprehensive wildlife survey and habitat analysis in old-growth Douglas-fir forests. Cal-Neva Wildlife Transactions 1981:106–121.

RAPHAEL, M. G., and R. H. BARRETT. 1984. Diversity and abundance of wildlife in late successional Douglas-fir forests. Pp. 352–360 *in* New Forests for a Changing World, Proceedings of the 1983 Convention of the Society of American Foresters, Portland, Oregon. Society of American Foresters, Bethesda, MD.

RAPHAEL, M. G., and K. V. ROSENBERG. 1983. An integrated approach to wildlife inventories in forested habitats. Pp. 219–222 *in* J. F. Bell and T. Atterbury (eds.), Renewable Resource Inventories for Monitoring Changes and Trends. College of Forestry, Oregon State University, Corvallis, OR.

RAPHAEL, M. G., and M. WHITE. 1984. Use of snags by cavity-nesting birds in the Sierra Nevada. Wildlife Monographs 86:1–66.

REGIER, H. A. 1978. A Balanced Science of Renewable Resources. Washington Sea Grant Publication. University of Washington Press, Seattle. 108 pp.

REYNOLDS, R. T., J. M. SCOTT, and R. A. NUSSBAUM. 1980. A variable circular-plot method for estimating bird numbers. Condor 82:309–313.

RHODES, M. J., T. J. CLOUD, JR., and D. HAAG. 1983. Habitat evaluation procedures for planning surface mine reclamation in Texas. Wildlife Society Bulletin 11:222–231.

RICE, J. 1981. Summarizing remarks: Sampling design. Pp. 450–451 *in* C. J. Ralph and J. M. Scott (eds.), Estimating Numbers of Terrestrial Birds. Studies in Avian Biology 6. Cooper Ornithological Society.

RICE, J., R. D. OHMART, and B. W. ANDERSON. 1981. Bird community use of riparian habitats: The importance of temporal scale in interpreting discriminant analysis. Pp. 186–196 *in* D. E. Capen (ed.), The Use of Multivariate Statistics in Studies of Wildlife Habitat. U.S. Department of Agriculture, Forest Service, General Technical Report RM–87. Rocky Mountain Forest Range and Experiment Station, Fort Collins, CO.

RICE, J., R. D. OHMART, and B. W. ANDERSON. 1983a. Habitat selection attributes of an avian community: A discriminant analysis investigation. Ecological Monographs 53:263–290.

RICE, J., R. D. OHMART, and B. W. ANDERSON. 1983b. Turnovers in species composition of avian communities in contiguous riparian habitats. Ecology 64:1444–1455.

RICE, J., B. W. ANDERSON, and R. D. OHMART. 1984. Comparison of the importance of different habitat attributes to avian community organization. Journal of Wildlife Management 48:895–911.

RICKLEFS, R. E. 1973. Fecundity, mortality, and avian demography. Pp. 366–435 *in* D. S. Farner (ed.), Breeding Biology of Birds. National Academy of Sciences, Washington, D.C.

RITTEL, H. 1972. On the planning crisis: Systems analysis of the "first and second generations." Bedriftsokonomen 8:380–396.

ROACH, B. A. 1974. Scheduled Timber Cutting for Sustained Yields of Wood Products and Wildlife. U.S. Department of Agriculture, Forest Service, General Technical Report NE–14. Northeastern Forest Experiment Station, Upper Darby, PA. 13 pp.

ROBBINS, C. S. 1970. An international standard for a mapping method in bird census work recommended by the International Bird Census Committee. Audubon Field Notes 24:722–726.

ROBBINS, C. S. 1978. Determining habitat requirements of nongame species. Transactions of the North American Wildlife and Natural Resources Conference 43:57–68.

ROBBINS, C. S. 1979. Effect of forest fragmentation on bird populations. Pp. 199–212 *in* R. M. DeGraaf (tech. coord.), Management of Northcentral and Northeastern Forests for Nongame Birds: Proceedings of the Workshop. U.S. Department of Agriculture, Forest Service, General Technical Report NC–51. North Central Forest Experiment Station, St. Paul, MN.

ROBEL, R. J. 1972. Body fat content of bobwhites in relation to food plantings in Kansas. Proceedings of the National Bobwhite Quail Symposium 1:139–149.

ROBEL, R. J., and N. A. SLADE. 1965. The availability of sunflower and ragweed seeds during fall and winter. Journal of Wildlife Management 29:202–206.

ROBEL, R. J., A. R. BISSET, A. D. DAYTON, and K. E. KEMP. 1979. Comparative energetics of bobwhites on six different foods. Journal of Wildlife Management 43:987–992.

ROBEL, R. J., A. R. BISSET, T. M. CLEMENT, JR., A. D. DAYTON, and K. L. MORGAN. 1979. Metabolizable energy of important foods of bobwhites in Kansas. Journal of Wildlife Management 43:982–987.

ROBEL, R. J., J. M. BRIGG, A. D. DAYTON, and L. C. HULBERT. 1970. Relationships between visual obstruction measurements and weight of grassland vegetation. Journal of Range Management 23:295–297.

ROBEL, R. J., R. M. CASE, A. R. BISSET, and T. M. CLEMENT, JR. 1974. Energetics of food plots in bobwhite management. Journal of Wildlife Management 38:653–664.

ROHLF, F. J., and R. R. SOKAL. 1981. Statistical Tables. Second edition. W. H. Freeman, San Francisco. 219 pp.

ROLLINGS, C. T. 1945. Habits, foods and parasites of the bobcat in Minnesota. Journal of Wildlife Management 9:131–145.

ROMESBURG, H. C. 1981. Wildlife science: Gaining reliable knowledge. Journal of Wildlife Management 45:293–313.

ROMNEY, V. E. 1945. The effect of physical factors upon catch of the beet leafhopper (*Eutettix tenellus* (Bak.)) by a cylinder and two sweep-net methods. Ecology 26:135–147.

ROOT, R. B. 1967. The niche exploitation patterns of the blue-gray gnatcatcher. Ecological Monographs 37:317–350.

ROTENBERRY, J. T. 1981. Why measure bird habitat? Pp. 29–32 *in* D. E. Capen (ed.), The Use of Multivariate Statistics in Studies of Wildlife Habitat. U.S. Department of Agriculture, Forest Service, General Technical Report RM–87. Rocky Mountain Forest and Range Experiment Station, Fort Collins, CO.

ROTENBERRY, J. T., and J. A. WIENS. 1978. Nongame bird communities in northwestern rangelands. Pp. 32–46 *in* R. M. DeGraaf (tech. coord.), Nongame Bird Habitat Management in the Coniferous Forests of the Western United States: Proceedings of the Workshop. U.S. Department of Agriculture, Forest Service, General Technical Report PNW–64. Pacific Northwest Forest and Range Experiment Station, Portland, OR.

ROTENBERRY, J. T., and J. A. WIENS. 1980. Habitat structure, patchiness, and avian communities in North American steppe vegetation: A multivariate analysis. Ecology 61:1228–1250.

ROTENBERRY, J. T., and J. A. WIENS. 1981. A synthetic approach to principal component analysis of bird/habitat relationships. Pp. 197–208 *in* D. E. Capen (ed.), The Use of Multivariate Statistics in Studies of Wildlife Habitat. U.S. Department of Agriculture, Forest Service, General Technical Report RM–87. Rocky Mountain Forest and Range Experiment Station, Fort Collins, CO.

ROTH, R. R. 1976. Spatial heterogeneity and bird species diversity. Ecology 57:773–782.

ROTH, R. R. 1979. Vegetation as a determinant in avian ecology. Pp. 162–174 *in* D. L. Drawe (ed.), Proceedings of the First Welder Wildlife Foundation Symposium. Bob and Bessie Welder Wildlife Foundation, Sinton, TX.

ROTHSTEIN, S. I. 1982. Successes and failures in avian egg and nestling recognition with comments on the utility of optimality reasoning. American Zoologist 22:547–560.

ROUGHGARDEN, J. 1983. Competition and theory in community ecology. American Naturalist 122:583–601.

RUDIS, V. A., and A. R. EK. 1981. Optimization of forest island spatial patterns: Methodology for analysis of landscape pattern. Pp. 241–256 *in* R. L. Burgess and D. M. Sharpe (eds.), Forest Island Dynamics in Man-Dominated Landscapes. Springer-Verlag, New York.

RUSSELL, R. M., D. A. SHARPNACK, and E. L. AMIDON. 1975. Wildland Resource Information System: User's Guide. U.S. Department of Agriculture, Forest Service, General Technical Report PSW–10. Pacific Southwest Forest and Range Experiment Station, Berkeley, CA. 36 pp.

RUTHERFORD, W. H. 1964. The Beaver in Colorado: Its Biology, Ecology, Management and Economics. Technical Publication Number 17. Colorado Game, Fish and Parks Department, Denver, CO. 49 pp.

SALWASSER, H. 1982. California's wildlife information system and its application to resource decisions. Cal-Neva Wildlife Transactions 1982:34–39.

SALWASSER, H., and J. C. TAPPEINER II. 1981. An ecosystem approach to integrated timber and wildlife habitat management. Transactions of the North American Wildlife and Natural Resources Conference 46:473–487.

SALWASSER, H., J. C. CAPP, H. BLACK, JR., and J. F. HURLEY. 1980. The California wildlife habitat relationships program: an overview. Pp. 369–378 *in* R. M. DeGraaf (tech. coord.), Management of Western Forests and Grasslands for Nongame Birds: Proceedings of the Workshop. U.S. Department of Agriculture, Forest Service, General Technical Report INT–86. Intermountain Forest and Range Experiment Station, Ogden, UT.

SALWASSER, H., I. D. LUMAN, and D. DUFF. 1982. Integrating wildlife and fish into public land forest management. Proceedings of the Annual Conference of the Western Association of Fish and Wildlife Agencies 62:293–299.

SALWASSER, H., C. K. HAMILTON, W. B. KROHN, J. F. LIPSCOMB, and C. H. THOMAS. 1983. Monitoring wildlife and fish: Mandates and their implications. Transactions of the North American Wildlife and Natural Resources Conference 48:297–307.

SAMSON, F. B. 1980. Island biogeography and the conservation of nongame birds. Transactions of the North American Wildlife and Natural Resources Conference 45:245–251.

SANDER, I. L. 1977. Manager's Handbook for Oaks in the North Central States. U.S. Department of Agriculture, Forest Service, General Technical Report NC–37. North Central Forest Experiment Station. St. Paul, MN. 35 pp.

SAS INSTITUTE. 1982a. SAS User's Guide: Basics. SAS Institute, Cary, NC. 923 pp.

SAS INSTITUTE. 1982b. SAS User's Guide: Statistics. SAS Institute, Cary, NC. 584 pp.

SAS INSTITUTE. 1983. S.U.G.I. Supplemental Library User's Guide. SAS Institute, Cary, NC. 402 pp.

SAVARD, J. P. 1978. Birds in Metropolitan Toronto: Distribution Relationships with Habitat Features, and Nesting Sites. M.S. thesis. University of Toronto, Toronto, Ontario. 221 pp.

SCHAMBERGER, M., and A. FARMER. 1978. The habitat evaluation procedures: Their application in project planning and impact evaluation. Transactions of the North American Wildlife and Natural Resources Conference 43:274–283.

SCHAMBERGER, M., and W. B. KROHN. 1982. Status of the habitat evaluations procedures. Transactions of the North American Wildlife and Natural Resources Conference 47:154–164.

SCHAMBERGER, M., A. H. FARMER, and J. W. TERRELL. 1982. Habitat Suitability Index Models: Introduction. FWS/OBS–82/10. U.S. Department of Interior, Fish and Wildlife Service, Washington, D.C. 2 pp.

SCHNABEL, Z. E. 1938. Estimator of the total fish population of a lake. American Mathematics Monthly 45:348–352.

SCHOENER, T. W. 1971. Theory of feeding strategies. Annual Review Ecology and Systematics 2:369–404.

SCHOENER, T. W. 1976. The species-area relation within archipelagoes: Models and evidence from island land birds. Pp. 629–642 *in* H. J. Frith and J. H. Calaby (eds.), Proceedings of the XVI International Ornithological Congress. Australian Academy of Science, Canberra, Australia.

SCHONEWALD-COX, C. M., S. M. CHAMBERS, B. MACBRYDE, and W. L. THOMAS. 1983. Genetics and Conservation: A Reference for Managing Wild Animal and Plant Populations. Benjamin/Cummings, Menlo Park, CA. 722 pp.

SCHROEDER, R. L. 1982. Habitat Suitability Index Models: Pine Warbler. FWS/OBS–82/10.28. U.S. Department of Interior, Fish and Wildlife Service, Office of Biological Services, Washington, D.C. 8 pp.

SCHROEDER, R. L. 1983. Habitat Suitability Index Models: Pileated Woodpecker. FWS/OBS–82/10.39. U.S. Department of Interior, Fish and Wildlife Service, Office of Biological Services, Washington, D.C. 15 pp.

SCHROEDER, R. L., and P. J. SOUSA. 1982. Habitat Suitability Index Models: Eastern Meadowlark. FWS/OBS–82/10.29. U.S. Department of Interior, Fish and Wildlife Service, Office of Biological Services, Washington, D.C. 9 pp.

SCHULTZ, J. C. 1983. Habitat selection and foraging tactics of caterpillars in heterogeneous trees. Pp. 61–89 *in* R. F. Denno and M. S. McClure (eds.), Impact of Variable Host Quality upon Herbivorous Insects. Academic, New York.

SCOTT, V. E., and J. L. OLDEMEYER. 1983. Cavity-nesting bird requirements and responses to snag cutting in ponderosa pine. Pp.

19–23 *in* J. W. Davis, G. A. Goodwin, and R. A. Ockenfels (tech. coords.), Snag Habitat Management: Proceedings of the Symposium. U.S. Department of Agriculture, Forest Service, General Technical Report RM–99. Rocky Mountain Forest and Range Experiment Station, Fort Collins, CO.

Scott, E., R. Holm, and R. E. Reynolds. 1958. A study of the phosphorus requirements of young bobwhite quail. Poultry Science 37:1425–1428.

Seagle, S. W. 1983. Habitat Availability and Animal Community Characteristics. Ph.D. dissertation. University of Tennessee, Knoxville, TN. 88 pp.

Seber, G. A. F. 1982. The Estimation of Animal Abundance and Related Parameters. Second edition. Macmillan, New York. 654 pp.

Seitz, W. K., C. L. Kling, and A. H. Farmer. 1982. Habitat evaluation: A comparison of three approaches on the northern Great Plains. Transactions of the North American Wildlife and Natural Resources Conference 47:82–95.

Selye, H. 1950. The Physiology and Pathology of Exposure to Stress. Acta, Montreal, Quebec. 1025 pp.

Serafin, J. A. 1974. Studies on the riboflavin, niacin, pantothenic acid and choline requirements of young bobwhite quail. Poultry Science 53:1522–1532.

Severinghaus, W. D. 1981. Guild theory as a mechanism for assessing environmental impact. Environmental Management 5:187–190.

Sharpe, D. M., F. W. Stearns, R. C. Burgess, and W. C. Johnston. 1981. Spatio-temporal patterns of forest ecosystems in man-dominated landscapes of the eastern United States. Pp. 109–116 *in* S. P. Tjallingii and A. A. de Veer (eds.), Perspectives in Landscape Ecology. Center for Agricultural Publication and Documentation, Wageningen, Netherlands.

Sheffield, R. M. 1981. Multiresource Inventories: Techniques for Evaluating Nongame Bird Habitat. U.S. Department of Agriculture, Forest Service, Research Paper SE–218. Southeastern Forest Experiment Station, Asheville, NC. 28 pp.

Shelford, V. E. 1913. Animal Communities in Temperate America. Bulletin of the Geographic Society of Chicago Number 5. 368 pp.

Sheppard, J. L., D. L. Wills, and J. L. Simonson. 1982. Project applications of the Forest Service Rocky Mountain Region wildlife and fish habitat relationships system. Transactions of the North American Wildlife and Natural Resources Conference 47:128–141.

Shields, W. M. 1979. Avian census techniques: An analytical review. Pp. 23–51 *in* J. G. Dickson, R. N. Conner, R. R. Fleet, J. C. Kroll, and J. A. Jackson (eds.), The Role of Insectivorous Birds in Forest Ecosystems. Academic, New York.

Shipes, D. A., T. T. Fendley, and H. S. Hill. 1980. Woody vegetation as food items for South Carolina coastal plain beaver. Proceedings of the Annual Conference of the Southeastern Association of Fish and Wildlife Agencies 33:202–211.

Short, H. L. 1983. Wildlife Guilds in Arizona Desert Habitats. Technical Note 362. U.S. Department of Interior, Bureau of Land Management, Washington, D.C. 258 pp.

Short, H. L., and K. P. Burnham. 1982. Technique for structuring wildlife guilds to evaluate impacts on wildlife communities. Special Scientific Report—Wildlife Number 244. U.S. Department of Interior, Fish and Wildlife Service, Washington, D.C. 34 pp.

Shortliffe, E. H. 1976. Computer-Based Medical Consultations: MYCIN. Elsevier, New York. 264 pp.

Shugart, H. H., Jr. 1981. An overview of multivariate methods and their application to studies of wildlife habitat. Pp. 4–10 *in* D. E. Capen (ed.), The Use of Multivariate Statistics in Studies of Wildlife Habitat. U.S. Department of Agriculture, Forest Service, General Technical Report RM–87. Rocky Mountain Forest and Range Experiment Station, Fort Collins, CO.

Shugart, H. H., Jr. 1984. A Theory of Forest Dynamics. Springer-Verlag, New York. 278 pp.

Shugart, H. H., Jr., and D. James. 1973. Ecological succession of breeding bird populations in northwestern Arkansas. Auk 90:62–77.

Shugart, H. H., Jr., and B. C. Patten. 1972. Niche quantification and the concept of niche pattern. Pp. 248–327 *in* B. C. Patten (ed.), Systems Analysis and Simulation in Ecology. Volume 2. Academic, New York.

Shugart, H. H., Jr., and D. C. West. 1977. Development of an Appalachian deciduous forest succession model and its application to assessment of the impact of the chestnut blight. Journal of Environmental Management 5:161–179.

Shugart, H. H., Jr., and D. C. West. 1980. Forest succession models. BioScience 30:308–313.

Shugart, H. H., Jr., and D. C. West. 1981. Long-term dynamics of forest ecosystems. American Scientist 69:647–652.

Shugart, H. H., Jr., T. R. Crow, and J. M. Hett. 1973. Forest succession models: A rationale and methodology for modeling forest succession over large regions. Forest Science 19:203–212.

Shugart, H. H., Jr., R. D. Dueser, and S. H. Anderson. 1974. Influence of habitat alterations on bird and small mammal populations. Occasional Papers of the Missouri Academy of Science 3:92–96.

Siegel, S. 1956. Nonparametric Statistics for the Behavioral Sciences. McGraw-Hill, New York. 312 pp.

Simberloff, D. S. 1974. Equilibrium theory of island biogeography and ecology. Annual Review Ecology and Systematics 5:161–182.

Simberloff, D. S. 1976. Species turnover and equilibrium island biogeography. Science 194:572–578.

Simberloff, D. S. 1983. Competition theory, hypothesis testing, and other community ecological buzz words. American Naturalist 122:626–635.

Simberloff, D. S., and L. G. Abele. 1976. Island biogeography theory and conservation practice. Science 191:285–286.

Simberloff, D. S., and L. G. Abele. 1982. Refuge design and island biogeographic theory: Effects of fragmentation. American Naturalist 120:41–50.

Simberloff, D., and L. B. Abele. 1984. Conservation and obfuscation: Subdivision of reserves. Oikos 42:399–401.

Simon, H. A. 1977. The New Science of Management Decision. Prentice-Hall, Englewood Cliffs, NJ. 175 pp.

Slagsvold, T. 1977. Bird song activity in relation to breeding cycle, spring weather, and environmental phenology. Ornis Scandinavica 8:197–222.

Smith, K. G. 1977. Distribution of summer birds along a forest moisture gradient in an Ozark watershed. Ecology 58:810–819.

Smith, T. M., H. H. Shugart, Jr., and D. C. West. 1981a. FORHAB: A forest simulation model to predict habitat structure for nongame bird species. Pp. 114–123 *in* D. E. Capen (ed.), The Use of Multivariate Statistics in Studies of Wildlife Habitat. U.S. Department of Agriculture, Forest Service, General Technical Report RM–87. Rocky Mountain Forest and Range Experiment Station, Fort Collins, CO.

Smith, T. M., H. H. Shugart, Jr., and D. C. West. 1981b. Use of forest simulation models to integrate timber harvest and nongame bird management. Transactions of the North American Wildlife and Natural Resources Conference 46:501–510.

Snee, R. D. 1971. Design and analysis of mixture experiments. Journal of Quality Technology 3:159–169.

Sokal, R. R., and R. J. Rohlf. 1969. Biometry: The Principles and

Practice of Statistics in Biological Research. W. H. Freeman, San Francisco. 776 pp.

Sokal, R. R., and R. J. Rohlf. 1973. Introduction to Biostatistics. W. H. Freeman, San Francisco. 368 pp.

Soulé, M. E., and B. A. Wilcox (eds.). 1980. Conservation Biology: An Evolutionary-Ecological Perspective. Sinauer Associates, Sunderland, MA. 395 pp.

Sousa, P. J. 1982. Habitat Suitability Index Models: Veery. FWS/OBS-82/10.22. U.S. Department of Interior, Fish and Wildlife Service, Office of Biological Services, Washington, D.C. 12 pp.

Sousa, P. J. 1983. Habitat Suitability Index Models: Field Sparrow. FWS/OBS-82/10.62. U.S. Department of Interior, Fish and Wildlife Service, Office of Biological Services, Washington, D.C. 14 pp.

Southwood, T. R. E. 1966. Ecological Methods. Methuen, London. 391 pp.

Southwood, T. R. E. 1977. Habitat, the templet for ecological strategies? Journal of Animal Ecology 46:337–365.

Spencer, W. D., R. H. Barrett, and W. J. Zielinski. 1983. Marten habitat preferences in the northern Sierra Nevada. Journal of Wildlife Management 47:1181–1186.

Stage, A. R. 1973. Prognosis Model for Stand Development. U.S. Department of Agriculture, Forest Service, Research Paper INT–137. Intermountain Forest and Range Experiment Station, Odgen, UT. 32 pp.

Stauffer, D. F., and L. B. Best. 1980. Habitat selection by birds of riparian communities: Evaluating effects of habitat alterations. Journal of Wildlife Management 44:1–15.

Stauffer, H. B. 1982. A sample size table for forest sampling. Forest Science 28:777–784.

Steel, R. G., and J. H. Torrie. 1960. Principles and Procedures of Statistics. McGraw-Hill, New York. 481 pp.

Steele, R. O., S. V. Cooper, D. M. Ondov, D. W. Roberts, and R. D. Pfister. 1983. Forest Habitat Types of Eastern Idaho-Western Wyoming. U.S. Department of Agriculture, Forest Service, General Technical Report INT–144. Intermountain Forest and Range Experiment Station, Ogden, UT. 122 pp.

Steen, H. K. 1976. The U.S. Forest Service—A History. University of Washington Press, Seattle. 356 pp.

Stefik, M., J. Aikins, R. Balzar, J. Benoit, L. Birnbaum, F. Hayes-Roth, and E. Sacerdoti. 1983. Basic concepts for building expert systems. Pp. 59–88 *in* F. Hayes-Roth, D. A. Waterman, and D. B. Lenat (eds.), Building Expert Systems. Addison-Wesley, Reading, MA.

Steinhorst, R. K. 1979. Analysis of niche overlap. Pp. 263–278 *in* L. Orloci, C. R. Rao, and W. M. Stiteler (eds.), Multivariate Methods in Ecological Work. International Co-operative Publishing House, Fairland, MD.

Stewart, R. E., and H. A. Kantrud. 1971. Classification of Natural Ponds and Lakes in the Glaciated Prairie Region. Resource Publication 92. U.S. Department of Interior, Fish and Wildlife Service, Washington, D.C. 57 pp.

Stjernberg, T. 1979. Breeding biology and population dynamics of the scarlet rosenfinch, *Carpodacus erythrinus*. Acta Zoologica Fennica 157:1–88.

Strauss, R. E. 1979. Reliability estimates for IvLev's electivity index, the forage ratio, and a proposed linear index of food selection. Transactions of the American Fisheries Society 108:344–352.

Strecker, E. L., and J. T. Emlen. 1953. Regulatory mechanisms in house-mouse populations: The effect of limited food supply on a confined population. Ecology 34:375–385.

Strong, D. R. 1979. Biogeographic dynamics of insect-host plant communities. Annual Review of Entomology 24:89–119.

Strong, D. R, L. A. Szyska, and D. S. Simberloff. 1979. Tests of community-wide character displacement against a null hypothesis. Evolution 33:897–913.

Sugihara, G. 1980. Minimal community structure: An explanation of species abundance patterns. American Naturalist 116:770–787.

Sugihara, G. 1981. $S = CA^z$, $z \approx 1/4$: A reply to Connor and McCoy. American Naturalist 117:790–793.

Sumner, I., and J. S. Dixon. 1953. Birds and Mammals of the Sierra Nevada, with Records from Sequoia and Kings Canyon National Parks. University of California Press, Berkeley, CA. 484 pp.

Sundström, K. E. 1927. Ökologisch-geographische Studien über die Vogelfauna der Gegend von Ekenes. Acta Zoologica Fennica 3:1–170.

Svendsen, G. E. 1980. Seasonal changes in feeding patterns of beaver in southeastern Ohio. Journal of Wildlife Management 44:285–290.

Svensson, S. 1978. Size and isolation of natural reserves: Some applications of ecological theory. Anser, Supplement 3:225–234.

Swaine, M. D., and P. Greig-Smith. 1980. An application of principal components analysis to vegetation change in permanent plots. Journal of Ecology 68:33–41.

Sweeney, J. M., and H. S. Bhullar. 1983. A computer tool for integrated resource management. Pp. 295–297 *in* J. F. Bell and T. Atterbury (eds.), Renewable Resource Inventories for Monitoring Changes and Trends. College of Forestry, Oregon State University, Corvallis, OR.

Sweeney, J. M., C. R. Wenger, and N. S. Yoho. 1981. Bobwhite Quail Food in Young Arkansas Loblolly Pine Plantations. University of Arkansas Agricultural Experiment Station Bulletin 852. Fayetteville, AR. 18 pp.

Swift, B. L., J. S. Larson, and R. M. DeGraaf. 1984. Relationship of breeding bird density and diversity to habitat variables in forested wetlands. Wilson Bulletin 96:48–59.

Szaro, R. C., and R. P. Balda. 1979. Bird Community Dynamics in a Ponderosa Pine Forest. Studies in Avian Biology 3. Cooper Ornithological Society. 66 pp.

Szaro, R. C., and R. P. Balda. 1982. Selection and Monitoring of Avian Indicator Species: An Example from a Ponderosa Pine Forest in the Southwest. U.S. Department of Agriculture, Forest Service, General Technical Report RM–89. Rocky Mountain Forest and Range Experiment Station, Fort Collins, CO. 8 pp.

Taber, R. D., and K. J. Raedeke. 1980. Environmental Assessment Report: The Roosevelt Elk of the Olympic National Forest. College of Forest Resources, University of Washington, Seattle. 109 pp.

Tagashi, K., and F. Takahashi. 1977. Preferential feeding of the last instar larvae of *Dendrolimus spectabilis* Butler (Lepidoptera: Lasiocampidae) on the old needles of *Pinus thunbergii* Parl. in the field. Japanese Journal of Ecology 27:159–162.

Tatsuoka, M. M. 1971. Multivariate Analysis: Techniques for Educational and Psychological Research. John Wiley and Sons, New York. 310 pp.

Taylor, S. M. 1965. The Common Birds Census: Some statistical aspects. Bird Study 12:268–286.

Temple, S. A. 1981. Applied island biogeography and the conservation of endangered island birds in the Indian Ocean. Biological Conservation 20:147–161.

Temple, S. A. 1983. Why have certain songbirds declined in a fragmented eastern deciduous forest? P. 93 *in* Abstracts of Presented Posters and Papers, 101st Stated Meeting of the American Ornithologists' Union, New York.

Terborgh, J. W. 1974. Preservation of natural diversity: The problem of extinction prone species. BioScience 24:715–722.

Terborgh, J. W. 1975. Faunal equilibria and the design of wildlife preserves. Pp. 369–380 *in* F. G. Golley and E. Medina (eds.), Tropical Ecological Systems: Trends in Terrestrial and Aquatic Research. Springer-Verlag, New York.

Terborgh, J. W. 1976. Island biogeography and conservation: Strategy and limitations. Science 193:1029–1030.

Terborgh, J. W., and B. Winter. 1980. Some causes of extinction. Pp. 119–133 *in* M. E. Soulé and B. A. Wilcox (eds.), Conservation Biology: An Evolutionary-Ecological Perspective. Sinauer Associates, Sunderland, MA.

Terres, J. K. 1956. Death in the night. Audubon 58:18–20.

Thomas, J. W. (ed.). 1979. Wildlife Habitats in Managed Forests: The Blue Mountains of Oregon and Washington. U.S. Department of Agriculture, Forest Service, Agriculture Handbook Number 553. U.S. Government Printing Office, Washington, D.C. 512 pp.

Thomas, J. W. 1982. Needs for and approaches to wildlife habitat assessment. Transactions of the North American Wildlife and Natural Resources Conference 47:35–46.

Thomas, J. W., R. Miller, C. Maser, R. Anderson, and B. Carter. 1978. The relationship of terrestrial vertebrates to plant communities and their successional stages. Pp. 281–303 *in* A. Marmelstein (ed.), Classification, Inventory and Analysis of Fish and Wildlife Habitat. FWS/OBS-78/76. U.S. Department of Interior, Fish and Wildlife Service, Office of Biological Services, Washington, D.C.

Thomas, J. W., C. Maser, and J. E. Rodiek. 1979. Edges. Pp. 48–59 *in* J. W. Thomas (ed.), Wildlife Habitats in Managed Forests: The Blue Mountains of Oregon and Washington. U.S. Department of Agriculture, Forest Service, Agriculture Handbook Number 553. U.S. Government Printing Office, Washington, D.C.

Tiainen, J. 1983. Dynamics of a local population of the willow warbler, *Phylloscopus trochilus,* in southern Finland. Ornis Scandinavica 14:1–15.

Tilghman, N. G., and D. H. Rusch. 1981. Comparisons of line-transect methods for estimating breeding bird densities in deciduous woodlots. Pp. 202–208 *in* C. J. Ralph and J. M. Scott (eds.), Estimating Numbers of Terrestrial Birds. Studies in Avian Biology 6. Cooper Ornithological Society.

Titus, K. 1984. Uniformity in Relative Habitat Selection by *Buteo lineatus* and *B. platypterus* in Two Temperate Forest Regions. Ph.D. dissertation. University of Maryland, Catonsville, MD. 95 pp.

Titus, K., and J. A. Mosher. 1981. Nest-site habitat selected by woodland hawks in the central Appalachians. Auk 98:270–281.

Tiwari, J. L., and J. E. Hobbie. 1976. Random differential equations as models of ecosystems: Monte Carlo simulations approach. Mathematical Biosciences 28:25–44.

Tiwari, J. L., J. E. Hobbie, J. P. Reed, D. W. Stanley, and M. C. Miller. 1978. Some stochastic differential equation models of an aquatic ecosystem. Ecological Modelling 4:3–27.

Tjernberg, M. 1984. Breeding habitat of red-breasted flycatcher, *Ficedula parva,* in Uppland, eastern Sweden, Vår Fågelvärld 43:275–282.

Toft, C. A., and P. J. Shea. 1983. Detecting community-wide patterns: Estimating power strengthens statistical inference. American Naturalist 122:618–625.

Tomlin, C. D. 1980. The Map Analysis Package. Papers in Spatial Information Systems Series. Yale University School of Forestry and Environmental Studies, New Haven, CT. 81 pp.

Townes, H. 1962. Design for a Malaise trap. Proceedings of the Entomological Society of Washington 64:253–262.

Trivedi, M. M., C. L. Wyatt, and D. R. Anderson. 1982. A multispectral approach to remote detection of deer. Photogrammetric Engineering and Remote Sensing 48:1879–1889.

Turner, F. B., and P. A. Medica. 1982. The distribution and abundance of the flat-tailed horned lizard (*Phrynosoma mcallii*). Copeia 1982:815–823.

Urich, D. L., and J. P. Graham. 1983. Applying habitat evaluation procedures (HEP) to wildlife area planning in Missouri. Wildlife Society Bulletin 11:215–222.

Välikangas, I. 1937. Qualitative und quantitative Untersuchungen über die Vogelfauna der isolierten Insel Suursaari (Hogland) im Finnischen Meerbusen. Die Landvogelfauna. Annales Academiae Scientiae Fenniae 45(5):1–236.

Van Balen, J. H. 1973. A comparative study of the breeding ecology of the great tit, *Parus major,* in different habitats. Ardea 61:1–93.

Van Horne, B. 1982. Niches of adult and juvenile deer mice (*Peromyscus maniculatus*) in seral stages of coniferous forest. Ecology 63:992–1003.

Van Horne, B. 1983. Density as a misleading indicator of habitat quality. Journal of Wildlife Management 47:893–901.

Van Horne, B., and R. G. Ford. 1982. Niche breadth calculation based on discriminant analysis. Ecology 63:1172–1174.

Van Melle, W. J. 1979. A domain-independent production-rule system for consultation programs. Pp. 923–925 *in* J. S. Gakkai (coord.), Proceedings of the Sixth International Joint Conference on Artificial Intelligence, Tokyo, Japan.

Van Melle, W. J. 1981. System Aids in Constructing Consultation Programs. University of Michigan Research Press, Ann Arbor, MI. 173 pp.

Vassallo, M. I., and J. C. Rice. 1982. Ecological release and ecological flexibility in habitat use and foraging of an insular avifauna. Wilson Bulletin 94:139–155.

Verner, J. 1980. Bird communities of mixed-conifer forests of the Sierra Nevada. Pp. 198–223 *in* R. M. DeGraaf (tech. coord.), Management of Western Forests and Grasslands for Nongame Birds: Proceedings of the Workshop. U.S. Department of Agriculture, Forest Service, General Technical Report INT-86, Intermountain Forest and Range Experiment Station, Ogden, UT.

Verner, J. 1981. Measuring responses of avian communities to habitat manipulation. Pp. 543–547 *in* C. J. Ralph and J. M. Scott (eds.), Estimating Numbers of Terrestrial Birds. Studies in Avian Biology 6. Cooper Ornithological Society.

Verner, J. 1983. An integrated system for monitoring wildlife on the Sierra National Forest. Transactions of the North American Wildlife and Natural Resources Conference 48:355–366.

Verner, J. 1984. The guild concept applied to management of bird populations. Environmental Management 8:1–14.

Verner, J. 1985. Assessment of counting techniques. Pp. 247–302 *in* R. F. Johnston (ed.), Current Ornithology. Volume 2. Plenum, New York and London.

Verner, J., and A. S. Boss (tech. coords.). 1980. California Wildlife and Their Habitats: Western Sierra Nevada. U.S. Department of Agriculture, Forest Service, General Technical Report PSW-37. Pacific Southwest Forest and Range Experiment Station, Berkeley, CA. 439 pp.

Verner, J., and L. V. Ritter. 1985. Comparison of transect and point counts in oak-pine woodlands of California. Condor 87:47–68.

Vickery, W. L., and T. D. Nudds. 1984. Detection of density: Dependent effects in annual duck censuses. Ecology 65:96–104.

Vickholm, M. 1983. Avointen reunojen vaikutus metsälinnustoon. M.S. thesis. University of Helsinki, Helsinki, Finland. 120 pp.

Vogt, R. C., and R. L. Hine. 1982. Evaluation of techniques for assessment of amphibian and reptile populations in Wisconsin. Pp. 201–217 *in* N. J. Scott (ed.), Herpetological Communities. Research Report 13. U.S. Department of Interior, Fish and Wildlife Service, Washington, D.C.

Von Haartman, L. 1960. The ortstreue of the pied flycatcher. Pp. 266–273 *in* G. Bergman, K. O. Donner, and L. von Haartman (eds.), Proceedings of the XII International Ornithological Congress. Tilgmannin Kirjapaino, Helsinki, Finland.

Von Haartman, L., O. Hildén, P. Linkola, P. Suomalainen, and R. Tenovuo. 1963–1972. Pohjolan linnut värikuvin. Otava, Helsinki, Finland. 1092 pp. + illus.

Wagner, G. R. 1982. Mind support systems. ICP Interface 7:19–22.

Wallace, A. R. 1869. The Malay Archipelago. Harper, New York. 638 pp.

Walsh, S. J. 1980. Coniferous tree species mapping using Landsat data. Remote Sensing of Environment 9:11–26.

Walters, C. J. 1974. An interdisciplinary approach to development of watershed simulation models. Technological Forecasting and Social Change 6:299–323.

Walters, C. J. 1977. Methodological problems in the modeling and analysis of ecological systems. Pp. 24–39 *in* H. R. Grumm (ed.), Analysis and Computation of Equilibria and Regions of Stability. International Institute for Applied Systems Analysis. Schloss, Laxenburg, A-2361, Austria.

Watson, A. M. (ed.). 1970. Animal Populations in Relation to Their Food Resources. Blackwell, Oxford, England. 477 pp.

Watt, A. S. 1925. On the ecology of British beechwoods with special reference to their regeneration. Part II. The development and structure of beech communities on the Sussex Downs. Journal of Ecology 13:212–226.

Watt, A. S. 1947. Pattern and process in the plant community. Journal of Ecology 35:1–22.

Watt, K. E. F. 1977. Why won't anyone believe us? Simulation 28:1–3.

Webb, W. L., D. F. Behrend, and B. Saisorn. 1977. Effect of logging on songbird populations in a northern hardwood forest. Wildlife Monographs 55:1–35.

Weber, B. J., M. L. Wolfe, G. C. White, and M. M. Rowland. 1984. Physiologic response of elk to differences in winter range quality. Journal of Wildlife Management 48:248–253.

Weinstein, D. A., and H. H. Shugart, Jr. 1983. Ecological modeling of landscape dynamics. Pp. 29–45 *in* H. Mooney and M. Gordon (eds.), Disturbance and Ecosystems. Springer-Verlag, New York.

Weisberg, S. 1980. Applied Linear Regression. John Wiley and Sons, New York. 283 pp.

Weise, C. M., and J. R. Meyer. 1979. Juvenile dispersal and development of site-fidelity in the black-capped chickadee. Auk 96:40–55.

Weiss, S. M., and C. A. Kulikowski. 1984. A Practical Guide to Designing Expert Systems. Rowman and Allanheld, Totowa, NJ. 174 pp.

Weller, M. W., and C. S. Spatcher. 1965. Role of Habitat in the Distribution and Abundance of Marsh Birds. Iowa State University Agriculture and Home Economics Experiment Station Report 43. Ames, IA. 31 pp.

Wells, S. M., R. M. Pyle, and N. M. Collins. 1983. The IUCN Invertebrate Red Data Book. IUCN, Gland, Switzerland. 632 pp.

Wengert, A., A. A. Dyer, and H. A. Deutsch. 1979. The "Purposes" of the National Forests: A Historical Re-interpretation of Policy Development. Colorado State University, Fort Collins, CO. 44 pp. + 4 appendices.

West, D. C., H. H. Shugart, Jr., and D. B. Botkin. 1981. Forest Succession: Concepts and Application. Springer-Verlag, New York. 517 pp.

Westman, W. E. 1980. Gaussian analysis: Identifying environmental factors influencing bell-shaped species distributions. Ecology 61:733–739.

Whitcomb, B. L., R. F. Whitcomb, and D. Bystrak. 1977. Island biogeography and "habitat islands" of eastern forest. Part III. Long-term turnover and effects of selective logging on the avifauna of forest fragments. American Birds 31:17–23.

Whitcomb, R. F. 1977. Island biogeography and "habitat islands" of eastern forest. American Birds 31:3–5.

Whitcomb, R. F., C. S. Robbins, J. F. Lynch, B. L. Whitcomb, M. K. Klimkiewicz, and D. Bystrak. 1981. Effects of forest fragmentation on avifauna of the eastern deciduous forest. Pp. 125–205 *in* R. L. Burgess and D. M. Sharpe (eds.), Forest Island Dynamics in Man-Dominated Landscapes. Springer-Verlag, New York.

White, D. H., and D. James. 1978. Differential use of fresh water environments by wintering waterfowl of coastal Texas. Wilson Bulletin 90:99–111.

White, P. S. 1979. Pattern, process, and natural disturbance in vegetation. Botanical Review 45:229–299.

Whitmore, R. C. 1975. Habitat ordination of passerine birds of the Virgin River Valley, southwestern Utah. Wilson Bulletin 87:65–74.

Whitmore, R. C. 1981. Structural characteristics of grasshopper sparrow habitat. Journal of Wildlife Management 45:811–814.

Whittaker, R. H. 1952. A study of summer foliage insect communities in the Great Smoky Mountains. Ecological Monographs 22:1–44.

Whittaker, R. H. 1953. A consideration of climax theory: The climax as a population and pattern. Ecological Monographs 23:41–78.

Whittaker, R. H. 1967. Gradient analysis of vegetation. Biological Review 42:207–264.

Wiens, J. A. 1969. An Approach to the Study of Ecological Relationships Among Grassland Birds. Ornithological Monographs Number 8. American Ornithologists' Union. 93 pp.

Wiens, J. A. 1976. Population responses to patchy environments. Annual Review of Ecology and Systematics 7:81–120.

Wiens, J. A. 1977. On competition and variable environments. American Scientist 65:590–597.

Wiens, J. A. 1981a. Single-sample surveys of communities: Are the revealed patterns real? American Naturalist 117:90–98.

Wiens, J. A. 1981b. Scale problems in avian censusing. Pp. 513–521 *in* C. J. Ralph and J. M. Scott (eds.), Estimating Numbers of Terrestrial Birds. Studies in Avian Biology 6. Cooper Ornithological Society.

Wiens, J. A. 1983. Avian community ecology: An iconoclastic view. Pp. 355–403 *in* A. H. Brush and G. A. Clark, Jr. (eds.), Perspectives in Ornithology. Cambridge University Press, New York.

Wiens, J. A., and J. T. Rotenberry. 1981a. Censusing and the evaluation of avian habitat occupancy. Pp. 522–532 *in* C. J. Ralph and J. M. Scott (eds.), Estimating Numbers of Terrestrial Birds. Studies in Avian Biology 6. Cooper Ornithological Society.

Wiens, J. A., and J. T. Rotenberry. 1981b. Habitat associations and community structure of birds in shrubsteppe environments. Ecological Monographs 51:21–41.

Wilcove, D. S. 1985. Nest predation in forest tracts and the decline of migratory songbirds. Ecology 66:1211–1214.

Wilcove, D. S., and R. F. Whitcomb. 1983. Gone with the trees. Natural History Magazine (September): 82–91.

Wilcox, B. A. 1978. Supersaturated island faunas: A species-age relationship for lizards on post-Pleistocene land-bridge islands in the Gulf of California. Science 199:996–998.

Wilcox, B. A. 1980a. Insular ecology and conservation. Pp. 95–117 *in* M. E. Soulé and B. A. Wilcox (eds.), Conservation Biology: An Evolutionary-Ecological Perspective. Sinauer Associates, Sunderland, MA.

WILCOX, B. A. 1980b. Some Aspects of the Biogeography and Evolutionary Ecology of Vertebrates in the Gulf of California. Ph.D. dissertation. University of California, San Diego, CA. 184 pp.

WILCOX, B. A. 1984. Concepts in conservation biology: Applications to the management of biological diversity. Pp. 155–172 *in* J. L. Cooley and J. H. Cooley (eds.), Natural Diversity in Forest Ecosystems: Proceedings of the Workshop. Institute of Ecology, Athens, GA.

WILCOX, B. A., and D. D. MURPHY. 1985. Conservation strategy: The effects of fragmentation on extinction. American Naturalist 125:879–887.

WILLIAMS, B. K. 1981. Discriminant analysis in wildlife research: Theory and applications. Pp. 59–71 *in* D. E. Capen (ed.), The Use of Multivariate Statistics in Studies of Wildlife Habitat. U.S. Department of Agriculture, Forest Service, General Technical Report RM–87. Rocky Mountain Forest and Range Experiment Station, Fort Collins, CO.

WILLIAMS, B. K. 1983. Some observations on the use of discriminant analysis in ecology. Ecology 64:1283–1291.

WILLIAMS, G. L., K. R. RUSSELL, and W. K. SEITZ. 1978. Pattern recognition as a tool in the ecological analysis of habitat. Pp. 521–531 *in* A. Marmelstein (ed.), Classification, Inventory, and Analysis of Fish and Wildlife Habitat. FWS/OBS–78/76. U.S. Department of Interior, Fish and Wildlife Service, Office of Biological Services, Washington, D.C.

WILLIAMS, W. T., and W. STEPHENSON. 1973. The analysis of three-dimensional data (sites × species × times) in marine ecology. Journal of Experimental Marine Biology and Ecology 11:207–227.

WILLIAMSON, K. 1969. Habitat preferences of the wren on English farmland. Bird Study 16:53–59.

WILLIAMSON, M. 1975. The design of wildlife preserves. Nature 256:519.

WILLIAMSON, M. 1981. Island Populations. Oxford University Press, Oxford, England. 286 pp.

WILLIAMSON, M. 1983. The land-bird community of Skokholm: Ordination and turnover. Oikos 41:378–384.

WILLIAMSON, P. 1971. Feeding ecology of the red-eyed vireo (*Vireo olivaceus*) and associated foliage gleening birds. Ecological Monographs 41:129–152.

WILLIS, E. O. 1974. Populations and local extinctions of birds on Barro Colorado Island, Panama. Ecological Monographs 44:153–169.

WILSON, E. O., and E. O. WILLIS. 1975. Applied biogeography. Pp. 522–534 *in* M. L. Cody and J. M. Diamond (eds.), Ecology and Evolution of Communities. Harvard University Press, Cambridge, MA.

WILSON, H. R., M. W. HOLLAND, JR., and R. H. HARMS. 1972. Dietary calcium and phosphorus requirements for bobwhite chicks. Journal of Wildlife Management 36:965–968.

WILSON, M. V. 1981. A statistical test for the accuracy and consistency of ordinations. Ecology 62:8–12.

WINSTON, P. H. 1977. Artificial Intelligence. Addison-Wesley, Reading, MA. 444 pp.

WINSTON, P. H. 1982. Learning new principles from precedents and exercises. Artificial Intelligence 19:321–350.

WINTERNITZ, B. L. 1976. Temporal change and habitat preference of some montane breeding birds. Condor 78:383–393.

WOODS, S. E. 1981. The Squirrels of Canada. National Museums of Canada, Ottawa, Ontario. 199 pp.

WOOLFENDEN, G. E., and S. A. ROHWER. 1969. Breeding birds in a Florida suburb. Bulletin of the Florida State Museum 13:1–83.

WOOLHOUSE, M. E. J. 1983. Theory and practice of the species-area effect, applied to the breeding birds of British woods. Biological Conservation 27:315–332.

WOOTEN, C. 1981. Avian Community Composition and Habitat Associations in an Upland Deciduous Forest in Northwestern Arkansas. M.S. thesis. University of Arkansas, Fayetteville, AR. 93 pp.

WRIGHT, S. J. 1981. Intra-archipelago vertebrate distributions: The slope of the species-area relation. American Naturalist 118:726–748.

WYKOFF, W. R., N. L. CROOKSTON, and A. R. STAGE. 1982. User's Guide to the Stand Prognosis Model. U.S. Department of Agriculture, Forest Service, General Technical Report INT–133. Intermountain Forest and Range Experiment Station, Ogden, UT. 112 pp.

ZAR, J. H. 1974. Biostatistical Analysis. Prentice-Hall, Englewood Cliffs, NJ. 620 pp.

ZAR, J. H. 1984. Biostatistical Analysis. Second edition. Prentice-Hall, Englewood Cliffs, NJ. 718 pp.

ZIELINSKI, W. J., W. D. SPENCER, and R. H. BARRETT. 1983. Relationship between food habits and activity patterns of pine martens. Journal of Mammalogy 64:387–396.

ZIVNUSKA, J. A. 1961. The multiple problems of multiple use. Journal of Forestry 59:555–560.

ZUBOY, J. R. 1981. A new tool for fishery managers: The Delphi technique. North American Journal of Fish Management 1:55–59.

Index